Cellular response to biomaterials

Related titles:

Surfaces and interfaces for biomaterials
(ISBN 978-1-85573-930-7)
Given such problems as rejection, the interface between an implant and its human host is a critical area in biomaterials. *Surfaces and interfaces for biomaterials* presents our current level of understanding on the nature of a biomaterial surface, the adaptive response of the biomatrix to that surface, techniques used to modify biocompatibility, and state-of-the-art characterisation techniques to follow the interfacial events at that surface.

Tissue engineering using ceramics and polymers
(ISBN 978-1-84569-176-9)
Tissue engineering is rapidly developing as a technique for the repair and regeneration of diseased tissue in the body. This book reviews how ceramic and polymeric biomaterials are being used in tissue engineering. The first part of the book reviews the nature of ceramics and polymers as biomaterials together with techniques for using them such as building tissue scaffolds, transplantation techniques, surface modification and ways of combining tissue engineering with drug delivery and biosensor systems. The second part of the book discusses the regeneration of particular types of tissue from bone, cardiac and intervertebral disc tissue to skin, liver, kidney and lung tissue.

Molecular interfacial phenomena of polymers and biopolymers
(ISBN 978-1-85573-928-4)
This book combines three fundamental areas of interest to the science and engineering community, namely material science, nanotechnology and molecular engineering. Although there have been various results published in this field, there has yet to be a fully comprehensive review. *Molecular interfacial phenomena of polymers and biopolymers* covers key research on molecular mechanisms and thermodynamic behaviour of (bio)polymer surfaces and interfaces, from theoretical and experimental perspectives.

Details of these and other Woodhead Publishing materials books can be obtained by:

- visiting our web site at www.woodheadpublishing.com
- contacting Customer Services (e-mail: sales@woodheadpublishing.com; fax: +44 (0) 1223 893694; tel: +44 (0) 1223 891358 ext. 130; address: Woodhead Publishing Limited, Abington Hall, Granta Park, Great Abington, Cambridge CB21 6AH, England)

If you would like to receive information on forthcoming titles, please send your address details to: Francis Dodds (address, tel. and fax as above; e-mail: francis.dodds@woodheadpublishing.com). Please confirm which subject areas you are interested in.

Cellular response to biomaterials

Edited by
Lucy Di Silvio

CRC Press
Boca Raton Boston New York Washington, DC

WOODHEAD PUBLISHING LIMITED
Cambridge   New Delhi

Published by Woodhead Publishing Limited, Abington Hall, Granta Park
Great Abington, Cambridge CB21 6AH, England
www.woodheadpublishing.com

Woodhead Publishing India Pvt Ltd, G-2, Vardaan House, 7/28 Ansari Road, Daryaganj, New Delhi – 110002, India

Published in North America by CRC Press LLC, 6000 Broken Sound Parkway, NW, Suite 300, Boca Raton, FL 33487, USA

First published 2009, Woodhead Publishing Limited and CRC Press LLC
© 2009, Woodhead Publishing Limited
The authors have asserted their moral rights.

This book contains information obtained from authentic and highly regarded sources. Reprinted material is quoted with permission, and sources are indicated. Reasonable efforts have been made to publish reliable data and information, but the authors and the publishers cannot assume responsibility for the validity of all materials. Neither the authors nor the publishers, nor anyone else associated with this publication, shall be liable for any loss, damage or liability directly or indirectly caused or alleged to be caused by this book.

Neither this book nor any part may be reproduced or transmitted in any form or by any means, electronic or mechanical, including photocopying, microfilming and recording, or by any information storage or retrieval system, without permission in writing from Woodhead Publishing Limited.

The consent of Woodhead Publishing Limited does not extend to copying for general distribution, for promotion, for creating new works, or for resale. Specific permission must be obtained in writing from Woodhead Publishing Limited for such copying.

Trademark notice: Product or corporate names may be trademarks or registered trademarks, and are used only for identification and explanation, without intent to infringe.

British Library Cataloguing in Publication Data
A catalogue record for this book is available from the British Library.

Library of Congress Cataloging in Publication Data
A catalog record for this book is available from the Library of Congress.

Woodhead Publishing ISBN 978-1-84569-358-9 (book)
Woodhead Publishing ISBN 978-1-84569-547-7 (e-book)
CRC Press ISBN 978-1-4200-9373-5
CRC Press order number WP9373

The publishers' policy is to use permanent paper from mills that operate a sustainable forestry policy, and which has been manufactured from pulp which is processed using acid-free and elemental chlorine-free practices.
Furthermore, the publishers ensure that the text paper and cover board used have met acceptable environmental accreditation standards.

Project managed by Macfarlane Book Production Services, Dunstable, Bedfordshire, England (e-mail: macfarl@aol.com)

Typeset by SNP Best-set Typesetter Ltd, Hong Kong
Printed by TJ International Limited, Padstow, Cornwall, England

Contents

Contributor contact details		xv
Introduction L Di Silvio, King's College London, UK		xxi
Part I	**Cell responses to biomaterials using polymers and ceramics**	**1**
1	Biocompatible three-dimensional scaffolds for tendon tissue engineering using electrospinning L Bosworth and S Downes, The University of Manchester, UK	3
1.1	Introduction	3
1.2	Electrospinning	4
1.3	Electrospinning applications in tissue engineering	8
1.4	Scaffold requirements	10
1.5	Cell response	10
1.6	Porosity	11
1.7	Fibre orientation	12
1.8	Surface wettability	14
1.9	Functionalising fibres	15
1.10	Electrospinning for tendon regeneration	17
1.11	Electrospun fibre bundles	19
1.12	Future trends	23
1.13	Sources of further information and advice	23
1.14	References	23
2	Degradable polymers and polymer composites for tissue engineering S Deb, King's College London, UK	28
2.1	Introduction	28
2.2	Degradable polymers	29

2.3	Degradable composites	42
2.4	Fibre scaffolds	49
2.5	Injectable composites	51
2.6	Drug delivery from resorbable scaffolds	52
2.7	Conclusions	53
2.8	References and further reading	53

3	**Biocompatibility of degradable polymers for tissue engineering**	**61**
	N GURAV, Northwick Park Institute of Medical Research (NPIMR), UK and L DI SILVIO, King's College London, UK	
3.1	Introduction	61
3.2	Biocompatibility	63
3.3	Degradable polymers	66
3.4	Practical problems during biocompatibility testing	68
3.5	Degradation mechanisms and devising the appropriate tests	70
3.6	*In-vitro* biocompatibility testing	70
3.7	*In-vivo* biocompatibility testing	75
3.8	Immune response	76
3.9	Discussion	79
3.10	References	80

4	**Cellular response to the surface chemistry of nanostructured biomaterials**	**85**
	M A BARBOSA, M C L MARTINS and J N BARBOSA, INEB, Universidade do Porto, Portugal	
4.1	Introduction	85
4.2	Nanostructured surfaces and materials	86
4.3	Influence of surface chemistry on protein adsorption	93
4.4	Influence of surface chemistry on cell response	102
4.5	Future trends	106
4.6	References	107

5	**Biocompatibility and other properties of hydrogels in regenerative medicine**	**114**
	A SANNINO and M MADAGHIELE, University of Salento, Italy and L AMBROSIO, Institute of Composite and Biomedical Materials (IMCB-CNR), Italy	
5.1	Gels and hydrogels: definition and classification	114
5.2	Hydrogel swelling ratio	116
5.3	Microstructural parameters/design variables for polymeric networks	117

5.4	Hydrogels in biomedical fields: optimization of chemical/physical properties for selected applications	126
5.5	Conclusions	131
5.6	References	132
6	**Cellular response to bioceramics**	**136**
	A WOESZ and S M BEST, University of Cambridge, UK	
6.1	Introduction	136
6.2	A short history of the usage of ceramic materials in the human body	136
6.3	Types of ceramics used in the human body: bioinert, bioactive and bioresorbable	137
6.4	Biological and mechanical behaviour of various classes of materials	138
6.5	Dense versus porous ceramic materials	141
6.6	Cell reactions to ceramic materials	142
6.7	Summary and future trends	149
6.8	References	150
7	**Biocompatibility and other properties of phosphate-based glasses for medical applications**	**156**
	E A ABOU NEEL and J C KNOWLES, University College London, UK	
7.1	Historical overview	156
7.2	Glass structure theories	157
7.3	Glass formation	158
7.4	Chemistry of phosphate-based glasses	159
7.5	Properties of phosphate-based glasses	162
7.6	Biomedical applications of phosphate-based glasses	164
7.7	*In vitro* studies on phosphate-based glasses	165
7.8	*In vivo* studies on phosphate-based glasses	172
7.9	Phosphate-based glasses in composites	173
7.10	Phosphate-based glass fibres	173
7.11	Future trends using phosphate-based glasses	177
7.12	References	178
Part II	**Cell responses and regenerative medicine**	**183**
8	**Biodegradable scaffolds for tissue engineering**	**185**
	V SALIH, University College London, UK	
8.1	Introduction	185
8.2	What is a scaffold?	187

8.3	Importance of scaffold design and manufacture	188
8.4	Types of scaffold material	189
8.5	Scaffold design and processing	193
8.6	Advantages and limitations of scaffolds	196
8.7	Monitoring cell behaviour in scaffolds	197
8.8	Nanotechnology and scaffolds	198
8.9	Scaffold applications in gene therapy	199
8.10	Future trends	200
8.11	Conclusions	202
8.12	Acknowledgements	202
8.13	Sources of further information and advice	202
8.14	References	203

9 Developing smaller-diameter biocompatible vascular grafts 212
K D ANDREWS and J A HUNT, University of Liverpool, UK

9.1	Introduction	212
9.2	Vascular grafts with respect to natural blood vessels	213
9.3	Why development of small-calibre vascular grafts is needed	215
9.4	Initial scaffolds investigated for use as vascular grafts	217
9.5	Further development and modification to biomaterials used as vascular grafts	222
9.6	Electrostatically spun scaffolds as vascular grafts	230
9.7	Future trends	232
9.8	Conclusions	232
9.9	References	233

10 Improving biomaterials in tendon and muscle regeneration 237
V MUDERA, U CHEEMA, R SHAH and M LEWIS, University College London, UK

10.1	Tendons and their mechanical properties	237
10.2	Biomaterials used in tendon regeneration	238
10.3	Muscles and their mechanical properties	242
10.4	Biomaterials used in muscle regeneration	243
10.5	Conclusions	248
10.6	References	248

11 Biomaterials for the repair of peripheral nerves 252
S HALL, King's College London, UK

| 11.1 | Introduction | 252 |
| 11.2 | The structure of a peripheral nerve fibre: an overview | 254 |

11.3	The injury response in the peripheral nervous system (PNS)	257
11.4	Clinical need: alternatives to autografts	262
11.5	Nerve guides	263
11.6	Why work on short gaps?	276
11.7	The 'smart' conduit	277
11.8	References	278
12	**Stem cells and tissue scaffolds for bone repair**	**291**
	S SCAGLIONE and R QUARTO, University of Genova, Italy and P GIANNONI, Advanced Biotechnology Center (CBA), Italy	
12.1	Introduction	291
12.2	Biomaterials as biomimetic scaffolds	292
12.3	Adult stem cell sources	294
12.4	Targeting stemness within the bone marrow niche: microenvironmental evidence and co-cultures	300
12.5	Stem cell-induced immunomodulation: a shared characteristic?	305
12.6	Future trends	306
12.7	Conclusions	307
12.8	References	308
13	**Cellular response to osteoinductive materials in orthopaedic surgery**	**313**
	L DI SILVIO and P JAYAKUMAR, King's College London, UK	
13.1	Introduction	313
13.2	Osteoinduction and bone healing	316
13.3	Bone morphogenetic proteins (BMPs)	318
13.4	Bone graft biomaterials and bone healing	323
13.5	Osteoinductivity and bone healing	324
13.6	Osteoconduction and bone healing	325
13.7	Osseointegration and bone healing	326
13.8	Bone graft materials	326
13.9	Osteoinductive materials	326
13.10	Synthetic bone graft substitutes (SBGSs)	332
13.11	Bone graft biomaterials and remodelling	334
13.12	Discussion	335
13.13	References	336
14	**Cellular response to bone graft matrices**	**344**
	A B M RABIE and R W K WONG, University of Hong Kong, Hong Kong	
14.1	Introduction	344

14.2	Demineralized intramembranous bone matrix (DBM_{IM})	344
14.3	HMG-CoA reductase inhibitor: HMGRI-collagen matrix	353
14.4	Polymethoxylated flavonoid-collagen matrix	358
14.5	Phytoestrogen-collagen matrix	360
14.6	Conclusions	364
14.7	References	364
14.8	Appendix	368
15	**Cellular response to hydroxyapatite and Bioglass® in tissue engineering and regenerative medicine** J HUANG, University College London, UK	**371**
15.1	Introduction	371
15.2	Cellular responses to Bioglass® and hydroxyapatite particles	372
15.3	Cellular responses to Bioglass® and hydroxyapatite reinforced biocomposites	382
15.4	Interaction of human osteoblast (HOB) cells and biocomposites at the ultrastructural level	385
15.5	Summary	387
15.6	References	388
16	**Diamond-like carbon (DLC) as a biocompatible coating in orthopaedic and cardiac medicine** W MA, A J RUYS and H ZREIQAT, The University of Sydney, Australia	**391**
16.1	Introduction	391
16.2	Deposition and characterization of DLC coatings	393
16.3	Physical properties of DLC coatings	397
16.4	Biocompatibility of DLC	401
16.5	Corrosion behaviour of DLC	405
16.6	DLC for orthopaedic applications	406
16.7	DLC for selective blood-interfacing applications	408
16.8	Conclusions	411
16.9	Acknowledgements	412
16.10	References	412
Part III	**The effect of surfaces and proteins on cell response**	**427**
17	**Cell response to nanofeatures in biomaterials** A CURTIS and M DALBY, Glasgow University, UK	**429**
17.1	Introduction: why should cells respond to nanofeatures?	429

17.2	Making artificial nanofeatures: fabrication methods including embossing	430
17.3	The range of reactions to nanotopography	436
17.4	Cell signalling in response to nanofeatures	440
17.5	Reactions at the genome to morphological effects of topography on cells	446
17.6	Related topics	451
17.7	Future trends	452
17.8	References	453
18	**Cell response to surface chemistry in biomaterials**	**462**
	C A SCOTCHFORD, University of Nottingham, UK	
18.1	Introduction	462
18.2	The role of proteins in the cell response to surface chemistry	464
18.3	Cell response to model surfaces	465
18.4	Mechanisms for cell responses to surfaces	468
18.5	Exploitation of cell response to surface chemistry	471
18.6	Future trends	473
18.7	References	474
19	**Bioactive surfaces using peptide grafting in tissue engineering**	**479**
	M DETTIN, University of Padova, Italy	
19.1	Biomimetic materials and bioactive surfaces	479
19.2	Peptide mimicry for bioactive surface design	483
19.3	Delivery of bioactive peptides to tissue–surface interface	486
19.4	Criteria for bioactive surfaces design	489
19.5	*In vitro* and *in vivo* applications of bioactive surfaces	490
19.6	Future trends	502
19.7	References	502
20	***In vitro* testing of biomaterial toxicity and biocompatibility**	**508**
	S INAYAT-HUSSAIN and N F RAJAB, Universiti Kebangsaan Malaysia, Malaysia and E L SIEW, Melaka Biotechnology Corporation, Malaysia	
20.1	Introduction	508
20.2	Biocompatibility	509
20.3	Toxicity testing: *in vitro* methods	512
20.4	Cytotoxicity	513
20.5	Cell death	519
20.6	*In vitro* genotoxicity testing of biomaterial	525

20.7	Future trends	531
20.8	References	532

21	**The influence of plasma proteins on bone cell adhesion**	**538**
	Å ROSENGREN and S OSCARSSON, Mälardalen University, Sweden	
21.1	Introduction	538
21.2	Protein material interactions: flow-based protein adsorption analysis	540
21.3	Bicinchoninic acid (BCA) protein assay	541
21.4	Two-dimensional gel electrophoresis	543
21.5	Western immunoblotting	544
21.6	Protein characterization	544
21.7	Ellipsometry	546
21.8	Cell–substrate interactions: study methods	548
21.9	Protein–cell interactions	550
21.10	References	555

22	**Degradation of calcium phosphate coatings and bone substitutes**	**560**
	S OVERGAARD, Odense University Hospital, Clinical Institute and University of Southern Denmark, Denmark	
22.1	Introduction and demand for coatings and bone substitutes	560
22.2	Requirements for coatings and bone substitutes	562
22.3	Mechanical properties of coatings and bone substitutes	562
22.4	Degradation of coatings and bone substitutes	563
22.5	Discussion	566
22.6	Choice of bone substitute, and future trends	567
22.7	Conclusions	568
22.8	References	568

23	**Surface-modified titanium to enhance osseointegration in dental implants**	**572**
	L SARINNAPHAKORN and L DI SILVIO, King's College London, UK	
23.1	Introduction	572
23.2	Dental implant and osseointegration	573
23.3	Mechanism of peri-implant endosseous healing	574
23.4	The 'loss of osseointegration' phenomenon	574
23.5	Surface modification of dental implant	575
23.6	Commercial surface-modified titanium implants	580

23.7	Current treatment modalities	582
23.8	Future trends in dental implant surfaces	584
23.9	Conclusions	585
23.10	References	586
24	**Surface modification of titanium for the enhancement of cell response**	**589**
	R CHIESA, Politecnico di Milano, Italy	
24.1	Titanium for hard and soft tissue applications	589
24.2	Mechanical, physical and thermal modification treatments of titanium	593
24.3	Chemical and electrochemical modification treatments of titanium	594
24.4	The biocompatibility enhancement of titanium	602
24.5	Conclusions	605
24.6	References	605
Index		609

Contributor contact details

(* = main contact)

Editor

Dr Lucy Di Silvio
Department of Biomaterials and Biomimetics Research
King's College London Dental Institute
Floor 17, Guy's Tower, Guy's Campus
London Bridge
London SE1 9RT
UK
E-mail: Lucy.di_silvio@kcl.ac.uk

Chapter 1

Miss L. Bosworth* and Professor S. Downes
Department of Biomaterials
School of Materials Science
University of Manchester
Grosvenor Street
Manchester M1 7HS
UK
E-mail: lbosworth81@yahoo.co.uk
sandra.downes@manchester.ac.uk

Chapter 2

Dr Sanjukta Deb
Department of Biomaterials and Biomimetics Research
King's College London Dental Institute
Floor 17, Guy's Tower, Guy's Campus
London Bridge
London SE1 9RT
UK
E-mail: sanjukta.deb@kcl.ac.uk

Chapter 3

Dr Neelam Gurav*
Northwick Park Institute of Medical Research (NPIMR)
Northwick Park Hospital and St Marks Trust (Y Block)
Watford Road
Harrow HA1 3UJ
UK
E-mail: n.gurav@imperial.ac.uk

Dr Lucy Di Silvio
Department of Biomaterials and Biomimetics Research
King's College London Dental Institute
Floor 17, Guy's Tower, Guy's Campus
London Bridge
London SE1 9RT
UK

Chapter 4

M.A. Barbosa,* M.C. Lopes Martins and J. Novais Barbosa
INEB-Instituto de Engenharia Biomédica
Universidade do Porto
Rua do Campo Alegre, 823
4150-180 Porto
Portugal
E-mail: mbarbosa@ineb.up.pt

Chapter 5

A. Sannino and M. Madaghiele
Department of Engineering for Innovation
University of Salento
Via per Monteroni
73100 Lecce
Italy

L. Ambrosio*
Institute of Composite and Biomedical Materials (IMCB-CNR)
National Research Council
Piazzale Tecchio 80
80125 Naples
Italy
E-mail: ambrosio@unina.it

Chapter 6

Dr A. Woesz* and Dr S.M. Best
Cambridge Centre for Medical Materials
University of Cambridge
Pembroke Street
Cambridge CB2 3QZ
UK
E-mail: Alexander@woesz.com
smb51@hermes.cam.ac.uk

Chapter 7

E.A. Abou Neel and J.C. Knowles*
Division of Biomaterials and Tissue Engineering
University College London
Eastman Dental Institute
256 Gray's Inn Road
London WC1X 8LD
UK
E-mail: j.knowles@eastman.ucl.ac.uk

Chapter 8

V. Salih
Division of Biomaterials & Tissue Engineering
University College London
Eastman Dental Institute
256 Gray's Inn Road
London WC1X 8LD
UK
E-mail: v.salih@eastman.ucl.ac.uk

Chapter 9

Dr K.D. Andrews and
  Dr J.A. Hunt*
Clinical Engineering (UK CTE)
UK BioTEC
School of Clinical Sciences
University of Liverpool
Ground Floor
Duncan Building
Daulby Street
Liverpool L69 3GA
UK
E-mail: huntja@liverpool.ac.uk

Chapter 10

Dr V. Mudera,* Dr U. Cheema,
  Dr R. Shah and Dr M. Lewis
Tissue Repair and Engineering
  Centre
University College London
Institute of Orthopaedics and
  Musculoskeletal Science
Royal National Orthopaedic
  Hospital
Brockley Hill
Stanmore HA7 4LP
UK
E-mail: rmhkvim@ucl.ac.uk

Chapter 11

Professor Susan Hall
Department of Anatomy and
  Human Sciences
King's College London
London SE1 1UL
UK
E-mail: susan.standring@kcl.ac.uk

Chapter 12

Dr S. Scaglione
Department of Communication,
  Computer and System Sciences
  (DIST)
University of Genova
Viale Causa 13
16145 Genova
Italy
E-mail: silvia.scaglione@unige.it

Dr P. Giannoni
Stem Cell Laboratory
Advanced Biotechnology Center
  (CBA)
Largo Rosanna Benzi n°10
16132 Genova
Italy
E-mail: paolo.giannoni@
  cba-biotecnologie.it

Professor R. Quarto*
Department of Pharmaceutical and
  Food Chemistry and
  Technologies (DICTFA)
University of Genova
Largo Rosanna Benzi n°10
16132 Genova
Italy
E-mail: rodolfo.quarto@unige.it

Chapter 13

Dr Lucy Di Silvio and P Jayakumar
Department of Biomaterials and
  Biomimetics Research
King's College London Dental
  Institute
Floor 17, Guy's Tower, Guy's
  Campus
London Bridge
London SE1 9RT
UK
E-mail: Lucy.di_silvio@kcl.ac.uk

Chapter 14

Professor A.B.M. Rabie* and
  Dr R.W.K. Wong
Biomedical and Tissue Engineering
University of Hong Kong
2/F, Prince Philip Dental Hospital
34 Hospital Road, Saiyingpun
Hong Kong
E-mail: rabie@hkusua.hku.hk

Chapter 15

Dr J. Huang
Department of Mechanical
  Engineering
University College London
Torrington Place
London WC1E 7JE
UK
E-mail: jie.huang@ucl.ac.uk

Chapter 16

W.J. Ma, Professor A.J. Ruys and
  Dr H. Zreiqat*
Biomedical Engineering
School of AMME
Camperdown Campus 2006
The University of Sydney
Sydney
Australia
E-mail: hzreiqat@usyd.edu.au

Chapter 17

Professor A. Curtis* and
  Dr M.J. Dalby
Centre for Cell Engineering
Joseph Black Building
Glasgow University
University Avenue
Glasgow G12 8QQ
Scotland
E-mail: a.curtis@bio.gla.ac.uk

Chapter 18

Dr Colin Scotchford
School of Mechanical, Materials
  and Manufacturing Engineering
University of Nottingham
University Park
Nottingham NG7 2RD
UK
E-mail: Colin.Scotchford@
  nottingham.ac.uk

Chapter 19

Monica Dettin
Department of Chemical Process
  Engineering
University of Padova
Via Marzolo, 9
35131 Padova
Italy
E-mail: monica.dettin@unipd.it

Chapter 20

Professor S.H. Inayat-Hussain*
Environmental Health Programme
Faculty of Allied Health Sciences
Universiti Kebangsaan Malaysia
Jalan Raja Muda Abdul Aziz
50300 Kuala Lumpur
Malaysia
E-mail: salmaan@streamyx.com

E.L. Siew
Toxicology Laboratory
Melaka Biotechnology Corporation
Lot 7, MITC City
75450 Ayer Keroh, Melaka
Malaysia

Dr N.F. Rajab
Biomedical Science Department
Faculty of Allied Health Sciences
Universiti Kebangsaan Malaysia
Jalan Raja Muda Abdul Aziz
50300 Kuala Lumpur
Malaysia

Chapter 21

Å. Rosengren and S. Oscarsson*
Mälardalen University
Department of Surface
 Biotechnology
Box 325
63105 Eskilstuna
Sweden
E-mail: sven@svenoscarsson.com

Chapter 22

Professor Søren Overgaard
Department of Orthopaedic
 Surgery
Odense University Hospital
Faculty of Health Science
Clinical Institute
University of Southern Denmark
Denmark
E-mail: soeren.overgaard@ouh.
 regionsyddanmark.dk

Chapter 23

Dr Lucy Di Silvio and
 Lertrit Sarinnaphakorn*
Department of Biomaterials and
 Biomimetics Research
King's College London Dental
 Institute
Floor 17, Guy's Tower, Guy's
 Campus
London Bridge
London SE1 9RT
UK
E-mail: lertrit.sarinnaphakorn@
 kcl.ac.uk
Lucy.di_silvio@kcl.ac.uk

Chapter 24

Professor Roberto Chiesa
Dipartimento di Chimica, Materiali
 e Ingegneria Chimica 'G. Natta'
Politecnico di Milano
Via Mancinelli 7
20131 Milano
Italy
E-mail: roberto.chiesa@polimi.it

Introduction

L DI SILVIO, King's College London, UK

Medical devices are essential to modern medical practice. They are used in the treatment of disease, trauma and disability, thus saving or improving the quality of life for millions. Currently, damaged tissues are replaced by synthetic biomedical implants, which often fail as a result of not fully integrating with the host tissue, thus resulting in significant socio-economic costs and reduced quality of life to the patient.

Biomaterials is a field with over 50 years of innovation and development. Until recently, the design of biomaterials focused primarily on biocompatibility in determining the interaction of host tissues with an implant. The aim of a biocompatible implant or material is to obtain the desired effect of the device in the application for which it is intended. This involves not only optimization of desirable properties but also minimization of deleterious and damaging effects. This process includes the selection of the material design and manufacture of the device and the testing systems to verify biocompatibility.

Understanding how cells respond to biomaterials is critical in designing biologically inspired implants for biomedical applications. Current strategies aim to engineer biomaterials that have the ability to control cell response. Understanding how microenvironmental cues can regulate cell functions such as adhesion, proliferation and phenotype will assist in the understanding of cellular behaviour and provide novel approaches for tissue engineering and regeneration.

An 'appropriate' cellular response to implanted surfaces is essential for tissue regeneration and integration. Following implantation, materials are immediately coated with proteins from blood and interstitial fluids, and it is through this adsorbed layer and not the surface itself that cells initially respond. The chemistry, surface energy and micro-architecture of the material also influence the types of proteins that adsorb onto the surface, and cells have the ability to distinguish complex differences in surface structure and to respond accordingly.

Various biomaterials are being developed with surface modifications and specific properties, with the aim of inducing healing in a more physiological manner. Materials that can respond to changes in their surrounding environment are attractive because these changes, particularly *in vivo*, can be exploited to control parameters such as drug release, cell adhesiveness, mechanical properties or permeability. Other types include nanostructured macroporous materials designed to mimic tissue structure and properties, and to stimulate new tissue growth while degrading in the body.

The list of new materials being developed specifically for biological applications is very long. We have chosen to focus on both synthetic and natural biomaterials, examples of which include polymer composites, ceramics, hydrogels, phosphate glasses, ceramics and titanium, and demineralized bone matrix and collagen, respectively. In an effort to develop complex scaffolds that better match natural tissue structure, several efforts have focused on modifying or functionalizing surfaces to control protein interaction and to decorate the surface appropriately with signalling molecules. Biomaterials are also being designed to specifically and maximally enhance mesenchymal cell attachment, migration, proliferation and differentiation, which will markedly increase the value of these materials for tissue engineering.

It is now being recognized that the future of regenerative medicine will be influenced in a major way by increased understanding of the underlying principles of stem cell biology. Human stem cell research holds enormous potential for contributing to our understanding of fundamental human biology and offers the possibility of treatments and ultimately of cures for many diseases for which adequate therapies do not exist. In its fully developed form, tissue engineering involves three key elements: a specific living cell type (or several cell types), a material scaffold that provides the supporting architecture for the cells, and a growth stimulus for inducing cell differentiation. The goal is to provide materials that are capable of supporting tissue regeneration *in vivo*, often at sites compromised by infection and loss of structure. Tissue engineering requires a 3-D matrix that will not only guide tissue formation into the desired shape but also provide adequate transport of nutrients and growth factors to promote tissue growth and vascularization.

Future advances in tissue engineering will be dependent upon advances in biodegradable polymers, rapid prototyping, drug delivery, cell culture (sterilization and tissue storage), developing more sensitive and sophisticated stem cell methodologies, angiogenesis, and biomimetic strategies for recreating extracellular, matrix-like biomimetics. Also, advances in gene delivery and surgery will be needed to fully realize the potential of tissue engineering.

The success of biomaterials is reflected in their importance to modern medical therapies, the economic potential of the market, and the steady growth of the field over the last 50 years. There is no doubt that tissue engineering will change the face of biomaterials, as scaffolds play a central role and are necessary in order to provide structure and allow delivery of cells and induction factors to the target site.

This book describes a wide range of biomaterials, both degradable and non-degradable, for a number of applications. Part I specifically focuses on cellular interactions with polymers and ceramics, and describes biocompatibility, encompassing biosafety and also functionality with respect to tissue engineered systems. It also describes the influence of surface chemistry and degradation products of degradable scaffolds on cellular response. Part II addresses cell response with specific reference to tissues – vascular, muscle, tendon, nerves and bone. It also describes tissue engineering systems using stem cells. Part III is concerned with more challenging advances in surface modification, including chemical and topographical and also peptide functionalization of biomaterials in order to enhance cell responses and osseointegration.

In addition to explaining how biomaterials influence cells, a greater understanding of how these new classes of materials interact at a cellular level can assist in improving device integration and function, for example by controlling direct responses in stem cells, and factors influencing and their differentiation into mature, functional cell types.

This book provides an interesting and up-to-date insight into cellular responses to biomaterials and it is hoped that it will act as a catalyst for the challenging 'Brave New World' of tissue engineering.

Part I
Cell responses to biomaterials using polymers and ceramics

1
Biocompatible three-dimensional scaffolds for tendon tissue engineering using electrospinning

L BOSWORTH and S DOWNES,
The University of Manchester UK

Abstract: This chapter first discusses the importance of applying ideal electrospinning parameters for creating fibres of micron to nanometre diameter and their applications in tissue engineering. Examples of biopolymers and structural aspects of the scaffold are considered. Particular attention is drawn to the different fibre collection techniques for creating bundles of electrospun fibres and how these could be used for regenerating tendon by mimicking the morphological and mechanical properties of the natural tissue.

Key words: electrospinning, biopolymers, tendon regeneration, tissue engineering, scaffold.

1.1 Introduction

Tissue engineering has become an important and emerging research field and utilises the concept that biomaterials can be designed and engineered to encourage living cells to repair and restore damaged tissues. There have been notable advances in tissue engineering in cartilage, bone and skin. Biomaterials have an important role, usually as scaffolds for cells cultured *in vitro* prior to implantation in patients or as implantable devices to encourage cell infiltration, growth and differentiation. The current priority is to utilise scientific and engineering principles to design and fabricate support structures, using degradable polymers that can be seeded with living cells. The aim is that the appropriate cells will produce growth factors and extracellular matrix molecules, at the same time as the polymer scaffold degrades. In this chapter, the methods to produce 3-D scaffolds using electrospinning techniques are evaluated. Details of the technology, the parameters that can be altered to control the structure and biomechanical properties of the electrospun fibres, methods to produce a hierarchical structure resembling the natural tissue and *in vitro* assessment of the scaffolds are considered. This chapter focuses on tendon damage or degeneration, which presents a clinical problem due to degenerative disease and trauma affecting thousands of people every year in the UK alone.

One of the approaches in tissue engineering is to mimic the tissue being replaced. Fabricated scaffolds whose structures and dimensions are similar to those of the tissue's extracellular matrix (ECM) further aid simulation by providing the cells with an environment not too dissimilar from their natural habitat, thus ensuring the implant is analogous in terms of its physical and structural properties to the original tissue.

A number of techniques are emerging to aid development of artificial ECM constructs, including the following:

- *Self-assembly* – independent arrangement of small molecules capable of producing structures with fibre diameters of several nanometres (Whitesides and Grzybowski, 2002).
- *Phase separation* – sponge-like scaffolds obtained by the separation of a polymer solution and extraction of the solvent (Nam and Park, 1999).
- *Electrospinning* – application of an electrostatic force to a polymer solution producing nanofibres of controllable dimensions.

This chapter focuses on the use of electrospinning to produce scaffold constructs. This technique uses simple apparatus to quickly manufacture large quantities of fibres. The structure of ECM is comparable to that of electrospun fibres in terms of fibre diameter and composition (Pham *et al.*, 2006). The distribution of fibre diameters obtained is dependent upon the spinning parameters employed; this can be controlled to replicate the diameter range for collagen fibre bundles, 50–500 nm. Control of the electrospinning process ultimately makes this method a suitable technique for tissue engineering.

Other advantages of electrospinning are as follows:

- Fibres of nanoscopic diameters provide high surface area to volume ratio, allowing significant cell attachment (Lee *et al.*, 2005).
- Control of process parameters allows resulting fibre diameters to be adjusted to match original tissue dimensions (Wnek *et al.*, 2003).
- Control of fibre orientation enables mechanical properties of fibrous scaffolds to be tailored more appropriately (Matthews *et al.*, 2002).
- A diverse range of materials can be electrospun.

1.2 Electrospinning

Despite the recent expansion in the use of electrospinning, this simple technique has been employed for a variety of purposes since it was first patented by Formhals in 1934. This straightforward technique is a

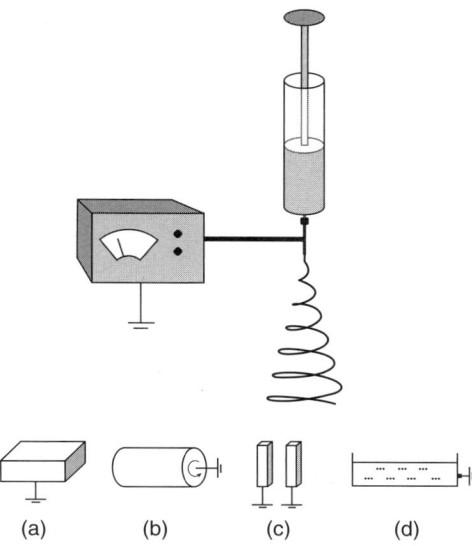

1.1 Basic electrospinning setup with collectors: (a) stationary plate, (b) rotating mandrel, (c) fixed plates, (d) liquid reservoir.

cost-effective method for fabricating long, continuous fibres. Additionally, the ability to vary the materials used and the fibre diameter produced offers versatility to suit a range of requirements. Recently, electrospinning has found significant use in the tissue engineering field, in particular for creating temporary scaffolds mimicking the tissue they are intended to restore.

The technique has a simple apparatus and utilises a high voltage to complete an electric circuit (Fig. 1.1). A high voltage power supply is connected to a needle-tipped syringe containing a polymeric solution with an applied flow rate. The earthed target collector is positioned at a known distance from the needle-tip. Application of a sufficiently high voltage causes the polymeric solution to charge and formation of a Taylor cone is observed. Expulsion of the polymer as a charged jet occurs once the charge intensity of the solution is sufficient to overcome the surface tension and viscoelastic forces of this Taylor cone (Doshi and Reneker, 1993). Owing to the charges present, the polymer jet stretches and thins as it travels towards the collector. Upon impact on the collector the charge of the fibres dissipates and the electrical circuit is completed.

1.2.1 Process parameters

Production and morphology of fibres is governed by a number of parameters, which can be split into three main groups (Huang *et al.*, 2003; Chronakis, 2005):

- *Polymeric solution properties* – viscosity, concentration, molecular weight, conductivity, surface tension.
- *Electrospinning setup* – voltage, solution flow rate, needle-tip to collector distance.
- *Ambient conditions* – external temperature and humidity, air velocity within the electrospinning setup.

Polymeric solution properties

Solution viscosity
The viscosity of the solution is directly affected by the concentration of polymer present. If the polymer concentration is high, greater quantities of polymer chains are present, increasing the number of chain entanglements with solvent molecules and ultimately raising the solution viscosity. A polymer's molecular weight also affects solution viscosity. A polymer of low molecular weight reduces the number of solvent/polymer entanglements because of the shorter chain length, and hence decreases solution viscosity. Fibre production is heavily dependent on the concentration of the solution being electrospun. Generally, if the concentration is too low, bead formation as opposed to fibre production is observed; if too high, pumping of the solution will be difficult and fabricated fibres are mostly micrometres in diameter. A solution of polycaprolactone (PCL) dissolved in acetone at a concentration of 5% w/v produces fibres with an average diameter of 200 nm when spun under constant parameters. Doubling the quantity of PCL with the same spinning parameters, however, more than doubles the average fibre diameter obtained, measuring 900 nm (Fig. 1.2(a)).

Solution conductivity
The solvent chosen to dissolve the polymer has a significant role in the level of conductivity present within the solution, which directly affects the fibre morphology generated from the electrospinning process. Solvents with high dielectric constants cause the emitted polymer jet to experience increased longitudinal force brought about by the higher accumulation of charge present within the polymeric solution (Wannatong *et al.*, 2004). Consequently the polymer jet experiences a greater degree of charge repulsion, leading to an increased level of stretching and elongation, resulting in fibres of finer diameter (Fong *et al.*, 1999).

Surface tension
The surface tension of the polymeric solution must be overcome in order for the electrospinning process to be initiated. The polymeric solution's viscosity directly affects its surface tension; high viscosity reduces the surface

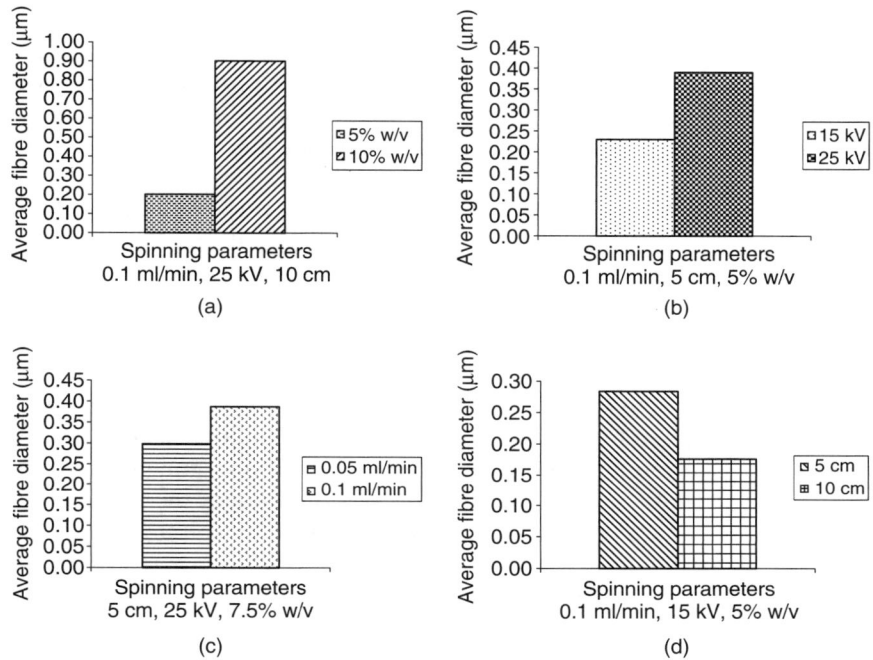

1.2 Average fibre diameters for electrospun polycaprolactone at different spinning parameters: (a) concentration, (b) voltage, (c) flow rate, (d) needle-tip to collector distance.

tension due to significant entanglement between solvent molecules and polymer chains preventing molecule clustering (Shawon and Sung, 2004).

Electrospinning setup

Voltage
The applied voltage has a direct effect on fibre morphology. High voltage causes the polymer jet to be emitted with rapid acceleration. This limits the jet's flight time and hence decreases the amount of stretching and solvent evaporation prior to collector impact (Ramakrishna *et al.*, 2005). The resulting fibres may be thicker and contain solvent, although flight time is also dependent upon the tip to collector distance. Raising the voltage applied to a solution of PCL/acetone by 10 kV follows this theory, giving a slight increase in average fibre diameter (Fig. 1.2(b)).

Solution flow rate
The applied flow rate determines the quantity of polymeric solution available to the electrospinning process. Generally, high flow rates yield fibres

of larger diameter. This is because the greater volume of solution pumped out may not have sufficient time for solvent evaporation and adequate stretching of the jet prior to contact with the collector. Doubling the flow rate for a PCL/acetone solution gave rise to a slight increase in average fibre diameter for the spinning parameters, as shown in Fig. 1.2(c).

Needle-tip to collector distance
Varying the distance between the needle-tip and the collector has an important role in determining fibre outcome. Short deposition distance reduces the polymer jet flight time, limiting the rate of solvent evaporation and polymer stretching and often resulting in the fabrication of thick, merged fibres. A minimum distance to allow significant drying and stretching of the jet is required for the production of long, fine fibres (Reneker et al., 2000). For the spinning parameters applied in Fig. 1.2(d), increasing the distance allowed a PCL/acetone solution longer time before impacting on the collector, thus permitting greater stretching and thinning of the polymer jet and resulting in fibre formation of finer diameter.

Ambient conditions

Humidity
The humidity surrounding the electrospinning process has been found to have a significant effect on fibre morphology, in terms of surface porosity (Casper et al., 2004). Studies on electrospun polystyrene fibres by Casper et al. (2004) found that rising levels of humidity led to an increase in pore size, number and distribution over the fibre surface.

Temperature
Investigations by Mit-uppatham et al. (2004) found that increasing the polymeric solution temperature resulted in a faster rate of fibre deposition and that fibres were of smaller diameter. These effects were attributed to the reduced viscosity of the polymeric solution caused by decreased entanglements between polymer and solvent as a result of polymer molecule expansion (Mit-uppatham et al., 2004).

1.3 Electrospinning applications in tissue engineering

A wide range of polymers – synthetic and natural – can be electrospun, and are currently being researched in a number of tissue engineering applications:

- A combination of collagen and PCL electrospun fibre matrices has successfully supported the growth and proliferation of dermal fibroblasts,

and shown promise as a dermal substitute for skin regeneration (Venugopal *et al.*, 2005).
- For peripheral nerve regeneration, poly(L-lactide-*co*-glycolide) (PLGA) electrospun tubes have been able to function as nerve guidance channels, and an *in vivo* rat study proved that nerve regeneration was successful after one month in 5 out of 11 rats implanted (Bini *et al.*, 2006).
- A study by Mo *et al.* (2004) demonstrated adhesion and proliferation of smooth muscle and endothelial cells on randomly oriented nanofibres of poly(L-lactide-*co*-ε-caprolactone) (PLLA-CL).
- An *in vitro* study on cardiomyocytes seeded onto PCL based nanofibrous meshes showed that beating started after just three days of culture, creating contractile cardiac grafts (Shin *et al.*, 2003).
- Electrospun polyurethane (PU) membranes fabricated by Khil *et al.* (2003) were used as wound dressings, and were found to promote epithelialisation and dermal organisation due to the unique structural properties of the nanofibrous structure.
- Nanocomposites of gelatin and hydroxyapatite were successfully electrospun by Kim *et al.* (2005), their purpose being to mimic human bone matrix, and were found to significantly improve bone-derived cellular activity.
- Biodegradable poly(L-lactide acid) (PLLA) electrospun fibres may potentially be used in drug delivery applications. Zeng *et al.* (2003) demonstrated the slow release of encapsulated drugs when degradation of the polymer fibres occurred.
- The structure and morphology of electrospun poly(hydroxybutyrate valerate) (PHBV) was proven to promote adhesion and growth of chondrocytes when compared to PHBV cast films (Chen and Wu, 2005).

1.3.1 Examples of biopolymers

Polycaprolactone (PCL)

PCL is a fully FDA-approved biocompatible, bioresorbable, semicrystalline, aliphatic polyester. An advantage of this biopolymer, dependent upon its application, is the slower rate of degradation compared to other biopolymers. Degradation of PCL occurs via hydrolysis of the ester group, and degradation products are metabolised via the tricarboxylic acid (TCA) cycle, and/or removed directly by renal secretion (Kweon *et al.*, 2003).

Poly(lactide-co-glycolide) (PLGA)

Similarly, PLGA is a FDA-approved copolymer of lactic acid and glycolic acid. Depending on the ratio of the two polymers, a range of different

PLGA forms can be created. This biopolymer is an amorphous, aliphatic polyester, and hydrolysis of the present ester groups causes its degradation. The rate of degradation is, however, dependent on the ratio of lactic acid to glycolic acid – a higher quantity of glycolic acid yields a shorter degradation time.

Poly(hydroxybutyrate) (PHB)

A polyhydroxyalkanoate belonging to the class of polyesters has a slow degradation rate and is made via a biological fermentation process (Pompe et al., 2007). PHB has a high level of crystallinity, with mechanical properties similar to those of polypropylene (Doyle et al., 1991).

1.4 Scaffold requirements

Scaffolds are synthetic constructs designed to adopt the function of naturally occurring tissues. To be successful, scaffolds must recreate the basic three-dimensional framework and properties of the original tissue to assume its function and stimulate appropriate regeneration. Recent developments include the incorporation of cell seeding onto bioresorbable scaffolds so that new tissue develops parallel to its degradation.

Constructs must be able to temporarily provide mechanical support comparable to that of the original tissue until sufficient ingrowth from the body's natural tissue is established (Das et al., 2003). For load transfer to be sustained, the rate of scaffold elimination needs to match the rate of newly formed tissue. If these rates do not provide adequate transition, the stability of the implant and new tissue could be compromised and may result in failure. Bioresorption eliminates the need for further invasive surgery (Hutmacher, 2000). An important, yet easily overlooked, design parameter is the ease of modifying the scaffold into the desired shape and size as required by each patient (Sachlos et al., 2003).

Microscopically, the porosity and surface topography of scaffolds must be considered. Adequate porosity is necessary to allow migration of cells into the scaffold structure and to ensure diffusion of nutrients and metabolic waste. Scaffold topography should promote cell adhesion and proliferation, whilst preventing cell dedifferentiation (Sachlos et al., 2003).

1.5 Cell response

Cellular response to materials is dependent on the surrounding interactive environment of mechanical, physicochemical and topographical stimuli (Wong et al., 2004). Further studies are required on the effects of mechanical stimuli on cell performance. Cell–material adhesion is influenced by the presence of specific molecules, such as integrins and proteins, whose presence may improve the surface hydrophilicity (Hasirci et al., 2006). Topographical cues, such as surface ridges and roughness, not only aid cell alignment but can also affect cell morphology and phenotype (Hasirci et al., 2006). Accurate surface topography is essential for guiding cell orientation and ensuring correct tissue function.

Research by Teixeira et al. (2003) has proven that human cells are able to attach, proliferate and orient themselves on fibres whose diameters are considerably smaller than their own. Polymeric nanofibres in tissue engineered constructs make ideal environments for cells to be able to attach, proliferate and maintain their phenotype.

Aligned nanofibrous structures for potential use in blood vessel engineering have been proven to control the migration of smooth muscle cells along the axis of aligned nanofibres (Xu et al., 2004). The distribution and arrangement of the cell's cytoskeleton were similarly found to lie parallel to the oriented fibres. Research by Yang et al. (2005) found that aligned nanofibres provided sufficient contact guidance for parallel neural stem cell growth.

1.6 Porosity

In order to provide adequate cell–surface interactions, nutrient and waste transfer, and sufficient space for ECM generation, research by Agrawal and Ray (2001) found that scaffolds should have a minimum porosity of 90%.

An important structural advantage of electrospun fibres is the degree of porosity present among the fibres, conferring substantial pore interconnectivity. Fibres that are collected randomly tend to exhibit a high level of porosity (Bhattarai et al., 2004), whereas purposefully oriented fibres lose this interconnectivity. Whilst a high pore volume may weaken the scaffold, a more effective transfer of nutrient and metabolic waste between the scaffold and surrounding environment is achievable (Ma and Choi, 2001).

The size of the pore is critical for the infiltration of cells into the scaffold. If pore size is too small, cell movement into the material may be inhibited; however, owing to the fineness and fragility of the fibres it is thought that cells are able to migrate through the scaffold by effectively pushing them out of their path (Boland et al., 2004). If pore size is too large, attachment and proliferation of cells between fibres could be reduced. Inadequate pore size is likely to inhibit effective tissue formation.

Porosity within the fibre itself may also occur, as demonstrated by Bognitzki et al. (2001) in whose research spun fibres of poly-L-lactide (PLLA) dissolved in dichloromethane were riddled with pores. It was proposed that the volatility of the solvent caused its rapid evaporation from solvent-rich regions and resulted in intra-porous fibres.

1.7 Fibre orientation

The orientation of fibres deposited from the electrospinning process is dependent upon the mode of collection. Random, non-woven arrangements of fibres are created when the collector incorporated is a stationary plate. Networks of this type make ideal applications for wound-healing dressings because the high pore volume allows diffusion of oxygen molecules through the scaffold to aid healing whilst preventing larger foreign bodies, such as bacteria, from penetrating through and causing infection (Huang et al., 2003).

In certain tissue engineering applications uniaxial alignment of fibres is required as surface topographies of this type have been found to encourage cell alignment and migration along such fibres by 'contact guidance' (Clark et al., 1987). A number of different collector methods for fabricating aligned fibres are available. One of the most common is utilisation of a rotating mandrel. A problem associated with this method, however, is the speed of rotation. If the mandrel rotates too slowly compared to the speed of fibre emission, alignment will be inhibited and a random network will result; if rotation is too fast, fibre breakage will occur (Huang et al., 2003). Collection of fibres between two parallel earthed plates, also known as the 'gap method of alignment', is another technique which produces highly aligned fibres (Dzenis, 2004). A limitation to this mode of collection is the restriction in obtainable fibre length; there is a maximum distance that the two plates can be apart (Li et al., 2003).

Fibrous yarns containing aligned individual fibres that are grouped together can also be fabricated, and are achieved by spinning the polymer solution directly into an earthed liquid reservoir (Smit et al., 2005). The network of fibres collected on the liquid surface is drawn off and into the air. Alignment and grouping of fibres occurs due to the effect of surface tension between the fibres and liquid during the drawing process. Three-dimensional fibrous bundles are the end product after lifting the fibres off the liquid surface.

1.7.1 Fibre alignment

Alignment is determined by the angle between the fibre and the direction of alignment; the smaller the angle the greater the alignment. The range of mandrel rotating speeds clearly affects the alignment of fibres collected for electrospun PCL, as shown in Fig. 1.3. A rotating speed of 500 rpm resulted in the highest order of fibre alignment.

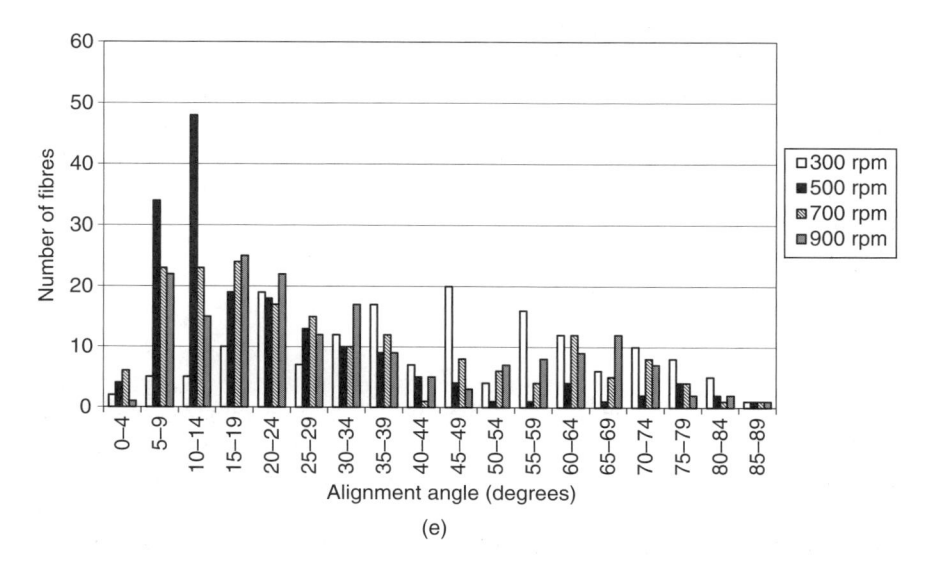

Biocompatible three-dimensional scaffolds 13

1.3 Electron micrographs depicting fibre orientation when collected on a rotating mandrel at speeds of (a) 300 rpm, (b) 500 rpm, (c) 700 rpm and (d) 900 rpm; (e) degree of alignment for these four speeds, small angles representing superior alignment.

14 Cellular response to biomaterials

1.8 Surface wettability

The hydrophilicity (or hydrophobicity) of a material's surface is vitally important for the adherence of cells. The wettability of a surface is determined by measuring the contact angle of a droplet of deionised water on the substrate surface. Both tissue culture plastic and glass coverslips are widely known for their significant wettability, making them suitable materials for the attachment of cells.

Post-electrospinning, the nanofibres of PCL are of an extremely hydrophobic nature, with a contact angle of approximately 85°. In order to make the surface ideal for cell attachment, pre-treatment of the fibres is necessary. Pre-treatments can include argon-plasma or the immersion of fibrous samples in cell culture media. The latter was the chosen method for these particular scaffolds. Immersion in cell culture media allows serum present proteins to adsorb onto the fibre surface; cells recognise these proteins via specific surface receptors, which enable them to adhere to the surface.

Nanofibrous scaffolds were vacuum dried at room temperature for 24 hours and then sterilised over two days through increasing concentrations of ethanol, followed by immersion in PBS for 24 hours and a further 24 hours in culture media prior to seeding of cells onto the structure's surface, as described by Mo et al. (2004).

As previously mentioned, PCL is known to be a biocompatible polymer and is approved for implantation by the FDA. The contact angle of water on a sheet of compression moulded (CM) PCL is approximately 55° (Fig. 1.4(b)) and is comparable to that for glass coverslips (Fig. 1.4(a)). However, when PCL is electrospun with acetone and the fibres are vacuum dried, the surface is highly hydrophobic (Fig. 1.4(c)). Pre-wetting the electrospun fibres with culture media, and further vacuum drying, demonstrates how the fibre surface can be altered as the contact angle is notably reduced (Fig. 1.4(d)). Submerging the scaffolds in culture media allows contained proteins to deposit onto the fibre surface, subsequently improving the surface's hydrophilicity. Measured contact angles are detailed in Table 1.1.

As stated previously, there is large scope for the use of electrospun nanofibres in the tissue engineering sector, as these structures provide an ideal

Table 1.1 Average water contact angles for samples investigated in Fig. 1.4

Material	Average contact angle [±S.D.]
Glass coverslip	49° [±1.4]
Compression moulded PCL	52° [±3.5]
Pre-treated electrospun PCL	86° [±5.3]
Post-treated electrospun PCL	33° [±3.5]

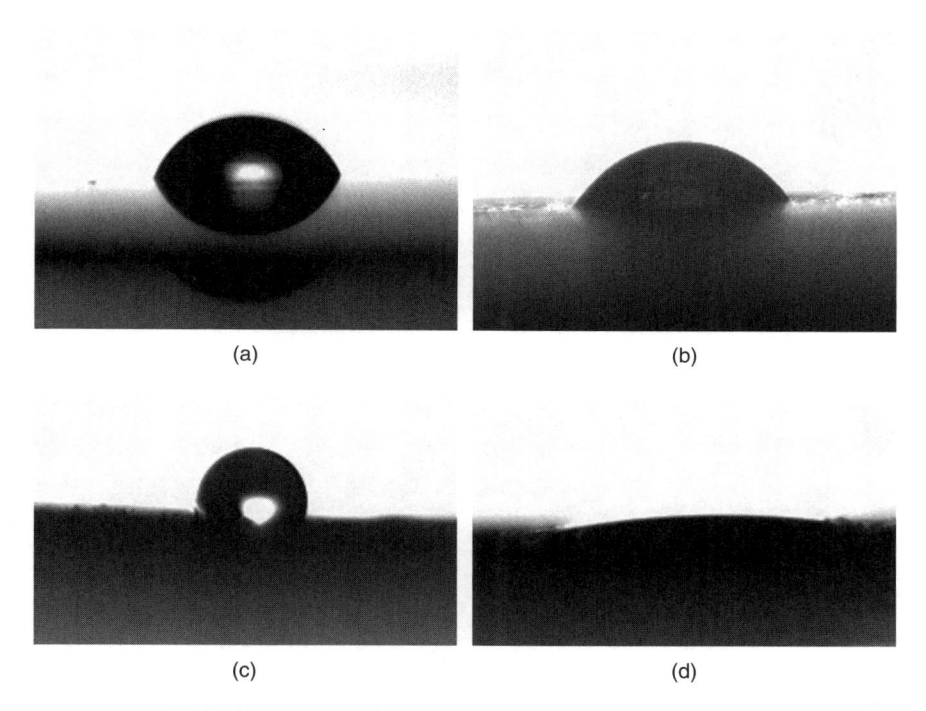

Biocompatible three-dimensional scaffolds 15

1.4 Digital images of deionised water droplets on the surface of (a) glass coverslip, (b) compression moulded PCL, (c) pre-treated electrospun PCL, (d) post-treated electrospun PCL.

platform for the adhesion and growth of cells by closely mimicking native ECM. Research has shown that a wide range of cell types are able to be successfully cultured on such fibrous structures. The micrographs in Fig. 1.5 demonstrate the adhesion and spreading of cells between the post-treated electrospun PCL fibres for fibroblasts, osteoblasts and tenocytes.

1.9 Functionalising fibres

Electrospun fibres can be functionalised by incorporating biomolecules onto the fibre surface to improve their adhesive qualities for cell attachment. Examples of such biomolecules include collagen, fibronectin, RGD peptides and bisphosphonates. These can be immobilised by coating the fibres (Kim and Park, 2006), by blending them with the polymer solution to be electrospun (Dalton *et al.*, 2006), or by dual electrospinning (Kidoaki *et al.*, 2005), whereby two or more solutions are simultaneously spun onto the same collector.

A study on the proliferation of smooth muscle cells on electrospun PCL by Venugopal *et al.* (2005) proved that cell adhesion and proliferation were

16   Cellular response to biomaterials

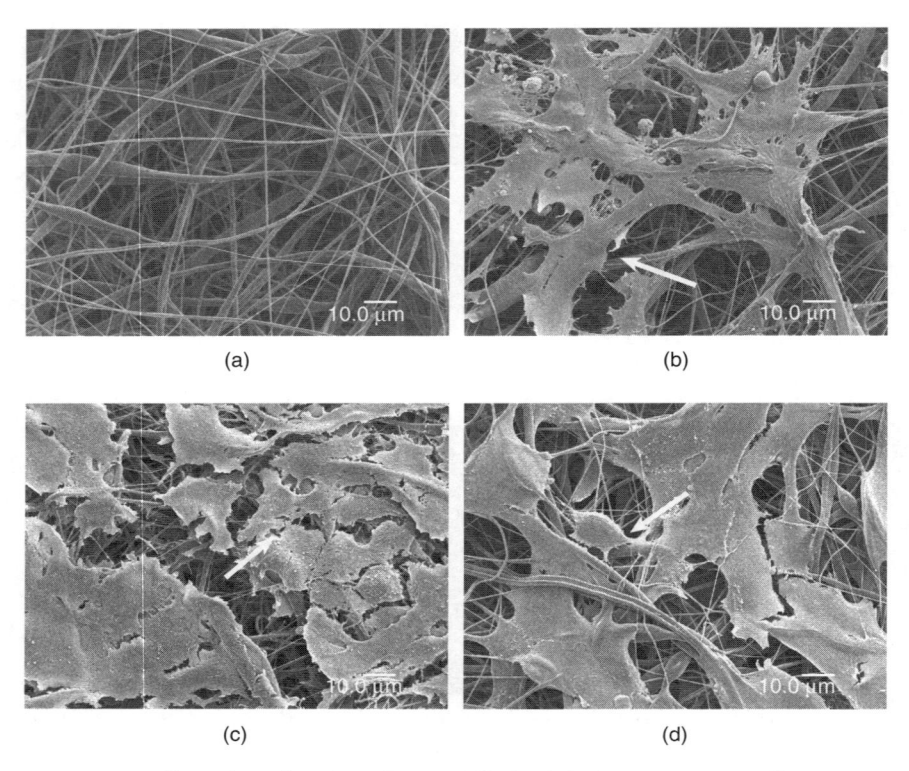

1.5 Scanning electron micrographs for (a) electrospun nanofibrous scaffold; and this scaffold with seeded (b) fibroblasts, (c) osteoblasts and (d) tenocytes. All cell micrographs are post 12-day culture. For clarity, arrows indicate individual cells.

increased when cells were seeded onto PCL fibres coated with collagen. Other benefits of collagen-coated fibres are to help maintain the cell's phenotype by preventing differentiation, and also to provide an initial temporary ECM, which may improve the chances of new tissue being formed (Venugopal *et al.*, 2005). One of the more important advantages of a PCL-collagen-coated scaffold is that while the scaffold is superior in terms of biocompatibility, it also maintains its mechanical integrity, both of which increase the chances of a new, sustainable, tissue being formed. The simplest way of incorporating collagen into the polymer fibre scaffold is to directly coat the fibres with collagen as shown in Fig. 1.6 (a and b). Coating PCL fibres with collagen greatly improved the hydrophilicity of the structure by reducing the contact angle by approximately 40° (Fig. 1.6 (c and d), Table 1.2).

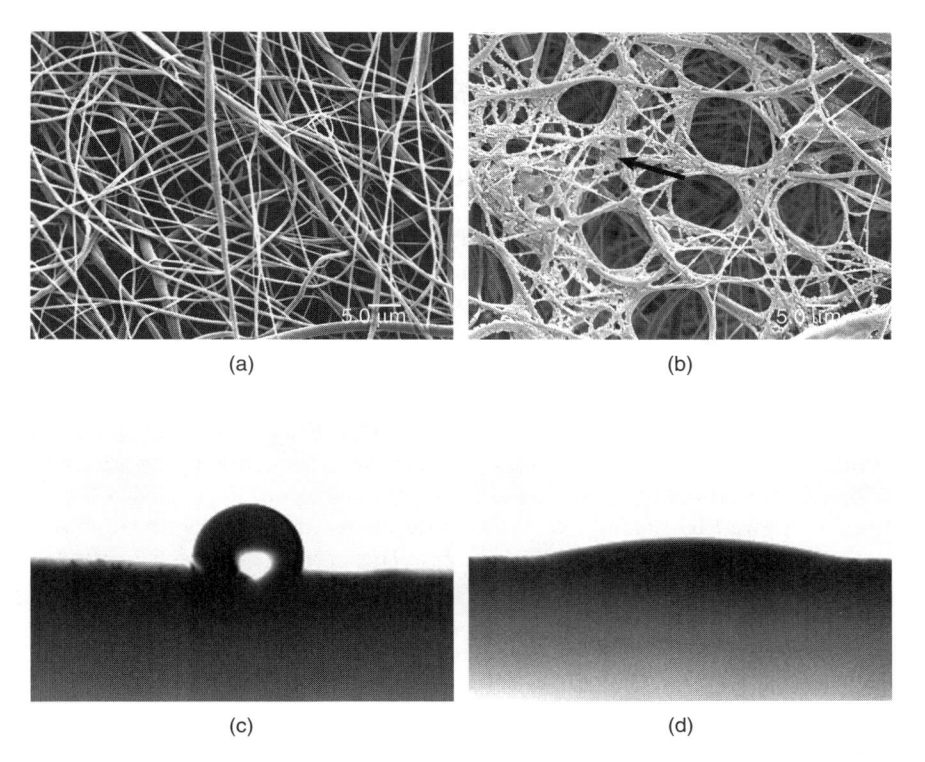

Biocompatible three-dimensional scaffolds 17

(a) (b)

(c) (d)

1.6 (a) A random network of electrospun PCL fibres; (b) the same network of electrospun PCL with sprayed collagen coating (the arrow demonstrates the roughened texture gained from this coating); (c and d) contact angles for the respective micrographs demonstrating the change in wettability with collagen surfaces.

Table 1.2 Average water contact angles for electrospun polycaprolactone (PCL) and PCL/collagen coated fibres as shown in Fig. 1.6

Material	Average contact angle [±S.D.]
Electrospun PCL	86° [±5.3]
Electrospun PCL with collagen	21° [±1.4]

1.10 Electrospinning for tendon regeneration

As mentioned, electrospinning has been incorporated in many tissue engineering studies, and research into tendon regeneration by using novel electrospun scaffolds is currently being investigated at the University of Manchester (Bosworth *et al.*, 2008).

Tendons are a type of connective tissue having great flexibility and elasticity, which enable forces generated by contracting muscles to cause movement of the attached bone (Viidik, 1996). Depending on their location and function within the body, tendons come in a variety of anatomical structures, with their fibrous ultrastructures exhibiting rounded cords, flattened ribbons, strap-like bands or fan shapes (Kannus, 2000). The versatility of the electrospinning process would enable fibres of different morphologies, such as rounded or flattened, and of various widths, to be produced.

Like other connective tissues, tendons are mainly composed of ECM (80% tissue volume) with few cells present (20% tissue volume). The main component of the ECM is collagen, the fibres of which primarily run along the length of the tendon axis; however, some of these fibres are oriented transversely to this (Sharma and Maffulli, 2005). The other important characteristic of tendon structure is its hierarchical arrangement – grouping of collagen molecules forms collagen fibrils, further grouping forms collagen fibres, followed by the subfascicle, fascicle layers and tertiary fibre bundle, before reaching the overall tendon unit. The cross-sectional size of each level varies between individuals and from tendon to tendon depending on its location and purpose. The average diameter for each hierarchical level is shown in Table 1.3. This hierarchical configuration allows the tendon to withstand high loads whilst maintaining its flexibility, and enables tendons to bend round bony prominences if necessary (Screen et al., 2004). The bundles are encompassed by a thin, loose, connective tissue layer, the endotenon, which lies within the tendon structure and binds the collagen fibre bundles together (Wang, 2006). The entire tendon is further covered by a thin sheet of connective tissue, the epitenon, which is contiguous with both the endotenon and the outer layer, the paratenon (Kannus, 2000). For sheath-free tendons, the paratenon, composed of loose areolar connective tissue, is the very outer layer of the structure (Sharma and Maffulli, 2005). The role of the paratenon is to reduce friction between the tendon and neighbouring tissue, ensuring unhindered movement. Certain tendons,

Table 1.3 Diameter range for hierarchical tendon structure

Hierarchical tendon structure	Diameter range
Collagen molecules	1.5 nm
Collagen fibrils	3.5 nm
Collagen fibres	10–20 nm
Primary fibre bundle (subfascicle)	50–500 nm
Secondary fibre bundle (fascicle)	10–50 μm
Tertiary fibre bundle	50–400 μm
Tendon unit	100–500 μm

Source: Screen et al. (2004).

such as those in the hands and feet, are further enclosed within a sheath because they undergo increased levels of mechanical stress. A thin film of fluid present between the paratenon and the sheath provides sufficient lubrication for ease of movement while the tendon is placed under high stress.

The great tensile strength that tendons provide can be attributed to the parallel alignment of collagen fibres, which are further strengthened by covalent crosslinks between them (Maffulli et al., 2000). While there are a number of different types of collagen, collagen type I is the main structural component present within all tendons and is responsible for tensile strength; however, collagen types II, III, IV and V are also present, though in far smaller quantities.

Following injury to the tissue, the tendon heals by formation of scar tissue, which is both structurally and mechanically inferior to healthy tendon tissue (Maffulli et al., 2000). Substandard tissue remains and causes ongoing morbidity. Owing to the tissue's poor healing qualities, there is sufficient cause to tailor a tissue engineered construct, which can temporarily fulfil the tendon function while the seeded cells secrete sufficient ECM to regenerate natural tendon tissue.

1.11 Electrospun fibre bundles

Fibre bundles can be fabricated by electrospinning a solution of PCL/acetone and collecting the fibres by one of the three methods shown in Fig. 1.1(b–d). The bundles themselves are composed of hundreds of PCL nanofibres, which mainly lie parallel to the longitudinal axis, providing the degree of alignment necessary to mimic the original tendon tissue (Fig. 1.7). Bundles created by the liquid reservoir technique yield the smallest diameters, while those removed from the thin mandrel produce the largest (Table 1.4). Fixed point bundles exhibit the greatest degree of variation in diameter (Table 1.4). In terms of bundle mechanical properties, the fixed point method resulted in the highest tensile strength (5.3 MPa) and the second highest modulus (16.9 MPa). Liquid reservoir mechanical properties demonstrated the greatest spread, highlighting the lack of reproducibility in this bundle collection type. The mechanical properties of thin mandrel bundles are the weakest, and this follows the work of Tan et al. (2005), whereby tensile testing of an electrospun, single PCL fibre was affected by its diameter. They reported that fibres of larger diameter resulted in a lower tensile strength.

Depending on their location and the expected demands to be withstood, the mechanical properties of the tendon are suitably adjusted. For example, the human Achilles' tendon when tested in vivo for five male subjects was found to yield a stress of 36.5 MPa and a Young's modulus of 788 MPa (Magnusson et al., 2003), while the human tibialis anterior tendon in vivo

20   Cellular response to biomaterials

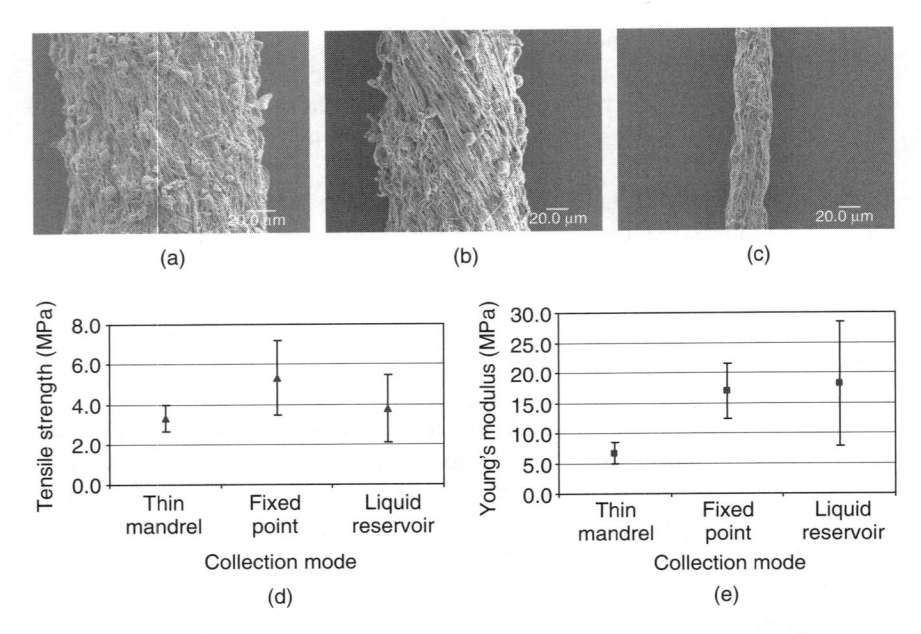

1.7 Predominantly aligned fibrous bundles fabricated from different collection techniques: (a) thin mandrel, (b) fixed point, and (c) liquid reservoir; (d) tensile strength and (e) Young's modulus of these fibre bundles.

Table 1.4 Average bundle diameter for collection modes

Bundle type	Average diameter (μm) [±S.D.]
Thin mandrel	189.83 [±15.85]
Fixed point	144.22 [±49.26]
Liquid reservoir	46.07 [±15.79]

has a Young's modulus of 1.2 GPa and a tensile strength of 25 MPa (Maganaris and Paul, 1999). However, the literature suggests that mechanical properties of tendons vary according to the method of testing employed, the nature of the test (either *in vivo* or *in vitro*), the age of the tissue, the tendon source (human or animal), sedentary or athletic patients, and also the sample gender. Whilst all methods are valid, no set value for each tendon can be given.

1.11.1 Advantages and disadvantages of fibre bundles

The different techniques used to fabricate fibrous bundles are each associated with strengths and weaknesses and are listed in Table 1.5.

Table 1.5 Comparison of advantages and disadvantages for each collection mode

Advantages	Disadvantages
Liquid reservoir • Very fine, long fibres can be drawn off the liquid surface. • Bundles with the smallest diameters are created using this method. • Bundles of this type have the highest Young's modulus.	• Technically difficult to draw fibres off the liquid surface. • Lack of uniformity in bundles drawn. • Fibres remain very fragile.
Fixed point • Bridged fibres can be easily removed, submerged in deionised water, and twisted together to form tight fibre bundles. • Bundles of this type measure the highest tensile strength.	• Limited distance for fibres to successfully bridge the gap. • Fibre deposition is not easy to control; resultant fibre bundles may contain a greater number of fibres compared to other bundles fabricated by this method.
Thin mandrel • Fibres can be easily removed from the mandrel edge, submerged in deionised water, and twisted together to form tight fibre bundles. • Bundles of long length are easily produced. • Bundles of this type are the most robust and easiest to manipulate.	• Bundles with largest diameter are fabricated. • Young's modulus and overall tensile strength are lowest for this collection type.

1.11.2 Cell attachment on fibre bundles

The main cell type present within tendons and which is responsible for the production of ECM is the tenocyte; these lie in columns among the collagen bundles. Based on this cellular arrangement, the degree of porosity present within the electrospun bundles is negligible as cells do not need to migrate into the scaffold structure, but merely align along the fibre direction. Studies on the seeding of tenocytes on electrospun fibre bundles demonstrated a degree of contact guidance on cell orientation. Tenocytes can be clearly seen to align and spread along the longitudinal fibre bundle axis (Fig. 1.8).

22   Cellular response to biomaterials

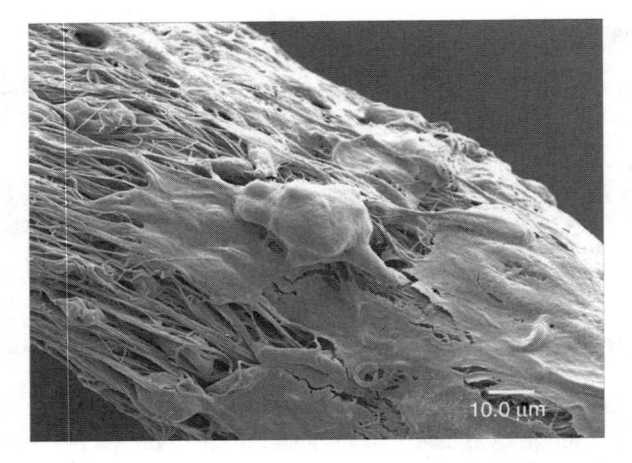

1.8 Tenocytes aligning along the fibre bundle axis.

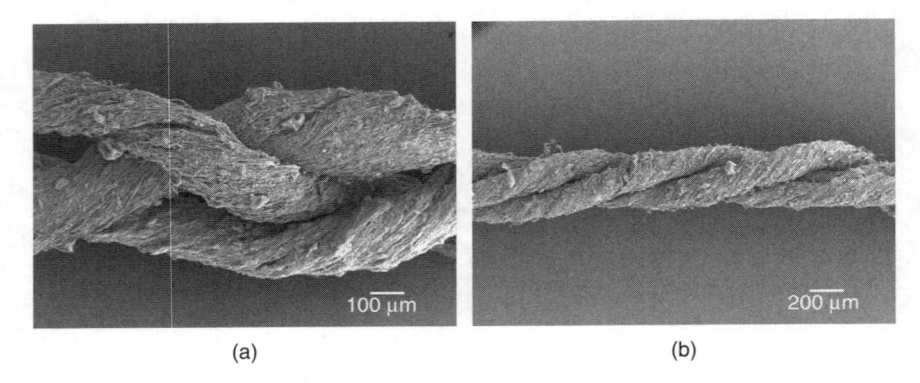

(a)   (b)

1.9 Rope-like structures of electrospun PCL bundles: (a) plaited fibre, (b) twisted fibre.

1.11.3 Rope-like structures

The scaffold proposed to act as a replacement structure for tendons is a number of these fibrous bundles grouped together in one of two ways. A plaited structure is one of these methods, and is formed by interweaving three fibrous bundles with each other (Fig. 1.9(a)). The other method creates a twisted fibre structure, whereby three fibrous bundles are twisted tightly together (Fig. 1.9(b)). The morphology of tendons is reminiscent of rope, as some rope structures display clear fibre bundling in their cross-sections. With this in mind, it is intended that further plaiting and twisting of electrospun bundles will be performed until overall dimensions are nearly identical to those of the tendon being replaced.

1.12 Future trends

There is an increasing interest in tissue engineering in clinically important areas, such as tendon, muscle and nerve repair. Tendons connect muscles to bones, thus allowing the transmission of forces and hence joint movement and making them essential in musculoskeletal biology and repair. Healthy tendons are fibroblastic and have a variety of anatomical structures; the fibres that form their ultrastructure can be rounded cords, strap-like bands or flattened ribbons. Tendons have a hierarchical structure; the future trend may be the formation of 3-D scaffolds using nanofibres or 'bundles' of fibres, thus mimicking the natural tissue. To date, it has not been possible to match simultaneous polymer degradation with ECM formation, nor has it been possible to produce scaffolds that maintain sufficient mechanical strength during the early healing phase and to then allow the newly formed tissue to take up the biomechanical load. Specifically, the scaffold constructs rarely mimic the morphology of the natural tissue at the ultrastructural level. This is a pivotal factor in the repair of structural tissues that have directional orientation and hierarchical structures such as tendons. An electrospinning approach may offer the ability for biomaterial engineers to produce scaffolds for tendon repair with controllable structures at both the nano level and the macro level, providing structural support and a delivery system for therapeutic cells.

1.13 Sources of further information and advice

An Introduction to Electrospinning and Nanofibers, Ramakrishna S, Fujihara K, Teo W-E, Lim T-C, Ma Z. World Scientific Publishing Co. Ltd, 2005. ISBN-10: 981-256-415-2

Polymeric Nanofibers, edited by Reneker D H, Fong H. American Chemical Society, Washington, DC, 2006. ISBN-10: 0-8412-3919-3

Medical Textiles and Biomaterials for Healthcare, edited by Anand S C, Kennedy J F, Miraftab M, Rajendran S. Woodhead Publishing Ltd, Cambridge, UK, 2006. ISBN-13: 978-1-85573-683-2

Human Tendons: Anatomy, Physiology and Pathology, Józsa L, Kannus P. Human Kinetics, Publishers, Champaign, IL, 1997. ISBN: 0-87322-484-1

1.14 References

Agrawal C M, Ray R B (2001), 'Biodegradable polymeric scaffolds for musculoskeletal tissue engineering', *Journal of Biomaterial Research*, 55 (2), 141–150.

Bhattarai S R, Bhattarai N, Yi H K, Hwang P H, Cha D I, Kim H Y (2004), 'Novel biodegradable electrospun membrane: scaffold for tissue engineering', *Biomaterials*, 25, 2595–2602.

Bini T B, Gao S, Wang S, Ramakrishna S (2006), 'Poly(l-lactide-co-glycolide) biodegradable microfibers and electrospun nanofibers for nerve tissue engineering: an in vitro study', *Journal of Materials Science*, 41, 6453–6459.

Bognitzki M, Czado W, Frese T, Schaper A, Hellwig M, Steinhart M, Greiner A, Wendorff J H (2001), 'Nanostructured fibres via electrospinning', *Advanced Materials*, 13 (1), 70–72.

Boland E D, Telemeco T A, Simpson D G, Wnek G E, Bowlin G L (2004), 'Utilising acid pretreatment and electrospinning to improve biocompatibility of poly(glycolic acid) for tissue engineering', *Journal of Biomedical Materials Research Part B: Applied Biomaterials*, 71B, 144–152.

Bosworth L, Clegg P, Downes S (2008), 'Electrospun nanofibres of polycaprolactone, and their use for tendon regeneration', *International Journal of Nano and Biomaterials*, 1(3), 263–279.

Casper C L, Stephens J S, Tassi N G, Chase D B, Rabolt J F (2004), 'Controlling surface morphology of electrospun polystyrene fibres: effect of humidity and molecular weight in the electrospinning process', *Macromolecules*, 37, 573–578.

Chen G-Q, Wu Q (2005), 'The application of polyhydroxyalkanoates as tissue engineering materials', *Biomaterials*, 26, 6565–6578.

Chronakis I S (2005), 'Novel nanocomposites and nanoceramics based on polymer nanofibres using electrospinning process – a review', *Journal of Materials Processing Technology*, 167, 283–293.

Clark P, Connolly P, Curtis A S G, Dow J A T, Wilkinson C D W (1987), 'Topographical control of cell behaviour – 1. Simple step cues', *Development*, 99, 439–448.

Dalton P D, Klinkhammer K, Salber J, Klee D, Möller M (2006), 'Direct in vitro electrospinning with polymer melts', *Biomacromolecules*, 7, 686–690.

Das S, Hollister S J, Flanagan C, Adewunmi A, Bark K, Chen C, Ramaswamy K, Rose D, Widjaja E (2003), 'Freeform fabrication of nylon-6 tissue engineering scaffolds', *Rapid Prototyping Journal*, 9 (1), 43–49.

Doshi J, Reneker D H (1993), 'Electrospinning process and applications of electrospun fibres', *Industry Applications Society Annual Meeting*, 3, 1698–1703.

Doyle C, Tanner E T, Bonfield W (1991), 'In vitro and in vivo evaluation of polyhydroxybutyrate and of polyhydroxybutyrate reinforced with hydroxyapatite', *Biomaterials*, 12, 841–847.

Dzenis Y (2004), 'Spinning continuous fibres for nanotechnology', *Science*, 304, 1917–1919.

Fong H, Chun I, Reneker D (1999), 'Beaded nanofibres formed during electrospinning', *Polymer*, 40, 4585–4592.

Formhals A (1934), US Patent 1, 975, 504.

Hasirci V, Vrana E, Zorlutuna P, Ndreu A, Yilgor P, Basmanav F B, Aydin E (2006), 'Nanobiomaterials: a review of the existing science and technology, and new approaches', *Journal of Biomaterials Science – Polymer Edition*, 17 (11), 1241–1268.

Huang Z M, Zhang Y Z, Kotaki M, Ramakrishna S (2003), 'A review on polymer nanofibers by electrospinning and their applications in nanocomposites', *Composites Science and Technology*, 63, 2223–2253.

Hutmacher D (2000), 'Scaffolds in tissue engineering bone and cartilage', *Biomaterials*, 21, 2529–2543.

Kannus P (2000), 'Structure of the tendon connective tissue', *Scandinavian Journal of Medicine and Science in Sports*, 10, 312–320.

Khil M-S, Cha D-I, Kim H-Y, Kim I-S, Bhattarai N (2003), 'Electrospun nanofibrous polyurethane as wound dressing', *Journal of Biomedical Materials Research Part B: Applied Biomaterials*, 67B (2), 675–679.

Kidoaki S, Kuen Kwon I, Matsuda T (2005), 'Mesoscopic spatial designs of nano- and microfibre meshes for tissue engineering matrix and scaffold based on newly devised multilayering and mixing electrospinning techniques', *Biomaterials*, 26, 37–46.

Kim C H, Khil M S, Kim H Y, Lee H U, Jahng K Y (2005), 'An improved hydrophilicity via electrospinning for enhanced cell attachment and proliferation', *Journal of Biomedical Materials Research Part B: Applied Biomaterials*, 78 (2), 283–290.

Kim T G, Park T G (2006), 'Biomimicking extracellular matrix: cell adhesive RGD peptide modified electrospun poly(D,L-lactic-co-glycolic acid) nanofiber mesh', *Tissue Engineering*, 12 (2), 221–233.

Kweon H, Yoo M K, Park I K, Kim T H, Lee H C, Lee H S, Oh J S, Akaike T, Cho C S (2003), 'A novel degradable polycaprolactone networks for tissue engineering', *Biomaterials*, 24, 801–808.

Lee C, Shin H, Cho I, Kang Y M, Kim I, Park K D, Shin J W (2005), 'Nanofibre alignment and direction of mechanical strain affect the ECM production of human ACL fibroblast', *Biomaterials*, 26, 1261–1270.

Li W-J, Danielson K G, Alexander P G, Tuan R S (2003), 'Biological response of chondrocytes cultured in three-dimensional nanofibrous poly(ε-caprolactone) scaffolds', *Journal of Biomedical Materials Research*, 67A (4), 1105–1114.

Ma P X, Choi J W (2001), 'Biodegradable polymer scaffolds with well defined interconnected spherical pore network', *Tissue Engineering*, 7 (1), 23–33.

Maffulli N, Ewen S, Waterston S, Reaper J, Barrass V (2000), 'Tenocytes from ruptured and tendinopathic Achilles tendons produce greater quantities of type III collagen than tenocytes from normal Achilles tendons: an in vitro model of human tendon healing', *American Journal of Sports Medicine*, 28 (4), 499–505.

Maganaris C N, Paul J P (1999), '*In vivo* human tendon mechanical properties', *Journal of Physiology*, 521.1, 307–313.

Magnusson S P, Hansen P, Aagaard P, Brønd J, Dyhre-Poulsen P, Bojsen-Moller J, Kjaer M (2003), 'Differential strain patterns of the human gastrocnemius aponeurosis and free tendon, *in vivo*', *Acta Physiologica Scandinavica*, 177 (2), 185–195.

Matthews J A, Wnek G E, Simpson D G, Bowling G L (2002), 'Electrospinning of collagen nanofibres', *Biomacromolecules*, 3, 232–238.

Mit-uppatham C, Nithitanakul M, Supaphol P (2004), 'Ultrafine electrospun polyamide-6 fibres: effect of solution conditions on morphology and average fibre diameter', *Macromolecular Chemistry and Physics*, 205, 2327–2338.

Mo X M, Xu C Y, Kotaki M, Ramakrishna S (2004), 'Electrospun P(LLA-CL) nanofiber: a biomimetic extracellular matrix for smooth muscle cell and endothelial cell proliferation', *Biomaterials*, 25, 1883–1890.

Nam Y S, Park T G (1999), 'Porous biodegradable polymeric scaffolds prepared by thermally induced phase separation', *Journal of Biomedical Materials Research*, 47 (1), 8–17.

Pham Q P, Sharma U, Mikos A G (2006), 'Electrospinning of polymeric nanofibres for tissue engineering applications: a review', *Tissue Engineering*, 12 (5), 1197–1211.

Pompe T, Keller K, Mothes G, Nitschke M, Teese M, Zimmerman R, Werner C (2007), 'Surface modification of poly(hydroxybutyrate) films to control cell–matrix adhesion', *Biomaterials*, 28, 28–37.

Ramakrishna S, Fujihara K, Teo W-E, Lim T-C, Ma Z (2005), *An Introduction to Electrospinning and Nanofibers*, World Scientific Publishing Co. Ltd.

Reneker D H, Yarin A L, Fong H, Koombhongse S (2000), 'Bending instability of electrically charged liquid jets of polymer solutions in electrospinning', *Journal of Applied Physics*, 87 (9), 4531–4547.

Sachlos E, Reis N, Ainsley C, Derby B, Czernuska, J (2003), 'Novel collagen scaffolds with predefined internal morphology made by solid freeform fabrication', *Biomaterials*, 24, 1487–1497.

Screen H R C, Lee D A, Bader D L, Shelton J C (2004), 'An investigation into the effects of the hierarchical structure of tendon fascicles on micromechanical properties', *Proceedings of the Institution of Mechanical Engineers: Journal of Engineering in Medicine*, 218 part H.

Sharma P, Maffulli N (2005), 'Basic biology of tendon injury and healing – clinical review', *The Surgeon*, 3 (5).

Shawon J, Sung C (2004), 'Electrospinning of polycarbonate nanofibres with solvent mixtures THF and DMF', *Journal of Materials Science*, 39, 4605–4613.

Shin M, Ishii O, Sueda T, Vacanti J P (2003), 'Contractile cardiac grafts using a novel nanofibrous mesh', *Biomaterials*, 25, 3717–3723.

Smit E, Buttner U, Sanderson R D (2005), 'Continuous yarns from electrospun fibres', *Polymer*, 46, 2419–2423.

Tan E, Ng S, Lim C (2005), 'Tensile testing of a single ultrafine polymeric fibre', *Biomaterials*, 26, 1453–1456.

Teixeira A I, Abrams G A, Bertics P J, Murphy C J, Nealey P F (2003), 'Epithelial contact guidance on well defined micro- and nanostructured substrates', *Journal of Cell Science*, 116, 1881–1892.

Venugopal J, Ma L L, Yong T, Ramakrishna S (2005), 'In vitro study of smooth muscle cells on polycaprolactone and collagen nanofibrous matrices', *Cell Biology International*, 29, 861–867.

Viidik A (1996), *Extracellular Matrix, Vol 1, Tissue Function*, Harwood Academic Publishers, Amsterdam.

Wang J H-C (2006), 'Mechanobiology of tendon', *Journal of Biomechanics*, 39, 1563–1582.

Wannatong L, Sirivat A, Supahol P (2004), 'Effects of solvents on electrospun polymeric fibres: preliminary study on polystyrene', *Polymer International*, 53, 1851–1859.

Whitesides G M, Grzybowski B (2002), 'Self-assembly at all scales', *Science*, 295, 2418–2421.

Wnek G E, Carr M E, Simpson D G, Bowlin G L (2003), 'Electrospinning of nanofibre fibrinogen structures', *Nano Letters*, 3 (2), 213–216.

Wong J Y, Leach J B, Brown X Q (2004), 'Balance of chemistry, topography, and mechanics at the cell–biomaterial interface: issues and challenges for assessing the role of substrate mechanics on cell response', *Surface Science*, 570, 119–133.

Xu C X, Inai R, Kotaki M, Ramakrishna S (2004), 'Aligned biodegradable nanofibrous structure: a potential scaffold for blood vessel engineering', *Biomaterials*, 25 (5), 877–886.

Yang F, Murugan R, Wang S, Ramakrishna S (2005), 'Electrospinning of nano/micro scale poly(L-lactic acid) aligned fibers and their potential in neural tissue engineering', *Biomaterials*, 26, 2603–2610.

Zeng J, Xu X, Chan X, Liany Q, Bian X, Yang L, Jing X (2003), 'Biodegradable electrospun fibres for drug delivery', *Journal of Controlled Release*, 92, 227–231.

2
Degradable polymers and polymer composites for tissue engineering

S DEB, King's College London, UK

Abstract: In the last two decades significant advances have been made in the development of biodegradable polymers for biomedical applications. Polymeric biomaterials that are degradable in nature are the preferred candidates, especially for tissue engineering applications, drug delivery and temporary therapeutic devices. In the field of biomaterials, the aim is to develop and characterise artificial materials for use in the human body to restore, repair and replace diseased tissue to enhance survival and quality of life. Specific applications demand the materials to provide the physical, biological and biomechanical properties required for efficient function. Biodegradable polymers can be either natural or synthetic. In general synthetic polymers can be tailored to confer a required range of properties necessary for a certain application with greater degree of control. Natural polymers are often associated with problems related to immunogenicity that may originate from the source. Degradable polymers form an important class of materials in medicine due to their ability to provide temporary support for regeneration of tissues, eliminating the need for a second surgery for removal of implants. Furthermore, they have the advantage that degradation rates can be controlled to regulate the healing process.

An overview of degradable polymers and some polymer composites in biomedical applications is provided in this chapter, including important aspects of the different polymers used, the mechanisms of degradation, and their applications in areas of tissue engineering and drug delivery. However, it will be prudent to mention that the degradable polymers and composites discussed in this chapter are by no means exhaustive.

Key words: biomaterials, degradable polymers, polymer composites, drug delivery, tissue engineering.

2.1 Introduction

Polymers consist of molecules that have high molar masses and are composed of a large number of repeating units. They occur in nature and can be found in living species such as proteins, collagen, DNA, etc. Synthetic polymers comprise a large group of materials that have become an everyday commodity in life around us, ranging from the common carrier bags to heavy plastics to medical devices. A number of natural and synthetic polymers are biodegradable in nature. Polymers that can be made to degrade via either biodegradation or enzymatic processes are rapidly gaining importance and increasingly replacing non-degradable polymers for biomedical applications. A device that is able to provide a temporary support during the healing process is desirable for clinicians [Hollinger and Battistone 1986] and, if its subsequent removal is not required, it can be of great advantage to the patient and clinician and to healthcare costs. Imagine an implant such as a bone screw or pin that is inserted to stabilise a fracture, which performs its function during the stabilisation process and eventually degrades within the body and is eliminated by natural pathways.

Polymers that can provide a temporary function are nevertheless required to fulfil a number of stringent criteria, prior to their use in the human body. Parameters such as their biocompatibility, resorbability, adequate mechanical support during the healing process, non-toxic response to degradation products, and ability to be eliminated by natural pathways in the body, are features of importance. Easy processing, sterilisability and adequate shelf life, are other important considerations that need be addressed.

Polymers for biomedical applications are designed to be either biostable or bioabsorbable. Biologically stable polymers provide a permanent support over time and ideally should perform during the lifetime of the patient, examples being poly(methylmethacrylate), poly(ethylene), etc., that are used as bone cements and the articulating surfaces of hip and knee joints respectively. Bioabsorbable polymers are designed to provide temporary support and should be absorbed in the body over a period of time, some examples being poly(lactide), poly(glycolide), poly(dioxanone) or the copolymers thereof that are used as degradable sutures, pins and staples.

Synthetic biomaterials intended for biomedical applications are being developed to restore or repair function of diseased or traumatised tissue in the human body and thus improve quality of life. Significant advances have been made in the last four decades in the development of biodegradable polymers for biomedical applications. More recently, polymer composites are being increasingly examined for biomedical applications, especially to function as scaffolds in the field of tissue engineering. There are obvious advantages associated with bioabsorbable devices for use in medical applications, in particular polymer composites comprising degradable polymers that can support three-dimensional tissue formation.

2.2   Degradable polymers

2.2.1   Synthetic polymers

Biodegradable polymers can be either natural or synthetic in origin. In general, synthetic polymers offer greater advantages than natural materials

Table 2.1 Examples of natural and synthetic biodegradable polymers

Natural polymers	Synthetic polymers
Albumin	Poly lactic acid:
Alginates	• Poly-L-lactic acid
Collagen (proteins)	• Poly-D,L-lactic acid
Chitin, chitosan (polysaccharides)	Poly glycolic acid
Fibrin	Poly-ε-caprolactone
	Poly-p-dioxanone
	Tri-methylene carbonate
	Poly anhydrides
	Poly ortho ester
	Poly urethanes
	Poly amino acids
	Poly hydroxy alanoates
	Poly phosphazenes

because they can be tailored to impart a wider range of properties and have a more predictable batch-to-batch uniformity in comparison to natural materials. Synthetic polymers also provide a more reliant source of raw materials, free from the concerns of immunogenicity. Examples of synthetic and naturally occurring polymers that have relevance in the field of biomaterials are presented in Table 2.1. The essential component in a degradable polymer is the presence of a hetero-atom within the backbone of the polymer. In general a polymer with a —C—C— backbone is stable and does not tend to degrade; however, the presence of anhydrides, esters, amide, etc., can confer biodegradable properties to the polymer. Biodegradability can therefore be engineered into polymers by the judicious introduction of chemical linkages such as anhydride, esters or amide bonds.

According to the American Society for Testing and Materials (ASTM) and the International Standards Organisation (ISO), degradable polymers are defined 'as those that undergo significant change in their chemical structure under specific environmental conditions. These changes result in a loss of physical and mechanical properties as measured by standard methods'. There are several types of degradable polymers; however, only the polymers that have been used for biomedical applications will be discussed in this chapter. Degradation of a polymeric device within the body can occur as a result of hydrolysis or enzymatic degradation. There are numerous factors that can affect the degradation of bioresorbable polymers and can be related to the bulk properties and the surface properties of the polymer itself; however, as the manufacturing conditions involved in producing the ultimate implant material are also affected by a number of parameters, it is important also to consider the second group of variables such as the effects of processing, sterilisation, size and shape of the implant that contribute to degradation.

Poly(α-hydroxyacids)

Polymers derived from α-hydroxy acids, namely poly(lactic acid) and poly(glycolic acid), have found the most extensive use, primarily as materials for sutures dating back to the early 1960s due to their superior biocompatibility and acceptable degradation profiles. These polymers remain popular for a variety of reasons, including the fact that both of these materials have properties that allow hydrolytic degradation. The degradation products of these polymers are eliminated by natural pathways: poly(glycolic acid) is ultimately converted to glycine that can also be eliminated by the tricarboxylic acid (TCA) cycle, being metabolised as water and carbon dioxide, and lactic acid that can be eliminated through the TCA cycle. The group of esters derived α-hydroxy acids yield polymers such as poly(glycolic acid) (Fig. 2.1a) and poly(lactic acid) (Fig. 2.1b), which are hydrolytically unstable. Degradable polyesters derived from monomers such as lactide, glycolide and caprolactone (Fig. 2.1c) are commonly used for clinical application. The degradation rates vary for each of the polymers; however, varying parameters such as molecular weight, crystallinity, copolymer, fabrication techniques, composite formation, etc., can vary the kinetics of degradation.

Poly(lactic acid), PLA, is one of the most significant members of the degradable polymer family, which can be produced from abundant, naturally occurring sources such as corns, sugars or beetroot. Poly(glycolic acid), PGA, and poly(lactic acid), PLA, belong to the family of polyesters and are obtained via ring opening polymerisation. They are semi-crystalline in nature and their molecular weights can be varied by the method of synthesis. PLA or poly(lactic) acid is derived from lactic acid, which is a chiral molecule and can therefore exist as D and L isomers, the L-isomer being the biological metabolite. Hence PLA is often referred to as PLLA/L-PLA (L-isomer), D-PLA (D-isomer) and PDLA/DL-PLA (mixture of D and L isomers), depending on the isomeric forms used. The additional methyl group renders the polymer more hydrophobic than PGA and also exhibits slower degradation

2.1 Structure of linear aliphatic polyesters: (a) PGA, (b) PLA, (c) PCL.

rates. PLLA is a thermoplastic material with a melting point T_m of about 180°C and a glass transition temperature T_g higher than room temperature [Vert et al. 1984]. Thus, PLLA can be processed into fibres, films and blocks by techniques that are applicable for poly(ethylene terephthalate) (PET) and various products with superior mechanical properties.

Poly(ε-caprolactone), PCL, is derived by the ring opening polymerisation of ε-caprolactone (Fig. 2.1). It is a biodegradable polyester with a low melting point (~60°C) and a glass transition temperature of –60°C. PCL is a hydrophobic, semi-crystalline polymer and has a relatively slow rate of degradation. PCL undergoes hydrolytic degradation and is assisted by enzymatic degradative processes.

The bulk and surface properties of these polymers are largely affected by the properties shown in Table 2.2 that also govern the degradation rate and overall degradation of the material. The factors that may affect the degradation of the device or implant that is fabricated from a degradable polymer also depend on the parameters shown in Table 2.2.

Polymers can be linear, branched, or cross-linked with other chains. The polymer chains can be either amorphous or crystalline and can typically be made up of both amorphous and crystalline regions. This 'semi-crystalline' structure affects the strength and absorption of these implants. The more crystalline the polymer, the more ordered the microstructure, resulting in less slippage between the neighbouring polymer chains, thus leading to a stronger construct with superior mechanical properties. This slippage of the chains is time dependent under load, hence they are viscoelastic in nature. Properties such as crystallinity and morphology are strongly dependent on the thermal processing and sterilisation techniques, which have a bearing on the rate of degradation, an important parameter in the design of bioabsorbable devices. In general, higher molecular weight polymers possess superior mechanical properties and slower degradation rates in comparison with their lower counterparts.

Table 2.2 Parameters that can affect the properties of the implant

Bulk and surface properties of the polymer	The fabricated implant from the polymer
Chemical structure and composition Molecular weight and molecular weight distribution Configuration and conformation Morphology Physical properties Presence of ionic species Surface charge, roughness	Size, shape and topography of the implant Processing conditions Sterilisation methods used Storage conditions Site of implantation

Morphology and crystallinity also play a major role in the degradation of the semi-crystalline polymers, the amorphous regions being more prone to degradation than the crystalline domains. The processing parameters also strongly influence the mechanical properties, and increasing crystallinity through processing, annealing and sterilisation is reflected in the mechanical properties such as tensile strength and Young's modulus as shown in Table 2.3 [Callister 1994]. PGA and PLA undergo hydrolytic degradation and are totally degradable with water and carbon dioxide, their ultimate degradation products. The degradation proceeds via the hydrolysis of the ester bonds and the degradation kinetics are significantly affected by the molecular weight, morphology and crystallinity.

Table 2.4 lists some properties of common bioabsorbable materials. Poly(lactides), poly(glycolides), poly(caprolactones) and other degradable polymers have been used clinically as absorbable sutures with a high degree of success. High molecular weight polymers of PLLA have also been used in the internal fixation of bone in the form of screws, pins and plates. In general, mass loss has been reported to be slow with no apparent acute or chronic inflammatory response; however, the degradation products being acidic, any build-up of such products is likely to cause an inflammation. PLLA screws have been typically reported to behave as an inert material

Table 2.3 Mechanical properties of some selected degradable polymers

Polymer	Tensile modulus (GPa)	Ultimate tensile strength (MPa)	Flexural modulus (GPa)	Strain to break (%)
PLLA	3–4	50–70	4–5	4
PGA	6–7	60–100	–	1–20
Melt spun PLLA fibres	6.5–9.3	390–1800		
Solution spun PLLA	9.6–16	560–2300		
PDLLA films/discs	1.9–2.4	28–50 (film)		

Table 2.4 Properties of common bioabsorbable materials

Polymer	Melting point (°C)	Glass transition temperature (°C)	Modulus (GPa)	Loss of strength (months)	Loss of mass (months)
PGA	225–230	35–40	12.8	1–2	6–12
PLLA	173–178	60–65	4.8	6	–
PDLA	Amorphous	55–60	1.9	1–2	12–16
PDS	NA	–10 to 0	1.5	1–2	6–12

Poly(dioxanone)

Poly(dioxanone) belongs to the poly(ether-ester) family, which is commonly used as biodegradable suture material. Polydioxanones can be unsymmetrically substituted yielding poly(1,4-dioxanone-2,5-diones), poly(1,3-dioxane-2-one) and poly(1,4-dioxane-2,3-dione). It is a crystalline polymer that is obtained via the ring opening polymerisation of *p*-dioxanone in the presence of heat and organometallic catalyst, as shown in Fig. 2.2 [Doddi *et al.* 1976]. Polydioxanone has a low glass transition temperature ranging between 10°C and 0°C; the ether linkage within the polymeric backbone imparts flexibility to the polymer. The polymer has approximately 37–55% crystallinity and low-temperature processing is necessary to prevent depolymerisation to the monomer. Polydioxanone suture was first marketed by Ethicon and has been shown to be non-toxic. Monofilaments of PDS are absorbed within six months and mechanical strength decreases by 50% within three weeks, hence it is advantageous for applications where rapid degradation is required [Sabino *et al.* 2000].

Degradation of PDS proceeds via hydrolysis, and changes in elastic modulus, pH and molecular weight monitored over time suggest that hydrolysis occurs at the ester bonds. The crystallinity is found to increase initially, suggesting that the amorphous regions undergo more rapid degradation, followed by the crystalline regions. Elastic modulus decreases as does the molecular weight with time. It is important to note that the geometry of the polymer, thickness and shape govern the rate of degradation along with molecular weight and crystallinity, hence comparison between different studies is difficult. Copolymers with *p*-dioxanone, lactides and glycolides yield a family of polymers which also find use in drug delivery systems. PDS® has been investigated for arterial regeneration in rabbits, for closure of abdominal wounds, for fixation of facial fractures and for orbital floor [Lizuka *et al.* 1991].

2.2 Synthesis of poly(dioxanone).

Polyhydroxyalkanoates

Polyhydroxyalkanoates (PHAs) are aliphatic polyesters obtained from microorganisms, of which polyhydroxybutyrate (PHB) is the most abundant. They are composed of 3-hydroxyacid monomer units and exist as a small number of cytoplasmic granules per cell. Copolymers of 3-hydroxybutyrate (PHB) and 3-hydroxyvalerate (PHV) monomer units have modified physical properties and can be processed, thus the polymers and copolymers thereof have been widely investigated as biomaterials. PHB is a linear homopolymer biosynthesised by various strains of bacteria by condensation of D(-)-β-hydroxybutyric acid and used as an energy and carbon source [Anderson and Dawes 1990]. PHB seems to be biocompatible and biodegrades readily to carbon dioxide in bacteria; however, in humans, the reports are few and contradictory [Williams 2005]. The polymer is non-toxic and its monomer seems to be well tolerated in relatively high concentrations. Blends of PHB and PHV have been used in the form of films and scaffolds for biomedical applications. The composition of the blends can be manipulated to yield cytocompatible films with greater ductility [Zhao *et al.* 2003].

Poly(ethylene oxide), poly(amino acids) and poly (alkyl 2-cyanoacrylates)

Polyethylene oxide, PEO, has the repeat structural unit —CH_2CH_2O— and finds application in drug delivery. The material known as poly(ethylene glycol), PEG, possesses a similar repeating unit; however, in addition it has hydroxyl groups at each end of the molecule. Typically, high molecular weight polyethylene oxide polymers are used as biomaterials, whereas PEG is frequently used with a low degree of polymerisation with *n* ranging from 12 to 400, although high molecular weight polymers are commercially available. PEO and PEG are highly biocompatible, which makes them attractive as biomaterials. Polyactive is a block copolymer composed of a polyethylene oxide (PEO) soft segment and a polybutylene terephthalate (PBT) hard segment that has been reported to show bone bonding properties.

An *in vitro* evaluation carried out by immersing the polymer in a metastable calcium phosphate solution, similar to that of physiological fluids, has shown the formation of a hydroxycarbonate apatite layer with calcification over time [Li *et al.* 1997]. The carboxylic acid groups produced during hydrolysis of Polyactive are thought to play an important role in nucleating hydroxyapatite. The mechanical properties of Polyactive can be modulated and an increase in the PBT content in the copolymer is reported to enhance the strength and modulus of the polymers [Sakkers *et al.* 1998]. Thus, Polyactive has been widely studied as a bone bonding biomaterial. The micro patterning of PEG-PLLA copolymers has recently been reported to promote the proliferation of human microvascular endothelial cells due to the topography of the grooves [Gao *et al.* 2008]; furthermore, PEO/PBT three-dimensional fibres have been examined for articular cartilage tissue engineering [Moroni *et al.* 2007].

Poly(amino acids) widely occur in nature and synthetic analogues have been investigated for biomedical applications. Poly(α-amino acids) are synthesised from aspartic acid and glutamic acid or can be fermented. Pure poly(α-amino acids) are highly crystalline, which results in slow rates of degradation, and the antigenicity of these polymers can make *in vivo* use more challenging. Poly(esteramides) are the most commonly investigated of this group of polymers.

Poly(alkyl 2-cyanoacrylate) is synthesised from cyanoacrylates and has excellent adhesive properties as a result of the strong bonds that can form with polar substrates including living tissue and skin [Piskin 1994; King and Kinney 1999; Oowaki *et al.* 2000]. Such substances are widely used as surgical adhesives [Coover *et al.* 1990]. Polycyanoacrylates are synthesised from cyanoacrylates as shown in Fig. 2.3.

Polyalkylcyanoacrylates are used in several biomedical applications. More recently, with increasing interest in the field of nanotechnology, these polymers have also been prepared in nanoparticulate sizes and utilised in drug delivery.

2.2.2 Degradation

Most degradable polymers are designed to degrade via a hydrolytic mechanism and the degradation of aliphatic polyesters occurs mainly via the bulk degradation through hydrolysis. Hydrolytic degradation occurs as polymers are in contact with tissue fluids or moisture and preferentially onsets at the amorphous regions of the polymeric network. Hydrolytic attack causes cleavage at the ester bonds, resulting in chain scission yielding low molecular weight species in the initial stages. Chain scission may also occur in the presence of non-specific esterases and carboxy peptidases, which can break down PGA into glycolic acid units. Both PLA and PCL can also be degraded

2.3 Polymerisation of methylcyanoacrylate to poly(methylcyanoacrylate).

enzymatically. For a polymer to be rendered degradable the availability of the hydrolysable groups is essential. The hydrophilicity and the structure of the polymer contribute to the rate of degradation. PLA is more hydrophobic than PGA by virtue of the additional —CH unit per repeating unit (see Fig. 2.1) which also increases the steric hindrance, thus the availability of the hydrolysable linkage is diminished, rendering it more stable.

Degradation has often been related to weight loss with time when exposed to simulated body fluids, water or solutions at different pH. However, it has often been observed that the polymers undergo catastrophic mechanical failure even with lower amounts of fluid uptake. It is thought that in semi-crystalline polymers, as the amorphous domains are more amenable to diffusion and thus degradation, these regions are hydrolysed easily and as these regions are responsible for holding the crystalline regions together, a catastrophic failure occurs. The mechanism of degradation is also autocatalytic in nature and with the hydrolysis of PLA/PGA, thus with higher molecular weight polymers there are fewer acid-catalysing groups present and the longer chain length requires chain scission to occur more drastically until the oligomeric units are small enough to diffuse through the matrix [Li and Vert 1995; Li et al. 1990]. The general trend observed in these semi-crystalline polymers during the degradation is an increase in crystallinity as most of the amorphous regions degrade preferentially, followed by the loss of the crystalline phase. As this process continues, the polymer loses its mechanical strength and fragmentation occurs. Further hydrolysis yields smaller fragments that can be assimilated by phagocytes, whilst the soluble monomeric anions such as glycolate and lactate dissolve in the intercellular fluid. The glycolic acid is converted to glycine, then serine and subsequently to pyruvic acid that enters the tricarboxylic acid (TCA) cycle and is eliminated as carbon dioxide and water. PLA is broken down into lactic acid, which is converted to pyruvic acid and eliminated as CO_2 and water. PCL degrades to hydrohexanoic acid, a metabolisable metabolite. The polymers eventually break down to lactic acid and glycolic acid, enter the Krebs cycle, and are further broken down into carbon dioxide and water and excreted through normal processes. Figure 2.4 summarises the degradation process.

The degradation process can be modulated by modification of the backbone of the polymer chain and thus provides the possibility of tailoring to regulate bulk degradation rates. A systematic *in vivo* study by Chawla and Chang [1985] with four different molecular weight PLA polymers implanted in rats showed conclusively that the lower molecular weight polymers had a faster rate of degradation, associating this observation to difference in morphology and crystallinity. Although extrapolating *in vitro* degradation studies to *in vivo* models is not always linear, *in vitro* degradation studies confirm that degradation rates vary and thus the tailoring of copolymers to best justify a particular application is possible.

38   Cellular response to biomaterials

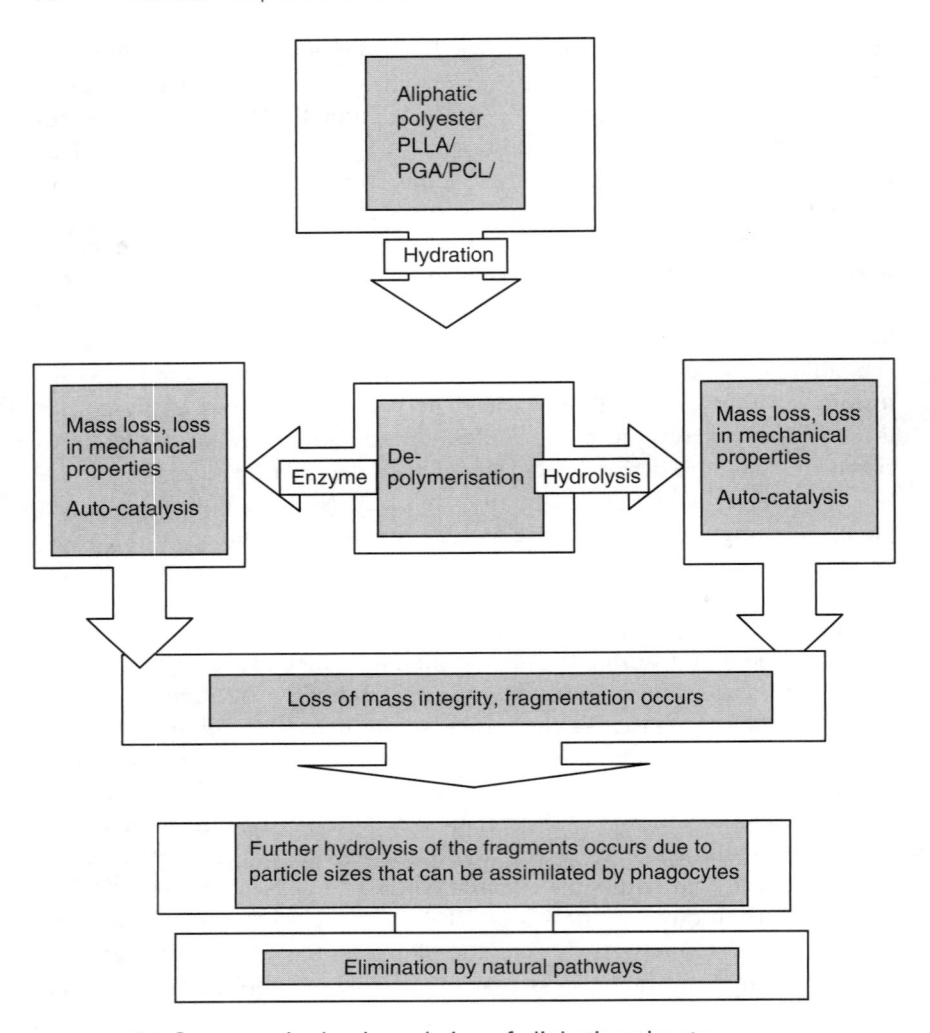

```
          ┌─────────────┐
          │  Aliphatic  │
          │  polyester  │
          │   PLLA/     │
          │   PGA/PCL/  │
          └─────────────┘
              Hydration
                 ↓
┌──────────────┐  ┌──────────┐  ┌──────────────┐
│ Mass loss,   │  │   De-    │  │ Mass loss,   │
│ loss in      │←─│polymeri- │─→│ loss in      │
│ mechanical   │Enzyme sation Hydrolysis mechanical │
│ properties   │  │          │  │ properties   │
│ Auto-catalysis│ └──────────┘  │ Auto-catalysis│
└──────────────┘                └──────────────┘
         ↓                              ↓
    ┌──────────────────────────────────────┐
    │ Loss of mass integrity, fragmentation occurs │
    └──────────────────────────────────────┘
                      ↓
    ┌──────────────────────────────────────┐
    │ Further hydrolysis of the fragments occurs due to │
    │ particle sizes that can be assimilated by phagocytes │
    └──────────────────────────────────────┘
                      ↓
         ┌────────────────────────────┐
         │ Elimination by natural pathways │
         └────────────────────────────┘
```

2.4 Sequence in the degradation of aliphatic polyesters.

Recently, it was hypothesised that surface erosion could be modified by tuning the hydrophobicity of a polymer backbone using a modular approach to polymer synthesis wherein judiciously designed building blocks of varying lipophilicities are linked together using hydrophobic spacers. Based on this rationale, Xu *et al.* [2006] reported the synthesis of a library of polymers by designing macromonomer diols with differing lipophilicities and then chemically reacting them with diacid spacers with thermal and physical properties distinct from pure poly(L-lactic acid) (PLLA) and enhanced surface erosion behaviour. Thus, the design of devices for any biomedical application necessitates the tailoring of polymers to suit the physical, mechanical and biological requirements.

2.2.3 Natural polymers in biomedical applications

Numerous biologically derived polymers have been explored as potential scaffolds for tissue engineering. Polymers include albumin, alginates, chitin/chitosan, cellulose, collagen, starch, fibrin, etc. These natural materials exhibit good cytocompatibility; however, controlling degradation rates, variation of properties from batch to batch and inadequate mechanical properties pose some limitations. Nevertheless, the ability to modify the chemical structures and influence properties such as hydrophilicity, swellability and cytocompatibility renders them a versatile group of polymers for medical application.

Alginates

Alginic acid is an insoluble polysaccharide and the sodium, potassium or ammonium salts are the alginates. Alginate is a water-soluble linear polysaccharide extracted from brown seaweed and is composed of alternating blocks of 1-4 linked α-L-guluronic acid (G) and β-D-mannuronic acid (M) residues (see Fig. 2.5). Alginates are not random copolymers and the patterns of the two residues D and M differ according to the source of the algae, each of which has different conformational preferences and behaviour; for example, the M/G ratio of alginate from *Macrocystis pyrifera* is about 1.6 whereas that from *Laminaria hyperborea* is about 0.45.

The solubility and water retention capacity of alginates depend on the pH, and precipitation occurs at about pH 3.5. The molecular weight also affects water uptake and calcium alginates with low molecular weights or fewer than 500 residues, exhibit higher water uptake. Sodium alginate and most other alginates from monovalent metals are soluble in water and yield

2.5 The M and G units of alginic acid.

40   Cellular response to biomaterials

solutions of high viscosity, thus making them particularly useful as thickening agents.

Alginates thus have long been used in the pharmaceutical industry as thickening or gelling agents, as colloidal stabilisers and as blood expanders. Alginates are used to take impressions of the oral tissue due to their ability to undergo a transformation from the sol state to a gel state through ionotropic gelation in the presence of many multivalent ions such as Ca^{2+}. The crosslinking can be carried out under very mild conditions at low temperatures and high humidity. Alginates have been widely investigated for cartilage and bone regeneration either as a scaffold or as carriers for biologically active molecules and drug delivery [Smidsrod and Draget 1996; Sriamornsak et al. 2006; Eiselt et al. 2000; Alsberg et al. 2001].

Chitosan

Chitin is the second most abundant natural polymer in the world after cellulose. Chitosan is is obtained by the deacetylation of chitin, which is the structural element of the shells of crustaceans such as crab and shrimps. The degree of deacetylation (%DA) and the molecular weight determine the properties of chitosan. The degree of deacetylation can be determined from NMR data and ranges between 60 and 100% in commercial chitosan. Chitosan is a linear copolymer of glucosamine and *N*-acetyl glucosamine in a β-1-4 linkage, and is positively charged and soluble in acidic to neutral solution with a charge density dependent on pH and the %DA value.

The DA which signifies the mole fraction of the *N*-acetylated units is a structural parameter and influences charge density, crystallinity, solubility and proneness to enzymatic degradation. Chitosan (Fig. 2.6) readily binds to negatively charged surfaces such as mucosal membranes and hence acts as a bioadhesive. As a polysaccharide of natural origin, chitosan has many useful features such as non-toxicity, biocompatibility, biodegradability and

2.6 Structure of chitosan.

antimicrobial properties [Rinuado 2006]. It exhibits excellent biological properties such as biodegradation in the human body, and immunological, antibacterial and wound-healing activity, and it is used as microcapsule implants for controlled release in drug delivery [Meera and Emilia 2006]. Chitosan is one of the most promising biopolymers for tissue engineering and possible orthopaedic applications. Chitosan has been proven to be safe for use as a pharmaceutical excipient. It has properties that make it suitable for drug delivery applications. As chitosan shows great promise in controlled release, this material is being increasingly exploited for developing nano and micro particles for drug delivery, gene therapy and regenerative medicine [Kim et al. 2008a; Springate et al. 2008].

Albumin

Albumins are water-soluble proteins, which are heat sensitive and coagulate. Egg white contains albumin and the most commonly known albumin is 'serum albumin' present in human blood. It is the most abundant of the blood plasma proteins, which act as transport proteins for several steroid hormones and fatty acids. Albumin, being a naturally occurring protein in the human body, has also been considered for biomedical applications. Functionalised albumin has been covalently linked to polymers such as polyethylene, polycarbonates, polypropylene, etc., and surface modification has been effected to enhance blood compatibility. The grafted albumin molecules are retained on the surface even on exposure to blood for extended periods and function as good oxygenators. Hydrogels, injectable drug delivery carriers and novel tissue adhesives have also been developed using albumin as a component [Gonen-Wadmany et al. 2007; Kakinoki et al. 2007].

Collagen

Collagen is another abundant protein and the main component of connective tissues in mammals. It has a long fibrous structure and is found as 'collagen fibrils', which are tough bundles of fibrillar proteins. They possess good tensile strength and are a component of cartilage, tendons, ligaments, bone and teeth. Collagen is also responsible for the elasticity and strength of the skin and its degradation leads to wrinkles and accompanying ageing. There are at least 28 types of known collagen [from Wikipedia, the free encyclopedia]. Collagen Type I is the most abundant in the human body, forms a structural component of bone and teeth and is present in scar tissue and in tendons. Collagen has been known for its use as a 'glue' since ancient times. More recently it has been widely used in soft tissue augmentation. Collagen has been used to develop biological patches and for purposes of tendon augmentation and is used in sports medicine, namely Zimmer Collagen Repair, TissueMend (TEI Biosciences, Boston, MA). Collagens on their own or in combination with silicones, glycosaminoglycans, fibroblasts, growth factors and other substances are employed in the management of burns and as artificial skin substitutes [MacNeil 2008].

Hyaluronic acid

Hyaluronan or hyaluronic acid is a non-sulfated glycosaminoglycan distributed widely throughout connective, epithelial and neural tissues. Hyaluronan is a polymer consisting of disaccharides with D-glucuronic acid and D-N-acetylglucosamine linked together via alternating beta-1,4 and beta-1,3 glycosidic bonds. It is one of the main components of the extracellular matrix and is also present in the articular cartilage as a coating on chondrocytes. Hyaluronan is a major component of the synovial fluid and is found to increase the viscosity of the fluid. The high molecular mass hyaluronic acid (HA), which plays an important role in many biological processes such as in tissue hydration, proteoglycan organisation in the extracellular matrix, and tissue repair, has found application in several clinical treatments and cosmetic use. Degradation of hyaluronic acid is enzyme mediated and the responsible enzymes are hyaluronidases. The degradation products are oligosaccharides and low molecular weight hyaluron. It has been used in the treatment of osteoporosis and is generally administered as a course of injections. Hyaluronic acid is one of the most promising biomaterials and has been investigated extensively as a biomaterial and for tissue engineering; this topic has been eloquently reviewed by Allison & Grande-Allen [2006] and further details are not within the scope of this chapter.

2.3 Degradable composites

2.3.1 Scaffolds for tissue engineering

Bioabsorbable surgical devices have received much attention over the last few decades and degradable polymers have contributed significantly to medicine and surgery. However, damaged tissues that are currently replaced by synthetic biomedical implants often fail to integrate with the host tissue. In recent years, the major research thrust has been towards tissue engineering strategies to engineer tissue replacement of damaged or diseased tissues. Since the last decade, tissue engineering is considered as the most attractive alternative to surgical procedures. Tissue engineering/regenerative medicine is a rapidly emerging multidisciplinary field, which is expected to improve the health and quality of life for millions of people worldwide. Tissue engineering combines the principles of engineering with biology and strives towards the development of biological substitutes. There are three key elements in tissue engineering and they involve (i) cells, (ii) cell signalling factors and (iii) scaffolds. The scaffolds are generally derived from existing biomaterials or novel biomaterials that are designed to direct the organisation, growth and differentiation of cells for forming functional tissue by providing both physical and chemical cues that eventually degrade within the living system without generating any side effects and are eliminated by natural processes.

The common principle in tissue engineering involves the isolation of specific cells that are then grown on a scaffold with appropriate signalling factors under controlled cell culture conditions and are subsequently placed at the target site. The other approach directly places the scaffold *in vivo* with an aim to generate '*de novo*' tissues. As a consequence the development of ideal scaffolds for the engineering of different tissues is a topic of enormous interest. Current challenges include development of scaffolds that are compatible with both the mechanical and biological properties of the 'said' tissue and more crucially support the vascularisation and integration of the construct to the host tissue. Thus, materials used for scaffold fabrication should not elicit any unresolved inflammatory response on implantation and should be cytocompatible and non-immunogenic. As the scaffold is required to facilitate cellular ingrowth, it needs either to be porous or to allow the transport of nutrients through diffusion and space for the cells to grow. Another important requirement is the ability of the construct to integrate with the host tissue, thus vascularisation is vital. From a materials standpoint, depending on the tissue in question and its specific biomechanical requirements, the physical, mechanical and chemical properties of the material are important. Furthermore, the formability, sterilisability and cost are variables that need to be considered in developing a scaffold. The rate of degradation of the scaffold is a critical issue, thus degradable scaffolds with different rates of degradation are desirable. However, as polymers on their own may lack the mechanical integrity or the requisite properties for scaffold formation, degradable composites are being investigated as alternatives to polymer scaffolds.

Scaffolds for guided tissue regeneration involve creating 3-D structures that are able to support seeding of cells. Central to the formation of new tissue is the role of the scaffold, which should either induce the formation of tissue or act as a carrier or template for implanted cells or other growth inducing agents. So an ideal scaffold should be fabricated from a biomaterial that is not only cytocompatible but also adequately porous to allow cellular infiltration [Hollister 2005]. Depending on the tissue, the scaffold may need to provide mechanical support of varying loads. The addition of biodegradable polymers to ceramics/glasses and cements can modify the physical and biological properties; some common examples include polymers such as poly(lactides)/poly(glycolides), poly(propylene fumarate), poly(caprolactones), carbohydrates, collagen, fibrin, chitosan with bioactive glasses, calcium phosphates and hydroxyapatites. Such polymer-composite materials have been researched widely as scaffolds and only a few examples are discussed in this chapter. Various tissue analogues such as skin, cartilage, bone, liver and nerve have been created using this method; however, to date the most successful and clinically applied engineered tissue remains to be skin.

2.3.2 Composites

Composite materials are made from at least two or more constituent materials that have significantly different physical and mechanical properties that remain separate and distinct on a macroscopic level and are not soluble in each other. There are at least two phases in a composite, namely a continuous phase and a dispersed phase. Typically, composite materials are designed to provide a combination of properties that is not achievable with a single-phase material and it is usually the dispersed phase that is responsible for often displaying structural features over a large range of dimensional scales, enhancing the properties. Composite materials have gained popularity for use as high performance materials that demand specific properties. Fibre reinforced polymers, carbon reinforced polymers, epoxy resin–fibre composites, ceramic matrix composites such as hydroxyapatite–collagen, ceramic and metal, concrete, dental composites, etc., are a few of the most widely used composites in day-to-day life.

Biodegradability, biocompatibility, mechanical properties and the interfacial bond between the two phases of a composite are the important considerations in developing a composite for biomedical applications. Modern composites are aimed at mimicking nature, for example bone and teeth are composites containing hydroxyapatite and collagen and have unique properties. Thus, biocomposites are generally modelled on the tissue that is being replaced, for example Cerosium® was one of the first biomedical composites designed to mimic the elastic properties of bone. Cerosium®, a composite of alumina and epoxy resin, was developed with an idea to imbibe the mechanical and biological compatibility required to replace bone. The material, however, failed due to biodegradation and the rupture of interfacial bonds.

One of the most successful composites modelled on this theory was the polyethylene–hydroxyapatite composite to create a bone analogue [Bonfield 1988a, 1988b]. This material was known as HAPEX® and a number of composites of particulate HA with polyethylene were prepared with volume fractions ranging from 0.1 to 0.6 HA. The Young's modulus of the composites ranged from approximately 1 to 8 GPa with an upper limit of 0.4

volume fraction of HA in the composite. The fracture toughness of the composite was very similar to that of cortical bone; however, *in vivo* studies showed that lower concentrations of HA compromised the bioactivity, with the formation of a fibrous capsule around the implant. Furthermore, chemical coupling to enhance the interfacial adhesion between the matrix and filler was attempted, though it has not been clinically used [Deb *et al.* 1995]. The HA–PE composite was evaluated clinically for sub-orbital floor reconstruction and as a bone substitute for ear, nose and throat surgery, i.e. for non-load-bearing applications. The FDA approval led to the material being used as a sub-orbital floor reconstruction and middle ear prosthesis. However, with the advent of tissue engineering, biodegradable scaffolds are finding a new place in regenerative medicine.

Composites and more specifically combinations of polymers and ceramics are highly favoured for scaffold materials in tissue engineering, in particular for bone tissue engineering [Mano *et al.* 2004; Kim *et al.* 2004; Niiranen *et al.* 2004; Rezwan 2006; Misra *et al.* 2008; Habraken *et al.* 2007]. Degradability of polymer-composite scaffold is an important issue in tissue engineering and ideally the degradation should match the rate of the neotissue formation; however, in polymer composites the biodegradation is rather more complex, with at least two phases present, each with a different rate of degradation that may also be affected by the amount of stress transfer, the effect of the components on the degradability, if any, on each other, loading conditions, environment, surface area and porosity. More specifically, the tissue engineering of bone would benefit from a scaffold that is able to mimic the biological and mechanical properties of bone and essentially be porous with interconnected networks that would allow the ingrowth of cells and transport of nutrients, whilst possessing the necessary regulatory and osteoinductive factors. The material processing for composite scaffolds with interconnecting pores is thus relevant to produce viable structures for bone tissue engineering. Several methods have been reported in the literature for composite manufacture. Some of the common methods are melt processing, solvent casting and particle leaching, solid free form, thermally induced phase separation, electrospinning, scaffold coating and microsphere sintering. Other approaches include the biomimetic synthesis of scaffolds that may be directly mineralised into the backbone of polymers via nucleation sites by virtue of the phosphorus-containing anionic functionality.

Bioactive ceramics are usually the materials of choice for the formulation of composite scaffolds due to their ability to confer bioactivity to the polymer composite. A bioactive glass exhibits a time-dependent modification of its surface and forms a biologically active hydroxy carbonate apatite (HCA) layer on implantation that provides a bond with biological tissues. Hench [1991] first reported the ability of Bioglass® to form a biological bond with bone. This ability to bond to bone is known as 'bioactivity' and can be assessed by *in vitro* methods that utilise the interaction of the glass with simulated body fluid (SBF), a protein-free medium with ionic concentrations similar to those in human blood plasma. The interaction of bioactive glasses with SBF leads to the formation of a HCA layer, which is similar to the mineral phase of bone. Bioglass® has since been used for several clinical applications such as treatment of periodontal disease, bone filler and implants with a good degree of success. Not surprisingly, bioactive glasses are promising candidates for composite formation for scaffolds for bone tissue engineering.

Poly(lactic acid) and poly(glycolic acid) composites

The most widely studied polymer as a matrix for composites is poly(lactic acid) either as a homopolymer or as the stereocopolymers or racemic mixtures. Composites of PLA have ranged from partially resorbable to totally resorbable with inclusions ranging from carbon fibres [Zimmerman et al. 1991] to calcium phosphates. Poly(lactic acid), poly(glycolic acid), their copolymers, poly(caprolactone) and racemic mixtures of the different forms of PLA are some of the polymers that have been used for the development of composite scaffolds. The mechanism of degradation has been discussed in Section 2.2.2 and in general the degradation rates decrease in the following order: PGA > PDLLA > PLLA > PCL. As discussed earlier, the rate of degradation is dependent on multiple factors and thus specific degradation rates of different polymers are difficult to summarise due to the variation in sample size, shape, molecular weight, ageing conditions, etc. Poly(DL lactic acid), PDLA, have been combined with 45S5 Bioglass® particles to fabricate optimised foams for both soft and hard tissue engineering [Boccaccini et al. 2002, 2003; Roether et al. 2002; Hench and Polak 2002].

Different methods, such as the thermally induced phase separation process, have been applied to prepare highly porous PDLLA foams filled with 10 wt% Bioglass® particles that have yielded stable and homogeneous layers of Bioglass® particles on the surface of the PDLLA/Bioglass® composite foams. A slurry dipping technique was also reported to yield composites with a porous network with *in vitro* bioactivity in simulated body fluid (SBF) showing the formation of hydroxyapatite (HA) on the surface of the PDLLA/Bioglass® composites [Verrier et al. 2004]. PDLA/Bioglass® foam scaffolds have been reported to provide an appropriate microenvironment for bovine annulus fibrosus cell culture which enhances cell proliferation and promotes the production of sGAG, collagen type I and collagen type II. These findings provide preliminary evidence for the use of PDLLA/Bioglass® composite scaffolds as cell-carrier materials for future treatments of intervertebral discs [Wilda et al. 2007].

Polyactive®, a copolymer based on poly(ethylene oxide) and poly(butylene terephthalate) (PEO/PBT), has been reported to be an osteoconductive polymer that has been mixed with hydroxyapatite to yield bioactive composites [van Blitterswijk *et al.* 1993; Jansen *et al.* 1995; Liu *et al.* 1998]. Ameer *et al.* [2002] reported a biodegradable composite system consisting of a PGA non-woven mesh capable of entrapping cells within a support of pre-defined shape, to potentially facilitate cell delivery into a target site (e.g., meniscal tears in the avascular zone).

Collagen, alginate and gelatin based composites

Collagen and hydroxyapatite are the two main components of bone, which is essentially a composite. There has been a trend to combine these two components to derive bone replacement materials with a view to mimicking the natural composite. Collagen–HA composites [Murugan and Ramakrishna 2005; Olszta *et al.* 2007] have been widely studied and Collagraft [Cornell *et al.* 1991], Bio-Oss [Terheyden *et al.* 2001] and Healos [Tay *et al.* 1998] are examples of commercially available collagen-based bone grafts for clinical use. Collagraft is made up of calcium phosphates and collagen and is available as a paste or in the form of strips (Zimmer, Warsaw, IN, and Collagen Corporation, Palo Alto, CA) that is used clinically in non-load-bearing or low-load-bearing applications. Bio-Oss is another composite that is made up of bovine collagen and deproteinated bovine bone and is used for repair of craniofacial defects. Healos® is an osteoconductive matrix constructed of crosslinked collagen fibres that are fully coated with hydroxyapatite; when combined with autogenous bone marrow aspirate, it provides an excellent environment for osteoprogenitor cell attachment, proliferation and differentiation and is mainly used as an osteoconductive matrix. Collagen has also been used in combination with rhBMP-2 as a bone substitute. The self-assembly of HA coatings on collagen matrices has been shown to form in the presence of citric acid and SBF; however, the mechanical properties are not adequate [Rhee and Tanaka 1998].

In order to enhance the mechanical properties of the mineralised collagen, a glutaraldehyde-crosslinked porous HA/collagen nanocomposite was reported by Chang *et al.* [2001]. Porous scaffolds from hydroxyapatite–poly(vinylalcohol)–gelatin nanocomposites have been also reported with a good control on the organisation of the HA particles in the matrix [Nayar and Sinha 2004]. Kim *et al.* [2005] used a biomimetic approach to form gelatin–hydroxyapatite nanocomposites by biomimetically precipitating HA on gelatin through freeze-drying that yielded a porous structure with less crystalline HA deposits; these nanocomposites are shown to be significantly favoured by human osteoblastic cells.

A biomimetic method of synthesising carbonated hydroxyapatite that was further combined with sodium alginate and crosslinked was used for developing a hydrogel-like composite for bone tissue engineering, which was further modified to function as an injectable scaffold with stem cell encapsulation [Rebeling et al. 2006]. The method provides a facile method of stem cell encapsulation in simulated body conditions [Rebeling et al. 2007].

Starch based composites

Starch is a naturally occurring polysaccharide that is a mixture of amylose and amylopectin. Starch can be found in plants and the corn crop is a major source of starch. Starch has adequate thermal stability for melt blending with synthetic plastics and has been used in the production of novel degradable composites. Starch based polymers present an enormous potential as biomaterials and have been emerging as potential candidates as scaffolds for the tissue engineering of bone, cartilage, bone fixation devices, bony defect filling materials, drug delivery vehicles and hydrogels for biomedical applications. Blends of starches and synthetic polymers have been extensively studied as biodegradable or partially biodegradable plastics as both an environmentally friendly plastic and a renewable resource; however, the overall degradability of the final composite is still questionable.

Starch based blends have been extensively studied by Reis and Cunha [2000] as biomaterials [Gomes et al. 2002; Elvira et al. 2002]. Starch blends with ethylene vinyl alcohol copolymers, cellulose acetate, polycaprolactone and poly(lactic acid) have been studied as a range of biomaterials with different rates of degradation for different applications [Reis and Cunha 2000; Pereira et al. 1998]. The starch blends have been shown to exhibit good biocompatibility by several *in vitro* and *in vivo* studies and, due to the ability to manipulate the physical and mechanical properties of such blends, the combination with bioactive ceramic fillers has also been investigated. The inclusion of bioactive ceramic fillers not only confers bioactivity but also tends to enhance mechanical properties that are necessary for any load-bearing applications. Reis and Cunha reported a SEVA-C/hydroxyapatite processed composite and used shear controlled injection moulding and also combined this technique with twin screw extrusion to yield highly oriented and mechanically strong composites. Bioglass®, a bioactive glass that is known for its ability to exhibit bone bonding, was also used as reinforcing filler with starch as the matrix to yield composites, with the tensile modulus being dependent on the amount of the glass filler. A tensile modulus of 3.5 GPa was reported with as little as 10% of Bioglass® filler that also exhibited bioactive characteristics [Lenor et al. 2002] whereas the bioactivity was only apparent in SEVA-C/HA composites with at least 30% of HA present in the matrix.

Chitosan based composites

Chitosan is a biodegradable, non-antigenic and biofunctional polymer that has excellent pre-requisites as a material for tissue regeneration. It is useful as a biomaterial due to its low cost and wide availability, and in addition, its hydrophilic surface promotes cell adhesion, proliferation and differentiation that is known to evoke a minimal foreign body reaction on implantation [Martino *et al.* 2005; Amaral *et al.* 2006]. Despite the fact that chitosan is a widely acceptable tissue compatible material [Rinaudo 2006; Kim *et al.* 2008b; Ta *et al.* 2008; Dang and Leong 2006; Habraken *et al.* 2007], it is mechanically weak and dimensionally unstable due to swelling, thus unable to maintain a predefined shape. Hence a number of blends with other polymers and composites with bioactive ceramics have been investigated to enhance the properties.

Hydroxyapatite/chitosan/alginate composites with uniform pore sizes that show the growth of the apatitic structure post nucleation of hydroxyapatite on the scaffold have been reported to function as scaffolds for bone tissue engineering. Biomimetic routes of synthesis of nanocomposites of HA/polygalacturonic/chitosan have been reported by Verma *et al.* [2008] which exhibit significant enhancement in elastic moduli and compressive strength over composites of chitosan and HA or polygalacturonic/HA. This observation was attributed to strong interfacial interaction between the polygalacturonic and chitosan domains, thus indicating the process of biomineralisation. The cellular response to various chitosan composites reports that it is not cytotoxic and is able to support the growth and differentiation of goat marrow stromal cells, thus the biological properties render these materials as attractive tissue engineering scaffolds [Oliveira *et al.* 2006].

2.4 Fibre scaffolds

Connective tissues such as tendons, ligaments and cartilage are important in the body due to their specific mechanical function. For example, cartilage is a dense tissue and comprises chondrocytes that produce large amounts of extracellular matrix made up of collagen fibre; the tendon is a strong band of fibrous tissue that usually connects the muscle to bone, and ligaments connect one bone to another. These tissues are oriented fibres with anisotropic properties and each of them has a load-bearing function that is prone to damage; endogenous repair processes do not restore function. The function of these structures is largely dependent on their architectural form, and as such, scaffold organisation is an important design parameter in generating tissue analogues. Biological regeneration using cartilage tissue engineering can provide new treatment options for articular cartilage defects [Stoop 2008]. Although there are several methods of spinning fibres, it is not possible to cover them in their entirety. This section presents a short overview on electrospinning methods for fibre formation due to its current resurgence in the production of fibres and its application in tissue engineering.

2.4.1 Electrospinning

Electrospinning is a technique that has been available for over 100 years or so. It involves the fabrication of fine fibres using an electrical charge. The basic setup requires a spinneret with a needle in which a polymer, sol-gel or composite solution can be loaded and is then driven by a pump to the tip of the needle. The application of a voltage then causes the droplet to stretch and, depending on the viscosity of the solution, an electrified jet is formed. The jet can be elongated and, as it moves, the solvent is lost, leading to the deposition of solid fibres. The fibres produced by this method typically can have diameters from a few microns to nanometres. The ease of fabrication and the ability to create fibres of different diameters has generated an interest in this process for developing scaffolds for tissue engineering and drug delivery carriers. Scaffolds for tissue engineering require the fabrication of 3-D constructs that can support cellular growth and proliferation.

While the electrospinning process itself is old, the concept of electrospun scaffolding for biomedical applications appears to have first emerged in 1978. Annis *et al.* [1978] reported a vascular prosthesis made from polyurethane using electrospinning techniques. Aligned scaffolds mimic fibre-reinforced tissue organisation by acting as a micro patterned three-dimensional grid for neo-tissue formation. Electrospinning techniques have been used to fabricate such scaffolds with FDA-approved degradable polymers such as the clinically used poly(α-hydroxy esters). Li *et al.* [2006a] used electrospinning techniques to fabricate fibres with uniform structures that were morphologically similar to the extracellular matrix. Highly porous three-dimensional polymeric scaffolds with a range of diameters could be produced using this technique and the mechanical properties could be altered by changing the polymer/copolymer composition. According to this study a PGA:PLGA yielded the stiffest polymers and the PLLA:PCL polymer fibres were most compliant. Similarly, the degradation pattern was dependent on the composition of the copolymer, as has been shown earlier, and of PGA, PDLLA and PLGA the more amorphous in nature had a faster rate of degradation in comparison to PLLA or PCL fibre scaffolds. The PLLA and PCL based fibres also exhibited good cellular response, making them good candidates for tissue engineering. However, due to the high density or packing of these fibres, slow rates of cellular infiltration have often been a problem; thus increasing the porosity of the fibrous scaffolds is one method of enhancing cell ingress that would still maintain the anisotropy of the matrix.

Baker *et al.* [2008] recently combined poly(ε-caprolactone) and poly(ethylene oxide) using a co-spinning technique from two different spinnerets to form dual composite fibre aligned scaffolds with PEO forming the fibres that would be removed; being water soluble rendered higher porosity, resulting in enhanced cell ingress. Seeding of mesenchymal stem cells in these scaffolds showed that cells were present throughout the matrix, in contrast with PCL where only fibres on the periphery had cells growing. Thus, this strategy may be utilised to enhance *in vitro* and *in vivo* formation and maturation of functional constructs for fibrous tissue engineering.

It is also well known that cells are sensitive to topography. This has an important bearing on cell adhesion and proliferation. Electrospinning techniques have also been applied for fabricating scaffolds for bone tissue engineering. The techniques either process the fibre first to mimic the extracellular matrix followed by mineralisation [Liao *et al.* 2008] or use a mix of the two components. Liao *et al.* used electrospun nanofibres from collagen and PLGA respectively to form the ECM that was subsequently mineralised using optimised conditions, and showed that the open porous structure did support mineralisation, though it was found to be more uniform in the collagen matrix in comparison to PLGA. This finding was attributed to the functional groups present on the surface of the nanofibres, and the presence of the carboxyl and carbonyl groups in collagen enhanced the mineralisation. The authors also indicated that the mineralisation process predominantly induced the formation of nanosized carbonated hydroxyapatite (CHA) during collagen mineralisation, while nanosized hydroxyapatite (HA) was formed during PLGA mineralisation. A composite of PCL containing chopped glass fibres showed that hydroxyapatite crystals were rapidly deposited on the surface post immersion in simulated body fluids, indicating good levels of bioactivity. Furthermore, the cell viability of murine-derived osteoblastic cells improved in the composite scaffolds in comparison to simple PCL fibre scaffolds [Lee *et al.* 2008]. PCL fibres have been successfully spun into fibres using electrospinning methods. In a bid to tailor these scaffolds for bone tissue engineering, coating of such fibres with calcium phosphates has also been reported.

2.5 Injectable composites

Injectable cements or composites that consist of polymers and ceramics that are capable of self-hardening in tissue matrices are increasingly being researched. Considering that bone defects can be of various sizes, shapes and contours, materials that can be moulded into the defects with minimally invasive techniques are effective in minimising scars and operation time and in improving healing time. Bone defects occur in a variety of clinical situations and their reconstruction is necessary for the rehabilitation of the patient.

A number of systems have been reported, such as calcium phosphate cements in combination with gelatin, alginates and chitosan [Liu et al. 2006]. One of the main drawbacks of calcium phosphate cements is its brittleness, low strength and slow integration with bone due to its low porosity. Mouldable scaffolds capable of *in situ* self-hardening with higher porosities have been reported by Xu *et al.* [2008] with CPC reinforced with absorbable fibres, chitosan and mannitol as porogen. The macropores present in this nanocomposite created a matrix suitable for cell infiltration and supported adhesion and proliferation of osteoblast-like cells, thus rendering it a viable scaffold for bone tissue engineering. The relatively high-strength and osteoconductive CPC composites may be an effective delivery vehicle for osteoinductive growth factors, antibiotics and other molecules necessary to promote bone regeneration. Potential dental and craniofacial uses of an improved CPC include mandibular and maxillary ridge augmentation, since CPC could be moulded to the desired shape and set to form a scaffold for bone ingrowth.

2.6 Drug delivery from resorbable scaffolds

In drug delivery, degradable polymers hold the promise of zero-order release, or continuous release of a drug that is independent of time. Current devices have not achieved this gold standard, but they still aid in delivering appropriate amounts of drugs with less reliance on patient compliance. In principle, biodegradation of degradable polymers, calcium phosphates and ceramic composites can be tailored to resorb at different rates and thus used as drug delivery matrices. The means of delivery of any active factor from a carrier material to the surrounding environment can be classified as a drug delivery vehicle. Conventional drugs such as methotrexate or gentamicin can be incorporated within degradable vehicles or scaffolds developed to function as temporary support for tissue regeneration may also be used to release growth factors, cells, proteins or drugs. Growth factors such as bone morphogenetic proteins, transforming growth factors, vascular endothelial cells, stem cells, osteoblast-like cells and others have been introduced in polymer composite scaffolds to impart osteoconductive properties and aid vascularisation. The presence of these biological moieties promotes tissue growth and the ceramic and glass components can bind strongly to these biological molecules with a modulated release pattern.

2.7 Conclusions

The advent of hard tissue repair using biomaterials started with bioinert materials and a number of them are still in clinical use, namely hydroxyapatite and related materials due to their close resemblance to natural bone. Other bioinert implant materials include stainless steel, titanium, poly(ethylene), etc., that are used to fabricate permanent implants such as hip, knee and dental prostheses. Degradable polymers were subsequently used to fabricate absorbable sutures and pins, screws and plates for fracture fixation.

In the past few years increasing efforts have been made in developing and understanding bioactive glasses and ceramics, especially for bone repair due to their bone bonding properties. The merit in combining degradable polymers with resorbable bioactive glasses was soon recognised and a host of polymer-composites with different physical, mechanical and biological properties have emerged. With the advent of tissue engineering and its importance in regenerative medicine, numerous scaffolds either with resorbable polymers on their own or in combination with resorbable glasses and their specific cellular interactions are of growing interest. The requirements for making composites and scaffolds for implantable devices are complex and specific to the structure and function of the tissue of interest. The composites and scaffolds serve as both physical support and adhesive substrates for isolated or host cells during *in vitro* cell culture and subsequent *in vivo* implantation. The scaffolds are used to deliver cells at the desired sites in the body and to progress the process of tissue development. Thus, properties such as biocompatibility, resorbability, rate of degradation, porosity, pore size, shape and mechanical properties of the composite need to be assessed prior to *in vivo* use.

There are numerous methods of fabricating such composites and some of them have been discussed; however, the ultimate method of production should be cost-effective, efficient and viable for creating heterogeneous and homogeneous scaffolds of varying sizes and shapes. Although there is a vast amount of literature available on fabrication of scaffolds and their specific cell interactions, an ideal scaffold that will provide a well vascularised and integrated tissue engineered construct is yet to become a reality.

Current research now focuses on the introduction of stem cells within scaffolds. Owing to the ability of stem cells to differentiate into different lineages, it is expected that specific combinations of scaffolds and cells may lead to the growth of biological tissues and organs.

2.8 References and further reading

Allison D D and Grande-Allen K J, Review. Hyaluronan: a powerful tissue engineering tool, *Tissue Eng.*, **12**(8), 2131–2140, 2006.

Alsberg V, Anderson K, Albeiruti A, Franceshi R T and Mooney D J, Cell interactive alginate hydrogels for bone tissue engineering, *J. Dent. Res.*, **80**, 2025–2029, 2001.

Amaral I F, Sampaio P and Barbosa M A, Three-dimensional culture of human osteoblastic cells in chitosan sponges: the effect of the degree of acetylation, *J. Biomed. Mater. Res.*, **76**, 335, 2006.

Ameer G A, Mahmood T A and Langer R, A biodegradable composite scaffold for cell transplantation, *J. Orthopaedic Res.*, **20**, 16–19, 2002.

Anderson A J and Dawes E A, Review. Occurrence, metabolism, metabolic role, and industrial uses of bacterial polyhydroxyalkanoates, *Microbiol. Rev.*, **54**(4), 450–472, 1990.

Annis D, Bornat A, Edwards R O, Higham A, Loveday B and Wilson J, An elastomeric vascular prosthesis, *Trans. American Society for Artificial Internal Organs*, **24**, 209, 1978.

Ayres C E, Bowlin G L, Pizinger R, Taylor L T, Keen C A and Simpson D G, Incremental changes in anisotropy induce incremental changes in the material properties of electrospun scaffolds, *Acta Biomater.*, **3**(5), 651–661, 2007.

Baker B M and Mauck R L, The effect of nanofiber alignment on the maturation of engineered meniscus constructs, *Biomaterials*, **28**(11), 1967–1977, 2007.

Baker B M, Gee A O, Metter R B, Nathan A S, Marklein R A, Burdick J A and Mauck R L, The potential to improve cell infiltration in composite fiber-aligned electrospun scaffolds by the selective removal of sacrificial fibers, *Biomaterials*, **29**(15), 2348–2358, 2008.

Boccaccini A R, Roether J A, Hench L L, Maquet V and Jérôme R, A composites approach to tissue engineering, *Ceram. Eng. Sci. Proc.*, **23**, 805–816, 2002.

Boccaccini A R, Notingher I, Maquet V and Jérôme R, Bioresorbable and bioactive composite materials based on polylactide foams filled with and coated by Bioglass® particles for tissue engineering applications, *J. Mater. Sci. Mater. Med.*, **14**, 443–450, 2003.

Bonfield W, Composites for bone replacement, *J. Biomed. Eng.*, **10**, 522–526, 1988a.

Bonfield W, Hydroxyapatite-reinforced polyethylene as an analogous material for bone replacement. In: Ducheyene P and Lemons J F (eds), *Bioceramics: Materials Characteristics versus in vivo Behaviour*, Vol 253, Annals of the New York Academy of Science, 1988b, pp. 173–177.

Callister W D, *Materials Science and Engineering. An Introduction*, 3rd edition, New York: Wiley, 1994, pp. 473–482.

Chang M C, Ikoma T, Kikuchi M and Tanaka J, Preparation of a porous hydroxyapatite/collagen nanocomposite using glutaraldehyde as a crosslinkage agent, *J. Mater. Sci. Lett.*, **20**, 1199–1201, 2001.

Chawla A S and Chang T M S, In-vivo degradation of poly(lactic acid) of different molecular weights, *Biomater. Med. Dev. Art. Org.*, **13**(3 & 4), 153–162, 1985–86.

Chen Q Z, Thompson I D and Boccaccini A R, 45S5 Bioglass®-derived glass–ceramic scaffolds for bone tissue engineering, *Biomaterials*, **27**, 2414–2425, 2006.

Cheung H Y, Lau K T, Lu T P and Hui D, A critical review on polymer-based bio-engineered materials for scaffold development, *Composites: Part B*, **38**, 291–300, 2007.

Chu C C, Degradation phenomena of two linear aliphatic polyester fibres used in medicine and surgery, *Polymer*, **26**, 591–594, 1985.
Coover H W, Dreifus D W and O'Connor J T, Cyanoacrylate adhesives. In: Skeist I (ed.), *Handbook of Adhesives*, 3rd edition, New York: Van Nostrand Reinhold, 1990, pp. 463–477.
Cornell C N, Lane J M, Chapman M, Merkow R, Seligson D, Henry S, Gostilo R and Vincent K, Multicenter trial of Collagraft as bone graft substitute, *J. Orthop. Trauma*, **5**(1), 1–8, 1991.
Courtney T, Sacks M S, Stankus J, Guan J and Wagner W R, Design and analysis of tissue engineering scaffolds that mimic soft tissue mechanical anisotropy, *Biomaterials*, **27**(19), 3631–3638, 2006.
Dang J M and Leong K W, Natural polymers for gene delivery and tissue engineering, *Advanced Drug Delivery Reviews*, **58**(4), 487–499, 2006.
Deb S, Wang M, Tanner K E and Bonfield W, Hydroxyapatite–polyethylene composites: effect of grafting and surface treatment of hydroxyapatite, *Proc. 12th European Conference on Biomaterials*, Porto, Portugal, 1995, pp. 10–13.
DiSilvio L, Dalby M J and Bonfield W, In vitro response of osteoblasts to hydroxyapatite reinforced polyethylene composites, *J. Mater. Sci.: Mater. in Med.*, **9**, 845–848, 1998.
Doddi N, Versfelt C C and Wasserman D, Synthetic absorbable devices of polydioxanone, US Patent 4,052,988, 1976.
Eiselt P, Yeh J, Latvala R K, Shea L D and Mooney D J, Porous carriers for biomedical applications based on alginate hydrogels, *Biomaterials*, **21**, 1921–1927, 2000.
Eling B, Gogolewski S and Pennings A J, Biodegradable materials of poly(lactic acid).1. Melt spun and solution spun fibres, *Polymer*, **23**, 1587–1593, 1982.
Elvira C, Mano J F, San Román J and Reis R L, Starch based biodegradable hydrogels with potential biomedical applications as drug delivery systems, *Biomaterials*, **23**(9), 1955–1966, 2002.
Fambri L, Pegoretti A, Fenner A, Incardona S D and Migliaresi C, Biodegradable fibres of poly(lactic acid) produced by melt spinning, *Polymer*, **38**, 79–85, 1997.
Gao D, Kumar G, Co C and Ho C C, Formation of capillary tube-like structures on micropatterned biomaterials, *Adv. Exp. Med. Biol.*, **614**, 199–205, 2008.
Gomes M E, Godinho J S, Tchalamov D, Cunha A M and Reis R L, Alternative tissue engineering scaffolds based on starch: processing methodologies, morphology, degradation and mechanical properties, *Mater. Sci. Eng.: C*, **20**, 19–26, 2002.
Gonen-Wadmany M, Oss-Ronen L and Seliktar D, Protein–polymer conjugates for forming photopolymerizable biomimetic hydrogels for tissue engineering, *Biomaterials*, **28**, 3876–3886, 2007.
Guilak F, Butler D L and Goldstein S A, Functional tissue engineering: the role of biomechanics in articular cartilage repair, *Clin. Orthop. Relat. Res.*, **391**, S295–S305, 2001.
Habraken W J E M, Wolke J G C and Jansen J A, Ceramic composites as matrices and scaffolds for drug delivery in tissue engineering, *Advanced Drug Delivery Reviews*, **59**(4–5), 234–248, 2007.
Hench L L, Bioceramics: from concept to clinic, *J. Am. Ceram. Soc.*, **74**, 1487–1510, 1991.

Hench L L and Polak J M, Third-generation biomedical materials, *Science*, **295**, 1014–1017, 2002.

Hollinger J O and Battistone G C, Biodegradable bone repair materials: synthetic polymers and ceramics, *Clin. Orthop.*, **207**, 290–305, 1986.

Hollister S J, Porous scaffold design for tissue engineering, *Nature Mater.*, **4**, 518–524, 2005.

Iroh J O, PCL Poly(epsilon-caprolactone). In: Mark J E (ed.), *Polymer Data Handbook*, Oxford: Oxford University Press, 1999, pp. 361–362.

Jansen J A, de Ruijter J E, Janssen P T M and Paquay Y G C J, Histological evaluation of a biodegradable Polyactive®/hydroxyapatite membrane, *Biomaterials*, **16**, 819–827, 1995.

Jin H-H, Lee C-H, Lee W-K, Lee J-K, Park H-C and Yoon S-Y, In-situ formation of the hydroxyapatite/chitosan-alginate composite scaffolds, *Mater. Lett.*, **62**(10–11), 1630–1633, 2008.

Kakinoki S, Taguchi T, Saito H, Tanaka J and Tateishi T, Injectable in situ forming drug delivery system for cancer chemotherapy using a novel tissue adhesive: characterization and in vitro evaluation, *European Journal of Pharmaceutics and Biopharmaceutics*, **66**, 383–390, 2007.

Kim H W, Knowles J C and Kim H E, Hydroxyapatite/poly([epsilon]-caprolactone) composite coatings on hydroxyapatite porous bone scaffold for drug delivery, *Biomaterials*, **25**, 1279–1287, 2004.

Kim H W, Kim H E and Salih V, Stimulation of osteoblast responses to biomimetic nanocomposites of gelatin–hydroxyapatite composites for tissue engineering scaffolds, *Biomaterials*, **26**, 5221–5230, 2005.

Kim I-Y, Seo S-J, Moon H-S, Yoo M-K, Park I-Y, Kim B-C and Cho C-S, Chitosan and its derivatives for tissue engineering applications, *Biotechnology Advances*, **26**(1), 1–21, 2008b.

Kim J-H, Kim Y-S, Park K, Kang E, Lee S, Nam H Y, Kim K, Park J H, Chi D Y, Park R-W, Kim I-S, Choi K and Kwon I C, Self assembled glycol chitosan nanoparticles for the sustained and prolonged delivery of angiogenic small peptide drugs in cancer therapy, *Biomaterials*, **29**, 1920–1930, 2008a.

King M E and Kinney A Y, Tissue adhesives: a new method of wound repair, *Nurse Pract.*, **24**, 66–73, 1999.

Lee H-H, Yu H-S, Jang J-H and Kim H-W, Bioactivity improvement of poly(ε-caprolactone) membrane with the addition of nanofibrous bioactive glass, *Acta Biomaterialia*, **4**(3), 622–629, 2008.

Lenor I B, Sousa R A, Vunha A M, Zhong Z, Greenspan D and Reis R L, Novel starch thermoplastic/Bioglass® composites: mechanical properties, degradation behaviour and in vitro bioactivity, *J. Mater. Sci. Mater. in Med.*, **13**, 1–7, 2002.

Li M, Mondrinos M J, Gandhi M R, Ko F K, Weiss A S and Lelkes P I, Electrospun protein fibers as matrices for tissue engineering, *Biomaterials*, **26**(30), 5999–6008, 2005.

Li P, Bakker D and van Blitterswijk C A, The bone-bonding polymer Polyactive 80/20 induces hydroxycarbonate apatite formation in vitro, *J. Biomed. Mater. Res.*, **34**(1), 79–86, 1997.

Li S and Vert M, Biodegradation of aliphatic polyesters. In: Scott G and Gilead D (eds), *Degradable Polymers Principles and Applications*, London: Chapman & Hall, 1995, pp. 43–87.

Li S, Garreau H and Vert M, Structure–property relationships in the case of the degradation of massive poly(α-hydroxy acids) in aqueous media. Part 3: Influence of the morphology of poly(L-lactic acid), *J. Mater. Sci. Mater. Med.*, **1**, 198–206, 1990.

Li W J, Mauck R L and Tuan R S, Electrospun nanofibrous scaffolds: production, characterization, and applications for tissue engineering and drug delivery, *J. Biomed. Nanotech.*, **1**(3), 259–275, 2005.

Li W-J, Cooper J A Jr, Mauck R L and Tuan R S, Fabrication and characterization of six electrospun poly(α-hydroxy ester)-based fibrous scaffolds for tissue engineering applications, *Acta Biomaterialia*, **2**(4), 377–385, 2006a.

Li W J, Jiang Y J and Tuan R S, Chondrocyte phenotype in engineered fibrous matrix is regulated by fiber size, *Tissue Eng.*, **12**(7), 1775–1785, 2006b.

Li W J, Mauck R L, Cooper J A, Yuan X and Tuan R S, Engineering controllable anisotropy in electrospun biodegradable nanofibrous scaffolds for musculoskeletal tissue engineering, *J. Biomech.*, **40**(8), 1686–1693, 2007.

Liao S, Murugan R, Chan C K and Ramakrishna S, Processing nanoengineered scaffolds through electrospinning and mineralization suitable for biomimetic bone tissue engineering, *J. Mech. Behavior of Biomed. Mats*, **1**(3), 252–260, 2008.

Liu H, Li H, Cheng W, Yang Y, Zhu M and Zhou C, Novel injectable calcium phosphate/chitosan composites for bone substitute materials, *Acta Biomaterialia*, **2**, 557–565, 2006.

Liu Q, deWijn J R, Bakker D, van Toledo M and van Blitterswijk C A, Polyacids as bonding agents in hydroxyapatite polyester-ether (Polyactive™ 30/70) composites, *J. Mater. Sci.: Mater. Med.*, **1**(9), 23–30, 1998.

Lizuka T, Mikkonen P, Paukku P and Lindqvist C, Reconstruction of orbital floor with polydioxanone plate, *Int. J. Oral Maxillofacial Surg.*, **20**, 83–87, 1991.

Lu L C and Mikos A G, Poly(lactic acid). In: Mark J E (ed.), *Polymer Data Handbook*, Oxford: Oxford University Press, 1999, pp. 527–633.

MacNeil S, Biomaterials for tissue engineering of skin, *Materials Today*, **11**(5), 26–35, 2008.

Mano J F, Sousa R A, Boesel L F, Neves N M and Reis R L, Bioinert, biodegradable and injectable polymeric matrix composites for hard tissue replacement: state of the art and recent developments, *Compos. Sci. Technol.*, **64**, 789–817, 2004.

Martino A D, Sittinger M and Risbud M V, Chitosan: A versatile biopolymer for orthopaedic tissue-engineering, *Biomaterials*, **26**, 5983, 2005.

Meera G and Emilia A T, Polyionic hydrocolloids for the intestinal delivery of protein drugs: alginate and chitosan – a review, *J. Controlled Release*, **114**, 1–14, 2006.

Middleton J C and Tipton A J, Synthetic biodegradable polymers as orthopaedic devices, *Biomaterials*, **21**, 2335–2346, 2000.

Misra S K, Mohn D, Brunner T J, Stark W J, Philip S E, Roy I, Salih V, Knowles J C and Boccaccini A R, Comparison of nanoscale and microscale bioactive glass on the properties of P(3HB)/Bioglass® composites, *Biomaterials*, **29**, 1750–1761, 2008.

Moroni L, Hendriks J A, Schotel R, de Wijn J R and van Blitterswijk C A, Design of biphasic polymeric 3-dimensional fiber deposited scaffolds for cartilage tissue engineering applications, *Tissue Eng.*, **13**(2), 361–371, 2007.

Moutos F T, Freed L E and Guilak F, A biomimetic three-dimensional woven composite scaffold for functional tissue engineering of cartilage, *Nature Materials*, **6**, 162–167, 2007.

Mow V C, Ratcliffe A and Poole A R, Cartilage and diarthrodial joints as paradigms for hierarchical materials and structures, *Biomaterials*, **13**, 67–97, 1992.

Murugan R and Ramakrishna S, Development of nanocomposites for bone grafting, *Composites Sci. Techn.*, **65**(15–16), 2385–2406, 2005.

Nayar S and Sinha A, Systematic evolution of a porous hydroxyapatite-poly(vinylalcohol)-gelatin composite, *Colloids and Surfaces B: Biointerfaces*, **35**, 29–32, 2004.

Nerurkar N L, Elliott D M and Mauck R L, Mechanics of oriented electrospun nanofibrous scaffolds for annulus fibrosus tissue engineering, *J. Orthop. Res.*, **25**(8), 1018–1028, 2007.

Niiranen H, Pyhältö T, Rokkanen P, Kellomäki M and Törmälä P, In vitro and in vivo behavior of self-reinforced bioabsorbable polymer and self-reinforced bioabsorbable polymer/bioactive glass composites, *J. Biomed. Mater. Res. A*, **69A**, 699–708, 2004.

Oliveira J M, Rodrigues M T, Silva S S, Malafaya P B, Gomes M E, Viegas C A, Dias I R, Azevedo J T, Mano J F and Reis R L, Novel hydroxyapatite/chitosan bilayered scaffold for osteochondral tissue-engineering applications: Scaffold design and its performance when seeded with goat bone marrow stromal cells, *Biomaterials*, **27**(36), 6123–6137, 2006.

Olszta M J, Cheng X, Jee S S, Kumar R, Kim Y-Y, Kaufman M J, Douglas E P and Gower L B, Bone structure and formation: a new perspective, *Mater. Sci. Eng.: R: Reports*, **58**(3–5), 77–116, 2007.

Oowaki H, Matsuda S, Sakai N, Ohta T, Iwata H, Sadato A, Taki W, Hashimoto N and Ikada Y, Non-adhesive cyanoacrylate as an embolic material for endovascular neurosurgery, *Biomaterials*, **21**, 1039–1046, 2000.

Pereira C S, Vazquez B, Cunha A M, Reis R L and San Román J, New starch based thermoplastic hydrogels for use as bone cements or drug delivery carriers, *J. Mater. Sci. Mater. Med.*, **9**, 825–833, 1998.

Piskin E, Review. Biodegradable polymers as biomaterials, *J. Biomaterials Science, Polymer Edition*, **6**, 775–795, 1994.

Rebeling S, Di Silvio L and Deb S, Biomimetic synthesis and biological evaluation of carbonated HA and sodium alginate composites, *Proc. European Society of Biomaterials*, Nantes, France, 2006.

Rebeling S, Deb S and Di Silvio L, Biomimetic carbonated hydroxyapatite/alginate composites for use as an injectable cell-seeded bone graft, *Proc. European Society of Biomaterials*, Brighton, UK, 2007.

Reis R L and Cunha A M, New degradable load bearing biomaterials composed of reinforced starch based blends, *J. Appl. Med. Polym.*, **4**, 1–5, 2000.

Reneker D H and Chun I, Nanometre diameter fibres of polymer, produced by electrospinning, *Nanotechnology*, **7**, 216–223, 1996.

Rezwan K, Chen Q Z, Blaker J J and Boccaccini A R, Biodegradable and bioactive porous polymer/inorganic composite scaffolds for bone tissue engineering: *Biomaterials*, **27**(18), 3413–3431, 2006.

Rhee S H and Tanaka J, Hydroxyapatite coating on a collagen membrane by a biomimetic method, *J. Am. Ceram. Soc.*, **81**, 3029–3031, 1998.

Rinaudo M, Review. Chitin and chitosan: properties and applications, *Progress in Polymer Science*, **31**, 603–632, 2006.

Roether J A, Boccaccini A R, Hench L L, Maquet V, Gautier S and Jerome R, Development and in vitro characterisation of novel bioresorbable and bioactive composite materials based on polylactide foams and Bioglass® for tissue engineering applications, *Biomaterials*, **23**, 3871–3878, 2002.

Sabino M A, González S, Márquez L and Feijoo J L, Study of the hydrolytic degradation of polydioxanone PPDX, *Polymer Degradation and Stability*, **69**, 209–216, 2000.

Sakkers R J, de Wijn J R, Dalmeyer R A, van Blitterswijk C A and Brand R, Evaluation of copolymers of polyethylene oxide and polybutylene terephthalate (polyactive): mechanical behaviour, *J. Mater. Sci. Mater. Med.*, **9**(7), 375–379, 1998.

Seal B L, Otero T C and Panitch A, Polymeric biomaterials for tissue and organ regeneration, *Mater. Sci. Eng. R: Rep.*, **34**, 147–230, 2001.

Smidsrod O and Draget K I, Chemistry and physical properties of alginates, *Carbohydr. Eur.*, **14**, 6–13, 1996.

Sodergard A and Stolt M, Properties of lactic acid based polymers and their correlation with composition, *Prog. Polym. Sci.*, **27**, 1123–1163, 2002.

Springate C M, Jackson J K, Gleave M E and Burt H M, Clustering antisense complexed with chitosan for controlled intratumoral delivery, *Int. J. Pharmaceutics*, **350**, 53–64, 2008.

Sriamornsak P, Burton M A, Ross A and Kennedy P, Development of polysaccharide gel coated pellets for oral administration: 1 Physico-mechanical properties, *Int. J. Pharmaceutics*, **326**(1–2), 80–88, 2006.

Stoop R, Smart biomaterials for tissue engineering of cartilage, *Injury*, **39**(1), Supplement 1, 77–87, 2008.

Ta H T, Dass C R and Dunstan D E, Injectable chitosan hydrogels for localised cancer therapy, *J. Controlled Release*, **126**(3), 205–216, 2008.

Tao J, Abdel-Fattah W I and Laurencin C T, In vitro evaluation of chitosan/poly(lactic acid–glycolic acid) sintered microsphere scaffolds for bone tissue engineering, *Biomaterials*, **27**(28), 4894–4903, 2006.

Tay B K B, Le A X, Heilman M, Lotz J and David S, Use of a Collagen-Hydroxyapatite Matrix in Spinal Fusion: A Rabbit Model, *Spine*, **23**(21), 2276–2281, November 1, 1998.

Terheyden H, Knak C, Jepsen S, Palmie S and Rueger D R, Mandibular reconstruction with a prefabricated vascularized bone graft using recombinant human osteogenic protein-1: an experimental study in minative pigs. Part 1: Prefabrication, *International Journal of Oral and Maxillofacial Surgery*, **30**(5), 373–379, 2001.

Van Blitterswijk C A, deWijn J R, Leenders H, Brink J V D, Hesseling S C and Bakker D, A comparative study of the interactions of two calcium phosphates, PEO/PBT copolymer (Polyactive®) and a silicone rubber with bone and fibrous tissue, *Cells and Materials*, **3**, 11–22, 1993.

VandeVord P J, Matthew H W, DeSilva S P, Mayton L, Wu B and Wooley P H, Evaluation of the biocompatibility of a chitosan scaffold in mice, *J. Biomed. Mater. Res.*, **59**, 585, 2002.

Verma D, Katti K S, Katti D R and Mohanty B, Mechanical response and multilevel structure of biomimetic hydroxyapatite/polygalacturonic/chitosan nanocomposites, *Mater. Sci. Eng.: C*, **28**(3), 399–405, 2008.

Verrier S, Blaker J J, Maquet V, Hench L L and Boccaccini A R, PDLLA/Bioglass® composites for soft-tissue and hard-tissue engineering: an in vitro cell biology assessment, *Biomaterials*, **25**(15), 3013–3021, 2004.

Vert M, Christel P, Chabot F and Leray J, In: Hasting G W and Ducheyne P (eds), *Macromolecular Materials*, Boca Raton, FL: CRC Press, 1984, pp. 119–142.

Wilda H, Merry C L R, Blaker J J and Gough J E, Three dimensional culture of annulus fibrosus cells within PDLLA/Bioglass® composite foam scaffolds: assessment of cell attachment, proliferation and extracellular matrix production, *Biomaterials*, **28**(11), 2010–2020, 2007.

Williams D F, Review. The proving of polyhydroxybutyrate and its potential in medical technology, *Med. Device Technol.*, **16**(1), 9–10, 2005.

Xu H K, Weir M D and Simon C G, Injectable and strong nano-apatite scaffolds for cell/growth factor delivery and bone regeneration, *Dental Materials*, **24**(9), 1212–1222, 2008.

Xu X-J, Sy J C and Shastri V P, Towards developing surface eroding poly(α-hydroxy acids), *Biomaterials*, **27**, 3021–3030, 2006.

Yang S, Leong K F, Du Z and Chua C K, The design of scaffolds for use in tissue engineering. Part I. Traditional factors, *Tissue Eng.*, **7**, 679–689, 2001.

Zhao K, Deng Y, Chun Chen J and Chen G Q, Polyhydroxyalkanoate (PHA) scaffolds with good mechanical properties and biocompatibility, *Biomaterials*, **24**(6), 1041–1045, 2003.

Zimmerman M C, Alexander H, Parsons J R and Bajpai P K, The design and analysis of laminated degradable composite bone plates for fracture fixation. In: Vigo T L and Turbank A F (eds), *High Tech Fibrous Materials*, Washington, DC: American Chemical Society, 1991, Chapter 10, pp. 132–148.

3
Biocompatibility of degradable polymers for tissue engineering

N GURAV, Northwick Park Institute of
Medical Research (NPIMR), UK and L DI SILVIO,
King's College London, UK

Abstract: For several decades biodegradable polymers have been attracting interest as the materials for filling in defects or being used as stabilising materials. However, with the increasing pace in the developments in tissue engineering, the usage of these materials has also increased. The chapter presents some data on the biocompatibility testing of these materials and some of the challenges at the basic level. In determining the biocompatibility of materials and devices, we need to focus more specifically on the applications, which in turn will affect the type of cells and tests that will be used. We also need to determine the duration the material will be *in-vivo* and where appropriate consider breakdown products. The use of accelerated degradation using enzymes *in-vitro* can change the properties of material which can be used to 'mimic' degradation *in-vivo*. These and other aspects of biocompatibility will be discussed in this chapter.

Key words: biocompatibility testing, resorbable materials, biomaterials, *in-vitro* methods, *in-vivo* methods.

3.1 Introduction

Biocompatibility is central to biomaterials science and concerns the interaction of any material that is to be either placed in contact with or transplanted in the body. There are many definitions of biomaterials and biocompatibility but the one given by Williams remains widely accepted in science and medicine: a biomaterial is '*any substance (other than a drug) or combination of substances synthetic or natural in origin, which can be used for any period of time, as a whole or part of a system which treats, augments, or replaces tissue, organ, or function of the body*' (Williams 1987). Historically biomaterials have been classed as biocompatible on the basis that they do not cause any adverse or negative responses when implanted in the host.

The International Organisation for Standardisation (ISO) provides guidelines for *in-vitro* testing of biomaterials. It defines degradable materials as 'those which undergo a significant change in chemical structure resulting in a loss of physical and mechanical properties, as measured by standard methods'.

For almost four decades degradable materials had predominantly been used for filling in defects or as stabilising materials. However, with the new emerging field of tissue engineering they have been promoted to being quite important constituents for regenerating new tissues. Their requirements have changed significantly and thus the definition of a biomaterial now encompasses *'any material used in a medical device intended to interact with biological systems'* (Ratner and Bryant 2004). Biomaterials now also need to be able to support cell growth and allow differentiation into different lineages as well as not being cytotoxic or eliciting an immune response. Hence this existing biomaterial now needs to satisfy the challenges associated with tissue engineering to become the scaffold of choice.

With this new wave of biomaterial development the original definition of biocompatibility still needs to be fulfilled as defined by Williams, but over the years the definition has been redefined to encompass the constantly changing field of tissue engineering and repair (Williams 1989). To determine biocompatibility we need to focus more specifically on the application, hence the cell types that will come into contact with the material need to be considered. We also need to determine the duration the material will be *in-vivo* and where appropriate, consider breakdown products. Then finally a protocol needs to be developed to test this biocompatibility and ultimately the functionality of the biomaterial as it changes from the original construct to the degrading material colonised by the cells into integrated matrix containing host tissue. Finally, when there is no biomaterial remaining at the intended implant site the functionality of the new tissue as compared to the host tissue remains to be assessed. The effect of the degradation products throughout this process remains crucial and needs to be tested in several ways. Tests naturally need to become more sophisticated to accommodate the challenges this new discipline is throwing to the 'humble biomaterial'. Ultimately it is the range of tests which are carried out that will determine how the material will be developed into engineered constructs ready for implantation before any clinical trials. This selection will be based not only on their biocompatibility but on their ease of use, cost-effectiveness and availability.

This chapter will describe various aspects of biocompatibility testing and give some examples of the *in-vitro* and *in-vivo* tests used. We will focus on the most common, the aliphatic polyesters, and describe the mechanisms and changes leading to their degradation, and we will review the literature of the current status in the development of degradable tissue engineered constructs. The scientific literature is abundant with some excellent reviews on the issues of biocompatibility (Kirkpatrick *et al.* 2005; Williams 2008; Leach and Mooney 2008; Helmus *et al.* 2008; Thevenot *et al.* 2008). This chapter does not aim to reiterate these, but to give a practical perspective on how the whole process of testing these materials is brought about. In our experience over the last decade *in-vitro* biocompatibility testing remains a fastidious and time-consuming task, throwing up new challenges with each new biomaterial. It is hoped that the reader will gain an insight into the practical challenges of testing biocompatibility, particularly *in-vitro*, and how to go about achieving a solution.

3.2 Biocompatibility

Biocompatibility assessment is essential for pre-clinical testing of materials intended for use as implantable devices. The Medical Device Directive states that a material to be used for medical devices should comply with toxicity tests (Haustveit *et al.* 1984; Robertson 1998; Dobbs 2007). Before beginning any biocompatibility testing programme the Medicines and Healthcare products Regulatory Agency (MHRA) and the ISO 10993 standard should be consulted. The MHRA is responsible for ensuring the safety and integrity of medicines and medical devices. The International Organisation for Standardisation (ISO) is a worldwide federation of national standards bodies in which the ISO 10993 consists of a number of parts under the general title of 'Biological evaluation of medical devices'. The international standard ISO 10993 defines a series of guidelines that can help in selection of the most appropriate test for biocompatibility screening. It does not state exactly which tests need to be carried out, which is pertinent as the nature of the tests would depend on the material and application and it is impossible to state exact tests for such a large number of materials and devices. They are constantly updated and provide a useful guideline to establish a testing programme for biomaterials.

ISO 10993-1: Evaluation and testing
ISO 10993-2: Animal welfare requirements
ISO 10993-3: Tests for genotoxicity, carcinogenicity and reproductive toxicity
ISO 10993-4: Selection of tests for interactions with blood
ISO 10993-5: Tests for *in-vitro* cytotoxicity
ISO 10993-6: Tests for local effects after implantation
ISO 10993-7: Ethylene oxide sterilisation residuals
ISO 10993-9: Framework for identification and quantification of potential degradation products
ISO 10993-10: Tests for irritation and delayed-type hypersensitivity
ISO 10993-11: Tests for systemic toxicity
ISO 10993-12: Sample preparation and reference materials
ISO 10993-13: Identification and quantification of degradation products from polymeric medical devices

ISO 10993-14: Identification and quantification of degradation products from ceramics
ISO 10993-15: Identification and quantification of degradation products from metals and alloys
ISO 10993-16: Toxicokinetic study design for degradation products and leachables
ISO 10993-17: Establishment of allowable limits for leachable substances
ISO 10993-18: Chemical characterisation of materials
ISO/TS 10993-19: Physico-chemical, morphological and topographical characterisation of materials
ISO/TS 10993-20: Principles and methods for immunotoxicology testing of medical devices

The testing procedures available for the assessment of biocompatibility should give quantitative data as well as qualitative data using appropriate cell types. Assessment of biodegradable polymers, however, is complex due to the degradation of the materials during testing and the need to artificially age the materials in order to assess longer-term effects. Owing to the wide variety of methods for material formulation which have different molecular weights, the biocompatibility of many resorbable materials in the long term remains unclear. However, 'combinative techniques' which involve morphological analysis and molecular biology present a useful set of methods for the biocompatibility testing of materials (Kirkpatrick *et al.* 1997).

Biocompatible in the *in-vitro* methods situation means that the material or any leachable product from it does not cause cell death or impair cellular functions. Biomaterials used *in-vivo* should be biocompatible both in the short term and also in the long term during degradation of the material and the release of degradation products. These include the bioresorbable materials and bioactive materials. For example, a material that is intended for use as a drug delivery vehicle will require different properties from one that will be used for tissue replacement. A biomaterial designed for hip replacement will require high mechanical strength and have low corrosion and degradation, in addition to having good bone induction, protein adsorption and cell adhesive capabilities. However, a material used for vascular reconstruction will need to have good stability, low degradation and low protein and cell adsorption properties. Hence, each material to be tested should be considered in the context of its intended use.

When the researcher is presented with a biomaterial to test, a systematic approach should be adopted to determine its safety when used in the body. The first challenge is to get the material supplied or produced in sufficient quantities in order to obtain data with good statistical significance, and to prevent batch-to-batch variation. This can actually be harder than it appears, mainly because the material can be costly or difficult to make or the supplier does not have it in sufficient quantities. Certainly at an early developmental stage obtaining sufficient quantities of the biomaterial has always proved to be very challenging in university research laboratories, where perhaps cost is an issue compared to private companies. Once sufficient quantities of the raw polymer are available, handling, sterilisation and test protocols to be used need to be determined. The formulation process of the device may also change the physical and chemical characteristics of the material, thus testing at every stage of the developmental process needs to be carried out.

To begin testing a biomaterial for its biocompatibility, the first thing that needs to be determined is where the final implant or device is to be used. This will allow the determination of the duration of the implant and the cells that will be in direct contact with it. *In-vitro* tests are not fully confirmative of the performance of a biomaterial, but they do provide essential quantitative data on cell viability and proliferation as the material is placed in a biological environment over time. A material, however, can perform very badly in an *in-vitro* test but go on to perform very well in an *in-vivo* situation. This is usually due to the accumulation of toxic products in the *in-vitro* situation which in an *in-vivo* environment would be readily removed by the body. Interpretation of data from *in-vitro* tests can be quite subjective, more so than that of *in-vivo* data in certain circumstances, and hence it is important to know what these are before allowing the material to progress to the next stage of development. More sophisticated tests that utilise perfusion and continuous flow of cultures are being developed in order to mimic more physiological conditions (Sikavitsas *et al.* 2003; Pan *et al.* 2008; Fan *et al.* 2008).

In determining the suitability of a biomaterial for a specific application, the material and site of implantation play a major role. The reaction is a two-way process: the effect the material has on the body and the effect the body has on the material in question. Degradable biomaterials are constantly changing with time as the biomaterial undergoes bulk and surface degradation which leads to changes in the environment around the material. Non-degradable materials also undergo changes in their chemical and structural nature, but this change is not as dramatic and the time involved for physical and chemical changes to occur is longer.

Appropriate biocompatibility testing is an essential pre-requisite in the development of any implantable device. However, biocompatibility testing has associated with it numerous, often undocumented, problems which can have a significant bearing on the final outcome of the various *in-vitro* test results. A systematic approach should therefore be adopted when testing materials; the number and types of tests and specimen replicates should be considered for statistical significance.

3.3 Degradable polymers

Biodegradable polymers are categorised under several generic names, some of which include polyesters, polyurethanes, polyamides, polyureas, polyanhydrides, polyphosphazanes, polyacrylates and polycyanoacrylates. The polymers most studied and those under intensive research currently are the polyesters, which include PLA and PGA. The first synthetic absorbable material was developed in 1962 by the Cyanamid Corporation using polyglycolic acid which was marketed under the trade name Dexon in the late 1960s and is still in use (De Buren 1970; Alveryd and Jacobsson 1975; Kersey et al. 2006). There are now a variety of degradable polymers available for medical use. These include poly(DL-lactide) (DLPLA), poly(L-lactide) (LPLA), polyglycolide (PGA), poly(dioxanone) (PDO), poly(glycolide-co-trimethylene carbonate) (PGA-TMC), poly(L-lactide-co-glycolide) (PGA-LPLA), poly(DL-lactide-co-glycolide) (PGA-DLPLA), poly(L-lactide-co-DL-lactide) (LPLA-DLPLA) and poly(glycolide-co-trimethylene carbonate-co-dioxanone) (PDO-PGA-TMC).

There has been increasing interest in degradable polymer systems for use in biomedical applications such as drug delivery (Koosha et al. 1989), fracture repair (Ewers and Lieb-Skowron 1990), tissue remodelling (Freed et al. 1993; Gilbert et al. 1993; Puelacher et al. 1994) and soft tissue implants (Mooney et al. 1996). Degradable materials have certain advantages that make them desirable for some orthopaedic applications. Their degradation rates and tensile strength can be controlled by varying molecular weight and, for copolymers, varying the ratio of the components can also dramatically affect their degradation rates (Nakamura et al. 1989). The majority of degradable materials have been poorly characterised using *in-vitro* methods which have usually involved the assessment of fibroblast, and some osteoblast and hepatocyte cell models. In contrast, there has been considerable work reported using degradable polymers *in-vivo*, but the mechanisms of cell attachment and proliferation on these polymers are not fully understood due to the lack of appropriate *in-vitro* methods. Most importantly, the effect of the cellular activity on the degradation of the polymer and the effect of the degradation products on the cells are poorly understood.

There are numerous disciplines that are actively using degradable materials in their research. Many are more established than others. Each has its own objectives and aims and cell types that need to be evaluated. Polyesters of α-hydroxyacids and other degradable polymers have attracted and are continuing to attract interest for use as controlled drug delivery systems (Schakenraad et al. 1988; Huatan et al. 1995a, 1995b). Most common are the homo- and co-polymers of lactic and glycolic acids, although collagen-based hydrogels and gelatin in the form of microspheres have also been investigated for use in delivering antibiotics and growth factors (Eldridge et al.

1992; Di Silvio *et al.* 1994). These drug delivery systems offer the same advantages as the degradable materials used in other applications. There is no need for a second surgical intervention to remove the implant, allowing for reduced inconvenience to the patient, reduced health costs and reduced risk of infections and complications, as well as long-term drug sustainability.

The most common delivery vehicles are microparticulate drug delivery systems which are produced by solvent evaporation, solvent extraction phase separation or spray drying. Spray drying is the most commonly used method as it is fast and results in the formation of particles which are 1–15 µm in diameter and are easily scaled up. The biocompatibility of the drug delivery biomaterials is a little more complicated, as the material needs to be biocompatible and also possess the desired release kinetics.

PLA has been used as a delivery vehicle for calcitonin, which is a synthetic analogue of hypocalcaemic peptide hormone used to treat osteoporosis, Paget's disease and hypercalcaemia and has to be injected regularly over a long time period (Asano *et al.* 1993). The release of calcitonin from poly(DL-lactide) with molecular weights of 1400, 2000 and 4400 was measured *in-vitro*. The lowest molecular weight polymer released the entire drug within 3 days, whereas the 4400 molecular weight polymer released at a constant rate for 24 days.

There are increasing demands that new vaccines be without risk and free from side effects. As the isolated antigens are sometimes of low immunogenicity, there is a need for delivery vehicles which are safe and which enhance the antibody response. In a study, Eldridge *et al.* (1991) used microspheres of DL-L-lactide-co-glycolide as vehicles for staphylococcal enterotoxin B toxoid which enhances the level of toxin neutralising antibodies. They concluded that due to the biocompatibility of the copolymer and the flexibility in terms of the release kinetics, it has been approved for human use and is being investigated for mucosal immunisation because of its ability to protect the vaccine and enhance adsorption into associated tissues. In another study, progesterone was loaded in microspheres of poly(DL-lactide-co-glycolide) and it was found that variations in the progesterone loading affected the structural and thermal properties (Rosilio *et al.* 1991).

Poly(D,L-lactide) and poly(D,L-lactide-co-glycolide) microspheres containing Piroxicam were prepared by a spray drying method and *in-vitro* release studies were carried out. Release from the DL-PLA microspheres was slow: less than 20% of the drug had been released by day 10. The DL-PLGA microspheres, however, released about 50% of the drug in the first 5 hours of the study. It was suggested that release from the DL-PLA was by diffusion through the intact polymer barrier. The release rate from DL-PLGA was much faster and this was due to the immediate water absorption by the polymer due to the glycolic acid content which is more hydrophilic that PLA. This allows the microsphere to swell thus allowing for diffusion of the drug out through pores.

The major problem associated with biocompatibility testing of polymers is the long time periods associated with their degradation. As a result of this, the *in-vivo* results are slow to emerge and incorrect conclusions are often drawn regarding the biocompatibility of the polymers. However, this does not tackle the problem caused by the rapid degradation of the implant leading to severe inflammatory responses. To tackle the problems with long-term degradation, more studies are being carried out which use polymers or polymer degradation products that have been aged *in-vitro* prior to implantation into animal models (Mainil-Varlet *et al.* 1997a, 1997b; De Jong *et al.* 2005). We have shown that polymers (Fig. 3.1) can be degraded *in-vitro* and these can then be used to test biocompatibility at different stages of degradation (Gurav 1997).

The definition of biocompatibility as given by Williams (1987) is widely accepted as the general definition which is applied to all materials used for implantation purposes. However, the designation of 'biocompatibility' is a combination of biofunctionality and biosafety. For each current application and newly developing applications these need to be assessed. This is inevitably a long and difficult task, but one that is necessary, as currently a whole range of tests are performed by different groups and researchers which are not comparable due to the differences in the methodologies used. It is first necessary to determine the major requirements of each medical device and then accordingly choose a material that will fulfil the criteria. Applications generally fall into the headings: dental, orthopaedic, vascular implants, drug delivery, transplantation, nerve regeneration, and soft tissue implants. A major requirement for all these applications is that the materials must be non-immunogenic, that is, they should not induce gross inflammatory or immune responses.

3.4 Practical problems during biocompatibility testing

It is important to perform a basic physical analysis of the material being tested. Physico-chemical characteristics including swelling, water uptake and leachables should be assessed. The size of the test sample is important so that they can fit into standard sized tissue culture plastic dishes such as, 24, 48 and 96 well plates allowing for better reproducibility and ease of handling. Biocompatibility testing has practical problems experienced by many researchers but rarely documented, that can influence the test results and final conclusion. Examples of such problems include materials floating in culture, the uptake of reagent dyes by the test material, inability to view cells on opaque materials, interference of media additives with specific

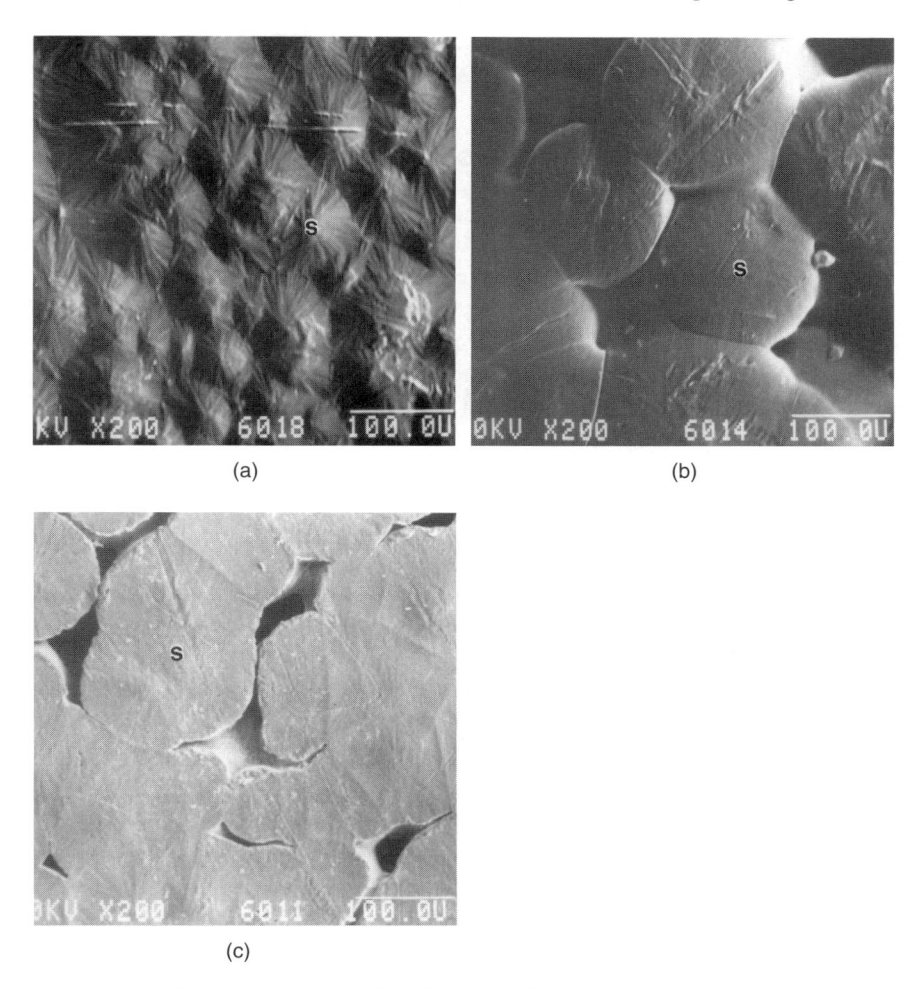

Biocompatibility of degradable polymers for tissue engineering 69

3.1 Surface of as-cast PCL films showing degradation by enzymes (s = spherulites): (a) as-cast PCL film; (b) PCL film stored in a trypsin solution for 6 weeks at 37°C; (c) surface of PCL film stored in papain for 6 weeks at 37°C.

tests and assays, and the limited number of samples made available for testing.

During the development stage materials can be expensive and supplies are limited, and the use of endpoint assays requires many samples in order to obtain statistically significant results. Some materials leach out toxins which affect the biocompatibility of the materials and hence they need to be washed before tests can be carried out. Tests on gels and porous materials may be more complex in nature due to infiltration of cells into the scaffold; this may be desirable or undesirable depending on the application. Removal or visualisation of these cells in a way that allows accurate determination of results can be difficult and may give rise to non-conclusive information.

3.5 Degradation mechanisms and devising the appropriate tests

Many factors, almost too numerous to list, influence the degradation of a material and hence predicting the exact outcome of a polymer in the *in-vivo* situation is made even more difficult. However, *in-vitro* we can determine the degradation pattern with a relative degree of accuracy. If the chemical structure and chemical composition are known, it is possible to predict the degradation in a given environment. Factors such as the shape, site of implantation, sterilisation and processing as well as storage will influence the material, so from the development stage to a stage where the material will be used *in-vivo* many factors will influence degradation rate.

The degradation rate of a polymer is influenced by factors such as crystallinity, initial molecular weight and the morphology of the specimen. Degradation of PLA and PGA homopolymers and copolymers is by hydrolysis, and copolymers have a wide range of degradation rates governed by the hydrophilic and hydrophobic nature of the copolymers as well as the crystallinity. Degradation is described as the hydrolysis of ester bonds via a bulk erosion process which is autocatalysed by the generation of carboxylic acid end groups (Vert *et al.* 1992, 1994a, 1994b). The main polymer chains undergo cleavage, resulting in smaller fragments (oligomers) which break up further to form monomers. The *in-vitro* degradation of high molecular weight aliphatic polyesters derived from glycolic acid and lactic acid is heterogeneous with a faster degradation rate in the centre compared to the surface (Leenslag *et al.* 1987; Ali *et al.* 1993a, 1993b; Vert *et al.* 1994a, 1994b). This can be explained by the fact that there is greater acid autocatalysis in the centre so that the degradation products from the centre of the polymer cannot be removed quickly, and there is a build-up of acidic degradation products which catalyse the further breakdown of the polymer. Degradation is also slower *in-vitro* as compared to *in-vivo*, most likely due to the role of enzymes (Therin *et al.* 1992). There is considerable evidence suggesting the role of enzymes in the degradation of polymers (Schakenraad *et al.* 1990; Chen *et al.* 2004; Zhao *et al.* 2008). In 1987 Smith *et al.* further demonstrated the degradation of non-resorbable polymers by a number of enzyme solutions which included trypsin, papain and chymotrypsin.

3.6 *In-vitro* biocompatibility testing

In-vitro assessment of materials initially concerns biosafety, that is, the exclusion of deleterious effects due to toxicity either directly or indirectly by toxic leachables. This initial screening can yield significant data about the material or device, saving valuable time and money prior to moving on to more complex biofunctionality tests, where the 'appropriateness' of the device for its intended site of use is tested. Although *in-vitro* testing cannot completely replace *in-vivo* tests, more sophisticated and sensitive tests are being developed in order to reduce animal testing. *In-vivo* testing is also more expensive and lacks control due to animal and species differences, and problems are frequently encountered with obtaining ethical permission to carry out tests. *In-vitro* testing using cell culture models has a number of advantages over *in-vivo* testing, in that the response of a variety of cells can be compared and evaluated over a shorter time period in a reproducible and controlled manner as compared to *in-vivo* tests. *In-vitro* testing is usually carried out before *in-vivo* testing and involves the culture of appropriate cell types on the material concerned.

In-vitro biocompatibility testing involves studying the behaviour of various cell types in response to a particular test agent, topography, chemistry or surface activity. It allows all aspects of cell metabolic activity to be evaluated; the simplest of these is the monitoring of cell viability and proliferation when in contact with a biomaterial. Any toxic effect on cells will cause alterations in cell membrane integrity, and this can be tested with a combination of dyes, for example trypan blue and erythromycin red, which are exclusion or inclusion dyes. Trypan blue stains dying cells blue and erythromycin red stains non-viable cells pink. A combination of fluorescein diacetate (FDA) and ethidium bromide (EB) is frequently used. FDA is taken up by the intact cells and gives a green fluorescence at the correct excitation wavelength, indicating the presence of viable cells. These tests can give a rapid estimation of cell viability. Coulter counters and cell sorters utilising fluorescent dyes can be used to distinguish dead cells from live ones to give a more accurate cell count.

In the evaluation of biocompatibility the choice of cells is important. If for example, the device is being developed for orthopaedic fixation, then osteoblasts will usually be cultured on the surface. In the case of material to be used for vascular reconstruction, blood compatibility will be investigated. Different material characteristics are needed, depending on the final application. For the two applications the material will have different properties and be expected to behave differently; for example, in orthopaedic fixation, the material will need to have good protein adsorption characteristics to aid cell adhesion and proliferation, while for the blood contact material it needs to have low protein adsorption and low cell attachment.

Although the time points usually used for *in-vitro* testing are short compared to *in-vivo* testing, *in-vitro* testing does yield information on early cell response, rate of cell adhesion, changes in cellular morphology and retention of phenotype. *In-vivo* testing is usually long term and can last several years, thus involving extra cost. The advantage of *in-vitro* testing for degradable polymers is that they can be artificially degraded, resulting in a reduction of time periods involved for subsequent *in-vivo* testing. Artificial degradation methods can be used, such as storage of polymers at high temperature, and in enzyme solutions which accelerates their degradation compared to *in-vivo*. This has been demonstrated by several groups of researchers who have artificially degraded polymers before implantation into animals. We have shown in a previous study that PLLA pre-degraded at elevated temperatures resulted in a reduction in follow-up times (Gurav 1997). Long-term biocompatibility testing is particularly important for degradable materials that are constantly changing their chemical and physical structures and releasing degradation products which may be detrimental to the cells adjacent to the implant.

Primary cells isolated from tissues are a good choice but obtaining sufficient cell numbers to carry out the numerous tests can be problematic, as well as obtaining normal human tissue for a particular cell type. Cell lines are therefore more commonly used, and sometimes more useful as they can replicate faster and have a well-defined phenotype. For other devices fibroblasts, epithelial cells and hepatocytes can be used, while studies of the immune response can utilise monocyte/macrophages and other cells of the immune system. *In-vitro* testing allows the standardisation of test methods and reduces the need for large numbers of experimental animals. It also allows qualitative determination of signs of cell damage, detected by light or electron microscopy. The morphology of cells varies on different materials (Fig. 3.2) and this change represents how the cell senses the material surface. The changes in surface structure caused by the degradation of polymers will also influence the cellular response to the materials.

Although quantitative methods are valuable for direct comparisons for biocompatibility testing, qualitative analysis of cells on materials is crucial for determining the morphology and behaviour of cells. Differences in the morphology of various cell types can be observed and cytotoxic effects which lead to dramatic changes in morphology of the cells can be examined at the light microscope level or in more detail at the electron microscope level. Cell spreading, for example, is a parameter of importance in studying biomaterials designed to be fully integrated with host tissue. Cytoskeletal reorganisation is an essential function of cells that have adhered to a surface, and precedes the function of cell proliferation. Dedifferentiation of cells is one of the disadvantages of *in-vitro* testing and precautions must be taken to monitor cell dedifferentiation.

The majority of *in-vitro* biocompatibility tests described in the literature involve the morphological assessment of cells on materials, usually by light or scanning electron microscopy. The time courses for cell exposure are

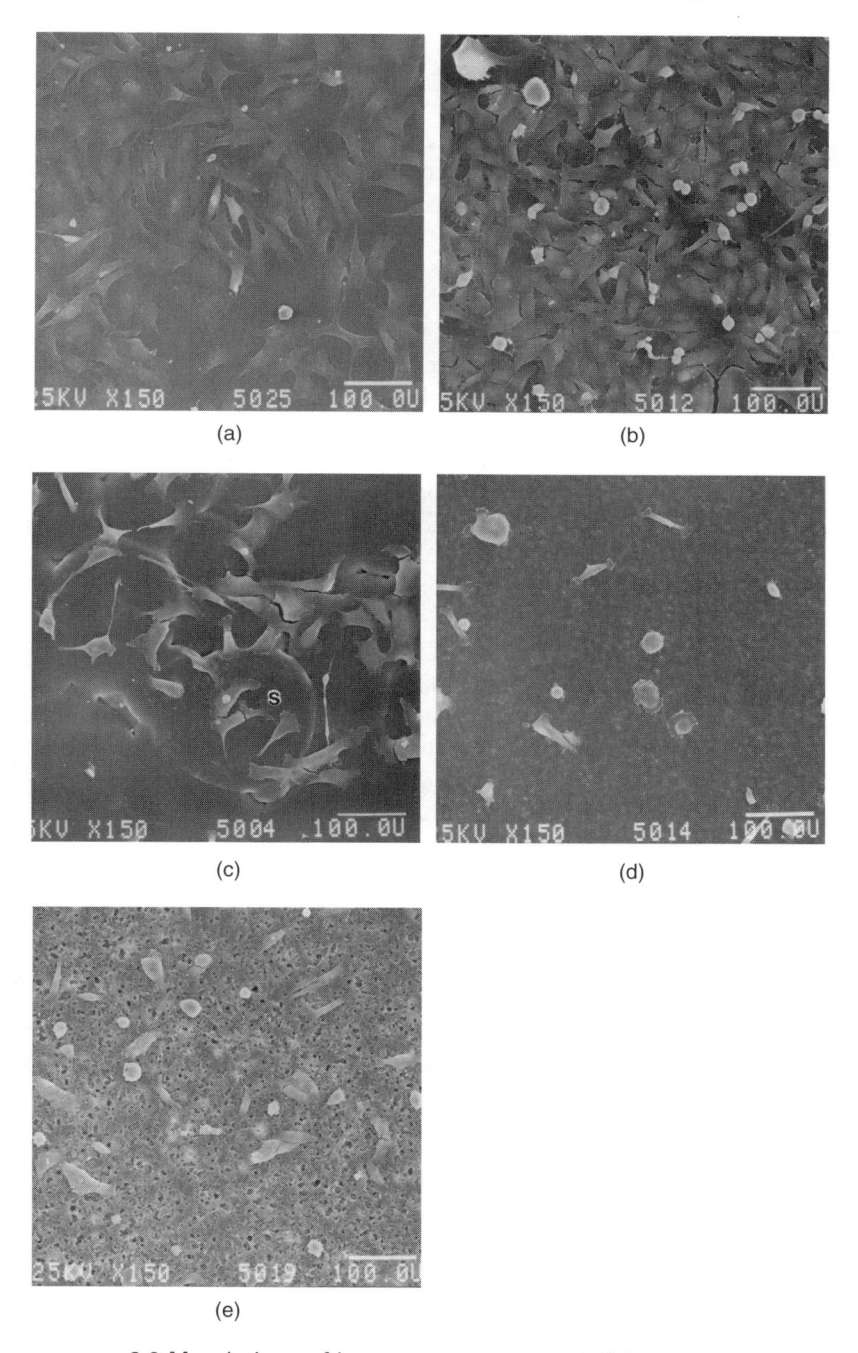

Biocompatibility of degradable polymers for tissue engineering 73

3.2 Morphology of human osteosarcoma (HOS) cells on the surfaces of as-cast degradable polymer films: (a) HOS cells on Thermanox used as a control surface; (b) HOS cells on PLA; (c) HOS cells on PCL (s = spherulite); (d) HOS cells on PHB; (e) HOS cells on PHB–PHV copolymer.

usually short term and at the most last for a few days. The types of cells used are often different in source and species. This can give conflicting results and make comparisons between different laboratories difficult. It has been demonstrated that different cells will behave differently on the same material surface. Differences were observed between periodontal ligament (PDL) fibroblasts and L929 cells (a mouse fibroblast cell line) when measuring toxicity of a dental material (Al-Nazhan and Spangberg 1990); the PDL fibroblast cells were less sensitive than the L929 cells, and ultrastructural differences between the two cell types were observed. Also different cell types, for example fibroblasts and osteoblasts, respond differently to a range of biomaterials (Hunter et al. 1995).

Despite the problems associated with in-vitro screening of materials, these tests can provide fast, reproducible data, which include appropriate quality controls such as negative (non-toxic) and positive (toxic) controls. Furthermore, other controls can be introduced to monitor cell culture conditions and material controls, thus providing additional useful data. This is not possible for in-vivo tests due to the differences in site and species and the size and type of materials used. Thus in-vitro tests can be used both to determine initial attachment of different cell types on various material surfaces and for carrying out more 'long-term' biocompatibility testing (Kirkpatrick 1998; Kirkpatrick et al. 1998; Rickert et al. 2006).

There is evidence to suggest that certain aspects of toxicity at the cellular level are not easily detectable in-vivo, due to the numerous other systemic factors that may be involved, whereas in-vitro systems allow the investigation of single effects without interference from the whole immune system. The range of applications for in-vitro tests is vast, but what is required is a standardised method of carrying out these tests. A standardised protocol needs to be developed that sets out tests which provide meaningful results, which can then be used for comparison within groups carrying out similar tests on the same materials. It is not only the cell types that vary, but also the test material preparation, treatment and cell seeding mode and density. For example, polylactic acid may be the material under investigation but the method of formulation and sterilisation may be different. New applications of materials require that they be fully characterised and have defined material properties, which is currently not the case. There is a lack of reliable data and of tests carried out and materials cannot be compared due to the variability in the samples (Engelberg and Kohn 1991).

Cell death is a major concern with regard to biocompatibility testing. If cells die on a device or a material under investigation, it is obvious that there is some toxic leachable or that the material itself has a detrimental effect and is killing the cells. There are, however, different types of cell death, notably necrosis and apoptosis. There are various methods available to test cell death and the mechanisms of cell death. There are two ways in which cells can die, either by necrosis (passively) or by apoptosis (actively) which involves signal transduction pathways. In necrosis, there is mitochondrial swelling and plasma membrane permeation. In the case of apoptosis, death involves breakdown of the cell into membrane-bound particles, referred to as apoptotic bodies. For example, expression of phosphotidylserine on the cell surface can be measured *in-vitro* using binding protein annexin A5 conjugated to fluorochromes.

More sophisticated methods have now been developed that look at all aspects of biofunctionality, beyond cytocompatibility. Specific aspects of inflammation, for example metal corrosion and uptake of metal products by polymorphonuclear (PMN) leucocytes, macrophage activation in the presence of orthopaedic implant materials, mechanisms of blood-material interactions using fluid shear systems, and molecular biological approaches in determining gene activity at both transcriptional and translational levels, are all areas that are being applied for *in-vitro* testing.

3.7 *In-vivo* biocompatibility testing

Degradable biomaterials are constantly changing with time as the biomaterial undergoes bulk and surface degradation which leads to changes in the local environment around the material. Non-degradable materials also undergo changes in their chemical and structural nature but this change is not as dramatic and the time involved for physical and chemical changes to occur is much longer than for degradable materials.

In-vivo testing usually involves the implantation of a test device (which has been sterilised by an appropriate method) into an animal model. *In-vivo* tests are also listed under the ISO 10993 guidelines: Part 3 – tests for genotoxicity, carcinogenicity and reproductive toxicity; Part 4 – selection of tests for interaction with blood; Part 6 – tests for local effects after implantation; Part 10 – tests for irritation and sensitisation; and Part 11 – tests for systemic toxicity. Such studies yield information on the long-term biocompatibility of materials and the effect of the material on the immune system. The problems associated with *in-vivo* biocompatibility testing are usually due to the difficulty in quantifying specific cellular and biological events. However, several methods have been developed which aim to isolate cell–biomaterial interactions *in-vivo*, one of which is the cage implant system (Marchant 1989). This allowed for the evaluation of the effect of cellular and humoral components of the exudates which surround polymeric materials following implantation. In this method, the biomaterial was placed inside a stainless steel cylindrical cage which was then implanted subcutaneously into a rat model. Aliquots of the exudates surrounding the implant were removed using a syringe and analysed. The types of cells present in the sample could then be identified.

In-vivo inflammatory potential of implants has been assessed by implanting the materials into experimental animals and examining the surrounding tissue histologically and biochemically. The presence of inflammatory cells has been demonstrated microscopically by immunostaining and by enzyme histochemistry. The surface area of the particle is also important in determining the inflammatory response (Gelb *et al.* 1994). This is applicable not only to wear particles from large non-degradable implants but also to degradable materials where the material is undergoing degradation giving rise to particles which elicit an inflammatory response. Degradation of PLA and PGA yielded molecular weight products of 10 000–20 000 and an increase in inflammation was observed (Spenlehauer *et al.* 1989). Although the long-term biocompatibility of some large degradable devices is still unclear, it can be said that the poly(esters) and their copolymers are considered biocompatible and have found use in a large number of applications (Claes and Ignatius 2002).

Bergsma *et al.* (1993) used SR-PLLA screws and plates to treat zygomatic fractures. The number average molecular weight M_n was 7.6×10^5 and 10 patients were treated. After three years some patients returned with a painful swelling at the site of implantation and on recalling the others a similar reaction was noted. Re-operation was carried out and the material and area were analysed over a period of 3.3 to 5.7 years. It was found that there was a fibrous capsule around the implant and on molecular weight analysis at both points the M_n was approximately 5000. Very little change in the M_n value of the polymer had taken place at the two points. It was hypothesised that this was due to the high crystallinity of the PLLA fragments which take a long time to degrade. Also, remnants of PLLA particles which were needle shaped were found in cells such as macrophages and fibroblasts around the implant. This may contribute to the slow degradation of the polymer as inside the cells they were not able to degrade, since the main degradation mechanism is hydrolysis and even though macrophages contain a large range of enzymes this does not have very much effect on the polymer.

3.8 Immune response

Biomaterials when implanted into the body are essentially foreign bodies, which are expected to elicit an immune response. The severity and nature of this response will depend on the material under investigation. Therefore the initial host response must encompass a repair process. Neutrophils are the first cells to reach the site of infection, attracted by chemotactic factors such as complement, with their predominant role being phagocytosis. The phagocytosed body is fused with a lysosome and destroyed by lytic enzymes. The neutrophil respiratory burst involves a sharp uptake of oxygen and results in the generation of toxic oxygen products, such as superoxide anions and hydroxyl radicals. If the material cannot be phagocytosed the neutrophil will release destructive substances into the extracellular environment in a process known as frustrated phagocytosis. Any of these parameters can be used as an indicator of neutrophil activation by biomaterials. Once at a site of the biomaterial, there are two ways in which neutrophils could act, either by respiratory burst or by the release of lysosomal enzymes (Delves and Roitt 2000a, 2000b; Delves *et al.* 2006).

Once it has been determined that the novel material device does not cause cell death, it can be tested further to assess effects on immune response. One of the first things encountered by a material following implantation in the body is blood, which contains cells of the immune system. Polymorphonuclear leukocytes (PMNs) and macrophages become phagocytic when cellular debris or foreign particles are present. When phagocytosis is initiated, the cell becomes metabolically active and there is a respiratory burst resulting in the production of hydrogen peroxide, oxygen free radicals and anions. These free radicals are effective in killing many types of bacteria and have been linked with initiating the degradation of some polymers (Schakenraad *et al.* 1990; Freyria *et al.* 1991; Anderson *et al.* 1999; Dawes *et al.* 2003; Chang *et al.* 2008). The role of the macrophage in the inflammatory process is very important because it is the longest surviving and most active of the inflammatory cells. Studies have shown that the activated macrophages are able to stimulate bone resorption when they phagocytose particles (Murray and Rushton 1990).

Adverse responses *in-vivo* include decreased cell viability and proliferation of cells around the site of implantation or any reaction not normal for that tissue. An inflammatory reaction can also occur due to the influx of inflammatory cells to the area. Although some inflammation is necessary for the wound healing process to proceed normally, if this inflammatory reaction continues it can lead to chronic inflammation and granuloma. Biocompatibility testing must therefore address all of these parameters both *in-vivo* and *in-vitro*. However, in the case of *in-vivo* testing, this would require the use of a large number of animals in order to obtain statistically significant data.

The monocyte/macrophage response to biomaterials has been the most extensively studied of all the cells of the immune system (Bonfield and Anderson 1993; Benahmed *et al.* 1996; Xia *et al.* 2004; Xia and Triffitt 2006) and has been implicated in causing increased degradation of the materials. Macrophages have been shown to infiltrate the implant site after implantation of a biomaterial and, once activated, macrophages have a much altered metabolism. This includes altered phagocytic ability and increased lysosomal enzyme release as well as secretion of cytokines and growth factors. Macrophage activation by materials can be assessed by measuring any of the above parameters. The effect of metals on macrophage viability has been determined by a number of researchers, but the effect of degradation products and the effect of degradable polymers on macrophages both *in-vivo* and *in-vitro* has not been as extensively studied.

Activated macrophages display altered morphology which can be identified microscopically. Macrophages cultured with a range of biomaterials were analysed by scanning electron microscopy for alterations in morphology, attachment and cell density (Miller and Anderson 1989; Miller *et al.* 1989). After 24 hours in culture the cells displayed an activated morphology; they were elongated or flattened with numerous filopodia for attachment. The highest density of morphologically active cells was seen on Dacron and polyethylene, while the lowest were on polydimethylsiloxane (PDMS). Supernatants from these cultures were assayed for interleukin-1 (IL1). A good correlation between IL1 secretion and morphological activation was seen.

Lymphocytes are the cells responsible for the specific immune response. Their reaction towards foreign bodies includes secretion of cytokines and antibodies. Activation is also seen as expression of surface antigens. Metal ions can be released into the body due to the biodegradation of metallic prostheses and other implants. Again, a large amount of work has been carried out on the effect of metal particles on these cells, but very little work has been done on the effect of degradation products from degradable polymers on these cells.

The normal healing process occurs in two phases (Williams 1989). The first phase is inflammation, followed by repair. Immediately following injury, blood enters the area and a fibrin clot forms, trapping red blood cells and activating platelets. The blood vessels in the local area dilate and white cells, mainly neutrophils initially, plus plasma proteins and other inflammatory mediators, diffuse out. The neutrophils phagocytose cellular debris generated by the injury and an acute inflammatory response is set up. Vascular regeneration begins with the formation of new capillaries in the wounded area. Fibroblasts lay down collagen in the form of scar tissue. The extent of injury determines the duration of inflammation and the time taken for the repair process.

If the source of irritation is not removed before repair begins, which is what happens in an implant situation, then the inflammation process continues and may delay the repair. The extent of this will depend on the nature of the implanted material. With a single, solid, non-biodegradable material the inflammatory response will be unaffected. As before, fibroblasts will lay down collagen, but it will not be able to infiltrate the implant. Instead, a fibrous capsule will form around the implant. This is known as classic fibrous encapsulation, and is rare. Usually there is a reaction between the implant and the tissue, involving cells of the immune system. In extreme cases this results in persistent inflammation and repair is never complete. In cases involving degradable materials the response is slightly different, as it is dependent on the breakdown products.

3.9 Discussion

Biocompatibility continues to be described as 'the ability of a material to perform with an appropriate host response in a specific application' (Williams 1987), though now those specific host responses are better defined. The problem associated with using the data available to form a complete picture on the biocompatibility of degradable polymers is the diversity of the tests performed. The time points are usually inappropriate, and usually not long enough to take into account the effect of late degradation products on the biocompatibility. The molecular weight, physical characteristics and changes in surface structure following sterilisation of the polymers are not established or stated. As the initial molecular weight influences the rate of degradation, it will affect the biocompatibility, and hence the immune response. It is therefore important to determine the molecular weight of a device immediately before implantation after it has been sterilised, as sterilisation processes are known to influence the molecular weight of certain polymers.

In the current light of tissue engineering, the cell of choice is the stem cell derived from any number of emerging sources. With this diversity in cell origin and differentiation potential, the methods of analysing biocompatibility need to become more quantitative, allowing for several parameters to be assessed simultaneously. There are a number of technologies available today which will allow scientists the ability to do this. Screening for a large number of gene expressions and proteins will be possible by microarray technologies and proteomics. With new techniques that allow the generation of material with nano topography and different surface charges, subtle changes will need to be measured in the cellular response. With this change in the biological capabilities there needs to be a change in the material technologies. Currently the majority of the materials that are in use as tissue engineering constructs or scaffold materials are based on PLA, PGA or copolymers of the two. There is a need for the generation of some new polymers that are designed with specific degradation mechanisms to match the desired implant and scaffold characteristics rather than conforming existing materials to the rapidly evolving field of tissue engineering that demands a better class of biomaterial (Vert 2005). It is hoped that this chapter has helped the reader to understand the 'fluid nature' of biocompatibility testing and in doing so lead their device to fulfil the criteria of biocompatibility as redefined by Williams: '*Biocompatibility refers to the ability of a biomaterial to perform its desired function with respect to a medical therapy, without eliciting any undesirable local or systemic effects in the recipient or beneficiary of that therapy, but generating the most appropriate beneficial cellular or tissue response in that specific situation, and optimising the clinically relevant performance of that therapy'* (Williams 2008).

3.10 References

Al-Nazhan S and Spangberg L (1990), 'Morphological cell changes due to chemical toxicity of a dental material: an electron microscopic study on human periodontal ligament fibroblasts and L929 cells', *J Endod*, 16(3), 129–34.

Ali SA, Doherty PJ and Williams DF (1993a), 'Mechanisms of polymer degradation in implantable devices. 2. Poly(DL-lactic acid)', *J Biomed Mater Res*, 27(11), 1409–18.

Ali SA, Zhong SP, Doherty PJ and Williams DF (1993b), 'Mechanisms of polymer degradation in implantable devices. I. Poly(caprolactone)', *Biomaterials*, 14(9), 648–56.

Alveryd A and Jacobsson SI (1975), 'Dexon for sutures and ligatures', *Acta Chir Scand*, 141(4), 256–8.

Anderson JM, Defife K, McNally A, Collier T and Jenney C (1999), 'Monocyte, macrophage and foreign body giant cell interactions with molecularly engineered surfaces', *J Mater Sci Mater Med*, 10(10/11), 579–88.

Asano M, Yoshida M, Omichi H, Mashimo T, Okabe K, Yuasa H, Yamanaka H, Morimoto S and Sakakibara H (1993), 'Biodegradable poly(DL-lactic acid) formulations in a calcitonin delivery system', *Biomaterials*, 14(10), 797–9.

Benahmed M, Bouler JM, Heymann D, Gan O and Daculsi G (1996), 'Biodegradation of synthetic biphasic calcium phosphate by human monocytes in vitro: a morphological study', *Biomaterials*, 17(22), 2173–8.

Bergsma EJ, Rozema FR, Bos RR and De Brujin WC (1993), 'Foreign body reactions to resorbable poly(L-lactide) bone plates and screws used for the fixation of unstable zygomatic fractures', *J Oral Maxillofac Surg*, 51(6), 666–70.

Bonfield TL and Anderson JM (1993), 'Functional versus quantitative comparison of IL-1 beta from monocytes/macrophages on biomedical polymers', *J Biomed Mater Res*, 27(9), 1195–9.

Chang DT, Jones JA, Meyerson H, Colton E, Kwon IK, Matsuda T and Anderson JM (2008), 'Lymphocyte/macrophage interactions: Biomaterial surface-dependent cytokine, chemokine, and matrix protein production', *J Biomed Mater Res A*, 87(3), 676–87.

Chen Y, Jia Z, Schaper A, Kristiansen M, Smith P, Wombacher R, Wendorff JH and Greiner A (2004), 'Hydrolytic and enzymatic degradation of liquid-crystalline aromatic/aliphatic copolyesters', *Biomacromolecules*, 5(1), 11–16.

Claes L and Ignatius A (2002), '[Development of new, biodegradable implants]', *Chirurg*, 73(10), 990–6.

Dawes EN, Clarke SA, Lamanuzzi N, Pinto E, Brooks RA and Rushton N (2003), 'The response of macrophages to particles of resorbable polymers and their degradation products', *J Mater Sci Mater Med*, 14(3), 271–5.

De Buren N (1970), '[Studies on a new absorbable synthetic suture material (Dexon)]', *Helv Chir Acta*, 37(6), 616–7.

De Jong WH, Eelco BJ, Robinson JE and Bos RR (2005), 'Tissue response to partially in vitro predegraded poly-L-lactide implants', *Biomaterials*, 26(14), 1781–91.

Delves PJ and Roitt IM (2000a), 'The immune system. First of two parts', *N Engl J Med*, 343(1), 37–49.

Delves PJ and Roitt IM (2000b), 'The immune system. Second of two parts', *N Engl J Med*, 343(2), 108–17.

Delves PJ, Martin SJ, Burton DR and Roitt IM (2006), *Roitt's Essential Immunology*, 11th edition, Wiley-Blackwell, Oxford.

Di Silvio L, Gurav N, Kayser MV, Braden M and Downes S (1994), 'Biodegradable microspheres: a new delivery system for growth hormone', *Biomaterials*, 15(11), 931–6.

Dobbs H (2007), 'Current developments and the Medical Device Directive – a regulator's perspective', *Drugs*, 10(4), 253–5.

Eldridge JH, Staas JK, Meulbroek JA, Tice TR and Gilley RM (1991), 'Biodegradable and biocompatible poly(DL-lactide-co-glycolide) microspheres as an adjuvant for staphylococcal enterotoxin B toxoid which enhances the level of toxin-neutralizing antibodies', *Infect Immun*, 59(9), 2978–86.

Eldridge JH, Staas JK, Tice TR and Gilley RM (1992), 'Biodegradable poly(DL-lactide-co-glycolide) microspheres', *Res Immunol*, 143(5), 557–63.

Engelberg I and Kohn J (1991), 'Physico-mechanical properties of degradable polymers used in medical applications: a comparative study', *Biomaterials*, 12(3), 292–304.

Ewers R and Lieb-Skowron J (1990), 'Bioabsorbable osteosynthesis materials', *Facial Plast Surg*, 7(3), 206–14.

Fan H, Liu H, Toh SL and Goh JC (2008), 'Enhanced differentiation of mesenchymal stem cells co-cultured with ligament fibroblasts on gelatin/silk fibroin hybrid scaffold', *Biomaterials*, 29(8), 1017–27.

Freed LE, Marquis JC, Nohria A, Emmanual J, Mikos AG and Langer R (1993), 'Neocartilage formation in vitro and in vivo using cells cultured on synthetic biodegradable polymers', *J Biomed Mater Res*, 27(1), 11–23.

Freyria AM, Chignier E, Guidollet J and Louisot P (1991), 'Peritoneal macrophage response: an in vivo model for the study of synthetic materials', *Biomaterials*, 12(2), 111–8.

Gelb H, Schumacher HR, Cuckler J, Ducheyne P and Baker DG (1994), 'In vivo inflammatory response to polymethylmethacrylate particulate debris: effect of size, morphology, and surface area', *J Orthop Res*, 12(1), 83–92.

Gilbert JC, Takada T, Stein JE, Langer R and Vacanti JP (1993), 'Cell transplantation of genetically altered cells on biodegradable polymer scaffolds in syngeneic rats', *Transplantation*, 56(2), 423–7.

Gurav N (1997), 'Biocompatibility testing of resorbable materials using improved in-vitro techniques', University College London.

Haustveit G, Torheim B, Fystro D, Eidem T and Sandvik M (1984), 'Toxicity testing of medical device materials tested in human tissue cultures', *Biomaterials*, 5(2), 75–80.

Helmus MN, Gibbons DF and Ebon D (2008), 'Biocompatibility: meeting a key functional requirement of next-generation medical devices', *Toxicol Pathol*, 36(1), 70–80.

Huatan H, Collett JH and Attwood D (1995a), 'The microencapsulation of protein using a novel ternary blend based on poly(epsilon-caprolactone)', *J Microencapsul*, 12(5), 557–67.

Huatan H, Collett JH, Attwood D and Booth C (1995b), 'Preparation and characterization of poly(epsilon-caprolactone) polymer blends for the delivery of proteins', *Biomaterials*, 16(17), 1297–303.

Hunter A, Archer CW, Walker PS and Blunn GW (1995), 'Attachment and proliferation of osteoblasts and fibroblasts on biomaterials for orthopaedic use', *Biomaterials*, 16(4), 287–95.

Kersey TL, Patel S and Thaller VT (2006), 'Old habits tie hard: an in vitro comparison of first-throw tension holding in Polyglycolic acid (Dexon S) and Polyglactin 910 (Coated Vicryl)', *Clin Experiment Ophthalmol*, 34(2), 152–5.

Kirkpatrick CJ (1998), 'New aspects of biocompatibility testing: where should it be going?', *Med Device Technol*, 9(7), 22–9.

Kirkpatrick CJ, Wagner M, Kohler H, Bittinger F, Otto M and Klein CL (1997), 'The cell and molecular biological approach to biomaterial research: a perspective', *J Mater Sci Mater Med*, 8(3), 131–41.

Kirkpatrick CJ, Bittinger F, Wagner M, Kohler H, Van Kooten TG, Klein CL and Otto M (1998), 'Current trends in biocompatibility testing', *Proc Inst Mach Eng*, Part H, 212(2), 75–84.

Kirkpatrick CJ, Peters K, Hermanns MI, Bittinger F, Krump-Konvalinkova V, Fuchs S and Unger RE (2005), 'In vitro methodologies to evaluate biocompatibility: status quo and perspective', *ITBM-RBM*, 26(3), 192–9.

Koosha F, Muller RH and Davis SS (1989), 'Polyhydroxybutyrate as a drug carrier', *Crit Rev Ther Drug Carrier Syst*, 6(2), 117–30.

Leach JK and Mooney DJ (2008), 'Synthetic extracellular matrices for tissue engineering', *Pharm Res*, 25(5), 1209–11.

Leenslag JW, Pennings AJ, Bos RR, Rozema FR and Boering G (1987), 'Resorbable materials of poly(L-lactide). VII. In vivo and in vitro degradation', *Biomaterials*, 8(4), 311–4.

Mainil-Varlet P, Curtis R and Gogolewski S (1997a), 'Effect of in vivo and in vitro degradation on molecular and mechanical properties of various low-molecular-weight polylactides', *J Biomed Mater Res*, 36(3), 360–80.

Mainil-Varlet P, Rahn B and Gogolewski S (1997b), 'Long-term in vivo degradation and bone reaction to various polylactides. 1. One-year results', *Biomaterials*, 18(3), 257–66.

Marchant RE (1989), 'The cage implant system for determining in vivo biocompatibility of medical device materials', *Fundam Appl Toxicol*, 13(2), 217–27.

Miller KM and Anderson JM (1989), 'In vitro stimulation of fibroblast activity by factors generated from human monocytes activated by biomedical polymers', *J Biomed Mater Res*, 23(8), 911–30.

Miller KM, Rose-Caprara V and Anderson JM (1989), 'Generation of IL-1-like activity in response to biomedical polymer implants: a comparison of in vitro and in vivo models', *J Biomed Mater Res*, 23(9), 1007–26.

Mooney DJ, Mazzoni CL, Breuer C, McNamara K, Hern D, Vacanti JP and Langer R (1996), 'Stabilized polyglycolic acid fibre-based tubes for tissue engineering', *Biomaterials*, 17(2), 115–24.

Murray DW and Rushton N (1990), 'Macrophages stimulate bone resorption when they phagocytose particles', *J Bone Joint Surg Br*, 72(6), 988–92.

Nakamura T, Hitomi S, Watanabe S, Shimizu Y, Jamshidi K, Hyon SH and Ikada Y (1989), 'Bioabsorption of polylactides with different molecular properties', *J Biomed Mater Res*, 23(10), 1115–30.

Pan H, Jiang H and Chen W (2008), 'The biodegradability of electrospun Dextran/PLGA scaffold in a fibroblast/macrophage co-culture', *Biomaterials*, 29(11), 1583–92.

Puelacher WC, Mooney D, Langer R, Upton J, Vacanti JP and Vacanti CA (1994), 'Design of nasoseptal cartilage replacements synthesized from biodegradable polymers and chondrocytes', *Biomaterials*, 15(10), 774–8.

Ratner BD and Bryant SJ (2004), 'Biomaterials: where we have been and where we are going', *Annu Rev Biomed Eng*, 6, 41–75.

Rickert D, Lendlein A, Peters I, Moses MA and Franke RP (2006), 'Biocompatibility testing of novel multifunctional polymeric biomaterials for tissue engineering applications in head and neck surgery: an overview', *Eur Arch Otorhinolaryngol*, 263(3), 215–22.

Robertson CW (1998), 'The medical device directive', *Health Estate*, 52(5), 26–7.

Rosilio V, Benoit JP, Deyme M, Thies C and Madelmont G (1991), 'A physicochemical study of the morphology of progesterone-loaded microspheres fabricated from poly(D,L-lactide-co-glycolide)', *J Biomed Mater Res*, 25(5), 667–82.

Schakenraad JM, Oosterbaan JA, Nieuwenhuis P, Molenaar I, Olijslager J, Potman W, Eenink MJ and Feijen J (1988), 'Biodegradable hollow fibres for the controlled release of drugs', *Biomaterials*, 9(1), 116–20.

Schakenraad JM, Hardonk MJ, Feijen J, Molenaar I and Nieuwenhuis P (1990), 'Enzymatic activity toward poly(L-lactic acid) implants', *J Biomed Mater Res*, 24(5), 529–45.

Sikavitsas VI, Bancroft GN, Holtorf HL, Jansen JA and Mikos AG (2003), 'Mineralized matrix deposition by marrow stromal osteoblasts in 3D perfusion culture increases with increasing fluid shear forces', *Proc Natl Acad Sci USA*, 100(25), 14683–8.

Smith R, Oliver C and Williams DF (1987), 'The enzymatic degradation of polymers in vitro', *J Biomed Mater Res*, 21(8), 991–1003.

Spenlehauer G, Vert M, Benoit JP and Boddaert A (1989), 'In vitro and in vivo degradation of poly(D,L lactide/glycolide) type microspheres made by solvent evaporation', *Biomaterials*, 10(8), 557–63.

Therin M, Christel P, Li S, Garreau H and Vert M (1992), 'In vivo degradation of massive poly(alpha-hydroxy acids): validation of in vitro findings', *Biomaterials*, 13(9), 594–600.

Thevenot P, Hu W and Tang L (2008), 'Surface chemistry influences implant biocompatibility', *Curr Top Med Chem*, 8(4), 270–80.

Vert M (2005), 'Aliphatic polyesters: great degradable polymers that cannot do everything', *Biomacromolecules*, 6(2), 538–46.

Vert M, Li S and Garreau H (1992), 'New insights on the degradation of bioresorbable polymeric devices based on lactic and glycolic acids', *Clin Mater*, 10(1–2), 3–8.

Vert M, Li SM and Garreau H (1994a), 'Attempts to map the structure and degradation characteristics of aliphatic polyesters derived from lactic and glycolic acids', *J Biomater Sci Polym Ed*, 6(7), 639–49.

Vert M, Mauduit J and Li S (1994b), 'Biodegradation of PLA/GA polymers: increasing complexity', *Biomaterials*, 15(15), 1209–13.

Williams DF (1987) *Definitions in Biomaterials; Proceedings of a Consensus Conference of the European Society for Biomaterials*, Chester, England, 3–5 March 1986. In *Progress in Biomedical Engineering*, Elsevier, Amsterdam, p. 54.

Williams DF (1989), 'A model for biocompatibility and its evaluation', *J Biomed Eng*, 11(3), 185–91.

Williams DF (2008), 'On the mechanisms of biocompatibility', *Biomaterials*, 29(20), 2941–53.

Xia Z and Triffitt JT (2006), 'A review on macrophage responses to biomaterials', *Biomed Mater*, 1(1), R1–R9.

Xia Z, Ye H, Choong C, Ferguson DJ, Platt N, Cui Z and Triffitt JT (2004), 'Macrophagic response to human mesenchymal stem cell and poly(epsilon-caprolactone) implantation in non-obese diabetic/severe combined immunodeficient mice', *J Biomed Mater Res A*, 71(3), 538–48.

Zhao Z, Yang L, Hua J, Wei J, Gachet S, El Ghzaoui A and Li S (2008), 'Relationship between enzyme adsorption and enzyme-catalyzed degradation of polylactides', *Macromol Biosci*, 8(1), 25–31.

4
Cellular response to the surface chemistry of nanostructured biomaterials

M A BARBOSA, M C L MARTINS and
J N BARBOSA,
INEB, Universidade do Porto, Portugal

Abstract: Cell–biomaterial interactions are largely governed by the chemical composition of the surface. Molecular design of substrates allows the development of biomaterials that lead to specific and desirable biological interactions with the surrounding tissues. Self-assembly in nature has inspired the development of a new generation of biomaterials. After implantation, biomaterials are rapidly covered with proteins. Adsorbed proteins are key mediators of cell behaviour.

In this chapter the process of self-assembly – in nature, in synthetic analogues of the extracellular matrix and in model surfaces (self-assembled monolayers, SAMs) – is overviewed. The application of SAMs as model surfaces to investigate cell–biomaterial interactions is discussed in detail.

Key words: cell–biomaterial interactions, self-assembled monolayers, nanostructured surfaces, protein adsorption, cell adhesion.

4.1 Introduction

Cell–biomaterial interactions are largely governed by the chemical composition of the substrate, although surface morphology has also been shown to play an important role. The molecular design of substrates is now possible, helping in guiding cell behaviour and shedding light into the mechanisms of cell–biomaterial interactions. The design of new biomaterials does not necessarily have to follow a molecular approach, but some of the most exciting developments have been achieved with this methodology. By learning with nature, and in particular with the naturally occurring self-assembling processes, one is now capable of designing biomaterials that mimic the extracellular matrix. Engineered self-assembling peptides, artificial proteins and derivatized polymers have been successfully developed based on molecular approaches. The fibrillar and non-fibrillar components of the extracellular matrix can now be produced with specific targeted functionalities to promote, prevent or control cell behaviour. This chapter initially addresses the basic aspects of self-assembly in nature and then reviews examples of nanostructured artificial scaffolds that mimic the extra-cellular matrix.

When a biomaterial is implanted in the human body the first cellular events that occur on its surface are largely determined by the nature and concentration of adsorbed proteins. It is generally considered that cell–biomaterial interactions are mediated by a layer of adsorbed plasma proteins that interact with integrins expressed on the cell membrane. Much of what is now known regarding the molecular interactions of proteins with surfaces has been gained through the use of model surfaces. Among these, self-assembled monolayers (SAMs) provide highly reproducible chemistries with immense possibilities of exposing selected functionalities. This chapter provides background information on the mechanisms of adsorption and detailed experimental data on the adsorption of plasma proteins, namely fibronectin and fibrinogen, to SAMs. The importance of selective protein adsorption in cell adhesion, and in particular strategies to promote it, are also reviewed.

Cellular responses to biomaterials surface chemistry have been investigated using a wide range of biomaterials. However, this chapter focuses on what is known about the behaviour of several cell types based on the use of SAMs.

4.2   Nanostructured surfaces and materials

4.2.1   Self-assembly in nature

Animals and plants exhibit an amazing diversity, but, however large or small, they result from the association of relatively similar minute building blocks. Cells are used to construct much more complex and larger structures than the cells themselves. These lower-order structures can be repeated millions of times to produce tissues and other higher-order structures. From entities that are only a few micrometres in size, structures orders of magnitude larger are built. This simple and ingenious strategy of aggregation of minute building blocks helps living organisms to survive by continuously repairing and regenerating their tissues, often at no major loss of function. At a smaller scale the same principle is adopted to build up cell membranes, cellular organelles and the medium that surrounds and supports cells: the *extra-cellular matrix* (ECM). For example, all proteins, in spite of their diversity in molecular weight and conformation – which influence their function – result from the combination of a limited number of amino acids, more precisely 20. About a dozen lipids, a few tens of sugars and a few nucleotides constitute nature's toolbox to fabricate all animals and plants. The fabrication of long polypeptide and polysaccharide chains follows the same principle of addition of small entities, originating complex polymeric molecules. The existence of several types of such molecules (e.g. proteins and sugars) and ions with various electrical charges and polarities in the biological milieu provides the driving force for their assembly into three-dimensional structures. The process of spontaneous association, or *self-assembly*, of the above lower-order entities is the basis of the structure for all biological organisms.

Self-assembly is governed by weak physical bonds, such as hydrogen bonds, electrostatic interactions, hydrophobic interactions and van der Waals forces. Water-mediated hydrogen bonds are particularly relevant in self-assembly of biological systems due to the ubiquitous presence of water and its interaction with all other molecules. In spite of the relative weakness of each individual bond, complex and strong structures (e.g. bone and ligaments) can be constructed from simple building blocks. This weakness guarantees also the ability of the biological system to adapt to changes in the environment without rupturing the stronger covalent bonds that ensure the integrity of the individual building blocks.

The lipid bilayer is a classical example of self-assembly. Several mathematical models have been proposed to simulate its constitution from single, disordered lipid molecules. The constructed models have to resemble the natural bilayers in several aspects, namely their physical properties. This includes the ability to deform under lateral stresses without rupturing, which requires a large degree of flexibility. This is necessary, for instance, when the cell deforms during migration. The bilayers should behave like an ordered fluid, rather than like a solid. Modelling can be done by assuming various degrees of complexity for the assembling lipids, ranging from very detailed, atomistic, model lipids to coarse-grained structures. One of the latter, composed of a head group and two tail groups, has been proposed [1]. In this work the authors have adopted a fluid-free model, since the fluid, together with more detailed model lipids, would significantly increase the computing time. The existence of a broad attraction potential between the tail groups drives the molecules to self-assembly, while maintaining a certain degree of cohesion between adjacent molecules. A random gas of lipid molecules first forms small clusters that then self-assemble into a lipid bilayer.

4.2.2 Mimicking the ECM – self-assembling polymers

Nature has inspired the development of designed self-assembling polymers capable of forming 3-D structures induced by changes in the medium, namely pH and ionic concentrations. A class of self-assembling peptides has been extensively investigated by Zhang for a variety of applications, including neural regeneration, encapsulation of chondrocytes, osteoblast differentiation and *in vitro* culture of hepatocytes. These peptides have been synthesized after the discovery of a simple repetitive sequence – AEAEAKAKAEAEAKAK – found in a yeast protein called zuotin. This peptide (EAK16-II) was the first of a family of peptides developed by Zhang. Basically, they contain 50% of charged (positive and negative) residues and form β-sheet structures in aqueous neutral solutions. On one side the chain is non-polar and on the other it is polar (Fig. 4.1). On this side the positively and negatively charged residues alternate. In solution the non-polar residues shield themselves from water, while the polar residues face the liquid and establish ionic bonds between themselves. Upon self-assembly, positive and negative residues alternate in the polar surface of the structure, like in a checkerboard. Nanofibres, with a diameter of *ca.* 10 nm, are formed during this process [2, 3]. These fibres absorb very large amounts of water, *ca.* 99.5%, and contain only 0.5% of protein. In these hydrogels the fibres are interwoven in a manner similar to an extracellular matrix, namely collagen. Porosity may range from 5 to 200 nm [4], promoting cell attachment in a 3-D environment. In microporous structures, pores usually have dimensions larger than those of cells (5–100 μm) and

4.1 EAK16 peptide chains are held together by hydrophobic (lysine–alanine interactions) and ionic bonds (glutamic acid–alanine interactions), forming a β-sheet structure [5].

the scaffold functions essentially as a 2-D structure. However, when the pores have much smaller dimensions, as happens with the above scaffolds, cells are capable of attaching in all directions, adopting a behaviour similar to that found in natural extracellular matrices (ECM).

Recombinant DNA engineering has been used to generate artificial extracellular matrix (aECM) proteins, namely for applications as small-diameter vascular grafts [6]. Proteins with two types of fibronectin domains have been produced: one with an RGD sequence from the tenth type III domain (Fig. 4.2a) and the other with the CS5 cell-binding domain (Fig. 4.2b). The T7-tag is included to increase expression levels and to favour protein detection. The hexahistidine tag was incorporated for the purpose of purification. The elastin-like domain provides for elasticity [7]. The proteins are produced in a process that involves cloning, bacterial growth, protein expression and purification. The yield reported was of the order of 600 mg of protein per 10 L of fermentation. Human umbilical vein endothelial cells (HUVECs) spread more rapidly on the former type of protein. The incorporation of lysine residues as crosslinking sites has been realized in order to provide for better mechanical properties [8]. Lysine was inserted either within the elastine domain or at the ends of a CS5 cell-binding protein. Cell adhesion and spreading was more pronounced when the lysine residues are bound at the end of the chains, indicating that amino acids outside the binding domain are important in cell–matrix interactions. The formation of a gel in near-neutral solutions occurs by self-assembly of the artificial proteins [9]. Interchain interactions occur via coiled-coil aggregates that form a network that retains the solvent within it.

In natural ECM the fibrillar components of the extracellular matrix are interwoven with non-fibrillar, highly hydrated glycosoaminoglycan (GAG) chains. These chains are covalently linked to proteins forming proteoglycans. Due to their strong negative charge GAGs attract large numbers of positive ions, resulting in a high osmotic pressure that is responsible for the

(a)
aECM 1:
M-MASMTGGQQMG-HHHHHHH-DDDDK(LD-YAVTGRGDSPASSKPIA((VPGIG)$_2$VPGKG(VPGIG)$_2$)$_4$VP)$_3$-LE
    T7 tag   His tag   Cleavage   RGD cell-binding   Elastin-like domain
                                 site   domain

(b)
aECM 3:
M-MASMTGGQQMG-HHHHHHH-DDDDK(LD-EEIQIGHIPREDVDYHLYPG((VPGIG)$_2$VPGKG(VPGIG)$_2$)$_4$VP)$_3$-LE
    T7 tag   His tag   Cleavage   CS5 cell-binding   Elastin-like domain
                                 site   domain

4.2 Amino acid sequence of aECM proteins: (a) protein with the RGD cell-binding domain; (b) protein with the CS5 cell-binding domain [6].

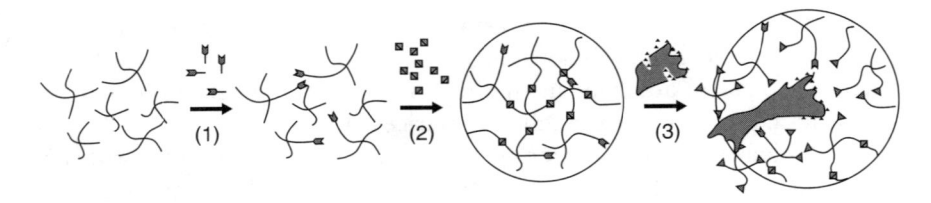

4.3 Multiarm PEGs, previously functionalized with VS, react first with mono-cysteine peptides (step 1) and then with bis-cysteine peptides (step 2). The former provide anchorage to cells while the latter act as crosslinkers between PEG arms and as substrates for MMP produced by cells during invasion of the matrix (step 3) [11].

absorption of large quantities of water. The remarkable hydrophilicity of GAGs is responsible for the good compression strength of the ECM, while fibrillar proteins, such as collagen, fibrin and elastin, provide resistance to tensile stresses. The non-fibrillar phase of ECM has been mimicked by derivatizing poly(ethylene glycol) (PEG) with mono- and bis-cysteine peptides [10, 11]. Multiarm PEG is initially functionalized with vinyl sulfone (VS) and then made to react with mono-cysteine adhesion peptides in high stoichiometric deficit (Fig. 4.3, step 1). This leaves a large proportion of branches free to bind to the bis-cysteine peptides, which act as crosslinkers between PEG chains (Fig. 4.3, step 2), as well as a substrate for matrix metalloproteases (MMP) during cell invasion (Fig. 4.3, step 3). The pendant receptor-binding mono-cysteine peptides enable cells to hold to the scaffold and migrate, while the MMP-binding bis-cysteine peptides provide mechanical strength before cell migration occurs.

4.2.3 Self-assembled monolayers (SAMs)

Self-assembled monolayers (SAMs), particularly those formed by the adsorption of organosilanes on hydroxylated surfaces and alkanethiols on gold, have been extensively used as models for fundamental studies of the interactions of surfaces with proteins and cells, because of the simple method of production associated with precise control of composition at the interface. Many reviews of SAM preparation, characterization and application are available [12–18]. In this chapter a brief description of these two types of SAMs is summarized.

SAMs are formed spontaneously by the adsorption of a surfactant with a specific affinity of its head group to a substrate [12, 13]. SAMs of organosilanes are prepared after immersion of hydroxylated surfaces (usually the native oxide of silicon, glass or titanium), into a diluted solution of organosilanes such as alkyltrichlorosilanes (X-$(CH_2)_n$-$SiCl_3$; $n > 3$) or alkyltriethoxysilanes (X-$(CH_2)_n$-Si-$(OCH_3)_3$; $n > 3$) in an organic solvent.

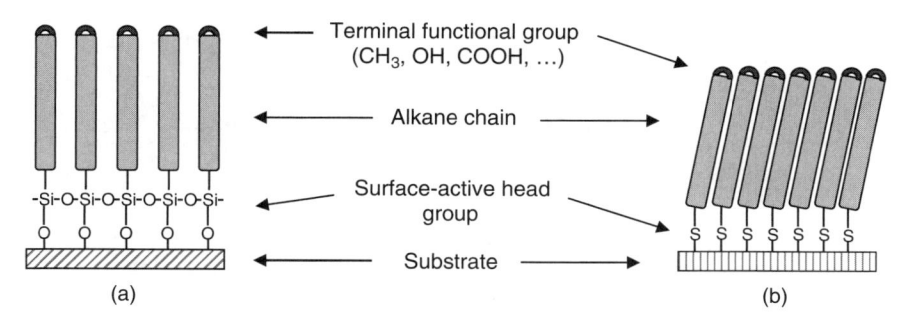

Cellular response to nanostructured biomaterials 91

4.4 Schematic diagram of SAMs of (a) alkylsiloxanes on hydroxylated surfaces, and (b) alkanethiolates on gold.

During adsorption, the silane group reacts with the OH groups present on the surface of the substrate and with trace water (which can be adsorbed to the hydrophilic substrate), to form a network with Si—O—Si bonds. The result is a monolayer in which the molecules are connected both to each other and to the surface by strong chemical bonds, exposing different functionalities, X (e.g., amine, acid or carboxyl) at the monolayer–liquid interface (Fig. 4.4a) [13, 16, 17]. One disadvantage of this type of SAMs is the limited range of functional groups available and the reproducibility, since they are very sensitive to reaction conditions, namely the amount of water.

SAMs of alkanethiols on gold are formed spontaneously after the immersion of a gold substrate into a diluted solution of long chain alkanethiol (X-$(CH_2)_n$SH; $n > 10$). The driving force for self-assembly is the strong specific interaction between the sulfur of the thiol and the gold surface. This will force the exposure of the functional group X (which can be alkyl, alcohol, carboxylic acid, ester, amide, oligo(ethylene glycol), etc.) at the monolayer–liquid interface (Fig. 4.4b). Van der Waals forces between the carbon chains orient and stabilize the monolayer. As a consequence, in general, the longer the chain length, the more ordered the monolayer [14, 17, 19, 20]. One disadvantage of alkanethiolate SAMs is their limited stability to oxidation [21].

The wettability of a surface can be precisely controlled using SAMs with different terminal functional groups (Table 4.1) [20, 22, 23]. SAMs that present polar terminal functional groups, such as carboxylic acid and hydroxyl, are wetted by water. Those that present non-polar, organic groups, such as methyl and trifluoromethyl, emerge dry from water.

The chemistry of the monolayer can be further controlled by the preparation of mixed SAMs from solutions with two or more different alkanethiols [24–27]. Spatial control over the chemistry of the surface can be obtained using surface patterning techniques such as photolithography and microcontact printing [28]. Photolithography consists in the exposure of a surface, covered with a patterned mask, to a UV light source of proper wavelength

Table 4.1 Contact angle of water on SAMs of alkanethiols on gold under cyclooctane and air[a]

Thiol	θ_{co} (deg)[b]	θ_{air} (deg)[c]
$HS(CH_2)_{10}CH_3$	165	112
$HS(CH_2)_{11}OPh$	156	85
$HS(CH_2)_2(CF_2)_9CF_3$	154	118
$HS(CH_2)_{11}CN$	146	63
$HS(CH_2)_{11}OMe$	106	85
$HS(CH_2)_{10}CONHMe$	94	76
$HS(CH_2)_{11}OH$	65	<15
$HS(CH_2)_{11}(OCH_2CH_2)_6OH$	52	38
$HS(CH_2)_{10}CONH_2$	20	<15

[a] Advancing contact angles were measured in triplicate. The values of all the replicates were within ±3° of the mean.
[b] Advancing contact angle of water under cyclooctane.
[c] Advancing contact angle of water under air.
Source: adapted from Sigal et al. [23].

[29]. Regarding SAMs of organosilanes, the surface could first be covered with a thin layer of a photoresist (UV-sensitive polymer). After UV irradiation, the exposed layer is washed out, creating a patterning on the surface that allows the formation of the organosilane monolayer. The remaining photoresist can then be removed and a different monolayer is formed on the complementary region. Concerning SAMs of alkanethiols on gold, a freshly prepared SAM is exposed to UV through a patterning mask. After this the oxidized regions (oxidized sulfur) can be easily washed and replaced by another alkanethiol [30, 31].

Microcontact printing is based on the process of ink stamp printing, where ink is the surfactant solution [32–34]. The patterning stamp is usually made of poly(dimethyl siloxane) (PDMS) produced after polymerization and curing on the top of a silicon wafer contained the desired pattern relief (grooves, pits, etc.). After being inked with the surfactant, the stamp is placed onto the surface. The surfactant is transferred to the surface only at the regions where the stamp contacts the surface [35, 36].

Of the several analytical techniques that can be applied to characterize the structure and the properties of SAMs, X-ray photoelectron spectroscopy (XPS), contact angle measurements, ellipsometry, infrared reflection absorption spectroscopy (IRAS), surface probe microscopies (such as AFM and STM) and electrochemical methods are the most frequently applied [17, 19].

4.3 Influence of surface chemistry on protein adsorption

4.3.1 Mechanisms of adsorption

Adsorption is the accumulation of a substance at an interface. This process occurs on all surfaces, both in air and in solution. The principles that govern adsorption in both media are essentially the same. Adsorption is often a dynamic process: species may adsorb and desorb from a surface, depending on the strength of the binding. Usually, we refer to physical and chemical adsorption, depending on the strength of the bond. Physical (or weak) adsorption is characterized by energies in the order of 20–40 kJ/mol, while for chemical adsorption the energies are in the range 100–400 kJ/mol. These are similar to the strength of covalent bonds (e.g. the binding energy for H—H is 436.4 kJ/mol). For comparison, the energy of hydrogen bonds is *ca.* 40 kJ/mol in water, while other van der Waals forces are much weaker (*ca.* 1 kJ/mol). In physical adsorption the adsorbate (i.e. the adsorbed species) is relatively free to diffuse along the surface, in contrast with what occurs in chemical adsorption. In this case the strength of adsorption (often to specific sites) renders the adsorbate immobile.

The amount (or surface concentration, Γ) of an adsorbed species is a function of its concentration in solution, c. A plot of Γ as a function of c is an adsorption isotherm. There are several types of adsorption isotherms. One of the most common in liquids is the Langmuir adsorption isotherm. It is represented by the equation

$$\theta = \frac{K_L}{1 + K_L c}$$

where θ is the coverage and K_L is the Langmuir constant. For high concentrations, monolayer coverage is reached. The problem with the application of this equation is that it assumes that adsorption is a reversible process, which does not always hold true for protein adsorption, as we will see below.

Apart from proteins a large number of substances compete for adsorption on a biomaterial surface. Water molecules are major competitors, and due to the strength of their intermolecular bonds proteins have to overcome a layer of water molecules before they successfully adsorb. As a result of the various forces prevailing at an interface a surface tension develops. At a triple interface (solid–liquid–vapour, Fig. 4.5) the interfacial tension (which has the same value as the interfacial energy) is given by the Young equation

$$\gamma_{sv} = \gamma_{sl} + \gamma_{lv} \cos\theta$$

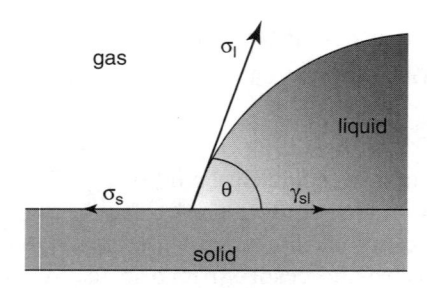

4.5 Solid–liquid–gas interface and interfacial tensions.

where γ_{sv}, γ_{sl} and γ_{lv} are the solid–vapour, solid–liquid and liquid–vapour interfacial tensions and θ is the contact angle.

The surface energy has a polar and a dispersive component:

$$\gamma = \gamma^p + \gamma^d$$

The polar component includes all polar interactions (dipole–dipole and dipole-induced dipole, i.e. a dipole developed in a molecule as a result of the approach of a charged species), whereas the dispersive component corresponds to the effect of instantaneous dipoles formed between non-polar molecules (short-lived dipoles formed in a molecule due to the instantaneous non-uniform distribution of electrons around the nucleus).

In studying protein adsorption to biomaterials the adhesion tension, τ, has been considered a more useful parameter than other variables [37]. τ is given by

$$\tau = \gamma_{sv} - \gamma_{sl} = \gamma_{lv} \cos\theta$$

Protein adsorption to surfaces has to overcome a barrier of adsorbed molecules, as mentioned earlier. However, the organization of these molecules depends on whether the surface is hydrophilic or hydrophobic. Generally speaking, a surface is considered hydrophobic if $\theta > 90°$ and hydrophilic if $\theta < 90°$. A less arbitrary definition, based on the actual organization of the adsorbed water layer, has been proposed [37]. A hydrophilic surface would be one capable of disrupting the structure of bulk water, whereas in the vicinity of a hydrophobic surface the water would essentially keep its bulk structure. Protein adsorption would occur on hydrophobic surfaces but not on hydrophilic surfaces, because adsorption would be energetically unfavoured on the latter.

The tendency for a protein to adsorb on a surface can be measure by calculating the difference between the adhesion tension of a solution of the protein of interest and of a solution without the protein (water or PBS), $\tau^{protein} - \tau^0$. When this difference is negative, protein adsorption is not favoured, whereas a positive value indicates the reverse. Figure 4.6 illus-

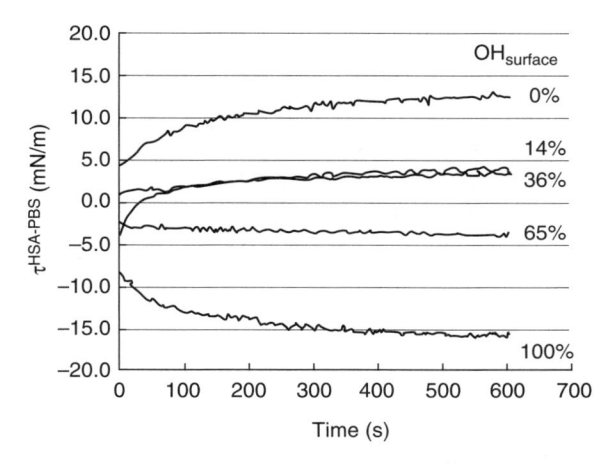

4.6 Difference between the adhesion tension of human serum albumin (HSA) solutions and PBS solutions measured on gold substrates functionalized with mixed SAMs of OH- and CH$_3$-terminated alkanethiols [26].

trates the application of this parameter to the adsorption of human albumin to self-assembled monolayers (SAMs) of mixed OH- and CH$_3$-terminated alkanethiols [26].

The higher adhesion tension observed for hydrophobic surfaces in Fig. 4.6 reflects the higher affinity of albumin for these surfaces. Generally, hydrophobic surfaces are reported to adsorb higher amounts of protein than hydrophilic surfaces. Protein adsorption requires some dehydration of both surface and protein, a process that increases the entropy of water. Since rupturing the water structure in the vicinity of hydrophilic surfaces is more difficult than on hydrophobic surfaces, protein adsorption is less favoured on the former. The driving force for adsorption associated with dehydration may overcome the repulsive forces between proteins and surfaces with like electrostatic charges. Apart from dehydration, other factors influence protein adsorption [38]:

- Redistribution of charges in the double layers around the protein and the sorbent surface when they overlap
- Dispersion forces between the protein and the material
- Structural rearrangements in the protein molecule.

As a result of structural rearrangements in solution a protein tends to protect its hydrophobic residues from the aqueous environment. However, small proteins may still expose 40–60% of their non-polar groups [39]. When a protein adsorbs on the surface of a material the extent of the structural rearrangements will vary. 'Hard' proteins will change their structure

'Soft' protein **'Hard' protein**

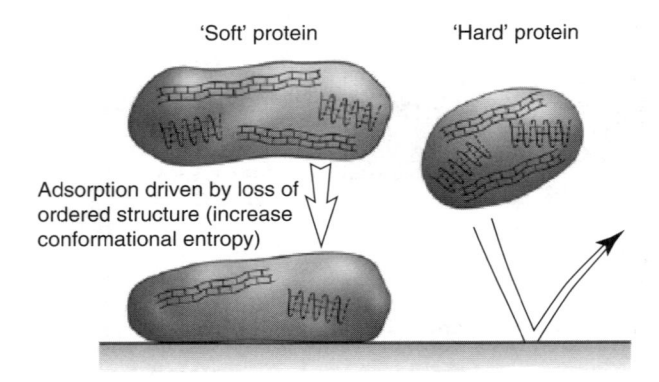

Adsorption driven by loss of ordered structure (increase conformational entropy)

Polar (hydrophilic), electrostatically repelling surface

4.7 Adsorption of 'soft' and 'hard' proteins on an electrostatically repelling surface. Soft proteins undergo structural changes that are capable of overcoming the repulsive forces, whereas hard proteins do not adsorb [38].

very little, whereas 'soft' proteins will undergo considerable changes that may alter their secondary structure; α-helix and β-sheet structures may be disrupted during this process. On electrostatically unfavourable surfaces a 'hard' protein will be repelled, whereas a 'soft' protein will be able to overcome the repulsive forces and adsorb, as shown in Fig. 4.7.

Although proteins undergo structural changes during adsorption, they do not completely unfold, i.e. they are not denatured. In some cases, namely for flexible proteins (e.g. albumin), this enables proteins to desorb, regaining their native conformation when they return to solution. This process may occur as a consequence of competition with other proteins. The nature and concentration of proteins in the medium determine which ones will win the race for the surface, forming the first layer of adsorbate. However, in the course of time, these proteins (or part of them) will exchange with others having a higher affinity for the surface, a well-known phenomenon referred to as the *Vroman effect*.

Counter-intuitive processes may occur during protein adsorption, particularly if one tends to reason in terms of either electrostatic or hydrophobic interactions alone. Table 4.2 illustrates predicted and observed interactions according to [38].

Proteins with a weak (labile) internal structure will adsorb to any surface, irrespective of their charge and the hydrophobic/hydrophilic nature of the substrate, whereas proteins with a stable structure would not adsorb on hydrophilic surfaces with similar electrical charges. However, protein adsorption may be mediated by adsorbed ions. Titanium oxide, which is negatively charged in PBS, adsorbs albumin, which is also negatively charged

Table 4.2 Scheme to predict whether ('Yes') or not ('no') proteins adsorb at surfaces[a]

Protein	Surface			
	Hydrophobic		Hydrophilic	
	+	−	+	−
Stable structure				
+	Yes	Yes	No	Yes
−	Yes	Yes	Yes	No
Labile structure				
+	Yes	Yes	Yes	Yes
−	Yes	Yes	Yes	Yes

[a]The '+' and '−' signs refer to the net electric charge of the protein and the surface.
Source: Norde [38].

at physiological pH (the isoelectric pH of albumin is 4.7–4.9) [40]. The explanation for this apparently unexpected result may reside in the adsorption of calcium ions, which would have a bridging effect in albumin adsorption. The chemical, structural and morphological characteristics of titanium oxide play an essential role in protein adsorption. For instance, adsorption of fibronectin, an adhesion protein, is favoured on titanium oxide obtained by chemical polishing, compared to an oxide with similar bulk composition (TiO_2) obtained by sputtering [41]. Elution from the former substrate is also more difficult. These differences could be related to the larger surface roughness of TiO_2 (14 times higher) obtained by chemical polishing or to the presence of a hydroxyl-rich layer on these substrates [42].

4.3.2 Protein adsorption on functionalized surfaces

After implantation, biomaterials are rapidly covered with proteins from blood and interstitial fluids. The composition of the adsorbed protein layer is a key mediator of cell behaviour, since cells depend on specific proteins for anchorage and extracellular instructions [43].

The type and amount of protein adsorbed to a surface, as well as its degree of unfolding and retention at the surface, vary according to the physical and chemical properties of the surface (such as wettability and charge) and the characteristics of the proteins (such as size, charge distribution and structure stability) [44].

SAMs are good models for fundamental studies regarding the effect of surface chemistry on protein adsorption, due to the possibility of precisely controlling the interface between the surface and the biological environment at a molecular scale [14–16].

The correlation between wettability of the surface and non-specific adsorption of proteins was well demonstrated using SAMs presenting non-charged terminal functional groups with different polarities (CH_3, OPh, CF_3, CN, OCH_3, $CONHCH_3$, OH, EG_nOH (n = 2–6), $CONH_2$) [23, 45]. In general, proteins adsorb preferentially [23] and strongly [45] on more hydrophobic surfaces (CH_3, OPh, CF_3, CN, OCH_3) in contrast to neutral and hydrophilic surfaces (OH, $CONH_2$ and EG_nOH) which exhibit reduced protein adsorption. Since protein adsorption involves the displacement of water molecules from the surface, hydrophilic surfaces present a substantial energy barrier to protein adsorption [37]. However, although to a lower degree, adsorption on more hydrophilic surfaces is also possible. The adsorption of certain proteins such as fibrinogen, γ-globulin [23], complement factor 3c (C3c) [46] and fibronectin [47] to the OH-terminated surfaces could be related to the hydrogen bonds formed between the surface and the side-end groups of polar protein amino acids [48].

A gradient of wettabilities can also be achieved using SAMs prepared with different percentages of uncharged polar (EG_nOH (n = 2–6) or OH) and unpolar (CH_3) terminal functional groups [26, 27, 49–52]. Protein adsorption (such as albumin [26], fibrinogen [27, 49] and fibronectin [51]) decreased with the incorporation of hydrophilic thiols in the monolayer, when adsorption was performed from pure solutions. Retention of adsorbed proteins is also lower on more hydrophilic surfaces [26, 49, 51].

When the adsorption was performed from serum, where many proteins compete for adsorption sites at the surface, a decrease of the total amount of adsorbed protein was not observed [50, 52]. Hirata et al. [50] described an increase of the adsorption of the complement protein C3b with the incorporation of OH groups on the monolayer, in contrast to the adsorption of albumin which still decreases with increased hydrophilicity of the surface. These processes can be explained since albumin (the first protein to adsorb at surfaces due to its high concentration and diffusion coefficient on serum and plasma) adsorbs on non-charged hydrophilic surfaces in a more reversible way and can be easily displaced and replaced by other proteins such as fibrinogen [26] or other cell adhesive proteins [52] that can bind tightly to this type of surfaces.

Surface charge can also influence protein adsorption by electrostatic interactions, particularly at lower ionic strength. Independently of protein overall charge at physiological pH, they can adsorb on both positively and negatively charged surfaces. This can be explained because proteins have localized regions or domains where positive and negative charges may be present and also because it is possible that some counterions can work as a bridge between the surface and the protein [53].

Some proteins, such as albumin and immunoglobulin G [48], adsorb preferentially, from pure solutions, on the negatively charged COOH rather than on the positively charged NH_2-terminated SAMs. However, Kidoaki and Matsuda [54] described how the strength of adhesion of these proteins is higher on NH_2 than on COOH. Nevertheless, adsorption was always higher on the hydrophobic CH_3-terminated SAM. Adsorption of the adhesive protein fibronectin [55] is higher on NH_2 than on COOH-terminated SAMs and the adsorbed amount of osteopontin is similar for both surfaces [56].

The negatively charged COOH- and OSO_3H-terminated SAMs have also been associated with the adsorption of the contact activation proteins high molecular weight kininogen (HMWK) and Factor XII (FXII), indicating that these surfaces are coagulation activators *in vitro* [46].

The behaviour of protein adsorption was also correlated with the size of the protein [23, 45]. Smaller proteins such as RNAse A and lysozyme (14 kDa) are very sensitive to the wettability of the surface, adsorbing only on the less wettable surfaces (SAMs terminated in CH_3, CF_3 and OPh). Larger proteins such as fibrinogen and γ-globulin (170–340 kDa) are less sensitive to the wettability of the surface, adsorbing to almost all the SAMs presented in Table 4.1, except to those terminated in EG_nOH. This could be explained by the high molecular weight proteins presenting higher multipoint attachments to the surface than smaller proteins. The intermediate size protein albumin (69 kDa) presents an intermediate sensitivity to the wettability of the surface, adsorbing to all the surfaces described in Table 4.1, except to the very wettable SAMs (OH, $CONH_2$ and EG_6OH).

Protein adsorption can also be related to the stability of the protein, since protein unfolding is likely to expose more points for protein–surface contact. Therefore, less stable proteins can bind strongly to a surface, decreasing the reversibility of adsorption [44].

Conformational changes of the adsorbed protein are dependent on the surface chemistry. Hydrophobic surfaces, such as CH_3-terminated SAMs, induce higher alteration of the organized secondary structure of the proteins after adsorption than the hydrophilic OH-terminated SAM [57]. The enhancement of the functionality of proteins such as fibrinogen and fibronectin after adsorption has been associated with alterations on their conformation. Structural changes in fibrinogen after adsorption may expose the dodecapeptide sequence of the fibrinogen γ chain [58, 59], which is essential for platelet adhesion and activation. Fibrinogen is only able to bind inactivated platelets in the adsorbed state [60]. Concerning fibronectin, while certain structural changes may reduce its functionality, others may enhance it. The soluble form of fibronectin shows reduced cell-binding ability [61].

Although adsorption of fibrinogen to CH_3-terminated SAMs induces elevated amounts of platelet adhesion and activation [49], the adsorption of fibronectin to this hydrophobic SAM prevents cell adhesion because the cell integrin binding domains are not accessible [47, 62].

Adsorption using low concentrations of protein solution may also favour spreading and denaturation of the adsorbed protein, because adsorption occurs slowly and proteins have time to unfold before the entire surface is covered [53, 63].

4.3.3 Selective protein adsorption

Besides models for fundamental studies, SAMs are also used in biosensors and other technologies (e.g. affinity chromatography), where the design of surfaces that can recognize and specifically bind certain biomolecules (proteins, peptides, nucleotides, etc.) is necessary [64]. These surfaces must possess specificity to a particular biomolecule, while simultaneously resisting non-specific adsorption of other molecules, such as proteins.

Selective protein surfaces can be synthesized by the immobilization of a ligand with specific affinity to a particular protein (antibody [65, 66], organic compounds [67, 68], etc.) into a surface that resists non-specific adsorption of other proteins and cell adhesion (non-fouling surfaces). These surfaces must also prevent the denaturation of the immobilized ligands to avoid the loss of their biological activity. In some applications, it is also desirable to maintain the native conformation of the adsorbed protein [26, 68].

SAMs terminated in oligo(ethylene glycol) (EG_nOH; $n = 3–6$) [56, 69], mannitol [70], semifluorinated groups [71] and oligo(phosphorylcholine) [72] have been described as resistant to protein adsorption and cell adhesion.

Based on the knowledge obtained with biosensors, SAMs can be very useful in the design of new biomaterials, where it is important to guide a cellular response by controlling the adsorption of specific proteins. One example is the design of surfaces with specific recognition for albumin [26, 68] (which is considered a *passivant* protein) or for thrombin [73] (the key enzyme of the coagulation cascade) in order to create anti-thrombogenic and anti-inflammatory biomaterials.

4.3.4 Characterization techniques

Measurements of the amount of adsorbed proteins require highly accurate techniques, since the adsorbed surface concentration is very low. Among a great variety of analytical techniques available, protein labelling with radio-isotopes, ellipsometry, surface plasmon resonance (SPR), quartz crystal microbalance (QCM) and infrared spectroscopy (IR) are those most frequently used to study protein adsorption on SAMs and will be briefly described here.

The radiolabelling technique is based on the incorporation of a radioactive nuclide (usually ^{125}I) into the molecular structure of the protein to be studied. The increase of radioactivity of the sample after contact with a solution containing the labelled protein allows the calculation of the amount of the protein adsorbed (detection limit 0.05 ng/cm^2). This technique can be used to study protein adsorption from a mixture of proteins, e.g. from blood or plasma. However, it does not give information about the conformation and biological status of the adsorbed protein [74, 75].

SPR, ellipsometry and QCM are very sensitive techniques that can be used for real time and *in situ* analysis of protein adsorption to SAMs or other thin films deposited on a flat substrate. Although SPR and QCM need special substrates, ellipsometry can be applied to any reflector and flat substrate. These techniques enable the direct monitoring of the adsorption kinetics without labelling the protein [76].

SPR is an optical technique that measures changes in the refractive index of the medium near a metal surface (usually a thin film of gold or silver deposited on a glass substrate). When the *p*-polarized light is reflected from the back of a glass prism interfacing with the thin metal film, an electromagnetic field component of the light (the evanescent wave) penetrates the metal layer. This evanescent wave is able to couple with the free oscillating electrons (plasmons) in the metal film at the specific angle of incidence. The SPR angle (angle at which resonance occurs) depends very strongly on the adsorbates on the metal film. The SPR shift due to protein adsorption is proportional to the surface concentration of the adsorbed protein [48, 76].

Ellipsometry is an optical technique that provides the refractive index and thickness of a thin film adsorbed into a reflective and flat surface. Through the refractive index of the protein and the increase in film thickness, the amount of adsorbed protein can be calculated [77, 78].

QCM uses the piezoelectric effect to measure changes in the fundamental frequency of vibration of a quartz crystal (thin quartz disc sandwiched between a pair of electrodes covered with the material to be tested) as protein adsorbs to it [79]. The change of the resonant frequency is proportional to the adsorbed mass on the crystal. QCM with dissipation monitoring (QCM-D) allows the measurement of changes in adsorbed mass, as well as the viscoelastic properties of the adsorbed protein layer on the crystal [80–82].

When combined with antibodies, SPR, ellipsometry and QCM-D can be used to quantify the concentration of a specific protein adsorbed from serum or plasma [76, 78]. It can also be used to detect conformational changes of adsorbed proteins, using antibodies directed at epitopes only present at conformationally changed proteins [76, 77].

Fourier-transformed infrared spectroscopy, using an attenuated total reflection accessory (ATR/FTIR) or a grazing angle accessory (infrared reflection absorption spectroscopy – IRAS), is a surface-sensitive technique that has been used to investigate the alteration of protein structure after adsorption. The amide group of proteins and polypeptides present characteristic vibrational modes that are sensitive to protein conformation. Each secondary structural element (helix, sheet, turns and random structures) gives rise to characteristic stretching frequencies in the amide I region (1700–1600 cm^{-1}). Hence, quantitative structural information is obtained from the band position and relative band areas [83–86].

4.4 Influence of surface chemistry on cell response

Biomaterial surface chemistry has profound consequences on cellular responses both *in vitro* and *in vivo*. The engineering of surfaces to control cell adhesion represents an active area of biomaterials research.

The effects of biomaterial surface properties on cellular responses are generally attributed to material-dependent differences in adsorbed protein species, concentration, and/or biological activity. In as short a time as can be measured after implantation in a living system, proteins are already observed on biomaterials surfaces. In seconds to minutes, a monolayer of proteins adsorbs to most surfaces. The protein adsorption event occurs well before cells arrive at the surface. Therefore, cells see primarily a protein layer, rather than the actual surface of the biomaterial [87, 88].

Adhesion of cells to biomaterials is an important prerequisite for the successful incorporation of implants or the colonization of scaffolds for tissue engineering. It is desirable to understand how implant surfaces should be composed to support the attachment, growth and function of cells. In addition, biocompatibility of implanted medical devices is determined by the host foreign body response to the surface of the implanted material. Monocytes migrate to the tissue/material interface and adhere to the surface of the implant, then differentiate into macrophages which, in turn, may fuse to form foreign body giant cells that have an important role in the stress cracking and oxidative damage of the implant [89, 90]. Once more, the chemical composition of the surface of a biomaterial will modulate the presence and activity of these cells.

Model substrates with well-controlled properties, in particular self-assembled monolayers (SAMs) of alkanethiols on gold, are very useful models to systematically investigate the effect of surface chemistry without changing other surface properties [13, 15]. The biological response to SAMs has been investigated using different chemically defined surfaces and several proteins and cell types.

Thrombus formation, caused by a cascade of protein adsorption events and platelet adhesion and activation, still remains an obstacle in the application of blood-contacting medical devices. Lin and Chuang [91] have investigated the adhesion of human platelets to SAMs terminated with CH_3, COOH, OH and SO_3H, and observed higher numbers of adherent platelets on the COOH-terminated SAM. The OH-terminated SAM was the surface that presented the lowest number of adherent platelets. Similar results were obtained by Sperling et al. [92] for the COOH-terminated SAMs, but high numbers of adherent platelets were also found on the CH_3-terminated surface. These authors have, in addition, investigated binary mixtures of these alkanethiols and observed that platelet adhesion was greatly enhanced in the case of mixed SAMs prepared with CH_3 and COOH groups, and inhibited in the case of mixtures with COOH- and OH-terminated alkanethiols. Mixed SAMs prepared from NH_2- and COOH-terminated alkanethiols, using various solution mole fractions of the NH_2-terminated alkanethiol, were employed for platelet adhesion studies [93]. SAMs with intermediate values of NH_2 mole fraction exhibited the least platelet adhesion values.

Rodrigues et al. [49] have investigated the adhesion of platelets to SAMs prepared with mixtures of OH- and CH_3-terminated alkanethiols. They have, in addition, assessed the effect of blood proteins on platelet adhesion and activation. Their results demonstrated a decrease of human fibrinogen adsorption with the increase of OH groups on the monolayer, the same being observed in the platelet adhesion and activation studies. Fibrinogen adsorption was related to high platelet adhesion, and adsorption of human serum albumin caused a *passivant* effect on platelet adhesion and activation in the case of SAMs containing OH-terminated alkanethiol.

Leukocyte adhesion to artificial surfaces is an important phenomenon in the evaluation of biomaterials, since adherent leukocytes are often related to the inflammatory response seen after implantation [94]. Adhesion of both human mononuclear and polymorphonuclear leukocytes to SAMs with the terminal functionalities of OH, COOH and CH_3 was investigated *in vitro* and the adhesion of the two different types of cells was higher for the CH_3-terminated SAMs; OH-terminated surfaces also presented high numbers of adherent leukocytes [95]. This study was also performed *in vivo* using a rodent air pouch model of inflammation [96]. The CH_3-terminated surface accounted for the lower adhesion density of cells, while the OH-covered surfaces presented the higher density of adherent leukocytes. These differences highlight the important role of proteins in the process of cell adhesion to artificial surfaces. Sperling et al. [92] investigated the adhesion of human leukocytes to SAMs with the same terminal functional groups and also with binary mixtures of these alkanethiols, which resulted in the adhesion of leukocytes being greatly enhanced by OH-group bearing surfaces, which is in contrast with the results observed in the studies with platelets. Tegoulia and Cooper [97] have studied the adhesion of human polymorphonuclear leukocytes to SAMs with different terminal functional groups and have also investigated the influence of fibrinogen adsorption. The adhesion was higher on the CH_3-terminated surface. Pre-incubation with fibrinogen decreased adhesion on all the surfaces tested.

Improved initial attachment of osteoblasts or osteoblast precursor cells to orthopaedic implant surfaces may lead to improved bone integration of the implant and longer-term stability. Primary human osteoblasts were cultured on SAMs of alkanethiols on gold with COOH and CH_3 terminal functional groups, and the number of adherent cells was higher on the carboxylic-terminated surface [31], being 10 times higher after 24 hours. In addition, osteoblasts were also cultured on SAMs that were patterned by photolithography techniques, and cells attached almost exclusively to the carboxylic-terminated SAMs [31]. In another study performed by the same authors [98] the effect of the adsorption of fibronectin and albumin was assessed. Attachment was higher on COOH-terminated SAMs precoated with fibronectin, being significantly reduced if this protein was omitted. On OH- and CH_3-terminated SAMs, increasing the proportion of albumin in the solution was sufficient to reduce cell attachment. Schweikl et al. [99] used self-assembled monolayers (SAMs) of various alkanethiols and alkylsilanes as model surfaces with the terminal functional groups COOH, NH_2, CH_3, CF_3 and poly(ethylene glycol) (PEG) to investigate proliferation of human MG-63 osteoblasts. High proliferation of osteoblasts occurred on NH_2-terminated and hydrophobic CH_3-terminated surfaces, while PEG-modified surfaces induced slightly lower cell numbers after an incubation period of 8 days. Cell proliferation was very slow on hydrophobic CF_3-terminated surfaces.

Attachment, spreading and growth of endothelial cells to synthetic surfaces is relevant in the vascularization of tissue engineered scaffolds. The effect of specific chemical functionalities on the growth of bovine aortic endothelial cells (BAEC) was investigated using SAMs of alkanethiols on gold with CH_3, OH, CO_2CH_3 and COOH as terminal functional groups, and the role of albumin and fibronectin was also evaluated [100]. Surface COOH groups promoted cell spreading and growth to a greater extent than the CH_3, OH or CO_2CH_3 groups. The best cell growth substrate demonstrated significantly higher fibronectin adsorption than the other substrates. Similar studies were performed by Mrksich et al. [101] using microcontact printing to pattern the attachment of bovine capillary endothelial cells to self-assembled monolayers of alkanethiols on gold. SAMs were patterned into regions terminated in methyl groups and tri(ethylene glycol) groups; methyl-terminated SAMs promoted hydrophobic adsorption of proteins and the tri(ethylene glycol) terminated ones resisted the adsorption of proteins.

Immersion of these patterned surfaces in a fibronectin solution resulted in adsorption of protein to the CH_3-terminated regions of the SAM; subsequent placement of these substrates in culture medium containing the cells resulted in attachment and spreading in the regions that present fibronectin.

In work performed by Liu *et al.* [56] osteopontin adsorption and subsequent bovine aortic endothelial cell adhesion were investigated using SAMs of alkanethiols on gold with CH_3, OH, COOH and NH_2 as terminal functional groups. The results show that the amounts of osteopontin adsorbed on the NH_2- and COOH-terminated SAMs are similar and close to a full monolayer. However, cell adhesion and spreading on the NH_2 surface is much higher than on the COOH SAM. These results suggest that the orientation and conformation of the protein on the positively charged NH_2 surface is more favourable for cell adhesion and spreading than on the negatively charged COOH surface. In another study [52], the adhesion of human umbilical vein endothelial cells (HUVECs) and the effect of albumin on cell adhesion were investigated using SAMs terminated with CH_3, OH, COOH and NH_2 groups, as well as SAMs prepared with binary mixtures of these alkanethiols. Few cells adhered to the OH- and CH_3-terminated SAMS. On CH_3/COOH, CH_3/NH_2 and CH_3/OH (only up to 85% OH groups) mixed SAMs, cells adhered and gradually spread in relation to the increase of surface concentrations of COOH, NH_2 and OH groups. Cells were then cultured on the mixed SAMs pretreated with albumin; cell adhesion was strongly decreased on hydrophobic surfaces; on the other hand, cells adhered to albumin-coated hydrophilic SAMs.

The effect of surface chemistry on nerve cell functions is a key issue in the understanding of cell adhesion and growth, nerve regeneration, biocompatibility and modelling of biomaterials performance. Shimizu *et al.* [102] investigated the effect of specific chemical functionalities on neurite outgrowth. They prepared SAMs with CH_3, COOH and NH_2 as terminal functional groups. Neurons with neurite outgrowth were observed predominantly on NH_2-terminated SAMs. Romanova *et al.* [103] used SAMs to evaluate neuronal growth and function, and observed that COOH-terminated surfaces support neuron attachment and growth even without an intermediate protein layer. Addition of a poly-L-lysine layer to the COOH-terminated surfaces significantly increases neurite outgrowth. Mixed monolayers of CH_3/COOH poorly support neuron adhesion and preclude neurite extension.

Material-driven control of stem cell behaviour and differentiation is a very interesting possibility. Curran *et al.* [104] have used silane-modified surfaces (OH, NH_2, COOH, CH_3 and SH) to control bone-marrow-derived mesenchymal stem cell adhesion and differentiation *in vitro*. The CH_3 and the control surfaces (glass) maintained mesenchymal stem cell phenotype; the SH- and NH_2-modified surfaces promoted and maintained osteogenesis. These surfaces did not support long-term chondrogenesis. The OH- and COOH-modified surfaces promoted and maintained chondrogenesis but did not support osteogenesis. This study demonstrates that different functional surface chemistries control the differentiation potential of human mesenchymal stem cells.

An appropriate cellular response to implanted surfaces is essential for tissue regeneration and integration. Implanted materials are immediately covered with proteins and through this adsorbed layer cells sense foreign surfaces. The nature of cell–surface interactions contributes to survival, growth and differentiation of cells.

4.5   Future trends

This chapter has covered various molecular aspects of cell–biomaterial interactions, including protein adsorption. In order to guide cell behaviour, which is particularly important in regenerative therapies, we should be able to design biomaterials that incorporate the precise biological and chemical cues that are essential for that purpose. The need to account, simultaneously, for cell attachment, guided differentiation, proliferation and migration appears to be an unaccomplishable task. The matrix should behave like the extracellular matrix (ECM) but with a complexity (or simplicity) compatible with our present (limited) technologies. Degradation of ECM-like matrices should be orchestrated by the cells, namely by production of metalloproteases, while the hybrid tissue should maintain the original tissue's mechanical and biological functionalities.

From the non-degradable biomaterials, which marked the early days of biomaterials engineering, we have now come to the era of fully degradable, hopefully at the 'right' rate, biomaterials. But is this a 'must', or can we compromise with an adequate (whatever that means) rate of degradation? Certainly we need more pliable materials, capable of adapting to changes in the environment (e.g. mechanical and chemical), and the best way to approach this goal is to understand how natural tissues have evolved during development. If our beliefs in protein-guided cell–biomaterials interactions are true, then we should devote a great effort to understanding how the pattern of adsorption *in vivo* may be different from the one we think holds true *in vitro*. Most of what we know at present on protein adsorption has been gained through experiments where we use a liquid, within which diffusion of species is completely different from what probably hides in the interstices of the ECM. This will probably have a great influence on how we will see protein adsorption and cell adhesion, as well as the paracrine mechanism of intracellular communication.

4.6 References

[1] Cooke IR, Deserno M. Solvent-free model for self-assembling fluid bilayer membranes: Stabilization of the fluid phase based on broad attractive tail potentials. *Journal of Chemical Physics*, 2005, 123(22): 224710.
[2] Gelain F, Horii A, Zhang SG. Designer self-assembling peptide scaffolds for 3-D tissue cell cultures and regenerative medicine. *Macromolecular Bioscience*, 2007, 7(5): 544–51.
[3] Zhang SG. Fabrication of novel biomaterials through molecular self-assembly. *Nature Biotechnology*, 2003, 21(10): 1171–8.
[4] Horii A, Wang X, Gelain F, Zhang S. Biological designer self-assembling peptide nanofiber scaffolds significantly enhance osteoblast proliferation, differentiation and 3-D migration. *PLoS ONE*, 2007, 2(2): e190.
[5] Zhang SG, Holmes T, Lockshin C, Rich A. Spontaneous assembly of a self-complementary oligopeptide to form a stable macroscopic membrane. *Proceedings of the National Academy of Sciences of the United States of America*, 1993, 90(8): 3334–8.
[6] Liu JC, Heilshorn SC, Tirrell DA. Comparative cell response to artificial extracellular matrix proteins containing the RGD and CS5 cell-binding domains. *Biomacromolecules*, 2004, 5(2): 497–504.
[7] Heilshorn SC, DiZio KA, Welsh ER, Tirrell DA. Endothelial cell adhesion to the fibronectin CS5 domain in artificial extracellular matrix proteins. *Biomaterials*, 2003, 24(23): 4245–52.
[8] Heilshorn SC, Liu JC, Tirrell DA. Cell-binding domain context affects cell behavior on engineered proteins. *Biomacromolecules*, 2005, 6(1): 318–23.
[9] Petka WA, Harden JL, McGrath KP, Wirtz D, Tirrell DA. Reversible hydrogels from self-assembling artificial proteins. *Science*, 1998, 281(5375): 389–92.
[10] Lutolf MP, Hubbell JA. Synthetic biomaterials as instructive extracellular microenvironments for morphogenesis in tissue engineering. *Nature Biotechnology*, 2005, 23(1): 47–55.
[11] Lutolf MP, Raeber GP, Zisch AH, Tirelli N, Hubbell JA. Cell-responsive synthetic hydrogels. *Advanced Materials*, 2003, 15(11): 888–92.
[12] Schreiber F. Structure and growth of self-assembling monolayers. *Progress in Surface Science*, 2000, 65(5–8): 151–256.
[13] Ulman A. Formation and structure of self-assembled monolayers. *Chemical Reviews*, 1996, 96(4): 1533–54.
[14] Love JC, Estroff LA, Kriebel JK, Nuzzo RG, Whitesides GM. Self-assembled monolayers of thiolates on metals as a form of nanotechnology. *Chemical Reviews*, 2005, 105(4): 1103–69.
[15] Ostuni E, Yan L, Whitesides GM. The interaction of proteins and cells with self-assembled monolayers of alkanethiolates on gold and silver. *Colloids and Surfaces B – Biointerfaces*, 1999, 15(1): 3–30.
[16] Mrksich M, Whitesides GM. Using self-assembled monolayers to understand the interactions of man-made surfaces with proteins and cells. *Annual Review of Biophysics and Biomolecular Structure*, 1996, 25: 55–78.
[17] Ulman A. *An Introduction to Ultrathin Organic Films: From Langmuir-Blodgett to Self-Assembly*, Academic Press, Boston, MA, 1991.

[18] Gooding JJ, Mearns F, Yang WR, Liu JQ. Self-assembled monolayers into the 21st century: Recent advances and applications. *Electroanalysis*, 2003, 15(2): 81–96.
[19] Chechik V, Stirling CJM. Gold-thiol self-assembled monolayers. In: Patai S, Rappoport Z, eds. *The chemistry of organic derivatives of gold and silver*: John Wiley & Sons, LTD 1999.
[20] Bain CD, Troughton EB, Tao YT, Evall J, Whitesides GM, Nuzzo RG. Formation of monolayer films by the spontaneous assembly of organic thiols from solution onto gold. *Journal of the American Chemical Society*, 1989, 111(1): 321–35.
[21] Schoenfisch MH, Pemberton JE. Air stability of alkanethiol self-assembled monolayers on silver and gold surfaces. *Journal of the American Chemical Society*, 1998, 120(18): 4502–13.
[22] Bain CD, Whitesides GM. Correlations between wettability and structure in monolayers of alkanethiols adsorbed on gold. *Journal of the American Chemical Society*, 1988, 110(11): 3665–6.
[23] Sigal GB, Mrksich M, Whitesides GM. Effect of surface wettability on the adsorption of proteins and detergents. *Journal of the American Chemical Society*, 1998, 120(14): 3464–73.
[24] Folkers JP, Laibinis PE, Whitesides GM. Self-assembled monolayers of alkanethiols on gold – comparisons of monolayers containing mixtures of short-chain and long-chain constituents with CH_3 and CH_2OH terminal groups. *Langmuir*, 1992, 8(5): 1330–41.
[25] Atre SV, Liedberg B, Allara DL. Chain-length dependence of the structure and wetting properties in binary composition monolayers of OH-terminated and CH_3-terminated alkanethiolates on gold. *Langmuir*, 1995, 11(10): 3882–93.
[26] Martins MCL, Ratner BD, Barbosa MA. Protein adsorption on mixtures of hydroxyl- and methyl-terminated alkanethiols self-assembled monolayers. *Journal of Biomedical Materials Research Part A*, 2003, 67A(1): 158–71.
[27] Prime KL, Whitesides GM. Self-assembled organic monolayers – model systems for studying adsorption of proteins at surfaces. *Science*, 1991, 252(5009): 1164–7.
[28] Falconnet D, Csucs G, Grandin HM, Textor M. Surface engineering approaches to micropattern surfaces for cell-based assays. *Biomaterials*, 2006, 27(16): 3044–63.
[29] Dulcey CS, Georger JH, Krauthamer V, Stenger DA, Fare TL, Calvert JM. Deep UV photochemistry of chemisorbed monolayers – patterned coplanar molecular assemblies. *Science*, 1991, 252(5005): 551–4.
[30] Ryan D, Parviz BA, Linder V, Semetey V, Sia SK, Su J, *et al*. Patterning multiple aligned self-assembled monolayers using light. *Langmuir*, 2004, 20(21): 9080–8.
[31] Scotchford CA, Cooper E, Leggett GJ, Downes S. Growth of human osteoblast-like cells on alkanethiol on gold self-assembled monolayers: The effect of surface chemistry. *Journal of Biomedical Materials Research*, 1998, 41(3): 431–42.
[32] Mrksich M, Whitesides GM. Patterning self-assembled monolayers using microcontact printing – a new technology for biosensors. *Trends in Biotechnology*, 1995, 13(6): 228–35.

[33] Xia YN, Whitesides GM. Soft lithography. *Annual Review of Materials Science*, 1998, 28: 153–84.
[34] Jackman RJ, Wilbur JL, Whitesides GM. Fabrication of submicrometer features on curved substrates by microcontact printing. *Science*, 1995, 269(5224): 664–6.
[35] Lehnert D, Wehrle-Haller B, David C, Weiland U, Ballestrem C, Imhof BA, *et al*. Cell behaviour on micropatterned substrata: Limits of extracellular matrix geometry for spreading and adhesion. *Journal of Cell Science*, 2004, 117(1): 41–52.
[36] Chen CS, Mrksich M, Huang S, Whitesides GM, Ingber DE. Micropatterned surfaces for control of cell shape, position, and function. *Biotechnology Progress*, 1998, 14(3): 356–63.
[37] Vogler EA. Structure and reactivity of water at biomaterial surfaces. *Advances in Colloid and Interface Science*, 1998, 74: 69–117.
[38] Norde W. My voyage of discovery to proteins in flatland...and beyond. *Colloids and Surfaces B – Biointerfaces*, 2008, 61(1): 1–9.
[39] Haynes CA, Norde W. Structures and stabilities of adsorbed proteins. *Journal of Colloid and Interface Science*, 1995, 169(2): 313–28.
[40] Lima J, Sousa SR, Ferreira A, Barbosa MA. Interactions between calcium, phosphate, and albumin on the surface of titanium. *Journal of Biomedical Materials Research*, 2001, 55(1): 45–53.
[41] Sousa SR, Moradas-Ferreira P, Barbosa MA. TiO_2 type influences fibronectin adsorption. *Journal of Materials Science – Materials in Medicine*, 2005, 16: 1173–8.
[42] Sousa SR, Moradas-Ferreira P, Saramago B, Melo LV, Barbosa MA. Human serum albumin adsorption on TiO_2 from single protein solutions and from plasma. *Langmuir*, 2004, 20(22): 9745–54.
[43] Wilson CJ, Clegg RE, Leavesley DI, Pearcy MJ. Mediation of biomaterial–cell interactions by adsorbed proteins: a review. *Tissue Engineering*, 2005, 11(1–2): 1–18.
[44] Dee KC, Puleo DA, Bizios R. *An Introduction to Tissue–Biomaterial Interactions*, John Wiley & Sons, Hoboken, NJ, 2002.
[45] Sethuraman A, Han M, Kane RS, Belfort G. Effect of surface wettability on the adhesion of proteins. *Langmuir*, 2004, 20(18): 7779–88.
[46] Lestelius M, Liedberg B, Tengvall P. In vitro plasma protein adsorption on omega-functionalized alkanethiolate self-assembled monolayers. *Langmuir*, 1997, 13(22): 5900–8.
[47] Keselowsky BG, Collard DM, Garcia AJ. Surface chemistry modulates fibronectin conformation and directs integrin binding and specificity to control cell adhesion. *Journal of Biomedical Materials Research Part A*, 2003, 66A(2): 247–59.
[48] Silin V, Weetall H, Vanderah DJ. SPR studies of the nonspecific adsorption kinetics of human IgG and BSA on gold surfaces modified by self-assembled monolayers (SAMs). *Journal of Colloid and Interface Science*, 1997, 185(1): 94–103.
[49] Rodrigues SN, Goncalves IC, Martins MCL, Barbosa MA, Ratner BD. Fibrinogen adsorption, platelet adhesion and activation on mixed hydroxyl-/methyl-terminated self-assembled monolayers. *Biomaterials*, 2006, 27(31): 5357–67.
[50] Hirata I, Hioko Y, Toda M, Kitazawa T, Murakami Y, Kitano E, *et al*. Deposition of complement protein C3b on mixed self-assembled monolayers carrying

surface hydroxyl and methyl groups studied by surface plasmon resonance. *Journal of Biomedical Materials Research Part A*, 2003, 66A(3): 669–76.
- [51] Capadona JR, Collard DM, Garcia AJ. Fibronectin adsorption and cell adhesion to mixed monolayers of tri(ethylene glycol)- and methyl-terminated alkanethiols. *Langmuir*, 2003, 19(5): 1847–52.
- [52] Arima Y, Iwata H. Effect of wettability and surface functional groups on protein adsorption and cell adhesion using well-defined mixed self-assembled monolayers. *Biomaterials*, 2007, 28(20): 3074–82.
- [53] Andrade JD, Hlady V. Protein adsorption and materials biocompatibility – a tutorial review and suggested hypotheses. *Advances in Polymer Science*, 1986, 79: 1–63.
- [54] Kidoaki S, Matsuda T. Adhesion forces of the blood plasma proteins on self-assembled monolayer surfaces of alkanethiolates with different functional groups measured by an atomic force microscope. *Langmuir*, 1999, 15(22): 7639–46.
- [55] Lee MH, Ducheyne P, Lynch L, Boettiger D, Composto RJ. Effect of biomaterial surface properties on fibronectin-alpha(5)beta(1) integrin interaction and cellular attachment. *Biomaterials*, 2006, 27(9): 1907–16.
- [56] Liu LY, Chen SF, Giachelli CM, Ratner BD, Jiang SY. Controlling osteopontin orientation on surfaces to modulate endothelial cell adhesion. *Journal of Biomedical Materials Research Part A*, 2005, 74A(1): 23–31.
- [57] Roach P, Farrar D, Perry CC. Interpretation of protein adsorption: Surface-induced conformational changes. *Journal of the American Chemical Society*, 2005, 127(22): 8168–73.
- [58] Tsai WB, Grunkemeier JM, Horbett TA. Variations in the ability of adsorbed fibrinogen to mediate platelet adhesion to polystyrene-based materials: a multivariate statistical analysis of antibody binding to the platelet binding sites of fibrinogen. *Journal of Biomedical Materials Research Part A*, 2003, 67A(4): 1255–68.
- [59] Weisel JW, Nagaswami C, Vilaire G, Bennett JS. Examination of the platelet membrane glycoprotein IIb–IIIa complex and its interaction with fibrinogen and other ligands by electron microscopy. *Journal of Biological Chemistry*, 1992, 267(23): 16637–43.
- [60] Savage B, Ruggeri ZM. Selective recognition of adhesive sites in surface-bound fibrinogen by glycoprotein-Iib-Iiia on nonactivated platelets. *Journal of Biological Chemistry*, 1991, 266(17): 11227–33.
- [61] Klebe RJ, Bentley KL, Schoen RC. Adhesive substrates for fibronectin. *Journal of Cell Physiology*, 1981, 109(3): 481–8.
- [62] McClary KB, Ugarova T, Grainger DW. Modulating fibroblast adhesion, spreading, and proliferation using self-assembled monolayer films of alkylthiolates on gold. *Journal of Biomedical Materials Research*, 2000, 50(3): 428–39.
- [63] Norde W, Favier JP. Structure of adsorbed and desorbed proteins. *Colloids and Surfaces*, 1992, 64(1): 87–93.
- [64] Davis F, Higson SPJ. Structured thin films as functional components within biosensors. *Biosensors & Bioelectronics*, 2005, 21(1): 1–20.
- [65] Herrwerth S, Rosendahl T, Feng C, Fick J, Eck W, Himmelhaus M, *et al.* Covalent coupling of antibodies to self-assembled monolayers of carboxy-

functionalized poly(ethylene glycol): Protein resistance and specific binding of biomolecules. *Langmuir*, 2003, 19(5): 1880–7.
[66] Kenseth JR, Harnisch JA, Jones VW, Porter MD. Investigation of approaches for the fabrication of protein patterns by scanning probe lithography. *Langmuir*, 2001, 17(13): 4105–12.
[67] Martins MCL, Naeemi E, Ratner BD, Barbosa MA. Albumin adsorption on Cibacron Blue F3G-A immobilized onto oligo(ethylene glycol)-terminated self-assembled monolayers. *Journal of Materials Science: Materials in Medicine*, 2003, 14: 945–54.
[68] Gonçalves IC, Martins MCL, Barbosa MA, Ratner BD. Protein adsorption on 18-alkyl chains immobilized on hydroxyl-terminated self-assembled monolayers. *Biomaterials*, 2005, 26(18): 3891–9.
[69] Prime KL, Whitesides GM. Adsorption of proteins onto surfaces containing end-attached oligo(ethylene oxide) – a model system using self-assembled monolayers. *Journal of the American Chemical Society*, 1993, 115(23): 10714–21.
[70] Luk YY, Kato M, Mrksich M. Self-assembled monolayers of alkanethiolates presenting mannitol groups are inert to protein adsorption and cell attachment. *Langmuir*, 2000, 16(24): 9604–8.
[71] Klein E, Kerth P, Lebeau L. Enhanced selective immobilization of biomolecules onto solid supports coated with semifluorinated self-assembled monolayers. *Biomaterials*, 2008, 29(2): 204–14.
[72] Chen SF, Liu LY, Jiang SY. Strong resistance of oligo(phosphorylcholine) self-assembled monolayers to protein adsorption. *Langmuir*, 2006, 22(6): 2418–21.
[73] Gouzy MF, Sperling C, Salchert K, Pompe T, Streller U, Uhlmann P, et al. In vitro blood compatibility of polymeric biomaterials through covalent immobilization of an amidine derivative. *Biomaterials*, 2004, 25(17): 3493–501.
[74] Yu XJ, Brash JL. Measurement of protein adsorption to solid surfaces in relation to blood compatibility using radioiodine labelling methods. In: Dawids S (ed.), *Test Procedures for the Blood Compatibility of Biomaterials*, Kluwer Academic Publishers, Dordrecht, the Netherlands, 1993: 287–330.
[75] Horbett TA. Techniques for protein adsorption studies. In: Williams DF (ed.), *Techniques of Biocompatibility Testing*, CRC Press, Boca Raton, FL, 1986: 183–214.
[76] Green RJ, Frazier RA, Shakesheff KM, Davies MC, Roberts CJ, Tendler SJB. Surface plasmon resonance analysis of dynamic biological interactions with biomaterials. *Biomaterials*, 2000, 21(18): 1823–35.
[77] Elwing H. Protein adsorption and ellipsometry in biomaterial research. *Biomaterials*, 1998, 19(4–5): 397–406.
[78] Tengvall P, Lundstrom I, Liedberg B. Protein adsorption studies on model organic surfaces: an ellipsometric and infrared spectroscopic approach. *Biomaterials*, 1998, 19(4–5): 407–22.
[79] Marx KA. Quartz crystal microbalance: a useful tool for studying thin polymer films and complex biomolecular systems at the solution–surface interface. *Biomacromolecules*, 2003, 4(5): 1099–120.
[80] Rodahl M, Hook F, Krozer A, Brzezinski P, Kasemo B. Quartz-crystal microbalance setup for frequency and Q-factor measurements in gaseous and liquid environments. *Review of Scientific Instruments*, 1995, 66(7): 3924–30.

[81] Hook F, Rodahl M, Brzezinski P, Kasemo B. Energy dissipation kinetics for protein and antibody–antigen adsorption under shear oscillation on a quartz crystal microbalance. *Langmuir*, 1998, 14(4): 729–34.
[82] Andersson M, Andersson J, Sellborn A, Berglin M, Nilsson B, Elwing H. Quartz crystal microbalance-with dissipation monitoring (QCM-D) for real time measurements of blood coagulation density and immune complement activation on artificial surfaces. *Biosensors & Bioelectronics*, 2005, 21(1): 79–86.
[83] Sethuraman A, Belfort G. Protein structural perturbation and aggregation on homogeneous surfaces. *Biophysical Journal*, 2005, 88(2): 1322–33.
[84] Chittur KK. FTIR/ATR for protein adsorption to biomaterial surfaces. *Biomaterials*, 1998, 19(4–5): 357–69.
[85] Bramanti E, Benedetti E. Determination of the secondary structure of isomeric forms of human serum albumin by a particular frequency deconvolution procedure applied to Fourier transform IR analysis. *Biopolymers*, 1996, 38(5): 639–53.
[86] Vedantham G, Sparks HG, Sane SU, Tzannis S, Przybycien TM. A holistic approach for protein secondary structure estimation from infrared spectra in H_2O solutions. *Analytical Biochemistry*, 2000, 285(1): 33–49.
[87] Eriksson C, Nygren H. Polymorphonuclear leukocytes in coagulating whole blood recognize hydrophilic and hydrophobic titanium surfaces by different adhesion receptors and show different patterns of receptor expression. *Journal of Laboratory and Clinical Medicine*, 2001, 137(4): 296–302.
[88] Tang LP, Eaton JW. Mechanism of acute inflammatory response to biomaterials. *Cells and Materials*, 1994, 4(4): 429–36.
[89] Brodbeck WG, Voskerician G, Ziats NP, Nakayama Y, Matsuda T, Anderson JM. In vivo leukocyte cytokine mRNA responses to biomaterials are dependent on surface chemistry. *Journal of Biomedical Materials Research Part A*, 2003, 64A(2): 320–9.
[90] MacEwan MR, Brodbeck WG, Matsuda T, Anderson JM. Monocyte/lymphocyte interactions and the foreign body response: in vitro effects of biomaterial surface chemistry. *Journal of Biomedical Materials Research Part A*, 2005, 74A(3): 285–93.
[91] Lin JC, Chuang WH. Synthesis, surface characterization, and platelet reactivity evaluation for the self-assembled monolayer of alkanethiol with sulfonic acid functionality. *Journal of Biomedical Materials Research*, 2000, 51(3): 413–23.
[92] Sperling C, Schweiss RB, Streller U, Werner C. In vitro hemocompatibility of self-assembled monolayers displaying various functional groups. *Biomaterials*, 2005, 26(33): 6547–57.
[93] Chuang WH, Lin JC. Surface characterization and platelet adhesion studies for the mixed self-assembled monolayers with amine and carboxylic acid terminated functionalities. *Journal of Biomedical Materials Research Part A*, 2007, 82A(4): 820–30.
[94] Brunstedt MR, Anderson JM, Spilizewski KL, Marchant RE, Hiltner A. In vivo leukocyte interactions on pellethane surfaces. *Biomaterials*, 1990, 11(6): 370–8.
[95] Barbosa JN, Barbosa MA, Aguas AP. Adhesion of human leukocytes to biomaterials: an in vitro study using alkanethiolate monolayers with different

chemically functionalized surfaces. *Journal of Biomedical Materials Research Part A*, 2003, 65A(4): 429–34.
[96] Barbosa JN, Barbosa MA, Aguas AP. Inflammatory responses and cell adhesion to self-assembled monolayers of alkanethiolates on gold. *Biomaterials*, 2004, 25(13): 2557–63.
[97] Tegoulia VA, Cooper SL. Leukocyte adhesion on model surfaces under flow: Effects of surface chemistry, protein adsorption, and shear rate. *Journal of Biomedical Materials Research*, 2000, 50(3): 291–301.
[98] Scotchford CA, Gilmore CP, Cooper E, Leggett GJ, Downes S. Protein adsorption and human osteoblast-like cell attachment and growth on alkylthiol on gold self-assembled monolayers. *Journal of Biomedical Materials Research*, 2002, 59(1): 84–99.
[99] Schweikl H, Muller R, Englert C, Hiller KA, Kujat R, Nerlich M, *et al.* Proliferation of osteoblasts and fibroblasts on model surfaces of varying roughness and surface chemistry. *Journal of Materials Science – Materials in Medicine*, 2007, 18(10): 1895–905.
[100] Tidwell CD, Ertel SI, Ratner BD, Tarasevich BJ, Atre S, Allara DL. Endothelial cell growth and protein adsorption on terminally functionalized, self-assembled monolayers of alkanethiolates on gold. *Langmuir*, 1997, 13(13): 3404–13.
[101] Mrksich M, Dike LE, Tien J, Ingber DE, Whitesides GM. Using microcontact printing to pattern the attachment of mammalian cells to self-assembled monolayers of alkanethiolates on transparent films of gold and silver. *Experimental Cell Research*, 1997, 235(2): 305–13.
[102] Shimizu N, Naka Y, Takei H. Neurite outgrowth of neurons immobilized on patterned self-assembled monolayers. *Abstracts of Papers of the American Chemical Society*, 2002, 224: U235–6.
[103] Romanova EV, Oxley SP, Rubakhin SS, Bohn PW, Sweedler JV. Self-assembled monolayers of alkanethiols on gold modulate electrophysiological parameters and cellular morphology of cultured neurons. *Biomaterials*, 2006, 27(8): 1665–9.
[104] Curran JM, Chen R, Hunt JA. The guidance of human mesenchymal stem cell differentiation in vitro by controlled modifications to the cell substrate. *Biomaterials*, 2006, 27(27): 4783–93.

5
Biocompatibility and other properties of hydrogels in regenerative medicine

A SANNINO and M MADAGHIELE,
University of Salento, Italy and L AMBROSIO,
Institute of Composite and Biomedical
Materials (IMCB-CNR), Italy

Abstract: The word 'gel' refers to a wide class of materials displaying a high capability of absorbing and retaining a liquid medium. Hydrogels are defined as a particular type of macromolecular gels, formed via chemical or physical stabilization of the polymer chains into a three-dimensional network, for which the absorbed liquid is water or a water solution. The hydrophilic nature of hydrogels gives them unique properties in terms of biocompatibility, rubbery mechanical properties similar to those of soft tissues, and mild-gelling conditions, that are suitable for drug delivery and cell transplantation strategies. Moreover, the strong sensitivity to certain environmental stimuli, displayed by some hydrogels in terms of reversible swelling/deswelling phase transitions, makes them particularly attractive as smart materials for use in a wide range of applications.

In this chapter, we introduce the theory describing the sorption thermodynamics of chemical hydrogels, discussing the relationships between their microstructural parameters and the resulting macroscopic properties. Various approaches to evaluate the most relevant network parameters affecting the hydrogel swelling capability and mechanical stiffness will be discussed. It is worth noting that the same theory, traditionally formulated for chemical hydrogels, might be applied to describe the behaviour of physical hydrogels as well.

Key words: macromolecular hydrogels, degree of crosslinking, sorption thermodynamics, sorption kinetics, regenerative medicine.

5.1 Gels and hydrogels: definition and classification

A remarkable capability of swelling, when placed in contact with specific liquid media, is the distinctive feature of a wide class of materials which are commonly designated as 'gels'. Owing to its particular nature, the gel, or swollen state, shows a hybrid behaviour, with structural properties similar to those of a solid but diffusive properties matching those of a liquid. Tanaka [1] defined the gel state as an intermediate state of matter between solid and liquid, and, before him, Jordan-Lloyd [2] stated that 'the colloidal state, of the gel, is one which is easier to recognize than to define'. A number of natural and synthetic materials can form gel-like structures under appropriate conditions. A first classification distinguishes between macromolecular (i.e., polymer-based) and non-macromolecular gels (e.g., silica gels). In the following, we focus on macromolecular gels, due to their large use in the field of biomaterials, with the aim of illustrating their design principles and structure–properties relationships.

Macromolecular gels are polymeric networks, stabilized via chemical or physical interactions among the polymer chains, which are able to retain large amounts of liquid in their mesh structure. The dissolution of the polymer in the solvent is prevented by the junctions, i.e. the crosslinking sites, existing among the macromolecules and forming the network. Polymer gels are classified into chemical and physical gels, with regard to the mechanisms underlying gel formation. Chemical gels are formed by means of covalent bonds among the macromolecules, that lead to a three-dimensional, highly stable polymer network. Gel formation, in this case, is not reversible. Conversely, in physical gels the network stabilization occurs by means of secondary interactions, such as Van der Waals forces, hydrophobic associations, hydrogen bonding, ionic interactions, crystallite formation, and physical entanglements. As such, physical gels might flow under mechanical loading and undergo a thermally or chemically triggered dissolution (gel-sol transition) [3]. This fundamental distinction between stable, stiff networks and injectable, reversible ones, for chemical and physical gels respectively, suggests how structural parameters of the polymer network may affect the macroscopic behaviour of the resulting gel, thus determining the range of applications for which the gel is suitable or not.

Hydrogels are defined as a particular type of macromolecular gels, for which the absorbed liquid is water or a water solution. The hydrophilic nature of chemical and physical hydrogels gives them unique properties in terms of biocompatibility, rubbery mechanical properties similar to those of soft tissues, and mild-gelling conditions, that are ideal for drug delivery and cell transplantation strategies [4, 5].

A basic classification of hydrogels relates to the type of electrostatic charges that are tethered on their macromolecular backbone, and distinguishes ionic or polyelectrolyte hydrogels from neutral or non-ionic ones.

With regard to their elastic mechanical behaviour, hydrogel networks are assumed to be either affine or phantom. In the former model, the network structure is affine with respect to the elastic deformation under applied stress, i.e. the end-to-end length of a chain scales linearly with the linear extension of the hydrogel. In the latter model, polymer chains are assumed to be immaterial or 'phantom', thus they can cross each other and move independently from the applied stress. Both models clearly represent two limit cases. For the reader's convenience, the chemical classification of gels and hydrogels discussed above is summarized in Fig. 5.1.

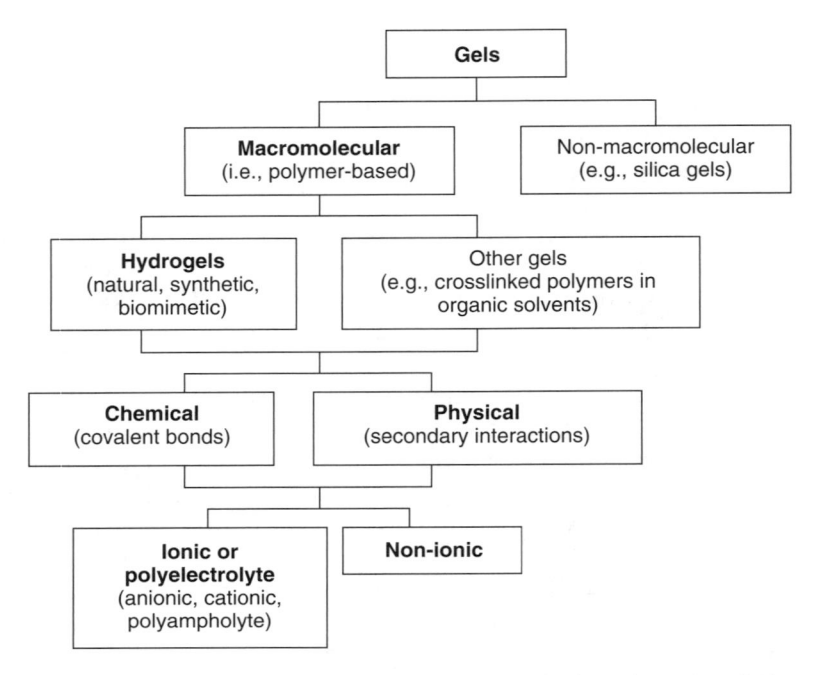

5.1 Schematic classification of gel and hydrogels, based on chemical parameters.

In this chapter, we discuss the relationships between microstructural parameters and macroscopic properties of hydrogel networks. We aim at underscoring the importance of such relationships for correctly designing hydrogel-based devices for specific applications. In this perspective, the microstructural parameters of the hydrogel network can be regarded as design variables, which can be adjusted *ad hoc* to satisfy selected requirements.

5.2 Hydrogel swelling ratio

When designing a hydrogel, the network ability to absorb and retain water is the first and most important feature to be evaluated. 'Smart' hydrogels may respond to small variations of certain environmental stimuli (e.g., pH, ionic strength, solvent composition) by means of remarkable swelling/shrinking phase transitions [1].

The hydrogel swelling capacity can be quantified by means of its mass swelling ratio or volume swelling ratio, Q_m and Q respectively, defined as follows:

$$Q_m = \frac{W_s - W_d}{W_d} = \frac{M_1}{M_2} \qquad 5.1$$

Biocompatibility and other properties in regenerative medicine 117

$$Q = \frac{V_s}{V_d} = \frac{V_1+V_2}{V_2} = 1+Q_m\frac{\rho_2}{\rho_1} \qquad 5.2$$

where W_s and W_d are the weights of the network in the swollen and dry states respectively, V_s and V_d are the corresponding volumes, M_1 and M_2 indicate the masses of the solvent (i.e. water) and the polymer respectively, V_1 and V_2 their volumes and ρ_1 and ρ_2 their densities. The polymer volume fraction in the swollen state can be easily determined as:

$$V_{2,s} = \frac{1}{Q} \qquad 5.3$$

5.3 Microstructural parameters/design variables for polymeric networks

The distinction between chemical and physical hydrogels, as well as that between ionic and neutral ones (Fig. 5.1), suggest that different properties are displayed by these hydrogel types. However, the type of bonds forming the polymeric network, either covalent or physical, and the presence of fixed charges on the macromolecular backbone are only two of the structural parameters affecting the macroscopic behaviour of hydrogels. Polymer–solvent interaction, degree of crosslinking, mesh size, and micro- and macroporous structure are additional variables to be taken into account for an optimal design of hydrogel-based devices, as they may affect the hydrogel's swelling capability, mechanical stiffness, sensitivity to environmental stimuli, and mass transport properties. Provided that the difference between chemical and physical gels lies in the ability of the latter to be injected and dissolved, as opposed to the chemical and mechanical stability of the former, the role of the other design variables is discussed in the following, in terms of thermodynamics of the gel state and swelling kinetics. The close interplay among the different design variables and some of the most significant hydrogel properties are summarized in Table 5.1.

5.3.1 Polymer–solvent interaction

In order for a gel to absorb a high amount of water or water solution, the polymer making up its network has to be highly hydrophilic. The hydrogel state in fact can be considered as a particular type of polymer–water solution which shows an elastic behaviour rather than a viscous one. The thermodynamic theory of polymer solutions [6] leads to the following expression for the free energy change associated with the mixing process between the solvent and the polymer network:

$$\Delta G_{mix} = kT\lfloor n_1 \ln(1-V_{2,s}) + \chi_{1,2} n_1 V_{2,s} \rfloor \qquad 5.4$$

Table 5.1 How network parameters affect macroscopic behaviour of hydrogels. The symbol ✓ highlights the variables to be modulated in relation to specific hydrogel properties

	Network parameters/design variables					
	Chemical				Physical	
	Type of bonds	Polymer–solvent interaction	Degree of crosslinking	Fixed charges	Mesh size (nanopores)	Porosity (micro/macropores)
Injectability (flow properties)	✓		✓			
Swelling capacity	✓	✓	✓	✓		✓
Swelling/shrinking kinetics						✓
Sensitivity to external stimuli	✓	✓		✓		
Mechanical stiffness	✓		✓			
Molecular permeability and diffusion			✓		✓	

where k is the Boltzmann constant, T the absolute temperature, n_1 the number of solvent molecules and $\chi_{1,2}$ the Flory–Huggins polymer–solvent interaction parameter, which is positive or negative for endothermic or exothermic mixing, respectively. According to the theory of phase equilibria, in good solvents (i.e., polymer and solvent miscible over the entire composition range), $\chi_{1,2}$ is less than 0.5. Experimentally, such a parameter is found to be a function of both temperature and polymer concentration, according to the following:

$$\chi_{1,2} = \chi_a + \chi_b V_{2,s} + \chi_c V_{2,s}^2 + \cdots \qquad 5.5$$

with χ_a, χ_b, etc. being functions of the temperature. This finding introduces the role played by the polymer–solvent interaction parameter in the design of thermosensitive hydrogels. Such hydrogels can undergo a reversible volume phase transition when the temperature approaches the upper or lower critical solution temperature (UCST and LCST, respectively). Most polymers increase their water solubility as the temperature increases (i.e., the $\chi_{1,2}$ parameter decreases), thus they might form positive temperature-sensitive hydrogels that shrink upon cooling below the UCST. However, polymers which decrease their water solubility as the temperature increases (i.e., $\chi_{1,2}$ increases) might be synthesized by combining hydrophilic and hydrophobic groups (e.g., ethyl, methyl and propyl groups) in the polymer chain. For example, negative temperature-responsive hydrogels, which shrink when heating above the LCST, can be obtained from polymers such as poly(N-isopropylacrylamide) (PNIPAAm), which form chemical hydrogels [1], and polypropylene oxide–polyethylene oxide–polypropylene oxide (PPO–PEO–PPO) block copolymers, which form physical hydrogels undergoing a reversible sol-gel transition around 37°C (i.e. upon injection *in vivo*) [7]. At low temperatures, hydrogen bonds between hydrophilic polymer groups and water molecules are stable. However, as temperature increases, polymer chains interact hydrophobically, losing their bound water and thus resulting in network shrinking. The LCST can be adjusted by controlling the ratio of hydrophobic to hydrophilic groups in the polymer chain [8].

5.3.2 Degree of crosslinking

Any polymeric network is characterized by having a certain fraction of macromolecular chains engaged in crosslinkages and a given fraction of free chain ends or terminal chains. By chain we mean the segment or structural unit of the primary polymer molecule extending from one crosslinking site to the next one. Terminal chains are linked to the network by only one crosslinking site. In order for the three-dimensional network to form, the interchain linkage is required to have a functionality $f \geq 3$, i.e. at least three chains need to be joined together at each crosslinking site (Fig. 5.2).

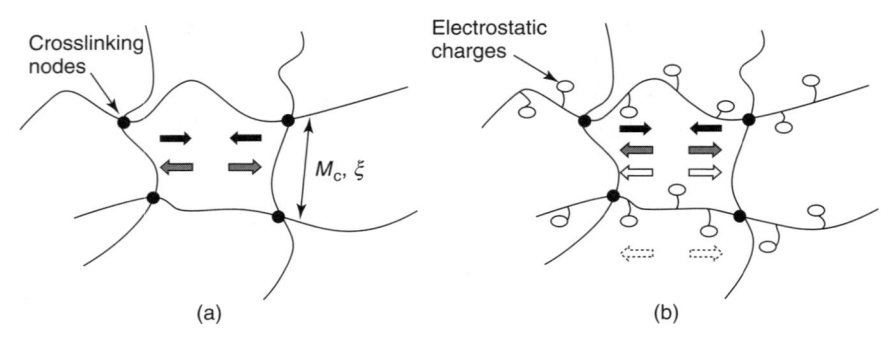

5.2 (a) Forces governing the swelling equilibrium of non-polyelectrolyte hydrogels. Black arrows indicate the force of retraction acting on the tetrafunctional polymer network, induced by the crosslinking bonds and opposing chain expansion. Grey arrows indicate the force of mixing deriving from polymer–solvent interaction and favouring the swelling. Molecular weight between crosslinks M_c and mesh size ξ are also evidenced. (b) Forces governing the swelling equilibrium of polyelectrolyte hydrogels. The presence of electrostatic charges anchored to the network yields two additional contributions to hydrogel expansion, with respect to non-polyelectrolyte hydrogels. White arrows indicate the ionic contribution to swelling due to a Donnan-type equilibrium established between the network and the solvent. Dotted arrows indicate the electrostatic repulsion between charges of the same sign, which is practically negligible for low charge densities, i.e. for high swelling ratios.

The degree of crosslinking (d.c.) of the network is defined as the density of junctions joining the chains into a permanent structure:

$$\text{d.c.} = \frac{2v}{fV} \qquad 5.6$$

In equation (5.6), v is the number of units engaged in crosslinks ($2v/f$ is the number of crosslinks) and V is the volume of the polymer network after crosslinking (i.e., the relaxed volume). If we define N_0 as the total number of structural units composing the network, then v is given by:

$$v = N_0 \frac{M_0}{M_c} = \frac{V}{vM_c} \qquad 5.7$$

where M_0 is the number average molecular weight of a structural unit, M_c is the number average molecular weight between two consecutive crosslinks, and v is the specific volume of the polymer.

The total number of crosslinked units is expressed in terms of crosslink density, which represents the moles of crosslinked chains per unit volume of network:

$$\rho_x = \frac{v}{V} = \frac{1}{vM_c} \qquad 5.8$$

Defining N as the number of primary linear polymer molecules forming the network, then:

$$N = N_0 \frac{M_0}{M_n} = \frac{V}{vM_n} \qquad 5.9$$

where M_n is the number average molecular weight of the primary molecules.

In an ideal, perfect network, no terminal chains are present, i.e. the network is infinite, and all chains are elastically effective, or active in deformation ($v = v_e$). Conversely, any real network contains $2N$ terminal chains, so that:

$$v_e = v - 2N \qquad 5.10$$

By combining equations (5.8), (5.9) and (5.10), we yield the value of the elastically effective crosslink density:

$$\rho_{xe} = \frac{v_e}{V} = \frac{1}{vM_c}\left(1 - \frac{2M_c}{M_n}\right) = \frac{1}{v}\left(\frac{1}{M_c} - \frac{2}{M_n}\right) \qquad 5.11$$

The above equation overestimates the real value of ρ_{xe}, since it is based on the assumption that terminal chains are the only imperfection of the network (e.g., no loops are present). In the following the effect of the crosslink density on both the hydrogel sorption capability and the mechanical stiffness is explained according to the classical thermodynamic theory of the gel state. It is worth noting that, although originally formulated for chemical hydrogels, the theory might be applied to describe the behaviour of physical hydrogels as well.

Evaluation of the degree of crosslinking through swelling measurements

The swelling capability of a chemically crosslinked, non-ionic hydrogel can be expressed in terms of the specific polymer–solvent interaction and the degree of crosslinking of the polymer network. The Flory–Rehner thermodynamic theory states that, when a hydrogel is allowed to swell, it is subjected to two counteracting forces, i.e. the force of mixing, which contributes positively to the swelling and is related to the hydrophilic nature of the polymer, and the entropic, retractive response of the polymer network, which opposes the deformation of the chains to a more elongated state and is directly related to the degree of crosslinking (Fig. 5.2). In terms of change in free energy associated with the mixing of the pure solvent and the pure polymer network, we can write:

$$\Delta G_{tot} = \Delta G_{mix} + \Delta G_{el} \qquad 5.12$$

where ΔG_{mix} is the variation in free energy due to mixing of the solvent molecules with the polymer chains, and ΔG_{el} is the variation associated with the elastic force established within the network upon swelling. If differentiating the above equation with respect to the number of solvent molecules, at constant temperature and pressure, we yield the equivalent expression in terms of variation in chemical potential of the solvent:

$$\Delta \mu_1 = \mu_1 - \mu_{1,0} = \Delta \mu_{mix} + \Delta \mu_{el} \qquad 5.13$$

where μ_1 is the chemical potential of the solvent inside the gel and $\mu_{1,0}$ is the chemical potential of the pure solvent. By dividing $\Delta \mu_1$ by the molar volume of the solvent, we yield the osmotic pressure π, which is established within the hydrogel upon swelling:

$$\pi = \pi_{mix} + \pi_{el} \qquad 5.14$$

As long as the pressure π is positive, the network swells. When the dry network is placed in contact with the solvent, the osmotic pressure driving the solvent inside the gel, π_{mix}, is maximum, since the polymer concentration in the gel is high. As the solvent molecules penetrate the network, the polymer concentration in the gel and the pressure π_{mix} decrease, while the elastic retraction force of the polymer chains, expressed in terms of pressure π_{el}, increases. At equilibrium, the chemical potential of the solvent inside the gel will be equal to that of the solvent outside the gel, i.e. $\Delta \mu_1 = 0$ and $\pi = 0$. Based on the classical theory of rubber elasticity and the thermodynamic theory of polymer solutions, it is possible to calculate the theoretical expressions for $\Delta \mu_{mix}$ and $\Delta \mu_{el}$, or equivalently for π_{mix} and π_{el}. In cases where the crosslinking reaction occurs in the presence of the solvent, the equilibrium condition for neutral gels leads to the following expression for M_c [9]:

$$\frac{1}{M_c} = \frac{2}{M_n} - \frac{\left(\frac{v}{V_1}\right)[\ln(1 - V_{2,s}) + V_{2,s} + \chi_{1,2} V_{2,s}^2]}{V_{2,r}\left[\left(\frac{V_{2,s}}{V_{2,r}}\right)^{1/3} - \frac{2}{f}\left(\frac{V_{2,s}}{V_{2,r}}\right)\right]} \qquad 5.15$$

where V_1 is the molar volume of the solvent, $V_{2,s}$ is the polymer volume fraction at equilibrium swelling, $V_{2,r}$ is the polymer volume fraction soon after crosslinking and before swelling, i.e. in the relaxed state, and $\chi_{1,2}$ is the Flory–Huggins polymer–solvent interaction parameter. The above equation is valid for loosely crosslinked networks, where the number of repeating units within each chain is large enough so that the chains can be represented by a Gaussian distribution. Furthermore, the expression for $\Delta \mu_{el}$, from which equation (5.15) is derived, is based on the assumption of affine network deformation, which is valid only when the polymer chains are extended less than half their fully stretched length, i.e. for moderate degrees of swelling. More complex equations have been developed for highly crosslinked networks, where the polymer chains contain a low number of monomers between crosslinks [10], and for non-affine deformations [11, 12].

Evaluation of the degree of crosslinking through mechanical measurements

Once the equilibrium swelling ratio is measured, for relatively simple hydrogel networks the average molecular weight between crosslinks and the corresponding crosslink density can be calculated according to equation (5.15) or its modifications. However, an approach to directly estimate the elastically effective crosslink density is provided by the theory on the entropic elasticity of rubbers, which relates the degree of crosslinking of a rubber-like crosslinked polymer to its macroscopic mechanical properties. Assuming that the deformation of the chains is affine and that the volume of the polymer does not change upon uniaxial deformation, Flory [6] derived a relationship between the uniaxial stress and the uniaxial deformation of a polymer network:

$$\sigma = RT\rho_{xe}\left(\alpha - \frac{1}{\alpha^2}\right) = G\left(\alpha - \frac{1}{\alpha^2}\right) \qquad 5.16$$

where σ is the stress, R is the universal gas constant, T is the absolute temperature, $\alpha = L/L_i$ is the deformation ratio, with L the actual thickness of the deformed sample and L_i the initial thickness of the sample ($\alpha > 1$ for elongation and $\alpha < 1$ for compression, respectively), and G is the shear modulus of the polymer network.

If the polymer network is swollen isotropically in a solvent [6], and the crosslinking reaction occurs in the presence of the solvent [9], the modulus G in the above equation can be expressed as follows:

$$G = RT\rho_{xe}V_{2,s}^{1/3}V_{2,r}^{-1/3} = RT\frac{V_e}{V_0}V_{2,s}^{1/3}V_{2,r}^{2/3} \qquad 5.17$$

where V_0 is the volume of the dry polymer network. Therefore the modulus G, and the corresponding effective crosslink density, can be measured from uniaxial elongation or compression tests. Dynamic mechanical analysis might be used to estimate ρ_{xe} as well [13].

The chemically effective degree of crosslinking, which is different from the elastically effective one due to the imperfections of a real polymer network, quantifies the covalent bonds originated from the crosslinking reaction and can be measured via chemical analyses [14].

5.3.3 Mesh size

Any polymer network is identified not only by the degree of crosslinking, but also by the correlation length ξ, also referred to as mesh size, which defines the linear distance between two adjacent crosslinks. For isotropically swollen hydrogels:

$$\xi = V_{2,s}^{-1/3}\left(\overline{r_0^2}\right)^{1/2} = V_{2,s}^{-1/3} l (C_n N)^{1/2} = V_{2,s}^{-1/3} l \left(C_n \frac{2M_c}{M_r}\right)^{1/2} \qquad 5.18$$

In the above equation, $\left(\overline{r_0^2}\right)^{1/2}$ is the root-mean-squared end-to-end distance between adjacent crosslinks in the unperturbed state, l is the bond length along the polymer chain, C_n is the characteristic ratio of the polymer, N is the number of bonds per chain, and M_r is the molecular weight of the polymer repeating unit. Typical mesh sizes for hydrogels used in biomedical applications range from 5 to 100 nm in the swollen state [15, 16]. The mesh size is a linear measure of the free space among polymer chains which is available for diffusion. Such a space is sometimes regarded as a 'pore', although of nanometric size.

5.3.4 Pendant fixed charges (polyelectrolytes)

Crosslinked polyelectrolyte chains in a solvent can ionize, thus giving rise to free, mobile counterions and a charged network. For ionic hydrogels, the total osmotic pressure π arising from swelling (equation 5.14) presents two additional contributions, respectively π_{Coul} and π_{ion}, both ascribed to the presence of fixed charges on the polymer backbone (Fig. 5.2). The former contribution accounts for the electrostatic repulsion established between charges of the same sign tethered on the polymer network, which causes the polymer chains to stretch to a more elongated state than that found in a neutral network, thus favouring the swelling. The second term accounts for the different concentration of mobile counterions found between the hydrogel and the external solution, with more counterions present in the gel, which induces more solvent to enter the network. This effect is explained in terms of a Donnan type equilibrium [6] established between the gel, that acts as a semipermeable membrane for the counterions, and the surrounding solution. The counterions are indeed trapped into the gel to ensure macroscopic electrical neutrality. The gel network acts simultaneously as a solute, a membrane semipermeable to itself and a pressure generating device [6]. For weakly charged networks, the electrostatic term π_{Coul} is negligible compared to the ionic one; conversely, for high charge densities, the electrostatic term becomes preponderant, also due to the counterion condensation, which reduces the effective ionization of the polymer chains and thus the ionic osmotic pressure.

Biocompatibility and other properties in regenerative medicine 125

In the ideal Donnan theory the pressure π_{ion} is given by:

$$\pi_{ion} = RT\left(\frac{i^2 c_2^2}{4I}\right) \qquad 5.19$$

In the above equation, R is the universal gas constant, T is the absolute temperature, i is the degree of ionization of the polymer, c_2 is the concentration of ionizable repeating groups on gel chains, and I is the ionic strength of the surrounding solution, which is defined as follows:

$$I = \frac{1}{2}\sum_i z_i^2 c_i \qquad 5.20$$

where z_i is the charge on ion i and c_i is the ion concentration. The concentration of ionizable polymer, c_2, is given by:

$$c_2 = \frac{V_{2,s}}{\upsilon M_r} \qquad 5.21$$

where M_r is the molecular weight of the polymer repeating unit and υ the polymer specific volume.

The degree of ionization i will depend on the dissociation constant of the ionizable network groups and the pH of the external solution. In case of anionic and cationic hydrogels, we can express i respectively as:

$$i_{an} = \frac{10^{-pK_a}}{10^{-pH} + 10^{-pK_a}} \qquad 5.22$$

$$i_{cat} = \frac{10^{-pK_b}}{10^{pH-14} + 10^{-pK_a}} \qquad 5.23$$

where K_a and K_b are the dissociation constants for the acid and the base, respectively. By combining equations (5.14), (5.19), (5.21) and (5.22) we yield the following expression for the equilibrium swelling of an anionic hydrogel with a tetrafunctional network:

$$V_1 \left(\frac{10^{-pK_a}}{10^{-pH} + 10^{-pK_a}}\right)^2 \left(\frac{V_{2,s}^2}{4I\upsilon^2 M_r^2}\right)$$
$$= [\ln(1-V_{2,s}) + V_{2,s} + \chi_{1,2}V_{2,s}^2] +$$
$$V_{2,r}\left(\frac{V_1}{\upsilon M_c}\right)\left(1 - \frac{2M_c}{M_n}\right)\left[\left(\frac{V_{2,s}}{V_{2,r}}\right)^{1/3} - \frac{V_{2,s}}{2V_{2,r}}\right] \qquad 5.24$$

Ionic hydrogels thus display a swelling capacity that is dependent on the ionic strength and the pH of the surrounding solution. Generally, when increasing the ionic strength of the external solution, ionic hydrogels yield a lower value of swelling ratio Q, due to the reduced difference in concentration of mobile ions between inside and outside the gel. With regard to the pH sensitivity, anionic hydrogels tend to deprotonate and swell when external pH is higher than pK_a of their ionizable groups, whereas cationic hydrogels protonate and swell when external pH is lower than pK_b of their ionizable groups. Further stimuli that can affect the swelling of polyelectrolyte hydrogels are electrical and magnetic fields [7, 17].

Polyampholyte hydrogels are polyelectrolyte networks carrying acid as well as basic functional groups, i.e. anionic and cationic charges respectively. Many biopolymers, such as proteins and nucleic acids, are polyampholyte. Weakly charged polyampholyte networks display a behaviour which is strongly dependent on the environmental pH. At the isoelectric point (pI), i.e. the pH at which the network has an equal number of positive and negative charges, the swelling is minimum due to electrostatic attraction between charges. When the pH is lower or higher than pI, the network has a net positive or negative charge respectively, behaving like a cationic or anionic hydrogel.

5.3.5 Micro- and macro-porosity

In cases where a hydrogel is required to respond quickly to certain environmental stimuli, the swelling/shrinking rates are of particular importance. Due to the nanometric dimension of the polymer mesh, the swelling of hydrogels is usually a slow, diffusion-driven process [18]. Since the time required for diffusion is proportional to the square of the characteristic length of the gel, the swelling rate is usually enhanced by reducing the size of the hydrogel-based device in powder or granular form. An alternative, size-independent approach to enhance the swelling kinetics consists of producing porous hydrogels, with interconnected micro- or macro-pores. Porous hydrogels typically display a higher swelling capability compared to non-porous ones, as long as their pores are small enough to retain the liquid medium by means of capillarity forces [19–21]. More importantly, the presence of an interconnected porous structure within the hydrogel improves significantly the swelling/deswelling kinetics, since a convective motion of fluid is established within the pores, in addition to the diffusional mechanism of mass transport taking place within the polymer mesh [21, 22]. The swelling rate results enhanced also due to the higher surface area per unit volume of the porous hydrogel.

5.4 Hydrogels in biomedical fields: optimization of chemical/physical properties for selected applications

Owing to their large water content (usually higher than 30% by weight [4]) as well as the absence of toxic monomers, hydrogels are highly biocompatible. A number of independent investigations show that many natural and synthetic hydrogels are well tolerated by cells, i.e. do not compromise cell viability and do not give rise to severe inflammatory responses *in vivo*. Biocompatibility, together with the 'smart' nature of hydrogels, makes them optimal candidates for a number of biomedical applications.

This section deals with some of those applications, particularly highlighting how to modulate selected network parameters to design hydrogels suitable for specific bio-applications. In particular, the complex phenomena coupling chemical–physical network parameters and hydrogel sorption thermodynamics and kinetics are briefly discussed for each of the applications treated in the following.

5.4.1 Devices for controlled drug delivery

The short half-life displayed *in vivo* by many drugs and proteins has prompted the need for delivery devices able to release such bioactive molecules at selected sites and for prolonged times. Hydrogels are particularly appealing for these applications, due to their ability to swell and shrink in a reversible manner, under the influence of physiologically relevant variables, such as temperature, pH and ionic strength. Swelling/shrinking phase transitions affect only the mesh size of the polymer network, thus regulating the diffusion release of loaded molecules [23]. Due to the strong pH variations from neutral to acidic, and from acidic to neutral, found when transitioning respectively from the mouth to the stomach, and from the stomach to the intestine, most polyelectrolyte hydrogels investigated find application as devices for oral drug delivery.

Swelling capacity and swelling kinetics can be particularly important. For example, if aiming at a sustained drug release inside the stomach (e.g., for the treatment of ulcers), an ideal hydrogel should possess at least two requisites: first, a high water sorption capacity inside the stomach (i.e., at low pH), in order for the hydrogel to reach a size large enough to prevent an early passage to the intestine through the pylorus; and second, a fast swelling response, as stomach emptying is not controllable and occurs about every 2 hours. To this purpose, Chen and coworkers [22] developed 'superporous' acrylate-based hydrogels, with a pore size in the order of tens of microns, that swell much faster than conventional superabsorbent hydrogels.

Moreover, mechanical properties of hydrogels, which are modulated by their degree of crosslinking and pore volume fraction, have been found to be critical, as hydrogels should withstand the compressive and frictional forces encountered in the stomach, for a relatively long time. In particular, superporous hydrogels with satisfactory mechanical elasticity and strength have been developed by creating interpenetrating polymer networks (IPNs)

[24]. Hydrogels for oral drug delivery are usually eliminated from the body through the faeces, thus gel biodegradability is not a necessary requirement. However, for topical delivering of drugs or proteins at different sites, hydrogel biodegradation, which may occur *in vivo* via either enzymatic digestion, hydrolysis or dissolution [4], becomes fundamental in order to avoid long-term foreign body reactions and/or the need for surgical removal. *In situ*-forming hydrogels are particularly advantageous for local drug delivery, as they might be delivered *in vivo* in a minimally invasive manner, i.e. by injecting a liquid solution that can polymerize either spontaneously under physiological conditions (reversible, physical gels) or upon UV exposure (photopolymerizable, chemical gels [5]). It is worth noting that, when the hydrogel is degradable *in vivo*, the drug release kinetics are also affected by the degradation rate.

As stated above, the degree of crosslinking of the polymer network, in addition to controlling the network mesh size, controls the hydrogel stiffness, which may be particularly important in cases where the drug release is meant to be stimulated by the mechanical compression to which the hydrogel is subjected *in vivo* [25].

5.4.2 Body water retainers

Superabsorbent hydrogels hold promise for the treatment of pathological conditions for which elimination of excess water from the body is required, e.g. for the treatment of diuretic-resistant, or intractable, oedemas. To this purpose, polyanionic cellulose-based hydrogels have been investigated, which show strong sensitivity to environmental ionic strength and pH, high swelling capability at neutral pH, and good biocompatibility [26, 27]. Such hydrogels are envisaged to be administered orally, to remain shrunk in the stomach and then to start absorbing in the intestine, where pH is about 6–7, before being expelled. The water sorption capability can be maximized by acting on the following variables: polyelectrolyte composition, degree of crosslinking, microporosity and, interestingly, introduction of molecular spacers in the polymer network, which appear to make the crosslinking junctions longer and more flexible, thus yielding higher expansions of the network [28]. As already mentioned, porosity enhances also the swelling kinetics, thus this variable should be addressed for the development of fast responsive body water retainers.

5.4.3 Artificial muscles, actuators and valves

The remarkable swelling/shrinking phase transitions, to which some thermosensitive and polyelectrolyte hydrogels are subjected when changing the environmental conditions or when applying electrical or magnetic fields,

Biocompatibility and other properties in regenerative medicine 129

suggest their possible use as artificial muscles, actuators and valves. Indeed the gels can be regarded as motors, able to convert chemical energy to mechanical energy. Depending on the particular experimental conditions, some hydrogels may undergo swelling or shrinking, or swelling on one side and deswelling on the other, which results in bending [17]. In addition to environmental sensitivity and macroscopic swelling/deswelling transitions, swelling kinetics are very important for a rapid response to selected stimuli. Gels in the form of small fibres have been employed for the first, functional poly(acrylonitrile)-based artificial muscle. Porous hydrogels, regardless of their size, are likely to act as fast responsive actuators, although porosity may greatly limit their mechanical properties. By tuning the hydrogel stiffness, the degree of crosslinking is a key variable in the design of actuators meant to work under mechanical loading.

5.4.4 Scaffolds for regenerative medicine

A scaffold is a temporary platform or template that hosts cells, either seeded *in vitro* or recruited *in vivo*, and guides them towards the reconstruction of a tissue or organ. The scaffold acts as an insoluble regulator of cell function, providing cells with molecular cues (i.e., scaffold composition and surface chemistry) and mechanical cues (i.e., scaffold stiffness), similar to those of the natural extracellular matrix, as well as preferential paths for cell migration and tissue deposition (i.e., pore structure). Hydrogels are particularly advantageous for application in regenerative medicine for the following reasons:

- Their processing conditions are usually mild [4, 29], thus allowing the simultaneous incorporation of cells. In this way exogenous cells can be seeded homogeneously throughout the matrix and then delivered to the graft site.
- Being highly hydrated, they possess mechanical properties matching those of many soft tissues.
- Even though they do not readily promote cellular adhesion (cell adhesion onto hydrophilic surfaces is usually very low [4], with the exception of collagen gels), their bulk or surface chemistry can be easily modified by incorporating proteins and small peptide sequences into the macromolecular backbone. This strategy can be used not only to promote cell attachment and deliver desired chemical signals to the cells [30, 31], but also to design selective materials, able to interact only with specific cell types [32], as well as to produce enzyme-sensitive scaffolds with controlled degradation rate [30] (Fig. 5.3).

In particular, in the last years self-assembling polypeptide hydrogels [33], more than others, have received increasing attention as potential

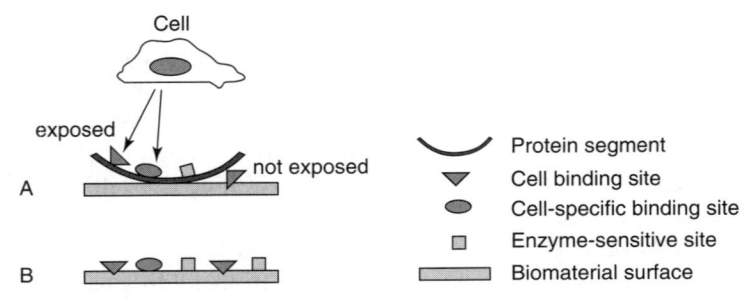

5.3 Cell adhesion onto biomaterials. Cell–material interactions are mediated by extracellular matrix proteins, which deliver specific information to the cells, including generic binding domains (e.g., peptide sequence RGD), cell-specific binding domains (e.g., peptide sequence VAPG for smooth muscle cells) and enzyme-sensitive moieties. The biomaterial's surface can adsorb such proteins from physiological fluids (A). However, cell response will depend on the conformational changes to which the proteins are subjected when adsorbed, i.e. some protein sites may not be exposed or available to cells. An uncontrolled adsorption of proteins may even lead to protein denaturation and, as a result, to a foreign-body reaction [31]. In the case of hydrogels, for which protein adsorption is very low, peptide sequences affecting cell behaviour, as well as genes encoding for them, may be linked to the macromolecular network (B), so that cell–material interactions can be induced and accurately controlled.

regenerative scaffolds, due to their nano- and micro-architecture, which resemble very closely those of the mammalian extracellular matrix, as well as to their spontaneous setting under physiological conditions.

However, the use of hydrogels in regenerative medicine is greatly limited to the treatment of small defects and/or tissues with a low metabolic activity (e.g., cartilage), since the only diffusion of nutrients to cells is not sufficient to ensure cell survival in the central area of large grafts (i.e., thicker than 200 µm) [34].

Degree of crosslinking and porosity are key variables affecting the biodegradation rate, the mass transport properties, and the mechanical stiffness of the hydrogel scaffold. An increase in the degree of crosslinking, in addition to increasing the hydrogel stiffness, leads to a shorter mesh size of the polymer network, thus down-regulating the diffusional mass transport and decreasing the degradation rate. For collagen-based scaffolds, it has been confirmed that there exists an optimal scaffold crosslink density for peripheral nerve regeneration [35]. This finding might be explained in terms of premature scaffold resorption in the case of a too low degree of crosslinking (i.e., high degradation rate) and retarded remodelling, with a possible foreign body recognition of the implant in the case of too high a degree of crosslinking (i.e., low degradation rate). With regard to the role played by an interconnected porosity, primarily cell attachment and infiltration are facilitated when the open pores are large enough to accommodate the cells (>10 μm). In analogy to what has been observed for crosslinking density, an optimal pore size has been reported for the regeneration of specific tissues [36]. It is likely that, while too small pores are not able to host the cells, too large pores yield a low surface area and, as a result, poor cell binding [37]. Interconnected porosity also facilitates survival of cells within the scaffold, due to the enhanced mass transport kinetics to and from the cells established within the open pores. A number of manufacturing methods for producing porous hydrogels, including gas foaming, freeze-drying and particulate leaching, have been developed [21, 38–42]. Porous fibre-based scaffolds have been investigated as well. The manufacturing techniques allow the modulation of the hydrogel pore volume fraction and pore size, thus hydrogels tailored to the regeneration of specific tissues can be produced. For instance, esters of hyaluronic acid in the form of non-woven fibres have recently shown potential as 3-D scaffolds for intestine tissue engineering [43].

However, when designing hydrogel templates, it should be kept in mind that porosity lowers the hydrogel stiffness, and usually increases the degradation rate, due to the higher surface area/volume ratio. Moreover, as the scaffold degrades, its mechanical strength and stiffness tend to decrease. For applications in which the scaffold is meant to temporarily replace load-bearing tissues, the mechanical properties of hydrogel-based scaffolds should be appropriately controlled, and this might be achieved by developing composite devices. For example, hyaluronan hydrogel-based scaffolds, reinforced with polylactic acid (PLA) fibres, have been recently designed for meniscus regeneration, and their regenerative behaviour, tested in a sheep model, appeared promising [44].

5.5 Conclusions

Hydrogels are polymeric networks in which a certain amount of water solution is retained, as a result of complex thermodynamic equilibria established among several counteracting factors, such as polymer–solvent interaction, elastic retraction, ion-driven swelling and electrostatic repulsion. Environmental changes (e.g., in terms of temperature, ionic strength, pH, electrical and magnetic fields) may lead hydrogels to sharp swelling–deswelling transitions, whose kinetics depend on hydrogel size and microporosity.

The hydrogel structure–function relationships discussed in this chapter open a wide range of design possibilities, since hydrogel properties can be properly tuned to the envisaged application by changing selected network parameters (Table 5.1). Based on a recent and innovative theory, which considers the cytoplasm itself as a gel, with water being retained by a network of protein macromolecules [45], we believe that cell biologists may also take advantage of interpreting hydrogel macroscopic properties in terms of microstructural parameters.

With particular focus on biomedical applications, it is worth noting that the old-fashioned concept of biocompatibility is being rapidly substituted by the novel ideas of biomimicry and bioactivity, introduced recently by regenerative medicine and tissue engineering. The versatility displayed by hydrogel networks, with regard to the modulation of their properties, makes us confident that hydrogels are ideal platforms for the creation of biomimetic and bioactive scaffolds able to induce the regeneration of several tissues and organs.

5.6 References

1. Tanaka T. Gels. *Sci Am* 1981, 244(1): 124–136, 138.
2. Jordan-Lloyd D. The problem of gel structure. In: Alexander J, editor, *Colloid Chemistry*. New York: Chemical Catalogue Company, 1926, pp. 767–782.
3. Te Nijenhuis K. On the nature of crosslinks in thermoreversible gels. *Polym Bull* 2007, 58: 27–42. doi: 10.1007/s00289-006-0610-7
4. Drury JL, Mooney DJ. Hydrogels for tissue engineering: scaffold design variables and applications. *Biomaterials* 2003, 24(24): 4337–4351. doi:10.1016/S0142-9612(03)00340-5
5. Nguyen KT, West JL. Photopolymerizable hydrogels for tissue engineering applications. *Biomaterials* 2002, 23(22): 4307–4314. doi:10.1016/S0142-9612(02)00175-8
6. Flory PJ. *Principles of Polymer Chemistry*. Ithaca, NY: Cornell University Press, 1953.
7. Qiu Y, Park K. Environment-sensitive hydrogels for drug delivery. *Adv Drug Deliv Rev* 2001, 53(3): 321–339. doi:10.1016/S0169-409X(01)00203-4
8. Bokias G, Staikos G, Iliopoulos I. Solution properties and phase behaviour of copolymers of acrylic acid with *N*-isopropylacrylamide: the importance of the intrachain hydrogen bonding. *Polymer* 2000, 41(20): 7399–7405. doi:10.1016/S0032-3861(00)00090-2
9. Peppas NA, Merrill EW. Crosslinked poly(vinyl alcohol) hydrogels as swollen elastic networks. *J Appl Polym Sci* 1977, 21: 1763–1770. doi: 10.1002/app.1977.070210704
10. Peppas NA, Moynihan HJ, Lucht LM. The structure of highly crosslinked poly(2-hydroxyethyl methacrylate) hydrogels. *J Biomed Mater Res* 1985, 19(4): 397–411. doi: 10.1002/jbm.820190405
11. James HM, Guth E. Statistical thermodynamics of rubber elasticity. *J Chem Phys* 1953, 21(6): 1039–1049. doi:10.1063/1.1699106
12. Hermans JJ. Statistical thermodynamics of swollen polymer networks. *J Polym Sci* 1962, 59(167): 191–208. doi: 10.1002/pol.1962.1205916715
13. Sannino A, Pappadà S, Madaghiele M, Maffezzoli A, Ambrosio L, Nicolais L. Crosslinking of cellulose derivatives and hyaluronic acid with water-soluble

carbodiimide. *Polymer* 2005, 46: 11206–11212. doi:10.1016/j.polymer.2005.10.048

14. Lenzi F, Sannino A, Borriello A, Porro F, Capitani D, Mensitieri G. Probing the degree of crosslinking of a cellulose based superabsorbing hydrogel through traditional and NMR techniques. *Polymer* 2003, 44: 1577–1588. doi:10.1016/S0032-3861(02)00939-4

15. Cruise GM, Scharp DS, Hubbell JA. Characterization of permeability and network structure of interfacially photopolymerized poly(ethylene glycol) diacrylate hydrogels. *Biomaterials* 1998, 19(14): 1287–1294. doi:10.1016/S0142-9612(98)00025-8

16. Lin CC, Metters AT. Hydrogels in controlled release formulations: network design and mathematical modeling. *Adv Drug Deliv Rev* 2006, 58(12–13): 1379–1408. doi:10.1016/j.addr.2006.09.004

17. Tanaka T, Nishio I, Sun ST, Ueno-Nishio S. Collapse of gels in an electric field. *Science* 1982, 218(4571): 467–469. doi: 10.1126/science.218.4571.467

18. Tanaka T, Sato E, Hirokawa Y, Hirotsu S, Peetermans J. Critical kinetics of volume phase transition of gels. *Phys Rev Lett* 1985, 55(22): 2455–2458. doi: 10.1103/PhysRevLett.55.2455

19. Esposito F, Del Nobile MA, Mensitieri M, Nicolais L. Water sorption in cellulose-based hydrogels. *J Appl Polym Sci* 1996, 60(13): 2403–2407. doi: 10.1002/(SICI)1097-4628(19960627)60:13<2403::AID-APP12>3.0.CO;2-5

20. Sannino A, Madaghiele M, Conversano F, Mele G, Maffezzoli A, Netti PA, et al. Cellulose derivative–hyaluronic acid-based microporous hydrogels cross-linked through divinyl sulfone (DVS) to modulate equilibrium sorption capacity and network stability. *Biomacromolecules* 2004, 5(1): 92–96. doi: 10.1021/bm0341881

21. Sannino A, Netti PA, Madaghiele M, Coccoli V, Luciani A, Maffezzoli A, et al. Synthesis and characterization of macroporous poly(ethylene glycol)-based hydrogels for tissue engineering application. *J Biomed Mater Res A* 2006, 79(2): 229–236. doi: 10.1002/jbm.a.30780

22. Chen J, Park H, Park K. Synthesis of superporous hydrogels: hydrogels with fast swelling and superabsorbent properties. *J Biomed Mater Res* 1999, 44(1): 53–62. doi: 10.1002/(SICI)1097-4636(199901)44:1<53::AID-JBM6>3.0.CO;2-W

23. Peppas NA. Hydrogels and drug delivery. *Curr Opin Colloid Interface Sci* 1997, 2: 531–537.

24. Qiu Y, Park K. Superporous IPN hydrogels having enhanced mechanical properties. *AAPS Pharm Sci Tech* 2003, 4(4): E51.

25. Lee KY, Peters MC, Anderson KW, Mooney DJ. Controlled growth factor release from synthetic extracellular matrices. *Nature* 2000, 408(6815): 998–1000. doi:10.1038/35050141

26. Sannino A, Esposito A, Nicolais L, Del Nobile MA, Giovane A, Balestrieri C, et al. Cellulose-based hydrogels as body water retainers. *J Mater Sci Mater Med* 2000, 11(4): 247–253. doi: 10.1023/A:1008980629714

27. Sannino A, Esposito A, De Rosa A, Cozzolino A, Ambrosio L, Nicolais L. Biomedical application of a superabsorbent hydrogel for body water elimination in the treatment of edemas. *J Biomed Mater Res A* 2003, 67(3): 1016–1024. doi: 10.1002/jbm.a.10149

134 Cellular response to biomaterials

28. Esposito A, Sannino A, Cozzolino A, Quintiliano SN, Lamberti M, Ambrosio L, *et al.* Response of intestinal cells and macrophages to an orally administered cellulose-PEG based polymer as a potential treatment for intractable edemas. *Biomaterials* 2005, 26(19): 4101–4110. doi:10.1016/j.biomaterials.2004.10.023
29. Williams CG, Malik AN, Kim TK, Manson PN, Elisseeff JH. Variable cytocompatibility of six cell lines with photoinitiators used for polymerizing hydrogels and cell encapsulation. *Biomaterials* 2005, 26(11): 1211–1218. doi:10.1016/j.biomaterials.2004.04.024
30. Seliktar D. Extracellular stimulation in tissue engineering. *Ann N Y Acad Sci* 2005, 1047: 386–394. doi: 10.1196/annals.1341.034
31. Ratner BD, Bryant SJ. Biomaterials: where we have been and where we are going. *Annu Rev Biomed Eng* 2004, 6: 41–75. doi:10.1146/annurev.bioeng.6.040803.140027
32. Gobin AS, West JL. Val-ala-pro-gly, an elastin-derived non-integrin ligand: smooth muscle cell adhesion and specificity. *J Biomed Mater Res A* 2003, 67(1): 255–259. doi: 10.1002/jbm.a.10110
33. Zhang S. Fabrication of novel biomaterials through molecular self-assembly. *Nat Biotechnol* 2003, 21(10): 1171–1178. doi:10.1038/nbt874
34. Muschler GF, Nakamoto C, Griffith LG. Engineering principles of clinical cell-based tissue engineering. *J Bone Joint Surg Am* 2004, 86-A(7): 1541–1558.
35. Harley BA, Spilker MH, Wu JW, Asano K, Hsu HP, Spector M, *et al.* Optimal degradation rate for collagen chambers used for regeneration of peripheral nerves over long gaps. *Cells Tissues Organs* 2004, 176(1–3): 153–165. doi: 10.1159/000075035
36. Yang S, Leong KF, Du Z, Chua CK. The design of scaffolds for use in tissue engineering. Part I. Traditional factors. *Tissue Eng* 2001, 7(6): 679–689. doi: 10.1089/107632701753337645
37. O'Brien FJ, Harley BA, Yannas IV, Gibson LJ. The effect of pore size on cell adhesion in collagen-GAG scaffolds. *Biomaterials* 2005, 26(4): 433–441. doi:10.1016/j.biomaterials.2004.02.052
38. Stokols S, Tuszynski MH. The fabrication and characterization of linearly oriented nerve guidance scaffolds for spinal cord injury. *Biomaterials* 2004, 25(27): 5839–5846. doi:10.1016/j.biomaterials.2004.01.041
39. Stachowiak AN, Bershteyn A, Tzatzalos E, Irvine DJ. Bioactive hydrogels with an ordered cellular structure combine interconnected macroporosity and robust mechanical properties. *Adv Mater* 2005, 17(4): 399–403. doi: 10.1002/adma.200400507
40. Lee WK, Ichi T, Ooya T, Yamamoto T, Katoh M, Yui N. Novel poly(ethylene glycol) scaffolds crosslinked by hydrolyzable polyrotaxane for cartilage tissue engineering. *J Biomed Mater Res A* 2003, 67(4): 1087–1092. doi: 10.1002/jbm.a.10570
41. Ford MC, Bertram JP, Hynes SR, Michaud M, Li Q, Young M, *et al.* A macroporous hydrogel for the coculture of neural progenitor and endothelial cells to form functional vascular networks *in vivo*. *Proc Natl Acad Sci USA* 2006, 103(8): 2512–2517. doi: 10.1073/pnas.0506020102
42. Bryant SJ, Cuy JL, Hauch KD, Ratner BD. Photo-patterning of porous hydrogels for tissue engineering. *Biomaterials* 2007, 28(19): 2978–2986. doi: 10.1016/j.biomaterials.2006.11.033

43. Esposito A, Mezzogiorno A, Sannino A, De Rosa A, Menditti D, Esposito V, *et al.* Hyaluronic acid based materials for intestine tissue engineering: a morphological and biochemical study of cell-material interaction. *J Mater Sci Mater Med* 2006, 17(12): 1365–1372. doi: 10.1007/s10856-006-0612-x
44. Chiari C, Koller U, Dorotka R, Eder C, Plasenzotti R, Lang S, *et al.* A tissue engineering approach to meniscus regeneration in a sheep model. *Osteoarthritis Cartilage* 2006, 14(10): 1056–1065. doi:10.1016/j.joca.2006.04.007
45. Pollack GH. *Cells, Gels and the Engines of Life.* Seattle, WA: Ebner & Sons, 2001.

6
Cellular response to bioceramics

A WOESZ and S M BEST, University of Cambridge, UK

Abstract: Over the last few decades bioceramics have improved the quality of life of millions of people. The term '*bioceramic*' encompasses a wide range of materials used in skeletal repair and sometimes in soft tissue repair. Applications of bioceramics are varied, ranging from repair or replacement of damaged bones or joints to artificial heart valves and even keratoprostheses. Equally the types of ceramics used are diverse (ranging from fully dense high strength aluminium oxide to low density porous calcium phosphates), as are the reactions these ceramics trigger at the implantation site.

In this chapter we will introduce the most important ceramic materials used in the human body and present their main applications. The main part of the chapter will concentrate on the possible reactions of the surrounding organic tissue to the implant material, which depends primarily on the material used (including its chemical composition, porosity and surface roughness) but also on the type of tissue, the age and health status of the patient, any accompanying medical treatment, mechanical loading and many other factors.

Key words: bioceramics, implants, bone replacement, cell reaction, alumina, calcium phosphates, bioglass, glass-ceramic, mechanical properties, biological properties, bioinert, bioactive, bioresorbable.

6.1 Introduction

With the increasing average age of the population the need for the replacement and restoration of diseased tissue has increased significantly. Nowadays this need for implants is met with a vast number of materials for use in an even larger number of different applications. Among these materials, ceramics now dominate in a number of areas; they have become gold standards for several applications. In this chapter we will give an introduction to the world of bioceramics and review the reactions that these ceramics trigger in the surrounding tissue.

6.2 A short history of the usage of ceramic materials in the human body

The first time that usage of ceramics in the human body was reported in a scientific journal was in 1892, when Dreesmann described the filling of bone defects with Plaster of Paris (calcium sulphate) in eight patients. In 1920 a report on the application of tricalcium phosphate for similar use in animals was published (Albee and Morrison 1920). During the following decades, numerous studies on the applicability of these resorbable materials were carried out, partly because it was believed that the dissolving biomaterials would supply the growing tissue with essential minerals (Hulbert et al. 1987).

The earliest suggestion for the use of a bioinert ceramic, aluminium oxide, in the human body was in 1932; however, it was not actually applied in orthopaedics until 1963 (Smith 1963) and in 1964 in dental applications (Sandhaus 1967). These first experiments were soon to be followed by studies on porous alumina (Hulbert et al. 1970) and the development of the most frequent application in the human body since then, the alumina hip prostheses, which was first reported in 1971 by Boutin.

About the same time the bone-bonding properties of Bioglass®, a mixture of SiO_2, Na_2S, CaO and P_2O_5, were reported by Hench and co-workers (Hench and Paschall 1973), the first successful application in an animal study was published in 1975 (Piotrowski et al. 1975). In parallel the use of hydroxyapatite in various applications – bone, tooth and orthopaedic implant (Monroe et al. 1971) and periodontal treatment (Nery et al. 1975) – was suggested, conducted soon afterwards and found to be successful.

Since these early investigations, ceramics have gained major importance in biomedicine and have become the gold standard for the treatment of some indications. Apart from the above-mentioned joint replacements, nowadays they are used in numerous applications, such as bone screws, for alveolar ridge and for craniomaxillofacial reconstruction, for the replacement of ossicular bones and as keratoprostheses, to mention just a few.

6.3 Types of ceramics used in the human body: bioinert, bioactive and bioresorbable

In this section we will give a short introduction into the definition of the terms used to describe the reaction an implant, or more generally a foreign body, triggers in the human body. According to Hench and coworkers (Hench and Wilson 1993; Hench 1998; Dubok 2000), and now generally accepted within the biomaterials community, materials can be classified according to four different body reactions:

- **Bioinert:** a material is considered to be 'bioinert' when it has minimal interaction with the surrounding tissue after implantation, when there is minimal chemical reaction occurring at the interface. The reaction of the human body to an 'inert' implant is the formation of a fibrous capsule covering the implant (for details see Section 6.6.1).

- **Bioactive:** bioactive materials trigger a chemical reaction between the implant and the surrounding tissue, and the materials undergo time-dependent changes. These chemical changes at the surface lead to the formation of a layer of carbonated apatite, similar to the mineral component of bone, which provides binding sites for the newly forming or repairing tissue. Detailed descriptions of the mechanisms can be found in Sections 6.6.2, 6.6.3 and 6.6.4.
- **Bioresorbable:** this term refers to materials which undergo time-dependent chemical changes not only on their surface, but which are either dissolved or resorbed with time and subsequently replaced by the advancing surrounding organic tissue. More details can be found in Section 6.6.3.
- **Toxic:** this term refers to materials that cause cell death in the surrounding tissue, either locally or remotely, by transport of dissolution products.

The reaction of the natural tissue to an implant is multifactorial and these factors can either be implant-related (examples include surface roughness and mechanical loading) or tissue-related (including effects such as tissue type, age and health of the patient and blood supply to the implant site). Therefore the biological reaction to an implant cannot be assumed to be independent of the application site or the patient.

Furthermore it may not always be easy to distinguish specific types of tissue reactions to a particular material type due to variations in composition, preparation and processing conditions. In some cases there may be a smooth transition between one behaviour type and the next (compare Table 6.1).

In addition to the terms used to describe the reaction of natural tissue to implant material, two additional expressions connected with bone replacement should be introduced. These are osteoconductivity and osteoinductivity. A bone replacement material is considered to be osteoconductive when it has the ability to induce bone tissue ingrowth and bone formation after implantation. An osteoinductive material triggers the same reaction, the formation of bone, even when they are implanted in non-osseous sites (Ben-Nissan 2003).

6.4 Biological and mechanical behaviour of various classes of materials

Metallic implants are used where high mechanical strength is essential. Stainless steel, cobalt chrome steels and titanium alloys are the most commonly used implant metals; their strength in tension and compression is high, as is their stiffness. The strength is important in any load-bearing

Table 6.1 Biological reaction and main mechanical properties of implant materials. Note that the symbol + denotes high, not necessarily advantageous values of the respective property

Material class	Examples	'Bioinert'	Bioactive	Bioresorbable	Tensile strength	Compressive strength	Stiffness	Fracture toughness
Metals	stainless steel, titanium	↕			+	+	+	+
Ceramics	alumina, zirconia	↕			○	+	+ ○	−
Ceramics	hydroxyapatite, bioglass, A/W glass-ceramic		↕		−	−		−
Ceramics	tricalcium phosphate, biphasic calcium phosphate			↕	−	−	○	−
Polymers	UHMWPE, PMMA	↕	↕		○	○	○	+
Polymers	PLA, PGA			↕	− ○	− ○	− ○	+ ○
Composites	various							

application, but stiffness can be a problem, as will be elaborated in Section 6.5. Orthopaedic alloys have high resistance to corrosion and wear, although in some applications wear can cause significant problems. It is for this reason that 'bioinert' ceramics began to be applied, since they have very high compressive strength and considerable tensile strength, and their frictional and wear resistance is unsurpassed by any other orthopaedic material, which makes them ideal in articulation applications.

Generally, the bioactive ceramics currently in use in orthopaedics have much lower tensile and compressive strength; their applications are either non-load-bearing or minimal load-bearing sites. Their main advantage over the aforementioned materials is their ability to trigger a chemical bonding with living tissue, which allows a mechanical connection and prevents loosening and failure of the implant. These ceramics are often used as coatings on load-bearing metallic implants. Bioresorbable ceramics find their main applications in the filling of defects, where they are not mechanically loaded and successively replaced by the advancing surrounding natural tissue.

Bioinert polymers have comparably low mechanical strength and stiffness; their high fracture toughness and therefore high reliability nevertheless make them a good choice in many applications. They are, for example, used as a counterpart for the alumina joint replacements. Bioresorbable polymers, which have similar mechanical properties, are used as sutures, for defect filling and in low load-bearing applications.

The high melting points of most ceramics prevent them from being processed by melt casting; their inherent brittleness inhibits shaping by plastic deformation. Almost all ceramics are therefore shaped by processes which comprise the production of a powder, pressing of the powder into shape (green compact) and subsequent heat treatment (sintering) at elevated temperatures. Shrinkage occurs during sintering, as most pores in the green compact are removed. Again, the brittleness and high hardness of most ceramics limit the possibilities to adjust the shape after sintering; therefore attention has to be turned to the shape of the green compact and the amount of shrinkage.

Since brittle materials like ceramics do not exhibit mechanisms which prevent a crack from running through the whole part after it has been started, the strength of such materials is determined by the size of the largest crack or flaw within them. As soon as a critical stress concentration at the crack tip is reached, the crack starts to grow and the part fails. Stress concentration at the crack tip depends on the shape of the crack and its size and on the stresses applied. According to linear elastic fracture mechanics (Griffith 1921), the maximum stress a brittle material with Young's modulus E and surface energy G_c, which contains a crack of length a, can bear is given by:

$$\sigma_{cr} = Y\sqrt{\frac{EG_c}{\pi a}}$$

with *Y* being a factor depending on geometry. If a small grain size can be produced in the ceramic then the flaw and/or pore sizes are also likely to be small; therefore much emphasis has to be put in keeping the grains small during sintering.

6.5 Dense versus porous ceramic materials

Ceramic materials are used in dense and porous form, and both of these forms have relative advantages and disadvantages. For example, a femoral head replacement made from alumina or zirconia is required to contain minimal porosity after sintering, as the mechanical load-bearing capacity is the most important parameter for this application, whereas the filling of a supercritically sized defect after removal of a bone tumour ideally requires a porous scaffold such that the surrounding healthy bone can grow into the cavities and eventually fill the implant.

Superior mechanical strength is one of clear advantages of dense ceramics over porous ones, as the strength of a material is significantly reduced with increasing porosity. The correlation between strength and porosity has to be considered at two orders of magnitude. For very small porosities, as described above, the size of the largest crack (or pore) determines the strength of a brittle material, whereas for high porosities Gibson and Ashby (1997) have approximated the dependence of the strength of a brittle, random foam on the porosity to be as follows:

$$\sigma_{cr} \approx 0.2Y \frac{K_c}{\sqrt{\pi a}} \left(\frac{\rho^*}{\rho_s}\right)^{3/2}$$

with K_c being the fracture toughness, ρ^* the density of the foam and ρ_s the density of the solid it is made of. Besides the porosity and the dense material's properties, the architecture strongly influences the performance of a porous material (Woesz *et al.* 2004).

On the other hand, the first and most important benefit gained when using porous ceramics is the potential for mechanical interlocking with the surrounding natural tissue (Chang *et al.* 1996), a mechanism often referred to as biological fixation. The ingrowth of bone into a porous implant can stabilise and strengthen the material itself, as described in Vuola *et al.* (1998). In order to ensure survival of the natural tissue within the pores of the implant, vascularisation is essential to supply the cells with nutrients. Various studies have been reported on the influence of pore size on bone cell ingrowth, but as yet there does not seem to be a consensus on the optimal pore size: in Chang et al. (2000) 50 μm and higher are mentioned to be optimal, a minimum of 200 μm is advised by Klawitter (1979), Kuboki et al. (1998) mentioned 300–400 μm as an optimum, and in Gauthier et al. (1998) 565 μm is found to be better than 300 μm. With an optimal pore size alone, bone ingrowth and proliferation are not assured, since pore interconnectivity and material characteristics also play an important role (Ducheyne and Qiu 1999).

Another advantage accompanying porosity is resorbability. The osteoclast cells attach to the surface of the material to be resorbed and dissolve it by producing a local acidic environment (Teitelbaum 2000). The resulting resorption lacunae have a depth of approximately 50 to 100 μm. This suggests a feature thickness of 200 μm (twice the depth of the lacunae) as maximum for resorbable, but not soluble, materials, if one wants the graft to be removed and replaced by natural bone material.

One of the main problems when using porous instead of dense reactive materials, especially resorbable materials, is the increase in surface area, which can lead to higher concentrations of products at the implantation site, sometimes exceeding the maximum concentration for cell viability (Hench 1980).

6.6 Cell reactions to ceramic materials

The first and inevitable event after an implant is placed into the body is the adsorption of protein on its surface. This mechanism takes place within seconds after contact of the implant's surface with the blood and other body fluids (Baier 1977; Vroman et al. 1977). Any further interaction of the biological surroundings with the implant surface happens via this proteinaceous interface, therefore it is of considerable importance (Williams 1992). It is known to influence the corrosion and degradation processes of the implant, and most importantly it determines the attachment of cells onto the implant's surface. Furthermore, the interaction between the initial protein layer and the proteins produced by the cells in the course of time is of high significance, too, as we will see later. The possibility to pre-coat implant surfaces with specific proteins should be mentioned here, as it is a way of altering the interaction and therefore altering the reaction of the body tissue to such a surface.

The tissue damage which is inevitable during implantation releases a cascade of breakdown products, which triggers a reaction in the surrounding cells and vessels and attracts cells from the blood and surrounding tissue (Gross et al. 1988). After organisation and repair of the damaged tissue, remodelling of the initially formed tissue takes place. In the case of bioactive and bioresorbable implants this remodelling also includes the interface, whereas the fibrous interfaces of implants from bioinert materials do not change with time, apart from increasing in thickness, if the implant moves relatively to the surrounding tissue.

According to Hench (1998), the reactivity of an implant material is quite well correlated with the time dependence of its bonding to surrounding organic tissue and also with the thickness of the fibrous capsule (in case of a bioinert material) or the reaction layer (in case of a bioactive material). Inert materials show almost no reactivity and do not bond with natural tissue; a fibrous capsule is formed around them. With increasing bioreactivity, the time it takes for tissue to bond with the material decreases, as does the thickness of the reaction layer. Bioresorbable materials can be considered as having the highest bioreactivity; therefore natural tissue is supposed to bond fastest. These materials have a 'negative' reaction layer thickness, as they are dissolved and gradually replaced by the advancing living tissue.

6.6.1 Alumina

High-density, high-purity alumina is extensively used in the human body. The best known and best investigated area of application is most probably the use of alumina in hip joint replacement (Learmonth *et al.* 2007), where a ceramic ball placed onto a metallic stem, which is placed into the femur, and a ceramic cup, fitted into the acetabulum of the hip, can be made of alumina. The excellent biocompatibility and strength as well as excellent corrosion and wear resistance and low friction of alumina make this the material of choice of many surgeons (Dearnley 1999; Slonaker and Goswami 2004; Rahaman *et al.* 2007).

Dense high-purity alumina has very high reported hardness (1900 HV: Kusaka *et al.* 1999), resulting in excellent tribological properties; it has an extremely high melting point (2054°C) and very high strength ($\sim$5000 N/mm^2: Ashby and Jones 1980), leading to high form stability. Its high chemical stability results in its biocompatibility, as very few ions are released into the physiological environment after implantation. In fact, there do not appear to have been any reports of negative effects of substances dissolved from an aluminium oxide implant (Li and Hastings 1998).

Besides the balls and cups in total hip joint replacement, alumina is used in many other applications in the human body. Its usage in knee prostheses and as bone screws, the latter also made of single crystal alumina, as well as in alveolar ridges and for maxillofacial reconstruction, have been reported. Also ossicular bone substitutes (middle ear implants) and keratoprostheses are produced from alumina. In dentistry the use of alumina is even more widespread.

The excellent mechanical, tribological and chemical properties discussed so far are valid for high-purity, fine-grain alumina. National and international standards such as the DIN, FDA and ISO standards therefore limit the amount of impurities and the grain size for implants used in the human body (Dorlot 1992; Willmann 1994).

As mentioned above, if a bioinert material such as alumina is placed in the human body, very little chemical interaction takes place and therefore there is virtually no direct mechanical bonding between the implant's surface and the surrounding tissue. If the implant is kept in place by means other than chemical bonding, by a mechanical fit or by biological fixation (via the ingrowth of bone into pores and roughness of the implant's surface), a relatively thin fibrous coating is produced by the natural tissue, which does not degrade the implant's functioning. But if mechanical fixation is insufficient, and therefore movement between implant and organic tissue occurs, the thickness of the fibrous tissue interface is generally expected to increase with time, subsequently leading to loosening and failure of the implant.

6.6.2 Calcium phosphates

Hydroxyapatite (HA) is a calcium phosphate-based bioceramic which has chemical similarities to the mineral component of bone and for this reason it has been of great interest for many years (De Groot 1993). Stoichiometric HA has a chemical composition of $Ca_{10}(PO_4)_6(OH)_2$, where the ratio between calcium ions and phosphate groups equals 1.67. The HA found in bone, enamel, dentin and other natural sources generally does not have this ratio, since some of the PO_4 groups are replaced by carbonate ions and other impurities. Thus bone mineral is more correctly referred to as calcium-deficient carbonated apatite (LeGeros 2002) and its chemical formula can be written as $(Ca, Na, Mg)_{10}(PO_4, HPO_4, CO_3)_6(OH)_2$. Bone has been reported to bond to HA; hence HA is generally regarded as bioactive and, according to some sources, even osteoconductive (Yuan et al. 1999). Because it develops a strong bond with bone, no fibrous capsule is formed on the bone/ceramic interface (Doremus 1992; Dorozhkin and Epple 2002).

Hydroxyapatite is produced in wet chemical precipitation reactions, via solid-state reactions and by a hydrothermal exchange reaction. However, calcium phosphate-based ceramics generally have relatively poor mechanical strength and their inherent brittleness limits their applicability to non-major load-bearing applications. Applications of HA include coatings of orthopaedic and dental implants, craniomaxillofacial surgery, alveolar ridge augmentation, the filling of space in bone after tumour resection, and as coatings in knee and hip joint replacements. For the space filling applications HA is provided in various forms – as powder, in granular and block form, with and without macropores (Woesz et al. 2005).

A family of calcium phosphates exists and these may be categorised according to their calcium to phosphate ratio. Tricalcium phosphate

($Ca_3(PO_4)_2$, TCP) has higher reactivity and higher solubility than HA. It is prepared by precipitation and sintering or by solid-state reactions at high temperatures and has a Ca to P ratio of 1.5. TCP has four different polymorphs, two of which are relevant in the field of bioceramics, α and β. These two polymorphs have different mechanical properties and dissimilar solubility (Gibson et al. 1996). They have been used as bone replacement material, but their solubility is very high, which is a strong restriction for most applications (Oonishi et al. 1997). Biphasic calcium phosphate (BCP). has been developed to overcome this limitation (Daculsi et al. 1999, 2003). It is a mixture of tricalcium phosphate and hydroxyapatite with varying composition and thus a Ca to P ratio varying between 1.5 and 1.67. It is prepared by sintering precipitated calcium-deficient apatite and has been used in macroporous form for more than 10 years in dental applications and for bone substitution, although its mechanical stability is low in porous form (Gauthier et al. 2001).

Calcium phosphates are, as mentioned above, bioactive. The phenomenon of bioactivity is thought to be connected with the precipitation of a layer of carbonated apatite on the surface of the implant. In the case of hydroxyapatite, early investigations found that bone, as a composite of collagen and nanoparticles of hydroxyapatite, is growing directly at the implant surface, without an interlayer of unmineralised tissue, but also without a previously precipitated layer of pure mineral. Transmission electron microscopy (TEM) studies have shown that the distance between the implant surface and the adjacent bone is smaller than 200 Å (Tracy and Doremus 1984). Another TEM study of the bone/implant interface presented in Jarcho et al. (1977) revealed that the bone formed directly at the hydroxyapatite surface is highly oriented perpendicular to the surface and changes its orientation to parallel at a distance between 500 and 1000 Å from the surface. De Groot (1988) compared the bonding strength of bone on titanium and hydroxyapatite, the former being less than 1 MPa, the latter being between 60 and 70 MPa, depending on the type of HA. Similar results were obtained in a push-out test presented in Korn et al. (1997), where the authors detected a significantly higher shear strength of the bone implant interface in HA, compared to bioinert ceramics and stainless steel.

Recent publications state that in fact dissolution and precipitation of ions are happening at the implant's surface; for the dense, stoichiometric HA, with its relatively low solubility, these mechanisms are difficult to measure. However, TEM studies have suggested that even dense stoichiometric HA undergoes a process of partial dissolution of calcium and phosphate ions, resulting in an increased concentration in the vicinity of the implant's surface, which subsequently leads to the precipitation of a layer of carbonated apatite, intermixed with proteins (Vanblitterswijk et al. 1985; Porter et al. 2005). This layer is thought to act as a substrate for the proliferation and differentiation of osteoblasts, thus impeding the attachment of fibroblasts and the formation of a fibrous capsule around the implant (Rokusek et al. 2005). With increasing reactivity the formation of this apatite-like layer becomes faster and more pronounced, and the quality of the bond between bone and implant increases (Ducheyne et al. 1990; Ducheyne and Qiu 1999).

One possibility to increase the reactivity and therefore the bioactivity of hydroxyapatite is the replacement of phosphate ions by carbonate ions (Gibson and Bonfield 2002; Barralet et al. 2002, 2003). Another is the substitution of phosphate by silicate ions (Gibson et al. 1999; Porter et al. 2004; Patel et al. 2005; Vallet-Regi and Arcos 2005; Thian et al. 2006a, 2006b; Pietak et al. 2007). High reactivity is also found in biphasic calcium phosphate and tricalcium phosphate; however, the limiting factor in this case is the speed of dissolution or resorption of the material, since the rates of implant dissolution and bone ingrowth need to be suitably matched.

6.6.3 Bioactive glasses

As introduced earlier, bioactivity is defined as the ability of a material to encourage bonding of living tissue to the implant's surface. Bioglass®, which was initially proposed (1967), investigated and presented (1974) by Hench and coworkers, was one of the earliest artificial materials to encourage bone bonding. Since then many researchers have investigated alternative compositions, but generally bioactive glasses comprise SiO, CaO, Na_2O and P_2O_5. In an extensive investigation Hench and coworkers defined and assessed the 'bioactivity index' of a range of bioactive glasses of different composition. It is defined as the time it takes for 50% of the surface of the material to bond to bone (Hench 1980).

In fact it is only in a small region of the ternary system of SiO, CaO and Na_2O that bone bonding was actually detected. Too high a SiO_2 content triggered the formation of a fibrous capsule around the implant (bioinert), whereas other compositions dissolved in the physiological surroundings or did not form glasses in the first place. Within the compositional area for materials exhibiting bioactivity, the bioactivity index increased, with a maximum at a composition of 45 wt% SiO, 24.4 wt% CaO, 24.5 wt% Na_2O and 6 wt% P_2O_5 (the amount of P_2O_5 was kept constant in this study). This composition was called Bioglass 45S5® and is probably the most thoroughly investigated bioactive glass (Hench and Paschall 1973; Hench 1980, 1998; Hench and Wilson 1984; De Aza et al. 2007).

The main advantage of Bioglass® is its rapid connection with bone, which takes place by a well-understood process consisting of 12 stages (Hench 1998). Immediately after implantation, the surface of any implant is exposed to a mixture of water, proteins and other biomolecules, which forms a biolayer within just a few milliseconds. In the case of bioglass this first step is followed by the dissolution of sodium ions from the surface of the glass, which are replaced by hydrogen ions. This happens during a few minutes after implantation and is controlled by diffusion, therefore being inversely dependent on the square root of time ($t^{-1/2}$). Thereafter the dissolution of silica and formation of silanols at the glass–liquid interface takes place. As a surface reaction this is linear with time. The next step comprises the condensation and repolymerisation of a silicon oxide rich layer on the surface of the implant, which is highly porous on a microscopic scale, followed by migration of calcium and phosphate ions to the surface, where they form an amorphous layer. Crystallisation of this calcium- and phosphate-rich layer into carbonated hydroxyapatite, which is intermixed with organic molecules, is complete 3–6 hours after implantation. Growth in thickness, necessary for sufficient mechanical compliance, takes 12–24 hours. The chemical composition and morphology of this layer are similar to the natural bone mineral and therefore it offers conditions which encourage the cells to attach 24–72 hours after implantation (Hench 1991; Shirtliff and Hench 2003). Additionally the layer incorporates organic molecules such as blood proteins and collagen, which might represent additional binding sites.

Following the formation of the bone-like apatite layer, the normal healing and regeneration processes take place. However, the time during which inflammatory reactions take place seems to be minimised by the use of Bioglass® (Hench and Paschall 1973), allowing the attachment of stem cells and their differentiation to happen faster than with other biomaterials.

Since the first clinical use of Bioglass® for the reconstruction of the bony ossicular chain, which increased the success rate of such treatments because Bioglass bonds not only to bone but also to unmineralised tissue (Lobel 1986), the material has mainly been used for cavity fillings after tooth extraction and for other treatments of bony defects in the oral cavity and as a coating material for metallic implants. Load-bearing applications are restricted due to the very low mechanical strength of Bioglass. However, research is going on and new fields of application such as drug delivery and the use of antibacterial effects are being examined (Leppäranta *et al.* 2008; Vallet-Regi *et al.* 2008; Zhao *et al.* 2008).

6.6.4 Glass-ceramics

Bioactive glass-ceramics represent a comparatively young field of research in the biomaterials community, but the materials are interesting since they offer two advantages over other bioceramics: they can easily and cheaply be shaped into complex shapes, by well-known and widespread techniques like casting, blowing, pressing and rolling; and because of their high relative density and subsequent crystallisation, they have superior mechanical properties compared to their amorphous precursor materials. Nucleation and crystallisation of specific phases within these materials, however, occur by mechanisms which are complex and often hard to control, and these need careful investigation and thorough understanding. It is particularly important to find a balance between the rate of nucleation (amount of nuclei formed per unit time), the growth rate of the nuclei (crystal growth rate) and the tendency of crystals to merge. These mechanisms have maxima at different temperatures; therefore the temperature treatment normally comprises two steps, one which aims, at lower temperatures, to form the maximum number of nuclei, and a second step, at higher temperatures, which is necessary to grow crystals fast enough to prevent coarsening. Heating rates which are too high limit this process, as they lead to temperature gradients within the parts and cracking due to thermal stresses (De Aza et al. 2007).

Many glass-ceramics used are based on compositions similar to the bioglasses introduced in Section 6.6.3, but also other compositions have been investigated and are used in clinical applications. The first glass-ceramic developed (by Bröhmer and co-workers in 1973) was Ceravital® (this expression, however, is nowadays used for a wide range of different compositions). Ceravital® has a bioactivity about half that of Bioglass 45S5 and is used to replace the ossicular chain in the middle ear, an application which includes only minimal loads on the implant material, so that the comparatively low mechanical strength of Ceravital® (in the range of that of dense HA) is not a problem.

Apatite/wollastonite glass-ceramics (A/W glass-ceramics) seem to be more promising. This material was developed by Kokubo and co-workers (Kokubo et al. 1986) and is commercially called Cerabone A/W®. Due to its good mechanical properties and machinability, its main applications are reconstruction of vertebrae, spinal discs and iliac crest as well as filling of bone defects.

A/W glass-ceramic is formed from MgO, CaO, SiO_2 and P_2O_5, with addition of a small amount of CaF_2. These are ground to a very small particle size, followed by isostatic pressing in the desired shape and a temperature treatment, which leads to densification and precipitation of the apatitic (oxyfluoroapatite, $Ca_{10}(PO_4)_6(O,F_2)$) and the wollastonite phase (β-$CaSiO_3$) (Kokubo et al. 1986). If the appropriate processing heat treatment is performed, a dense, fine-grained glass-ceramic with very good mechanical properties is obtained (Kokubo et al. 1985, 1987; Nakamura et al. 1985). In fact, the strength of Cerabone A/W exceeds the strength of human cortical bone. Its bioactivity index is roughly one quarter of that of Bioglass 45S5.

Unlike some glass-ceramics, casting of A/W glass-ceramic into a mould is not possible without significant loss of mechanical strength, as the crystallisation of the apatite and wollastonite phases starts at the mould wall and leads to a non-advantageous crystal size and orientation within the part.

A/W glass-ceramics bind to bone and maintain a high bonding strength over long periods of time. As opposed to Bioglass, no silicon-rich layer on the implant surface is formed, but the development of a calcium- and phosphate-rich layer is observed (Kokubo 1990). It consists of a carbonate containing hydroxyapatite with small crystal size and its thickness reached 7 µm after 10 days in simulated body fluid and remained constant thereafter. In Kokubo *et al.* (1990) the authors compared the behaviour of A/W glass-ceramics with that of non-bioactive ceramics of similar composition, on which they did not observe the formation of an apatite-rich layer. The conclusion that the formation of this layer is a prerequisite for bioactivity in these materials is therefore confirmed.

The structure and composition of the apatite-rich layer on the surface of an A/W glass-ceramic implant are similar to those of the apatite in natural bone; therefore it is permissible to assume that this surface will preferentially be inhabited by osteoblasts instead of fibroblasts, thus preventing the formation of a fibrous capsule as in the case of non-bioactive materials, which do not exhibit the formation of the apatitic layer.

Other glass-ceramics have been and are being developed, some with very promising mechanical and biological properties (Kitsugi *et al.* 1986, 1993; Kokubo *et al.* 2000; Ignatius *et al.* 2005; Almeida and Fernandes 2006).

6.7 Summary and future trends

Bioceramics are a very versatile group of materials, and due to their unique combination of properties (high strength and unsurpassed low friction of alumina or bone- and tissue-bonding ability of bioglasses, for example) they have gained a wide variety of applications in the human body. However, the potential for improvement is still large and, as the average age of humankind is predicted to increase further, thus amplifying the need for replacement and restoration of damaged tissue, these improvements will all the more be desired. In particular the ideal combination of high mechanical strength and bioactivity has not yet been accomplished in a single material. One of the most promising ways towards that kind of material are composites from tough biopolymers and particles of bioceramics. Of special interest seem to be nanoparticles, as with decreasing particle size the size of the flaws becomes limited, and the material eventually approaches its theoretical strength (Gao *et al.* 2003). Bone itself is in fact a composite of a biopolymer and nanoparticles of carbonated apatite. While each component alone would fail to meet the mechanical requirements of the skeleton, nature overcomes the limitation in choice of materials (Fratzl and

Weinkamer 2007) by using nanometre-sized particles and thereby combines the toughness of the protein collagen and the strength of the nanometre-sized mineral particles (Woesz 2008). The future of bioactive implants therefore appears to be in the design and development of materials that more closely match the structures and compositions of natural tissues.

6.8 References

Albee, F H and Morrison, H F (1920). 'Studies in bone growth: Triple CaP as a stimulus to osteogenesis'. *Annals of Surgery* **71**: 32–39.

Almeida, N A F and Fernandes, M H V (2006). 'Effect of glass ceramic crystallinity on the formation of simulated apatite layers'. *Materials Science Forum* **514–516**: 1039–1043.

Ashby, M F and Jones, D R H (1980). *Engineering Materials – An Introduction to Their Properties and Applications*. Oxford, UK: Pergamon Press.

Baier, R E (1977). 'The organization of blood components near interfaces'. *Annals of the New York Academy of Sciences* **283**: 17–36.

Barralet, J, Knowles, J C, Best, S and Bonfield, W (2002). 'Thermal decomposition of synthesised carbonate hydroxyapatite'. *Journal of Materials Science – Materials in Medicine* **13**(6): 529–533.

Barralet, J E, Fleming, G J P, Campion, C, Harris, J J and Wright, A J (2003). 'Formation of translucent hydroxyapatite ceramics by sintering in carbon dioxide atmospheres'. *Journal of Materials Science* **38**(19): 3979–3993.

Ben-Nissan, B (2003). 'Natural bioceramics: from coral to bone and beyond'. *Current Opinion in Solid State & Materials Science* **7**(4–5): 283–288.

Boutin, P (1971). '[Experimental study of alumina and its use in surgery of the hip]'. *Presse Medicale* **79**(14): 639–640.

Chang, B S, Lee, C K, Hong, K S, Youn, H J, Ryu, H S, Chung, S S and Park, K W (2000). 'Osteoconduction at porous hydroxyapatite with various pore configurations'. *Biomaterials* **21**(12): 1291–1298.

Chang, Y S, Oka, M, Kobayashi, M, Gu, H O, Li, Z L, Nakamura, T and Ikada, Y (1996). 'Significance of interstitial bone ingrowth under load-bearing conditions: A comparison between solid and porous implant materials'. *Biomaterials* **17**(11): 1141–1148.

Daculsi, G, Weiss, P, Bouler, J M, Gauthier, O, Millot, F and Aguado, E (1999). 'Biphasic calcium phosphate/hydrosoluble polymer composites: A new concept for bone and dental substitution biomaterials'. *Bone* **25**(2): 59s–61s.

Daculsi, G, Laboux, O, Malard, O and Weiss, P (2003). 'Current state of the art of biphasic calcium phosphate bioceramics'. *Journal of Materials Science – Materials in Medicine* **14**(3): 195–200.

De Aza, P N, De Aza, A H, Pena, P and De Aza, S (2007). 'Bioactive glasses and glass-ceramics'. *Boletín de la Sociedad Española de Cerámica y Vidrio* **46**(2): 45–55.

De Groot, K (1988). 'Effect of porosity and physiochemical properties on the stability, resorption and strength of calcium phosphate ceramics'. In *Bioceramics: Material Characteristics versus in Vivo Behavior*, Ducheyne, P and Lemons, J A (eds). New York: The New York Academy of Sciences, **523**: 227–233.

De Groot, K (1993). 'Clinical applications of calcium phosphate biomaterials'. *Ceramics International* **19**: 363–366.

Dearnley, P A (1999). 'A review of metallic, ceramic and surface-treated metals used for bearing surfaces in human joint replacements'. *Proceedings of the Institution of Mechanical Engineers Part H – Journal of Engineering in Medicine* **213**(H2): 107–135.

Doremus, R H (1992). 'Bioceramics'. *Journal of Materials Science* **27**(2): 285–297.

Dorlot, J M (1992). 'Long-term effects of alumina components in total hip prostheses'. *Clinical Orthopaedics and Related Research* **(282)**: 47–52.

Dorozhkin, S V and Epple, M (2002). 'Biological and medical significance of calcium phosphates'. *Angewandte Chemie – International Edition* **41**(17): 3130–3146.

Dreesmann, H (1892). 'Veber Knochenplombierung'. *Beiträge zur Klinische Chirurgie* **9**: 804–810.

Dubok, V A (2000). 'Bioceramics – Yesterday, today, tomorrow'. *Powder Metallurgy and Metal Ceramics* **39**(7–8): 381–394.

Ducheyne, P and Qiu, Q (1999). 'Bioactive ceramics: the effect of surface reactivity on bone formation and bone cell function'. *Biomaterials* **20**(23–24): 2287–2303.

Ducheyne, P, Beight, J, Cuckler, J, Evans, B and Radin, S (1990). 'Effect of calcium phosphate coating characteristics on early postoperative bone tissue ingrowth'. *Biomaterials* **11**(8): 531–540.

Fratzl, P and Weinkamer, R (2007). 'Nature's hierarchical materials'. *Progress in Materials Science* **52**: 1263–1334.

Gao, H J, Ji, B H, Jager, I L, Arzt, E and Fratzl, P (2003). 'Materials become insensitive to flaws at nanoscale: Lessons from nature'. *Proceedings of the National Academy of Sciences of the United States of America* **100**(10): 5597–5600.

Gauthier, O, Bouler, J M, Aguado, E, Pilet, P and Daculsi, G (1998). 'Macroporous biphasic calcium phosphate ceramics: influence of macropore diameter and macroporosity percentage on bone ingrowth'. *Biomaterials* **19**(1–3): 133–139.

Gauthier, O, Goyenvalle, E, Bouler, J M, Guicheux, J, Pilet, P, Weiss, P and Daculsi, G (2001). 'Macroporous biphasic calcium phosphate ceramics versus injectable bone substitute: a comparative study 3 and 8 weeks after implantation in rabbit bone'. *Journal of Materials Science – Materials in Medicine* **12**(5): 385–390.

Gibson, I R and Bonfield, W (2002). 'Novel synthesis and characterization of an AB-type carbonate-substituted hydroxyapatite'. *Journal of Biomedical Materials Research* **59**(4): 697–708.

Gibson, I R, Akao, M, Best, S M and Bonfield, W (1996). 'Phase transformation of tricalcium phosphates using high temperature X-ray diffraction'. *Proc. 9th International Symposium on Ceramics in Medicine*, Otsu, Japan, Elsevier Science.

Gibson, I R, Best, S M and Bonfield, W (1999). 'Chemical characterization of silicon-substituted hydroxyapatite'. *Journal of Biomedical Materials Research* **44**(4): 422–428.

Gibson, L J and Ashby, M F (1997). *Cellular Solids*. Cambridge, UK: Cambridge University Press.

Griffith, A A (1921). 'The phenomena of rupture and flow in solids'. *Philosophical Transactions of the Royal Society of London, Series A* **221**: 163–198.

Gross, U, Schmitz, H-J and Strunz, V (1988). 'Surface activities of bioactive glass, aluminum oxide and titanium in a living environment'. In *Bioceramics: Materials Characteristics versus in Vivo Behavior*, Ducheyne, P and Lemons, J A (eds). New York: The New York Academy of Sciences, **523**: 211–226.

Hench, L L (1980). 'Biomaterials'. *Science* **208**(4446): 826–831.

Hench, L L (1991). 'Bioceramics – from concept to clinic'. *Journal of the American Ceramic Society* **74**(7): 1487–1510.

Hench, L L (1998). 'Bioceramics'. *Journal of the American Ceramic Society* **81**(7): 1705–1728.

Hench, L L and Paschall, H A (1973). 'Direct chemical bond of bioactive glass-ceramic materials to bone and muscle'. *Journal of Biomedical Materials Research* **7**(3): 25–42.

Hench, L L and Wilson, J (1984). 'Surface-active biomaterials'. *Science* **226**(4675): 630–636.

Hench, L L and Wilson, J (1993). 'Chapter 1: Introduction'. In *An Introduction to Bioceramics*, Hench, L L and Wilson, J (eds). Singapore, New Jersey, London, Hong Kong: World Scientific Publishing, **1**: 386.

Hulbert, S F, Young, F A, Mathews, R S, Klawitter, J J, Talbert, C D and Stelling, F H (1970). 'Potential of ceramic materials as permanently implantable skeletal prostheses'. *Journal of Biomedical Materials Research* **4**: 433–456.

Hulbert, S F, Bokros, J C, Hench, L L, Wilson, J and Heimke, G (1987). 'Ceramics in clinical applications, past, present and future'. In *High Tech Ceramics*, Vincenzini, P (ed.). Amsterdam: Elsevier Science Publishers: 3–27.

Ignatius, A, Peraus, M, Schorlemmer, S, Augat, P, Burger, W, Leyen, S and Claes, L (2005). 'Osseointegration of alumina with a bioactive coating under load-bearing and unloaded conditions'. *Biomaterials* **26**(15): 2325–2332.

Jarcho, M, Kay, J F, Gumaer, K I, Doremus, R H and Drobeck, H P (1977). 'Tissue, cellular and subcellular events at a bone-ceramic hydroxyapatite interface'. *Journal of Bioengineering* **1**(2): 79–92.

Kitsugi, T, Yamamuro, T, Nakamura, T, Higashi, S, Kakutani, Y, Hyakuna, K, Ito, S, Kokubo, T, Takagi, M and Shibuya, T (1986). 'Bone bonding behavior of 3 kinds of apatite containing glass-ceramics'. *Journal of Biomedical Materials Research* **20**(9): 1295–1307.

Kitsugi, T, Yamamuro, T, Nakamura, T, Kotani, S, Kokubo, T and Takeuchi, H (1993). 'Four calcium phosphate ceramics as bone substitutes for non-weight-bearing'. *Biomaterials* **14**(3): 216–224.

Klawitter, J (1979). *A Basic Investigation of Bone Growth in Porous Materials*. Clemson, SC: Clemson University.

Kokubo, T (1990). 'Surface chemistry of bioactive glass-ceramics'. *Journal of Non-Crystalline Solids* **120**(1–3): 138–151.

Kokubo, T, Ito, S, Shigematsu, M, Sakka, S and Yamamuro, T (1985). 'Mechanical properties of a new type of apatite-containing glass ceramic for prosthetic application'. *Journal of Materials Science* **20**(6): 2001–2004.

Kokubo, T, Ito, S, Sakka, S and Yamamuro, T (1986). 'Formation of a high-strength bioactive glass ceramic in the system $MgO–CaO–SiO_2–P_2O_5$'. *Journal of Materials Science* **21**(2): 536–540.

Kokubo, T, Ito, S, Shigematsu, M, Sakka, S and Yamamuro, T (1987). 'Fatigue and lifetime of bioactive glass ceramic A-W containing apatite and wollastonite'. *Journal of Materials Science* **22**(11): 4067–4070.

Kokubo, T, Ito, S, Huang, Z T, Hayashi, T, Sakka, S, Kitsugi, T and Yamamuro, T (1990). 'Ca, P-rich layer formed on high-strength bioactive glass-ceramic A-W'. *Journal of Biomedical Materials Research* **24**(3): 331–343.

Kokubo, T, Kim, H-M, Kawashita, M and Nakamura, T (2000). 'Novel ceramics for biomedical applications'. *Journal of the Australian Ceramic Society* **36**(1): 37–46.

Korn, D, Soyez, G, Elssner, G, Petzow, G, Brès, E F, d'Hoedt, B and Schulte, W (1997). 'Study of interface phenomena between bone and titanium and alumina surfaces in the case of monolithic and composite dental implants'. *Journal of Materials Science – Materials in Medicine* **8**(10): 613–620.

Kuboki, Y, Takita, H, Kobayashi, D, Tsuruga, E, Inoue, M, Murata, M, Nagai, N, Dohi, Y and Ohgushi, H (1998). 'BMP-induced osteogenesis on the surface of hydroxyapatite with geometrically feasible and nonfeasible structures: Topology of osteogenesis'. *Journal of Biomedical Materials Research* **39**(2): 190–199.

Kusaka, J, Takashima, K, Yamane, D and Ikeuchi, K (1999). 'Fundamental study for all-ceramic artificial hip joint'. *Wear* **225**: 734–742.

Learmonth, I D, Young, C and Rorabeck, C (2007). 'The operation of the century: total hip replacement'. *Lancet* **370**: 1508–1519.

LeGeros, R Z (2002). 'Properties of osteoconductive biomaterials: Calcium phosphates'. *Clinical Orthopaedics and Related Research* **(395)**: 81–98.

Leppäranta, O, Vaahtio, M, Peltola, T, Zhang, D, Hupa, L, Hupa, M, Ylänen, H, Jukka, I S, Matti, K V and Eerola, E (2008). 'Antibacterial effect of bioactive glasses on clinically important anaerobic bacteria in vitro'. *Journal of Materials Science – Materials in Medicine* **19**: 547–551.

Li, J and Hastings, G W (1998). 'Oxide bioceramics: inert ceramics in medicine and dentistry'. In *Handbook of Biomaterials Properties*, Black, J and Hastings, G (eds). London: Chapman & Hall.

Lobel, K (1986). 'Ossicular replacement prosthesis'. In *Clinical Performance of Skeletal Prostheses*, Hench, L L and Wilson, J (eds). New York: Chapman & Hall.

Monroe, E A, Votava, W, Bass, D B and McMullen, J (1971). 'New calcium phosphate ceramic material for bone and tooth implants'. *Journal of Dental Research* **50**(4): 860–861.

Nakamura, T, Yamamuro, T, Higashi, S, Kokubo, T and Itoo, S (1985). 'A new glass-ceramic for bone-replacement – evaluation of its bonding to bone tissue'. *Journal of Biomedical Materials Research* **19**(6): 685–698.

Nery, E B, Lynch, K L, Hirthe, W M and Mueller, K H (1975). 'Bioceramic implants in surgically produced infrabony defects'. *Journal of Periodontology* **46**(6): 328–347.

Oonishi, H, Kushitani, S, Yasukawa, E, Iwaki, H, Hench, L L, Wilson, J, Tsuji, E I and Sugihara, T (1997). 'Particulate bioglass compared with hydroxyapatite as a bone graft substitute'. *Clinical Orthopaedics and Related Research* **(334)**: 316–325.

Patel, N, Brooks, R A, Clarke, M T, Lee, P M T, Rushton, N, Gibson, I R, Best, S M and Bonfield, W (2005). 'In vivo assessment of hydroxyapatite and silicate-substituted hydroxyapatite granules using an ovine defect model'. *Journal of Materials Science – Materials in Medicine* **16**(5): 429–440.

Pietak, A M, Reid, J W, Stott, M J and Sayer, M (2007). 'Silicon substitution in the calcium phosphate bioceramics'. *Biomaterials* **28**(28): 4023–4032.

Piotrowski, G, Hench, L L, Allen, W C and Miller, G J (1975). 'Mechanical studies of bone bioglass interfacial bond'. *Journal of Biomedical Materials Research* **9**(4): 47–61.

Porter, A, Patel, N, Brooks, R, Best, S, Rushton, N and Bonfield, W (2005). 'Effect of carbonate substitution on the ultrastructural characteristics of hydroxyapatite implants'. *Journal of Materials Science – Materials in Medicine* **16**(10): 899–907.

Porter, A E, Rea, S M, Galtrey, M, Best, S M and Barber, Z H (2004). 'Production of thin film silicon-doped hydroxyapatite via sputter deposition'. *Journal of Materials Science* **39**(5): 1895–1898.

Rahaman, M N, Yao, A H, Bal, B S, Garino, J P and Ries, M D (2007). 'Ceramics for prosthetic hip and knee joint replacement'. *Journal of the American Ceramic Society* **90**(7): 1965–1988.

Rokusek, D, Davitt, C, Bandyopadhyay, A, Bose, S and Hosick, H L (2005). 'Interaction of human osteoblasts with bioinert and bioactive ceramic substrates'. *Journal of Biomedical Materials Research Part A* **75A**(3): 588–594.

Sandhaus, S (1967). 'Bone implants and drills and taps for bone surgery'. British Patent 1083769.

Shirtliff, V J and Hench, L L (2003). 'Bioactive materials for tissue engineering, regeneration and repair'. *Journal of Materials Science* **38**(23): 4697–4707.

Slonaker, M and Goswami, T (2004). 'Review of wear mechanisms in hip implants: Paper II – Ceramics IG004712'. *Materials & Design* **25**(5): 395–405.

Smith, L (1963). 'Ceramic plastic material as bone substitute'. *Archives of Surgery* **87**: 653–661.

Teitelbaum, S L (2000). 'Bone resorption by osteoclasts'. *Science* **289**(5484): 1504–1508.

Thian, E S, Huang, J, Best, S M, Barber, Z H, Brooks, R A, Rushton, N and Bonfield, W (2006a). 'The response of osteoblasts to nanocrystalline silicon-substituted hydroxyapatite thin films'. *Biomaterials* **27**(13): 2692–2698.

Thian, E S, Huang, J, Vickers, M E, Best, S M, Barber, Z H and Bonfield, W (2006b). 'Silicon-substituted hydroxyapatite (SiHA): A novel calcium phosphate coating for biomedical applications'. *Journal of Materials Science* **41**(3): 709–717.

Tracy, B M and Doremus, R H (1984). 'Direct electron-microscopy studies of the bone–hydroxyapatite interface'. *Journal of Biomedical Materials Research* **18**(7): 719–726.

Vallet-Regi, M and Arcos, D (2005). 'Silicon substituted hydroxyapatites. A method to upgrade calcium phosphate based implants'. *Journal of Materials Chemistry* **15**(15): 1509–1516.

Vallet-Regi, M, Colilla, M and Izquierdo-Barba, I (2008). 'Bioactive mesoporous silicas as controlled delivery systems: Application in bone tissue regeneration'. *Journal of Biomedical Nanotechnology* **4**: 1–15.

Vanblitterswijk, C A, Grote, J J, Kuypers, W, Blokvanhoek, C J G and Daems, W T (1985). 'Bioreactions at the tissue/hydroxyapatite interface'. *Biomaterials* **6**(4): 243–251.

Vroman, L, Adams, A L, Klings, M, Fischer, G C, Munoz, P C and Solensky, R P (1977). 'Reactions of formed elements of blood with plasma-proteins at interfaces'. *Annals of the New York Academy of Sciences* **283**: 65–76.

Vuola, J, Taurio, R, Göransson, H and Asko-Seljavaara, S (1998). 'Compressive strength of calcium carbonate and hydroxyapatite implants after bone-marrow-induced osteogenesis'. *Biomaterials* **19**: 223–227.

Williams, D F (1992). 'Biofunctionality and biocompatibility'. In *Medical and Dental Materials*, Williams, D F (ed.). Weinheim, New York, Basel, Cambridge: VCH Publishers, **14**: 1–28.

Willmann, G (1994). 'Alumina ceramic looks back on 20 years of use in medical applications'. *Biomedizinische Technik* **39**(4): 73–78.

Woesz, A (2008). 'Porous scaffolds with controlled architecture'. In *Virtual Prototyping & Bio Manufacturing in Medical Applications*, Bidanda, B and Bartolo, P (eds). New York: Springer Science + Business Media: 171–206.

Woesz, A, Stampfl, J and Fratzl, P (2004). 'Cellular solids beyond the apparent density – an experimental assessment of mechanical properties'. *Advanced Engineering Materials* **6**(3): 134–138.

Woesz, A, Rumpler, M, Stampfl, J, Varga, F, Fratzl-Zelman, N, Roschger, P, Klaushofer, K and Fratzl, P (2005). 'Towards bone replacement materials from calcium phosphates via rapid prototyping and ceramic gelcasting'. *Materials Science & Engineering C – Biomimetic and Supramolecular Systems* **25**(2): 181–186.

Yuan, H P, Kurashina, K, de Bruijn, J D, Li, Y B, de Groot, K and Zhang, X D (1999). 'A preliminary study on osteoinduction of two kinds of calcium phosphate ceramics'. *Biomaterials* **20**(19): 1799–1806.

Zhao, L Z, Yan, X X, Zhou, X F, Zhou, L, Wang, H N, Tang, H W and Yu, C Z (2008). 'Mesoporous bioactive glasses for controlled drug release'. *Microporous and Mesoporous Materials* **109**: 210–215.

7
Biocompatibility and other properties of phosphate-based glasses for medical applications

E A ABOU NEEL and J C KNOWLES,
University College London, UK

Abstract: This chapter discusses phosphate-based glasses as biodegradable substitutes with a specific and controllable bioactivity. It starts with a general description of these glasses, highlighting the glass structure theories, formation, chemistry, terminology and properties. It also explores what these glasses can offer in terms of biomedical applications as bone tissue substitutes and antibacterial devices, with a particular focus placed on the *in vitro* and *in vivo* studies conducted on both monoliths and glass fibres. The application of these glasses as reinforcing agents for various composites is also considered. The chapter concludes with a discussion of the future of these glasses as biomaterials and highlights possible avenues of potential application.

Key words: phosphate-based glasses, bone tissue substitutes, antibacterial devices, *in vitro* and *in vivo* studies, future of these glasses.

7.1 Historical overview

The word glass was derived from *glaesum* or *vitrum*, a late-Latin term used to refer to a lustrous and transparent substance. The first appearance of glass dates back to 4000 BC when glazed stone beads and several coloured artefacts of excellent glass inlay work were found in Egypt.[1] Not until around 1500 BC were there significant improvements in the art and technology of glasses, and this was then followed by further developments in glass science by Faraday, Zeiss, Abbe and Scott. The primary interest of these scientists was in glasses for optical use. Nowadays, glasses are widely used for everyday applications, and the idea of using a soluble glass as a biomaterial has emerged within the last two decades.[2]

Phosphate compositions exhibit a pronounced tendency to form glasses upon cooling of their melts. The first reported phosphate composition was a binary glass composed of $Na_2O-P_2O_5$, which was discovered over a century ago under the name of Graham's salt or sodium hexametaphosphate. This glass found potential use in widespread industrial applications due to its ability to form complexes with either alkali or alkali earth metals and to exhibit a colloidal activity in solutions.[3,4] In recent decades, numerous phosphate-based glass formulations ranging among binary, ternary, quaternary and more complex compositions have been developed.

Phosphate-based glasses were developed for a wide range of technological applications such as sensors, solid-state batteries, laser devices, and airtight seals for metals with high coefficient of thermal expansion.[5] They were also used as matrices for vitrifying nuclear waste products due to their high waste loading ability, low processing temperature, and relatively high chemical durability.[6-9]

7.2 Glass structure theories

Goldsmith (1926) was the pioneer in studying glass structure and setting the parameters required for any oxide to form a glass. His theory stated that any oxide, R_mO_n, having ionic radius ratio of the cation, R, to oxygen ion, O, in the range of 0.2 to 0.4 would form a glass. This radius ratio tends to produce a tetrahedral arrangement of oxygen around the cation, and it was believed that this arrangement is necessary for glass formation.[10] However, there was a limitation to this theory, in that BeO with a radius ratio of 0.3 could not form glass. This led to the evolution of 'continuous random network theory' by Zachariasen (1932) that was considered the basis for the most widely used model for glass structure. Zachariasen described glass as amorphous inorganic polymers where the atoms are arranged in an extended three-dimensional network as in a crystal, but without periodicity and symmetry; however, the network is not entirely random. Lack of periodicity is responsible for the gradual breakdown of the glass network on heating, and the lack of symmetry is responsible for the isotropic properties of the glass, i.e. the properties are the same in all directions since the atomic arrangement will be the same in all directions, unless external fields of sufficient intensity are present.[11]

Zachariasen stated four rules for an oxide to form a glass with energy comparable to a crystal: (1) an oxygen atom must be linked to not more than two central atoms A, since a higher coordination number will prevent the formation of a non-periodic network; (2) the number of oxygen atoms surrounding atom A must be small, specifically either 3 or 4 since most of the glass available at that time contained network cations in triangular or tetrahedral coordination; (3) oxygen polyhedra must share only corners, not edges or faces, with each other to avoid the regular lining up of the polyhedra when building the glass network; and (4) at least three corners should be shared, for each tetrahedron to form three-dimensional networks.

Zachariasen finally modified the previous rules, and stated that any oxide glass may be formed (1) if the sample contains a high percentage of cations

B which are surrounded by oxygen tetrahedra or triangles, i.e. glass-forming cations; (2) if these tetrahedra or triangles share corners with each other; and (3) if some oxygen atoms are linked to only two such cations, and do not form further bonds with any other cations A. The cation B is essential for the formation of an open continuous irregular network structure; the presence of this opened structure is to obtain electrical stability of the framework (the sum of cations and anions must balance). These holes are filled with the cation A as the framework is being formed, since the addition of cation A should not cause a great increase in the potential energy. Cation A should have a large radius and a small charge opposite to that of cation B to reduce the repulsive forces between them.

The non-existence of a glass form of TiO_2 or Al_2O_3 could not be explained by the Zachariasen theory. Hagg (1935) studied the glass-forming process from the cooling process point of view; he believed that glasses consisted of chains or two-dimensional sheets. This theory explained why SiO_2 but not TiO_2 was able to form a glass.[12]

7.3 Glass formation

Glass formation usually takes place by rapid cooling of the molten mass past the crystallisation temperature and the solidification starts at the glass transition temperature. This process is usually described on the basis of change in enthalpy or volume as a function of temperature as shown in Fig. 7.1.

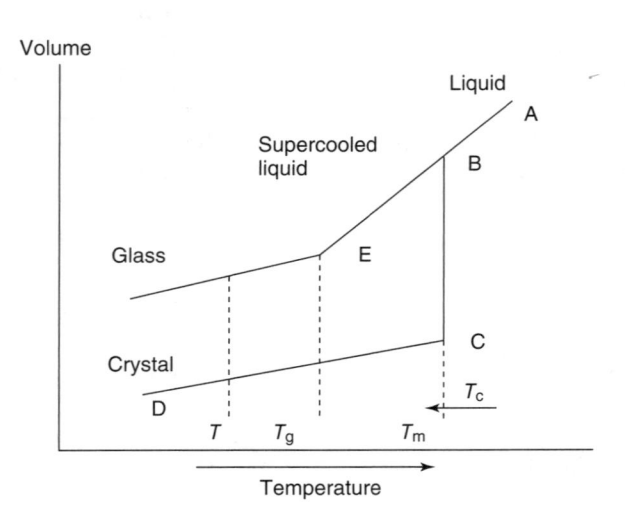

7.1 Volume change as a function of temperature during glass formation.

A continuous decrease in the temperature of the molten mass from point A in Fig. 7.1 results in a continuous reduction in volume along the line AB. Then slow cooling results in an abrupt decrease in volume, and crystallisation at the melting point will occur along BC. Further cooling results in a gradual change in volume with the formation of crystalline phases along the line CD. On the other hand, rapid cooling gives no chance for crystallisation to take place; instead a supercooled liquid is obtained, and the volume change will follow the line BE. As the melt continues to cool to room temperature, the viscosity increases so that the atoms cannot be arranged in a long range ordered arrangement (symmetrical units of grouped atoms) with the resultant formation of a glass. The temperature between that of the supercooled liquid and that of the frozen solid (glass) is called the glass transition temperature, T_g.

7.4 Chemistry of phosphate-based glasses

In studying the phosphate glass structure, three important elements should be considered: glass formers, glass modifiers, and the interaction between them.

Glass or network formers can be either primary or conditional glass formers. Primary network formers include SiO_2, B_2O_3 and P_2O_5 which are able to form glasses by themselves. Conditional glass formers, sometimes called intermediate oxides, however, are not able to form glasses under normal conditions, but under certain circumstances they can replace the network forming oxides. These oxides include GeO_2, Bi_2O_3, As_2O_3, Sb_2O_3, TeO_2, Al_2O_3, Ga_2O_3 and V_2O_5, and when they are incorporated into a glass, they can produce changes in colour and conductivity.[13]

Glass modifiers are not part of the glass network, but they act to balance the glass network. They are generally alkali and alkali earth metal oxides such as Na_2O, K_2O, MgO and CaO.[11,13,14] The addition of these metal oxides is essential for stability of the phosphate glasses since pure phosphate (P_2O_5) glass is chemically unstable due to its hygroscopic nature.[15,16]

The interaction between the network former and modifier affects the organisation of the glass network and therefore its properties.[5] The structure of phosphate glass should be considered in more detail before studying the interaction between the network former and modifier.

Generally, phosphate glasses are inorganic polymers composed entirely of $(PO_4)^{3-}$ tetrahedra that are regarded as the backbone of the glass as shown in Fig. 7.2. Each tetrahedron is composed of one phosphorus ion (P^{5+}) that is charge balanced with four O^{2-} ions. For charge compensation, one oxygen atom will share its two electrons with P^{5+} and is called a terminal double bonded oxygen (DBO). The other three oxygens share only one of their two electrons; they are still free to combine with other P^{5+}, in which

7.2 The phosphate anion tetrahedra.

case they can form bridging oxygens (BO)*, or with metal ions, where they can form non-bridging oxygen (NBO). The bridging oxygen is covalently bonded to two glass former atoms, while the non-bridging oxygens are ionically bonded only on one side; therefore, the binding energy of a BO is higher than that of a NBO.[17]

The structure of the tetrahedra is classified according to the number of BO per tetrahedron, and is denoted by Q^i terminology, where Q refers to the phosphorus atom bonded to four oxygen atoms forming a tetrahedron, and i refers to the number of bridging oxygens per tetrahedron starting from zero to a maximum of three atoms. For example, Q^3 tetrahedra (PO_4) possess three covalent bridging oxygen bonds to the neighbouring tetrahedra as in vitreous P_2O_5; accordingly, the Q^3 tetrahedron is a neutral unit and is known as a branching unit. Q^2 tetrahedra, $(PO_4)^-$, possess two covalent bridging oxygen bonds with the neighbouring tetrahedra; therefore, Q^2 carries one negative charge and is known as a middle unit. Q^1 tetrahedra, $(PO_4)^{2-}$, units possess one covalent bridging oxygen bond with the neighbouring tetrahedra and carry two negative charges; Q^1 is known as an end unit. Q^0, $(PO_4)^{3-}$, is an isolated tetrahedron unit with no bridging oxygen to the neighbouring tetrahedra; therefore, it carries three negative charges and known as an orthophosphate unit.[18-20] Therefore, the number of negative charges each unit carries depends on the number of NBOs it has, as in Fig. 7.3.

Generally, polymerisation of phosphates and formation of three-dimensional networks occurs in the absence of metallic cations; the addition of metallic oxide, however, results in the inhibition of polymerisation and leads to the formation of chain structures through a process called 'depolymerisation'.

One model that was proposed for the interaction between the glass network and the modifying oxides is the depolymerisation model. According to this model, the addition of modifying oxides leads to the cleavage of P—O—P links, i.e. depolymerisation of the glass network, which starts from

---

*Bridging oxygens are oxygen ions taking part in P—O—P bonds (by which the P tetrahedra join to other tetrahedra), while non-bridging oxygens are assigned to —P=O or M—O—P bonds (where M is a cation).[7]

Q³ tetrahedron (ultraphosphate)   Q² tetrahedron (metaphosphate)   Q¹ tetrahedron (pyrophosphate)   Q⁰ tetrahedron (orthophosphate)

7.3 Nomenclature and representation of PO$_4$ tetrahedra with different polymerisations.[18]

vitreous P$_2$O$_5$ with the creation of negatively charged NBO at the expense of BO. The negatively charged NBO will coordinate with the modifier cations for optimisation of the coordination number of metal ions.[19,21–23] The depolymerisation scheme according to Kirkpatrick and Brow follows the sequence Q³ → Q² → Q¹ → Q⁰ as the amount of modifier oxide (e.g. monovalent metal oxides) increases as follows:[18]

$$2Q^3 + R_2O \rightarrow 2Q^2$$

$$2Q^2 + R_2O \rightarrow 2Q^1$$

$$2Q^1 + R_2O \rightarrow 2Q^0$$

where R$_2$O are monovalent metal oxides. Accordingly, the amount of metal oxides in the glass (x) and hence the oxygen to phosphorus ratio sets the number of linkages of each tetrahedron via bridging oxygen to other tetrahedra, and hence the dominance of any Q species.

7.4.1 Ultraphosphate glass structure

In the ultraphosphate region, where $0 \leq x \leq 0.5$, the glass network will be dominated by Q² and Q³ units where x is the mole fraction of metal oxide. These tetrahedra appear to be randomly linked, at least in alkali ultraphosphate glasses. In glasses with P$_2$O$_5$ content greater than 0.75 mol, the phosphate network most resembles that found in vitreous P$_2$O$_5$ glass. With increasing modifier content, there is loss of extended-range order associated with vitreous P$_2$O$_5$, as the fraction of tetrahedra increases. The composition at which this transition occurs depends on the coordination number and valence of the modifying oxides.[19]

7.4.2 Metaphosphate glass structure

In the metaphosphate region, where $x = 0.5$, the glass network will be based on Q² units that form indefinitely long chains and/or rings which are linked by an ionic linkage between the modifying cations and NBO. Q^2 species are the most susceptible species to degradation because the crosslink density for the metaphosphate glass is zero.[20] Brow claimed that the properties of metaphosphate glasses are more affected by M—O—P (inter-chain) bonds (M refers to metal ions) than by P—O—P bonds.[19]

7.4.3 Polyphosphate glass structure

In the polyphosphate region, where $x > 0.5$, the glass network will be based on Q^2 units terminated by Q^1 units. In the polyphosphate region two possibilities can occur: firstly, at $x = 0.67$ the network is based on phosphate dimer, two Q^1 tetrahedra units linked by one bridging oxygen, and is called a pyrophosphate unit; secondly, at $x = 0.75$, the network is based on isolated Q^0 units.[19,20]

Another model that was proposed for the interaction between modifying oxides and the glass network is called 'repolymerisation'.[21] This model is applied for the glass in the ultraphosphate region, and is used to explain the packing density and M–O coordination number, where M refers to metal. In this model, it was assumed that all terminal oxygen atoms, including not only NBO in the Q^2 middle unit but also DBO in the Q^3 branching units, tend to coordinate with the metal oxides. This coordination leads to reorganisation of the network structure so that each PO_4 unit is finally connected to four phosphorus or metal atoms with the metal atoms included in M—O—P bridges. At the point where all terminal oxygen atoms are involved in M—O—P bridges, the structure is stabilised. For this model, it was supposed that the double bonded oxygens (DBOs) carry a slightly negative charge that should be compensated by the opposite charge of P atoms. The P atoms are not the first inter-tetrahedral neighbour of DBO because the direction of the P=O bonds tends to approach the centre of adjacent rings in P_4O_{10} or P_2O_5; therefore, a defect in the glass structure will result. This defect is eliminated by the presence of a real neighbour of modifying oxides. Hoppe concluded that depolymerisation is the predominating model affecting phosphate glass structure.[21]

7.5 Properties of phosphate-based glasses

Phosphate-based glasses are generally characterised by low melting and transition temperatures, and high coefficients of thermal expansion that extend their applications in a variety of technological fields.[18,24,25] They also have a unique property in that they are degradable and this property can be readily and predictably manipulated by changing the glass composition.[26–33] The glass composition can be adjusted to obtain degradation times ranging from hours to several months or longer. The degradation can be reduced by the incorporation of cations with high electrostatic field strength (Zn^{2+}, Pb^{2+}) to increase the covalency of M—O—P bonds, or cations with high valence (Al^{3+} and Fe^{3+}), or substitution of some of the oxygen atoms by nitrogen.[34]

7.5.1 Glass transition

All amorphous (non-crystalline or semi-crystalline) materials will yield at the glass transition temperature (T_g) during heating. This temperature is the main characteristic transformation temperature of the amorphous phase. The glass transition events occur when a hard, solid, amorphous material or component undergoes transformation to a soft, rubbery, liquid phase during heating. The reverse transformation also occurs during cooling. The classic T_g is observed as an endothermic stepwise change in the heat flow or heat capacity, and for polymers it is characterised by a change in the temperature-dependent properties during heating or cooling. T_g is heating rate dependent: the higher the heating rate, the higher the T_g.

7.5.2 Crystallisation

Crystallisation is a two-stage process of nucleation and then crystal growth. The nucleation can be spontaneously formed in the melt (homogeneous) or around existing impurities (heterogeneous). The crystal growth usually occurs at a temperature higher than that of nucleation and cannot proceed without nuclei formation; in such a case, a glass will be formed. Both stages of crystallisation (nucleation and crystal growth) occur at temperatures higher than the glass transition.

7.5.3 Degradation

In relation to time, the degradation process of phosphate-based glasses involves two stages: the first stage is slow degradation where the process varies exponentially with time, and the second stage is uniform degradation that varies linearly with time. The first stage is usually associated with the diffusion of H_2O molecules into the glass until the surface layer becomes totally surrounded with H_2O, while the second stage usually happens when this totally hydrated layer becomes separated from the partially hydrated layer still attached to the glass and then leached into the surrounding solution. The transition from the first to the second stage depends on the glass composition. The degradation is usually affected by several factors such as glass composition, pH, temperature, solubility/saturation effect, and surface area of the glass to solution volume ratio. The degradation of phosphate-based glasses is more sensitive to the glass composition; the more alkali the glass contains, the lower the durability. Moreover, the glass degradation is accelerated in acidic or basic conditions, but is more pronounced in an acidic environment. Phosphate glasses consume H$^+$ from acidic solution or OH$^-$ from basic solution; this consumption is linear over time after a few minutes of degradation, and this corresponds to the first stage of degradation. The rate of consumption increases with increasing degradation. Regardless of pH, the degradation generally increases with temperature. The solubility effect means that the glass degradation is inhibited by saturation of the solution, with the low solubility ions leached from the glass such as Ca^{2+}.[35,36] The surface area to volume ratio controls the leachant pH and hence the degradation. Gao et al. (2004) classified these factors according to the release kinetics into internal and external factors.[37] The internal factors include the composition and the thermal history of the glass, while the external factors include the medium pH, temperature, and concentration of PO$_4^{3-}$ and Ca^{2+}.

Bunker et al. (1984) stated that three types of reactions could be accounted for by the degradation mechanisms of phosphate glasses: acid–base, hydrolysis, and hydration reactions.[35] The acid–base reaction is responsible for the disruption of the ionic interaction between the glass chains, while hydrolysis involves the cleavage of P—O—P bonds and is highly dependent on pH. The other mechanism includes simple hydration of the entire phosphate chains, and subsequent separation of these chains into the surrounding medium.

7.6 Biomedical applications of phosphate-based glasses

Phosphate-based glasses in which the cations, which can be one or more of any electropositive element, may comprise up to about 0.65 mol of the glass, are generally referred to as controlled release glasses. This class of glasses is completely soluble in water and leaves no solid residue. Their degradation is an erosion-controlled process that follows zero-order release over the life of the material.[30] They can be produced in different forms such as powder, granules, fibre, cloth, tubes, and cast blocks of various shapes.[38] These glasses have been under development in the Standard Telecommunication Laboratories since the early 1970s, initially as a controlled source of inorganic ions.[39]

Polyphosphate glass provides a source of phosphate that support the growth of recombinant *Escherichia coli* to a density 40% higher than that obtained with typical fermentation media. The high solubility of polyphosphate together with the absence of precipitate formation when mixed with the fermentation media supports its use for this purpose.[40]

Soluble glasses containing trace elements such as copper, cobalt and selenium were manufactured under the trade name of Cosecure®. They were designed for oral administration in the form of a rumen bolus to ruminant animals for the treatment of trace element deficiencies.[41–43]

Copper-releasing phosphate glasses were also used as molluscicides to control the snail hosts of schistosomiasis. The glass composition and the physical form can be changed in a reproducible manner to suit the chemistry of the water body being treated. Moreover, most of the released Cu is in non-toxic or weakly toxic form as copper polyphosphate complex that acts as secondary releasing complexes.[44]

Silver-releasing phosphate glasses were used clinically to control long-term infection in indwelling catheters. A cartridge with silver-containing glass was inserted in-line between the catheter and the urine collection bag. When it was in position, this insert was bathed with urine as it flowed from the bladder into the collection bag. The silver ions released inhibit bacterial proliferation.[38] Also they can be potentially used for treatment of vesico-ureteral flux and urinary incontinence.[45]

7.7 *In vitro* studies on phosphate-based glasses

It has been shown that phosphate-based glasses have potential use as hard tissue substitutes or antimicrobial delivery vehicles.

As hard tissue substitutes, they possess similarity to the inorganic component of bone. In such cases, these glasses could have properties that make them a suitable candidate for synthetic orthopaedic graft materials.[2] Furthermore, incorporation of fluoride ions, which play an active role in stabilising the apatite, was also developed.[2] Moreover, they have a unique property of degradation that can be altered to vary from a few hours to several weeks according to the end applications. There is increasing interest in degradable scaffolds to replace lost or damaged tissues; theoretically, they would eventually be replaced by the natural tissue while they are degrading. They can also be doped with a variety of metal oxides to modify their physical properties,[46] or to induce a specific function.[31] Furthermore, the ionic environment, caused by leaching of ions from these glasses during their degradation, has an impact on the biological response of cells. For example, Ca^{2+} ions have been implicated in stimulating osteoblast-like cell proliferation and differentiation and phosphate ions act as an extracellular 'pool' responsible for the release of Cbfa-1, an important bone marker, from bone cells.[47]

As antimicrobial delivery vehicles, phosphate-based glasses offer potential alternatives to the current methods available for the treatment of infections where they can be used as a localised antibacterial delivery system via the inclusion of ions known for their antibacterial effects such as copper and silver. Such materials could therefore be placed at a site of infection, with the aim of releasing antibacterial ions as the glass degrades, which may be useful in wound-healing applications. Such applications could also function in the prevention of implant- or biomaterial-related infections which are one of the main causes of revision surgery, or even in negating the need for systemically administered antibiotics, a current prophylaxis.[48]

For both applications, a number of glass systems, which may be binary, ternary or quaternary, have been developed by incorporation of various metal oxides such as Fe_2O_3, CuO, Al_2O_3 and TiO_2 into the parent glass. Comprehensive studies have been carried out to give an overview of the correlation between the basic glass structure and how it affects the glass bulk and surface properties and hence its biocompatibility.

7.7.1 Binary systems

Binary sodium phosphate glasses (Na_2PO_4H–$NaPO_4H_2$), developed by Gough et al. (2002, 2003), demonstrated a minimal level of macrophage activation evident from low amounts of peroxide and interleukin-1β release.[49,50] Moreover, early primary craniofacial osteoblast attachment and spreading was also obtained on those glasses. Upon long-term culture up to 28 days, the craniofacial osteoblasts exhibited cytoskeletal characteristics and a level of collagen synthesis similar to those of the positive control. The biocompatibility of these glasses was related to their degradation and the ions being released, as it was difficult for cells to attach and to form a physical anchorage to highly degrading glasses as they had a labile surface.

7.7.2 Ternary system(s)

Uo et al. (1998) developed ternary glasses based on the P_2O_5–CaO–Na_2O system; the cytocompatibility of this system was assessed by a direct contact cytotoxicity assay using dental pulp cells.[36] The results showed that the samples containing 0.50 mol P_2O_5 showed low cytotoxicity that decreased with increasing P_2O_5 content due to a change in pH from neutral at 0.50 mol to acidic at 0.60 mol or more. Therefore, they related the cytotoxicity to the glass degradation, and hence the associated pH changes and ion concentration in the media.

Franks et al. (2000) developed a ternary glass system based on the composition of $0.45P_2O_5$–$xCaO$–$(0.55 - x)Na_2O$, where x was between 0.08 and 0.40 mol.[51] Initial work investigated the degradation and ion release of these glasses. The findings suggested that Ca^{2+} and its interaction with the glass network were suggested to be the dominant factors in the glass degradation, and an inverse relationship existed between calcium oxide content and the degradation rate. The biological response of this glass system was analysed by Salih et al. (2000) to assess its potential applications for bone regeneration.[26] Two human osteoblast cell lines, MG63 and HOS (TE85), were incubated in the glass extracts with different concentrations (neat, 1:4, 1:16, 1:64 dilution) for 2 and 5 days. MTT assay was used to study cell growth, and ELISA was used to measure the expression of antigens such as bone sialoprotein, osteonectin and fibronectin which play a vital role in bone metabolism and integrity. The results showed that low-solubility glasses enhanced bone cell growth and antigen expression at all tested dilutions. The highly soluble glasses, however, significantly reduced cell proliferation and downregulated antigen expression especially with neat and 1:4 dilutions at 5 days. The authors suggested that these results were related to ions released from the glass during degradation and the resultant pH changes. They also suggested that with low dissolution rate glasses, greater amounts of Ca^{2+} are released; Ca^{2+} is known to have an essential role in cell activation mechanisms affecting both cell growth and function. However, with highly soluble glasses, a sharp increase in pH associated with high release rates of Na^+ and phosphorus ions $(PO_4)^{2-}$ may have a deleterious effect on cells.

Owing to the higher degradation and unfavourable cellular response associated with high sodium content, i.e. highly degrading glasses, Franks (2000) also developed another ternary system by complete replacement of Na_2O in the above-mentioned system with potassium oxide (K_2O).[2] This new system was based on $0.45P_2O_5-xCaO-(0.55-x)K_2O$ composition, where x was between 0.16 and 0.32 mol. It should be noted that glasses with CaO content outside the above-mentioned range were difficult to prepare due to crystallisation on casting. It was observed that the $P_2O_5-CaO-K_2O$ system dissolved faster than the $P_2O_5-CaO-Na_2O$ system. Therefore, the authors did not conduct biocompatibility studies on this glass system.

Bitar et al. (2004) examined the short-term response of two typical cellular components of a hard/soft tissue interface such as the periodontal ligament/mandible and patellar tendon/tibia.[52] Human oral osteoblasts, oral fibroblasts and hand flexor tendon fibroblasts were co-cultured on phosphate glasses with different degradation rates ($0.50P_2O_5-xCaO-(0.50-x)Na_2O$ where x equals 0.30 to 0.48 mol). Quantitative and morphological assessment of cell adhesion and proliferation for all cell types was assessed. Immunolabelling was also used to establish phenotypying of both osteoblasts and fibroblasts. The results showed that glass discs with less than 0.40 mol CaO support little or no cell adhesion and survival. This behaviour was related to the high solubility of the surface layer of these glasses; therefore, it is difficult for cells to attach to a labile surface and to form a physical anchorage as observed by Gough et al.[50] The authors concluded that ternary glass compositions with high CaO (0.46 and 0.48 mol) supported high numbers of adherent and viable cells as indicated by DNA content, and they also maintain cellular function as indicated by phenotypic gene expression up to 7 days.

From these studies, it was clear that the degradation, and hence ions released from degrading glasses and pH of the surrounding environments, have been implicated as factors affecting the biocompatibility of these glasses. Therefore, addition of metal oxides known for their effect on the glass degradation was attempted as dopants into the parent ternary P_2O_5–CaO–Na_2O glasses to produce a new series of quaternary systems available for potential biomedical applications.

7.7.3 Quaternary systems/dopants

Bone-related glasses

Moving from ternary to quaternary glasses, a new system based on $0.45P_2O_5$–$(0.20, 0.24, 0.28$ and $0.32)$CaO–$(0.35 - x)$Na$_2$O–xK$_2$O, where x was between 0.0 and 0.25 mol, was developed by Knowles et al.[27] This system was synthesised by the partial substitution of K_2O for Na_2O to study the effect of substitution of a monovalent ion with another of different ionic radius. The degradation of this system was affected by CaO as well as K_2O content. An anomaly in degradation was observed at high CaO content, where weight gain was observed prior to weight loss. This anomaly was explained by the mixed alkali effect,[53] which was prominent in this quaternary system for two reasons: firstly, the larger ionic radius of the potassium ion (K^+) than that of Na^+ was expected to produce more disruption to the glass network, and to slow down the diffusion of K^+ in and out of the glass; and secondly, K^+ is heavier than Na^+, therefore the total weight loss associated with complete leaching of K^+ is larger than that with Na^+. The MTT assay showed that the K^+ had a positive effect on cell proliferation only at high amounts of K_2O (0.20 mol) regardless of the associated increase in degradation.

Another quaternary system based on $0.45P_2O_5$–$(0.32 - x)$CaO–$0.23Na_2O$–xMgO, where x was between 0 and 0.22 mol, was developed by Franks et al.[54] This system was formed by the partial substitution of Ca^{2+} in the ternary glass with magnesium ion (Mg^{2+}) that has the same valence but a different ionic radius. This study looked at the overall degradation characteristics and the effect of released ions on cell proliferation. The results showed that the trend of the degradation process lost its exponential nature and became more linear with time with decreasing CaO content. This emphasised the influential role of CaO on the degradation process. Generally, the degradation rate was decreased by substitution of CaO with MgO, although Mg^{2+} has the same valence as Ca^{2+}, thus the ionic radius has a role in the degradation process. The MTT assay was used to assess the effect of glass extracts at different dilutions as mentioned previously on proliferation of human osteoblast cell line (MG63) for 2 and 5 days. The results were normalised to the control cells incubated in normal medium. The result showed that glasses with little or no MgO showed a slight decrease in cell proliferation only after 2 days; however, after 5 days all tested glass compositions showed equal or greater cell proliferation than control cells.

Quaternary glasses based on zinc oxide as a dopant were also attempted by Salih et al. (2007) to promote osteoblast cell adhesion for potential application in bone tissue engineering.[55] The compositions investigated were $0.50P_2O_5$–$(0.40 - x)$CaO–$0.10Na_2O$–xZnO where x is between 0 and 0.20 mol. Attachment of osteoblast-like cells was assessed morphologically by scanning electron microscopy, and the effect of the glass extract (neat and 10% diluted) on cell proliferation up to 7 days was determined by Cyquant® assay. The results showed that after 24 hours of culture, the cells attached to all glass compositions, but they still maintained a rounded morphology, suggesting lack of spreading on the glass surfaces. Moreover, cell proliferation increased with increasing ZnO content up to 0.05 mol, but never reached those levels exhibited by cells grown on the positive control.

Macroporous resorbable phosphate glass constructs (Na_2O–CaO–P_2O_5–TiO_2) formed by foaming the glass particle slurry with H_2O_2 followed by subsequent sintering was also developed by Navarro et al.[56] These constructs also showed no cytotoxic effect on osteoblast-like cells with both direct and indirect contact methods.

Bulk glasses incorporating Ti^{4+} ions in the form of TiO_2 as dopant were also developed by Abou Neel et al. (2007) using the conventional melt quenching process.[33,57] The authors suggested that incorporation of Ti^{4+} ions could provide a dual action with Ca^{2+} ions to further improve the biological response of these glasses. The addition of TiO_2 could also boost the possibility of using a sol-gel route to prepare glasses with bioactive molecules known for their potent modulating effects. This glass system was based on $0.50P_2O_5$–$(0.20 - 0.15)Na_2O$–0.30CaO–$(0 - 0.05$ mol$)TiO_2$ composition. A more detailed study of cell proliferation, gene expression and bioactivity of MG63 cells on these glasses was also considered. Cell proliferation and gene expression (core binding protein factor alpha 1 (Cbfa1), alkaline phosphatase (ALP), collagen type I alpha subunit I (COLIAI) and osteonectin (Sparc)) were reproducibly enhanced on the TiO_2-containing glass surfaces, particularly those with 0.03 and 0.05 mol. The authors suggested that this enhancement may be associated with the low degradation of these two compositions with the maintenance of appropriate pH favoured by osteoblasts or may be related to the release of Ti^{4+} ions from them. Therefore, among the different glasses examined in the present study, the 0.05 mol TiO_2 glass induced the most favourable cellular response. Consequently, 0.05 mol TiO_2 glass represents the optimal TiO_2 content.

Complementary to the previous study, work investigating the effect of doping the optimal composition of titanium phosphate-based glasses (0.50P_2O_5–0.30CaO–0.15Na_2O–0.05TiO_2) with zinc oxide has also been carried out by Abou Neel et al.[58] The authors suggested that ZnO may act as adjunct to TiO_2 in enhancing the biological properties of these glasses. This study considered the effect of ZnO incorporation on the thermal properties, degradation, ion release, surface and biological properties. The results showed that incorporation of ZnO at 0.01, 0.03 and 0.05 mol at the expense of CaO was effective in controlling the bulk and surface properties of the glass. Glasses incorporating both modifying oxides maintained high viability of MG63 cells up to 7 days in a comparable manner to both parent glass with 0.05 mol TiO_2 and the Thermanox® positive control. This can be correlated to the release of beneficial ions such as Ca^{2+}, P, Ti^{4+} and a suitable level of Zn^{2+} coupled with the increase in hydrophilicity, which is thought to be associated with enhanced protein adsorption and adhesion of anchorage-dependent cells such as osteoblast, fibroblast and endothelial cells on the surface of biomaterials.[59]

Other quaternary glasses incorporating Ca, Co, Zn and Fe were developed by Kesisoglou et al. (2002) and in vitro immune response to these glass extracts was studied by looking at both direct and indirect effects on T lymphocytes.[60] Although all these extracts were capable of modulating an inflammatory response, they were unable to activate the immunological response directly.

Antimicrobial glasses

Antimicrobial glass systems incorporating either Cu^{2+} or Ag^+ ions were successfully developed by Mulligan et al. (2003) for potential application in treatment of oral infections.[61,62] These glasses could be placed at the site of infection, for example, in periodontal pockets to treat the infection with ions being released as the glass degrades. They had common features of containing a P_2O_5 fixed at 0.45 mol, and the two antibacterial ions, Cu^{2+} or Ag^+, fixed at 0, 0.01, 0.05, 0.10 and 0.15 mol. For each glass system, the calcium to sodium oxide (Ca/Na_2O) ratio was varied to normalise the degradation over compositions to give nominally the same release profile for the starting materials. Consequently, the overall effect on bacteria would be reflected by the presence or absence of antibacterial ions and their amounts. The effect of both glass systems on the viability of a *Streptococcus sanguis* biofilm* using constant depth film fermenter (CDFF) was evaluated under a simulated oral environment and compared to antimicrobial ion-

*Biofilm (dental plaque) can be described as a diverse microbial community attached to the tooth surface and encased in a matrix comprising bacterial and host products.[61,62]

free glass and HA discs as controls. It was observed that by 24 hours, there was a significant reduction in viable counts compared to the controls, which was attributed to the release of antimicrobial ions, and the reduction was higher for glasses containing higher levels of antimicrobial ions. However, a recovery of the cell count was observed at 48 hours but was still significantly lower than that of the controls and remained relatively constant between 48 hours and 8 days. The recovery was attributed to two reasons: firstly, the formation of a sacrificial layer formed by the dead bacterial cells that acted as a barrier against further penetration of antimicrobial ions into the biofilm; and secondly, the differentiation of bacteria into another phenotype that was resistant. The results also showed that Ag^+ ions are more potent antimicrobial ions than Cu^{2+}.

Further work on antimicrobial phosphate glasses was also carried out by Ahmed et al. (2006) who developed another silver-containing phosphate glass system with $0.50P_2O_5–0.30CaO–xAg_2O–(0.20 − x)Na_2O$ where x is 0–0.15.[63] A disc diffusion assay was used to screen the antibacterial activity of these glasses against various microorganisms including *Staphylococcus aureus, Escherichia coli, Bacillus cereus, Pseudomonas aeruginosa,* methicilline resistant *Staphylococcus aureus* (MRSA) and *Candida albicans*. The results showed that phosphate glasses and in particular those containing 0.03 and 0.05 mol Ag_2O were more effective than the remaining compositions in inhibiting bacterial growth. Therefore, a further study was carried out on these two compositions for their effects on the growth and viability of *Staphylococcus aureus, Escherichia coli* and *Candida albicans*. It was observed that the 0.03 mol Ag_2O composition was the optimal composition to mount a potent antibacterial effect against the test microorganisms. It was bactericidal for *Staphylococcus aureus* and *Escherichia coli* and significantly reduced the growth of *Candida albicans*. These findings were correlated with the excellent long-term release of Ag ions from this composition into the surrounding medium.

Another study, carried out by Valappil et al. (2007), studied silver-containing phosphate glasses (0.10, 0.15 and 0.20 mol) as a step forward to test their effect on the formation of the highly resistant *S. aureus* biofilms.[64] Ag ions were found to be an effective bactericidal agent against *S. aureus* biofilms, and the rate of their release was 0.42–1.22 µg/(mm^2 h) depending on the glass composition, and this accounted for the variation in their bactericidal effect. Due to its cytotoxic effect, a dead layer, approximately 20 µm in thickness, of non-viable bacterial cells was found attached to the glass surface and covered by an upper layer of viable cells as evidenced by confocal microscopy. This dead layer was caused by the presence of +1 oxidation state Ag, which is known to be a highly effective bactericidal agent compared to other oxidation states (+2 or +3) as confirmed by ^{31}P NMR and HEXRD.

Not only silver and copper but also gallium (Ga) was considered as a dopant for phosphate glasses for its antibacterial effect.[65] Novel quaternary gallium-doped phosphate-based glasses ($0.45P_2O_5$–$0.16CaO$–$(0.39-x)Na_2O$ –xGa_2O_3, where x was 0.01, 0.03 and 0.05 mol) were synthesised, and their bactericidal activities were tested against both Gram negative (*Escherichia coli* and *Pseudomonas aeruginosa*) and Gram positive (*Staphylococcus aureus*, methicillin-resistant *Staphylococcus aureus*, and *Clostridium difficile*) bacteria. The results confirmed that the controlled delivery of Ga from 0.01 mol was sufficient to mount a potent bactericidal effect, which makes these glasses potentially promising new therapeutics for pathogenic bacteria including the superbugs MRSA and *C. difficile*.

7.8 *In vivo* studies on phosphate-based glasses

Moving a step forward from *in vitro* to *in vivo* biocompatibility studies by filling a 3 mm defect in rat calvarium was performed by Abou Neel *et al.* (2007) to test the biocompatibility of TiO_2-containing phosphate glasses (0.01, 0.03 and 0.05 mol).[57] This study was conducted up to 5 weeks when an osteoid tissue, i.e. early bone formation, was expected to occur. It was observed that after 1 week of implantation, only soft tissue can be detected at the bone particles and in between the particles. This was expected, since the beginning of new bone formation normally occurs after 2 weeks postoperatively and is also probably associated with surgical trauma. However, after 5 weeks of implantation, the newly formed bone tissue was only seen at the bone–particle interface for the 0.05 mol TiO_2 doped glasses as shown in Fig. 7.4. This indicated that the new bone formation was dependent on the glass composition, and doping phosphate glasses with 0.05 mol TiO_2 enhanced the bone healing capacity. This finding further confirmed the highly significant gene up-regulation obtained for cells grown on 0.05 mol TiO_2 glass surfaces.

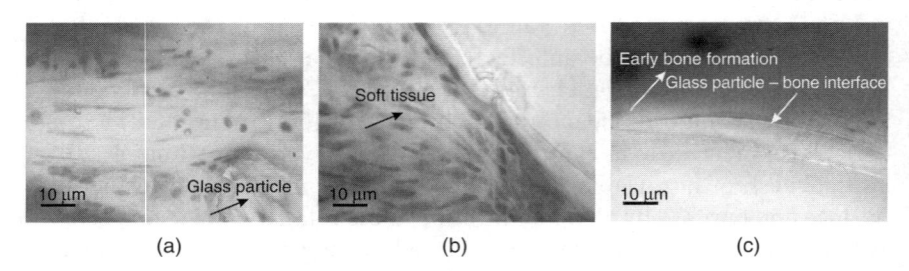

7.4 Histology images of glass particles doped with (a) 0.01, (b) 0.03, and (c) 0.05 mol TiO_2 after implantation in rats calvarium bone defect for 5 weeks at ×20 magnification (stain: Cole's haematoxylin eosin).[57]

7.9 Phosphate-based glasses in composites

Phosphate-based glasses have been used as reinforcing agents for many degradable polymers such as tyrosine based polymers,[66] acrylic acid based drug delivery systems,[67] polycaprolactone (PCL)[68,69] and polylactic acid (PLA)[56] for potential biomedical applications in hard tissue repair and reconstruction. The rationale for using such a combination is to produce a completely degradable composite with improved mechanical properties, and to improve the biological response of the material by incorporation of these bioactive glasses that release ions which are favourable for cell attachment and proliferation. By developing a totally degradable composite, the problems associated with conventional metallic implants of stress shielding and the need for secondary surgical intervention for implant removal against long-term foreign body reaction could then be eliminated.

7.10 Phosphate-based glass fibres

Fibres have potential applications in the engineering of soft tissue such as muscle and ligament due to their chemistry and the fibre morphology which could mimic the fibrous form of these tissues.[13] It has been suggested that glass fibres can orient the muscle cells along their long axis to grow and form myotubes, particularly the three-dimensional mesh arrangements that provide the best configuration for supporting cell attachment and proliferation.[70,71] Generally, glass fibrous meshes with open mesh morphology allow for diffusion of nutrient and waste products in and out of the scaffolds, and would allow for ingrowth of vasculature and hence the tissue. They would also provide the necessary structural support without compromising the porosity.[72–74] Recently, it was also suggested that the fibres could also act as a nerve conduit, since they act as contact guidance for cells to orient, proliferate and grow.[75,76]

Phosphate glasses within certain compositional ranges, i.e. those with more than 0.045 mol P_2O_5, can be easily drawn into fibres due to their polymeric nature. The fibre drawing ability or spinnability is related to the ability of the longer chains to entangle with other chains. Entanglements of these chains allow for continuous filaments to be formed instead of clusters or droplets. Milberg and Daly (1963) assumed that the metaphosphate chains are composed of long-chain molecules with the chain axes having a preference to be parallel to the long axis of the fibre, and that all chains have a random orientation around their respective axes.[77] In perfectly oriented fibres, all chain axes are parallel to the fibre axis, and the rotational disorder of the chains around their axes corresponds to the cylindrical symmetry of the fibre. Other authors have suggested that the drawing operation preferentially selects the strong bonds to be pulled because they are able to withstand the pulling stress. The continuity of fibres is related to the ability of these bonds to be aligned along the long axis of the fibre, whereas weak bonds can be extended for only a short distance along the strong bonds and are not able to form continuous fibres.[78]

It has been reported that glass fibres have higher strength properties and higher reactivity than bulk glass. Choueka *et al.* (1995) attributed this improvement in mechanical properties to inherent stresses developed during fibre production and rapid cooling, which forces some bonds to be frozen at angles and lengths in a non-equilibrium position and some other bonds to be broken.[66] However, De Diego *et al.* (2000) revealed the high strength properties of fibres to be due to the enhancement of the chemical bonding at the surface, and to the decreased possibility of phase separation during fibre production.[73]

High heat was the conventional route used for fibre drawing where fragments of the starting glass are remelted, and fibres are drawn onto a rotating collection drum.[66,73] Adjustment of the melt temperature is necessary for obtaining a suitable viscosity for fibre drawing, as it was not feasible for glass with low melt viscosity to be drawn into fibres.[70] Additionally, this temperature should be above the glass crystallisation temperature, otherwise the fibre drawing will be difficult,[79] or the bioactivity may be reduced.[80]

It has been reported that Fe_2O_3 doped phosphate glass fibres (PGF) have an intriguing ability to form capillary-like channels which can be simply and easily produced during their degradation as shown in Fig. 7.5. The degrada-

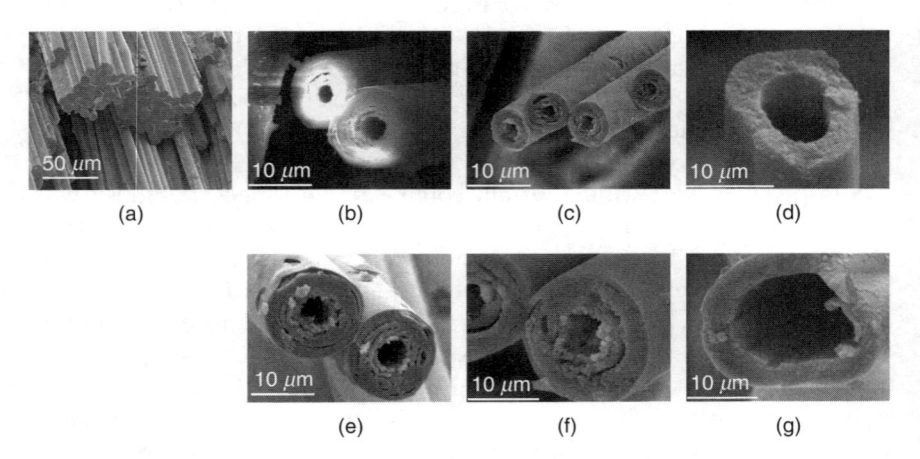

7.5 Scanning electron microscopy of glass fibres with diameter of 32 ± 7 µm: (a) as prepared, (b and e) after being degraded in deionised water at 37°C for 6 months, (c and d) and (f and g) after 18 months for 0.03 and 0.05 mol Fe_2O_3 respectively, under low and high magnification.[81]

tion mechanism of PGF is a combination of surface hydration and internal hydrolysis reaction. Initially, the hydration of the outer layer of PGF occurred, which did not leach into the surrounding solution, thus acting as a barrier. In the long term, bulk degradation took place by hydrolysis of the long P—O—P, Q^2 species of the parent fibres, into short P—O—P chains, Q^1 and Q^0 units that prevailed in the structure of these channels, as evidenced from both FTIR and Raman spectroscopy.[81]

The degradation mechanism proposed in this study has also been supported by the fact that the overall fibre diameter remained essentially unchanged. The preferential leaching out of the core fibres, moreover, was suggested to be associated with the formation of an outer skin layer with different chemical (degradation) properties than that of the core. This is believed to be associated with the variation in both tensional force distribution and cooling rate created across the fibres during their fabrication. The inhomogeneity in the produced fibres was previously manifested by an anisotropic X-ray scattering,[77] and variation in density and the average atomic mass across the fibre.[82]

The ability of microtube formation through the degradation of these glass fibres could potentially be applied in a number of areas including drug delivery and cell transportation, e.g. to act as a conduit during nerve healing by transporting nerve cells. Moreover, they can be used in combination with either natural or synthetic polymers to help the in-growth of vascularisation and the diffusion of nutrient and waste through three-dimensional scaffolds for soft and hard tissue engineering, e.g. muscle, ligament, tendon and bone. It is also possible for the same construct to have fibres with different degradation rates, so that the rapidly degrading fibres could provide *in situ* channels for the rapid growth of blood vessels, and the relatively slower degrading fibres could allow for the alignment of cells to form the target tissue.

7.10.1 *In vitro* studies on phosphate-based glass fibres

Ternary glass systems

Iron-containing phosphate glass fibres ($(0.56 - 0.73)P_2O_5$–CaO–$(0.05 - 0.225\,\text{wt})Fe_2O_3$) were used as reinforcing agents for the development of bioabsorbable composites for potential orthopaedic applications. A cortical plug method was used to test the biocompatibility of these glasses; the results showed that no inflammation was observed up to 5 weeks. Also, incorporation of iron oxide into the fibres enhanced their tensile strength.[83] Glass fibres with $0.629P_2O_5$–$0.219Al_2O_3$–$0.152ZnO$ supported proliferation and differentiation of human masseter muscle derived cells.[71]

Quaternary glass systems

Glass fibres with $0.50P_2O_5$–$0.30CaO$–$(0.20 - x)Na_2O$–xFe_2O_3, where x was between 0.01 and 0.05 mol, were produced. A dramatic improvement in an immortal muscle precursor cell line attachment with the highest cell density was observed for fibres with 0.04 and 0.05 mol Fe_2O_3.[70] Also, it was observed that glass fibres with 0.03 mol Fe_2O_3 demonstrated significant biocompatibility with both primary human osteoblasts and fibroblasts, supporting a clear proliferation pattern and permitting a well-spread morphology.[52]

Studies incorporating unidirectionally aligned 0.03 mol Fe_2O_3 containing PGF for *in situ* development of channels in 3-D dense collagen scaffolds, and the assessment of diffusion throughout this construct by ultrasound imaging and SEM, has also been attempted.[84] The free movement of coated micro-bubble agents, acting as a contrast agent, confirmed that the channels were continuous in nature and of 30–40 µm diameter (approximately the same as the fibre diameter) as shown in Fig. 7.6. Moreover, this construct maintained excellent viability of human oral fibroblasts after 24 hours in culture, and the cells showed a tightly packed spindle-shaped appearance forming a three-dimensional network and spreading over the collagen matrix and the glass fibres with no preference for either phase.

PGF with varying amounts of copper oxide (CuO) were developed for potential use in wound-healing applications. PGF with 0, 0.01, 0.05 and 0.10 mol CuO were produced with different diameters and characterised in terms of structural and antibacterial properties. The effect of two fibre diameters on short-term (3 hours) attachment and killing against *Staphylococcus epidermidis* was investigated and related to their rate of

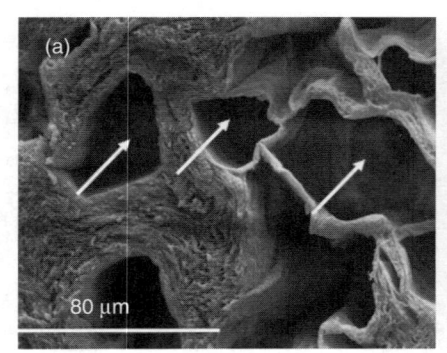

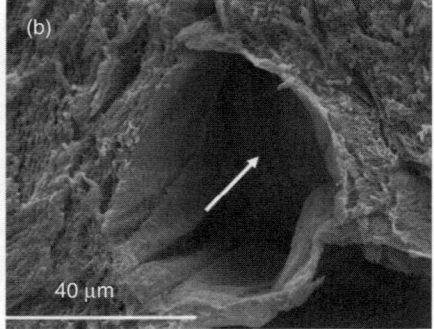

7.6 (a) SEM of a cross-section through the PGF-collagen scaffold showing a cluster of channels left in the matrix as the fibres degraded; (b) close-up of a channel.[84]

degradation in deionised water, as well as copper ion release measured using ion chromatography. The results showed that there was a significant decrease in the rate of degradation with increasing CuO content and with increase in the fibre diameter. Over 6 hours, both the amount and rate of copper ions released increased with CuO content, as well as a reduction in fibre diameter, thus increasing the surface area to volume ratio. There was a decrease in the number of viable *staphylococci* both attached to the CuO-containing fibres and in the surrounding environment.[31]

7.11 Future trends using phosphate-based glasses

- Zinc phosphate glasses could be potentially applied for the treatment of chronic inflammatory diseases such as Crohn's disease and rheumatoid arthritis which are characterised by decreased Zn^{2+} levels in the blood.
- Phosphate-based glass fibres could be used as a vehicle for cell delivery in inaccessible areas such as the periodontal ligament for the treatment of advanced periodontitis.
- Phosphate-based glass fibres with antimicrobial properties could be prepared in mesh form to be potentially used as a wound dressing for the treatment of severe burns, leg ulcers, pressure sores and infected surgical wounds, providing protection against ingress of microorganisms. In addition, when they dissolve they release antibacterial ions which could help to combat infection. The highly degradable nature of these fibres is beneficial for the release of the antimicrobial agents, and these meshes would be used on a temporary basis. They could also be incorporated into calcium phosphate bone cements used for the treatment of orthopaedic problems such as vertebral defects. The antibacterial ions released from the bone cement into the surrounding tissue around the replacement device could help to reduce the incidence of post-operative infection.
- The ability of microtube formation through the degradation of these glass fibres could potentially be applied in a number of areas including drug delivery and cell transportation, e.g. to act as a conduit during nerve healing by transporting nerve cells. Moreover, they can be used in combination with either natural or synthetic polymers to help the ingrowth of vascularisation and the diffusion of nutrient and waste through three-dimensional scaffolds for soft and hard tissue engineering, e.g., muscle, ligament, tendon and bone. It is also possible for the same construct to have fibres with different degradation rates, so that the rapidly degrading fibres could provide *in situ* channels for the rapid growth of blood vessels, and the relatively slower degrading fibres could allow for the alignment of cells to form the proposed tissue.

7.12 References

1. Varshneya A K (1994), *Fundamentals of Inorganic Glasses*, Elsevier Science, New York, p. 1.
2. Franks K (2000), 'The structure and properties of soluble phosphate based glasses', PhD thesis, University of London.
3. Van Wazer J R, Holst K A (1950), 'Structure and properties of the condensed phosphates. I. Some general considerations about phosphoric acids', *Journal of the American Chemical Society* 72, 639–643.
4. Van Wazer J R (1950), 'Structure and properties of the condensed phosphates. II. A theory of the molecular structure of sodium phosphate glasses', *Journal of the American Chemical Society* 72, 644–646.
5. Schneider J, Oliveira S L, Nunes L A O, Bonk F, Panepucci H (2005), 'Short-range structure and cation bonding in calcium–aluminium metaphosphate glasses', *Inorg Chem* 44, 423–430.
6. Ray C A, Fang X, Karabulut G K, Day D E (1999), 'Effect of melting temperature and time on iron valence and crystallisation of iron phosphate glasses', *Journal of Non-Crystalline Solids* 249, 1–16.
7. Marasinghe G K, Karabulut M, Ray C S, Day D E, Shuh D K, Allen P G, Saboungi M L, Grimsditch M, Haeffner D (2000), 'Properties and structure of vitrified iron phosphate nuclear wasteforms', *Journal of Non-Crystalline Solids* 263–264, 146–154.
8. Mesko M G, Day D E (1999), 'Immobilization of spent nuclear fuel in iron phosphate glass', *Journal of Nuclear Materials* 273, 27–36.
9. Fang X, Ray C S, Marasinghe G K, Day D E (2000), 'Properties of mixed Na_2O and K_2O iron phosphate glasses', *Journal of Non-Crystalline Solids* 263–264, 292–298.
10. Goldsmith V M (1926), 'Investigation concerning structure and properties of crystals', *Skr Nor Vidensk Akad KI-1: Mat Naturvidensk* K1 8, 130.
11. Zachariasen W H (1932), 'The atomic arrangement in glass', *American Chemical Society* 54(3), 3841–3851.
12. Hagg G (1935), 'The vitreous state', *J Chem Phys* 3, 42–45.
13. Knowles J C (2003), 'Phosphate based glasses for biomedical applications', *J Mater Chem* 13, 2395–2401.
14. Greaves G N, Smith W, Giulotto E, Pantos E (1997), 'Local structure, microstructure and glass properties', *Journal of Non-Crystalline Solids* 222, 13–24.
15. Lockyer M W C, Holland D, Howes A P, Dupree R (1995), 'Magnetic-angle spinning nuclear magnetic resonance study of the structure of some $PbO–Al_2O_3–P_2O_5$ glasses', *Solid State Nuclear Resonance* 5, 23–34.
16. Sales B C, Boatner L A, Ramey J O (2000), 'Chromatographic studies of the structures of amorphous phosphates: a review', *Journal of Non-Crystalline Solids* 263–264, 155–166.
17. Khattak G D, Mekki A, Wenger L E (2004), 'Local structure and redox state of copper in tellurite glasses', *Journal of Non-Crystalline Solids* 337, 174–181.
18. Kirkpatrick R J, Brow R K (1995), 'Nuclear magnetic resonance investigation of the structures of phosphate and phosphate-containing glasses: a review', *Solid State Nuclear Magnetic Resonance* 5, 9–21.
19. Brow R K (2000), 'Review: the structure of simple phosphate glasses', *Journal of Non-Crystalline Solids* 263–264, 1–28.

20. Walter G, Vogel J, Hoppe U, Hartmann P (2001), 'The structure of CaO–Na$_2$O–MgO–P$_2$O$_5$ invert glass', *Journal of Non-Crystalline Solids* 296, 212–223.
21. Hoppe U (1996), 'A structural model for phosphate glasses', *Journal of Non-Crystalline Solids* 195, 138–147.
22. Walter G, Hoppe U, Baade T, Kranold R, Stachel D (1997), 'Intermediate range order in MeO–P$_2$O$_5$ glasses', *Journal of Non-Crystalline Solids* 217, 299–307.
23. Hoppe U, Walter G, Kranold R, Stachel D (2000), 'Structural specifics of phosphate glasses probed by diffraction methods: a review', *Journal of Non-Crystalline Solids* 263–264, 29–47.
24. Shih P Y, Yung S W, Chin T S (1998), 'Thermal and corrosion behaviour of P$_2$O$_5$–Na$_2$O–CuO glasses', *Journal of Non-Crystalline Solids* 224, 143–152.
25. Shih P Y, Yung S W, Chin T S (1999), 'FTIR and XPS studies of P$_2$O$_5$–Na$_2$O–CuO glasses', *Journal of Non-Crystalline Solids* 244, 211–222.
26. Salih V, Franks K, James M, Hastings G W, Knowles J C (2000), 'Development of soluble glasses for biomedical use. Part 2: The biological response of human osteoblast cell lines to phosphate-based soluble glasses', *Journal of Material Science: Materials in Medicine* 11, 615–620.
27. Knowles J C, Franks K, Abrahams I (2001), 'Investigation of the solubility and ion release in the glass system K$_2$O–Na$_2$O–CaO–P$_2$O$_5$', *Biomaterials* 22, 3091–3096.
28. Ahmed I, Lewis M, Olsen I, Knowles J C (2004), 'Phosphate glasses for tissue engineering: Part 1. Processing and characterisation of a ternary-based P$_2$O$_5$–CaO–Na$_2$O glass system', *Biomaterials* 25, 491–499.
29. Parsons A J, Evans M, Rudd C D, Scotchford C A (2004), 'Synthesis and degradation of sodium iron phosphate glasses and their *in vitro* cell response', *J Biomed Mater Res* 71A, 283–291.
30. Gao H, Tan T, Wang D (2004), 'Effect of composition on the release kinetics of phosphate controlled release glasses in aqueous medium', *Journal of Controlled Release* 96, 21–28.
31. Abou Neel E A, Ahmed I, Pratten J, Nazhat S N, Knowles J C (2005), 'Characterisation of antimicrobial copper releasing degradable phosphate glass fibres', *Biomaterials* 26, 2247–2254.
32. Abou Neel E A, Ahmed I, Blaker J J, Bismarck A, Boccaccini A R, Lewis M P, Nazhat S N, Knowles J C (2005), 'Effect of iron on the surface, degradation and ion release properties of phosphate-based glass fibres', *Acta Biomaterialia* 1, 553–563.
33. Abou Neel E A, Knowles J C (2007), 'Physical and biocompatibility studies of novel titanium dioxide doped phosphate-based glasses for bone tissue engineering applications', *Journal of Materials Science: Materials in Medicine* DOI 10.1007/s10856-007-3079-5.
34. Delahaye F, Montagne L, Palavit G, Touray J C, Baillif P (1998), 'Acid dissolution of sodium-calcium metaphosphate glasses', *Journal of Non-Crystalline Solids* 242, 25–32.
35. Bunker B C, Arnold G W, Wilder J A (1984), 'Phosphate glass dissolution in aqueous solutions', *Journal of Non-Crystalline Solids* 64, 291–316.
36. Uo M, Mizuno M, Kuboki Y, Makishima A, Watari F (1998), 'Properties and cytotoxicity of water soluble Na$_2$O–CaO–P$_2$O$_5$ glasses', *Biomaterials* 19, 2277–2284.

37. Gao H, Tan T, Wang D (2004), 'Dissolution mechanism and release kinetics of phosphate controlled release glasses in aqueous medium', *Journal of Controlled Release* 96, 29–36.
38. Gilchrist T, Healy D M, Drake C (1991), 'Controlled silver-releasing polymers and their potential for urinary tract infection control', *Biomaterials* 12, 76–78.
39. Drake C F (1985), 'Continuous and pulsed delivery of bioactive materials using composite system based on inorganic glasses', *Consultation on Immunomodulation*, Bellagio, Italy, 16–18 April.
40. Curless C, Baclaski J, Sachdev R (1996), 'Phosphate glass as a phosphate source in high cell density *Escherichia coli* fermentations', *Biotechnol Prog* 12(1), 22–25.
41. Telfer S B, Illingworth D V, Anderson P J B, Zervas G, Carlos G (1985), 'Effect of soluble-glass boluses on the copper, cobalt and selenium status of sheep', *Biochemical Society Transactions* 13, 529.
42. Allen W M, Drake C F, Tripp M (1984), 'Use of controlled release system for supplementation during trace element deficiency – The administration of boluses of controlled release glass (CRG) to cattle and sheep', TEMA 5, 29 June–4 July 1984, Aberdeen.
43. Allen W M, Sansom B F, Drake C F, Moore P R (1983), 'Recent developments in the treatment of metabolic diseases', in Ruchebusch Y, Toutain P-L, Korltz G D, *Veterinary Pharmacology and Toxicology, Proceedings from the 2nd European Association for Veterinary Pharmacology and Toxicology*, Toulouse, 13–17 September, MTP Press, Boston, MA, pp. 183–191.
44. O'Sullivan T N, Smith J D, Thomas J D, Drake C (1991), 'Copper molluscicides for control of Schistosomiasis. 2. Copper phosphate controlled release glass', *Environ Sci Technol* 25, 1088–1091.
45. Cartmell S H, Dorthy P J, Hunt J A, Healy D M, Gilchrist T (1998), 'Soft tissue response to glycerol-suspended controlled-release glass particulate', *Journal of Material Science: Material in Medicine* 9, 773–777.
46. Parsons A J, Burling L D, Scotchford C A, Walker G S, Rudd C D (2006), 'Properties of sodium-based ternary phosphate glasses produced from readily available phosphate salts', *Journal of Non-Crystalline Solids* 352, 5309–5317.
47. Fujita T, Izumo N, Fukuyama R, Meguro T, Nakamuta H, Kohno T, Koida M (2001), 'Phosphate provides Runx2/Cbfa1 in bone cells', *Biochem Biophys Res Commun* 280, 348–352.
48. Gristina A G (1987), 'Biomaterial centered infections: microbial adhesion versus tissue integration', *Science* 237, 588–1595.
49. Gough J E, Christian P, Scotchford C A, Rudd C D, Jones I A (2002), 'Synthesis, degradation, and *in vitro* cell responses of sodium phosphate glasses for craniofacial bone repair', *J Biomed Mater Res* 59(3), 481–489.
50. Gough J E, Christian P, Scotchford C A, Jones I A (2003), 'Long-term craniofacial osteoblast culture on a sodium phosphate and a calcium/sodium phosphate glass', *J Biomed Mater Res* 66A, 233–240.
51. Franks K, Abrahams I, Knowles J C (2000), 'Development of soluble glasses for biomedical use. Part 1: *In vitro* solubility measurement', *Journal of Material Science: Materials in Medicine* 11, 609–614.
52. Bitar M, Salih V, Mudera V, Knowles J C, Lewis M (2004), 'Soluble phosphate glasses: *In vitro* studies using human cells of hard and soft tissue origin', *Biomaterials* 25, 2283–2292.

53. Chakradhar R P S, Ramesh K P, Rao J L, Ramakrishna J (2003), 'Mixed alkali effect in borate glasses – electron paramagnetic resonance and optical absorption studies in Cu^{2+} doped xNa_2O–$(30-x)K_2O$–$70B_2O_3$ glasses', *J Phys: Condens Matter* 15, 1469–1486.
54. Franks K, Salih V, Knowles J C (2002), 'The effect of MgO on the solubility behaviour and cell proliferation in a quaternary soluble phosphate based glass system', *Journal of Material Science: Materials in Medicine* 13, 549–556.
55. Salih V, Patel A, Knowles J C (2007), 'Zinc containing phosphate-based glasses for tissue engineering', *Biomed Mater* 2, 1–11.
56. Navarro M, Valle S D, Martinez S, Zeppetelli S, Ambrosio L, Planell J A, Ginebra M P (2004), 'New macroporous calcium phosphate glass ceramic for guided bone regeneration', *Biomaterials* 25, 4233–4241.
57. Abou Neel E A, Mizoguchi T, Ito M, Bitar M, Salih V, Knowles J C (2007), '*In vitro* bioactivity and gene expression by cells cultured on titanium dioxide doped phosphate-based glasses', *Biomaterials* 28, 2967–2977.
58. Abou Neel E A, O'Dell L A, Chrzanowski W, Smith M E, Knowles J C, 'Control of surface free energy in titanium doped phosphate based glasses by co-doping with zinc', submitted to *Biomedical Materials Research: Part B*.
59. Webster T J, Ergun C, Doremus R H, Siegel R W, Bizios R (2000), 'Specific proteins mediate enhanced osteoblast adhesion on nanophase ceramics', *J Biomed Mater Res* 51, 475–483.
60. Kesisoglou A, Knowles J C, Olsen I (2002), 'Effect of phosphate-based glasses on T lymphocytes in vitro', *Journal of Material Science: Materials in Medicine* 13, 1189–1192.
61. Mulligan A M, Wilson M, Knowles J C (2003), 'The effect of increasing copper content in phosphate-based glasses on biofilms of *Streptococcus sanguis*', *Biomaterials* 24(10), 1797–1807.
62. Mulligan A M, Wilson M, Knowles J C (2003), 'Effect of increasing silver content in phosphate-based glasses on biofilms of *Streptococcus sanguis*', *J Biomed Mater Res* 67A, 401–412.
63. Ahmed I, Ready D, Wilson M, Knowles J C (2006), 'Antibacterial effect of silver-doped phosphate-based glasses', *J Biomed Mater Res* 78A, 618–625.
64. Valappil S P, Pickup D M, Carroll D L, Hope C K, Pratten J, Newport R J, Smith M E, Wilson M, Knowles J C (2007), 'Effect of silver content on the structure and antibacterial activity of silver-doped phosphate-based glasses', *Antimicrobial Agents and Chemotherapy* 51, 4453–4461.
65. Valappil S P, Ready D, Abou Neel E A, Pickup D M, Chrzanowski W, O'Dell L A, Newport R J, Smith M E, Wilson M, Knowles J C (2008), 'Antimicrobial gallium-doped phosphate-based glasses', *Advanced Functional Materials* 18, 732–741.
66. Choueka J, Charvert J L, Alexander H, Oh Y H, Joseph G, Blumenthal N C, LaCourse W C (1995), 'Effect of annealing temperature on the degradation of reinforcing fibres for absorbable implants', *J Biomed Mater Res* 29, 1309–1315.
67. Fernández M, Méndez J A, Vázques B, San Román J, Ginebra M P, Gil F J, Manero J M, Planell J A (2002), 'Acrylic-phosphate glasses composites as self-curing controlled delivery systems of antibiotics', *Journal of Material Science: Materials in Medicine* 13, 1251–1257.
68. Corden T, Jones I A, Rudd C D, Christian P, Downes S, McDougall K E (2000), 'Physical and biocompatibility properties of poly-epsilon-caprolactone

produced using in situ polymerization; a novel manufacturing technique for long fiber composite materials', *Biomaterials* 21, 713–724.
69. Prabhakar R L, Brocchini S, Knowles J C (2005), 'Effect of glass composition on the degradation properties and ion release characteristics of phosphate glass-polycaprolactone composites', *Biomaterials* 26, 2209–2218.
70. Ahmed I, Collins C A, Lewis M, Olsen I, Knowles J C (2004), 'Processing, characterisation and biocompatibility of iron-phosphate glass fibres for tissue engineering', *Biomaterials* 25, 3223–3232.
71. Shah R, Sinanan A C M, Knowles J C, Hunt N P, Lewis M P (2005), 'Craniofacial muscle engineering using a 3-dimensional glass fibre construct', *Biomaterials* 26, 1497–1505.
72. Mooney D J, Mazzoni C L, Breuer C, McNamara K, Hern D, Vacanti J P, Langer R (1996), 'Stabilized polyglycolic acid fibre-based tubes for tissue engineering', *Biomaterials* 17, 115–124.
73. De Diego M A, Coleman N J, Hench L L (2000), 'Tensile properties of bioactive fibers for tissue engineering applications', *J Biomed Mater Res* 53, 199–203.
74. Domingues R Z, Clark A E, Brennan A B (2001), 'Sol-gel bioactive fibrous mesh', *J Biomed Mater Res* 55, 468–474.
75. Lee S J, Lee J H, Lee H B (1999), 'Interaction of fibroblast cells onto fibers with different diameter', *Korea Polymer Journal* 7(2), 102–107.
76. Hatcher B M, Seegert C A, Brennan A B (2003), 'Polyvinylpyrrolidone modified bioactive glass fibres as tissue constructs: *in vitro* mesenchymal stem cell response', *J Biomed Mater Res* 66A, 840–849.
77. Milberg M E, Daly M C (1963), 'Structure of oriented sodium metaphosphate glass fibres', *Journal of Chemical Physics* 39(11), 2966–2973.
78. Murgatroyd J B (1948), 'The delayed elastic effect in glass fibres and the constitution of glass in fibre form', *Journal of the Society of Glass Technology* 32, 291–300.
79. Clupper D C, Gough J E, Hall M M, Clare A G, La Course W C, Hench L L (2003), '*In vitro* bioactivity of S520 glass fibres and initial assessment of osteoblast attachment', *J Biomed Mater Res* 67A, 285–294.
80. Orifice R L, Hench L L, Clark A E, Brennan A B (2001), 'Novel sol-gel bioactive fibres', *J Biomed Mater Res* 55, 460–467.
81. Abou Neel E A, Young A M, Nazhat S N, Knowles J C (2007), 'A facile synthesis route to prepare microtubes from phosphate glass fibres', *Advanced Materials* 19, 2856–2862.
82. Potter R M, Mattson S M (1991), 'Glass fibre dissolution in a physiological saline solution', *Glastech Ber* 64, 16.
83. Lin S T, Krebs S L, Kadiyala S, Leong K W, LaCourse W C, Kumar B (1994), 'Development of bioabsorbable glass fibres', *Biomaterials* 15(13), 1057–1061.
84. Nazhat S N, Abou Neel E A, Kidane A, Ahmed I, Hope C, Kershaw M, Lee P D, Stride E, Saffari N, Knowles J C, Brown R A (2007), 'Controlled microchannelling in dense collagen scaffolds by soluble phosphate glass fibres', *Biomacromolecules* 8, 543–551.

Part II
Cell responses and regenerative medicine

8
Biodegradable scaffolds for tissue engineering

V SALIH, University College London UK

Abstract: This chapter addresses the concerted research shift in regenerative medicine methods from using solely medical devices and tissue grafts to a more explicit approach. It is an approach that utilises specific biodegradable synthetic or natural scaffolds in order to replace tissue function. The types of scaffold used to enable cells to function in an appropriate manner to produce the required extracellular matrix and ultimately a tissue of a desired geometry, size and composition are considered here.

Key words: regenerative medicine, natural and synthetic scaffolds, three-dimensional scaffolds, cell and tissue engineering.

8.1 Introduction

This chapter will review the use of three-dimensional biomaterial scaffolds as a potential future therapy for cell engineering and *de novo* tissue regeneration. This will include examples currently available to and under development within the biomaterials community. The philosophy of scaffolds in tissue engineering will be described, as well as their potential advantages and their obvious disadvantages. The interactions of cells within scaffolds will be examined as well as the importance of physico-chemical properties of the materials. Furthermore, the maturation and maintenance of a suitable extracellular matrix plays an important role in the potential success of any 3-D scaffold. A summary of current trends and future challenges for the next generation of inventors and developers of scaffolds for tissue engineering is provided along with a useful list of key reference books, societies, professional bodies, interest groups and websites which will act as a practical and valuable database for keeping interested parties informed of developments in 3-D scaffold research and development.

There has been a clear and defined hypothetical shift in regenerative medicine from using solely medical devices and tissue grafts, to a more explicit approach that utilises specific biodegradable synthetic or natural scaffolds combined with cells and/or biological molecules, in order to create

Table 8.1 The shift in research emphasis of materials used for implantation over 50+ years

1950–1975	bio**MATERIALS**
1975–2000	**BIOMATERIALS**
2000–present	**BIO**materials

Source: adapted from Anderson (2006).

a functional replacement tissue in a diseased or damaged tissue site. Every era in medical research over the past 50 years, involving the use of biomaterials in order to replace tissue function, has been distinct and identified by particular materials and developmental successes. For example, in the 1950s, there was a predominant use of metal implants and associated devices. Throughout the 1970s and 1980s, there was a significant increase in the use of polymers and synthetic materials, and more recently there has been a distinct and concentrated effort in the design and use of both natural and degradable scaffolds. This is summarised by the scheme outlined in Table 8.1.

There has been an evolution from the use of biomaterials to simply replace non-functioning tissue to that of utilising specific materials which will nurture, in three dimensions, a fully functioning and structurally acceptable tissue. Thus, the simple need to accomplish the replacement of a functioning joint using a fully metal prosthesis during the pioneering days of Sir John Charnley in the 1960s has been markedly enhanced to concentrate on biological aspects of the damaged or diseased tissue to be replaced by repaired, or better still, totally regenerated tissue. There was a very naive belief that materials were typically 'inert' and Anderson quite rightly suggests that this is a misleading interpretation, as it became clear that materials could indeed change physically and chemically following implantation.[1] Certainly from a biological perspective, no material should be considered inert!

It has become quite apparent, therefore, that the choice of scaffold is crucial to enable the cells in question to function in an appropriate manner to produce the required extracellular matrix, and thus tissue, of a desired geometry and size and normal functional capability.

There are, however, legislative and ethical hurdles which need to be cleared, e.g. consideration and implementation of the Human Tissue Act (2004), in order to ensure the continued acceleration of this exciting field of research. Clearly, the potential health benefits to individuals and health services are vast. Stern challenges lie ahead when one considers that, for all the research groups in many different countries and the numbers of materials being investigated, very few examples of scaffolds are in clinical use at present. Some of the issues that need to be addressed comprise the native tissues and organs and the variety of tissue structures. Do scientists need to consider two-dimensional culture techniques as possibly being obsolete as more and more groups switch to three-dimensional bioreactor culture? Furthermore, the question of critical-sized defects comes into play, i.e. do we consider microscale vs. nanoscale, various cell types, culture systems including perfusion of scaffolds, vascularity, and waste and nutrient consumption biomonitoring? These are important features of any future improvement of metrology within tissue engineering systems and the development of the next generation of scaffolds.

Researchers within the field are striving to improve understanding of cell/material interactions and the ability to control the host response, and to enable the standardisation of functional assays/protocols by limiting types of cells in use and by standardising isolation methods. This can be assessed by using the same markers of biological function and developing a national (and ultimately international) group of reference materials. The adoption of standards is vital as it will lead to innovation, reduced development, processing and manufacturing costs, and more products to market, and will highlight previous intellectual property problems in industry by encompassing validation and accurate measurement modalities. In due course, these considerations need to involve numerous stakeholders, namely the major research councils and government departments (i.e. DTI, DoH) as well as the main standards authorities and regulatory bodies (ASTM, ISO, BSI, NIBSC, MHRA). The long-term impact will help direct and form development of local ideas to be accepted nationally and internationally. These measures will provide 'best practice' methods in order to create safe, high-quality performance-related products. Educating regulatory authorities and end-users about the technical standards can only lead to long-tem benefits to the health sector and, ultimately, to the end-users, i.e. the patients.

8.2 What is a scaffold?

In the simplest terms, a scaffold may be described as a support structure with viable cells coupled with a suitable culture environment to support the development of functional tissue. For the purposes of this chapter, a scaffold for tissue engineering purposes will be considered as a permanently placed or temporary, three-dimensional porous and permeable natural or synthetic biomaterial that is biocompatible to a particular mammalian cell type with respect to its bulk form as well as its degraded constituent form. Moreover, the scaffold should ideally possess appropriate mechanical properties for the developing tissue(s) in question and possess suitable physical and chemical properties to allow cell adhesion, migration, proliferation, and differentiation to the cell's mature phenotype with simultaneous extracellular matrix production and maintenance. Although extensive, these seem reasonable and sensible demands of a material when one considers the need for tissue or organ replacement.

The concept of understanding and developing improved scaffolds is inherent in the need to fully comprehend how we have moved on from the small-scale monolayer principle to the larger three-dimensional-scale effort in growing cell constructs of a suitable size for clinical use. Thus, a suitable and controlled geometric cell scaffold would ultimately produce a viable tissue construct.[2]

Furthermore, the scaffold must be multifunctional. Prior to implantation, during the *in vitro* preparation phase of any cell construct, the scaffold must host the cells of interest and any extracellular matrices or additional growth/differentiation agents.[3] This is just the beginning, as mature constructs must be viable, and to do this they need to allow adequate perfusion of nutrients and waste and, if applicable, they need to allow appropriate vascularisation.[4-6] As well as the choice of adequate material – a decision often made with many assumptions and at the expense of fundamental biological understanding of the tissue requirements – there is an essential need for at least an appreciation of the physiochemical properties of the material. These properties will ultimately affect the basic requirements of cell survival, growth, proliferation and matrix reorganisation, but from a functional point of view they need to preserve appropriate gene and therefore protein function.

8.3 Importance of scaffold design and manufacture

The requirements for suitable scaffolds applied to tissue engineering are numerous. Hutmacher has elegantly summarised the most essential properties that need to be considered for three-dimensional scaffolds.[7] Of vital importance are adequate biocompatibility and vascularisation.[8,9] The scaffold needs to be non-immunogenic[10,11] and, if it is derived from animal tissues, it must be free from any agents involved in potential disease transmission.[7] Furthermore, there are considerations of geometry, sterilisation of products and ease of handling at the point of clinical use. All these factors are determined largely, or in part, by the material properties.

Determining the possibility of novel designer scaffolds being able to enhance the treatment of damaged or diseased tissues necessitates that the scaffolds can be manufactured and tested appropriately for material function well before their suitability for tissue regeneration. However, this latter point is one of the biggest obstacles, and complex three-dimensional designs require advanced manufacturing techniques. Porous scaffolds may be manufactured in a variety of ways. The most commonly used systems include woven or non-woven preparation utilising fibres,[12,13] a variety of sintered polymer particles[14–16] and solvent casting of polymer films.[17–19] Phase separation has also been a relatively successful method which allows for subtle changes in manufacture leading to profound differences in material properties.[20–23]

Porosity is a key factor in any design of scaffolds. It is essential that the pores are patent with an interconnecting geometry allowing for a large surface area to volume ratio which will benefit thorough cell attachment and migration through the material. Clearly, the size of pores and level of interconnectivity depend on the cell and tissue requirements under investigation, and a high porosity is likely to adversely affect mechanical properties – an important consideration particularly in the early phases of engineering bone tissue *in vivo*.[24,25] Pore size must be considered, as numerous studies have clearly shown that certain cell types grow optimally within certain pore sizes; up to 5 μm for neovascularisation, 20 μm for hepatocyes, 50–100 μm for osteoid ingrowth, and pore sizes of considerably greater than 500 μm for adequate survival of transplanted cells.[26,27]

Surface properties will fundamentally play a vital role in cell behaviour:[28] cell adhesion, migration and proliferation in the first instance, followed by maturation and differentiation and ultimately by gene and protein function.[29–33]

Mechanical properties should be sufficient to allow for cell in-growth and the maintenance of the appropriate extracellular matrix and this will subsequently depend on the intended application. For example, numerous groups are attempting to mimic the architecture and mineral phase of bone – probably the most commonly researched skeletal tissue with respect to biomaterial scaffold development. Perhaps unsurprisingly, there are inherent problems and issues when one compares the modulus and strength of the scaffold material with the natural tissue. Quite often, there can be differences of several orders of magnitude in the material physical properties. Moreover, there are often anisotropic material properties of natural tissues to be considered and this can be a veritable challenge when designing scaffolds.

8.4 Types of scaffold material

The choice of material is an important consideration for any scaffold design. The selection of appropriate materials for scaffold design and manufacture will undoubtedly determine the success of the scaffold for its intended application. Furthermore, it must have several desirable characteristics as an absolute minimum before it can be deemed a potential scaffold matrix; these include biocompatibility, appropriate surface and bulk chemistry, certain physical properties and biodegradability.[34]

Once the type of scaffold and its intended use have been established (this depends on the medical issue to be resolved, the type of tissue repair or regeneration to be created), there are currently two main approaches to utilising the scaffold:[35]

- Acellular scaffolds are predominantly aimed at the initiation of repair of native host cell populations *in situ*. Alternatively the scaffolds may consist just of a nude matrix, or of a matrix containing biological growth factors or other mediators that can also lead to the 'recruitment' of stem cells, which may undergo appropriate differentiation into the mature cells needed to develop a competent repair tissue.
- Cell-based approaches may involve the use of adult stem cells, haemopoietic stem cells, or primary differentiated cells recruited from predetermined sites. Quite often, cells are isolated by explantation of the native tissue or via enzymatic treatment, and if immature cells are used, they may either be left to differentiate *in vivo* or be induced to do so *in vitro*, prior to implantation using a cocktail of differentiation factors depending on the type of tissue required.

One of the most common approaches[35] is the combination of cells with a scaffold matrix, or the cells may be applied in isolation.[36] This latter approach, utilising a 'combined' technology, is the most popular of the three alternatives. In these scenarios single or even co-cultured cell populations are contained within a matrix that is functionalised with a biologically relevant molecule. Clearly, the maturation of the cell scaffold can be partially controlled *in vitro* prior to any implantation procedure or, indeed, immediately implanted post-seeding with cell differentiation occurring *in vivo*.

The importance of biomaterial surface structure, chemistry and scaffold configuration plays a vital role at this stage and it is during this period that we see how even minor, quite simple manipulation to a biomaterial can have a profound and marked impact on the surrounding cell function and tissue matrices.[37,38]

If a scaffold containing cells (or 'cell construct') is to be used, the main function then is to support and maintain the conduction of tissue throughout its geometry. Prospective materials with these characteristics have been sought and these include natural polymers, synthetic polymers, ceramics, metals, and various combinations of these. The actual types of material currently being utilised as scaffold materials may be categorised according to their derivation, i.e. synthetically derived materials and materials derived from natural sources (Table 8.2).

Since the 1950s numerous biocompatible materials such as metals, ceramics and polymers have been used for surgical implantations. Both metals and ceramics have contributed to considerable advances in implant medicine, particularly with respect to orthopaedic tissue replacements. Typical

Table 8.2 Commonly used materials of natural and synthetic origin used for scaffolds in tissue engineering

Synthetic polymers	Materials of natural origin
Polyglycolic acid (PGA)	Collagen
Polylactic acid (PLA)	Fibrin
Poly(lactic-co-glycolic acid) (PLGA)	Fibronectin
Polycaprolactone (PCL)	Silk (Fibroin)
Poly(propylene fumarates) (PPF)	Gelatin
Polyanhydrides	Hyaluronate
Poly(acrylic acid) PAA	Glycosaminoglycans
Poly(ethylene glycol) PEG	Chitin
Ceramics (including hydroxapatite, bioactive glasses, phosphate-based glasses, calcium phosphates)	Chitosan
	Dextran
	Agarose
	Alginates
	Starch
	Polyhydroxyalkanoates
	Coral
	Demineralised/deproteinised bone

implant metals currently used comprise stainless steels and cobalt-based and titanium-based alloys.[39] Perhaps the most extensively used ceramics include calcium phosphate, alumina, zirconia and, more recently, Bioglass®.[40] Based on these materials, there is a whole range of endoprostheses available that have contributed to a marked improvement of the quality of life of numerous patients worldwide over a number of decades. Polymeric materials have also received appropriate attention from the scientific and medical communities.[41] Natural polymers, such as collagens, glycosaminoglycan, starch, chitin and chitosan, have been used to repair nerves, skin, cartilage and bone.[42]

Ceramics have been widely used in the tissue engineering and regeneration fields.[43] This is not surprising as perhaps the most commonly treated tissue in such applications is bone, and ceramics can easily be formulated to have similar geometry, structure and, to some extent, physical and mechanical properties to native bone, such as the inorganic component, as well as having osteoconductive properties.[44] However, ceramics are certainly limited in their use amongst a variety of tissues and cannot therefore be expected to be appropriate for the support of cells from soft connective tissues, for example.

Ceramics can be derived from natural sources (e.g., coralline hydroxyapatite, HA) or be created, such as synthetic HA or β-tricalcium phosphate (β-TCP). Owing to their numerous qualities of ease of manufacture and osteoconductive and osteoinductive properties, they have been obviously considered for bone tissue engineering applications. Numerous researchers

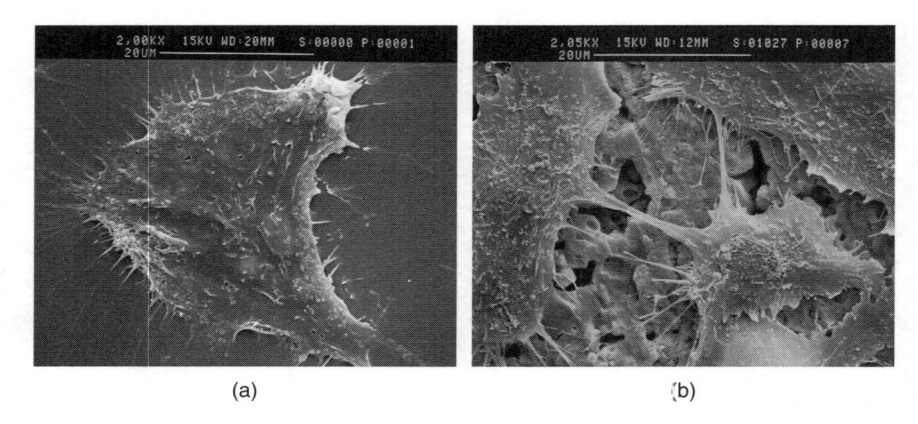

8.1 SEM of human osteoblast-like cells: (a) on control glass surface, and (b) on HA containing 5% glass. Note the flattened, well-spread morphology of cells on both surfaces and the cytoplasmic extensions. Note the pitting caused by processing of the material surface of HA5%.

have shown that by using ceramics with or without bone-derived cells, acceptable *in vitro* and *in vivo* results regarding bone regeneration could be achieved.[45–51] For example, a novel phosphate-glass formulation incorporated to a synthetic hydroxyapatite gives an appropriate composite for bone cells *in vitro* (see Fig. 8.1).

Very often soft tissues possess quite different types of cell receptors, respond to soluble biological mediators and a variety of cytokines, and have distinctly different mechanical property requirements compared with hard connective tissues. Thus, a different approach is required for soft tissue. In such cases, the use of natural polymers has seen a large research effort to create novel scaffolds for tissue repair and regeneration. The ideology here is that such polymers have normal functional properties *in situ* and a common trend is to mimic nature with such natural polymers and molecules – it makes sense! Thus, natural polymers display quite uniquely how all the properties displayed by biological materials and systems are exclusively determined by the physical–chemical properties of the monomers and their sequence.[52] A fundamental understanding of the molecular and functional aspects of natural materials can clearly lead to the development of natural three-dimensional scaffolds with predictable structure and suitable function at the micro- and mesoscale.[53–55]

Irrespective of the tissue or organ involved, there are many concepts that can be extrapolated from natural phenomena and are therefore applicable in the tissue engineering field to repair or regenerate a variety of tissues. In many cases, the matrices and scaffolds would ideally be made of biodegradable polymers whose properties closely resemble those of the extracellular matrix (ECM), a soft, tough and elastomeric proteinaceous network that provides mechanical stability and structural integrity to tissues and organs.[56] Within the natural biomaterials available to researchers we can find the commonly used species such as collagen,[57–60] fibrinogen,[61–63] chitosan,[64–66] starch,[67–70] hyaluronic acid[71–74] and poly(hydroxybutyrate).[75–77]

Synthetic biodegradable polymers are seemingly more commonly used within the biomedical engineering field and are widely reported as being successful in a variety of applications. Notably these materials have an adaptable chemistry and furthermore their processability varies according to structure. Thus any direct comparison with natural polymers cannot be ascertained. The most widely used are poly(α-hydroxy acids),[78–82] poly(ϵ-caprolactone),[83–85] poly(propylene fumarates),[86–89] poly(carbonates)[90–92] and poly(anhydrides).[93–95] The reader is referred to several excellent reviews which give further detailed information on natural and synthetic polymers and their use and applications in scaffold materials.[96–100]

8.5 Scaffold design and processing

There are several requirements in the design of scaffolds for tissue engineering. Many of these requirements are complex and not yet fully understood. In addition to being biocompatible both in bulk and in degraded form, these scaffolds should possess appropriate mechanical properties to provide the correct stress environment for the neo-tissues. Also, the scaffolds should exhibit the appropriate surface structure and chemistry for cell attachment. In addition they should be porous and permeable to permit the migration and permeation of cells and nutrients, respectively.

A range of techniques has been developed to process and fabricate synthetic and natural polymers into a variety of porous scaffolds and structures. As in many industrial applications, polymers are often preformed and distributed in monolith or pellet form.[101] For the purposes of this chapter, the conventional scaffold fabrication methods are limited to those that create structures that have an interpenetrating (i.e. continuous) and uninterrupted porous geometry. The most commonly used of these techniques are described in brief detail below.

8.5.1 Solvent-casting particulate-leaching

This technique involves producing a solution of polymer dissolved in a solvent and adding mineral particles (such as sodium chloride) of a specific diameter to produce a uniform suspension.[102,103] The solvent is allowed to evaporate, leaving behind a polymer matrix with salt particles embedded throughout. The composite is then immersed in water where the salt leaches out to produce a porous structure. Alternatively, the dispersion may be freeze-dried in order to evaporate the solvent. A lamination technique using chloroform as the binder was proposed to shape such scaffolds into 3-D structures.[104]

8.5.2 High-pressure gas foaming

A biodegradable polymer such as PLGA is saturated with carbon dioxide (CO_2) at high pressures.[105] The solubility of the gas in the polymer is then decreased rapidly by bringing the CO_2 pressure back to atmospheric level. This results in nucleation and growth of gas bubbles, or cells, with sizes ranging between 100 and 500 µm in the polymer.

8.5.3 Fibre mesh bonding

Fibres, produced by textile technology, have been used to make non-woven scaffolds from PGA and PLLA.[106] The lack of structural stability of these non-woven scaffolds, however, often resulted in significant deformation due to contractile forces of the cells that had been seeded on the scaffold. This led to the development of a fibre bonding technique to increase the mechanical properties of the scaffolds.[107] This is achieved by dissolving PLLA in methylene chloride and casting over the PGA mesh. The solvent is allowed to evaporate and the construct is then heated above the melting point of PGA. Once the PGA–PLLA construct has cooled, the PLLA is removed by dissolving in methylene chloride again. This treatment results in a mesh of PGA fibres joined at various 'cross-joints', and interconnected networks can be manufactured in this way, albeit with minimal control of porosity.[106]

8.5.4 Phase separation

Phase separation is a thermodynamic separation of a polymer solution into a polymer-rich component and a polymer-poor/solvent-rich component. A biodegradable synthetic polymer is dissolved in molten phenol or naphthalene and biologically active molecules may be incorporated.[108–110] The temperature is then lowered to produce a liquid–liquid phase separation and quenched to form a two-phase solid. The solvent is removed by sublimation to give a porous scaffold with bioactive molecules incorporated in the structure.

8.5.5 Injection moulding

This is a melt-based technique that depends on a raw polymer being interacted with a blowing agent. The developed methodology consists of a standard conventional injection moulding process, on which a solid blowing agent based on carboxylic acids is used to generate the foaming of the bulk of the moulded part. The proposed route allows for the production of scaffolds with a compact skin and a porous core, with promising mechanical properties.[111] As a result of the process, the blowing agent is degraded, releasing water and carbon dioxide, and this action causes pores to remain in the 3-D structure. This methodology can also be utilised to create a microstructured polymer scaffold, which favours the three-dimensional cultivation of cells within an array of cubic microcontainers.[112]

8.5.6 Freeze drying

This type of fabrication method does not rely on the use of solid porogens. Instead a synthetic polymer, such as PLGA or PLA, is dissolved in a suitable solvent (dichloromethane or glacial acetic acid). Water is then added to the polymeric solution, the two liquids are mixed and the result is an emulsion. In order to separate the two resultant phases, the emulsion is cast into a mould and very quickly frozen by mean of immersion into liquid nitrogen, for example. The now frozen emulsion is subsequently freeze-dried under appropriate conditions in order to remove the dispersed water and solvent. This process yields a solidified and porous polymeric structure which can be stored dry prior to use.[113] Similarly, collagen scaffolds have been made in this fashion by freezing a dispersion or solution of collagen and then freeze-drying.[114] By utilising subtle changes in the fabrication process, physical characteristics such as pore size can be controlled to some extent.[115,116] These collagen scaffolds are then crosslinked by either physical or chemical means to reduce the solubility, antigenicity and degradation rate.

8.5.7 Solid freeform fabrication

Despite the recent advancements in the fabrication of scaffolds in the last decade, manufacturing and processing techniques are still some distance from matching the natural structure and geometries of native tissues. The complexity of mammalian native tissues means that complex scaffold architecture designs need to be generated. Some improvements in these areas are currently being offered by technological advancements in scaffold design and creation. Complex scaffolds generated using conventional computer-aided design are successful in blueprint, but very often the available techniques cannot 'build' the scaffold and thus translate the concept from a design level to a clinically usable material. Thus, scaffold architectures are being built using a relatively new technique known as solid freeform fabrication (SFF, also referred to as Rapid Prototyping, RP) where the material is created layer by layer.[100] In this way, customised shapes with defined, reproducible and controlled internal morphology can be produced. There are some excellent reviews which evaluate the application of SFF[101,117,118] and suggest that this method can optimise the design and manufacture of tissue-engineering scaffolds.

8.6 Advantages and limitations of scaffolds

8.6.1 Advantages

The advantages seem obvious when one considers the reproducible manufacturing techniques using, in particular, natural materials as well as the numerous commercial polymers available. The main advantages of such materials are their generally low immunogenic potential, the potential bioactive behaviour and the capability of typically favourable interactions with the host's tissue. In the case of plant materials such as chitin and starch, their source is potentially unlimited.

Although materials of natural origin present some drawbacks, including the difficulties in controlling the variability from batch to batch, low mechanical properties or limited processability which utilise some potentially toxic solvents, their advantages clearly outweigh the drawbacks. Their degradability, biocompatibility, relative low costs of sourcing, retrieval, processing and ready availability, and the close associations with cell extracellular matrices and intrinsic cellular interactions, make them attractive candidates for biomedical applications. Furthermore, investigations of cell–scaffold interactions *in vitro* can inform the coherent formulation of scaffold composition and structure for improved performance in tissue engineering and regenerative medicine applications.

8.6.2 Limitations

One of the key issues and concerns, however, with the use of degradable synthetic polymers is the release of by-products which are normally acidic, and this is important both *in vitro* and *in vivo*. Lactic acid is released from materials such as PLLA during degradation, and this marked reduction in the local pH further accelerates the degradation rate due to autolysis,[119,120] resulting in a highly acidic microenvironment. Clearly, this may adversely affect cellular function, although many cells typically exist in a slightly acidic milieu.

Cells attached to scaffolds are often faced with several days or even weeks of culturing *in vitro* before the tissue is suitable for implantation. This can lead to considerable changes in the physical and chemical composition of the scaffold prior to use *in vivo*. During this period, particles released during polymer degradation can affect numerous cellular processes, and once *in vivo*, simultaneously elicit an inflammatory response. Furthermore, these materials are often too inflexible or difficult to handle, or even collapse on handling by the operator. Moreover, many materials used in tissue engineering lack specific reactive groups within their surface chemistry for the optimum covalent attachment of cells, which thrive *in vivo* on an extracellular matrix comprising mainly the extracellular proteins of collagen, elastin, glycoproteins, proteoglycans, laminin and fibronectin.[121] Thus, for these materials the major challenge of surface engineering is the introduction of various growth factors or matrix protein groups to the surface and this type of work has recently seen some excellent reviews.

Many existing 3-D scaffolds for tissue engineering and application *in vivo* prove less than ideal for clinical applications because they lack suitable mechanical strength, interconnected channels, controlled porosity and suitable reactive surfaces. Furthermore, metals and ceramics have two major disadvantages for tissue engineering applications. Firstly, they are not biodegradable (except biodegradable bioceramics such as β-tricalcium phosphate), and secondly, their processability is very limited. Metals often exhibit a strength and modulus many orders of magnitude greater than bone tissue, for example, and this can develop to a 'stress-shielding' effect, very often causing implant failure due to micromotion and/or infection between the implant surface and native tissue. Ceramics are brittle and thus present a low mechanical stability, which prevents their use in the regeneration of larger bone defects. Furthermore, due to natural cell physiology *in vivo* during bone resorption, such as osteoclastic activity, ceramic degradation/dissolution rates would be difficult to predict. This may present a considerable problem, because if degradation is excessively quick it will forfeit the already limited mechanical stability of the construct.

8.7 Monitoring cell behaviour in scaffolds

It should be a major aim in tissue engineering scaffold research to develop measurement tools and methods to characterise the surface interactions between cells and biologically active materials and thus advance the discovery of biomaterials to *facilitate* the development of regenerative therapies and advanced drug/gene delivery systems. Key researchers in this field have to ask questions such as 'can *in vitro* methods predict *in vivo* results?' and 'can we identify the "key interactions" between cells and material surfaces?' In order to improve the physicochemical and biological properties of the materials recommended for potential clinical use, we must consider (at the very least!) cell adhesion, migration, proliferation and differentiation, cell function (expression, localisation and regulation of proteins) as well as material surface phenomena.

198 Cellular response to biomaterials

To visualise and quantify such important material–cell mechanisms and interactions, we need to utilise a fitting armoury of techniques. These may include the following, though this by no means represents an exhaustive list: cell–surface interactions such as electron microscopy (EM), X-ray photo electron spectroscopy (XPS), the use of quartz crystal microbalance (QCM), Raman spectroscopy, Fourier-transform infrared spectroscopy (FTIR), nuclear magnetic resonance (NMR) and micro-computer tomography (micro-CT). Researchers also need to consider protein deposition and this can be evaluated via atomic force spectroscopy as well as QCM. At the cellular level, gene and protein expression can be determined and visualised using quantitative real-time polymerase chain reaction (qRT-PCR), histology, immunochemistry, light and confocal microscopy, biological plate assays and gel electrophoresis. Measurement of cell numbers and sizes is conventionally determined by simple indirect counting methods, flow cytometry (FCM) and the use of cell imaging software. Typically, several methods are often employed to determine details concerning numerous factors and information concerning the cell–material interface. Interpretation of the information revealed by these techniques should be carefully analysed before steadfast conclusions are made, as many of the techniques described above rely on intensely processed samples prior to evaluation.

8.8 Nanotechnology and scaffolds

Nanomaterials are defined as materials composed of one or more engineered nano-components, a nano-component having at least one of its dimensions between 1 nm and 100 nm. These nano-components, and their associated interactions, are engineered to impart the unique properties of these materials. Examples of such novel nano-devices include nanotubes, nanoparticles, nanostructured materials, and a host of designer molecules.

Nanotechnology has rapidly become the new science of the current decade and is fast developing into many aspects of medicine, including tissue engineering and scaffold design. The application of nanotechnology to scaffold cell-based therapies will potentially revolutionise tissue engineering. Several very recent reviews explore the current methodologies available.[122–124] New design and manufacture methodologies are required, and within our own group the engineering of novel nanofibre phosphate-based glasses produced via a sol-gel route is being developed.[125] Notable changes that occur with the shift from micro- to nanometre reinforcement are the high surface area to volume ratio and the significant improvements in mechanical performance (stiffness and tensile strength). The core rationale for this work is based on comparing micron with nano-sized Bioglass®

particulate filled composites.[75] At 10 wt% filler, a doubling of Young's modulus was achieved by reducing the particle size from tactoid to nanometre scale. Furthermore, nanoparticulates at 20 wt% doubled the protein adsorption. This interesting concept is being further developed by increasing the reinforcement aspect ratio, projecting further improvements in mechanical performance. Other work has yielded glass nanofibres down to 84 nm via sol-gel and electrospinning from a $70SiO_2$–$25CaO$–$5P_2O_5$ silicate-based glass.[126] One of the striking features of these fibres was the ability to become mineralised in an unprecedentedly short period of time. Their bioactivity and osteogenic potential were assessed *in vitro*, and the fibres showed almost complete dissolution at 24 hours with high levels of precipitation which were shown to be apatite based.

8.9 Scaffold applications in gene therapy

The advent of genomic techniques and the successful identification of entire genomes for a number of species have led to an inventive approach in order to manipulate the local cell environment for directing cell function. Thus, the concept of gene delivery utilising scaffold materials has identified several design parameters for both the scaffold material and vector which must be duly satisfied. Two excellent reviews highlight the types of developments that have inspired gene and DNA delivery in tissue engineering.[127,128]

The development of effective and safe protocols for the delivery of genes to alleviate a wide range of genetic abnormalities is of major importance. Current procedures for the introduction of healthy genes into the cells and tissues of affected individuals have relied primarily on the use of virus-derived 'carriers' for delivering the therapeutic gene, but their questionable safety is now a main public concern. Moreover, alternative synthetic particulate gene-delivery systems such as cationic liposomes, polymers and dendrimers are deemed to be inefficient and have had only very limited success *in vivo*. This is for a number of reasons, including (1) toxicity/instability of the particulate delivery system and enzymatic degradation of the gene before or after cell uptake, (2) short cell/tissue exposure time and targeting, and (3) poorly controlled separation of gene and particulate delivery device within the cell and limited nuclear targeting/uptake.[129–132]

A safer and more attractive alternative to the above viral and particulate systems is the co-precipitation of genes with calcium phosphate. Although such precipitation has been used with some success for gene transfer *in vitro* for many years, its application in gene therapy has been limited by a number of problems, including (1) inability to protect the gene from degradative enzymes, (2) the sensitivity of precipitate structure/

chemistry to small changes in pH or the presence of other components,[131] and (3) difficulties with targeting *in vivo*. Such parameters significantly affect cellular uptake and stability within the cell and thus the overall functional efficiency of transferred genes. Recently, however, gene-containing calcium phosphate nanoparticles (GeneCaP) have been produced using reverse microemulsions. These nanoparticles gave comparable transfection efficiency *in vitro* to the commercial agent Superfect and also demonstrated some gene expression *in vivo*.[133,134] Significantly, the entrapped gene was protected from degradative enzymes, probably because the water-soluble DNA remains near the centre of the microemulsion droplet while the insoluble calcium phosphate forms a protecting shell at the oil–water interface.

Other recent studies (in the absence of DNA) have shown that the size and shape (e.g., nanosphere, sheet, rod or needle) of calcium phosphate precipitates prepared using microemulsions can vary widely, which would significantly affect their cellular uptake and transport. Additionally, upon varying pH, calcium and phosphate can precipitate in different ratios, e.g. hydroxyapatite ($Ca_{10}-x(PO_4)_6-xOH_2-x$) in neutral, or dicalcium phosphate (brushite, $CaHPO_4$) in acid conditions. These chemical changes have marked effects on calcium phosphate solubility and therefore its potential ease of separation from the gene in the cell.

As noted above, targeting and sustained delivery have been major difficulties in the delivery *in vivo* of DNA-containing particles for gene therapy. An ideal particulate DNA delivery device should be biocompatible and initially fluid, but then able to set rapidly at the site of the target tissue. It should then degrade, enabling continuous controlled availability of the gene to cells in contact with the material surface.

8.10 Future trends

The characterisation, analysis and application of nanomaterials are current research topics that are among the most exciting areas emerging for the next generation of scaffold materials. Much work still remains to be done so as to increase the knowledge and usability of these new materials, but there is a vast interest in this area which has attracted academic curiosity as well as industrial financial support. As the demand for replacement human organs increases with ageing populations, and stricter controls are enforced on their supply, the need for alternative sources of reliable replacement tissue is already great.

Although most of the current design and fabrication methodologies occur at scales larger than 100 μm, future work should also endeavour to incorporate micrometre- and nanoscale features. Currently, integration of micrometre or tens of micrometre feature sizes occurs during postprocessing steps. Integration of micrometre- and nanoscale features into designed scaffolds could improve both mechanical properties through toughening mechanisms and tissue regeneration through improved control of cell adhesion.[100]

Tissue engineering offers a novel route for repairing damaged or diseased tissues by incorporating the patients' own healthy cells or donated cells into temporary housings or scaffolds. The structure and properties of the scaffold are critical to ensure normal cell behaviour and performance of the cultivated tissue. This new tissue engineering approach is only just beginning to be commercially exploited in products such as skin substitutes and bone substitutes incorporating ceramics. Once the technology has been sufficiently well developed, cells grown on a porous scaffold will be used to repair other tissues within the human body. In respect of this, co-cultures of two or more cell types would give the distinct advantage of creating near native-like tissue *in vitro* and potentially generate scaffolds for implantation for complex tissues, e.g. articular cartilage and the tendon–bone interface.[135]

Finally, of major importance would be the establishment of standards and test method procedures for characterising the structure of tissue scaffolds, improving the reliability of manufacturing processes, and enhancing confidence in performance *in vivo* through the development of reliable *in vitro* tests. The key to achieving these objectives is to identify appropriate metrology methods for determining the size and distribution of pores and the extent to which they are interconnected with the different types of scaffolds. These measures will be used to develop quality control procedures that will enable batch-to-batch variations to be assessed and minimised. Research such as this is being conducted by a consortium led by the National Physical Laboratory (UK), University of Brighton, and Queen Mary College, University of London.

Translation from laboratory to the clinics is an important consideration if the current generation of materials is to be successful and realised as working solutions to tissue regeneration issues. It is imperative that scaffolds are designed and developed using good manufacturing practice (GMP) conditions in a reproducible and quality-controlled fashion with economic efficiency. To move the current tissue engineering practices to the next developmental level, some manufacturing processes will be required to accommodate the incorporation of cells and/or biological molecules during the scaffold fabrication process. These approaches are working towards the possibility of current tissue engineered constructs, not only to have a controlled spatial distribution of cells and molecules, but also to possess a versatility of scaffold material and microstructure within one specifically designed and fabricated construct, for implantation in a realistically critical-sized anatomical site.[117]

8.11 Conclusions

The past half-century of research related to biomaterials scaffold design and characterisation for tissue replacement and augmentation has witnessed an evolutionary journey from simple, yet permanent, metallic implants to temporary and synthetic materials which have helped create the current 'tissue engineering' ideology. Today's concepts of scaffold- and matrix-based tissue engineering involve the combination of a scaffold with cells and/or biomolecules that promote the repair and/or regeneration of tissues. More recently, regenerative therapies have considered whole tissue architecture, the ultimate goal aimed at the creation of scaffolds that create a temporary 3-D matrix upon which cells and tissues can grow exclusively *in vitro* and/or *in vivo*.

Although many well-established and broadly defined methods are being universally utilised, there still remain several vital issues to project the next generation of scaffolds into the limelight. These include the control of cell growth in tissue engineered scaffolds to tissues greater than 1 mm in thickness in order to overcome the nutrient diffusion/consumption pathways which are currently limiting their application to critical-sized defects only. This remains a significant challenge. Moreover, the requirement for multiple architectures and geometric hierarchies which exist in many natural tissues needs to be further investigated.

8.12 Acknowledgements

The author would like to thank the editor, Dr Lucy Di Silvio, and Woodhead Publishing for their kind invitation to write this chapter.

8.13 Sources of further information and advice

8.13.1 Suggested books for further reading

1. Lanza RP, Langer R, Vacanti JP. *Principles of Tissue Engineering*. San Diego, CA: Academic Press, 2000.
2. Atala A, Lanza RP. *Methods in Tissue Engineering*. San Diego, CA: Academic Press, 2002.
3. Palsson B, Hubbell JA, Plonsey R, Bronzino JD. *Tissue Engineering*. In series *Principles and Applications in Engineering*. Boca Raton, FL: CRC Press, 2003.
4. Palsson B, Bhatia SN. *Tissue Engineering*. Upper Saddle River, NJ: Prentice Hall, 2003.
5. Saltzman WM. *Tissue Engineering: Principles for the Design of Replacement Organs and Tissues*. Oxford: Oxford University Press, 2004.

6. Ma PX, Elisseeff J (eds). *Scaffolding in Tissue Engineering*. Boca Raton, FL: CRC Press, 2005.

8.13.2 Useful websites/societies

- British Tissue Engineering Network, BriteNet (www.briten.org)
- European Society of Biomaterials, ESB (www.esbiomaterials.eu)
- The Biomaterials Network (www.biomat.net)
- Tissue Engineering and Regenerative Medicine International Society, TERMIS (http://www.termis.org/)
- Tissue & Cell Engineering Society (www.tces.org)
- United Kingdom Society for Biomaterials, UKSB (www.uksb.org.uk)
- World Biomaterials Congress, WBC (www.wbc2008.com)

8.13.3 Recommended journal titles

Acta Biomaterialia; Biomacromolecules; Biomaterials; Biomedical Materials: Materials for tissue engineering & regenerative medicine; European Cells and Material; Journal of Biomaterials Applications; Journal of Applied Biomaterials & Biomechanics; Journal of Tissue Engineering and Regenerative Medicine; Tissue Engineering Parts A, B (Reviews) and C (Methods); Journal of Applied Biomaterials & Biomechanics; Journal of Bioactive and Compatible Polymers; Journal of Biomedical Materials Research Parts A and B; Journal of Materials Science: Materials in Medicine; Journal of Applied Biomaterials (1990–1995); Journal of Biomechanical Engineering; Journal of Biomedical Materials Research Part B: Applied Biomaterials; Journal of Biomedical Nanotechnology; Journal of Long-Term Effects of Medical Implants; Journal of Nanobiotechnology; Journal of Tissue Engineering and Regenerative Medicine; Materials Science and Engineering C: Biomimetic and Supramolecular Medical Plastics and Biomaterials; Nanomedicine; Regenerative Medicine; Tissue Engineering; Trends in Biomaterials & Artificial Organs.

8.14 References

1. Anderson JM (2006) The future of biomedical materials. *J Mater Sci: Mater Med* 17: 1025–1028.
2. Brown RA, Phillips JB (2007) Cell responses to biomimetic protein scaffolds used in tissue repair and engineering. *Int Rev Cytol* 262: 75–150.
3. Agrawal CM, Ray RB (2001) Biodegradable polymeric scaffolds for musculo-skeletal tissue engineering. *J Biomed Mater Res* 55: 141–150.
4. Langer R, Vacanti JP (1993) Tissue engineering. *Science* 260(5110): 920–926.
5. Ko HC, Milthorpe BK, McFarland CD (2007) Engineering thick tissues – the vascularisation problem. *Eur Cell Mater* 14: 1–18.

6. Landman KA, Cai AQ (2007) Cell proliferation and oxygen diffusion in a vascularising scaffold. *Bull Math Biol* 69(7): 2405–2428.
7. Hutmacher DW (2001) Scaffold design and fabrication technologies for engineering tissues – state of the art and future perspectives. *J Biomater Sci: Polym Ed* 12(1): 107–124.
8. Bach AD, Arkudas A, Tjiawi J, Polykandriotis E, Kneser U, Horch RE, Beier JP (2006) A new approach to tissue engineering of vascularized skeletal muscle. *J Cell Mol Med* 10: 716–726.
9. Hutmacher DW, Schantz JT, Lam CX, Tan KC, Lim TC (2007) State of the art and future directions of scaffold-based bone engineering from a biomaterials perspective. *J Tissue Eng Regen Med* 1(4): 245–260.
10. Sarraf CE, Harris AB, McCulloch AD, Eastwood M (2003) Heart valve and arterial tissue engineering. *Cell Prolif* 36: 241–254.
11. Vats A, Tolley NS, Polak JM, Gough JE (2003) Scaffolds and biomaterials for tissue engineering: a review of clinical applications. *Clin Otolaryngol* 28(3): 165–172.
12. Ang KC, Leong KF, Chua CK, Chandrasekaran M (2006) Investigation of the mechanical properties and porosity relationships in fused deposition modelling-fabricated porous structures. *Rapid Prototyping Journal* 12(2): 100–105.
13. Srisuwan T, Tilkorn DJ, Wilson JL, Morrison WA, Messer HM, Thompson EW, Abberton KM (2006) Molecular aspects of tissue engineering in the dental field. *Periodontology 2000* 41(1): 88–108.
14. Ribeiro CC, Barrias CC, Barbosa MA (2006) Preparation and characterisation of calcium-phosphate porous microspheres with a uniform size for biomedical applications. *J Mater Sci: Mater Med* 17(5): 455–463.
15. Hao L, Savalani MM, Zhang Y, Tanner KE, Harris RA (2007) Selective laser sintering of hydroxyapatite reinforced polyethylene composites for bioactive implants and tissue scaffold development. *Proc Inst Mech Eng* 220(4): 521–531.
16. Kofron MD, Cooper JA Jr, Kumbar SG, Laurencin CT (2007) Novel tubular composite matrix for bone repair. *J Biomed Mater Res A* 82(2): 415–425.
17. Mikos AG, Thorsen AJ, Czerwonka LA, Bao Y, Langer R (1994) Preparation and characterisation of poly(L-lactic acid) foams. *Polymer* 35: 1068–1077.
18. Shimko DA, Shimko VF, Sander EA, Dickson KF, Nauman EA (2005) Effect of porosity on the fluid flow characteristics and mechanical properties of tantalum scaffolds. *J Biomed Mater Res B* 73(2): 315–324.
19. Quadrani P, Pasini A, Mattioli-Belmonte M, Zannoni C, Tampieri A, Landi E, Giantomassi F, Natali D, Casali F, Biagini G (2005) High-resolution 3D scaffold model for engineered tissue fabrication using a rapid prototyping technique. *Med Biol Eng Comp* 43(2): 196–199.
20. Nakamatsu J, Torres FG, Troncoso OP, Min-Lin Y, Boccaccini AR (2006) Processing and characterization of porous structures from chitosan and starch for tissue engineering scaffolds. *Biomacromolecules* 7(12): 3345–3355.
21. Blaker JJ, Maquet V, Jérôme R, Boccaccini AR, Nazhat SN (2005) Mechanical properties of highly porous PDLLA/Bioglass composite foams as scaffolds for bone tissue engineering. *Acta Biomater* 1(6): 643–652.
22. Blacher S, Maquet V, Jérôme R, Pirard JP, Boccaccini AR (2005) Study of the connectivity properties of Bioglass-filled polylactide foam scaffolds by image analysis and impedance spectroscopy. *Acta Biomater* 1(5): 565–574.

23. Day RM, Maquet V, Boccaccini AR, Jérôme R, Forbes A (2005) In vitro and in vivo analysis of macroporous biodegradable poly(D,L-lactide-co-glycolide) scaffolds containing bioactive glass. *J Biomed Mater Res A* 75(4): 778–787.
24. Vunjak-Novakovic G, Freed LE (1998) Culture of organized cell communities. *Adv Drug Deliv Rev* 33(1–2): 15–30.
25. Salgado AJ, Coutinho OP, Reis RL (2004) Bone tissue engineeering: state of the art and future trends. *Macromol Biosci* 4: 743–765.
26. Whang K, Healy E, Elenz DR *et al.* (1999) Engineering bone regeneration with bioabsorbable scaffolds with novel microarchitecture. *Tissue Eng* 5: 35-51.
27. Kelly A (2007) Why engineer porous materials? *Phil Trans A Math Phys Eng Sci* 364(1838): 5–14.
28. Brett PM, Harle J, Salih V, Mihoc R, Olsen I, Jones FH, Tonetti M (2004) Roughness response genes in osteoblasts. *Bone* 35(1): 124–133.
29. Baier RE (2006) Surface behaviour of biomaterials: the theta surface for biocompatibility. *J Mater Sci: Mater Med* 17(11): 1057–1062.
30. Yim EK, Leong KW (2005) Significance of synthetic nanostructures in dictating cellular response. *Nanomedicine* 1(1): 10–21.
31. Moroni L, de Wijn JR, van Blitterswijk CA (2008) Integrating novel technologies to fabricate smart scaffolds. *J Biomater Sci: Polym Ed* 19(5): 543–572.
32. Murugan R, Ramakrishna S (2007) Design strategies of tissue engineering scaffolds with controlled fiber orientation. *Tissue Eng* 13(8): 1845–1866.
33. Goldberg M, Langer R, Jia X (2007) Nanostructured materials for applications in drug delivery and tissue engineering. *J Biomater Sci: Polym Ed* 18(3): 241–268.
34. Cohen S, Bano MC, Cima LG *et al.* (1993) Design of synthetic polymeric structures for cell transplantation and tissue engineering. *Clin Mater* 13: 3–10.
35. Hunziker E, Spector M, Libera J, Gertzman A, Woo SL, Ratcliffe A, Lysaght M, Coury A, Kaplan D, Vunjak-Novakovic G (2006) Translation from research to applications. *Tissue Eng* 12(12): 3341–3364.
36. Brittberg M (1999) Autologous chondrocyte transplantation. *Clin Orthop Relat Res* 367 Suppl: S147–155.
37. Ellingsen JE, Thomsen P, Lyngstadaas SP (2006) Advances in dental implant materials and tissue regeneration. *Periodontol 2000* 41: 136–156.
38. Moradian-Oldak J, Wen HB, Schneider GB, Stanford CM (2006) Tissue engineering strategies for the future generation of dental implants. *Periodontol 2000* 41: 157–176.
39. Brunski JB (1996) Metals. In: Ratner BD, Hoffman AS, Schoen FJ, *et al.*, eds. *Biomaterials Science: An Introduction to Materials in Medicine*. New York: Academic Press, pp. 37–50.
40. Hench LL (1996) Ceramics, glasses, and glass-ceramics. In: Ratner BD, Hoffman AS, Schoen FJ, *et al.*, eds. *Biomaterials Science: An Introduction to Materials in Medicine*. New York: Academic Press, pp. 73–83.
41. Lin HR, Kuo CJ, Yang CY, Shaw SY, Wu YJ (2002) Preparation of macroporous biodegradable PLGA scaffolds for cell attachment with the use of mixed salts as porogen additives. *J Biomed Mater Res* 63(3): 271–279.
42. Harper EJ, Braden M, Bonfield W (2000) Mechanical properties of hydroxyapatite reinforced poly(ethylmethacrylate) bone cement after immersion in a physiological solution: influence of a silane coupling agent. *J Mater Sci: Mater Med* 11(8): 491–497.

43. LeGeros RZ (2002) Properties of osteoconductive biomaterials: calcium phosphates. *Clin Orthop Relat Res* 395: 81–98.
44. Hak DJ (2007) The use of osteoconductive bone graft substitutes in orthopaedic trauma. *J Am Acad Orthop Surg* 15(9): 525–536.
45. Srouji S, Kizhner T, Livne E (2006) 3D scaffolds for bone marrow stem cell support in bone repair. *Regen Med* 1(4): 519–528.
46. Cancedda R, Mastrogiacomo M, Bianchi G, Derubeis A, Muraglia A, Quarto R (2003) Bone marrow stromal cells and their use in regenerating bone. *Novartis Found Symp* 249: 133–143.
47. Ohgushi H, Miyake J, Tateishi T (2003) Mesenchymal stem cells and bioceramics: strategies to regenerate the skeleton. *Novartis Found Symp* 249: 118–127.
48. Ohgushi H, Caplan AI (1999) Stem cell technology and bioceramics: from cell to gene engineering. *J Biomed Mater Res* 48(6): 913–927.
49. Heymann D, Pradal G, Benahmed M (1999) Cellular mechanisms of calcium phosphate ceramic degradation. *Histol Histopathol* 14(3): 871–877.
50. Burwell RG (1985) The function of bone marrow in the incorporation of a bone graft. *Clin Orthop Relat Res* 200: 125–141.
51. Salih V, Georgiou G, Knowles JC, Olsen I (2001) Glass reinforced hydroxyapatite for hard tissue surgery – Part II: in vitro evaluation of bone cell growth and function. *Biomaterials* 22(20): 2817–2824.
52. Malafaya PB, Silva GA, Reis RL (2007) Natural-origin polymers as carriers and scaffolds for biomolecules and cell delivery in tissue engineering applications. *Adv Drug Deliv Rev* 59(4–5): 207–233.
53. Rodriguez-Cabello JC, Reguera J, Girotti A, Alonso M, Testera AM (2005) Developing functionality in elastin-like polymers by increasing their molecular complexity: the power of the genetic engineering approach. *Prog Polym Sci* 30: 1119–1145.
54. Brown RA, Wiseman M, Chuo CB, Cheema U, Nazhat SN (2005) Ultra-rapid engineering of biomimetic tissues: a plastic compression fabrication process for nano-micro structures. *Advan Func Mater* 15: 1762–1770.
55. Bitar M, Brown RA, Salih V, Kidane AG, Knowles JC, Nazhat SN (2008) Effect of cell density on osteoblastic differentiation and matrix degradation of biomimetic dense collagen scaffolds. *Biomacromolecules* 9(1): 129–135.
56. Guo XD, Zheng QX, Du JY, Yang SH, Wang H, Shao ZW, Sun EJ (2002) Molecular tissue engineering: concepts, status and challenge. *J Wuhan Univ Technol* 17: 30–34.
57. Sachlos E, Reis N, Ainsley C, Derby B, Czernuszka JT (2003) Novel collagen scaffolds with predefined internal morphology made by solid freeform fabrication. *Biomaterials* 24(8): 1487–1497.
58. Wahl DA, Sachlos E, Liu C, Czernuszka JT (2007) Controlling the processing of collagen-hydroxyapatite scaffolds for bone tissue engineering. *J Mater Sci: Mater Med* 18(2): 201–209.
59. Badylak SF (2007) The extracellular matrix as a biologic scaffold material. *Biomaterials* 28(25): 3587–3593.
60. Twardowski T, Fertala A, Orgel JP, San Antonio JD (2007) Type I collagen and collagen mimetics as angiogenesis promoting superpolymers. *Curr Pharm Des* 13(35): 3608–3621.
61. Linnes MP, Ratner BD, Giachelli CM (2007) A fibrinogen-based precision microporous scaffold for tissue engineering. *Biomaterials* 35: 5298–5306.

62. Chang J, Rasamny JJ, Park SS (2007) Injectable tissue-engineered cartilage using a fibrin sealant. *Arch Facial Plast Surg* 9(3): 161–166.
63. Alsberg E, Feinstein E, Joy MP, Prentiss M, Ingber DE (2006) Magnetically-guided self-assembly of fibrin matrices with ordered nano-scale structure for tissue engineering. *Tissue Eng* 12(11): 3247–3256.
64. Shi C, Zhu Y, Ran X, Wang M, Su Y, Cheng T (2006) Therapeutic potential of chitosan and its derivatives in regenerative medicine. *J Surg Res* 133(2): 185–192.
65. Di Martino A, Sittinger M, Risbud MV (2005) Chitosan: a versatile biopolymer for orthopaedic tissue-engineering. *Biomaterials* 26(30): 5983–5990.
66. Li Z, Gunn J, Chen MH, Cooper A, Zhang M (2008) On-site alginate gelation for enhanced cell proliferation and uniform distribution in porous scaffolds. *J Biomed Mater Res A* 86(2): 552–559.
67. Nakamatsu J, Torres FG, Troncoso OP, Min-Lin Y, Boccaccini AR (2006) Processing and characterization of porous structures from chitosan and starch for tissue engineering scaffolds. *Biomacromolecules* 7(12): 3345–3355.
68. Salgado AJ, Coutinho OP, Reis RL, Davies JE (2007) In vivo response to starch-based scaffolds designed for bone tissue engineering applications. *J Biomed Mater Res A* 80(4): 983–989.
69. Ghosh S, Gutierrez V, Fernández C, Rodriguez-Perez MA, Viana JC, Reis RL, Mano JF (2008) Dynamic mechanical behavior of starch-based scaffolds in dry and physiologically simulated conditions: Effect of porosity and pore size. *Acta Biomater* 4(4): 950–959.
70. Oliveira JT, Crawford A, Mundy JM, Moreira AR, Gomes ME, Hatton PV, Reis RL (2007) A cartilage tissue engineering approach combining starch–polycaprolactone fibre mesh scaffolds with bovine articular chondrocytes. *J Mater Sci: Mater Med* 18(2): 295–302.
71. Ji Y, Ghosh K, Shu XZ, Li B, Sokolov JC, Prestwich GD, Clark RA, Rafailovich MH (2006) Electrospun three-dimensional hyaluronic acid nanofibrous scaffolds. *Biomaterials* 27(20): 3782–3792.
72. Fraser SA, Crawford A, Frazer A, Dickinson S, Hollander AP, Brook IM, Hatton PV (2006) Localization of type VI collagen in tissue-engineered cartilage on polymer scaffolds. *Tissue Eng* 12(3): 569–577.
73. Gutowska A, Jeong B, Jasionowski M (2001) Injectable gels for tissue engineering. *Anat Rec* 263(4): 342–349.
74. Turner WS, Schmelzer E, McClelland R, Wauthier E, Chen W, Reid LM (2007) Human hepatoblast phenotype maintained by hyaluronan hydrogels. *J Biomed Mater Res B* 82(1): 156–168.
75. Misra SK, Mohn D, Brunner TJ, Stark WJ, Philip SE, Roy I, Salih V, Knowles JC, Boccaccini AR (2008) Comparison of nanoscale and microscale bioactive glass on the properties of P(3 HB)/Bioglass composites. *Biomaterials* 29(12): 1750–1761.
76. Chen QZ, Efthymiou A, Salih V, Boccaccini AR (2008) Bioglass-derived glass-ceramic scaffolds: study of cell proliferation and scaffold degradation in vitro. *J Biomed Mater Res A* 84(4): 1049–1060.
77. Valappil SP, Misra SK, Boccaccini AR, Keshavarz T, Bucke C, Roy I (2007) Large-scale production and efficient recovery of PHB with desirable material properties, from the newly characterised *Bacillus cereus* SPV. *J Biotechnol* 132(3): 251–258.

78. Valappil SP, Misra SK, Boccaccini AR, Roy I (2006) Biomedical applications of polyhydroxyalkanoates: an overview of animal testing and in vivo responses. *Exp Rev Med Dev* 3(6): 853–868.
79. Misra SK, Valappil SP, Roy I, Boccaccini AR (2006) Polyhydroxyalkanoate (PHA)/inorganic phase composites for tissue engineering applications. *Biomacromolecules* 7(8): 2249–2258.
80. Zinn M, Witholt B, Egli T (2001) Occurrence, synthesis and medical application of bacterial polyhydroxyalkanoate. *Adv Drug Deliv Rev* 53(1): 5–21.
81. Chen GQ, Wu Q (2005) The application of polyhydroxyalkanoates as tissue engineering materials. *Biomaterials* 26(33): 6565–6578.
82. Zhao K, Deng Y, Chun Chen J, Chen GQ (2003) Polyhydroxyalkanoate (PHA) scaffolds with good mechanical properties and biocompatibility. *Biomaterials* 24(6): 1041–1045.
83. Kim HW, Lee EJ, Kim HE, Salih V, Knowles JC (2005) Effect of fluoridation of hydroxyapatite in hydroxyapatite–polycaprolactone composites on osteoblast activity. *Biomaterials* 26(21): 4395–4404.
84. Swieszkowski W, Tuan BH, Kurzydlowski KJ, Hutmacher DW (2007) Repair and regeneration of osteochondral defects in the articular joints. *Biomol Eng* 24(5): 489–495.
85. Chen ZC, Ekaputra AK, Gauthaman K, Adaikan PG, Yu H, Hutmacher DW (2008) In vitro and in vivo analysis of co-electrospun scaffolds made of medical grade poly(epsilon-caprolactone) and porcine collagen. *J Biomater Sci: Polym Ed* 19(5): 693–707.
86. Mistry AS, Cheng SH, Yeh T, Christenson E, Jansen JA, Mikos AG (2008) Fabrication and in vitro degradation of porous fumarate-based polymer/alumoxane nanocomposite scaffolds for bone tissue engineering. *J Biomed Mater Res A* [Epub ahead of print].
87. Lee JW, Lan PX, Kim B, Lim G, Cho DW (2008) Fabrication and characteristic analysis of a poly(propylene fumarate) scaffold using micro-stereolithography technology. *J Biomed Mater Res B* [Epub ahead of print].
88. Christenson EM, Soofi W, Holm JL, Cameron NR, Mikos AG (2007) Biodegradable fumarate-based polyHIPEs as tissue engineering scaffolds. *Biomacromolecules* 8(12): 3806–3814.
89. Sitharaman B, Shi X, Tran LA, Spicer PP, Rusakova I, Wilson LJ, Mikos AG (2007) Injectable in situ cross-linkable nanocomposites of biodegradable polymers and carbon nanostructures for bone tissue engineering. *J Biomater Sci: Polym Ed* 18(6): 655–671.
90. Macario DK, Entersz I, Bolikal D, Kohn J, Nackman GB (2008) Iodine inhibits antiadhesive effect of PEG: Implications for tissue engineering. *J Biomed Mater Res B* 86(1): 237–244.
91. Welle A, Kröger M, Döring M, Niederer K, Pindel E, Chronakis IS (2007) Electrospun aliphatic polycarbonates as tailored tissue scaffold materials. *Biomaterials* 28(13): 2211–2219.
92. Meechaisue C, Dubin R, Supaphol P, Hoven VP, Kohn J (2006) Electrospun mat of tyrosine-derived polycarbonate fibers for potential use as tissue scaffolding material. *J Biomater Sci: Polym Ed* 17(9): 1039–1056.
93. Chevallay B, Herbage D (2000) Collagen-based biomaterials as 3D scaffold for cell cultures: applications for tissue engineering and gene therapy. *Med Biol Eng Comput* 38(2): 211–218.

94. Behravesh E, Yasko AW, Engel PS, Mikos AG (1999) Synthetic biodegradable polymers for orthopaedic applications. *Clin Orthop Relat Res* 367 Suppl: S118–129.
95. Burkoth AK, Burdick J, Anseth KS (2000) Surface and bulk modifications to photocrosslinked polyanhydrides to control degradation behavior. *J Biomed Mater Res* 51(3): 352–359.
96. Hutmacher DW, Schantz JT, Lam CX, Tan KC, Lim TC (2007) State of the art and future directions of scaffold-based bone engineering from a biomaterials perspective. *J Tissue Eng Regen Med* 1(4): 245–260.
97. Silva GA, Ducheyne P, Reis RL (2007) Materials in particulate form for tissue engineering. 1. Basic concepts. *J Tissue Eng Regen Med* 1(1): 4–24.
98. Sands RW, Mooney DJ (2007) Polymers to direct cell fate by controlling the microenvironment. *Curr Opin Biotechnol* 18(5): 448–453.
99. Lee J, Cuddihy MJ, Kotov NA (2008) Three-dimensional cell culture matrices: state-of-the-art. *Tissue Eng B* 14(1): 61–86.
100. Hollister SJ (2006) Porous scaffold design for tissue engineering. *Nat Mater* 4(7): 518–524.
101. Sachlos E, Czernuszka J (2003) Making tissue engineering scaffolds work. Review: the application of solid freeform fabrication technology to the production of tissue engineering scaffolds. *Eur Cell Mater* 5: 29–39.
102. Schruben DL, Gonzalez P (2000) Dispersity improvement in solvent casting: particle/polymer composite. *Polym Eng Sci* 40(1): 139–142.
103. Draghi L, Resta S, Pirozzolo MG, Tanzi MC (2005) Microspheres leaching for scaffold porosity control. *J Mater Sci: Mater Med* 16(12): 1093–1097.
104. Quadrani P, Pasini A, Mattiolli-Belmonte M, Zannoni C, Tampieri A, Landi E, Giantomassi F, Natali D, Casali F, Biagini G, Tomei-Minardi A (2005) High-resolution 3D scaffold model for engineered tissue fabrication using a rapid prototyping technique. *Med Biol Eng Comput* 43(2): 196–199.
105. Mooney DJ, Baldwin DF, Suh NP, Vacanti JP, Langer R (1996) Novel approach to fabricate porous sponges of poly(D,L-lactic-co-glycolic acid) without the use of organic solvents. *Biomaterials* 17(14): 1417–1422.
106. Lu L, Mikos AG (1996) The importance of new processing techniques in tissue engineering. *MRS Bull* 21(11): 28–32.
107. Jayawarna V, Smith A, Gough JE, Ulijn RV (2007) Three-dimensional cell culture of chondrocytes on modified di-phenylalanine scaffolds. *Biochem Soc Trans* 35 (Pt 3): 535–537.
108. Barnes CP, Sell SA, Boland ED, Simpson DG, Bowlin GL (2007) Nanofiber technology: designing the next generation of tissue engineering scaffolds. *Adv Drug Deliv Rev* 59(14): 1413–1433.
109. Weigel T, Schinkel G, Lendlein A (2006) Design and preparation of polymeric scaffolds for tissue engineering. *Exp Rev Med Dev* 3(6): 835–851.
110. Norman JJ, Desai TA (2006) Methods for fabrication of nanoscale topography for tissue engineering scaffolds. *Ann Biomed Eng* 34(1): 89–101.
111. Gomes ME, Ribeiro AS, Malafaya PB, Reis RL, Cunha AM (2001) A new approach based on injection moulding to produce biodegradable starch-based polymeric scaffolds: morphology, mechanical and degradation behaviour. *Biomaterials* 22(9): 883–889.
112. Giselbrecht S, Gietzelt T, Gottwald E, Guber AE, Trautmann C, Truckenmüller R, Weibezahn KF (2004) Microthermoforming as a novel technique for

manufacturing scaffolds in tissue engineering (CellChips). *IEE Proc Nanobiotechnol* 2004 151(4): 151–157.
113. Huang YC, Huang YY, Huang CC, Liu HC (2005) Manufacture of porous polymer nerve conduits through a lyophilizing and wire-heating process. *J Biomed Mater Res B* 74(1): 659–664.
114. Yannas IV, Burke JF, Gordon PL, Huang C, Rubenstein RH (1980) Design of an artificial skin. II. Control of chemical composition. *J Biomed Mater Res* 14(2): 107–132.
115. Sheu MT, Huang JC, Yeh GC, Ho HO (2001) Characterization of collagen gel solutions and collagen matrices for cell culture. *Biomaterials* 22(13): 1713–1719.
116. Madaghiele M, Sannino A, Yannas IV, Spector M (2008) Collagen-based matrices with axially oriented pores. *J Biomed Mater Res A* 85(3): 757–767.
117. Hutmacher DW, Cool S (2007) Concepts of scaffold-based tissue engineering – the rationale to use solid free-form fabrication techniques. *J Cell Mol Med* 11(4): 654–669.
118. Hutmacher DW (2000) Scaffolds in tissue engineering bone and cartilage. *Biomaterials* 21: 2529–2543.
119. Vert M, Mauduit J, Li S (1994) Biodegradation of PLA/PGA polymers: increasing complexity. *Biomaterials* 15: 1209–1213.
120. Taboas JM, Maddox RD, Krebsbach H, Hollister SJ (2003) Indirect solid free form fabrication of local and global porous, biomimetic and composite 3D polymer-ceramic scaffolds. *Biomaterials* 24: 181–194.
121. Badylak SF (2007) The extracellular matrix as a biologic scaffold material. *Biomaterials* 28(25): 3587–3593.
122. Ye Z, Mahato RI (2008) Role of nanomedicines in cell-based therapeutics. *Nanomed* 3(1): 5–8.
123. Nair LS, Laurencin CT (2008) Nanofibers and nanoparticles for orthopaedic surgery applications. *J Bone Joint Surg Am* 90 Suppl 1: 128–131.
124. Seidlits SK, Lee JY, Schmidt CE (2008) Nanostructured scaffolds for neural applications. *Nanomed* 3(2): 183–199.
125. Kim HW, Kim HE, Knowles JC (2006) Production and potential of bioactive glass nanofibres as a next-generation biomaterial. *Adv Funct Mater* 16: 1529–1535.
126. Godbey WT, Mikos AG (2001) Recent progress in gene delivery using non-viral transfer complexes. *J Control Release* 72(1–3): 115–125.
127. De Laporte L, Shea LD (2007) Matrices and scaffolds for DNA delivery in tissue engineering. *Adv Drug Deliv Rev* 59: 292–307.
128. Jang J-H, Houchin TL, Shea LD (2004) Gene delivery from polymer scaffolds for tissue engineering. *Exp Rev Med Dev* 1(1): 127–138.
129. Pouton CW, Seymour LW (2001) Key issues in non-viral gene delivery. *Adv Drug Deliv Rev* 46(1–3): 187–203.
130. Schmidt-Wolf GD, Schmidt-Wolf IG (2003) Non-viral and hybrid vectors in human gene therapy: an update. *Trends Mol Med* 9(2): 67–72.
131. Yang YW, Yang JC (1997) Calcium phosphate as a gene carrier: electron microscopy. *Biomaterials* 18(3): 213–217.
132. Spack EG, Sorgi FL (2001) Developing non-viral DNA delivery systems for cancer and infectious disease. *Drug Discov Today* 6(4): 186–197.

133. Bisht S, Bhakta G, Mitra S, Maitra A (2005) pDNA loaded calcium phosphate nanoparticles: highly efficient non-viral vector for gene delivery. *Int J Pharm* 288(1): 157–168.
134. Roy I, Mitra S, Maitra A, Mozumdar S (2003) Calcium phosphate nanoparticles as novel non-viral vectors for targeted gene delivery. *Int J Pharm* 250(1): 25–33.
135. Bitar M, C Knowles J, Lewis MP, Salih V (2005) Soluble phosphate glass fibres for repair of bone–ligament interface. *J Mater Sci: Mater Med* 16(12): 1131–1136.

9
Developing smaller-diameter biocompatible vascular grafts

K D ANDREWS and J A HUNT,
University of Liverpool, UK

Abstract: There is a serious clinical need for the development of a replacement blood vessel, particularly with regard to the smaller vasculature structures. This chapter examines the requirements for small-calibre vascular grafts (<6 mm diameter) and discusses the development to date in achieving patent replacement vessels. The different research methodologies are investigated, including the materials and cells used, modifications (both physical and chemical) made to the grafts and the ranging experimental conditions tested. The chapter includes a focus on the technique of electrostatic spinning which shows significant potential in the areas of controlled graft production and cell–material interactions.

Key words: vascular grafts, regenerative medicine, cell–material interactions, modifications, electrostatic spinning.

9.1 Introduction

There is a serious clinical need for the development of a replacement blood vessel: in 2005–2006, there were in excess of 225,000 surgical procedures performed in the UK alone, requiring vessel replacement, intervention or augmentation (HESonline). Success to date has been limited to large-calibre vessels, such as the thoracic and abdominal aortas, with many of the surgical procedures involving the use of Dacron® (polyethyleneterephthalate, PET) and expanded polytetrafluoroethylene (ePTFE) as standard. However, the successful application of artificial vessel replacements has, thus far, been limited to vessels of greater than 6 mm diameter (L'Heureux et al., 2007). Currently, when considering smaller-diameter vessels, autologous vessels such as the saphenous vein remain the best viable option. This, however, is not always possible due to the patient's vascular history (diseased vessels, previous harvest or amputation) and also introduces the additional complication of a second surgical procedure (Kannan et al., 2005a). Hence there is an obvious need for the development of a vessel replacement for these smaller structures (i.e., <6 mm in diameter).

This chapter focuses on the requirement for a small-calibre blood vessel replacement; due to the current success seen in the use of Dacron® and ePTFE for the larger-diameter vessels, these structures (i.e., vessels >6 mm diameter) have not been covered in any great detail.

9.2 Vascular grafts with respect to natural blood vessels

In order to replace or augment a blood vessel, a vascular graft or prosthesis is needed. A vascular graft has been defined as 'any transplanted or transposed tissue that is used to replace or by-pass a part of the vascular system' (Williams, 1999). In order to achieve a high degree of patency when implanted, the selected vascular graft must have certain properties; these should be determined by first considering the natural blood vessel.

9.2.1 The natural blood vessel

The artery is the vascular structure most in need of repair and replacement; it is also the most complex of the structures in the cardiovascular system. The wall of the artery consists of three separate layers (adventitia, media and intima), each comprising different structural and cellular components. Each of these layers plays a role in the functioning of the vessel. Figure 9.1 shows the natural structure and composition of an arterial vessel. The adventitia is the outer layer and consists of collagen-rich connective tissue (collagen and fibroblasts), with few elastic fibres; the media, or middle layer, is made up of smooth muscle cells, arranged circumferentially, and elastic fibres; the intima (inner layer) surrounds the lumen of the vessel and includes endothelial cells, normally aligned to match the direction of blood flow (Baguneid et al., 2006; Sarkar et al., 2007).

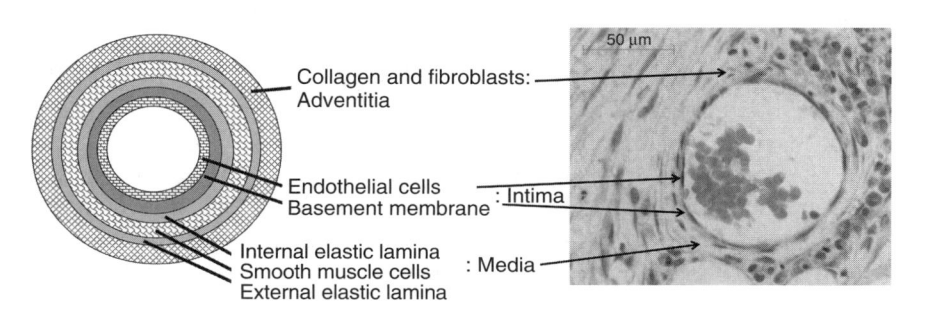

9.1 Structure of a natural arterial blood vessel.

9.2.2 Requirements of a replacement vessel, i.e. vascular graft

The properties of a blood vessel are highly complex, as is seen from the multi-layered nature of the arterial structure. In order for a vascular graft to achieve success when implanted, it must meet as many of these criteria as possible (discussed here in relation to the natural functional vessel) (Baguneid et al., 2006).

A vascular graft needs to be non-thrombogenic, non-immunogenic and resistant to infection (to be brought about by the endothelium, when in a confluent, quiescent, non-activated state); it requires appropriate mechanical properties (controlled in the natural vessel through the ratios of elastic and non-elastic tissues in the layers); the biological/cellular response induced should be non-inflammatory; it should be of appropriate dimensions for the intended location within the vasculature; it must enable the appropriate response, function and viability from involved cells (non-activated states, appropriately aligned, maintaining viability) (the grafts/vessels contact both the flowing blood and the surrounding tissue); it requires the ability to present physiological functions (constriction and/or relaxation); the graft must be non-toxic; it needs to be kink-resistant, demonstrate ease of handling and ability to suture (and to retain these sutures); and the manufacturing time and cost should be low. These requirements are summarised in Table 9.1.

Owing to the layered structure of the natural blood vessels, involving a number of cell types and fibrous tissue structures, the vessel as a whole contributes to its functioning. The endothelial cells promote a non-thrombogenic inner surface to the vessel, aligning in the direction of the blood flow to increase cell retention and remain in an unactivated state. The smooth muscle cells in the media layer are arranged in a circumferential direction, remaining in a non-proliferative state (proliferating smooth muscle cells can result in intimal hyperplasia, or narrowing of the artery). This media layer also contains elastin, an elastic component. The outer layer, the adventitia, comprises collagen and fibroblasts, with the fibroblasts producing connective tissue. The combination of the collagen and elastin present in the vessel as a whole produces an anisotropic viscoelastic nature in the structure, showing high elasticity at low pressures and increased stiffness (i.e., low distensibility) at high pressures.

Hence, in addition to meeting as many of the optimum properties as possible, there are three main key components that need to be included in any replacement blood vessel to produce a high patency rate. These are a collagen fibre analogue (component with a high tensile strength) to provide support; and an elastin fibre analogue (elastic component) to allow recoil and prevent aneurysm formation; and an endothelial analogue layer surrounding the lumen to prevent thrombosis (Baguneid et al., 2006; Isenberg et al., 2006b).

Table 9.1 Summary of requirements of an optimum vascular graft

General requirement	Specific ideal property of graft
Biological compatibility	• Non-thrombogenic • Non-immunogenic • Resistant to infection
Induce an appropriate healing response	• Non-inflammatory • Does not induce fibrous capsule formation or hyperplasia • Eventual integration of the graft within the body
Induce appropriate cellular responses	• Viable cells • Aligned appropriately (smooth muscle cells circumferentially; endothelial cells in the direction of flow) • Non-activated cells • Appropriate permeability of graft structure and material (water, cells, solute, nutrient, waste and gaseous exchange)
Demonstrate physiological functioning	• Constriction/relaxation of vessels • Appropriate peripheral resistance
Non-harmful to body	• Non-toxic
Appropriate mechanical properties	• Compliance matched to anastomosis of natural vessel • Resistant to creep • Resistant to fatigue • Physiologically appropriate strength and stiffness
Appropriate dimensions (for the intended location in the vasculature)	• Appropriate length • Appropriate diameter
Ease of handling	• Kink-resistant • Ability to suture easily • Ability to retain sutures • Easy to handle by surgical team, i.e. non-fragile
Suitable manufacturing	• Low cost • Low time period

9.3 Why development of small-calibre vascular grafts is needed

9.3.1 Clinical use to date of vascular grafts

The 'gold standard' for the replacement of large-diameter vessels (>6 mm) is still that of the autologous vessels; for example, the saphenous vein is used for lower limb arterial bypass and the internal mammary artery for coronary bypass. The preference for use is that of arterial grafts; venous grafts are prone to thrombosis and other complications, whilst arterial grafts show greater prostacyclin production, have a better blood supply and more appropriate dimensions. When these natural structures are not available, as is often the case, artificial alternatives are required (Kannan et al., 2005a). To this end, research has attempted to make available venous grafts more 'artery-like' (Wang et al., 2007).

Among the first artificial vessels, or vascular grafts, to be developed and used clinically on a wide scale were those of Dacron® and ePTFE (Baguneid et al., 2006). ePTFE (PTFE in its extruded and sintered form) is non-biodegradable, antithrombotic (due to its electronegative luminal surface) and relatively inert. Dacron® is woven or knitted polyester with a relatively high Young's modulus (stiffness) (Kannan et al., 2005a). Whilst having shown reasonable patency rates when used in large-diameter vascular applications, such as ePTFE for 7–9 mm diameter lower limb bypass grafts and Dacron® for general large-diameter vascular grafts, there is still much room for improvement when compared to the autologous vessels (Zdrahala, 1996a). Also, as previously discussed (Section 9.1), these synthetic vessels have not been used in lower-diameter clinical applications due to their high failure rates (Baguneid et al., 2006).

Reasons for failure and low patency rates of these grafts have included use of the prosthetic materials resulting in thrombus formation (particularly when used for narrow-diameter grafts and low-flow conditions); the synthetic materials have demonstrated an inherent increased risk of infection; they lacked a natural ability to heal, including the inability to spontaneously re-endothelialise *in vivo* (in human subjects – animal models do, however, show vastly improved rates of cellular regrowth); surgical placement of the grafts into the vasculature damaged the natural endothelium due to suturing at the anastomosis; the presence of a foreign material in the body could induce an inflammatory response resulting in the development of a fibrous capsule and restenosis; the materials showed limited durability and compliance mismatch resulting in intimal hyperplasia (Isenberg et al., 2006b; L'Heureux et al., 2007; Baguneid et al., 2006; Wang et al., 2007; Vara et al., 2005).

It was seen that, in relation to the natural properties of a blood vessel, there are several unresolved issues with the use of Dacron® and ePTFE, in addition to the problems of using autologous vessels. It was also seen that there have been additional complications incurred (whether considering artifical or natural vessels) for use at diameters lower than 6 mm. Hence, it was necessary to use and investigate other possibilities for use as vascular grafts.

9.3.2 Endothelialisation of vascular grafts – the regenerative medicine approach

In addition to further investigation into the material of choice for the vascular graft, it is also necessary to consider the protective layer of endothelial cells lining the lumen of the natural vessel. It is known that these cells, when confluent and unactivated, are responsible for the non-thrombogenic state of the blood vessel, preventing thrombosis and intimal hyperplasia. Therefore, it has been recognised that adding endothelial cells to the vascular grafts, normally before their implantation, would improve the patency rates (Kannan *et al.*, 2005a). To this end, endothelial cells must attach, and remain attached, to the luminal surface of the vascular graft, orienting in the direction of blood flow and remaining non-active; smooth muscle cells and/or fibroblasts must interact with the outer graft surface, orienting circumferentially in a non-active, non-proliferative state. The initiation of this approach, investigating both the choice of biomaterial and the behaviour of the cells when combined with the graft (commonly endothelial, but also smooth muscle cells and fibroblasts), was the beginning of the tissue engineering and regenerative medicine approach. By considering both the material and cellular behaviour, it was seen as possible to optimise the functioning of the vascular graft, and so enhance its patency (Wang *et al.*, 2007; Vara *et al.*, 2005; Baguneid *et al.*, 2006).

9.4 Initial scaffolds investigated for use as vascular grafts

There are four main categories of biomaterial that have been investigated for use as scaffolds for the tissue engineering/regenerative medicine based vascular grafts: synthetic, natural, decellularised, and cell-based.

9.4.1 Synthetic scaffolds

The synthetic biomaterials could be further categorised into two sections: permanent and degradable (Zdrahala, 1996a; Kannan *et al.*, 2005a, 2005b; Hoenig *et al.*, 2005).

The permanent biomaterials should not have been affected by the addition of cells or implantation (although themselves will influence/affect the cells), and should remain as a supporting scaffold for the lifetime of the vascular graft, staying in permanent contact with the cells. These materials must, therefore, have high durability and matched mechanical properties (to the anastomosis/ends of the adjoining natural vessel), should not suffer from creep or fatigue, and must show suitable cellular-material responses.

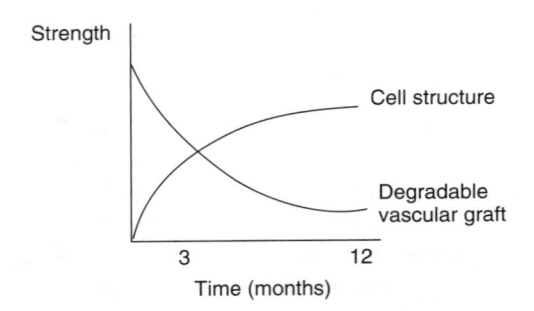

9.2 Schematic diagram to indicate the ideal: degradation of the vascular graft coordinated with the formation and increase in strength and support capability of the associated cell structure.

The degradable biomaterials were to be affected by the addition of cells, implantation, and the conditions over time. These biomaterials were intended to degrade gradually over a period of time, their rate of degradation coordinated with the formation of a self-supporting cell structure. The degradation products should not be harmful nor induce an inflammatory response. In the initial post-implantation stages, the biomaterial scaffolds should have matching mechanical properties (these alter with the degradation of the biomaterial, but should still be matched, with the continual growth and development of cellular involvement, to the natural vessel; Fig. 9.2). The scaffolds must also again have shown appropriate cellular-material interactions. Many research groups have preferred this (degradable) option for a vascular graft, with the eventual loss of the synthetic biomaterial leading to an improved patency rate and lower risk of infection and adverse immunological response. However, this scenario is seen to be more complex, with more variables and complications needing resolution.

One of the main synthetic biomaterials researched by numerous groups is that of polyurethane and its subsets (Baguneid *et al.*, 2006; Wang *et al.*, 2007; Zdrahala, 1996b). It was investigated in both permanent and degradable forms, depending on the exact type of the polyurethane being considered. The main reason for the keen interest in this polymer as a scaffold for vascular grafts was its mechanical properties, particularly its compliance which was better matched to the natural vessels than the previously investigated Dacron® or ePTFE.

Polyurethane consists of hard and soft segments, allowing control over the mechanical properties depending on the ratio of hard to soft (How and Clarke, 1984; How and Annis, 1987). The hard segments provide stiffness, rigidity and stability, while the soft allow flexibility. This allows the graft to be designed to match the compliance of its intended location within the vasculature, reducing the occurrence of intimal hyperplasia, and limiting damage to the cells and so their subsequent loss. It also allows a variety of fabrication techniques to be used, resulting in a range of possible structures and forms. However, the early forms of polyurethane investigated were prevalent to attack of the soft segments from oxidative and hydrolytic stresses induced by the conditions of implantation and cellular by-products. The next generation of polyurethanes investigated were made from either carbonate or ether groups, making them more stable and less likely to degrade, otherwise resulting in cellular loss and death. Cell death was also avoided by further chemical alteration, eliminating diisocyanates from the polymer instead substituting aliphatic diisocyanates. The originally used diisocyanates were found to degrade into toxic substances, causing adverse cell responses and carcinogenic reactions in animal models. Other modified polyurethanes examined included groups such as urea (Kannan et al., 2005a).

These early studies on polyetherurethane and polyetherurethane urea, and later work using poly(carbonate-urea) urethane and poly(ester urethane) urea (Annis, 1987; de Cossart et al., 1989; Salacinski et al., 2000; Stankus et al., 2006), have shown encouraging results with cellular attachment and some retention on the graft surfaces. There was minimal, if any, occurrence of intimal hyperplasia (Annis et al., 1978). Endothelial and smooth muscle cells were both seen to adhere and proliferate actively on the materials, with no adverse cell reasponses. However, despite encouraging results (largely seen in animal models where re-endothelialisation occurs naturally), cell experiments performed *in vitro* showed disappointing cell morphologies with a lack of spreading observed. Furthermore, perfusion experiments revealed contradictory cell retention, indicating that further work was needed for this group of materials to be used successfully as a replacement for Dacron® and ePTFE (Mohanty et al., 1992).

Although the compliance of the investigated polyurethanes was indicated to be a success (as shown by the lack of intimal hyperplasia occurring in the animal models used), the cellular interactions were not as successful, leading to the need for investigation of other new materials. The strategy was to maintain the level of compliance seen with the polyurethanes, but to further encourage a higher degree of cell interaction with the material (namely improved morphology and retention). The concept of the use of (bio-)degradable biomaterials was also now being considered. Research groups were recognising that the permanency of a material within the body held potential for long-term immunogenic responses. If the biomaterial forming the vascular graft gradually biodegraded (or was bioabsorbed), this would allow the development of a solely cell-based vascular graft. However, the cell growth had to be matched to this material degradation, in order to constantly maintain graft strength above an optimal value to prevent failure (collapse, cellular detachment) (Fig. 9.2).

Degradable polymers considered include polylactic acid (PLA), polyglycolic acid (PGA), polyhydroxyalkanoate (PHA), polydioxanone (PDS) and polycaprolactone/poly(ε-caprolactone) (PCL) (Boland *et al.*, 2004; Chong *et al.*, 2007; Kannan *et al.*, 2005a). PGA is a crystalline, hydrophilic polymer with a high degradation rate (approximately 2 weeks). PLA is a methylated form of PGA, making the polymer less hydrophilic, and it therefore has a slower rate of degradation. It was seen that as single components these polymers were not ideal, with degradation occurring too quickly (with the accompanying loss of strength) for PGA and, although the degradation rate was more suitable for *in vivo* use, the PLA was known to affect the local pH upon the production of the degradation products, hence affecting the cells. PGA was also reported to promote the cellular formation of fibrous capsules around the implants (Boland *et al.*, 2004; Kannan *et al.*, 2005a). PCL is a biodegradable, hydrophilic polyester with a slower rate of degradation than PLA. Cells could readily adhere to grafts made from this biomaterial maintaining their phenotype, although reports suggested that the mechanical properties and cellular interactions could be further improved (Chong *et al.*, 2007; Kannan *et al.*, 2005a).

9.4.2 Natural scaffolds

Synthetic biomaterials used in vascular grafts initially demonstrated variable results, with fibrous encapsulation, non-optimal degradation and non-matched mechanical properties. As previously discussed, this can produce adverse cellular responses such as cell detachment and intimal hyperplasia. Hence, to focus on the desired cellular interactions, natural biomaterials were investigated. These included collagen, elastin, fibrin, hyaluronan and polysaccharide-based biomaterials. These natural biomaterials were deemed as suitable for the manufacture of vascular grafts as they themselves comprised the extracellular matrix of the natural vessels. Collagen particularly was seen as an attractive option due to its inherent specific biological properties, capable of influencing 'natural' cellular interactions (Wu *et al.*, 2007). However, collagen and other natural biomaterials, in their 'virgin' state, showed limited physical and chemical properties, and furthermore were not easily modified. These issues readily manifested themselves in the grafts showing limited burst strength, leading to early failure, when implanted as a simple, single-component (material), tubular structure. The natural 3-D structure was not maintained or accurately reproduced in artificial fabrication, as was also the case with the naturally occurring numerous extracellular matrix proteins. Hence the natural *in vivo* environment for the growing cells was adversely altered (Kannan *et al.*, 2005b; L'Heureux *et al.*, 2007; Vara *et al.*, 2005).

Despite the non-ideal physical and chemical properties exhibited by these biomaterials, the cellular interactions produced were seen to be favourable. Polysaccharide-based grafts demonstrated neointima formation in animal models, with collagen deposition and circumferentially aligned smooth muscle cells; there was also no intimal hyperplasia or aneurysm formation (Chaouat *et al.*, 2006). Endothelial cells adhered and spread along individual fibres of silk fibroin, showing a non-activated nature and desirable growth patterns (Unger *et al.*, 2004). Hyaluronan and its derived polymers showed fibroblastic proliferation, viability and a uniform cell distribution; animal models demonstrated endothelialisation and regeneration (Figallo *et al.*, 2007; Lepidi *et al.*, 2006). Collagen, elastin and fibrin scaffolds also produced advantageous cellular interactions; endothelialisation readily occurred, expressing vWF; smooth muscle cells proliferated into a confluent layer; and extracellular matrix components were readily produced (Simionescu *et al.*, 2006; He *et al.*, 2005; Amiel *et al.*, 2006; Chan *et al.*, 2007; Isenberg *et al.*, 2006a).

9.4.3 Decellularised scaffolds

Decellularised scaffolds may be similar in composition to the natural scaffolds; however, decellularised structures are produced directly from vessels, or other naturally occurring and functioning composite tubular structures in the body (animal or human), including ureter, skin, pericardium, fascia or small intestine. The theory behind this strategy was to use a functioning 'tube', with its already present components and structure (i.e. the extracellular matrix layered, fibrous composition), with the natural cells removed (normally) through chemical means. This extracellular matrix contains similar proteins to natural blood vessels, inducing preferential cellular responses (Laflamme *et al.*, 2005). This approach held several advantages: the elastin components of the natural vessel were maintained, providing a stable structure; the structure could be prepared in advance, allowing for a potentially rapid clinical turnaround; and these materials demonstrated low immunogenicity compared to cellular tissue and synthetic scaffolds (Borschel *et al.*, 2005). Decellularised scaffolds could encourage regrowth of cells, particularly endothelial cells, promoting the desired phenotypes, morphology and near-confluency; however, they lacked the ability to further retain cells upon exposure to shear stresses and so potentially suffered from the same issues of thrombogenicity as the previously considered synthetic materials (Walles *et al.*, 2003; Mertsching *et al.*, 2005; Darby *et al.*, 2006; Kerdjoudj *et al.*, 2006).

Owing to the initially promising cell interactions and the retention of the elastin components of the original structures (so also promising mechanical properties), research has now moved to focus on pre-seeding the scaffolds with cells and/or crosslinking the graft surfaces to improve mechanical properties further and enhance cell retention. These methods need to be considered in terms of their toxicity and long-term effects upon the material and the surrounding tissue.

9.4.4 Cell-based scaffolds

In a further development to the principle of the natural structure-based grafts, methods of producing cell-based structures are being developed. These can be produced *in vitro* or *in vivo*, using laboratory or body conditions to culture and grow the vessel constructs. Methods reported include the use of a cell and matrix mixture poured into a mould; the culture of cell-sheets, which can then be wrapped around a temporary mandrel to form the desired structure, either alone or with sheets of matrix material as additional layers; and a temporary mandrel being implanted into the peritoneal cavity of an animal model for cells to form layers around (Isenberg *et al.*, 2006b). These methods have, to date, been cultured/grown only in animals and have not been extensively tested in humans with preliminary clinical trials ongoing (L'Heureux *et al.*, 2007). The initial results in these animal models have been promising, with no signs of intimal hyperplasia or aneurysm formation (Baguneid *et al.*, 2006; Vara *et al.*, 2005; L'Heureux *et al.*, 2007). The formed mechanical properties were similar to those naturally required (in animals) and different dimensions of vessels can be grown. However, the risk of adhesion in humans needs to be minimised for the use of the peritoneal method, and two surgical interventions were often required (although not always; recent research has improved on this aspect of the procedure through the use of dermal fibroblasts to form the cell sheets and the later additional use of endothelial cells or their precursors (L'Heureux *et al.*, 2007)) for the cellular procurement/'graft' implantation and the 'graft' retrieval/graft implantation (Hoenig *et al.*, 2005). Use of this group of vascular graft scaffolds is rare in 'routine' clinical situations; however, owing to the promising initial data reported, particularly for use as arteriovenous shunts (L'Heureux *et al.*, 2007), it has been deemed worthy of further investigation; indeed, interest in this area (for regenerative medicine in general, not confined within the vascular field) has recently increased as evidenced in the results of literature searches.

9.5 Further development and modification to biomaterials used as vascular grafts

9.5.1 Combinations of biomaterials and copolymers investigated

Initial research into biomaterials suitable for use as vascular grafts showed variable results with insufficient properties inducing undesired and even adverse cellular responses. However, it was noted that the biomaterials comprised a mixture of both the desirable and unwanted properties; hence, it was deduced that a combination of selected biomaterials could combine optimum properties to obtain ideal cellular–graft interactions, leading to the focus of many recent research studies.

To date, a large and increasing number of copolymers have been or are being investigated for their suitability as vascular grafts. Some of these combinations include poly(lactic glycolic) acid (PLGA) (Stitzel et al., 2006); collagen/PLGA (Jeong et al., 2007); poly(L-lactid-co-ε-caprolactone) (P(LLA-CL)) (Xu et al., 2004b); gelatin/PCL (Zhang et al., 2005); poly(glycerol-sebacate) (Motlagh et al., 2006); biopolyester blends such as poly(3-hydroxybutyrate-co-3-hydroxyvalerate) (Tang and Wu, 2006); copolyesters such as PEGT/PBT, based on poly(ethylene glycol) and poly(butylene terephthalate) (Wang et al., 2005; Ring et al., 2006); hyaluronan-gelatin hydrogel (Mironov et al., 2005); chitosan/PEO (poly(ethylene oxide)) (Duan et al., 2004); and polyurethane-polydimethylsiloxane (Soldani et al., 1992).

These combinations, and multiple others, have produced promising results. Control was possible over the final chemical composition of the scaffolds, through the variation of the ratio of the polymer constituents. As a result, the physical bulk and surface properties of the scaffolds could be controlled and so the cellular interactions could be influenced. Owing to the many different combinations possible, these cell responses were seen to be extremely varied; it was possible to systematically and continually alter the blend of polymers until the desired end product is reached. However, the following cell–scaffold interactions were among those examined: aliphatic-aromatic copolyesters enabled the development of a confluent, adherent monolayer of bone marrow cells in the scaffold lumen; polyurethane-polydimethylsiloxane produced good results in animal models, with mature, stable luminal interfaces, no intimal hyperplasia and collagen deposition; P(LLA-CL) exhibited well-adhered smooth muscle cells, expressing a spindle-like contractile phenotype; and gelatin/PCL blends enhanced the attachment of cells due to the material's more favourable wettability, also promoting cellular ingrowth in comparison to the single polymers. These results largely demonstrated improved and more desirable cellular interactions, combined with the potential to optimise the mechanical properties of the system.

9.5.2 Further modifications

In addition to the copolymer combinations of biomaterials investigated, further modifications to the main graft constituent have been examined. These modifications included coatings of additional material and chemical/

molecular attachment of biological and/or pharmaceutical agents (e.g. RGD sequences, heparin and antibiotics).

Material coatings have been receiving increased clinical interest; it was hypothesised that the coating of vascular grafts could overcome the inability of human subjects to regrow the necessary layer of endothelial cells on the luminal surface, enhancing cell attachment and spreading and so improving the patency rates. The coatings investigated include gelatin, fibronectin, collagen (varying types both in single form and in combinations), fibrin and laminin. These were selected due to their inclusion within, or similarity to, the basement membrane that the endothelial cells cover (in their natural *in vivo* environment) (Fig. 9.1) (Ma *et al.*, 2005a; Patel *et al.*, 2006; Raman Sreerekha and Krishnan, 2006). The coating could be across the material surface as a whole, following the material structure and topography, or in a specific pattern (for example, through the use of micropattern printing). The use of these materials as coatings was reported to largely improve the initial cell adhesion to the graft luminal surfaces, as well as the spreading and proliferation (He *et al.*, 2005; Shindo *et al.*, 1987; Kaehler *et al.*, 1989). Although there were seen to be potential issues with the retention of both the cells and the coating itself, recent studies have shown promising results when analysing the effect of both surface coating/material impregnation and the underlying graft substrate itself. Dacron® and ePTFE, but more interestingly newly investigated materials, showed the capability for optimisation of properties and cellular responses (Andrews *et al.*, 2008). Cell orientation was also affected, depending on the coating-topographical correlation; gelatin coating following the underlying fibres has been reported to orient endothelial cells along the fibres (Ma *et al.*, 2005a). Expression of surface adhesion proteins has been examined to determine the maintenance of endothelial cell phenotype; CD31, CD106 and CD54 were upregulated on coated material surfaces compared to the 'virgin' surfaces (Ma *et al.*, 2005a). Again, however, there was some evidence that the endothelialisation may have been comparable to or lower than the tissue culture plastic controls. Hence, the results appeared initially promising but were dependent on the material and nature of the coating (Fernandez *et al.*, 2005).

The coatings have also been examined for their ability to aid in the formation of a media, or smooth muscle cell, layer. Again results were initially promising, with biomaterial coatings such as gelatin and collagen improving cell adhesion and spreading. The material substrate effects were also again connected, with a range of induced porosities affecting the resultant cellular responses, and hence requiring optimisation (Choi and Noh, 2005).

The attachment of biological and/or pharmaceutical agents to the biomaterial grafts was considered to negate the potential inflammatory, foreign body and infection responses of their implantation. When considering the inclusion of the biological/pharmaceutical sequences, nano-scale arrangements of RGD and other molecules were intended to aid the attachment of cells in the appropriate morphology. It has been shown that the molecules can be successfully incorporated into and onto the graft biomaterials. Again, as with the material coatings, these incorporations can be across the material as a whole, following the underlying topography, or in a specific pattern. Molecules investigated include RGD and WQPPRARI sequences (RGDS, KQAGDV, VAPG, GRGDS) (Gauvreau and Laroche, 2005; Alobaid *et al.*, 2006; Mann and West, 2002); bioactive molecules (elastin-mimetic protein, water, heparin) (Ma *et al.*, 2005b; Jordan *et al.*, 2007; Noh *et al.*, 2006; Patel *et al.*, 2006); proteases (matrix metalloproteinases such as MMP-2 and inhibitors such as TIMP-2) (Cavallaro *et al.*, 2007); growth factors (granulocyte colony-stimulating factor) (Lahtinen *et al.*, 2007; Cho *et al.*, 2006); matrix molecules (adhesion fragments such as tropoelastin, fibrillin-1 and fibulin-5) (Stephan *et al.*, 2006; Knetsch *et al.*, 2006; Choi *et al.*, 2005); and antibiotics (Blanchemain *et al.*, 2007; Kim *et al.*, 2004). Initial results were once again mixed: protein surface adsorption produced high blood cell adhesion, indicating that *in vitro* cell treatment would first be required before implantation (Ma *et al.*, 2005b; Jordan *et al.*, 2007; Noh *et al.*, 2006; Patel *et al.*, 2006); antibiotic involvement was successfully completed with corresponding increases in permeability and strength of the vascular grafts (Blanchemain *et al.*, 2007; Kim *et al.*, 2004); and biological sequences such as RGD showed that initial cell adhesion did not depend upon the chemical structure or combination of the molecules but that the arrangements did enhance cell proliferation and spreading (Gauvreau and Laroche, 2005; Alobaid *et al.*, 2006; Mann and West, 2002). The migration of the cells was clearly augmented by the specific patterning of the sequences, also increasing the quantitative data shown for the cell adhesion and proliferation. When modified through crosslinking, RGD sequences showed particular success in improving cell retention when under shear stress, i.e. fluid flow, conditions (Gauvreau and Laroche, 2005; Alobaid *et al.*, 2006; Mann and West, 2002).

9.5.3 Structures

Studying the copolymer combinations and further modifications developed to optimise the cellular interactions showed that the cell responses were dependent upon a number of factors. It was seen that the biomaterials used to form the vascular grafts, as a whole and/or as a combination of layers, could theoretically be optimised in terms of properties and cellular interactions. The biomaterial selected to form the bulk structure was the best choice to optimise the graft properties (compliance and strength), with the surface layers (recent attention largely focusing on natural and biologically related components) capable of influencing the cell interactions (chiefly endothelial or smooth muscle cells). When investigating the surface effects upon the cells it was reported that the configuration of the material or its attachments could influence the cell behaviour. Hence, the underlying biomaterial structure was seen to be significant in controlling the cell responses. This also correlated with widely reported research performed into cell behaviour with respect to material topography and chemistry (Curtis and Wilkinson, 1998; Imbert *et al.*, 1998; Kottke-Marchant *et al.*, 1996).

The structures of traditionally used Dacron® and ePTFE are woven and porous respectively; the Dacron® prostheses consist of aligned woven or knitted yarn with spacings of 650–900 µm, the ePTFE with regularly spaced nodes or fibrils (approximately 30 µm apart) forming a microporous structure (Fig. 9.3) (Andrews *et al.*, 2008). The improvement in performance of these materials has been seen as necessary both at >6 mm diameter and significantly so at <6 mm. A significant aspect of the low patency rates has been due to the thrombogenicity of the materials, a factor related to flow-induced endothelial cell detachment.

The literature reports that cells respond to wide ranges of underlying material surface topography. This includes fibre and pore features, leading to differences between pore sizes/fibre spacings, fibre diameters, surface roughness, surface area (connected to porosity) and fibre orientation, varying on a micro- and/or nano-scale (Salem *et al.*, 2002; Chung *et al.*, 2003; Xu *et al.*, 2004a). These topographical features have been shown to influence cell adhesion, cell morphology, cell spreading, cell growth, orientation and function (Miller *et al.*, 2005; Daxini *et al.*, 2006). The cell behaviour was affected by this topography through several mechanisms, the topography altering the extent and manner in which extracellular matrix proteins were

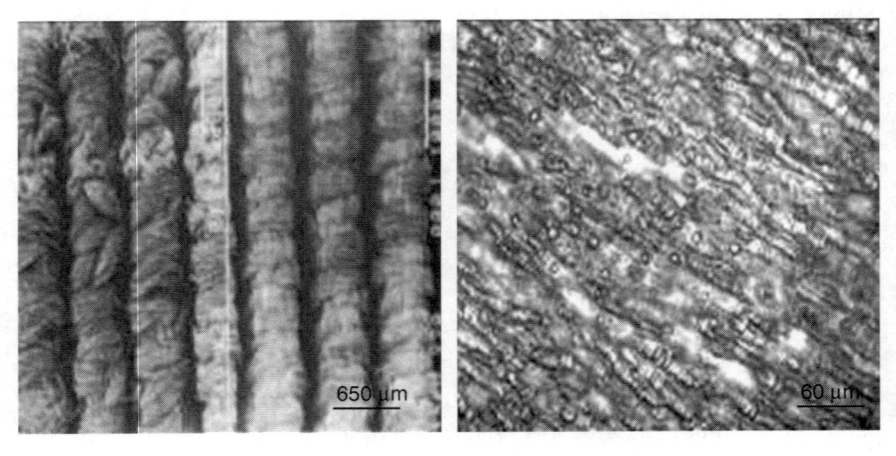

9.3 Microscopy images of Dacron® and ePTFE vascular grafts showing their structures.

adsorbed (the first stage of cell–material interaction); these features also resulted in 'contact guidance', orienting cells and concentrating cytoskeletal involvement, particularly focal adhesions (Salem et al., 2002). Due to these mechanisms, altering the topography of the underlying biomaterial could be used to influence the cell behaviour to aid and enhance or influence adhesion, retention, morphology and orientation. Provided the chosen biomaterials could be formed into a variety of topographies, through the fabrication technique(s), the features could be manipulated to produce cell behaviour optimal for the vascular graft application. These optimal cell interactions (as referred to in Section 9.3.2) included spread, confluent, non-activated, strongly adhered endothelial cells aligned in the direction of blood flow (i.e., longitudinally along the vessel); smooth muscle cells in a non-proliferative state, arranged circumferentially around the vessel; and non-activated fibroblasts also arranged in a circumferential direction, producing extracellular matrix and so connective tissue. The topography could also influence the ingrowth of the cells in the scaffold, and so the formation of a well-developed, functional vascular graft (particularly important when using degradable biomaterials as the scaffold).

Numerous reports in the literature have indicated that topographical features can be optimised for different cell types, with thresholds at both ends of the scale for producing improved cell behaviour, i.e. enhanced adhesion and retention. The majority of the research to date concerns well-defined porous or grooved features: fibroblasts could grow on a wide range of features, showing a cooperative mechanism of cell growth and spreading using their neighbouring cells to bridge gaps (pores up to 50 µm in size); endothelial cells were unable to perform this method of spreading and so showed enhanced cell behaviour on pores with smaller and more defined structures approximating to the average size of the cells themselves when spread (pore sizes of 18–60 µm being reported as optimum for endothelialisation, particularly on degradable materials, with cells unable to cover pores of greater than 80 µm and having material-dependent effects on pores of 15–30 µm; investigations concluded that features less than 18 µm induced inflammatory responses while features greater than 60 µm led to blood leakage from the grafts) (Kannan et al., 2005a); and smooth muscle cells responded to grooves of 40–160 µm width, switching from randomly aligned, fibroblast-like morphology to a spindle-shaped morphology, aligning along the grooves, approaching confluence and showing enhanced expression of smooth muscle α-actin (showing a shift to the desired non-proliferative contractile phenotype capable of producing organised extracellular matrix) (Sarkar et al., 2006; Shen et al., 2006). These topographical features could be produced in a range of biomaterials through a variety of techniques including photolithography, etching, moulding and particulate leaching.

Fibrous materials are currently receiving much increased attention as tissue engineering scaffolds. Their similarity to naturally occurring extracellular matrix, their possible production at either the nano- or micro-scale, and their organisation (as cell-seeded scaffolds) into 3-D networks have shown potential to form natural tissue analogues, forming guides for the desired hierarchical and organised (i.e., oriented and structured) cell–material interactions (Kidoaki et al., 2005). The surface topography features show similar results to those investigated on a 2-D level; however, their complex and more interactive nature requires further research into the precise effects induced. Nanostructures, both on surface and at 3-D levels, are also receiving increased attention due to this similarity to the natural extracellular matrix structures and protein arrangements.

9.5.4 Experimental conditions

In addition to the ability to alter the material, its modifications and structure, further variables can be examined to produce an optimum functional vascular graft. These variables have included the use of shear stress/flow, different cells, different seeding techniques and culture conditions.

When the vascular graft is implanted it is almost immediately subjected to the forces and stresses of blood flow. As the innermost layer of a tissue engineered/regenerative medicine vascular graft, the seeded endothelial cells are in direct contact with the flowing fluid and so are subjected to shear stresses. It is these shear stresses in the natural vessel that align the cells in the longitudinal direction. However, it is also these shear stresses that can be responsible for endothelial cell loss from the luminal surfaces resulting in thrombus formation and graft failure. It is also the pulsatile pressures resulting from the blood flow that reveal compliance mismatches in the natural vessel and the implanted graft, leading to adverse changes in the smooth muscle cells and intimal hyperplasia. To date, the majority of seeded endothelial cells have been investigated after static *in vitro* culture; recently research has begun to concentrate on the use of dynamic culture conditions and pre-conditioning of the cells under shear stress. *In vitro* and *ex vivo* testing of static seeded endothelial-lined vascular grafts at levels equivalent to 24 hours of arterial circulatory system shear stresses showed high incidence of cell loss (>50%) (Inoguchi et al., 2007). However, combinations of cells seeded and cultured under perfusion, pre-conditioning of endothelial cells with gradually increased shear stresses, and pre-conditioning of endothelial cells seeded onto RGD/biological molecule-material composite structures resulted in greater endothelial cell retention when exposed to physiological shear stresses (*in vitro*, *ex vivo* and *in vitro*) (Fernandez et al., 2006; Inoguchi et al., 2007; Lyman et al., 2006; Hahn et al., 2007; Kladakis and Nerem, 2004; Meinhart et al., 2005; Rademacher et al., 2001; Yang et al.,

2006). Pre-conditioning produced oriented endothelial cells, with elongated/ spread morphologies (Fernandez et al., 2006; Inoguchi et al., 2007; Lyman et al., 2006); smooth muscle cells responded to such conditions by altering their phenotype (with a compliance match to the desired non-proliferative phenotype), orientation, extracellular matrix production (depositing significantly higher levels of collagen) and growth factor release (Hahn et al., 2007). Hence, the application of shear stress was seen as capable of producing desirable pre-conditioning effects upon the tissue engineered grafts.

The conventional strategy for cell seeding the grafts with endothelial cells is a two-stage technique. In this method the cells are extracted from tissue and cultured for a prolonged period to increase the cell numbers before seeding. Culture for 2–4 weeks provides a high number of cells so that, even with cell loss upon exposure to shear stresses, there should theoretically be a reasonably confluent layer remaining. The clinical results for humans, however, do not support this theory, although high initial cell numbers are still sought. Once these cell numbers are reached, the cell-seeded grafts can, in turn, be cultured for endothelialisation to occur *in vitro* before implantation. Hence, this two-stage approach is time consuming, with the various stages leading to increased risk of infection, and requiring additional equipment and personnel costs. Consequently this approach is not suitable for all patients (emergency grafts are not possible with this method) and cannot be produced in all institutions. In order to overcome these issues, a single-stage procedure is currently under development. Here, the cells would be extracted during the same surgical procedure as the implantation of the grafts (the endothelial cells are harvested and immediately seeded onto the materials which are then implanted). This method immediately exposes the newly seeded cells to pulsatile blood flow and shear stresses, resulting in the loss of a high proportion of the cells. In order to negate this effect a technique known as 'sodding' is employed, where an extremely high seeding density is used to ensure some of the cells remain even after the shear stress effects. Owing to this large number of required cells, but the lack of an *in vitro* culture period to expand the numbers (in this single-stage process), other, more numerous and available, sources of endothelial cells are sought. To date, clinical success has only been notably achieved through the use of the two-stage technique (Hoenig et al., 2005; Vara et al., 2005).

The culture conditions for the tissue engineered grafts can include the static or dynamic nature of the pre-implantation environment, as already discussed, but can also involve the addition of growth factors and supplements into the culture medium. These additions can include ascorbic and retinoic acid to enhance collagen production. Due to the research into the single-stage seeding and 'sodding', alternative cell sources are being investigated with associated methods of cell extraction. Endothelial cells are normally harvested from venous or microvascular origin; alternative tissue sources being examined are subcutaneous and omental fat. Subcutaneous fat is the preferred source due to the ease with which it can be extracted (with a simple surgical procedure of liposuction, allowing a high yield of cells due to the surface area of obtained tissue that is available for enzymatic digestion). Centrifugation and percoll extraction are both used as possible methods of cell extraction from the tissue but result in reduced number of cells obtained, with poor expression of endothelial cell markers. Overpurifying the population of cells used may also result in no beneficial effect when seeded, as is also the case with respect to the continuous culture of the cells resulting in too high a passage number, the loss of the cells' proliferation potential, apoptotic cells and an over-expression of the pro-thrombotic protein vWF (Prasad Chennazhy and Krishnan, 2005; Hoenig et al., 2005; Vara et al., 2005).

The traditional cells used in vascular grafts are endothelial cells, smooth muscle cells and fibroblasts. Other cells are now also being investigated to overcome the issues of cell source and numbers, e.g. stem cells, CD14+ (monocytic) blood-derived cells, and more rarely myofibroblasts and chondrocytes (Roh et al., 2007; Krenning et al., 2007). The main cells being investigated are the endothelial cells, due to their importance in forming the non-thrombogenic barrier layer between the biomaterial scaffold and blood. A range of adult progenitor cells are now being considered for use as endothelial cells, including adipose-derived stem cells, muscle-derived stem cells, cardiac stem cells, endothelial progenitor cells (from bone marrow, peripheral blood and cord blood), haematopoietic stem cells, and mesenchymal stem cells (Sales et al., 2005; Aper et al., 2007). The use of these cells, singularly and in combination (cell type and source), is currently under investigation and development at the laboratory stage (clinical results are minimal and show limited success in animals). Genetically modified cells are also under investigation but are the subject of fierce debate due to inherent ethical issues.

9.6 Electrostatically spun scaffolds as vascular grafts

This chapter has discussed the necessary properties for an optimised functional tissue engineered/regenerative medicine approach to producing a vascular graft. Research has investigated combinations of biomaterial (all categories), in terms of chemistry, structure, modification and experimental conditions; physical and chemical properties have been examined in relation to influencing selected cellular behaviours.

Fibrous structures are receiving increased interest due to their similarity to naturally occurring extracellular matrix and their ability to be produced using a variety of fabrication techniques. One of these techniques is that of electrostatic spinning. This process has been recognised as having potential

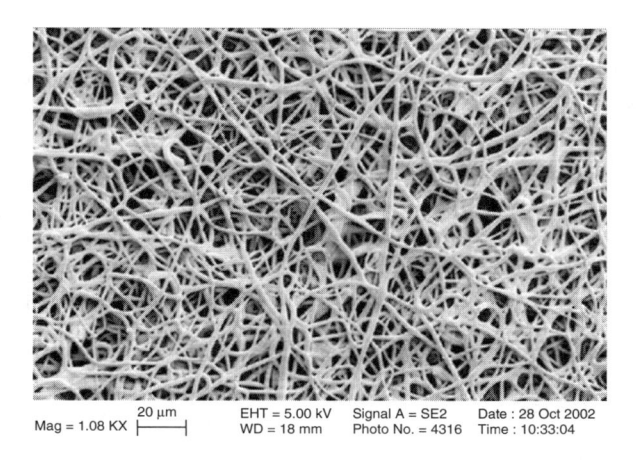

9.4 SEM image showing a typical electrostatically spun scaffold (here produced from polyurethane).

for producing scaffolds for a wide range of tissue engineering approach applications, particularly those in the cardiovascular area (Annis et al., 1978; Annis, 1987; How and Clarke, 1984; How and Annis, 1987). Scaffolds can be produced, using this method, from a wide range of polymers, both synthetic and natural. Research has shown that a variety of structures and properties could be spun in a controlled and predictable manner, with associated topographical features (Fig. 9.4) (Andrews et al., 2007a, 2007b). These features were capable of influencing the associated cellular behaviour at both early and late stages of culture (Andrews et al., 2008, 2007a, Andrews and Hunt, 2008).

The dynamic nature of the electrostatic spinning process, as compared to more static methods of scaffold fabrication, was determined to be ideally suited to producing a range of vascular graft-type constructs. The multiple spinning parameters present within the technique allow the manipulation of the polymeric fibres to form predictable structures, properties and topographical features. It was established that inter-fibre separation was affected through the alteration of flow rate, spray distance, grid and mandrel voltage; fibre diameter was affected by flow rate and mandrel voltage, void fraction by flow rate, fibre orientation by traverse speed and mandrel speed, and thickness by flow rate (Andrews et al., 2007b).

It was found that the cell interactions (to date having examined L929 murine immortalised fibroblasts, human embryonic lung fibroblasts and human umbilical vein endothelial cells) differ between the scaffolds (with their differing topographical features), the cell types and the culture periods. In general, less responsive cell types (i.e., the immortalised cells and to an extent the human fibroblasts) responded quickly to the scaffolds, adhering and spreading, resulting in greater cell coverage; with the more responsive and complex cell types investigated, there was seen to be a 'lag' period, after which the cells responded with more extremes of cell behaviour, such as cytoskeletal involvement and adhesion mechanisms (Andrews et al., 2008, 2007a, Andrews and Hunt, 2008).

Hence, this technique has been determined as worthy of further extensive research, particularly into the long-term interaction between appropriate vascular cells and the underlying scaffold structures and topographical features, and the effect upon these variables. Cellular ingrowth also requires further investigation.

9.7 Future trends

Although many important steps have been made towards forming a functioning vascular graft, based on the tissue engineering/regenerative medicine approach, there is still much research to be done. The knowledge obtained thus far needs to be furthered and combined to maximise the effects discovered. As previously discussed, the functioning of a natural vessel is due to many simultaneous factors, a principle that needs to be applied to the engineered vessel. The field needs not only to focus on the cellular interactions with the underlying biomaterial vascular graft, but also to determine their long-term effects upon this material and so its ability to perform physiologically.

The current work regarding decellularised tissues and cell-based grafts is promising, as are many of the 'modifications' that have been made to conventionally used grafts and procedures. This work needs to be continued, determining the full potentials of the strategies. Above all, any negative effect upon the body occurring as a result of chemical treatments or multiple surgical interventions requires minimising, with the time to implantation/treatment reduced.

9.8 Conclusions

There has been extensive research performed to date regarding the production of a vascular graft by adopting a regenerative medicine approach. Much knowledge has been acquired regarding the cellular interactions forming part of the vascular grafts. This knowledge needs to be combined with the biomaterials and physiological knowledge also available in order to optimise the functioning of the grafts within the body. Above all, if the field is to succeed in its production of a fully functioning, truly tissue engineered vessel, the time from concept to implantation needs to be reduced; only then will we be able to take full advantage of our knowledge of cellular responses and produce an 'off the shelf' vascular graft that is fully functional.

9.9 References

Alobaid, N., Salacinski, H. J., Sales, K. M., Ramesh, B., Kannan, R. Y., Hamilton, G. and Seifalian, A. M. (2006) *Eur J Vasc Endovasc Surg*, **32,** 76–83.
Amiel, G. E., Komura, M., Shapira, O., Yoo, J. J., Yazdani, S., Berry, J., Kaushal, S., Bischoff, J., Atala, A. and Soker, S. (2006) *Tissue Engineering*, **12,** 2355–2365.
Andrews, K.D. and Hunt, J.A. (2008) *Journal of Materials Science: Materials in Medicine*, **19,** 1601–1608.
Andrews, K. D., Hunt, J. A. and Black, R. A. (2007a) *Biomaterials*, **28,** 1014–1026.
Andrews, K. D., Hunt, J. A. and Black, R. A. (2007b) *Polymer International*, **57,** 203–210.
Andrews, K. D., Feugier, P., Black, R. A. and Hunt, J. A. (2008) *Journal of Surgical Research*, **149**(1), 39–46.
Annis, D. (1987) *Life Support Systems*, **5,** 47–52.
Annis, D., Bornat, A., Edwards, R. O., Higham, A., Loveday, B. and Wilson, J. (1978) *Trans Am Soc Artif Intern Organs*, **XXIV,** 209–213.
Aper, T., Schmidt, A., Duchrow, M. and Bruch, H.-P. (2007) *Eur J Vasc Endovasc Surg*, **33,** 33–39.
Baguneid, M. S., Seifalian, A. M., Salacinski, H. J., Murray, D., Hamilton, G. and Walker, M. G. (2006) *British Journal of Surgery*, **93,** 282–290.
Blanchemain, N., Haulon, S., Boschin, F., Marcon-Bachari, E., Traisnel, M., Morcellet, M., Hildebrand, H. F. and Martel, B. (2007) *Biomolecular Engineering*, **24,** 149–153.
Boland, E. D., Telemeco, T. A., Simpson, D. G., Wnek, G. E. and Bowlin, G. L. (2004) *J Biomed Mater Res B: Appl Biomater*, **71,** 144–152.
Borschel, G. H., Huang, Y.-C., Calve, S., Arruda, E. M., Lynch, J. B., Dow, D. E., Kuzon, W. M., Dennis, R. G. and Brown, D. L. (2005) *Tissue Engineering*, **11,** 778–786.
Cavallaro, G., Cucina, A., Randone, B., Polistena, A., Mosiello, G., Coluccia, P., De Toma, G. and Cavallaro, A. (2007) *Journal of Surgical Research*, **137,** 122–129.
Chan, B. P., Hui, T. Y., Chan, O. C. M., So, K.-F., Lu, W., Cheung, K. M. C., Salomatina, E. and Yaroslavsky, A. (2007) *Tissue Engineering*, **13,** 73–85.
Chaouat, M., Le Visage, C., Autissier, A., Chaubet, F. and Letourneur, D. (2006) *Biomaterials*, **27,** 5546–5553.
Cho, S.-W., Lim, J. E., Chu, H. S., Hyun, H.-J., Choi, C. Y., Hwang, K.-C., Yoo, K. J., Kim, D.-I. and Kim, B.-S. (2006) *J Biomed Mater Res*, **76A,** 252–263.
Choi, Y.-J. and Noh, I. (2005) *Current Applied Physics*, **5,** 463–467.
Choi, Y. J., Choung, S. K., Hong, C. M., Shin, I. S., Park, S. N., Hong, S. H., Park, H. K., Park, Y. H., Son, Y. and Noh, I. (2005) *J Biomed Mater Res*, **75A,** 824–831.
Chong, M. S. K., Lee, C. N. and Teoh, S. H. (2007) *Materials Science and Engineering*, **27C,** 309–312.
Chung, T.-W., Liu, D.-Z., Wang, S.-Y. and Wang, S.-S. (2003) *Biomaterials*, **24,** 4655–4661.
Curtis, A. and Wilkinson, C. (1998) *J Biomater Sci: Polym Ed*, **9,** 1313–1329.
Darby, C. R., Roy, D., Deardon, D. and Cornall, A. (2006) *Eur J Vasc Endovasc Surg*, **31,** 181–186.
Daxini, S. C., Nichol, J. W., Sieminski, A. L., Smith, G., Gooch, K. J. and Shastri, V. P. (2006) *Biorheology*, **43,** 45–55.
de Cossart, L., How, T. V. and Annis, D. (1989) *J Cardiovasc Surg*, **30,** 388–394.

Duan, B., Dong, C., Yuan, X. and Yao, K. (2004) *J Biomater Sci: Polym Ed*, **15,** 797–811.
Fernandez, P., Deguette, A., Pothuaud, L., Belleannee, G., Coste, P. and Bordenave, L. (2005) *Biomaterials*, **26,** 5042–5047.
Fernandez, P., Daculsi, R., Remy-Zolghadri, M., Bareille, R. and Bordenave, L. (2006) *Tissue Engineering*, **12,** 1–7.
Figallo, E., Flaibani, M., Zavan, B., Abatangelo, G. and Elvassore, N. (2007) *Biotechnol Prog*, **23,** 210–216.
Gauvreau, V. and Laroche, G. (2005) *Bioconjugate Chem*, **16,** 1088–1097.
Hahn, M. S., McHale, M. K., Wang, E., Schmedlen, R. H. and West, J. L. (2007) *Annals of Biomedical Engineering*, **35,** 190–200.
He, W., Ma, Z., Yong, T., Teo, W. E. and Ramakrishna, S. (2005) *Biomaterials*, **26,** 7606–7615.
HESonline (Hospital Episode Statistics) (2005–2006), www.hesonline.nhs.uk/
Hoenig, M. R., Campbell, G. R., Rolfe, B. E. and Campbell, J. H. (2005) *Arterioscler Thromb Vasc Biol*, **25,** 1128–1134.
How, T. V. and Annis, D. (1987) *Journal of Biomedical Materials Research*, **21,** 1093–1108.
How, T. V. and Clarke, R. M. (1984) *J Biomechanics*, **17,** 597–608.
Imbert, E., Poot, A. A., Figdor, C. G. and Feijen, J. (1998) *Biomaterials*, **19,** 2285–2290.
Inoguchi, H., Tanaka, T., Maehara, Y. and Matsuda, T. (2007) *Biomaterials*, **28,** 486–495.
Isenberg, B. C., Williams, C. and Tranquillo, R. T. (2006a) *Annals of Biomedical Engineering*, **34,** 971–985.
Isenberg, B. C., Williams, C. and Tranquillo, R. T. (2006b) *Circulation Research*, **98,** 25–35.
Jeong, S. I., Kim, S. Y., Cho, S. K., Chong, M. S., Kim, K. S., Kim, H., Lee, S. B. and Lee, Y. M. (2007) *Biomaterials*, **28,** 1115–1122.
Jordan, S. W., Haller, C. A., Sallach, R. E., Apkarian, R. P., Hanson, S. R. and Chaikof, E. L. (2007) *Biomaterials*, **28,** 1191–1197.
Kaehler, J., Zilla, P., Fasol, R., Deutsch, M. and Kadletz, M. (1989) *Journal of Vascular Surgery*, **9,** 535–541.
Kannan, R. Y., Salacinski, H. J., Butler, P. E., Hamilton, G. and Seifalian, A. M. (2005a) *J. Biomed Mater Res Part B: Appl Biomater*, **74B,** 570–581.
Kannan, R. Y., Salacinski, H. J., Sales, K., Butler, P. and Seifalian, A. M. (2005b) *Biomaterials*, **26,** 1857–1875.
Kerdjoudj, H., Boura, C., Marchal, L., Dumas, D., Schaff, P., Voegel, J.-C., Stoltz, J.-F. and Menu, P. (2006) *Bio-Medical Materials and Engineering*, **16,** S123-S129.
Kidoaki, S., Kuen Kwon, I. and Matsuda, T. (2005) *Biomaterials*, **26,** 37–46.
Kim, K., Luu, Y. K., Chang, C., Fang, D., Hsiao, B. S., Chu, B. and Hadjiargyrou, M. (2004) *J Control Release*, **98,** 47–56.
Kladakis, S. M. and Nerem, R. M. (2004) *Endothelium*, **11,** 29–44.
Knetsch, M. L. W., Aldenhoff, Y. B. J. and Koole, L. H. (2006) *Biomaterials*, **27,** 2813–2819.
Kottke-Marchant, K., Veenstra, A. A. and Marchant, R. E. (1996) *Journal of Biomedical Materials Research*, **30,** 209–220.
Krenning, G., Dankers, P. Y. W., Jovanovic, D., van Luyn, M. J. A. and Harmsen, M. C. (2007) *Biomaterials*, **28,** 1470–1479.

Laflamme, K., Roberge, C. J., Labonte, J., Pouliot, S., D'Orleans-Juste, P., Auger, F. A. and Germain, L. (2005) *Circulation*, **111**, 459–464.

Lahtinen, M., Blomberg, P., Baliulis, G., Carlsson, F., Khamis, H. and Zemgulis, V. (2007) *European Journal of Cardio-thoracic Surgery*, **31**, 383–390.

Lepidi, S., Abatangelo, G., Vindigni, V., Deriu, G. P., Zavan, B., Tonello, C. and Cortivo, R. (2006) *FASEB Journal*, **20**, 103–105.

L'Heureux, N., Dusserre, N., Marini, A., Garrido, S., de la Fuente, L. and McAllister, T. (2007) *Nature Clinical Practice: Cardiovascular Medicine*, **4**, 389–395.

Lyman, D. J., Stewart, S. F. C., Murray-Wijelath, J. and Wijelath, E. (2006) *J Biomed Mater Res Part B: Appl Biomater*, **77B**, 389–400.

Ma, Z., He, W., Yong, T. and Ramakrishna, S. (2005a) *Tissue Engineering*, **11**, 1149–1158.

Ma, Z., Kotaki, M., Yong, T., He, W. and Ramakrishna, S. (2005b) *Biomaterials*, **26**, 2527–2536.

Mann, B. K. and West, J. L. (2002) *J Biomed Mater Res*, **60**, 86–93.

Meinhart, J. G., Schense, J. C., Schima, H., Gorlitzer, M., Hubbell, J. A., Deutsch, M. and Zilla, P. (2005) *Tissue Engineering*, **11**, 887–895.

Mertsching, H., Walles, T., Hofmann, M., Schanz, J. and Knapp, W. H. (2005) *Biomaterials*, **26**, 6610–6617.

Miller, D. C., Haberstroh, K. M. and Webster, T. J. (2005) *Journal of Biomedical Materials Research*, **73A**, 476–484.

Mironov, V., Kasyanov, V., Shu, X. Z., Eisenberg, C., Eisenberg, L., Gonda, S., Trusk, T., Markwald, R. R. and Prestwich, G. D. (2005) *Biomaterials*, **26**, 7628–7635.

Mohanty, M., Hunt, J. A., Doherty, P. J., Annis, D. and Williams, D. F. (1992) *Biomaterials*, **13**, 651–656.

Motlagh, D., Yang, J., Lui, K. Y., Webb, A. R. and Ameer, G. A. (2006) *Biomaterials*, **27**, 4315–4324.

Noh, I., Choi, Y.-J., Son, Y., Kim, C.-H., Hong, S.-H., Hong, C.-M., Shin, I.-S., Park, S.-N. and Park, B.-Y. (2006) *Journal of Biomedical Materials Research*, **79A**, 943–953.

Patel, H. J., Su, S.-H., Patterson, C. and Nguyen, K. T. (2006) *Biotechnol Prog*, **22**, 38–44.

Prasad Chennazhy, K. and Krishnan, L. K. (2005) *Biomaterials*, **26**, 5658–5667.

Rademacher, A., Paulitschke, M., Meyer, R. and Hetzer, R. (2001) *Int J Artif Organs*, **24**, 235–242.

Raman Sreerekha, P. and Krishnan, L. K. (2006) *Artificial Organs*, **30**, 242–249.

Ring, A., Langer, S., Homann, H. H., Kuhnen, C., Schmitz, I., Steinau, H. U. and Drucke, D. (2006) *Burns*, **32**, 35–41.

Roh, J. D., Brennan, M. P., Lopez-Soler, R. I., Fong, P. M., Goyal, A., Dardik, A. and Breuer, C. K. (2007) *Journal of Pediatric Surgery*, **42**, 198–202.

Salacinski, H. J., Tai, N. R., Punshon, G., Giudiceandrea, A., Hamilton, G. and Seifalian, A. M. (2000) *Eur J Vasc Endovasc Surg*, **20**, 342–352.

Salem, A. K., Stevens, R., Pearson, R. G., Davies, M. C., Tendler, S. J. B., Roberts, C. J., Williams, P. M. and Shakesheff, K. M. (2002) *J Biomed Mater Res*, **61**, 212–217.

Sales, K. M., Salacinski, H. J., Alobaid, N., Mikhail, M., Balakrishnan, V. and Seifalian, A. M. (2005) *Trends in Biotechnology*, **23**, 461–467.

Sarkar, S., Lee, G. Y., Wong, J. Y. and Desai, T. A. (2006) *Biomaterials*, **27**, 4775–4782.

Sarkar, S., Schmitz-Rixen, T., Hamilton, G. and Seifalian, A. M. (2007) *Med Bio Eng Comput*, **45**, 327–336.
Shen, J. Y., Chan-Park, M. B., He, B., Zhu, A. P., Zhu, X., Beuerman, R. W., Yang, E. B., Chen, W. and Chan, V. (2006) *Tissue Engineering*, **12**, 2229–2240.
Shindo, S., Takagi, A. and Whittemore, A. D. (1987) *Journal of Vascular Surgery*, **6**, 325–332.
Simionescu, D. T., Lu, Q., Song, Y., Lee, J. S., Rosenbalm, T. N., Kelley, C. and Vyavahare, N. R. (2006) *Biomaterials*, **27**, 702–713.
Soldani, G., Panol, G., Sasken, H. F., Goddard, M. B. and Galletti, P. M. (1992) *Journal of Materials Science: Materials in Medicine*, **3**, 106–113.
Stankus, J. J., Guan, J., Fujimoto, K. and Wagner, W. R. (2006) *Biomaterials*, **27**, 735–744.
Stephan, S., Ball, S. G., Williamson, M., Bax, D. V., Lomas, A., Shuttleworth, C. A. and Kielty, C. M. (2006) *J Anat*, **209**, 495–502.
Stitzel, J., Liu, J., Lee, S. J., Komura, M., Berry, J., Soker, S., Lim, G., Van Dyke, M., Czerw, R., Yoo, J. J. and Atala, A. (2006) *Biomaterials*, **27**, 1088–1094.
Tang, X. J. and Wu, Q. Y. (2006) *J Mater Sci: Mater Med*, **17**, 627–632.
Unger, R. E., Peters, K., Wolf, M., Motta, A., Migliaresi, C. and Kirkpatrick, C. J. (2004) *Biomaterials*, **25**, 5137–5146.
Vara, D. S., Salacinski, H. J., Kannan, R. Y., Bordenave, L., Hamilton, G. and Seifalian, A. M. (2005) *Pathologie Biologie*, **53**, 599–612.
Walles, T., Herden, T., Haverich, A. and Mertsching, H. (2003) *Biomaterials*, **24**, 1233–1239.
Wang, L., Wu, Y., Chen, L., Gu, Y., Xi, T., Zhang, A. and Feng, Z.-G. (2005) *Current Applied Physics*, **5**, 557–560.
Wang, X., Lin, P., Yao, Q. and Chen, C. (2007) *World J Surg*, **31**, 682–689.
Williams, D.F. (1999) *The Williams Dictionary of Biomaterials*, Liverpool, Liverpool University Press.
Wu, H.-C., Wang, T.-W., Kang, P.-L., Tsuang, Y.-H., Sun, J.-S. and Lin, F.-H. (2007) *Biomaterials*, **28**, 1385–1392.
Xu, C., Yang, F., Wang, S. and Ramakrishna, S. (2004a) *J Biomed Mater Res*, **71A**, 154–161.
Xu, C. Y., Inai, R., Kotaki, M. and Ramakrishna, S. (2004b) *Biomaterials*, **25**, 877–886.
Yang, Z., Tao, J., Wang, J.-M., Tu, C., Xu, M.-G., Wang, Y. and Pan, S.-R. (2006) *Biochemical and Biophysical Research Communications*, **342**, 577–584.
Zdrahala, R. J. (1996a) *Journal of Biomaterials Applications*, **10**, 309–329.
Zdrahala, R. J. (1996b) *Journal of Biomaterials Applications*, **11**, 37–61.
Zhang, Y., Ouyang, H., Lim, C. T., Ramakrishna, S. and Huang, Z.-M. (2005) *J Biomed Mater Res: Part B*, **72B**, 156–165.

10
Improving biomaterials in tendon and muscle regeneration

V MUDERA, U CHEEMA, R SHAH and M LEWIS,
University College London, UK

Abstract: This chapter discusses current literature on use of synthetic and native biomaterials in tendon and muscle replacement and strategies for tissue engineering of these structures. An overview of different approaches to tendon and skeletal muscle *in vitro* and *in vivo* is also covered. Utilising polymer scaffolds has resulted in mechanically strong structures; however, where structure is function, mimicking structural architecture has been problematic. The use of native protein scaffolds and cell-generated matrices shows promise for future development of tissues with good structural and functional outcomes.

Key words: biomaterials, musculoskeletal tissue engineering, synthetic polymers, native collagen material, biomimetic materials, tissue architecture.

10.1 Tendons and their mechanical properties

Tendons are typically described as dense fibrous connective tissues that attach muscles to bones. They transmit mechanical forces generated by muscle to the skeleton and thus are crucial for movement. Performing primarily a mechanical function, the material properties of a tendon are determined by its structure, i.e. structure determines function. Tendons are complex materials composed primarily of water (55% of wet weight), proteoglycans (<1%), cells and Type I collagen (85% of dry weight) along with smaller amounts of other collagens such as collagens Type III, V, XII and XIV.

The mechanical properties of tendons are dominated by the fibrous protein collagen. They possess high tensile strength which is crucial in mediating the normal movement and stability of joints. It is important that the material properties of all connective tissues, including tendon, fulfil a variety of tensile, compressive, shear and torsional load-bearing functions by standard adaptations of the fibrous, anisotropic form of their collagen component.

Along with the matrix component of tendons, there is the cellular aspect. Compared to tissues such as liver and skeletal muscle, tendons have relatively low cellularity. Resident cells are fibroblasts, often termed 'tenocytes'.

Typically these cells are elongated and found aligned parallel to collagen fibres in tendon. Resident cells help maintain extracellular matrix components and also respond to mechanical load and injury in tendons. The mechanisms for this are not yet well elucidated.

10.1.1 Structure of tendon

The primary structure in the hierarchy of tendons is collagen polypeptides which are characterised by the presence of glycine at every third amino acid. Three polypeptides wind into a triple helix. Collagen monomers are assembled into fibrils which are then grouped into fibres. The bundles of collagen fibres along with fibroblasts (the resident cells) are grouped into fascicles. Tendons transmit load with minimal loss of energy and deformation but are not entirely inextensible. They are capable of withstanding up to 4% strain (elastic range), after which the viscous range commences and they are classified as viscoelastic (Goh *et al.* 2003).

10.2 Biomaterials used in tendon regeneration

Tendon reconstruction following damage due to trauma traditionally used biological auto grafts. However, factors such as donor site morbidity and limited source have limited extensive use of biological grafts. The use of biomaterials to replace damaged tendons has to fulfil specific criteria to enable long-term functional outcomes. Besides biocompatibility, mechanical properties have to be tailored to match the viscoelastic properties of tendon and this has proved challenging. Recent advances in tissue engineering have brought about a change in concept. The principal theme is to use a bioartificial scaffold as a framework which can be seeded with cells. Using mechanical and/or cytokine stimulation, cells should deposit a natural ECM to replace the degrading scaffold. Another approach is to start with a natural protein scaffold. However, there are associated problems with such approaches to tendon repair and regeneration, the primary one being that such protein scaffolds do not reach the required *in vivo* strength and the rate of tissue regeneration is too slow.

Traditional replacement of tendon structures *in vivo* has broadly adopted three strategies: (1) synthetic biocompatible non-degradable polymers, (2) synthetic biocompatible degradable polymers, and (3) natural proteins (predominantly collagen based).

The use of synthetic non-degradable polymer-based constructs (Gore-Tex and Dacron) has been reported with limited long-term results. Gore-Tex is a porous form of polytetrafluoroethylene with a microstructure characterised by nodes interconnected by fibrils. It has been used to aid repair of tendons *in vivo*, (Kollender *et al.* 2004). In this study, secondary reconstruction of the extensor mechanism using the middle third of the quadriceps tendon and the patellar retinaculum was augmented with Gore-Tex strips. Function was evaluated using a system which assigns numerical values (0–5) for each of six categories: pain, function, emotional acceptance, walking and gait ability, use of external supports, and patient satisfaction. Markedly increased muscle strength, especially of the hamstrings, was noted after surgery. After surgery, all patients regained active extension. All patient were ambulatory, and all patients reported no limitations in daily life activities. Overall function was estimated clinically to be good to excellent. (Kollender *et al.* 2004).

Dacron is another synthetic material that has been used successfully in the tissue engineering of tendons. Dacron, or polyethylene terephthalate, is a thermoplastic polymer resin of the polyester family. There are studies in which, following injury, implantation of silicone/Dacron tendon interposition prosthesis to reconstruct a 4-cm deficit in the flexor tendon not only gave good functional outcome, but does so up to 25 years post-implantation (Ketchum 2000). Such devices are effectively prostheses; however, as long as they retain their material properties and give replacement tissues good function, they will have applications in regenerative medicine. Current research is focused on developing biomaterials with suitable mechanical properties which are degraded and replaced by the host tissue *in vivo* (i.e., not prostheses), and tissue engineered replacements which when tested should give similar or improved clinical outcomes, compared with synthetic prosthetic materials.

The use of Gore-Tex and Dacron to repair and replace tendons and ligaments has shown that they induce (1) inflammatory reaction, (2) very small amounts of new collagen fibres laid down with poor orientation of collagen, and (3) adverse reactions to wear particles of these synthetic materials.

In recent times many synthetic biocompatible degradable polymers have been tested *in vivo*. These include PLA or poly(lactic acid) and PCL or poly-ε-caprolactone (Sato *et al.* 2000), PLLA or poly(L-lactic acid) (Laitinen *et al.* 1992), PGA or polyglycolic acid (Liu *et al.* 2006) and pHEMA or poly(2-hydroxyethylmethacrylate) (Davis *et al.* 1992).

Sato and colleagues conducted a study in which four bioabsorbable materials were used as scaffold implants in the reconstruction of extra-articular ligaments or tendons: (1) poly-*N*-acetyl-D-glucosamine (chitin), (2) poly-α-caprolactone (p-CL), (3) polylactic acid (PLA), and (4) chitin/p-CL composite (Sato *et al.* 2000). The tendon regeneration process was assessed using macroscopic, histologic, immunohistochemical, and mechanical (to failure) techniques. Both PLA and the chitin/p-CL composite tendons showed increased in-growth of fibrous tissue, suggesting that these materials are promising as artificial tendons (Sato, *et al.* 2000).

Studies conducted by Laitinen and colleagues suggest modest increases in the mechanical properties of an anterior cruciate ligament ovine model where PLLA was implanted (Laitinen et al. 1993). Up to 48 weeks post-implantation the maximum load increased to 21% (from 9% at week 6), and the axial rigidity was 48%, which were considered acceptable mechanical properties. These may seem relatively low, though good functional recovery with minor radiographic changes was reported alongside this. The authors suggested that strengthening of the reconstruction in the fascia lata-PLLA group was probably because the maturation of the fibroconnective tissue and orientation of the collagen fibres were higher using PLLA augmentation (Laitinen et al. 1993).

Recently Liu and colleagues used both autologous dermal fibroblasts and tenocytes seeded on polyglycolic acid (PGA) unwoven fibres to form cell-scaffold constructs (Liu et al. 2006). These were cultured *in vitro* for 7 days before *in vivo* implantation to repair a defect of flexor digital superficial tendon, and tested up to 26 weeks. The main aim was to test whether dermal fibroblasts could give similar results in terms of repair (tested for mechanical integrity and gross histology) as tenocytes. This is important as dermal fibroblasts provide a more accessible and abundant cell source. Cell sourcing is an important strategic consideration as taking skin biopsies and culture expanding dermal fibroblasts is far less invasive than having to biopsy a tendon and culture expand tenocytes. Tensile strength for PGA scaffolds seeded with both cell types was about 75% of natural tendon strength. By week 26, both engineered tendons exhibited histology similar to that of natural tendon, collagens became parallel throughout the tendon structure, and PGA fibres were completely degraded. A startling finding was the absence of collagen Type III in either of the engineered tendons. Collagen Type III is usually found in very early stages of tendon healing and is thought to compromise the mechanical strength. Natural tendon tissue usually does not contain collagen III; this type of collagen is a normal component of skin tissue. This finding indicates that dermal fibroblasts may undergo a phenotype change during *in vivo* tendon formation under mechanical loading (Liu et al. 2006).

Work by Davis et al. (1992) has focused on engineering artificial tendons from composites of water-swollen poly(2-hydroxyethylmethacrylate)/poly(caprolactone) blend (PHEMA/PCL) hydro gel matrix reinforced with poly(lactic acid) fibres. The fibres were filament wound to mimic the structural hierarchy of the natural tendon in a simplified manner. The engineered constructs were then implanted *in vivo* into the Achilles of rabbits, with good integration after 45 days, and more importantly, degradation of the fibre and matrix was observed followed by in-growth of new and differentiated collagenous material after 45 days (Davis et al. 1992).

Type I collagen is the most widely used scaffold material since it was observed that fibroblasts contract a collagen gel to form a tissue-like structure (Bell et al. 1979). However, the limitations of using native proteins, including collagen, is that at present the mechanical properties of *in vitro* cell-seeded collagen constructs are inferior compared to the native tissue they have to replace.

Current research is being directed towards understanding how best to increase mechanical properties of collagen-based TE devices (Awad et al. 2003, Butler et al. 2008). One major advantage of using cell/collagen constructs is the inherent cell-based modelling which occurs to increase strength and give aligned structure to such constructs. Awad and colleagues have focused their efforts on improving biomechanical properties of such constructs for tendon repair (Butler et al. 2008). By casting Type I collagen and mesenchymal stem cells in wells which contained a tensioned suture in the longitudinal axis (to mimic tendon alignment and mechanics) to aid gel attachment and contraction, they form relatively dense constructs, with both cellular and collagen components aligned along the principal axis of strain. Following *in vitro* maturation of these constructs, they were implanted in a rabbit tendon repair model and tested from 12 to 26 weeks. The authors have observed that reducing the cell/collagen ratio and changing the suture fixed points at the ends resulted in failure forces greater than peak *in vivo* forces that were measured. Through combining their original collagen gels with collagen sponges, they observed enhanced stiffness and maximum force strength of the implanted constructs, and noted that mechanical stimulation for up to 12 days of such constructs in bioreactors prior to implantation results in increased biomechanical properties (Butler et al. 2008).

A novel approach to tissue engineering tendons is culturing cells in a mechanically active, three-dimensional environment, using Type I collagen. Cells cultured in 3-D collagen gels express a more native state phenotype because they form a syncytial network which can be mechanically loaded (Garvin et al. 2003). Bio-artificial tendons or BATs have been developed by Banes and colleagues, by embedding avian flexor tendon cells into Type I collagen and setting over flexible bases, which were later subjected to mechanical loading. Results have indicated that expression of collagen genes I, III and XII as well as aggrecan, fibronectin and tenascin is consistent with expression levels of cells grown on collagen-bonded 2-D surfaces or in native avian flexor tendon (Garvin et al. 2003).

The approaches taken to overcome the challenges of tendon tissue engineering have been varied, from the use of synthetic polymer scaffolds through to native protein scaffolds, as well as relying to some degree on cell-generated ECM components to supplement scaffolds, each with varying degrees of success. Examples of the use of polymer scaffolds such as Gore-Tex and Dacron in clinical practice have shown promise; however, there remains a drive to engineer functional tendon, using native protein scaffolds. An additional outcome to this approach is the development of effective models to understand formation of tendon tissue, with the potential to replace much larger defects. As understanding of the process of regenerating and rebuilding tendon becomes more sophisticated, it is not difficult to envisage complex composite materials being used for tissue engineering tendon.

10.3 Muscles and their mechanical properties

It is understood that complex mechanical forces generated in the growing embryo play a crucial role in organogenesis. The differentiation and development of skeletal myoblasts *in vitro* is very different from that *in vivo*, as many aspects of the signalling are lost. This includes the missing nervous and vascular systems. However, the contribution of 3-D extracellular matrix through which alignment occurs, and the generation of mechanical tension, which is transmitted to skeletal muscle elements, are also key factors. Within developing muscle, complex patterns of mechanical loading are applied to the skeletal myoblasts and myofibres through the elongating skeleton and by foetal movements.

10.3.1 Structure and function of skeletal muscle

Skeletal muscle is a classical example of 'structure determining function'. Approximately 48% of body mass is made up of skeletal muscle (Saxena *et al.* 1999). It is responsible for the voluntary control and active movement of the body. Skeletal muscle consists of elongated, multinuclear muscle fibres encapsulated within connective tissue sheaths.

These 'composites' of muscle fibres and connective tissue are highly contractile structures with high cell density and cellular orientation. Adult fibres have specialised endpoint attachments to tendons (myotendinous junctions) that form an elaborate system for force transmission. As the muscle contracts and shortens, these forces are transmitted to the skeleton via the tendon and produce movement (Hanson & Huxley 1953).

The extracellular matrix (ECM) acts as a physical scaffold for the muscle fibres and has an essential role in the control and maintenance of cellular function. The ECM is composed of interstitial connective tissue and basal lamina that is in intimate contact with muscle fibres (Sanes *et al.* 1986). The basal lamina is composed of collagen IV, the glycoprotein laminin, ectactin and heparan sulphate proteoglycans (HSPGs), while the interstitial ECM (endo-, epi- and perimysium) is composed of collagen I, fibronectin and HSPGs (Lewis *et al.* 2001). As well as a structural role, many ECM components have a direct effect on the determination, movement, attachment, proliferation, alignment and fusion of cells committed to the myogenic lineage. For example, laminins present in the basal lamina act as major ligands for cell surface receptors involved in the transmission of force from the cell interior (Gullberg et al. 1998, Maley et al. 1995). Furthermore, proteoglycans are essential for binding growth factors to their receptors, and proteolytic fragments of fibronectin and laminin act as chemotactic signals for myogenic cells (Adams & Watt 1993, Bischoff 1997, Gullberg et al. 1998). Of vital importance to skeletal muscle regeneration are a population of mesenchymal fibroblasts that reside within this connective tissue milieu.

10.4 Biomaterials used in muscle regeneration

Biomaterials that are used for scaffold purposes may be constructed from synthetic or natural substances using a variety of techniques to impart desirable chemical, physical and biological characteristics. Biodegradable polyesters are frequently used and these include those derived from naturally occurring α-hydroxy acids (e.g., polyglycolic acid (PGA) and poly-L-lactic acid (PLLA)) and polycaprolactone (PCL). Other synthetic biomaterials that have been used include polyurethanes and polypropylene. Biological polymers have also been investigated and have focused exclusively on collagen and alginates. Recently, there has also been interest in the use of inorganic polymers such as soluble phosphate glasses. Shah et al. (2005) reported the use of phosphate-based dissolving glass fibres within cultures to provide topographic cues to which myoblasts align. This was found to promote differentiation and formation of parallel arrays of fibres, mimicking the natural structure of skeletal muscle in vitro.

The format or morphology of the material presented to the cells is key and various processing techniques have been devised. These include extrusion, moulding, solvent casting or solid free-form technology, utilised to construct the materials into fibres, porous sponges, tubular structures, hydrogel delivery systems and various other configurations (Mooney & Mikos 1999, Hutmacher 2001). Although these morphologies have been used to enable the seeded cells to adhere, proliferate and differentiate into muscle fibres, the majority do not represent the anatomical arrangement of native skeletal muscle. Materials formed as fibres have the scope to provide a biomimetic scaffold with a parallel alignment to enable the seeded cells to assume the form of native skeletal muscle fibres both in vitro and in vivo.

In terms of biocompatibility, biological polymers are ideal but they often suffer from very poor mechanical properties and are not always conducive to fibre formation. Nevertheless, investigations have shown very positive results with both collagen and alginates. Cylindrical gels of the former have been implanted with both C2C12 and primary chick MPCs and maintained in bioreactors. Histology of these constructs indicated nascent myofibres (myotubes), which were metabolically active (Shansky et al. 2006). Both native and peptide-modified alginate gels have been seeded with primary mouse and rat MPCs and investigated in both *in vitro* and *in vivo* environments. Myogenic differentiation was observed in both cell types seeded within both the native and peptide-modified gels *in vitro*. *In vivo* experiments using cell-seeded implants of native and peptide-modified gels also showed myogenic differentiation complementing the *in vitro* experiments (Hill et al. 2006, Kamelger et al. 2004).

For fibre-related morphologies, PGA has proved to be one of the first materials tested in tissue engineering applications due to its clinical acceptability, as it is used in dissolvable sutures (FDA approved). Primary rat MPC-loaded meshes consisting of 12 µm diameter fibres were implanted *in vivo* and allowed to integrate for up to 6 weeks, with the resulting histology indicating the formation of multinucleated syncytia reminiscent of myofibres (Saxena et al. 1999, 2001); (Kamelger et al. 2004). PLLA, a related polyester, can be formed into fibres by electro-spinning (Huang et al. 2006) or extrusion (Cronin et al. 2004), generating a wide range of fibre diameters (0.5 to 60 µm). Parallel arrays of fibres (mimicking the fibre arrangement of skeletal muscle) can be produced by tension alignment (Huang et al. 2006) or bundling (Cronin et al. 2004). Myotube formation and maturation was subsequently achieved following seeding of these scaffolds with C2C12 and primary human MPCs. Importantly, cellular response was only seen when the fibres were pre-coated with biological matrices such as matrigel, laminin, fibronectin and collagen (or its denatured product, gelatin) (Huang et al. 2006, Cronin et al. 2004). PCL is similarly well received clinically and C2C12 cell attachment and proliferation was supported on gravity-spun 80–150 µm diameter fibres in random arrays. Coating with a biopolymer (gelatin) was again a prerequisite (Williamson et al. 2006).

The potential disadvantage of these 'classical' degradable polyesters is that they are not particularly elastic while some newer materials can still be made into fibres but also possess highly improved elastomeric properties. Electro-spun DegraPol® (a degradable block elastomeric polyesterurethane) meshes (10 µm diameter fibres) have been seeded with C2C12 and L6 cell lines as well as human MPCs and shown to support adhesion, proliferation and differentiation into multinucleate myotubes *in vitro* (Riboldi et al. 2005). Once again, coating with matrigel, collagen or fibronectin was required to elicit the optimum biological effect. Thus, although non-biological organic polymers may be easy to manufacture to specification, a major drawback is the necessity to pre-coat the surfaces with a protein to enable cells to adhere. Interest has also focused on tissue engineering skeletal muscles with scaffolds made from materials that are elastomeric if not degradable. Attachment, proliferation, migration and differentiation of the G8 myogenic cell line was supported on such a spun-cast material (Mulder et al. 1998).

As can be appreciated, the fibre diameters are extremely important, with the optimum diameters appearing to be in the 10–100 µm cross-sectional area. What is less widely investigated is the ideal spacing between such fibres, although it has been suggested that 30 to 55 µm gaps between fibres was a prerequisite for nascent C2C12 muscle fibre (myotubes) formation on non-degradable polypropylene fibres; again laminin coating was vital (Neumann et al. 2003).

Despite the very encouraging results with non-biological organic polymers, there are some problems emerging as investigations develop. One of the major issues appears to be the mode of degradation of the polymers in that they degrade by hydrolysis and the breakdown products are acidic in nature, which may interfere with the regenerative process. Furthermore, as the degradation response is not iterative to the biological response, if the biological response lags behind, the constructs can fail without any control. Some attention has therefore been directed towards other materials that are biocompatible without the problems discussed above. Lin et al. (1994) developed bioabsorbable glass fibres composed of calcium iron phosphate. The solubility rate of such glasses is correlated with composition, and because it is highly linear with time, it is very predictable. These glasses are polymeric in nature and form fibres easily. Further modifications may be made by the addition of iron oxide, aluminium or magnesium, which have a strong effect on the glass network. Random meshes of heat-drawn phosphate-based glass fibres (10 µm diameter) have been shown to support the attachment, proliferation and myogenic differentiation of both human and murine MPCs; however, in common with the organic polymeric fibres, coating of the glass fibres with gelatin or matrigel was essential (Borrebaek 1966, Shah et al. 2005).

To study mechanical effects on skeletal myoblasts *in vitro*, three-dimensional cell culture has been employed to help mimic some of the actions of a 3-D environment on myotubes. The culture of cells within a 3-D environment is the next crucial advance in successful engineering of muscle, be it for tissue replacement *in vivo* or for development of pharmaceutical testing; this is a key link translating 2-D *in vitro* studies to functional biomimetic muscle tissue.

With regards to the specifics of engineering skeletal muscle, the effects of certain ECM components are well documented, and growing cultures in 3-D matrices which aid myogenesis will have benefits in maturation of the tissue model in question. In the presence of extracellular matrix in 2-D culture, differentiation of myoblasts is enhanced. Flasks are often coated in either gelatine or collagen to provide the extracellular matrix. The disruption of cell–matrix adhesion results in reduced myogenesis both *in vitro* and *in vivo*, thus highlighting the importance of cell–matrix interactions with laminin (Knudsen 1990). Disruption of adhesion of myoblasts to the matrix components, laminin and fibronectin, has resulted in reduced myogenesis (Knudsen 1990). In the presence of laminin, a matrix component, rodent muscle cells *in vitro* have increased cell motility, bipolar morphology and stimulated DNA synthesis (Goodman *et al.* 1989a). Increased levels of laminin result in enhanced myogenesis in newborn mouse cultures (Goodman *et al.* 1989b). Laminin and fibronectin are just two of the matrix components found under the basal lamina of skeletal muscle fibres. Following muscle damage, myogenic precursor cells (MPC) proliferate and differentiate. Although the exact roles for such matrix components are not completely understood, it is proposed that such cell–matrix adhesion promotes cell migration, whereas cell–cell adhesion molecules may promote myoblast interaction prior to and during cell fusion.

There are three main approaches to tissue engineering of skeletal muscle, in terms of biomaterials. Cells are seeded into either biodegradable polymer scaffolds or native matrix (i.e. collagen). Otherwise cells are left to produce their own ECM. Among the different groups which engineer muscle, there are known factors which affect the growth and myogenesis of these cultures. Dennis *et al.* have engineered 3-D skeletal muscle tissue 'myooids' from different primary muscle cultures (Dennis *et al.* 2001, Dennis & Kosnik 2000). Their sources of primary myoblasts included adult mouse, neonatal mouse and adult rat. Alongside this, co-cultures of mouse cell line myoblasts and fibroblasts have also been used in muscle cultures (Dennis *et al.* 2001, Dennis & Kosnik 2000). Co-cultures of the myoblast and the fibroblast cell lines have been used to mimic the cell combination found in the primary cultures. Also it is known that fibroblasts provide important cues, promoting myooid formation (Dennis *et al.* 2001). It seems likely that fibroblasts were important in part at least, to produce extracellular matrix to mechanically support the myooid itself. These cell systems were grown to form myooids (Dennis *et al.* 2001, Dennis & Kosnik 2000). Monolayer cultures of cells eventually formed their own extracellular matrix which detached from the substratum and spontaneously rolled up to form cylindrical structures, which is probably the function of fibroblast contraction of the collagen network. Within a two-week time frame the central portion of the structure was filled with fused differentiated myoblasts, and an outer ring of fibroblasts was present. The different contractile properties of the myooids were compared, both isometric and induced by electrical stimulation.

Successful engineering of skeletal muscle models has been possible by using both native materials and synthetic polymers (Levenberg *et al.* 2005, Powell *et al.* 2002). Early studies by Vandenburgh and colleagues (1991) used either collagen-coated substratum, over which cells were seeded, or

3-D culture of cells embedded in Matrigel (combination of ECM components, growth factors and glycosaminoglycans) in combination with collagen, or simply embedded in pure Type I collagen scaffold (Powell et al. 2002, Shansky et al. 2006). One of the benefits of using a 3-D matrix material which can be easily deformed (such as Type I collagen at low concentrations) is the ability of cells to actively remodel the biomaterial to aid with tissue architecture and, in the case of skeletal muscle, alignment. By attaching 3-D cell-seeded, Type I collagen at two ends, endogenous cell-generated forces align along the principal axis of strain, hence aligning fibres along this axis (Cheema et al. 2003). In this case cell-generated ECM components were crucial in the development of this model.

Some of the benefits of using native components of the ECM are that they actually enhance some of the processes required for maturation of any tissue model. In the case of skeletal muscle, collagen is known to enhance myogenesis. So a native material like collagen can provide biological benefits, in terms of aiding tissue maturation; it can be remodelled by cell-generated endogenous forces to give specific required architecture, i.e. alignment, which is also crucial to successful myogenesis, as well as good perfusive properties for both oxygen and nutrients. In terms of perfusive properties, studies have been carried out to establish the diffusion coefficient of Type I collagen scaffolds at defined densities for both glucose and oxygen, which are 1.3×10^{-6} cm^2 s^{-1} and 4.48×10^{-6} cm^2/s^{-1} respectively (Rong et al. 2006).

It is known that at the extremities of muscle, there is myotendinous junction formation. This transitional region is found to have collagen fibres of tendon inserted in foldings of the basal lamina. It is true that interstitial fibroblasts synthesise most of the tenascin (the myotendinous antigen) found in muscle tissue, and this serves as a marker for myotendinous junction development (Chiquet & Fambrough 1984). The ends of the mechanically loaded skeletal muscle cultures stained for tenascin (Vandenburgh et al. 1991). This corresponds well with the region of the muscle construct under the greatest amount of tension in the mechanical cell stimulator and indicates the possible role of mechanical loads in initiation of myotendinous development of muscle in the embryo *in vivo*.

There have also been examples of successful tissue engineering approaches in skeletal muscle constructs using polymer scaffolds, one of the most successful being a composite of 50% polylactic acid (PLLA) and 50% polylactic-glycolic acid (PLGA) (Levenberg et al. 2005). This material was highly porous (up to 93%) and pore sizes ranged from 212 to 600 µm (Miyagawa et al. 2002). This group has been able to successfully employ a 3-D multi-culture system consisting of myoblasts, embryonic fibroblasts and endothelial cells, with each cell type providing specific cues for the overall development of the construct (Levenberg et al. 2005, Caspi et al. 2007).

The myoblasts were the basis of the engineered muscle tissue; embryonic endothelial cells were able to induce an endothelial vessel network, as well as induce embryonic fibroblasts in the culture to become smooth muscle cells to help stabilise the vessel network (Levenberg et al. 2005). The material used was biodegradable, hence it is envisaged that following *in vivo* implantation the scaffold will gradually be replaced by cell-generated ECM components.

10.5 Conclusions

Current research strategies show promising and diverse approaches to tissue engineering muscle, right from the basic understanding of differentiating myoblasts through to creating biomimetic 3-D primitive neo-muscle. Future success will depend on controlling differentiation of muscle within 3-D constructs *in vitro* and *in vivo*. Increasing complexity in the form of perfusion systems and innervation need to be engineered before larger functional structures for implantation become a reality.

10.6 References

Adams, J. C. & Watt, F. M. 1993, 'Regulation of development and differentiation by the extracellular matrix', *Development*, vol. 117, no. 4, pp. 1183–1198.

Awad, H. A., Boivin, G. P., Dressler, M. R., Smith, F. N., Young, R. G. & Butler, D. L. 2003, 'Repair of patellar tendon injuries using a cell–collagen composite', *J. Orthop. Res.*, vol. 21, no. 3, pp. 420–431.

Bell, E., Ivarsson, B. & Merrill, C. 1979, 'Production of a tissue-like structure by contraction of collagen lattices by human fibroblasts of different proliferative potential in vitro', *Proc. Natl Acad. Sci. USA*, vol. 76, no. 3, pp. 1274–1278.

Bischoff, R. 1997, 'Chemotaxis of skeletal muscle satellite cells', *Dev. Dyn.*, vol. 208, no. 4, pp. 505–515.

Borrebaek, B. 1966, 'Increase in epididymal adipose tissue hexokinase activity induced by glucose and insulin', *Biochim. Biophys. Acta*, vol. 128, no. 1, pp. 211–213.

Butler, D. L., Juncosa-Melvin, N., Boivin, G. P., Galloway, M. T., Shearn, J. T., Gooch, C. & Awad, H. 2008, 'Functional tissue engineering for tendon repair: A multidisciplinary strategy using mesenchymal stem cells, bioscaffolds, and mechanical stimulation', *J. Orthop. Res.*, vol. 26, no. 1, pp. 1–9.

Caspi, O., Lesman, A., Basevitch, Y., Gepstein, A., Arbel, G., Habib, I. H., Gepstein, L. & Levenberg, S. 2007, 'Tissue engineering of vascularized cardiac muscle from human embryonic stem cells', *Circ. Res.*, vol. 100, no. 2, pp. 263–272.

Cheema, U., Yang, S. Y., Mudera, V., Goldspink, G. G. & Brown, R. A. 2003, '3-D in vitro model of early skeletal muscle development', *Cell Motil. Cytoskeleton*, vol. 54, no. 3, pp. 226–236.

Chiquet, M. & Fambrough, D. M. 1984, 'Chick myotendinous antigen. I. A monoclonal antibody as a marker for tendon and muscle morphogenesis', *J. Cell Biol.*, vol. 98, no. 6, pp. 1926–1936.

Cronin, E. M., Thurmond, F. A., Bassel-Duby, R., Williams, R. S., Wright, W. E., Nelson, K. D. & Garner, H. R. 2004, 'Protein-coated poly(L-lactic acid) fibers provide a substrate for differentiation of human skeletal muscle cells', *J. Biomed. Mater. Res. A*, vol. 69, no. 3, pp. 373–381.

Davis, P. A., Huang, S. J., Ambrosio, L., Ronca, D. & Nicolais, L. 1992, 'A biodegradable composite artificial tendon', *J. Mater. Sci.: Mater. Med.*, vol. 3, no. 5, pp. 359–364.

Dennis, R. G. & Kosnik, P. E. 2000, 'Excitability and isometric contractile properties of mammalian skeletal muscle constructs engineered in vitro', *In Vitro Cell Dev. Biol. Anim.*, vol. 36, no. 5, pp. 327–335.

Dennis, R. G., Kosnik, P. E., Gilbert, M. E. & Faulkner, J. A. 2001, 'Excitability and contractility of skeletal muscle engineered from primary cultures and cell lines', *Am. J. Physiol. Cell Physiol.*, vol. 280, no. 2, pp. C288–C295.

Garvin, J., Qi, J., Maloney, M. & Banes, A. J. 2003, 'Novel system for engineering bioartificial tendons and application of mechanical load', *Tissue Eng.*, vol. 9, no. 5, pp. 967–979.

Goh, J. C., Ouyang, H. W., Teoh, S. H., Chan, C. K. & Lee, E. H. 2003, 'Tissue-engineering approach to the repair and regeneration of tendons and ligaments', *Tissue Eng.*, vol. 9 Suppl 1, pp. S31–S44.

Goodman, S. L., Deutzmann, R. & Nurcombe, V. 1989a, 'Locomotory competence and laminin-specific cell surface binding sites are lost during myoblast differentiation', *Development*, vol. 105, no. 4, pp. 795–802.

Goodman, S. L., Risse, G. & von der Mark, K. 1989b, 'The E8 subfragment of laminin promotes locomotion of myoblasts over extracellular matrix', *J. Cell Biol.*, vol. 109, no. 2, pp. 799–809.

Gullberg, D., Velling, T., Lohikangas, L. & Tiger, C. F. 1998, 'Integrins during muscle development and in muscular dystrophies', *Front. Biosci.*, vol. 3, pp. D1039–D1050.

Hanson, J. & Huxley, H. E. 1953, 'Structural basis of the cross-striations in muscle', *Nature*, vol. 172, no. 4377, pp. 530–532.

Hill, E., Boontheekul, T. & Mooney, D. J. 2006, 'Designing scaffolds to enhance transplanted myoblast survival and migration', *Tissue Eng.*, vol. 12, no. 5, pp. 1295–1304.

Huang, N. F., Patel, S., Thakar, R. G., Wu, J., Hsiao, B. S., Chu, B., Lee, R. J. & Li, S. 2006, 'Myotube assembly on nanofibrous and micropatterned polymers', *Nano. Lett.*, vol. 6, no. 3, pp. 537–542.

Hutmacher, D. W. 2001, 'Scaffold design and fabrication technologies for engineering tissues – state of the art and future perspectives', *J. Biomater. Sci. Polym. Ed.*, vol. 12, no. 1, pp. 107–124.

Kamelger, F. S., Marksteiner, R., Margreiter, E., Klima, G., Wechselberger, G., Hering, S. & Piza, H. 2004, 'A comparative study of three different biomaterials in the engineering of skeletal muscle using a rat animal model', *Biomaterials*, vol. 25, no. 9, pp. 1649–1655.

Ketchum, L. D. 2000, 'Twenty-five-year follow-up evaluation of an active silicone/Dacron tendon interposition prosthesis: A case report', *J. Hand Surg. [Am.]*, vol. 25, no. 4, pp. 731–733.

Knudsen, K. A. 1990, 'Cell adhesion molecules in myogenesis', *Curr. Opin. Cell Biol.*, vol. 2, no. 5, pp. 902–906.

Kollender, Y., Bender, B., Weinbroum, A. A., Nirkin, A., Meller, I. & Bickels, J. 2004, 'Secondary reconstruction of the extensor mechanism using part of the quadriceps tendon, patellar retinaculum, and Gore-Tex strips after proximal tibial resection', *J. Arthroplasty*, vol. 19, no. 3, pp. 354–360.

Laitinen, O., Tormala, P., Taurio, R., Skutnabb, K., Saarelainen, K., Iivonen, T. & Vainionpaa, S. 1992, 'Mechanical properties of biodegradable ligament augmentation device of poly(L-lactide) in vitro and in vivo', *Biomaterials*, vol. 13, no. 14, pp. 1012–1016.

Laitinen, O., Pohjonen, T., Tormala, P., Saarelainen, K., Vasenius, J., Rokkanen, P. & Vainionpaa, S. 1993, 'Mechanical properties of biodegradable poly-L-lactide ligament augmentation device in experimental anterior cruciate ligament reconstruction', *Arch. Orthop. Trauma Surg.*, vol. 112, no. 6, pp. 270–274.

Levenberg, S., Rouwkema, J., Macdonald, M., Garfein, E. S., Kohane, D. S., Darland, D. C., Marini, R., van Blitterswijk, C. A., Mulligan, R. C., D'Amore, P. A. & Langer, R. 2005, 'Engineering vascularized skeletal muscle tissue', *Nat. Biotechnol.*, vol. 23, no. 7, pp. 879–884.

Lewis, M. P., Machell, J. R., Hunt, N. P., Sinanan, A. C. & Tippett, H. L. 2001, 'The extracellular matrix of muscle – implications for manipulation of the craniofacial musculature', *Eur. J. Oral Sci.*, vol. 109, no. 4, pp. 209–221.

Lin, S. T., Krebs, S. L., Kadiyala, S., Leong, K. W., LaCourse, W. C. & Kumar, B. 1994, 'Development of bioabsorbable glass fibres', *Biomaterials*, vol. 15, no. 13, pp. 1057–1061.

Liu, W., Chen, B., Deng, D., Xu, F., Cui, L. & Cao, Y. 2006, 'Repair of tendon defect with dermal fibroblast engineered tendon in a porcine model', *Tissue Eng.*, vol. 12, no. 4, pp. 775–788.

Maley, M. A., Davies, M. J. & Grounds, M. D. 1995, 'Extracellular matrix, growth factors, genetics: their influence on cell proliferation and myotube formation in primary cultures of adult mouse skeletal muscle', *Exp. Cell Res.*, vol. 219, no. 1, pp. 169–179.

Miyagawa, S., Sawa, Y., Taketani, S., Kawaguchi, N., Nakamura, T., Matsuura, N. & Matsuda, H. 2002, 'Myocardial regeneration therapy for heart failure: hepatocyte growth factor enhances the effect of cellular cardiomyoplasty', *Circulation*, vol. 105, no. 21, pp. 2556–2561.

Mooney, D. J. & Mikos, A. G. 1999, 'Growing new organs', *Sci. Am.*, vol. 280, no. 4, pp. 60–65.

Mulder, M. M., Hitchcock, R. W. & Tresco, P. A. 1998, 'Skeletal myogenesis on elastomeric substrates: implications for tissue engineering', *J. Biomater. Sci. Polym. Ed.*, vol. 9, no. 7, pp. 731–748.

Neumann, T., Hauschka, S. D. & Sanders, J. E. 2003, 'Tissue engineering of skeletal muscle using polymer fiber arrays', *Tissue Eng.*, vol. 9, no. 5, pp. 995–1003.

Powell, C. A., Smiley, B. L., Mills, J. & Vandenburgh, H. H. 2002, 'Mechanical stimulation improves tissue-engineered human skeletal muscle', *Am. J. Physiol. Cell Physiol.*, vol. 283, no. 5, pp. C1557–C1565.

Riboldi, S. A., Sampaolesi, M., Neuenschwander, P., Cossu, G. & Mantero, S. 2005, 'Electrospun degradable polyesterurethane membranes: potential scaffolds for skeletal muscle tissue engineering', *Biomaterials*, vol. 26, no. 22, pp. 4606–4615.

Rong, Z., Cheema, U. & Vadgama, P. 2006, 'Needle enzyme electrode based glucose diffusive transport measurement in a collagen gel and validation of a simulation model', *Analyst*, vol. 131, no. 7, pp. 816–821.

Sanes, J. R., Schachner, M. & Covault, J. 1986, 'Expression of several adhesive macromolecules (N-CAM, L1, J1, NILE, uvomorulin, laminin, fibronectin, and a heparan sulfate proteoglycan) in embryonic, adult, and denervated adult skeletal muscle', *J. Cell Biol.*, vol. 102, no. 2, pp. 420–431.

Sato, M., Maeda, M., Kurosawa, H., Inoue, Y., Yamauchi, Y. & Iwase, H. 2000, 'Reconstruction of rabbit Achilles tendon with three bioabsorbable materials: histological and biomechanical studies', *J. Orthop. Sci.*, vol. 5, no. 3, pp. 256–267.

Saxena, A. K., Marler, J., Benvenuto, M., Willital, G. H. & Vacanti, J. P. 1999, 'Skeletal muscle tissue engineering using isolated myoblasts on synthetic biodegradable polymers: preliminary studies', *Tissue Eng.*, vol. 5, no. 6, pp. 525–532.

Saxena, A. K., Willital, G. H. & Vacanti, J. P. 2001, 'Vascularized three-dimensional skeletal muscle tissue-engineering', *Biomed. Mater. Eng.*, vol. 11, no. 4, pp. 275–281.

Shah, R., Sinanan, A. C., Knowles, J. C., Hunt, N. P. & Lewis, M. P. 2005, 'Craniofacial muscle engineering using a 3-dimensional phosphate glass fibre construct', *Biomaterials*, vol. 26, no. 13, pp. 1497–1505.

Shansky, J., Creswick, B., Lee, P., Wang, X. & Vandenburgh, H. 2006, 'Paracrine release of insulin-like growth factor 1 from a bioengineered tissue stimulates skeletal muscle growth in vitro', *Tissue Eng.*, vol. 12, no. 7, pp. 1833–1841.

Vandenburgh, H. H., Swasdison, S. & Karlisch, P. 1991, 'Computer-aided mechanogenesis of skeletal muscle organs from single cells in vitro', *FASEB J.*, vol. 5, no. 13, pp. 2860–2867.

Williamson, M. R., Adams, E. F. & Coombes, A. G. 2006, 'Gravity spun polycaprolactone fibres for soft tissue engineering: interaction with fibroblasts and myoblasts in cell culture', *Biomaterials*, vol. 27, no. 7, pp. 1019–1026.

11
Biomaterials for the repair of peripheral nerves

S HALL, King's College London, UK

Abstract: Traumatic injury to a peripheral nerve not only provokes a localised response at the wound site but also triggers changes centrally within the spinal cord and brain and peripherally at the target organ. It is therefore perhaps not surprising that functional outcome after the (focal) clinical repair of such injuries, even in best-case scenarios, is frequently disappointing.

When clinical intervention is required, e.g. when the two stumps of a divided nerve cannot be sutured together without placing them under unacceptable tension, the 'gold standard' repair is a peripheral nerve autograft to bridge the gap between the nerve stumps. The microenvironment provided by an autograft is 'primed' to facilitate axonal regrowth, because it contains a population of axon-responsive Schwann cells lying in appropriately oriented laminin-rich basal lamina tubes. However, experience has shown that autografts are not ideal, for reasons which include fascicular and modality mismatch and donor site morbidity. Tissue-engineered conduits which mimic the microenvironment of an acutely transected peripheral nerve may provide an alternative to autografts. These bioprostheses must be able to direct and sustain axonal outgrowth, as well as providing haptotactic and chemotactic guidance cues for the regrowing axons and their associated cells as they grow towards the distal nerve stump. In addition, they must be fashioned from materials which are biocompatible and bioresorbable and which possess an appropriate compressive modulus.

This chapter reviews the experimental data informing the design of tissue-engineered alternatives to nerve autografts.

Key words: peripheral nerve, Wallerian degeneration, nerve grafts, conduits, tissue engineering.

11.1 Introduction

'Treatment of injuries to major nerve trunks ... remains a major and challenging reconstructive problem ...' (Lundborg and Rosen, 2007). The challenge confronts surgeons, scientists and patients.

Peripheral nerve injury is not uncommon. Estimates vary, but ~300,000 cases present annually in Europe alone (Lietz *et al.*, 2006; Kingham and Terenghi, 2006) and over 50,000 injuries require repair annually in the

United States (Bellamkonda, 2006). If plexus and root injuries are included, ~5% of all patients admitted to trauma centres in the United States present with nerve injuries which require surgical repair (Robinson, 2000). In peacetime, nerves are most likely to be damaged in road traffic accidents, penetrating trauma, falls and industrial accidents; significant numbers may also be cut, intentionally or inadvertently, during surgical interventions, e.g. during radical retropubic prostatectomy (Dean and Lue, 2005), or disrupted during complicated obstetric deliveries. In wartime, nerves are frequently involved in blast and ballistic injuries; indeed, current clinical practice is largely based upon principles established while managing these cases over the last century. '*Medical journals published during the last war* [WW1] *were filled with articles dealing with the diagnosis and treatment of nerve injuries.*' (Hoen, 1946).

When a nerve has been divided, physical continuity is restored either by suturing the nerve stumps together without placing the suture site under tension, or by bridging the gap with some kind of conduit (most commonly a nerve autograft). However, functional outcome is frequently unsatisfactory. Residual symptoms may include weakness, pain, dysaesthesia, cold intolerance and, when injuries affect the upper limb, lifelong impairment of hand function. There is an increased risk of secondary disability as a result of falls or fractures. Individuals may be unable to return to their previous employment or to any kind of profitable work; the economic burden to them, their carers and their dependents, and ultimately to the state, is considerable (Rosberg *et al.*, 2005; Lundborg and Rosen, 2007).

There are numerous reasons why repairing a focal injury may fail to restore function. They include the cascade of molecular and cellular events triggered by the injury, not only at the site of the lesion, but also in distant parts of the injured neurons and in their target organs; the extent to which neighbouring non-neural tissues are involved in the injury; the age, health and compliance of the patient; and the skill of the surgeon. Therapeutic intervention cannot (yet) manipulate responses which may occur many centimetres from the injury and which may influence functional outcome months or even years afterwards (Taylor, 2001; Zimmermann, 2001). Centrally, peripheral nerve injury induces reprogramming of synaptic connectivity in the cortex (Wall *et al.*, 2002; Lundborg, 2000b; Chen *et al.*, 2002; Rosen and Lundborg, 2004) and spinal cord (McLachlan *et al.*, 1993), and also triggers an immune response in affected dorsal root ganglia driven by macrophages, T lymphocytes and satellite cells, which may mediate the development and maintenance of chronic neuropathic pain (Scholz and Woolf, 2007). Peripherally, atrophy of chronically denervated end organs may preclude target reinnervation, even under optimal conditions (see Section 11.3.2).

Keeping more neurons alive, encouraging axons to cross long (>10 cm) inter-stump gaps, manipulating the cellular responses within an injured nerve, maximising the accuracy of target reinnervation and abrogating neuropathic pain are all major clinical goals which drive current research. Clinically, the gold standard for repairing divided nerves is an autograft, but there are good reasons to seek alternative procedures. Tubular prostheses (nerve conduits) fabricated from synthetic polymers are used extensively as experimental tools to promote axonal regrowth, and increasingly as implantable clinical devices.

Our understanding of the local injury response in the peripheral nervous system, PNS, provides a basis for the rational design of smart, tissue-engineered nerve conduits. This chapter will therefore begin with an overview of the structure of a normal peripheral nerve and what is known about the way in which it responds to traumatic injury. Subsequent sections will deal with alternatives to nerve autografts, concentrating particularly on the use of nerve conduits, in the laboratory and the clinic, charting their progress from hollow tube to tissue-engineered device. *Key points have been emboldened and italicised throughout the text.*

11.2 The structure of a peripheral nerve fibre: an overview

11.2.1 Connective tissue sheaths

Peripheral nerve fibres consist of variable numbers of bundles or fascicles covered by an epineurium composed of an outer paraneural layer of loose connective tissue enriched with adipocytes, and an inner interfascicular layer which is loosely attached to the perineurium. The paraneural layer may function as a shock absorber, protecting the entire nerve from recurrent compression and allowing it to glide within its tissue bed, while the interfascicular layer probably allows individual fascicles to slide independently within the main bundle.

As nerves pass from limb root to digital tips, the arrangement of fascicles changes continuously, sometimes over very short distances (Sunderland, 1945; Williams and Jabaley, 1986). Each fascicle contains axons and populations of non-neural cells, lying within an endoneurium and enclosed by a multi-layered, collagen-rich cellular sheath, the perineurium, which extends from the CNS/PNS transition zone to the termination of each nerve (Ushiki and Ide, 1986, 1990). The perineurium acts as a diffusion barrier between epineurium and endoneurium, and is thought to play a major role in determining the elasticity and tensile strength of an intact nerve. It maintains a positive intraneural pressure, best appreciated when a nerve is transected: the sudden loss of intraneural pressure causes the endoneurial contents to protrude beyond the cut ends of the nerve, producing a characteristic 'mushroom' appearance. The epineurium and perineurium are thicker at sites where fasciculi branch or pass under bands of connective tissue, or pass between the heads of muscles.

Blood vessels pass obliquely across the perineurium, running between the epineurium and the endoneurium (Bell and Weddell, 1984a, 1984b); perineurial sheaths may accompany blood vessels as they enter or leave a fascicle. The blood–nerve barrier and the perineurial barrier control the composition of the endoneurial milieu.

11.2.2 Axons and Schwann cells

Axons are the long peripheral processes of neurons whose cell bodies lie more centrally in either sensory or autonomic ganglia or in the spinal cord. They are associated along their entire length with Schwann cells. Broadly speaking, axons >2 μm diameter have a 1:1 relationship with their Schwann cells and are myelinated by these cells, whereas smaller axons are ensheathed in groups within Schwann cells and remain unmyelinated. Neighbouring Schwann cells overlap at nodes of Ranvier along myelinated axons, and their individual territories define myelinated internodes.

Schwann cells, once assumed to be cellular *tabulae rasae* under neuronal control, are now known to display myelinating/non-myelinating and sensory/motor phenotypes (Höke *et al.*, 2006). All axons and their associated Schwann cells are surrounded by tubes of laminin-rich basal lamina which are secreted by the Schwann cells and which extend continuously from the CNS/PNS interface to the target organs innervated by each axon (neuromuscular junctions, sensory corpuscles, etc.). In so doing, they isolate individual Schwann cell-axon units from their neighbours and from the endoneurial extracellular matrix.

The functional relationship between axons and their ensheathing Schwann cells is normally tightly regulated by reciprocal signalling (Jessen and Mirsky, 1998; Taveggia *et al.*, 2005). The way in which Schwann cells respond to the loss of that relationship, as a result of either segmental demyelination (when axonal integrity is maintained but myelin is removed) or axonal degeneration, plays a critical role in the injury response in the PNS. ***Manipulation of acutely denervated Schwann cells in order to facilitate axonal regrowth is a key strategy in the design of tissue-engineered conduits.***

11.2.3 Other endoneurial cells

The endoneurium also contains populations of fibroblasts, which secrete endoneurial collagen; endothelial cells which line blood vessels; resident macrophages, and mast cells. Although the normal functions of the macrophages and mast cells are not known, there is some evidence for the roles they play in nerve repair. Thus, resident macrophages initiate injury-induced myelin breakdown (see page 259) and mast cells are probably involved in changes in local nerve blood flow (Höke et al., 2001), and in the generation of pain and itching.

11.2.4 Biomechanical properties of peripheral nerves

Peripheral nerves are longitudinally anisotropic structures displaying time-dependent viscoelastic behaviour including stress relaxation and creep (Kendall et al., 1979). Their biomechanical properties reflect the structural organisation of their connective tissue sheaths and the packing of the axons and Schwann cells within the endoneurium (Millesi et al., 1995). The compressive pressures exerted by adjacent tendons, muscles and bones during normal bodily movements and postures expose peripheral nerves to combinations of tensile, shear and compressive stresses which they are designed to accommodate. Sustained high levels of strain (>6% elongation in the forearm, Grewal et al., 1996) produce physiological and structural changes which, depending upon the rate of lengthening, may have pathological consequences (Driscoll et al., 2002; Topp and Boyd, 2006).

Post-operative extraneural fibrosis not only may tether a regenerating nerve within its wound bed, restricting excursion during movement and exerting longitudinal traction on the repairing nerve, but also may exert prolonged compressive stress likely to compromise the microcirculation (Driscoll et al., 2002; Ju et al., 2006). Even moderate elevations in endoneurial fluid pressure can deform perineurial vessels from cylinders to ellipses, reducing the cross-sectional area of their lumens (Myers et al., 1986; Lundborg, 1988; Höke et al., 2001). Transient changes are accommodated with no discernible effect on function, whereas prolonged compression sufficient to produce ischaemia may produce demyelination or even degeneration.

Sustained local inflammation (such as that caused by the degradation of implanted nerve conduits, see Section 11.5.2), can induce mechanosensitivity to pressure or stretch in intact A and C fibres within the range of nerve stretch which occurs during normal limb movement (Dilley et al., 2005), and could contribute to persisting pain after repair.

Isolating repair sites from extraneural tissues by applying an anti-adhesive wrap, such as a hyaluronic acid–carboxymethylcellulose membrane, intraoperatively (Adanali et al., 2003) might offer a means of reducing extraneural scarring at sites where great mobility is required, such as the wrist, elbow or knee, but does not address the very considerable problem of reducing intraneural scarring (Petersen et al., 1996; Palatinsky et al.,

1997). The biomechanical properties of an injured nerve change over time as repair progresses and must be factored into the design of nerve conduits (Beel *et al.*, 1984; Temple *et al.*, 2003; Georgeu *et al.*, 2005; Tillett *et al.*, 2004).

11.3 The injury response in the peripheral nervous system (PNS)

Broadly speaking, the PNS displays three stereotyped local responses to noxious events. Ischaemia induced by mild pressure may produce transient parasthesiae but no obvious structural changes, and recovery is usually rapid. Sustained ischaemia, or attack by myelinotoxic agents, may induce a specific loss of myelin (primary or segmental demyelination), resulting in conduction slowing followed by conduction block. However, since axonal integrity is retained, recovery typically occurs within a few months, usually with little residual loss of function. Trauma which physically separates axons from their cell bodies produces Wallerian degeneration, WD. All affected axons in the extreme tip of the proximal stump and throughout the distal stump degenerate, irrespective of their calibre or functional modality: recovery is often slow and incomplete, and there may be considerable residual loss of function.

11.3.1 Wallerian degeneration

During the first two weeks after injury, the constitutive tissue response involves numerous events, some sequential, others consecutive. These include the production of debris, as axons and myelin sheaths are degraded; an increase in local blood flow; activation of resident macrophages, recruitment of exogenous macrophages and the removal of debris; activation of Schwann cells; and axonal sprouting from the proximal tips of the damaged axons (Fig. 11.1).

Axonal degradation

Until recently it was assumed that denervated axons withered away because they were no longer supported by their cell bodies. It is now known that disconnected axons auto-destruct, probably via a local caspase-independent process (Finn *et al.*, 2000; Raff *et al.*, 2002) which is triggered by events that occur before any morphological or electrophysiological changes can be detected, and which has been likened to 'lighting a fuse' (Tsao *et al.*, 1999). The speed with which this process is initiated suggests that therapeutic manipulation at this stage is an unrealistic clinical goal, since it is unlikely that axonal dissolution could be prevented once it has been triggered.

258   Cellular response to biomaterials

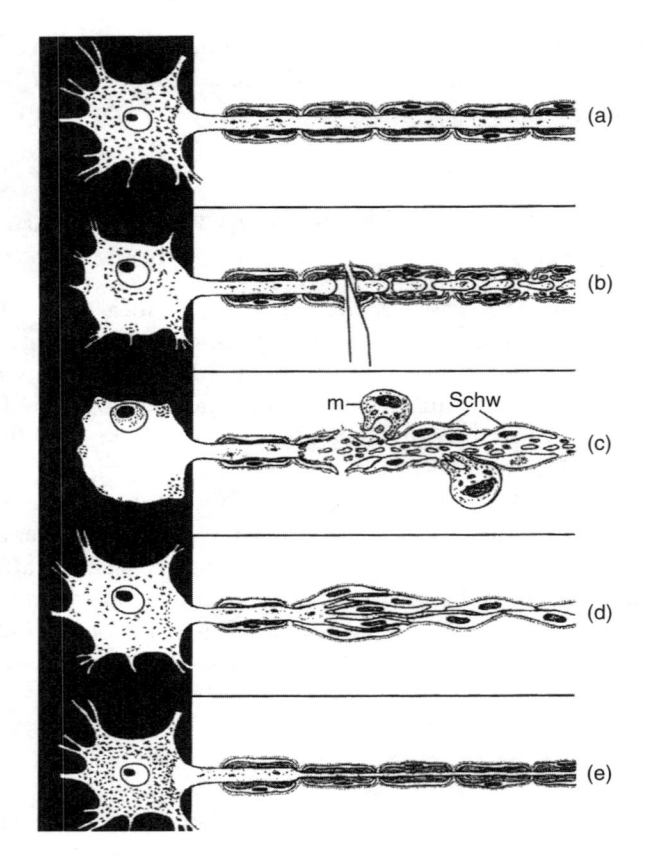

11.1 Wallerian degeneration: diagrammatic representation of the sequence of cellular events triggered by transection of a myelinated axon. (a) Normal myelinated axon associated with a longitudinal chain of Schwann cells and enclosed within a continuous basal lamina (dotted line). (b) Local injury produces dissolution of myelin sheaths and degeneration of axoplasm distally, and sealing of the tip of the proximal stump of the axon. (c) Schwann cell tube is invaded by macrophages (m) which breach the basal lamina; Schwann cells (Schw) distal to the injury proliferate; axon sprouts emerge from the proximal stump. (d) Axon sprouts elongate within the Schwann tube and associate with the Schwann cells therein. Since the injury shown caused minimal separation of the two ends of the damaged axon, there is no axonal misrouting. (e) Daughter Schwann cells remyelinate the regrowing axon. The new myelin sheaths are thinner and the internodal lengths are shorter than their proximal counterparts. Note that changes also occur in the cell body (the response shown here is stereotypical and is known as chromatolysis; readers should be aware that morphological changes in the neuronal cell body are not always so marked). (Reproduced with permission from Lundborg, G. (2005) Nerve Injury and Repair. Courtesy of Elsevier.)

Biomaterials for the repair of peripheral nerves 259

Macrophages are more than cellular dustbins

Resident endoneurial macrophages respond extremely rapidly to injury and are probably involved in the early stages of myelin breakdown (Mueller *et al.*, 2001). They are joined within 3–4 days by large numbers of haematogenously derived macrophages which access the endoneurium via a focally leaky blood–nerve barrier, enticed by locally generated chemokines (Dailey *et al.*, 1998; da Costa *et al.*, 1997). Endogenous and exogenous macrophages penetrate the Schwann cell basal laminae and phagocytose the cellular debris contained therein. They subsequently leave the tubes and lie free in the endoneurium, where debris processing continues until they are filled with lipid droplets, earning them the name of 'foam cells'. With time, foam cells either apoptose or migrate to draining lymph nodes.

Myelin degradation is essential because it removes components of the myelin sheaths such as myelin associated glycoprotein, MAG (Tang *et al.*, 1997) which would otherwise inhibit axonal (re)growth. Macrophages secrete a variety of chemokines, cytokines and neuroactive factors, some of which may be associated with the generation of neuropathic pain. They also secrete ApoE, which may scavenge lipids from axonal and myelin debris, and redeliver them to Schwann cells for reutilisation during regeneration. Experimentally abrogating the macrophage response delays the later stages of WD, but does not prevent axonal breakdown.

Acutely denervated Schwann cells

Within a few days of injury, denervated Schwann cells divide. Mitogens include axolemmal debris, macrophage-processed myelin and fragments of fibronectin exposed by the action of Schwann cell-derived MMP-3. Dedifferentiated daughter Schwann cells remain within the basal lamina tubes secreted by their parent cells, forming what are now usually called Schwann tubes but which were previously known as bands of Büngner. They behave as the 'presumed targets' of the regrowing axons until the actual targets (muscle or sensory end organ) have been reinnervated, and they secrete proteins which collectively facilitate axonal regrowth. Table 11.1 gives a representative sample of Schwann cell products which are upregulated during WD. Although they revert to more primitive phenotypes, motor and sensory Schwann cells retain their characteristic patterns of trophic factor expression (Höke *et al.*, 2006).

The period when denervated Schwann cells respond to axonal signals *in vivo* is relatively short (Hall, 1999), an experimental finding consistent with the clinical observation that there is ***a relatively narrow window of opportunity when surgical intervention is most likely to produce positive***

260 Cellular response to biomaterials

Table 11.1 Proteins upregulated in acutely denervated Schwann cells

- p75^{NTR}, low affinity neurotrophin receptor (Taniuchi *et al.*, 1986; Heumann *et al.*, 1987); growth factors NGF (Matsuoka *et al.*, 1991) and BDNF (Meyer *et al.*, 1992; Korsching, 1993); pro-protein convertases such as PC1, that may play a role in the processing of pro-BDNF (Marcinkiewicz *et al.*, 1998).
- *neu* receptors c-erbB2 and c-erbB3, involved in the regulation of Schwann cell differentiation and proliferation via cerbB-neuregulin signalling (Cohen *et al.*, 1992; Li *et al.*, 1997, 1998; Garratt *et al.*, 2000); the transcript encoding LP(A1), the receptor for an extracellular signalling phospholipid that regulates Schwann cell morphology and adhesion *in vitro* (Weiner *et al.*, 2001).
- Cell adhesion molecules N-CAM and L1 (Jessen *et al.*, 1987; Martini and Schachner, 1988), ninjurin 1 and 2 (Araki and Milbrandt, 2000) and galectin-1 (Hughes, 2001; Horie *et al.*, 1999); laminin (Kuecherer-Ehret *et al.*, 1990); J1/tenascin (Martini and Schachner, 1991); α5β1 fibronectin (Lefcort *et al.*, 1992); β4 integrin subunit (Feltri *et al.*, 1994); p200, a heparin-binding adhesive glycoprotein, implicated in Schwann cell-e.c.m. interactions (Chernousov *et al.*, 1999).
- Netrin-1, implicated in growth cone and axon guidance (Madison *et al.*, 2000); GAP-43 (Curtis *et al.*, 1992; Plantinga *et al.*, 1993).
- Cytokines and chemokines, e.g. interleukin (IL)-6 (Kurek *et al.*, 1996; Subang and Richardson, 2001), leukemia inhibitory factor, LIF, TNF-α and TNF-α converting enzyme (Shubayev and Myers, 2002); TGF-β1 (Scherer *et al.*, 1993; Einheber *et al.*, 1995) which may induce LIF expression (Matsuoka *et al.*, 1997).
- Erythropoeitin, a neuroprotective cytokine (Campana and Myers, 2001; Digicaylioglu *et al.*, 2004; Keswani *et al.*, 2004).
- Connexin 46, Cx46, a gap junction protein that may mediate junctional coupling in dedifferentiated Schwann cells, facilitating the rapid diffusion of signalling molecules between cells downstream of the injury site (Chandross, 1998).
- Hemopexin, an acute phase protein (Camborieux *et al.*, 1998).
- Matrix metalloproteinases MMP-2 and MMP-9 (Hughes *et al.*, 2002), whose many functions including remodelling of the blood nerve barrier; TIMP-1, the tissue inhibitor of some MMPs, which may protect basal laminae from dissolution during degeneration (La Fleur *et al.*, 1996).

results (Lundborg, 2000a). Since axon-responsiveness can be regained when chronically denervated cells are explanted *in vitro* (Li *et al.*, 1998), ***therapeutically prolonging the period of axon-responsiveness of endogenous denervated Schwann cells*** in long distal stumps, particularly where a graft or conduit has also been implanted, might facilitate axonal regrowth (see Sections 11.3.2 and 11.5.4).

11.3.2 Axonal outgrowth

Within hours of injury, the proximal stumps of severed axons begin to sprout from their tips and adjacent nodes of Ranvier and grow towards the lesion site. (They do not grow retrogradely into the endoneurium of the unaffected proximal stump because the microenvironment of an intact peripheral nerve is hostile to regrowing axons. This means that the conventional wisdom that the PNS is 'regeneration-friendly' is correct only when speaking of an acutely denervated distal stump or recently implanted nerve graft: Schwann cell proliferation and axonal sprouting and elongation are all inhibited in a normal nerve, but facilitated in an acutely repairing nerve.)

Role of the microenvironment

Depending upon the type of injury and the type of repair, axon sprouts may grow either within the original Schwann tube which enclosed their parent axon, or within a different tube in either a coapted distal stump or a graft, or into a non-neural environment such as the lumen of a nerve conduit. If the basal laminae of the original Schwann tubes are intact, the sprouts will encounter the activated, phenotypically correct progeny of the original Schwann cells. The possibility of pathfinding failures under these circumstances is potentially low and axons are likely to reinnervate their original targets if the injury is 'near-target'. However, axons are less likely to reach or to reinnervate a distant, chronically denervated target because the denervated Schwann cells and target organs will have either atrophied or become less responsive to axonal signals before the axons reach them. The concept that 'time is muscle' reflects the rapid destabilisation of the neuromuscular junction induced by nerve injury: significant changes occur in the gene profiles of nAChR subunits and myogenic regulatory factors in denervated skeletal muscle which rapidly become permanent (Ma *et al.*, 2007).

If nerve stumps have been coapted without tension, or an autograft has been implanted, axonal misrouting is inevitable because sprouts will encounter 'foreign' Schwann tubes distal to each suture line they cross. Denervated, dedifferentiated Schwann cells express distinct sensory and motor phenotypes (Höke *et al.*, 2006) and do not support phenotypically inappropriate ingrowing axon sprouts, which therefore soon atrophy, a phenomenon sometimes called 'pruning' (Brushart, 1993). These pathway differences, mediated at a molecular level, presumably explain why regeneration of motor axons is promoted by a motor graft and reinnervation of sensory pathways is promoted using a sensory graft (Nichols *et al.*, 2004; Lago *et al.*, 2007), and may undermine the effectiveness of a sural nerve graft when repairing a mixed nerve. Moreover, axonal misrouting probably mediates much of the functional reorganisation which occurs in the somatosensory cortex after nerve injury. **Manipulation of pathway selection is a key goal for a tissue-engineered nerve conduit.**

Minifascicles

When axons and Schwann cells grow out from a proximal nerve stump into non-neural territory, the morphology of the parent fascicle is not recreated. Instead, small bundles of axons, Schwann cells and fibroblasts form hundreds of 'minifascicles', each surrounded by a very thin perineurial sheath. The phenomenon involves the cellular outgrowth from the proximal stump exclusively, suggesting that interactions between perineurial cells, Schwann cells and axons are involved (Parmantier *et al.*, 1999). Minifascicular organisation disappears when regrowing axons gain access to the Schwann tubes of a distal stump.

Minifasciculation probably contributes to axonal dispersion across a gap, and may be exacerbated when the cellular outgrowth from the proximal stump is deflected by materials within the lumen of a conduit, such as inflammatory foci, degradation products and suture materials. Minifascicles which grow abortively into the epineurium and/or within the layers of the perineurium may produce painful, mechanosensitive neuromas around the lesion site.

11.4 Clinical need: alternatives to autografts

The microenvironment of an acutely denervated peripheral nerve facilitates axonal regrowth because it provides a vascularised segment of longitudinally oriented, laminin-rich basal lamina tubes filled with axon-responsive Schwann cells. The clinical gold standard for bridging an interstump gap therefore is a vascularised interfascicular or group fascicular nerve autograft, usually based on the sural nerve (a sensory nerve). Such a graft contributes all the cellular and acellular components of a peripheral nerve *except* neurons.

Somewhat counter-intuitively, an autograft does not guarantee a favourable outcome. Protocol-specific disadvantages include limited tissue volume (which may be significant when dealing with defects in large nerves in the upper limb) and fascicular and modality mismatches between host and donor nerves (all properties of the graft); and secondary donor site morbidity such as formation of a painful neuroma, potential infection, hyperesthesia or numbness within the distribution of the donor nerve (all consequences of harvesting the graft). Large-diameter grafts may become centrally or segmentally necrotic.

Any alternative to an autograft must provide an appropriate microenvironment for regrowing axons. Currently there are three options. The first is allografting in combination with low-grade immunosuppression (Grand *et al.*, 2002; Udina *et al.*, 2004; Jensen *et al.*, 2004), with the long-term prospect of inducing immune tolerance by costimulation blockade (Kvist *et al.*, 2007).

The second option is end-to-side (termino-lateral) neurorraphy, TLN, in which the distal end of an injured nerve is sutured to the side of an intact donor nerve, presumably creating a 'perineurial window' (Viterbo et al., 1992, 1994; Noah et al., 1997a, 1997b; Mennen, 1999; Rovak et al., 2000; Yan et al., 2002; Walker et al., 2004); both the first and second options use peripheral nerves, from a donor or the host respectively, with some success. A third, non-neural, option is to implant the proximal and distal stumps of the cut nerve into either end of a biological or synthetic tube (nerve guide) using materials which may be fabricated according to need (Fields et al., 1989; Hudson et al., 1999; Strauch, 2000; McDonald and Zochodne, 2003; Ijkema-Paassen et al., 2004): the protocol is often called entubulation or tubulisation.

11.5 Nerve guides

When a peripheral nerve is cut, the two stumps separate, producing an inter-stump gap. If that nerve consists of a few small fascicles and there is no other associated injury, it is likely that the gap will probably be of the order of one or two millimetres. Such a short gap will gradually be filled with minifascicles and connective tissue and it will be crossed successfully by many axons, particularly if the cut nerve stumps are held in reasonable alignment by surrounding tissues. However, when the injured nerve consists of numerous fascicles, and the inter-stump gap is a centimetre or more long, axons and Schwann cells may still cross the gap and access the distal stump, but many more of them usually form a neuroma (a tangled mass of axons and Schwann cells) around the proximal stump and do not cross the gap. Entubulation isolates regrowing axons within a nerve guide or conduit fashioned from a wide range of naturally occurring or synthetic materials, thereby optimising the probability that axons will reach the distal stump and also reducing the incidence of neuroma formation. Over the last 20 years, nerve guides have evolved from hollow tubes to complex tissue-engineered devices: a small number have transferred to the clinic, e.g. a copolymer of DL-lactic acid and poly ε-caprolactone Neurolac™ (Meek et al., 1999).

11.5.1 Natural nerve guides

Natural conduits which support axonal regrowth include a number of preparations based on peripheral nerves, e.g. 'pseudo-nerves' (Zhao et al., 1997) and acellular peripheral nerves (Hall, 1986b), preparations which retain longitudinal columns of Schwann cell basal lamina tubes, with or without their Schwann cells respectively; 'turnover' distal epineurial sheath tubes (Ayhan et al., 2000; Chatdokmaiprai et al., 2006); and epineurial tubes seeded with isogenic Schwann cells (Fansa et al., 2003). Other tissues which have been used include segments of tissues which are naturally tubular, e.g. inside-out small intestine (Wang et al., 1999), arteries (both inside-out and outside-in) and inside-out veins (Barcelos et al., 2003); flat sheets which may be rolled into a tube before or during implantation, e.g. omentum (Castaneda and Kinne, 2002) or amniotic membrane (Mohammad et al., 2000; Mligiliche et al., 2002); and fibrillar elements which may be teased apart without losing their structural integrity, e.g. tendon (Brandt et al., 1999a, 1999b). Mesothelia may also be preformed into a tube *in vivo* before implantation (Lundborg et al., 1981).

Some biological conduits are the natural prototypes of current tissue-engineered devices. They may be natural composites, e.g., denatured, and therefore acellular, striated muscle, which offers an external wall plus a lumen filled with longitudinally disposed laminin-rich sarcolemmal tubes, a feature which has been exploited both experimentally and clinically (Calder and Norris, 1993; Glasby et al., 1986; Hall and Enver, 1994; Meek et al., 2004), or tissues which are pre-loaded with Schwann cells, e.g. inside-out veins and arteries may retain an endogenous population of Schwann cells associated with adherent nervi vasorum.

Providing that they do not collapse, these natural guides are all capable of supporting varying degrees of axonal regrowth across a short gap in experimental animals, and some, e.g. acellular muscle, have been used clinically.

11.5.2 Synthetic nerve guides

Much of what is now known about the behaviour of axons and Schwann cells as they cross a short (≤ 1 cm) inter-stump gap is based on studies using 'regeneration chambers' made of non-resorbable polymers. The most frequently used material in these experiments has been silicone, poly(dimethylsiloxane), a synthetic non-biodegradable, impermeable and highly elastomeric polymer (Williams et al., 1983). Experiments using straight or inverted Y-shaped silicone-based non-resorbable conduits provided new data, or confirmed existing data, which are now being factored into the design of tissue-engineered devices, and which may be summarised as follows:

1. Axons and Schwann cells grow preferentially towards a segment of peripheral nerve when challenged with a selection of neural and non-neural 'lures', presumably responding to a diffusible concentration gradient of target-derived factors collectively called 'neurotropic factors' (Abernethy et al., 1994; Zheng and Kuffler, 2000).
2. The length of an inter-stump gap which axons and their associated Schwann cells cross *unaided in a tube* is limited (≤ 1 cm in adult rats, ~3 cm in primates).

3. It is not possible to increase significantly the rate at which axons cross a gap, although it is possible to influence the number and calibre of axons which cross.
4. The wound fluid accumulating within the lumen of a non-porous tube is enriched with a cocktail of bioactive factors (confirmed by *in vitro* assays). Many of these factors are secreted by activated Schwann cells and recruited inflammatory cells participating in the injury response (see Table 11.1): some have been incorporated into conduits in attempts to enhance axonal regrowth across short gaps, e.g. laminin, testosterone, ganglioside GM1 and catalase (Muller *et al.*, 1987); platelet-derived growth factor plus insulin-derived growth factor-1 (Wells *et al.*, 1997); brain-derived neurotrophic factor, BDNF (Utley *et al.*, 1996); laminin (Labrador *et al.*, 1998); nerve growth factor, NGF (Derby *et al.*, 1993; Wong and Oblinger, 1991); glial cell-derived neurotrophic factor, GDNF (Chen *et al.*, 2001); glial growth factor, GGF (Bryan *et al.*, 2000); alpha fibroblast growth factor, α-FGF (Cordeiro *et al.*, 1989); NGF, BDNF and NT-3 (Bloch *et al.*, 2001; Whitworth *et al.*, 1996; Sterne *et al.*, 1997); and inosine (Hadlock *et al.*, 1999).

In general, the addition of bioactive factors, whether delivered in biodegradable polymers, cross-linked to gel matrices within conduits or directly injected into the lumen of the conduit, promoted the number of axons crossing a short inter-stump gap. However, these results should be interpreted with caution for several reasons. Significantly, experimental controls are usually empty conduits and not autografts. Histomorphometric data *on their own* can be misleading: the correlation between the number of axons crossing a gap and subsequent functional outcome, as revealed by a battery of behavioural tests, is often poor (Munro *et al.*, 1998). Moreover, the effect of adding growth factors to conduits is usually transient, in that similar numbers of axons ultimately cross control tubes, but after a lag of several weeks (e.g. Rich *et al.*, 1989; Hollowell *et al.*, 1990). Most of the molecules delivered have very short half-lives: multiple-injection strategies have been suggested to overcome this problem (McDonald and Zochodne, 2003).

On the basis of encouraging experimental results, entubulation returned to the clinic, where the performance of implanted silicone tubes was found to be at least as successful as an autograft for bridging short inter-stump gaps (Lundborg *et al.*, 2004). However, although non-resorbability is a useful property in terms of retaining and concentrating neurotrophin-enriched fluid around the emerging front of axons and in preventing escape of axon sprouts into surrounding tissues, it became obvious that it can be a double-edged sword: those same axons may later be entrapped by fibrotic encapsulation of the implant. Indeed, some patients requested a second operation to remove their conduits (Merle et al., 1989), presumably because excessive fibrosis induced by a foreign body reaction to the persisting tube compressed the maturing nerves, causing pain and secondary axonal degeneration.

To address these issues, the second generation of nerve conduits were fabricated from naturally occurring and synthetic biodegradable polymers, used either independently or blended into novel materials which exploited the most useful features of each component (e.g. Li et al., 2007). An additional property of these materials, incorporated into many tissue engineered-devices, is that their controlled degradation allows site-specific, controlled and sustained release of growth factors.

The most frequently studied naturally occurring polymers are collagen (Archibald et al., 1995; Keilhoff et al., 2003; Stang et al., 2005a, 2005b; Chen et al., 2006b) and the polysaccharides alginate (Hashimoto et al., 2002) and chitosan (Wang et al., 2005), recently combined in an alginate/chitosan polyelectrolyte complex (Pfister et al., 2007a). Naturally occurring polymers are usually hydrophilic and potentially likely to interact well with cells, but they lack adequate mechanical strength and must be crosslinked before they can be used as the wall of a conduit, e.g. collagen can be crosslinked with poly-epoxy compounds, glutaraldehyde, UV or microwave irradiation (Ahmed et al., 2004): crosslinking not only confers stability but also permits modification of the rate of resorption. However, natural polymers tend to suffer from batch-to-batch variability and must be purified extensively to render them non-pathogenic and non-immunogenic.

Commonly used synthetic polymers include polyglycolic acid, PGA (Dellon and Mackinnon, 1988; Mackinnon and Dellon, 1990a), poly(DL-lactide) (Nyilas et al., 1983), poly(trimethylene carbonate) (Mackinnon and Dellon, 1990b), copolymers of poly(L-lactide) and polyurethane-based microvenous prostheses (Robinson et al., 1991), polylactides (Steuer et al., 1999), poly(L-lactide):poly(glycolide) (Luis et al., 2007), poly(DL-lactide-ε-caprolactone) Neurolac™ (Meek et al., 1999; Meek and Coert, 2002; Bertleff et al., 2005), and polyhydroxyalkanoates (Valappil et al., 2006) such as poly-3-hydroxybutyrate (Hazari et al., 1999; Young et al., 2002).

The batch consistency of synthetic polymers can be more rigorously controlled than that of bioartificial polymers. Materials can be fabricated to satisfy specific criteria, such as mechanical strength, flexibility, porosity and rate of degradation, e.g. the biodegradation rate of poly-D, L-lactic-glycolic acid co-polymer (PLGA) can be largely controlled by monomer ratio, molecular weight and crystallinity, and can be varied from weeks to months, whereas natural polymers tend to be degraded by proteases secreted by invading phagocytic cells. Irrespective of how a polymer is degraded, implanted materials should not trigger a massive macrophage infiltration in the vicinity of the regrowing axons (Ijkema-Paassen et al., 2004).

The rate of degradation of a nerve conduit should *at least* be synchronised with the rate of growth of the slowest axons it supports. If a conduit degrades too quickly, the delicate cable of regrowing axons it contains may be damaged or deflected before the axons reach the distal stump, whereas if degradation is too slow, that same cable may be compressed by the conduit and the maturing axons may degenerate. All axons available to cross a gap do not leave the proximal stump at the same time (so-called regeneration stagger) and so sufficient time should be allowed for all axons to have crossed the distal suture site and entered the distal stump before a short conduit degrades; heterochronous degradation along a proximo-distal gradient may be a design specification for long conduits. Whatever the length of the conduit, the **degradation products must not be cytotoxic, nor physically impede or deflect axonal growth, and the rate of degradation must not overload local mechanisms for removing the debris**. For example, commonly used aliphatic polyesters are degraded to acidic products and will therefore lower the pH of local tissues: if degradation is too fast, the pH will fall too quickly, affecting the viability of surrounding tissues and impairing vasculogenesis (Sung et al., 2004).

11.5.3 Conduit design

Physical characteristics, such as the chamber diameter, and the thickness, structure, permeability (Valentini et al., 1987) and porosity of the conduit wall, all influence the pattern of axonal regrowth. Walls must be mechanically strong enough to prevent suture dehiscence, and to resist external compressive forces, especially during limb or digit movement. They must be kink resistant and retain the ability to return to their original shape after bending. Walls with low compressive strength are likely to kink and collapse and obstruct or even occlude the lumen either before or during axonal outgrowth. Low modulus materials may be strengthened in a composite, e.g. poly(2-hydroxyethyl-methacrylate-co-methyl methacrylate) reinforced with custom-made coils of polycaprolactone (Fig. 11.2) (Katayama et al., 2006); chitosan strengthened with hydroxyapatite (Yamaguchi et al., 2003) or chitin (Yang et al., 2004). Biodegradable polymers also have a tendency to swell on implantation, reducing the internal diameter of the tube and potentially interfering with axonal outgrowth (den Dunnen et al., 1995).

Mechanical strength may be increased by increasing wall thickness, although there are obvious practical limitations to this strategy. Hardening a tube wall is likely to decrease its flexibility, whereas crimping (corrugating) or braiding tube walls will increase their flexibility (Fig. 11.3). Increasing the mismatch between the modulus of the tube and the surrounding tissues increases the probability of compression, constriction and secondary axonal

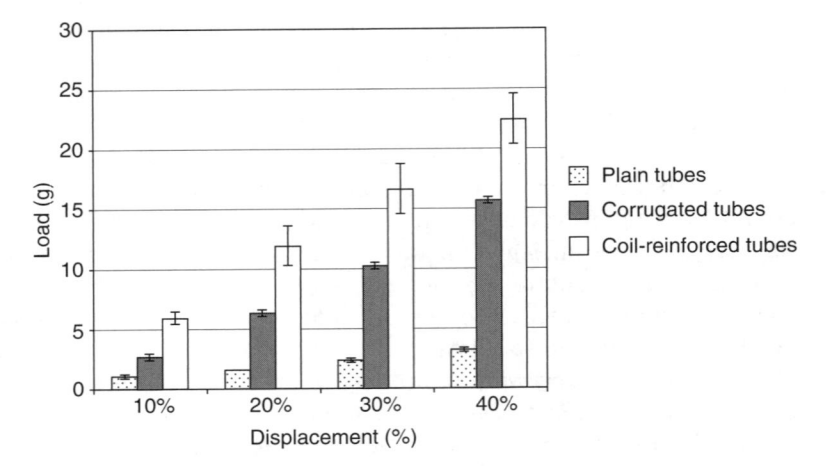

11.2 Load vs displacement graph reflects compressive strength of three differently prepared tubes of poly(2-hydroxyethyl methacrylate-co-methyl methacrylate) compared *in vivo*. Each bar is significantly different from the others ($p < 0.05$; $n = 3$, mean ± SD). (Reproduced with permission from Katayama, Y., Montenegro, R., Freier, T., Midha, R., Belkas, J. S. & Shoichet, M. S. (2006) Coil-reinforced hydrogel tubes promote nerve regeneration equivalent to that of nerve autografts. Biomaterials, 27, 505–18. Courtesy of Elsevier.)

degeneration: introducing some sort of matrix or scaffold into a hollow tube increases its overall compressive strength and so helps resist this compression (Fig. 11.4). Tensile strength and modulus both decrease with increasing porosity. The number, diameter and distribution of pores/channels in a tube wall may be optimised using computer-aided design or by manipulating the braiding angle and/or number of monofilaments in a fibre in a microbraided wall. Alternatively, pores may be randomly distributed, such as occurs when tube walls are fabricated from collagenous microbeads (Fig. 11.5) (Bender *et al.*, 2004). Thickness and porosity are important determinants of flexibility and of the speed of revascularisation: pores and holes sufficiently large to permit transmural entry of small blood vessels *into* a lumen are also large enough to convey minifascicles *out of* the lumen and into the surrounding connective tissue (Hall, unpublished observations).

Internal scaffolds

Numerous experiments have confirmed that vascularised minifascicles of axons and Schwann cells will grow across an 'empty' silicone tube of up to 1 cm length. To call such a tube empty is misleading, because soon after implantation the lumen fills with exudate derived from the cut ends of the nerves. The fibrinous clot which subsequently forms provides an irregular

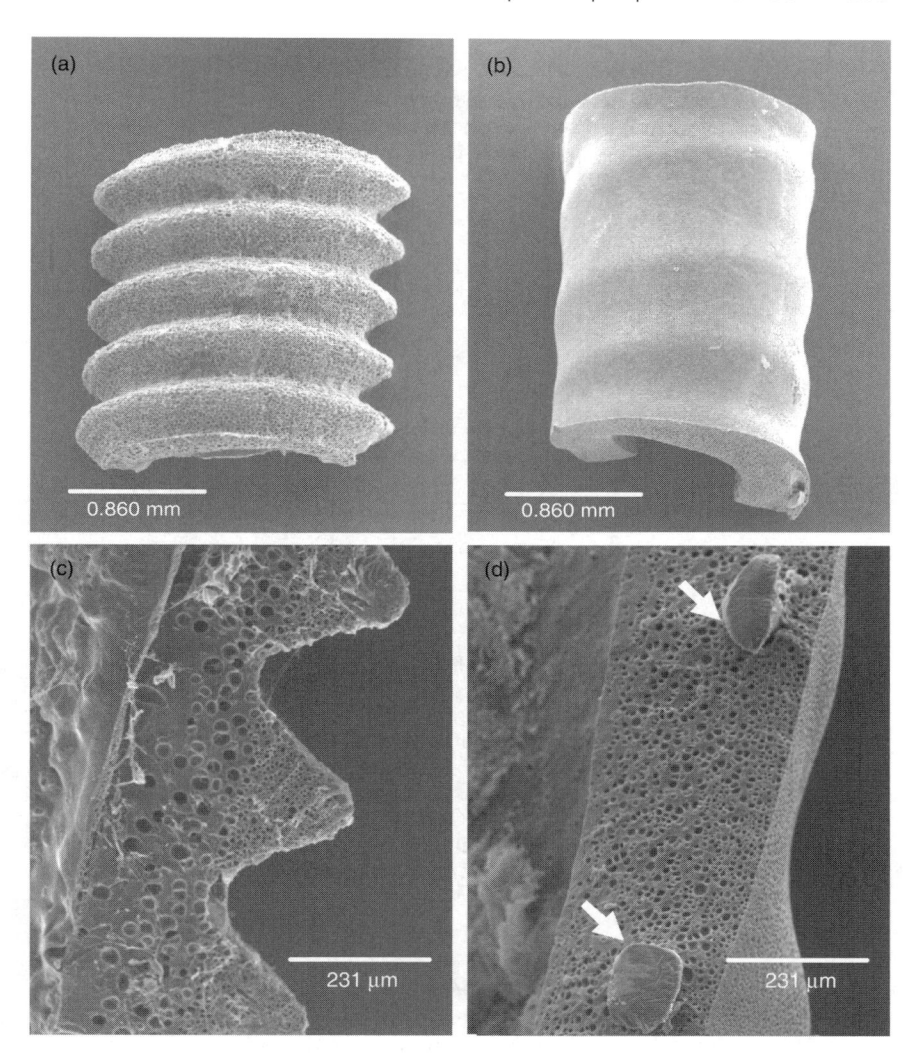

Biomaterials for the repair of peripheral nerves 269

11.3 Representative scanning electron micrographs of tubes of poly(2-hydroxyethyl methacrylate-co-methyl methacrylate) synthesised by SpinFX® technology: (a) corrugated tube, (b) coil-reinforced tube, (c) corrugated tube wall, (d) coil-reinforced tube wall (arrow indicates coil). (Reproduced with permission from Katayama, Y., Montenegro, R., Freier, T., Midha, R., Belkas, J. S. & Shoichet, M. S. (2006) Coil-reinforced hydrogel tubes promote nerve regeneration equivalent to that of nerve autografts. Biomaterials, 27, 505–18. Courtesy of Elsevier.)

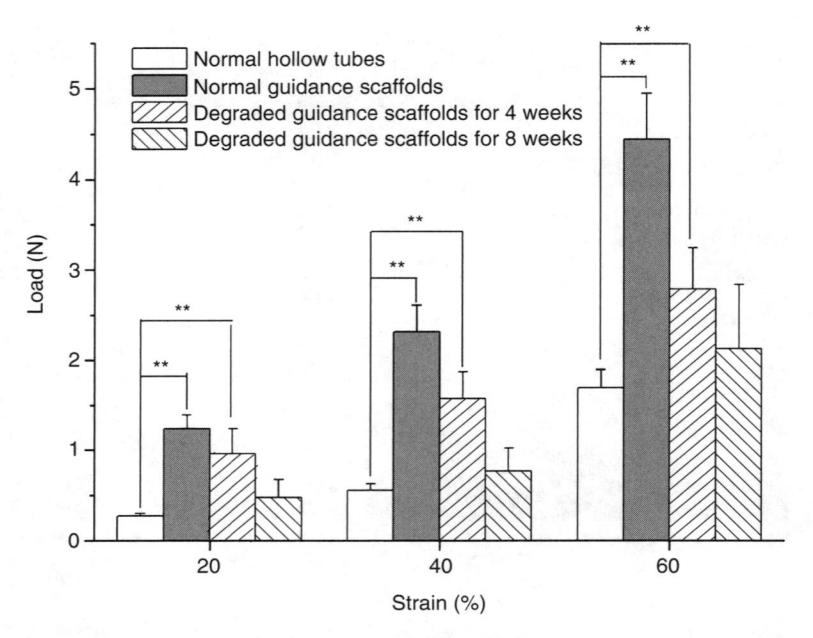

11.4 Adding a scaffold to a hollow conduit increases the compressive strength of the tube. Comparison of normal and degraded (for 4 or 8 weeks) hollow tubes containing guidance scaffolds with normal hollow tubes without inner matrices. **$p < 0.01$ relative to normal hollow tubes (mean ± S.D., $n = 6$ in each group). (Reproduced with permission from Wang, A., Ao, Q., Cao, W., Yu, M., He, Q., Kong, L., Zhang, L., Gong, Y. & Zhang, X. (2006) Porous chitosan tubular scaffolds with knitted outer wall and controllable inner structure for nerve tissue engineering. J Biomed Mater Res A. 79:36–46. Courtesy of John Wiley & Sons, Inc).

scaffold which is invaded by cells (perineurial cells, fibroblasts, Schwann cells, endothelial cells and macrophages) which grow out from the two nerve stumps, together with axon sprouts from the proximal stump. It is equally misleading to assume that this filling provides a microenvironment similar to that found in an acutely transected peripheral nerve. Indeed, the process is not as robust as is often assumed: there have been numerous reports that axons fail to cross short gaps bridged by empty, 'control' conduits (Strasberg *et al.*, 1999; Sierpinski *et al.*, 2008).

All but the extreme tips of regrowing axons (the growth cones) are segregated from the luminal contents not only by their associated, differentiating, Schwann cells but also by the perineurial cells which define the boundaries of each minifascicle, and by the basal laminae which surround the Schwann cells and perineurial cells. ***Therapeutically manipulating the behaviour of the growth cone will determine the path of the regrowing***

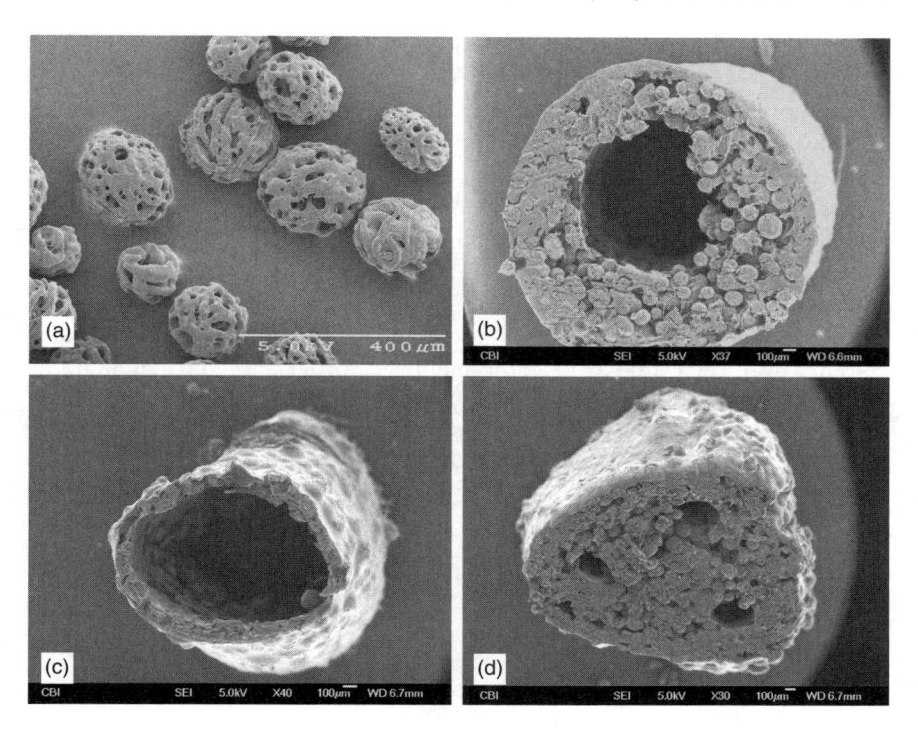

Biomaterials for the repair of peripheral nerves   271

11.5 Scanning electron micrographs of CultiSphers and PCL/CultiSpher composite nerve guides: (a) SEM of CultiSphers, macroporous, collagenous microbeads; (b) single-channelled PCL/CultiSpher nerve guide with a thick wall; (c) single-channelled PCL/CultiSpher nerve guide with a thin wall; (d) tri-channelled PCL/CultiSpher nerve guide (CultiGuide) fabricated by the wire mesh method using PVA-coated wire mandrels. (Reproduced with permission from Bender, M. D., Bennett, J. M., Waddell, R. L., Doctor, J. S. & Marra, K. G. (2004) Multi-channeled biodegradable polymer/CultiSpher composite nerve guides. Biomaterials, 25, 1269–78. Courtesy of Elsevier.)

axon, since wherever the growth cone goes, the rest of the attached axon must surely follow. Thus, growth cones entering either a distal stump or a graft must be persuaded to penetrate Schwann tubes of the appropriate functional modality, and growth cones entering a conduit must be offered appropriately cued substrates that offer directional guidance and facilitate adhesion and extension. Given that axons and Schwann cells co-migrate, it is perhaps surprising that little or no attention has thus far been paid to manipulating Schwann cell outgrowth from the nerve stumps.

During normal development, growth cones negotiate a path through a complex landscape, responding to soluble chemotropic factors and growth-permissive/inhibiting contact cues immobilised on the surface of local cells and in the surrounding extracellular matrix. During repair, a nerve conduit should offer a similarly dynamic 3-D landscape of cells and molecules to regrowing axons and their Schwann cells. It has been assumed that a nerve guide should contain a 3-D biomimetic scaffold reflecting the arrangement of Schwann tubes in an acutely denervated distal nerve stump (but see Bellamkonda, 2006). In the PNS, such a scaffold will be multi-functional: its topography will be '... *at least as important as* [its] *surface chemistry* ...' (Brown and Phillips, 2007). It should guide and accommodate the cellular outgrowth without compressing or deflecting it, provide an extensive surface area primed with topological and haptotactic cues to facilitate cell attachment, spreading and migration, and retain its mechanical and structural stability until such time as its components are programmed to biodegrade. An extensive literature describes ways of filling the lumen of a conduit to achieve these aims (Chen *et al.*, 2006a). Indeed, a recent paper reported a device that offered '... *a better alternative* ... *than conventional biodegradable conduits* ...' (Rutkowski *et al.*, 2004), which is perhaps a measure of how far the field has advanced since the early silicone tube experiments some 20 years ago, but also a reminder that the performance of any new device must be compared with that of an autograft and not of an empty tube.

Filling conduits with hydrogels containing extracellular matrix molecules such as laminin or collagen, which are known to promote cell adhesion and/or extension, and which are enriched in tissue exudates in regeneration chambers, produced variable results in early studies (Valentini *et al.*, 1987). The anticipated enhancement of axonal regrowth did not always occur, probably because the viscous gels formed plugs which physically impeded initial cellular outgrowth from the proximal stump, and deflected axon sprouts and their associated Schwann cells to the narrow interface between the inner aspect of the tube wall and the outer surface of the gel. The thickness of the neuronal cable which ultimately crossed the gap reflected the speed and extent to which the gels were degraded, since this produced additional space to accommodate more minifascicles. Predictably, diluting the gels proved more successful in facilitating axonal regrowth (Labrador *et al.*, 1998). A recent study using a keratin hydrogel as a provisional 'neuroinductive' matrix reported robust axonal regrowth across a short (4 mm gap) conduit which was attributed in part to activation of migrating Schwann cells by keratins (Sierpinski *et al.*, 2008).

The basal lamina tubes which persist in segments of acellular muscle or peripheral nerve offer longitudinally oriented channels within a supportive matrix (Hall, 1997). These natural conduits promote vigorous axonal regrowth across short gaps in rats, and so it was disappointing that acellular muscle grafts were only moderately successful when tested in the clinic (Calder and Norris, 1993): presumably they lacked sufficient mechanical strength to withstand longitudinal compression within the wound bed.

Longitudinal orientation can be built into scaffolds either directly, using channels or fibrils, or indirectly, e.g. by using cells to align tethered collagen gels (Phillips *et al.*, 2005). Minifascicles will grow *within* longitudinal channels of varying calibres created inside tubes of synthetic polymers, e.g. by depositing material on wire or PVA mandrels (Bender *et al.*, 2004; Bini *et al.*, 2005), or dip-moulding and low pressure injection moulding (Hadlock *et al.*, 2000). They will also grow in the spaces *between* longitudinally oriented bundles of fibrils, e.g. magnetically aligned collagen fibrils (Dubey *et al.*, 1999; Ceballos *et al.*, 1999) or rods of melt-derived bioactive glass, Bioglass® (Bunting *et al.*, 2005). Third generation tissue-engineered devices seeking to approximate ever more closely the macro- and micro-architecture of a peripheral nerve include multichannelled conduits with walls impregnated with bioactive growth factors and/or cells (Bender *et al.*, 2004), a tube with a microporous outer shell and a micropatterned inner lumenal surface selectively coated with laminin and pre-seeded with Schwann cells (Rutkowski *et al.*, 2004), and a tube with porous outer walls of knitted chitosan fibres and a highly porous inner scaffold with axially oriented 'macro' channels surrounded by interconnected micropores, which as yet has only been assessed *in vitro* (Wang *et al.*, 2006). (For further details of fabrication techniques see Huang and Huang, 2006.) The fourth generation of scaffolds will presumably mimic the architecture of a peripheral nerve at the nanoscale (Vasita and Katti, 2006; Schnell *et al.*, 2007; Willerth and Sakiyama-Elbert, 2007; Goldberg *et al.*, 2007), including perhaps multi-walled nanotubes (MWNTs) (Hu *et al.*, 2005).

Ensuring that pores and channels in scaffolds and gels are patent and accessible during the first weeks after implantation of a conduit is essential. Orifices may be occluded by clot, plugs of tissue debris or invading macrophages, and will therefore be invisible to outgrowing axon sprouts and Schwann cells. Whatever the protocol, impeding the initial outgrowth of axons produces a dense, disorganised mass of minifascicles which cap the end of the proximal stump, a phenomenon which delays and deflects subsequent outgrowth, and which could potentially produce a painful, mechanosensitive neuroma. The influence of the 'transparency factor' on the design of sieve electrodes (see Section 11.5.5) reinforces the importance of minimising the deflection of outgrowing axons from their original path as they search for a channel portal in a multichannel scaffold, or a pathway between aligned fibrils.

As well as offering guidance to outgrowing cells, an internal scaffold should be capable of replenishing what are sometimes loosely called 'neurotrophic factors' (growth-stimulating molecules capable of influencing neuronal survival and growth but with a short half-life), and delivering self-regulatory cells. (See Pfister *et al.*, 2007b, for a recent review of the design of conduits with integrated delivery systems for growth factors or growth factor-producing cells, and an assessment of growth factor delivery kinetics.)

11.5.4 Bridging a long gap

In order to cross an inter-stump gap, regrowing axons require an oriented substrate, a sustained supply of growth factors and a continuous supply of Schwann cells. All of these components are essential, but none is sufficient in isolation, e.g. regrowing axons will not leave a proximal stump and enter short grafts of acellular nerve or muscle, which are naturally packed with laminin-rich basal lamina tubes, unless accompanied by Schwann cells (Hall, 1986a).

The Schwann cells growing into the lumen of a short conduit support axonal regrowth across the entire extent of the gap. The consistent finding that axons fail to cross a long (>10 cm) gap in a tube has been attributed mainly to exhaustion of the Schwann cell pool when the axons are still some way from the distal stump. To compensate for the perceived glial shortfall, segments of peripheral nerve have been used either as 'stepping stones', sandwiched between grafts of acellular muscle (Calder and Green, 1995) or short lengths of silicone tubes (Maeda *et al.*, 1993), or as 'fillers' inside a vein autograft (Tang *et al.*, 1993) or a resorbable conduit (Francel *et al.*, 2003). More frequently, cultured Schwann cells have been added to a conduit, usually seeded within a hydrogel (Foidart-Dessalle *et al.*, 1997; Strauch *et al.*, 2001). A Schwann cell density of 80×10^6 cells/ml (approximately four times the concentration in normal adult nerve) has been reported to produce maximal enhancement of axonal regrowth in two different experimental studies (Mosahebi *et al.*, 2001; Guenard *et al.*, 1992). Axons will cross longer gaps (up to 6 cm) in the presence of exogenous Schwann cells, in larger experimental animals, but longer distances currently remain problematic. Ensuring that added cells not only survive but also adhere to the scaffold into which they have been introduced remains a major issue.

Obtaining sufficient numbers of autologous human Schwann cells is not a practical proposition in most clinics, on the grounds of technical difficulty and time required to expand cultures (Komiyama *et al.*, 2004). Stem cell technology may ultimately provide an alternative source of Schwann cells (Tohill and Terenghi, 2004; Hou *et al.*, 2006; Nie *et al.*, 2007). Novel cell delivery systems which are currently being developed in other tissues, e.g. pharmacologically active microcarriers, PAMs (Tatard *et al.*, 2007), might be a means of delivering such cells to specific areas within a device.

11.5.5 Sieve electrodes

One of the most exciting developments (quite literally) in recent years has been the use of micromachined sieve electrodes, which are inserted into conduits to create a neural interface which can be used to control a limb prosthesis. Sieve electrodes are usually made out of silicone (Wallman *et al.*, 2001) or polyimide (Zhao *et al.*, 1997; Navarro *et al.*, 1998): polyimide-based electrodes can be micromachined with a greater density of holes than is possible with silicone dice of the same area (Fig. 11.6). The geometric

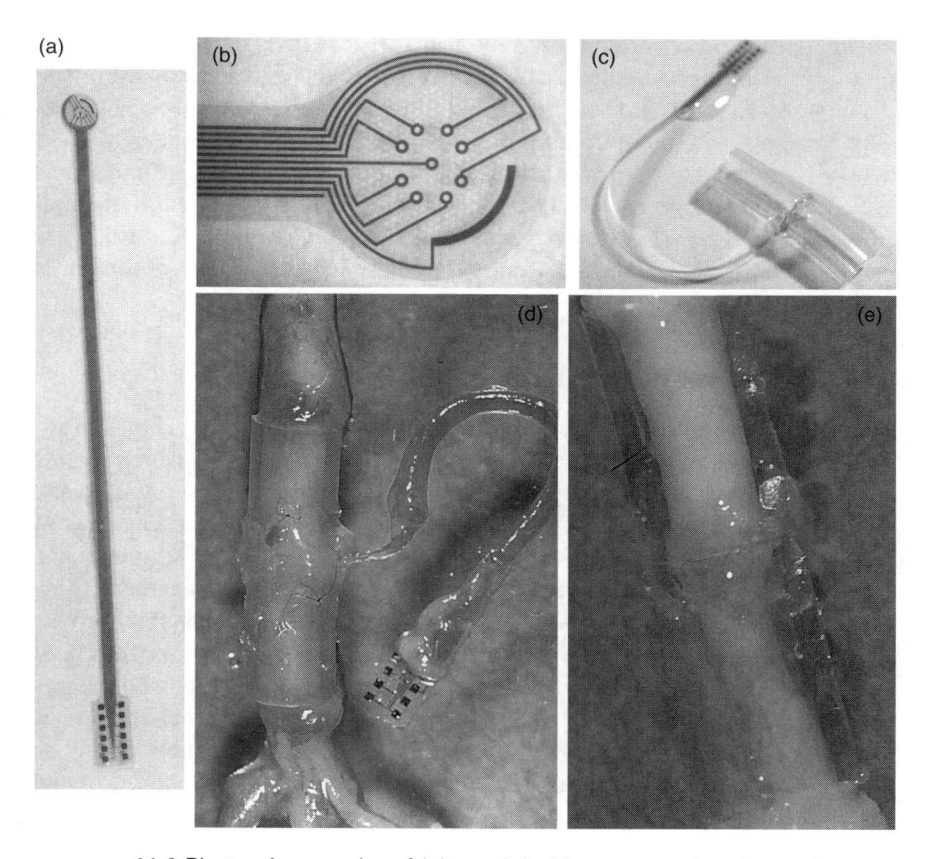

11.6 Photomicrographs of (a) a polyimide regenerative electrode, (b) enlarged view of the sieve portion of the regenerative electrode; (c) the silicone guide with the sieve electrode encased; (d) a sciatic nerve regenerated through the polyimide sieve electrode after removal at 6 months post-implantation; (e) view of the regenerated nerve after opening the silicone tube. In (d) and (e), the proximal side is at the top and the distal side is at the bottom. The regenerated nerve shows a conic-shaped enlargement at both sides of the sieve, most evident at the distal site. Note the thin fibrous tissue covering the polyimide ribbon. (Reproduced with permission from Lago, N., Ceballos, D., Rodriguez, F. J., Stieglitz, T. & Navarro, X. (2005) Long term assessment of axonal regeneration through polyimide regenerative electrodes to interface the peripheral nerve. Biomaterials, 26, 2021–31. Courtesy of Elsevier.)

design of the electrodes, particularly the diameters of the via holes (the holes through which axons regrow) and the transparency factor (the percentage of the total via hole area in relation to the actual surface area of the electrodes), influences functional outcome. The discussion concerning choice of materials for future sieve electrodes and the effect of varying transparency factors and via hole size touches on familiar themes (see above), such as the effects of internal fibrosis, controlling the extent of axonal deflection required to allow sprouts to access via holes, axonal compression if the holes in the scaffold are too small, and manipulating the dispersion of motor and sensory nerves (Wallman et al., 2001). For reasons which are not yet fully understood, large α motor neurons appear to have a lower capacity for growing through regenerative electrodes than other types of axons (Lago et al., 2005): this may simply reflect the fact that small sensory and sympathetic fibres may regenerate more rapidly during the early stages of repair, and so occupy the via holes before the motor fibres can access them.

11.6 Why work on short gaps?

Minifascicles can cross a short inter-stump gap without the aid of a graft, conduit or neuroactive factors, indeed '... *spontaneous regeneration in rats is highly effective ...*' (Keilhoff et al., 2005). It is therefore reasonable to ask why the great majority of experiments use conduits and tissue-engineered devices which bridge short (≤ 1 cm) inter-stump gaps. First, many of these experiments are designed to test the biocompatibility of walls and scaffolds and their rates of biodegradation, or the effectiveness and sustainability of drug delivery systems, often combining components previously tested *in vitro* before scaling up for longer gaps *in vivo*: the width of the gap is therefore of secondary importance in these proof-of-principle studies. Second, if it is possible to ensure a more robust outgrowth of larger calibre axons or facilitate an earlier penetration of the distal stump, even across a short gap, the chance that those axons will ultimately reach the periphery should be enhanced. Third, experimental models do not attempt to replicate clinical presentations. In the laboratory, a peripheral nerve is transected with a sharp blade at some point that is easily accessible to the scientist, the nerve stumps are sutured into the conduit immediately and there is minimal disturbance of the wound bed or surrounding tissues. In the clinic, '... *immediate nerve repairs are nothing but exceptions ...*' (Sulaiman et al., 2002) and surrounding tissues are almost certainly damaged, conditions which will adversely impact axonal regrowth (Glasby et al., 1998). It is therefore reasonable to assume that axonal regrowth across even short gaps would benefit from the support offered by tissue-engineered devices: a protocol which enhances the quality of regeneration by delivering the maximum number of axons across even a short gap in the minimum time is to be welcomed.

11.7 The 'smart' conduit

Distilling the extensive literature which has accumulated over the last 30 years, the design specifications of an ideal nerve conduit, as currently envisaged, may be summarised as follows. The device should provide a biocompatible, vascularised, bioresorbable scaffold to support outgrowing axons and their associated Schwann cells. Its walls must be sufficiently rigid to hold sutures during surgery and to prevent compression of the luminal contents, but have a Young's modulus close to that of a peripheral nerve so that the conduit is sufficiently flexible to accommodate movements of the host limb or digit (particularly if implanted across a joint). The material of which the conduit is constructed should be sufficiently permeable/porous to permit radial diffusion of facilitatory molecules (and possibly passage of endothelial and other cells), and yet not so porous that it becomes weak and collapses, occluding the lumen in the first few days after implantation. Its components must degrade at rates compatible with that of the slowest regrowing axons, so that the latter are neither exposed to the wound bed too early, nor secondarily compressed by persisting, unyielding walls (which would necessitate secondary surgery to remove the conduit). The degradation products must be non-cytotoxic, non-inflammatory and non-immunogenic. The outer layer of the conduit should be made of a material which neither adheres to the wound bed nor elicits an inflammatory response, on the grounds that epineurial fibrosis would tether the conduit and limit sliding of the nerve in its tissue bed during movement, events which are likely to compromise axonal regrowth, and which may also generate painful dysesthesiae. Transparency aids positioning of nerve ends during surgery.

More complex, tissue-engineered conduits should approximate the microenvironment of an acutely denervated peripheral nerve. Their construction should therefore include the capacity for the controlled and sustained delivery of bioactive agents (identified from experimental studies as key regulators of the injury response), along appropriate concentration gradients. Their internal architecture should provide a provisional scaffold of longitudinally oriented channels or fibres offering topological and haptotactic cues to guide the regrowing axons and populations of non-neural cells and to promote cell adhesion and outgrowth. The device should be surgically practicable: an off-the-shelf formulation should be available for short gaps, but guides requiring the addition of isogenous Schwann cells or stem cells will require pre-operative work-up.

11.8 References

Abernethy, D. A., Thomas, P. K., Rud, A. & King, R. H. (1994) Mutual attraction between emigrant cells from transected denervated nerve. *J Anat*, 184 (Pt 2), 239–49.
Adanali, G., Verdi, M., Tuncel, A., Erdogan, B. & Kargi, E. (2003) Effects of hyaluronic acid-carboxymethylcellulose membrane on extraneural adhesion formation and peripheral nerve regeneration. *J Reconst Microsurg*, 19, 29–36.
Ahmed, M. R., Venkateshwarlu, U. & Jayakumar, R. (2004) Multilayered peptide incorporated collagen tubules for peripheral nerve repair. *Biomaterials*, 25, 2585–94.
Araki, T. & Milbrandt, J. (2000) Ninjurin2, a novel homophilic adhesion molecule, is expressed in mature sensory and enteric neurons and promotes neurite outgrowth. *J Neurosci*, 20, 187–95.
Archibald, S. J., Shefner, J., Krarup, C. & Madison, R. D. (1995) Monkey median nerve repaired by nerve graft or collagen nerve guide tube. *J Neurosci*, 15, 4109–23.
Ayhan, S., Yavuzer, R., Latifoglu, O. & Atabay, K. (2000) Use of the turnover epineurial sheath tube for repair of peripheral nerve gaps. *J Reconstr Microsurg*, 16, 371–8.
Barcelos, A. S., Rodrigues, A. C., Silva, M. D. & Padovani, C. R. (2003) Inside-out vein graft and inside-out artery graft in rat sciatic nerve repair. *Microsurgery*, 23, 66–71.
Beel, J. A., Groswald, D. E. & Luttges, M. W. (1984) Alterations in the mechanical properties of peripheral nerve following crush injury. *J Biomech*, 17, 185–93.
Bell, M. A. & Weddell, A. G. (1984a) A descriptive study of the blood vessels of the sciatic nerve in the rat, man and other mammals. *Brain*, 107 (Pt 3), 871–98.
Bell, M. A. & Weddell, A. G. (1984b) A morphometric study of intrafascicular vessels of mammalian sciatic nerve. *Muscle Nerve*, 7, 524–34.
Bellamkonda, R. V. (2006) Peripheral nerve regeneration: an opinion on channels, scaffolds and anisotropy. *Biomaterials*, 27, 3515–8.
Bender, M. D., Bennett, J. M., Waddell, R. L., Doctor, J. S. & Marra, K. G. (2004) Multi-channeled biodegradable polymer/CultiSpher composite nerve guides. *Biomaterials*, 25, 1269–78.
Bertleff, M. J., Meek, M. F. & Nicolai, J. P. (2005) A prospective clinical evaluation of biodegradable neurolac nerve guides for sensory nerve repair in the hand. *Hand Surgery [Am]*, 30, 513–8.
Bini, T. B., Gao, S., Wang, S. & Ramakrishna, S. (2005) Development of fibrous biodegradable polymer conduits for guided nerve regeneration. *J Mater Sci Mater Med*, 16, 367–75.
Bloch, J., Fine, E. G., Bouche, N., Zurn, A. D. & Aebischer, P. (2001) Nerve growth factor- and neurotrophin-3-releasing guidance channels promote regeneration of the transected rat dorsal root. *Exp Neurol*, 172, 425–32.
Brandt, J., Dahlin, L. B., Kanje, M. & Lundborg, G. (1999a) Spatiotemporal progress of nerve regeneration in a tendon autograft used for bridging a peripheral nerve defect. *Exp Neurol*, 160, 386–93.
Brandt, J., Dahlin, L. B. & Lundborg, G. (1999b) Autologous tendons used as grafts for bridging peripheral nerve defects. *J Hand Surg [Br]*, 24, 284–90.

Brown, R. A. & Phillips, J. B. (2007) Cell responses to biomimetic protein scaffolds used in tissue repair and engineering. *Int Rev Cytol*, 262, 75–150.

Brushart, T. M. E. (1993) Motor axons preferentially reinnervate motor pathways. *J Neurosci*, 13, 2730–38.

Bryan, D. J., Holway, A. H., Wang, K. K., Silva, A. E., Trantolo, D. J., Wise, D. & Summerhayes, I. C. (2000) Influence of glial growth factor and Schwann cells in a bioresorbable guidance channel on peripheral nerve regeneration. *Tissue Eng*, 6, 129–38.

Bunting, S., Di Silvio, L., Deb, S. & Hall, S. (2005) Bioresorbable glass fibres facilitate peripheral nerve regeneration. *J Hand Surg [Br]*, 30, 242–7.

Calder, J. S. & Green, C. J. (1995) Nerve-muscle sandwich grafts: the importance of Schwann cells in peripheral nerve regeneration through muscle basal lamina conduits. *J Hand Surg*, 20B, 423–8.

Calder, J. S. & Norris, R. W. (1993) Repair of mixed peripheral nerves using muscle autografts: a preliminary communication. *Br J Plast Surg*, 46, 557–64.

Camborieux, L., Bertrand, N. & Swerts, J. P. (1998) Changes in expression and localization of hemopexin and its transcripts in injured nervous system: a comparison of central and peripheral tissues. *Neuroscience*, 82, 1039–52.

Campana, W. M. & Myers, R. R. (2001) Erythropoietin and erythropoietin receptors in the peripheral nervous system: changes after nerve injury. *FASEB J*, 15, 1804–6.

Castaneda, F. & Kinne, R. K. (2002) Omental graft improves functional recovery of transected peripheral nerve. *Muscle Nerve*, 26, 527–32.

Ceballos, D., Navarro, X., Dubey, N., Wendelschafer-crabb, G. & Kennedy, W. R. (1999) Magnetically aligned collagen gel filling a collagen nerve guide improves peripheral nerve regeneration. *Exp Neurol*, 158, 290–300.

Chandross, K. (1998) Nerve injury and inflammatory cytokines modulate gap junctions in the peripheral nervous system. *Glia*, 24, 21–31.

Chatdokmaiprai, C., Suwansingh, W. & Worapongpaiboon, S. (2006) The turnover distal epineurial sheath tube for repair of peripheral nerve gaps. *J Med Assoc Thai*, 89, 663–9.

Chen, M. B., Zhang, F. & Lineaweaver, W. C. (2006a) Luminal fillers in nerve conduits for peripheral nerve repair. *Ann Plast Surg*, 57, 462–71.

Chen, M. H., Chen, P. R., Chen, M. H., Hsieh, S. T., Huang, J. S. & Lin, F. H. (2006b) An in vivo study of tricalcium phosphate and glutaraldehyde crosslinking gelatin conduits in peripheral nerve repair. *J Biomed Mater Res B Appl Biomater*, 77, 89–97.

Chen, R., Cohen, L. G. & Hallett, M. (2002) Nervous system reorganization following injury. *Neuroscience*, 111, 761–73.

Chen, Z. Y., Chai, Y. F., Cao, L., Lu, C. L. & He, C. (2001) Glial cell line-derived neurotrophic factor enhances axonal regeneration following sciatic nerve transection in adult rats. *Brain Res*, 902, 272–6.

Chernousov, M. A., Scherer, S. S., Stahl, R. C. & Carey, D. J. (1999) p200, a collagen secreted by Schwann cells, is expressed in developing nerves and in adult nerves following axotomy. *J Neurosci Res*, 56, 284–94.

Cohen, J. A., Yachnis, A. T., Arai, M., Davis, J. G. & Scherer, S. S. (1992) Expression of the neu proto-oncogene by Schwann cells during peripheral nerve development and Wallerian degeneration. *J Neurosci Res*, 31, 622–34.

Cordeiro, P. G., Seckel, B. R., Lipton, S. A., D'amore, P. A., Wagner, J. & Madison, R. (1989) Acidic fibroblast growth factor enhances peripheral nerve regeneration in vivo. *Plast Reconstr Surg*, 83, 1013–9; discussion 1020–1.

Curtis, R., Stewart, H. J., Hall, S. M., Wilkin, G. P., Mirsky, R. & Jessen, K. R. (1992) GAP-43 is expressed by nonmyelin-forming Schwann cells of the peripheral nervous system. *J Cell Biol*, 116, 1455–64.

Da Costa, C. C., Van der Laan, L. J., Dijkstra, C. D. & Bruck, W. (1997) The role of the mouse macrophage scavenger receptor in myelin phagocytosis. *Eur J Neurosci*, 9, 2650–7.

Dailey, A. T., Avellino, A. M., Benthem, L., Silver, J. & Kliot, M. (1998) Complement depletion reduces macrophage infiltration and activation during Wallerian degeneration and axonal regeneration. *J Neurosci*, 18, 6713–22.

Dean, R. C. & Lue, T. F. (2005) Neuroregenerative strategies after radical prostatectomy. *Rev Urol*, 7 Suppl 2, S26–32.

Dellon, A. L. & Mackinnon, S. E. (1988) An alternative to the classical nerve graft for the management of the short nerve gap. *Plast Reconstr Surg*, 82, 849–56.

Den Dunnen, W. F., Van der Lei, B., Robinson, P. H., Holwerda, A., Pennings, A. J. & Schakenraad, J. M. (1995) Biological performance of a degradable poly(lactic acid-epsilon-caprolactone) nerve guide: influence of tube dimensions. *J Biomed Mater Res*, 29, 757–66.

Derby, A., Engleman, V. W., Frierdich, G. E., Neises, G., Rapp, S. R. & Roufa, D. G. (1993) Nerve growth factor facilitates regeneration across nerve gaps: morphological and behavioral studies in rat sciatic nerve. *Exp Neurol*, 119, 176–91.

Digicaylioglu, M., Garden, G., Timberlake, S., Fletcher, L. & Lipton, S. A. (2004) Acute neuroprotective synergy of erythropoietin and insulin-like growth factor I. *Proc Natl Acad Sci U S A*, 101, 9855–60.

Dilley, A., Lynn, B. & Pang, S. J. (2005) Pressure and stretch mechanosensitivity of peripheral nerve fibres following local inflammation of the nerve trunk. *Pain*, 117, 462–72.

Driscoll, P. J., Glasby, M. A. & Lawson, G. M. (2002) An in vivo study of peripheral nerves in continuity: biomechanical and physiological responses to elongation. *J Orthop Res*, 20, 370–5.

Dubey, N., Letourneau, P. C. & Tranquillo, R. T. (1999) Guided neurite elongation and schwann cell invasion into magnetically aligned collagen in simulated peripheral nerve regeneration. *Exp Neurol*, 158, 338–50.

Einheber, S., Hannocks, M. J., Metz, C. N., Rifkin, D. B. & Salzer, J. L. (1995) Transforming growth factor-beta 1 regulates axon/Schwann cell interactions. *J Cell Biol*, 129, 443–58.

Fansa, H., Dodic, T., Wolf, G., Schneider, W. & Keilhoff, G. (2003) Tissue engineering of peripheral nerves: Epineurial grafts with application of cultured Schwann cells. *Microsurgery*, 23, 72–7.

Feltri, M. L., Scherer, S. S., Nemni, R., Kamholz, J., Vogelbacker, H., Scott, M. O., Canal, N., Quaranta, V. & Wrabetz, L. (1994) Beta 4 integrin expression in myelinating Schwann cells is polarized, developmentally regulated and axonally dependent. *Development*, 120, 1287–301.

Fields, R. D., Le Beau, J. M., Longo, F. M. & Ellisman, M. H. (1989) Nerve regeneration through artificial tubular implants. *Prog Neurobiol*, 33, 87–134.

Finn, J. T., Weil, M., Archer, F., Siman, R., Srinivasan, A. & Raff, M. C. (2000) Evidence that Wallerian degeneration and localized axon degeneration induced by local neurotrophin deprivation do not involve caspases.PG. *J Neurosci*, 20, 1333–41.

Foidart-Dessalle, M., Dubuisson, A., Lejeune, A., Severyns, A., Manassis, Y., Delree, P., Crielaard, J. M., Bassleer, R. & Lejeune, G. (1997) Sciatic nerve regeneration through venous or nervous grafts in the rat. *Exp Neurol*, 148, 236–46.

Francel, P. C., Smith, K. S., Stevens, F. A., Kim, S. C., Gossett, J., Gossett, C., Davis, M. E., Lenaerts, M. & Tompkins, P. (2003) Regeneration of rat sciatic nerve across a LactoSorb bioresorbable conduit with interposed short-segment nerve grafts. *J Neurosurg*, 99, 549–54.

Garratt, A. N., Britsch, S. & Birchmeier, C. (2000) Neuregulin, a factor with many functions in the life of a schwann cell. *Bioessays*, 22, 987–96.

Georgeu, G. A., Walbeehm, E. T., Tillett, R., Afoke, A., Brown, R. A. & Phillips, J. B. (2005) Investigating the mechanical shear-plane between core and sheath elements of peripheral nerves. *Cell Tissue Res*, 320, 229–34.

Glasby, M. A., Gschmeissner, S. E., Huang, C. L. & De Souza, B. A. (1986) Degenerated muscle grafts used for peripheral nerve repair in primates. *J Hand Surg [Br]*, 11, 347–51.

Glasby, M. A., Fullerton, A. C. & Lawson, G. M. (1998) Immediate and delayed nerve repair using freeze-thawed muscle autografts in complex nerve injuries. Associated arterial injury. *J Hand Surg [Br]*, 23, 354–9.

Goldberg, M., Langer, R. & Jia, X. (2007) Nanostructured materials for applications in drug delivery and tissue engineering. *J Biomater Sci Polym Ed*, 18, 241–68.

Grand, A. G., Myckatyn, T. M., Mackinnon, S. E. & Hunter, D. A. (2002) Axonal regeneration after cold preservation of nerve allografts and immunosuppression with tacrolimus in mice. *J Neurosurg*, 96, 924–32.

Grewal, R., Xu, J., Sotereanos, D. G. & Woo, S. L. (1996) Biomechanical properties of peripheral nerves. *Hand Clin*, 12, 195–204.

Guenard, V., Kleitman, N., Morrissey, T. K., Bunge, R. P. & Aebischer, P. (1992) Syngeneic Schwann cells derived from adult nerves seeded in semipermeable guidance channels enhance peripheral nerve regeneration. *J Neurosci*, 12, 3310–20.

Hadlock, T., Sundback, C., Koka, R., Hunter, D., Cheney, M. & Vacanti, J. (1999) A novel, biodegradable polymer conduit delivers neurotrophins and promotes nerve regeneration. *Laryngoscope*, 109, 1412–6.

Hadlock, T., Sundback, C., Hunter, D., Cheney, M. & Vacanti, J. P. (2000) A polymer foam conduit seeded with Schwann cells promotes guided peripheral nerve regeneration. *Tissue Eng*, 6, 119–27.

Hall, S. (1997) Axonal regeneration through acellular muscle grafts. *J Anat*, 190 (Pt 1), 57–71.

Hall, S. M. (1986a) The effect of inhibiting Schwann cell mitosis on the re-innervation of acellular autografts in the peripheral nervous system of the mouse. *Neuropathol Appl Neurobiol*, 12, 401–14.

Hall, S. M. (1986b) Regeneration in cellular and acellular autografts in the peripheral nervous system. *Neuropathol Appl Neurobiol*, 12, 27–46.

Hall, S. M. (1999) The biology of chronically denervated Schwann cells. *Ann N Y Acad Sci*, 883, 215–233.

Hall, S. M. & Enver, K. (1994) Axonal regeneration through heat pretreated muscle autografts. An immunohistochemical and electron microscopic study. *J Hand Surg [Br]*, 19, 444–51.

Hashimoto, T., Suzuki, Y., Kitada, M., Kataoka, K., Wu, S., Suzuki, K., Endo, K., Nishimura, Y. & Ide, C. (2002) Peripheral nerve regeneration through alginate gel: analysis of early outgrowth and late increase in diameter of regenerating axons. *Exp Brain Res*, 146, 356–68.

Hazari, A., Johansson-Ruden, G., Junemo-bostrom, K., Ljungberg, C., Terenghi, G., Green, C. & Wiberg, M. (1999) A new resorbable wrap-around implant as an alternative nerve repair technique. *J Hand Surg [Br]*, 24, 291–5.

Heumann, R., Korsching, S., Bandtlow, C. & Thoenen, H. (1987) Changes of nerve growth factor synthesis in nonneuronal cells in response to sciatic nerve transection. *J Cell Biol*, 104, 1623–31.

Hoen, T. I. (1946) The repair of peripheral nerve lesions. *Am J Surg*, 72, 489–495.

Höke, A., Sun, H. S., Gordon, T. & Zochodne, D. W. (2001) Do denervated peripheral nerve trunks become ischemic? The impact of chronic denervation on vasa nervorum. *Exp Neurol*, 172, 398–406.

Höke, A., Redett, R., Hameed, H., Jari, R., Zhou, C., Li, Z. B., Griffin, J. W. & Brushart, T. M. (2006) Schwann cells express motor and sensory phenotypes that regulate axon regeneration. *J Neurosci*, 26, 9646–55.

Hollowell, J. P., Villadiego, A. & Rich, K. M. (1990) Sciatic nerve regeneration across gaps within silicone chambers: long-term effects of NGF and consideration of axonal branching. *Exp Neurol*, 110, 45–51.

Horie, H., Inagaki, Y., Sohma, Y., Nozawa, R., Okawa, K., Hasegawa, M., Muramatsu, N., Kawano, H., Horie, M., Koyama, H., Sakai, I., Takeshita, K., Kowada, Y., Takano, M. & Kadoya, T. (1999) Galectin-1 regulates initial axonal growth in peripheral nerves after axotomy. *J Neurosci*, 19, 9964–74.

Hou, S. Y., Zhang, H. Y., Quan, D. P., Liu, X. L. & Zhu, J. K. (2006) Tissue-engineered peripheral nerve grafting by differentiated bone marrow stromal cells. *Neuroscience*, 140, 101–10.

Hu, H., Ni, Y., Mandal, S. K., Montana, V., Zhao, B., Haddon, R. C. & Parpura, V. (2005) Polyethyleneimine functionalized single-walled carbon nanotubes as a substrate for neuronal growth. *J Phys Chem B*, 109, 4285–9.

Huang, Y. C. & Huang, Y. Y. (2006) Biomaterials and strategies for nerve regeneration. *Artif Organs*, 30, 514–22.

Hudson, T. W., Evans, G. R. & Schmidt, C. E. (1999) Engineering strategies for peripheral nerve repair. *Clin Plast Surg*, 26, 617–28, ix.

Hughes, P. M., Wells, G. M., Perry, V. H., Brown, M. C. & Miller, K. M. (2002) Comparison of matrix metalloproteinase expression during Wallerian degeneration in the central and peripheral nervous systems. *Neuroscience*, 113, 273–87.

Hughes, R. C. (2001) Galectins as modulators of cell adhesion. *Biochimie*, 83, 667–76.

Ijkema-Paassen, J., Jansen, K., Gramsbergen, A. & Meek, M. F. (2004) Transection of peripheral nerves, bridging strategies and effect evaluation. *Biomaterials*, 25, 1583–92.

Jensen, J. N., Tung, T. H., Mackinnon, S. E., Brenner, M. J. & Hunter, D. A. (2004) Use of anti-CD40 ligand monoclonal antibody as antirejection therapy in a murine peripheral nerve allograft model. *Microsurgery*, 24, 309–15.

Jessen, K. R. & Mirsky, R. (1998) Origin and early development of Schwann cells. *Microsc Res Tech*, 41, 393–402.

Jessen, K. R., Mirsky, R. & Morgan, L. (1987) Myelinated, but not unmyelinated axons, reversibly down-regulate N-CAM in Schwann cells. *J Neurocytol*, 16, 681–8.

Ju, M. S., Lin, C. C., Fan, J. L. & Chen, R. J. (2006) Transverse elasticity and blood perfusion of sciatic nerves under in situ circular compression. *J Biomech*, 39, 97–102.

Katayama, Y., Montenegro, R., Freier, T., Midha, R., Belkas, J. S. & Shoichet, M. S. (2006) Coil-reinforced hydrogel tubes promote nerve regeneration equivalent to that of nerve autografts. *Biomaterials*, 27, 505–18.

Keilhoff, G., Stang, F., Wolf, G. & Fansa, H. (2003) Bio-compatibility of type I/III collagen matrix for peripheral nerve reconstruction. *Biomaterials*, 24, 2779–87.

Keilhoff, G., Pratsch, F., Wolf, G. & Fansa, H. (2005) Bridging extra large defects of peripheral nerves: possibilities and limitations of alternative biological grafts from acellular muscle and Schwann cells. *Tissue Eng*, 11, 1004–14.

Kendall, J. P., Stokes, I. A., O'hara, J. P. & Dickson, R. A. (1979) Tension and creep phenomena in peripheral nerve. *Acta Orthop Scand*, 50, 721–5.

Keswani, S. C., Buldanlioglu, U., Fischer, A., Reed, N., Polley, M., Liang, H., Zhou, C., Jack, C., Leitz, G. J. & Höke, A. (2004) A novel endogenous erythropoietin mediated pathway prevents axonal degeneration. *Ann Neurol*, 56(6), 815–26.

Kingham, P. J. & Terenghi, G. (2006) Bioengineered nerve regeneration and muscle reinnervation. *J Anat*, 209, 511–26.

Komiyama, T., Nakao, Y., Toyama, Y., Vacanti, C. A., Vacanti, M. P. & Ignotz, R. A. (2004) Novel technique for peripheral nerve reconstruction in the absence of an artificial conduit. *J Neurosci Methods*, 134, 133–40.

Korsching, S. (1993) The neurotrophic factor concept: a reexamination. *J Neurosci*, 13, 2739–48.

Kuecherer-Ehret, A., Graeber, M. B., Edgar, D., Thoenen, H. & Kreutzberg, G. W. (1990) Immunoelectron microscopic localization of laminin in normal and regenerating mouse sciatic nerve. *J Neurocytol*, 19, 101–9.

Kurek, J. B., Austin, L., Cheema, S. S., Bartlett, P. F. & Murphy, M. (1996) Up-regulation of leukaemia inhibitory factor and interleukin-6 in transected sciatic nerve and muscle following denervation. *Neuromuscul Disord*, 6, 105–14.

Kvist, M., Lemplesis, V., Kanje, M., Ekberg, H., Corbascio, M. & Dahlin, L. B. (2007) Immunomodulation by costimulation blockade inhibits rejection of nerve allografts. *J Peripher Nerv Syst*, 12, 83–90.

La Fleur, M., Underwood, J. L., Rappolee, D. A. & Werb, Z. (1996) Basement membrane and repair of injury to peripheral nerve: defining a potential role for macrophages, matrix metalloproteinases, and tissue inhibitor of metalloproteinases-1. *J Exp Med*, 184, 2311–26.

Labrador, R. O., Buti, M. & Navarro, X. (1998) Influence of collagen and laminin gels concentration on nerve regeneration after resection and tube repair. *Exp Neurol*, 149, 243–52.

Lago, N., Ceballos, D., Rodriguez, F. J., Stieglitz, T. & Navarro, X. (2005) Long term assessment of axonal regeneration through polyimide regenerative electrodes to interface the peripheral nerve. *Biomaterials*, 26, 2021–31.

Lago, N., Rodriguez, F. J., Guzman, M. S., Jaramillo, J. & Navarro, X. (2007) Effects of motor and sensory nerve transplants on amount and specificity of sciatic nerve regeneration. *J Neurosci Res*, 85, 2800–12.

Lefcort, F., Venstrom, K., Mcdonald, J. A. & Reichardt, L. F. (1992) Regulation of expression of fibronectin and its receptor, alpha 5 beta 1, during development and regeneration of peripheral nerve. *Development*, 116, 767–82.

Li, H., Terenghi, G. & Hall, S. M. (1997) Effects of delayed re-innervation on the expression of c-erbB receptors by chronically denervated rat Schwann cells in vivo. *Glia*, 20, 333–47.

Li, H., Wigley, C. & Hall, S. M. (1998) Chronically denervated rat Schwann cells respond to GGF in vitro. *Glia*, 24, 290–303.

Li, X. K., Cai, S. X., Liu, B., Xu, Z. L., Dai, X. Z., Ma, K. W., Li, S. Q., Yang, L., Sung, K. L. & Fu, X. B. (2007) Characteristics of PLGA-gelatin complex as potential artificial nerve scaffold. *Colloids Surf B Biointerfaces*, 57, 198–203.

Lietz, M., Dreesmann, L., Hoss, M., Oberhoffner, S. & Schlosshauer, B. (2006) Neuro tissue engineering of glial nerve guides and the impact of different cell types. *Biomaterials*, 27, 1425–36.

Luis, A. L., Rodrigues, J. M., Lobato, J. V., Lopes, M. A., Amado, S., Veloso, A. P., Armada-da-Silva, P. A., Raimondo, S., Geuna, S., Ferreira, A. J., Varejao, A. S., Santos, J. D. & Mauricio, A. C. (2007) Evaluation of two biodegradable nerve guides for the reconstruction of the rat sciatic nerve. *Biomed Mater Eng*, 17, 39–52.

Lundborg, G. (1988) Intraneural microcirculation. *Orthop Clin North Am*, 19, 1–12.

Lundborg, G. (2000a) A 25-year perspective of peripheral nerve surgery: evolving neuroscientific concepts and clinical significance. *J Hand Surg [Am]*, 25, 391–414.

Lundborg, G. (2000b) Brain plasticity and hand surgery: an overview. *J Hand Surg [Br]*, 25, 242–52.

Lundborg, G. & Rosen, B. (2007) Hand function after nerve repair. *Acta Physiol (Oxf)*, 189, 207–17.

Lundborg, G., Dahlin, L. B., Danielsen, N. P., Hansson, H. A. & Larsson, K. (1981) Reorganization and orientation of regenerating nerve fibres, perineurium, and epineurium in preformed mesothelial tubes – an experimental study on the sciatic nerve of rats. *J Neurosci Res*, 6, 265–81.

Lundborg, G., Rosen, B., Dahlin, L., Holmberg, J. & Rosen, I. (2004) Tubular repair of the median or ulnar nerve in the human forearm: a 5-year follow-up. *J Hand Surg [Br]*, 29, 100–7.

Ma, J., Shen, J., Garrett, J. P., Lee, C. A., Li, Z., Elsaidi, G. A., Ritting, A., Hick, J., Tan, K. H., Smith, T. L., Smith, B. P. & Koman, L. A. (2007) Gene expression of myogenic regulatory factors, nicotinic acetylcholine receptor subunits, and GAP-43 in skeletal muscle following denervation in a rat model. *J Orthop Res*, 25(11), 1498–505.

Mackinnon, S. E. & Dellon, A. L. (1990a) Clinical nerve reconstruction with a bioabsorbable polyglycolic acid tube. *Plast Reconstr Surg*, 85, 419–24.

Mackinnon, S. E. & Dellon, A. L. (1990b) A study of nerve regeneration across synthetic (Maxon) and biologic (collagen) nerve conduits for nerve gaps up to 5 cm in the primate. *J Reconstr Microsurg*, 6, 117–21.

Madison, R. D., Zomorodi, A. & Robinson, G. A. (2000) Netrin-1 and peripheral nerve regeneration in the adult rat. *Exp Neurol*, 161, 563–70.

Maeda, T., Mackinnon, S. E., Best, T. J., Evans, P. J., Hunter, D. A. & Midha, R. T. (1993) Regeneration across 'stepping-stone' nerve grafts. *Brain Res*, 618, 196–202.

Marcinkiewicz, M., Savaria, D. & Marcinkiewicz, J. (1998) The pro-protein convertase PC1 is induced in the transected sciatic nerve and is present in cultured Schwann

cells: comparison with PC5, furin and PC7, implication in pro-BDNF processing. *Brain Res Mol Brain Res*, 59, 229–46.
Martini, R. & Schachner, M. (1988) Immunoelectron microscopic localization of neural cell adhesion molecules (L1, N-CAM, and myelin-associated glycoprotein) in regenerating adult mouse sciatic nerve. *J Cell Biol*, 106, 1735–46.
Martini, R. & Schachner, M. (1991) Complex expression pattern of tenascin during innervation of the posterior limb buds of the developing chicken. *J Neurosci Res*, 28, 261–79.
Matsuoka, I., Meyer, M. & Thoenen, H. (1991) Cell-type-specific regulation of nerve growth factor (NGF) synthesis in non-neuronal cells: comparison of Schwann cells with other cell types. *J Neurosci*, 11, 3165–77.
Matsuoka, I., Nakane, A. & Kurihara, K. (1997) Induction of LIF-mRNA by TGF-beta 1 in Schwann cells. *Brain Res*, 776, 170–80.
McDonald, D. S. & Zochodne, D. W. (2003) An injectable nerve regeneration chamber for studies of unstable soluble growth factors. *J Neurosci Methods*, 122, 171–8.
McLachlan, E. M., Janig, W., Devor, M. & Michaelis, M. (1993) Peripheral nerve injury triggers noradrenergic sprouting within dorsal root ganglia. *Nature*, 363, 543–6.
Meek, M. F. & Coert, J. H. (2002) Clinical use of nerve conduits in peripheral-nerve repair: review of the literature. *J Reconstr Microsurg*, 18, 97–109.
Meek, M. F., Den Dunnen, W. F., Schakenraad, J. M. & Robinson, P. H. (1999) Long-term evaluation of functional nerve recovery after reconstruction with a thin-walled biodegradable poly (DL-lactide-epsilon-caprolactone) nerve guide, using walking track analysis and electrostimulation tests. *Microsurgery*, 19, 247–53.
Meek, M. F., Varejao, A. S. & Geuna, S. (2004) Use of skeletal muscle tissue in peripheral nerve repair: review of the literature. *Tissue Eng*, 10, 1027–36.
Mennen, U. (1999) End-to-side nerve suture – a technique to repair peripheral nerve injury. *S Afr Med J*, 89, 1188–94.
Merle, M., Dellon, A. L., Campbell, J. N. & Chang, P. S. (1989) Complications from silicon-polymer intubulation of nerves. *Microsurgery*, 10, 130–3.
Meyer, M., Matsuoka, I., Wetmore, C., Olson, L. & Thoenen, H. (1992) Enhanced synthesis of brain-derived neurotrophic factor in the lesioned peripheral nerve: different mechanisms are responsible for the regulation of BDNF and NGF mRNA. *J Cell Biol*, 119, 45–54.
Millesi, H., Zoch, G. & Reihsner, R. (1995) Mechanical properties of peripheral nerves. *Clin Orthop*, 314, 76–83.
Mligiliche, N., Endo, K., Okamoto, K., Fujimoto, E. & Ide, C. (2002) Extracellular matrix of human amnion manufactured into tubes as conduits for peripheral nerve regeneration. *J Biomed Mater Res*, 63, 591–600.
Mohammad, J., Shenaq, J., Rabinovsky, E. & Shenaq, S. (2000) Modulation of peripheral nerve regeneration: a tissue-engineering approach. The role of amnion tube nerve conduit across a 1-centimeter nerve gap. *Plast Reconstr Surg*, 105, 660–6.
Mosahebi, A., Woodward, B., Wiberg, M., Martin, R. & Terenghi, G. (2001) Retroviral labeling of Schwann cells: in vitro characterization and in vivo transplantation to improve peripheral nerve regeneration. *Glia*, 34, 8–17.

Muller, H., Williams, L. R. & Varon, S. (1987) Nerve regeneration chamber: evaluation of exogenous agents applied by multiple injections. *Brain Res*, 413, 320–6.

Mueller, M., Wacker, K., Ringelstein, E. B., Hickey, W. F., Imai, Y. & Kiefer, R. (2001) Rapid response of identified resident endoneurial macrophages to nerve injury. *Am J Pathol*, 159, 2187–97.

Munro, C. A., Szalai, J. P., Mackinnon, S. E. & Midha, R. (1998) Lack of association between outcome measures of nerve regeneration. *Muscle Nerve*, 21, 1095–7.

Myers, R. R., Murakami, H. & Powell, H. C. (1986) Reduced nerve blood flow in edematous neuropathies: a biomechanical mechanism. *Microvasc Res*, 32, 145–51.

Navarro, X., Calvet, S., Rodriguez, F. J., Stieglitz, T., Blau, C., Buti, M., Valderrama, E. & Meyer, J. U. (1998) Stimulation and recording from regenerated peripheral nerves through polyimide sieve electrodes. *J Peripher Nerv Syst*, 3, 91–101.

Nichols, C. M., Brenner, M. J., Fox, I. K., Tung, T. H., Hunter, D. A., Rickman, S. R. & Mackinnon, S. E. (2004) Effects of motor versus sensory nerve grafts on peripheral nerve regeneration. *Exp Neurol*, 190, 347–55.

Nie, X., Zhang, Y. J., Tian, W. D., Jiang, M., Dong, R., Chen, J. W. & Jin, Y. (2007) Improvement of peripheral nerve regeneration by a tissue-engineered nerve filled with ectomesenchymal stem cells. *Int J Oral Maxillofac Surg*, 36, 32–8.

Noah, E. M., Williams, A., Fortes, W. & Terzis, J. K. (1997a) A new animal model to investigate axonal sprouting after end-to-side neurorrhaphy. *J Reconstr Microsurg*, 13, 317–25.

Noah, E. M., Williams, A., Jorgenson, C., Skoulis, T. G. & Terzis, J. K. (1997b) End-to-side neurorrhaphy: a histologic and morphometric study of axonal sprouting into an end-to-side nerve graft. *J Reconstr Microsurg*, 13, 99–106.

Nyilas, E., Chiu, T. H., Sidman, R. L., Henry, E. W., Brushart, T. M., Dikkes, P. & Madison, R. (1983) Peripheral nerve repair with bioresorbable prosthesis. *Trans Am Soc Artif Intern Organs*, 29, 307–13.

Palatinsky, E. A., Maier, K. H., Touhalisky, D. K., Mock, J. L., Hingson, M. T. & Coker, G. T. (1997) ADCON-T/N reduces in vivo perineural adhesions in a rat sciatic nerve reoperation model. *J Hand Surg [Br]*, 22, 331–5.

Parmantier, E., Lynn, B., Lawson, D., Turmaine, M., Namini, S. S., Chakrabarti, L., Mcmahon, A. P., Jessen, K. R. & Mirsky, R. (1999) Schwann cell-derived Desert hedgehog controls the development of peripheral nerve sheaths. *Neuron*, 23, 713–24.

Petersen, J., Russell, L., Andrus, K., Mackinnon, M., Silver, J. & Kliot, M. (1996) Reduction of extraneural scarring by ADCON-T/N after surgical intervention. *Neurosurgery*, 38, 976–83; discussion 983–4.

Pfister, L. A., Papaloizos, M., Merkle, H. P. & Gander, B. (2007a) Hydrogel nerve conduits produced from alginate/chitosan complexes. *J Biomed Mater Res A*, 80, 932–7.

Pfister, L. A., Papaloizos, M., Merkle, H. P. & Gander, B. (2007b) Nerve conduits and growth factor delivery in peripheral nerve repair. *J Peripher Nerv Syst*, 12, 65–82.

Phillips, J. B., Bunting, S. C., Hall, S. M. & Brown, R. A. (2005) Neural tissue engineering: a self-organizing collagen guidance conduit. *Tissue Eng*, 11, 1611–7.

Plantinga, L. C., Verhaagen, J., Edwards, P. M., Hol, E. M., Bar, P. R. & Gispen, W. H. (1993) The expression of B-50/GAP-43 in Schwann cells is upregulated in

degenerating peripheral nerve stumps following nerve injury. *Brain Res*, 602, 69–76.
Raff, M. C., Whitmore, A. V. & Finn, J. T. (2002) Axonal self-destruction and neurodegeneration. *Science*, 296, 868–71.
Rich, K. M., Alexander, T. D., Pryor, J. C. & Hollowell, J. P. (1989) Nerve growth factor enhances regeneration through silicone chambers. *Exp Neurol*, 105, 162–70.
Robinson, L. R. (2000) Traumatic injury to peripheral nerves. *Muscle Nerve*, 23, 863–73.
Robinson, P. H., Van Der Lei, B., Hoppen, H. J., Leenslag, J. W., Pennings, A. J. & Nieuwenhuis, P. (1991) Nerve regeneration through a two-ply biodegradable nerve guide in the rat and the influence of ACTH4–9 nerve growth factor. *Microsurgery*, 12, 412–9.
Rosberg, H. E., Carlsson, K. S., Hojgard, S., Lindgren, B., Lundborg, G. & Dahlin, L. B. (2005) Injury to the human median and ulnar nerves in the forearm–analysis of costs for treatment and rehabilitation of 69 patients in southern Sweden. *J Hand Surg [Br]*, 30, 35–9.
Rosen, B. & Lundborg, G. (2004) Sensory re-education after nerve repair: aspects of timing. *Handchir Mikrochir Plast Chir*, 36, 8–12.
Rovak, J. M., Cederna, P. S., Macionis, V., Urbanchek, M. S., Van Der Meulen, J. H. & Kuzon, W. M., Jr. (2000) Termino-lateral neurorrhaphy: the functional axonal anatomy. *Microsurgery*, 20, 6–14.
Rutkowski, G. E., Miller, C. A., Jeftinija, S. & Mallapragada, S. K. (2004) Synergistic effects of micropatterned biodegradable conduits and Schwann cells on sciatic nerve regeneration. *J Neural Eng*, 1, 151–7.
Scherer, S. S., Kamholz, J. & Jakowlew, S. B. (1993) Axons modulate the expression of transforming growth factor-betas in Schwann cells. *Glia*, 8, 265–76.
Schnell, E., Klinkhammer, K., Balzer, S., Brook, G., Klee, D., Dalton, P. & Mey, J. (2007) Guidance of glial cell migration and axonal growth on electrospun nanofibers of poly-epsilon-caprolactone and a collagen/poly-epsilon-caprolactone blend. *Biomaterials*, 28, 3012–25.
Scholz, J. & Woolf, C. J. (2007) The neuropathic pain triad: neurons, immune cells and glia. *Nat Neurosci*, 10, 1361–1368.
Shubayev, V. I. & Myers, R. R. (2002) Endoneurial remodeling by TNFalph- and TNFalpha-releasing proteases. A spatial and temporal co-localization study in painful neuropathy. *J Peripher Nerv Syst*, 7, 28–36.
Sierpinski, P., Garrett, J., Ma, J., Apel, P., Klorig, D., Smith, T., Koman, L. A., Atala, A. & Van Dyke, M. (2008) The use of keratin biomaterials derived from human hair for the promotion of rapid regeneration of peripheral nerves. *Biomaterials*, 29(1), 118–28.
Stang, F., Fansa, H., Wolf, G. & Keilhoff, G. (2005a) Collagen nerve conduits–assessment of biocompatibility and axonal regeneration. *Biomed Mater Eng*, 15, 3–12.
Stang, F., Fansa, H., Wolf, G., Reppin, M. & Keilhoff, G. (2005b) Structural parameters of collagen nerve grafts influence peripheral nerve regeneration. *Biomaterials*, 26, 3083–91.
Sterne, G. D., Brown, R. A., Green, C. J. & Terenghi, G. (1997) Neurotrophin-3 delivered locally via fibronectin mats enhances peripheral nerve regeneration. *Eur J Neurosci*, 9, 1388–96.

Steuer, H., Fadale, R., Müller, E., Müller, H. W., Planck, H. & Schlosshauer, B. (1999) Biohybride nerve guide for regeneration: degradable polylactide fibers coated with rat Schwann cells. *Neurosicence Letter*, 277, 165–8.

Strasberg, J. E., Strasberg, S., Mackinnon, S. E., Watanabe, O., Hunter, D. A. & Tarasidis, G. (1999) Strain differences in peripheral-nerve regeneration in rats. *J Reconstr Microsurg*, 15, 287–93.

Strauch, B. (2000) Use of nerve conduits in peripheral nerve repair. *Hand Clin*, 16, 123–30.

Strauch, B., Rodriguez, D. M., Diaz, J., Yu, H. L., Kaplan, G. & Weinstein, D. E. (2001) Autologous Schwann cells drive regeneration through a 6-cm autogenous venous nerve conduit. *J Reconstr Microsurg*, 17, 589–95; discussion 596–7.

Subang, M. C. & Richardson, P. M. (2001) Synthesis of leukemia inhibitory factor in injured peripheral nerves and their cells. *Brain Res*, 900, 329–31.

Sulaiman, O. A. R., Voda, J., Gold, B. G. & Gordon, T. (2002) FK506 increases peripheral nerve regeneration after chronic axotomy but not after chronic schwann cell denervation. *Exp Neurol*, 175, 127–37.

Sunderland, S. (1945) The intraneural topography of the radial, median, and ulnar nerves. *Brain*, 68, 243–99.

Sung, H. J., Meredith, C., Johnson, C. & Galis, Z. S. (2004) The effect of scaffold degradation rate on three-dimensional cell growth and angiogenesis. *Biomaterials*, 25, 5735–42.

Tang, J. B., Gu, Y. Q. & Song, Y. S. (1993) Repair of digital nerve defect with autogenous vein graft during flexor tendon surgery in zone 2. *J Hand Surg [Br]*, 18, 449–53.

Tang, S., Shen, Y. J., Debellard, M. E., Mukhopadhyay, G., Salzer, J. L., Crocker, P. R. & Filbin, M. T. (1997) Myelin-associated glycoprotein interacts with neurons via a sialic acid binding site at ARG118 and a distinct neurite inhibition site. *J Cell Biol*, 138, 1355–66.

Taniuchi, M., Clark, H. B. & Johnson, E. M., Jr. (1986) Induction of nerve growth factor receptor in Schwann cells after axotomy. *Proc Natl Acad Sci U S A*, 83, 4094–8.

Tatard, V. M., Sindji, L., Branton, J. G., Aubert-Pouessel, A., Colleau, J., Benoit, J. P. & Montero-Menei, C. N. (2007) Pharmacologically active microcarriers releasing glial cell line – derived neurotrophic factor: Survival and differentiation of embryonic dopaminergic neurons after grafting in hemiparkinsonian rats. *Biomaterials*, 28, 1978–88.

Taveggia, C., Zanazzi, G., Petrylak, A., Yano, H., Rosenbluth, J., Einheber, S., Xu, X., Esper, R. M., Loeb, J. A., Shrager, P., Chao, M. V., Falls, D. L., Role, L. & Salzer, J. L. (2005) Neuregulin-1 type III determines the ensheathment fate of axons. *Neuron*, 47, 681–94.

Taylor, B. K. (2001) Pathophysiologic mechanisms of neuropathic pain. *Curr Pain Headache Rep*, 5, 151–61.

Temple, C. L., Ross, D. C., Dunning, C. E., Johnson, J. A. & King, G. J. (2003) Tensile strength of healing peripheral nerves. *J Reconstr Microsurg*, 19, 483–8.

Tillett, R. L., Afoke, A., Hall, S. M., Brown, R. A. & Phillips, J. B. (2004) Investigating mechanical behaviour at a core-sheath interface in peripheral nerve. *J Peripher Nerv Syst*, 9, 255–62.

Tohill, M. P. & Terenghi, G. (2004) Stem cell plasticity and therapy for injuries of the peripheral nervous system. *Biotechnol Appl Biochem*, 40 (Pt 1), 17–24.

Topp, K. S. & Boyd, B. S. (2006) Structure and biomechanics of peripheral nerves: nerve responses to physical stresses and implications for physical therapist practice. *Phys Ther*, 86, 92–109.

Tsao, J. W., George, E. B. & Griffin, J. W. (1999) Temperature modulation reveals three distinct stages of Wallerian degeneration. *J Neurosci*, 19, 4718–26.

Udina, E., Gold, B. G. & Navarro, X. (2004) Comparison of continuous and discontinuous FK506 administration on autograft or allograft repair of sciatic nerve resection. *Muscle Nerve*, 29, 812–22.

Ushiki, T. & Ide, C. (1986) Three-dimensional architecture of the endoneurium with special reference to the collagen fibril arrangement in relation to nerve fibers. *Arch Histol Jpn*, 49, 553–63.

Ushiki, T. & Ide, C. (1990) Three-dimensional organization of the collagen fibrils in the rat sciatic nerve as revealed by transmission- and scanning electron microscopy. *Cell Tissue Res*, 260, 175–84.

Utley, D. S., Lewin, S. L., Cheng, E. T., Verity, A. N., Sierra, D. & Terris, D. J. (1996) Brain-derived neurotrophic factor and collagen tubulization enhance functional recovery after peripheral nerve transection and repair. *Arch Otolaryngol Head Neck Surg*, 122, 407–13.

Valappil, S. P., Misra, S. K., Boccaccini, A. R. & Roy, I. (2006) Biomedical applications of polyhydroxyalkanoates: an overview of animal testing and in vivo responses. *Expert Rev Med Devices*, 3, 853–68.

Valentini, R. F., Aebischer, P., Winn, S. R. & Galletti, P. M. (1987) Collagen- and laminin-containing gels impede peripheral nerve regeneration through semipermeable nerve guidance channels. *Exp Neurol*, 98, 350–6.

Vasita, R. & Katti, D. S. (2006) Nanofibers and their applications in tissue engineering. *Int J Nanomedicine*, 1, 15–30.

Viterbo, F., Trindade, J. C., Hoshino, K. & Mazzoni Neto, A. (1992) Latero-terminal neurorrhaphy without removal of the epineural sheath. Experimental study in rats. *Rev Paul Med*, 110, 267–75.

Viterbo, F., Palhares, A. & Franciosi, L. F. (1994) Restoration of sensitivity after removal of the sural nerve. A new application of latero-terminal neurorraphy. *Rev Paul Med*, 112, 658–60.

Walker, J. C., Brenner, M. J., Mackinnon, S. E., Winograd, J. M. & Hunter, D. A. (2004) Effect of perineurial window size on nerve regeneration, blood-nerve barrier integrity, and functional recovery. *J Neurotrauma*, 21, 217–27.

Wall, J. T., Xu, J. & Wang, X. (2002) Human brain plasticity: an emerging view of the multiple substrates and mechanisms that cause cortical changes and related sensory dysfunctions after injuries of sensory inputs from the body. *Brain Res Brain Res Rev*, 39, 181–215.

Wallman, L., Zhang, Y., Laurell, T. & Danielsen, N. (2001) The geometric design of micromachined silicon sieve electrodes influences functional nerve regeneration. *Biomaterials*, 22, 1187–93.

Wang, A., Ao, Q., Cao, W., Yu, M., He, Q., Kong, L., Zhang, L., Gong, Y. & Zhang, X. (2006) Porous chitosan tubular scaffolds with knitted outer wall and controllable inner structure for nerve tissue engineering. *J Biomed Mater Res A*, 79, 36–46.

Wang, K. K., Cetrulo, C. L., Jr. & Seckel, B. R. (1999) Tubulation repair of peripheral nerves in the rat using an inside-out intestine sleeve. *J Reconstr Microsurg*, 15, 547–54.

Wang, X., Hu, W., Cao, Y., Yao, J., Wu, J. & Gu, X. (2005) Dog sciatic nerve regeneration across a 30-mm defect bridged by a chitosan/PGA artificial nerve graft. *Brain*, 128, 1897–910.

Weiner, J. A., Fukushima, N., Contos, J. J., Scherer, S. S. & Chun, J. (2001) Regulation of Schwann cell morphology and adhesion by receptor-mediated lysophosphatidic acid signaling. *J Neurosci*, 21, 7069–78.

Wells, M. R., Kraus, K., Batter, D. K., Blunt, D. G., Weremowitz, J., Lynch, S. E., Antoniades, H. N. & Hansson, H. A. (1997) Gel matrix vehicles for growth factor application in nerve gap injuries repaired with tubes: a comparison of biomatrix, collagen, and methylcellulose. *Exp Neurol*, 146, 395–402.

Whitworth, I. H., Brown, R. A., Dore, C. J., Anand, P., Green, C. J. & Terenghi, G. (1996) Nerve growth factor enhances nerve regeneration through fibronectin grafts. *J Hand Surg [Br]*, 21, 514–22.

Willerth, S. M. & Sakiyama-elbert, S. E. (2007) Approaches to neural tissue engineering using scaffolds for drug delivery. *Adv Drug Deliv Rev*, 59, 325–38.

Williams, H. B. & Jabaley, M. E. (1986) The importance of internal anatomy of the peripheral nerves to nerve repair in the forearm and hand. *Hand Clin*, 2, 689–707.

Williams, L. R., Longo, F. M., Powell, H. C., Lundborg, G. & Varon, S. (1983) Spatial-temporal progress of peripheral nerve regeneration within a silicone chamber: parameters for a bioassay. *J Comp Neurol*, 218, 460–70.

Wong, J. & Oblinger, M. M. (1991) NGF rescues substance P expression but not neurofilament or tubulin gene expression in axotomized sensory neurons. *J Neurosci*, 11, 543–52.

Yamaguchi, I., Itoh, S., Suzuki, M., Osaka, A. & Tanaka, J. (2003) The chitosan prepared from crab tendons: II. The chitosan/apatite composites and their application to nerve regeneration. *Biomaterials*, 24, 3285–92.

Yan, J.-G., Matloub, H. S., Sanger, J. R., Zhang, L.-L., Riley, D. A. & Jaradeh, S. S. (2002) A modified end-to-side method for peripheral nerve repair: Large epineurial window helicoid technique versus small epineurial window standard end-to-side technique. *J Hand Surg*, 27, 484–92.

Yang, Y., Gu, X., Tan, R., Hu, W., Wang, X., Zhang, P. & Zhang, T. (2004) Fabrication and properties of a porous chitin/chitosan conduit for nerve regeneration. *Biotechnol Lett*, 26, 1793–7.

Young, R. C., Wiberg, M. & Terenghi, G. (2002) Poly-3-hydroxybutyrate (PHB): a resorbable conduit for long-gap repair in peripheral nerves. *British J Plastic Surg*, 55, 235–40.

Zhao, Q., Drott, J., Laurell, T., Wallman, L., Lindstrom, K., Bjursten, L. M., Lundborg, G., Montelius, L. & Danielsen, N. (1997) Rat sciatic nerve regeneration through a micromachined silicon chip. *Biomaterials*, 18, 75–80.

Zheng, M. & Kuffler, D. P. (2000) Guidance of regenerating motor axons in vivo by gradients of diffusible peripheral nerve-derived factors. *J Neurobiol*, 42, 212–9.

Zimmermann, M. (2001) Pathobiology of neuropathic pain. *Eur J Pharmacol*, 429, 23–37.

12
Stem cells and tissue scaffolds for bone repair

S SCAGLIONE and R QUARTO,
University of Genova, Italy and P GIANNONI,
Advanced Biotechnology Center (CBA), Italy

Abstract: This chapter discusses how cells and biomaterials interact to regenerate a functional tissue. Natural processes of bone repair are sufficient to restore the skeletal integrity for most fractures. However, the bone auto-regenerative potential cannot handle 'critical size' lesions. Therefore, manipulation of natural mechanisms to regenerate larger bone segments is required in reconstructive surgery. Biomaterials are not simple bio-inert 'cell carriers', but they have the task of driving tissue regeneration, being 'informative' for cells. The challenge is obtaining biomaterials mimicking a specific microenvironment and inducing cells to differentiate in a predetermined manner regenerating the desired tissue according to physiological pathways.

Key words: stem cells, biomimetic materials, bone repair, microenvironment.

12.1 Introduction

Several major progressions and improvements have been introduced in the field of bone regenerative medicine during the last few years, as innovative alternatives to current therapies, which still present many limitations. Natural processes of bone repair restore the skeletal integrity of most fractures; however, in 'critical sized' lesions the natural repair process is not sufficient. In these cases, therefore, some form of manipulation of the natural healing mechanisms is required for tissue regeneration and reconstructive surgery.

Very often in this context biomaterials are intended as mere cell delivery vehicles. However, biomaterials are not only simple bio-inert 'cell carriers' since they have the difficult task of driving tissue regeneration *in vivo* and, most importantly, of being 'informative' for cells, which means to direct cell differentiation fate providing a correct microenvironment and proper physical–chemical inputs. Biomaterial scaffolds have to be considered a critical component of the experimental design, since they have to promote cell adhesion, proliferation and differentiation as well as to encourage vascular invasion and ultimately, mimicking the microenvironment of a regenerating bone, new bone formation. Scaffold chemical composition is of crucial importance for the osteoconductive properties and the resorbability of the scaffold. The most intriguing concept in modern scaffold design is biomimetic scaffolds where a material is able to mimic the host tissue and pre-existing environment, thus aimed at inducing cells to differentiate in a predetermined manner and to regenerate according to physiological pathways. This approach would avoid the long-term stability problems associated with biomaterials such as mechanical loading and host reactions, since implanted grafts would be gradually resorbed and substituted by newly formed bone tissue. Besides their chemical composition, the other critical parameter to improve the efficiency of biomaterials to be used in bone tissue engineering is the overall structure and architecture.

12.2 Biomaterials as biomimetic scaffolds

In recent years, biomaterials design has evolved from the original prosthetic approach, favouring mechanical strength and durability, to a new generation of intelligent bio-functional engineered materials able to influence cellular activity such as proliferation, differentiation and morphogenesis. This tissue engineering (TE) approach postulates that novel biomaterials should provide the suitable physical–chemical inputs, *informative* for cells, addressing their response towards repair processes rather than just being a simple inert 'cell carrier'. In this view the mechanical strength becomes less important, being mostly limited to handling properties and workability.

According to these considerations, scaffold chemical composition is of crucial importance for the conductive properties and the resorbability of the graft. In addition, the structure should also be considered to mimic a specific pre-existing microenvironment. This would, in turn, induce cells to differentiate in a predetermined manner and to regenerate the desired tissue according to physiological pathways.

In the field of bone tissue engineering, a wide range of biomaterials, whose composition is such that they mimic natural bone, have been tested to stimulate ossification and to improve the osteogenic potential of osteoprogenitor cells. Calcium and phosphate ions are important components during the mineralization phase of the ossification process. Materials composed of calcium phosphate such as hydroxyapatite (HA) and tricalcium phosphate (TCP) are attractive candidates for bone substitutes (Boyde *et al.*, 1999; Dong *et al.*, 2001, 2002; Flautre *et al.*, 1999; Gauthier *et al.*, 1998, 2005; Kon *et al.*, 2000; Livingston *et al.*, 2003; Bruder *et al.*, 1998; Marcacci *et al.*, 1999). They are also particularly advantageous for bone tissue engineering applications as they induce neither immune nor inflammatory responses in recipient organisms (Erbe *et al.*, 2001; Livingston *et al.*, 2002; El-Ghannam, 2005).

HA is a natural component of bone tissue and therefore has been considered the ideal material to build bone substitutes. HA coatings have been shown to improve the outcome of prosthetic implants, improving the interaction between natural bone and implanted device. Porous HA ceramics support bone formation by marrow MSCs *in vitro* and *in vivo*. However, its brittleness and poor resorbability limits its application in the regeneration and repair of bone defects.

On the other hand, biocompatible polymers have also been regarded as candidates for bone substitutes (Bonzani *et al.*, 2007; Jiang *et al.*, 2006; Ren *et al.*, 2005; Williams *et al.*, 2005; Wu *et al.*, 2006). However, a number of practical problems still persist, such as the difficulty in controlling the *in vivo* degradation of bio-resorbable polymers, low efficiency of cell seeding, cytotoxicity of the breakdown products produced during scaffold degradation, and poor mechanical properties, not comparable with those of natural hard tissues.

To overcome these limitations, ceramic/polymer composite materials have been explored (Leung *et al.*, 2006; El-Amin *et al.*, 2006; Kim *et al.*, 2006; Kretlow and Mikos, 2007; Ren *et al.*, 2008). When used in blends with other polymers, HA particles exposed on the surface of scaffolds favour focal contact formation of osteoblasts. A bone-like mineral film, consisting mainly of calcium apatite, when layered onto the surface of polymeric-based substrates, does not achieve the same effect as when HA is incorporated into the bulk material. Other forms of calcium phosphate-containing material have been assessed for osteoconductivity.

Interestingly, heterogeneous composite scaffolds consisting of two distinct but integrated layers have been proposed to induce cells towards different lineages (i.e. cartilage and bone) and possibly generate osteochondral grafts (Martin *et al.*, 2007). Within these informative biomaterials, cells may recognize the different surfaces of the graft as pre-existing bone/cartilage tissue (mimesis) and deposit bone/cartilage extracellular matrix accordingly to their specific localization in the scaffold. This evidence further confirms that the use of an appropriate biomaterial may potentially address progenitor cells to alternative differentiation pathways, within the same lesion, if different tissues are to be repaired. Indeed osteochondral lesions may represent the ideal clinical target for this approach. Besides their chemical composition, the other critical parameter to improve the biomaterial efficiency is the overall structure and architecture (Chang *et al.*, 2000; De Oliveira *et al.*, 2003; Gauthier *et al.*, 1998; Karageorgiou and Kaplan, 2005; Mankani *et al.*, 2001; Lu *et al.*, 1999; Navarro *et al.*, 2004).

An important advancement in this field has been represented by the introduction of synthetic porous scaffolds. In these cases in fact the internal architecture can be intelligently designed and the density, pore shape, pore size and pore interconnection pathway of the material can be predetermined. The result is that the surface available for tissue regeneration as well as for cell delivery can be very broad. The increase in surface also has an effect on scaffold resorbability. Moreover, a high porosity level is necessary for the *in vivo* bone tissue ingrowth, since it enhances migration and proliferation of osteoblasts and mesenchymal cells, and matrix deposition in the empty spaces. Macroporosity has a strong impact on osteogenic outcomes and the pore interconnection pathway plays an important role as well. An incomplete pore interconnection or a limiting calibre of the interconnections could represent an important constraint to the overall biological system by limiting blood vessel invasion (Mastrogiacomo *et al.*, 2006). Therefore, bioceramics with high porosity and appropriate interconnection pathway should allow the neodeposited tissue to infiltrate and to fill the scaffold.

12.3 Adult stem cell sources

Stem cells undergo self-renewal in an undifferentiated state, retaining their differentiation potential towards multiple lineages. Both embryonic and adult stem cells have their advantages and disadvantages. Teratoma formation, immune rejection and – last but not least – ethical issues have to be resolved before embryonic stem cells (ES) can be of practical use for clinical application. Therefore adult tissues represent the current available stem cell sources. Their use eases the preparation of tissue-engineered substitutes in terms of availability, immuno-compatibility and applicability of differentiation protocols. However, adult stem cells do not perform as effectively *ex vivo*. Several factors contribute to this low efficiency: variability between donors, even between sampling sites in the same donors, age and life-habits. Historically bone marrow, due to its use in transplant procedures post-irradiation (McGovern *et al.*, 1959), was the first source of mesenchymal stem cells for skeletal tissue engineering purposes.

12.3.1 Bone marrow stromal cells

Colonies of adherent cells are formed in low-density marrow cultures (Kuznetsov and Robey, 2005). Single precursor cells of clonal colonies are referred to as colony forming unit-fibroblasts (CFU-f). The generated strains are virtually homogeneous (with the exception of mouse cells that also contain haematopoietic precursors), lack basic characteristics of macrophages, of leukocytes and of mature stromal elements, display a fibroblastic morphology, and can be expanded manyfold (Fig. 12.1). Bone

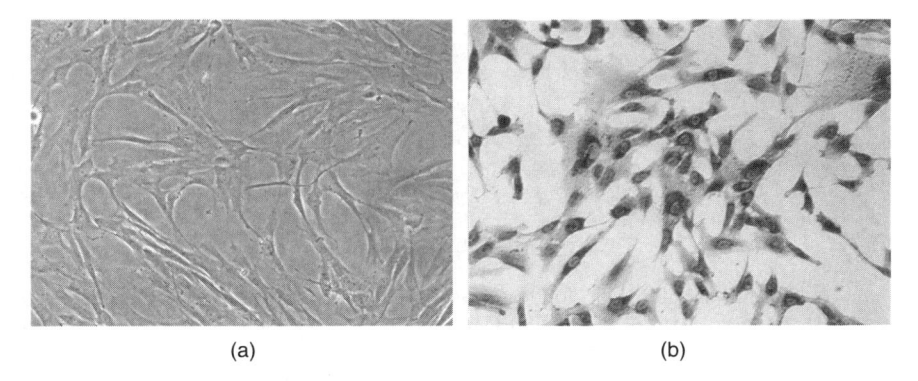

12.1 Basic morphology of adult stem cells. Cultured adult stem cells, independently of the tissue source of origin, share a common fibroblastic phenotype, displaying an elongated shape ((a): brightfield) and positivity to classical markers, such as type I collagen ((b): nuclear counterstaining with hematoxylin). The images presented were acquired with the same enlargements.

marrow stromal cells (BMSC) constitutively express telomerase activity *in vivo* (Gronthos *et al.*, 2003). They also share some characteristics with endothelial and smooth muscle cells. The general term 'bone marrow stromal cell' has been promoted to distinguish them from fibroblasts of other connective tissues. BMSC were first isolated using adherence to plastic; the presence of the STRO-1 antigen was first used to discriminate this subset of marrow cells (Simmons and Torok-Storb, 1991). A rapidly evolving research field has contributed to include several additional markers in the BMSC phenotypic profile, which comprises positivity to CD29, CD44, CD90, CD105 and CD146 and stands for a negative selection for haematopoietic surface markers such as CD34 and CD45 (Sacchetti *et al.*, 2007). BMSC also exhibit expression of vascular cell adhesion molecule-1. Alpha 4 integrin is also present, although to a lesser extent than in adipose tissue-derived stem cells.

By varying the culture conditions, BMSC can differentiate into virtually all mesoderm-derived cell types, suggesting that the BMSC population encompasses progenitor cells, similar to the haematopoietic stromal cells. Skeletal tissues that can be derived are cartilage, bone, haematopoietic stroma, and fibrous tissue (Fig. 12.2). More recently, additional phenotypes could be derived from BMSC, such as tenocytes and myogenic and skeletal muscle cells (Wakitani *et al.*, 1995; Gronthos *et al.*, 2000; Ferrari *et al.*, 1998; Awad *et al.*, 1999) including phenotypes that arise from different embryonic tissues, such as those of cardiomyocytes (Fukuda, 2001; Wang *et al.*, 2000), neurons (Woodbury *et al.*, 2000) and endothelial cells (Reyes *et al.*, 2001). However, this cell plasticity needs a broader and deeper comprehension

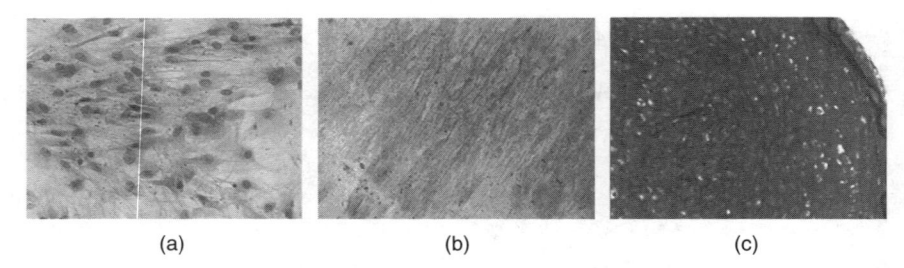

12.2 Examples of induced differentiation of cultured human bone marrow stromal cells. If cultured in the appropriate media, human adult bone marrow stromal cells express osteogenic markers such as bone sialoprotein ((a): nuclear counterstaining with hematoxylin) and produce a mineralized matrix ((b): Von Kossa staining). Upon changing the culture media and shifting to a 3-D micromass culture system, the same cells differentiate towards a chondrogenic phenotype, producing a cartilaginous metachromatic matrix ((c): Toluidine Blue staining).

to be fully exploited for tissue engineering purposes. Indeed, the cells' regenerative properties become limited upon expansion *ex vivo*: human BMSC in monolayer cultures lose the ability to differentiate towards the adipocyte and chondrocyte lineages after a certain number of cell duplications and depending on the culture conditions (Digirolamo *et al.*, 1999). The decreased differentiation potential may be ascribed to a dilution effect promoted by an increased committed progeny resulting from an asymmetrical division of the multipotent progenitor cells in culture (Rambhatla *et al.*, 2001). Contemporarily one lineage may be preferentially undertaken, depending upon the culture conditions (Lee *et al.*, 2003); actually subselection is one of the most applied strategies for tissue engineering purposes in order to achieve large number of committed cells from expanded stromal precursors.

Loss of the differentiation properties may also depend on the natural cell ageing process, which involves telomere shortening. Cell replication would be affected, possibly impairing subsets of progenitors committed to specific lineages. In the case of murine BMSC, deficiency of the telomerase inhibits the cells' differentiation into adipocytes and chondrocytes (Liu *et al.*, 2004). Conversely, overexpression of the telomerase gene sustains multipotency in human BMSC for more than 50 population doublings in culture (D'Ippolito *et al.*, 2004).

12.3.2 Microvascular-associated and circulating stem cells

Evidence that non-bone-related tissues may nest precursor cells able to produce bone under the proper stimuli has been provided ever since the early work of Friedenstein (1976). A relevant source of spontaneous osteogenic cells is localized in the vessels' basal membranes. These cells, called pericytes, may originate in mineralized nodules *in vivo* and are positive to several osteogenic markers such as alkaline phosphatase, bone sialoprotein, osteonectin, osteopontin and osteocalcin. They are able to differentiate along the osteogenic, chondrogenic and adipogenic pathways, supporting the existence of a microvessel-associated progenitor cell (Bianco *et al.*, 2001). Pericyte-derived cells can also form spontaneously differentiated myotubes associated with the expression of myogenic markers, restricted to the differentiated cells. When transplanted into severe combined immune deficient-X-linked, muscular dystrophy mice (scid-mdx), pericyte-derived cells colonize host muscle and generate numerous fibres expressing human dystrophin. These precursors, associated with microvascular walls, are distinguishable from the satellite cells and may represent an agglomeration of embryonic 'mesoangioblasts' present after birth, with a wider range of differentiation potentialities (Dellavalle *et al.*, 2007).

For tissue engineering purposes a readily available and easily accessible cell source is the key for a possible technology transfer from the bench side to the clinic, once the cell-mediated approach has been proven to function. In this respect, the finding that adherent clonogenic cells could be derived from circulating blood is relevant. Indeed those cells, retrieved from different species (human, guinea pig, rabbit and mouse) (Kuznetsov *et al.*, 2001), display a wide range of colony-forming efficiency, the lowest in human, where the CFU-f cells reaches only 0.025 cells per million of the total nucleated population. The derived cells, however, present both a fibroblastic and a polygonal morphology, although both cell types are similarly immunoreactive to most osteogenic, fibroblastic and smooth muscle cell markers. Absence of endothelial, haematopoietic cell and macrophage markers separates those circulating stem cells from the BMSC. They are also negative for Stro-1, endoglin and Muc-18, all of which are expressed in human BMSC. Additionally they are positive to the adipocyte marker CEBPα and are able to undergo adipogenic differentiation upon proper stimulation (Diascro *et al.*, 1998; Kuznetsov *et al.*, 2001). In guinea pig strains of circulating stem cells, only the fibroblast-like ones originated bone when transplanted *in vivo*, in some cases supporting a complete haematopoietic marrow compartment development within it. This is suggestive of a correlation between the phenotype and the cell stem potential (Kuznetsov *et al.*, 2001).

These findings demonstrated the presence of osteogenic precursors in peripheral blood, albeit with a highly variable frequency, raising several questions related to their origin, destination and role in reparative events.

12.3.3 Adipose tissue-derived stem cells

Adipose tissue is an easily accessible source of mesenchymal stem cells and is available in abundant amounts, compared with bone marrow. Its stromal vascular fraction is a rich source of adipose tissue-derived stem cells (ADSC). ADSC are multipotent: *in vitro*-expanded cells differentiate into adipocytes, chondrocytes, osteoblasts and myocytes. Flow cytometry and immunohistochemistry show that human adipose tissue-derived stromal cells display a protein expression phenotype similar to that of human bone marrow stromal cells. Expressed proteins include CD9, CD10, CD13, CD29, CD34, CD44, CD49(d), CD49(e), CD54, CD55, CD59, CD105, CD106 and CD146. Unlike human bone marrow-derived stromal cells, ADSC are not positive for the STRO-1 antigen. If cultured under adipogenic conditions, they uniquely express CEBPα and PPARδ, two transcriptional regulators of adipogenesis (Gronthos *et al.*, 2001).

For bone tissue applications ADSC are subjected to expansion procedures; thus some studies have compared the functional properties of freshly isolated ADSC with previously expanded cultures. The major immunophenotype difference consisted in high positivity for CD34, CD117 and HLA-DR of the freshly isolated ADSC. In contrast, expression of CD105 and especially CD166 was relatively low. After osteogenic stimulation Runx-2 and Collagen I gene expression were significantly increased in both cell populations, but with different kinetics. In general, however, the reported differences are of minor relevance (Varma *et al.*, 2007).

ADSC are known to respond to mechanical loading and osteogenic induction by the administration of 1,25-dihydroxyvitamin D3. Cells respond with an increased cyclooxygenase-2 gene expression, a key enzyme in prostaglandin (PG) synthesis. Local production and availability of PGE2 might be beneficial in promoting osteogenic differentiation of ADSC, resulting in enhanced bone formation for bone tissue engineering (Knippenberg *et al.*, 2005, 2007). ADSC have been combined, at a high density, with honeycomb collagen scaffolds in 3-D culture to promote their osteogenic potential. Cells were shown to express Cbf-a1 and specific staining confirmed the occurrence of calcification *in vitro*. Furthermore, subcutaneous transplantation of osteogenically induced cell-loaded scaffold in an ectopic nude mouse model stained positively for osteocalcin presence and bone deposition (Kakudo *et al.*, 2008). However, combined bone allograft and ADSC have not proven successful in all *in vivo* experimental settings, as assessed in a rabbit model (Follmar *et al.*, 2007).

Some improvement may come from the use of genetically engineered ADSC, as indicated by a canine ulnar bone defect model using BMP-2-expressing cells (Li *et al.*, 2007). As pluripotent progenitor cells, ADSC can also differentiate into other mesenchymal tissues, such as cartilage; several studies have been performed to assess the influence of the cell source on the cartilaginous extracellular matrix properties and components (Xu et al., 2007b; Mehlhorn et al., 2006). In this respect ADSC seem a potential source for full-thickness cartilage repair in animal models (Dragoo et al., 2007). However, comparative studies on stem cells isolated from donor-matched adipose tissue and bone marrow (Afizah et al., 2007) still indicate that BMSCs are, so far, more suitable than ADSC for chondrogenesis.

12.3.4 Cord blood stem cells

The recent identification of mesenchymal stem cells in cord-blood has revealed an unexploited source for non-haematopoietic stem cell-based therapeutic strategies for the replacement of injured or diseased connective tissue (Cetrulo, 2006). Umbilical cord blood multilineage cells (UCBC) are slower to establish in culture and display a lower precursor frequency and a lower level of bone antigen expression. Moreover, the lack of a constitutive expression of neural antigens is suggestive of a more primitive population than bone marrow cells (Goodwin et al., 2001). Early exposure of mononuclear UCBC cells to medium conditioned by osteoblastic cells in the presence of osteogenic supplements and human plasma markedly increased the frequency of stromal cell growth. Contemporarily it also enhanced the rate of osteogenic differentiation and spreading on calcium phosphate scaffolds (Hutson et al., 2005).

Subsets of cells derived from the mononuclear fraction of UCBC have been maintained in continuous culture for more than 10 passages. These cell populations express adhesion molecules CD13+, CD29+ and CD44+, but not antigens of haematopoietic differentiation; their exposure to osteogenic agents enhanced the expression of alkaline phosphatase and the appearance of hydroxyapatite nodules. Expanded UCBC maintained responsiveness to adipogenic agents. In addition, when treated with basic fibroblast growth factor and human epidermal growth factor, the cells underwent changes consistent with cells of neural origin, expressing neural-specific markers. Thus, similarly to bone marrow, cord blood contains a population of cells that can be expanded in culture and have the potential to express the phenotype of multiple lineages. These findings suggest that UCBC may act as an interesting source of osteoprogenitor cells, having a promising impact on the development of autologous tissue-engineered skeletal tissue constructs. Indeed, these cells have been tested in critical-sized femoral defects in a rat model of xenotransplant (Jager et al., 2007). Human cord blood-derived stem cells showed significant engraftment in bone marrow, surviving within a collagen-TCP scaffold for up to 4 weeks, with an increased local bone formation in a nude rat's femoral defect. However, in a study performed in rabbits, the outcomes of full-thickness cartilage defects repair were confronted using human UCBC, chondrocytes, BMSC or fibroblasts. Among the cell types used, the cord-derived cells performed worst, showing absence of hyaline cartilage, of tissue integration and of subchondral bone remodelling (Yan and Yu, 2007).

12.3.5 Placental stem cells

Cells with a BMSC-like potency have also been detected in the human placenta (Miao *et al.*, 2006). Placental tissues were trypsin-digested and the resulting cells were shown to possess a multilineage differentiation potential similar to BMSC in terms of morphology and cell-surface antigen expression. They also showed endothelial and neurogenic differentiation capabilities under appropriate conditions. Interestingly, the stromal network derived from the placenta has become a scaffold for soft tissue augmentation (Flynn *et al.*, 2006). Maintenance of the existing vascular network and ECM architecture render this tissue a valuable tool for plastic and reconstructive surgery, particularly for the adhesion of primary human adipose precursor cells (Table 12.1).

12.4 Targeting stemness within the bone marrow niche: microenvironmental evidence and co-cultures

The most intriguing concept in modern scaffolds is obtaining materials able to mimic a specific pre-existing environment, inducing cells to differentiate in a predetermined manner and to regenerate the desired tissue according to physiological pathways. The specific mechanisms by which bone marrow stromal cells leave their primary sites, move through and remodel matrices during their homing are not well understood. The microenvironment in which the mesenchymal progenitors reside is the post-natal marrow, where they form the outer coating of the sinusoid walls and the branching of the extravascular meshwork in which the myelopoiesis takes place. Once recovered from their natural niche, cells from various species may present different abilities to survive plating *in vitro*, although a majority will develop BMSC colonies. When transplanted in a highly vascularized environment, such as in the kidney capsule, BMSC can give rise to a haematopoietic stroma. However, in the absence of bone formation, no haematopoiesis is seen, a clear hint of the strong interaction between the cellular types involved (Kuznetsov and Robey, 2005). In this respect the need for a more comprehensive environment is evident, if the goal to be attained is the maintenance of stem cell characteristics.

Heterogeneous BMSC populations are often used without any selection for cell-mediated tissue engineering approaches. Initial experimental

Table 12.1 Osteogenic and non-osteogenic cell phenotypes

Cells*	Markers										
	CD29	CD34	CD44	CD45	CD49a	CD63	CD90	CD105	CD146	Stro-1	ALP
SF	n.d.	n.d.	n.d.	n.d.	+	+	+	+	low/dim	n.d.	–
MF	n.d.	n.d.	n.d.	n.d.	+	+	+	+	low/dim	n.d.	+
HTB	n.d.	n.d.	n.d.	n.d.	+	+	+	+	low/dim	n.d.	–
PC	n.d.	n.d.	n.d.	n.d.	+	+	+	+	low/dim	n.d.	+/–
FD	n.d.	n.d.	n.d.	n.d.	+	+	+	+	low/dim	n.d.	+
BMSC	n.d.	–	n.d.	–	n.d.	+	+	+	high/bright	n.d.	+/–
MACSC	+	–	+	–	n.d.	n.d.	n.d.	–	–	–	n.d.
ADSC	+	+	+	n.d.	+	n.d.	n.d.	+	low/dim	–	n.d.

*SF: skin fibroblasts; MF: muscle fibroblasts; HTB: human trabecular bone cells; PC: periosteal cells; FD: fibrotic bone marrow of fibrous dysplasia of bone; BMSC: bone marrow stromal cells; MACSC: microvascular-associated circulating stem cells; ADSC: adipose tissue-derived stem cells.

Source: adapted from Kuznetsov et al. (2001) and Sacchetti et al. (2007).

studies have tackled the task of isolating lineage-specific progenitors and evaluating the differentiation properties of the resulting expanded colonies. In spite of the differences among donor species, overall results indicate that although BMSC are clonogenic, proliferate and expand rapidly, only subsets of CFU-f derived cells give rise to real multi-potential progenitors. Among those, varied degrees of differentiation capacities can be found, enhanced and maintained. However, the maintenance of the stem cell properties seems to reside within the cooperative interaction of many cell types.

A significant example of this cooperation comes from the pluripotent embryonic stem (ES) cells. ES cells have complete potential for all the primary germ layers, such as ectoderm, mesoderm and endoderm; however, the cellular and molecular mechanisms that control their lineage-restricted differentiation are not completely understood. Interestingly, micromass culture systems have been used to induce the differentiation of ES cells into chondrocytes, the cartilage-producing cells, by co-culturing ES cells with limb bud progenitors. A high percentage of differentiated cells exhibited typical morphological characteristics of chondrocytes and expressed cartilage matrix genes such as collagen type II and proteoglycans. Expression of these specific markers was paralleled by a decrease in ES cell-specific transcription factor Oct-4 (Sui *et al.*, 2003), suggesting that signals from the progenitor cells are sufficient to induce ES cells into the chondrogenic lineage (Khillan, 2006). Similar results, although in a different co-culture setting, were obtained by using primary chondrocytes to induce human embryonic stem cells (hES) to differentiate towards the chondrocyte lineage. Co-cultures of hES and chondrocytes were established using well inserts, with control comprising hES grown alone or with fibroblasts. After 28 days, and removal of the chondrocyte inserts, hES differentiation was assessed by morphology, immunocytochemistry and reverse transcription polymerase chain reaction. Human ES, co-cultured or grown alone, were also implanted into SCID mice on a poly-D,L-lactide scaffold, harvested 35 days later and assessed in the same way. Co-cultures of hES and chondrocytes formed colonies and increased secretion of sulphated glycosaminoglycans (GAG). In addition, co-cultured hES expressed Sox 9 and collagen type II, unlike control hES alone or co-culture with fibroblasts. The implanted constructs contained significantly more type II and I collagen and GAG than the hES grown alone (Vats *et al.*, 2006). A similar crosstalk between uncommitted mesenchymal progenitors and differentiated cells was evidenced in a co-culture system applied to bone marrow mesenchymal stem cells and synovial cells or synovial fluid, where the latter induced the expression of chondrogenic markers in the marrow-derived cells in both an ovine (Chen *et al.*, 2005) and an equine model (Hegewald *et al.*, 2004).

To provide means of reconstituting the ideal physiological milieu in which different cell types play their role to maintain the progenitors' stem characteristics, new and more complex approaches are being developed. Different cell types are brought together and co-culture models are providing new insights to the mutual requirements of the different cells resident within the same tissue. Encompassing a collagen vitrified gel membrane in a nylon frame for easy handling, Takezawa and co-workers (Takezawa et al., 2007) were able to set a permeable scaffold for the contemporary growth human microvascular endothelial cells (HMVECs) and human dermal fibroblasts (HDFs) or HT-29 (a human colon carcinoma cell line) cells. The different cell types were cultured on opposite surfaces of the collagen vitrigel membrane. Histomorphological observations revealed the formation of three-dimensional crosstalk models composed of HMVECs and HDFs or HMVECs and HT-29 cells. Such models could be applied to research not only of paracrine factors, but also to epithelial– or endothelial–mesenchymal transitions.

The same environment in which BMSC are nested may affect their fate. Initially, BMSC differentiation may be affected by the nutritional environment, as demonstrated in closed culture systems. In this case nutrients would pass through a diffusion chamber and, in transplanted BMSC, cartilage would form in the inner core of the chamber, whereas bone would deposit a filter interface. Thus nutrients and soluble mediators diffusion has to be considered as factors deeply affecting the outcomes of differentiation protocols applied to mesenchymal progenitors. Additionally, the observation that successful applications of cell-based therapies in tissue engineering are currently limited to tissues a few millimetres thick led to the suggestion that introduction of a preformed *in vitro* vascular network may be a useful strategy for engineered tissues (Choong et al., 2006). *In vivo* capillary networks deliver oxygen and nutrients to thicker (>2 mm) tissues, but this network is totally absent in the engineered tissues. Griffith and collaborators (Griffith et al., 2005) developed a system for generating capillary-like networks within a thick fibrin matrix; in their system human umbilical vein endothelial cells were embedded in fibrin gels at a known distance from a monolayer of human dermal fibroblasts. Surprisingly, capillary network formation was shown as inversely correlated with the distance between the two layers of cells and did not depend on oxygen diffusion. Indeed cooperation between the two cell types seems to be the key for neo-vascularization. The issue of vascularization was also addressed by Choong et al. (2006) by investigating the development of a biomaterial surface that supports endothelial cell attachment and proliferation. Their work demonstrated that the presence of human bone marrow endothelial cell line (HBMEC-60) specifically enhanced human bone marrow-derived fibroblast (HBMF) proliferation and differentiation and that this effect was not observed with co-cultures with skin fibroblasts. Therefore co-culturing endothelial cells with HBMFs could be a promising system for bone tissue-engineering applications.

Current haematopoietic culture systems mainly utilize two-dimensional devices with limited ability to promote self-renewal of early progenitors. A possible improvement can be achieved by creating three-dimensional culture environments that, if properly engineered, may regulate stem cell proliferation and differentiation, mimicking the haematopoiesis *in vivo*. To simulate the marrow microenvironment and expand cord progenitors, human cord blood (CB) cells were cultured in 3-D non-woven matrix of polyethylene terephthalate (PET) fabric with defined microstructure. Compared to two-dimensional (2-D) CD34(+) cell culture, 3-D culture produced 30–100% higher total cells and progenitors without exogenous cytokines. By the addition of thrombopoietin and flt-3/flk-2 ligand the total cell number, the CD34(+) cell number and the colony-forming unit (CFU) number were significantly increased, indicating a haematopoiesis pathway that promoted progenitor production (Li *et al.*, 2001). Indeed, in the light of reproducing a living stroma and its microenvironment for the maintenance of the stem cell phenotypic characteristics, it is of relevance to note that cultures of 400 cGy-irradiated bone marrow CD34+ cells with endothelial cells, under non-contact conditions, supported the differential recovery of both viable progenitor cells and primitive SCID-repopulating cells. Conversely, exposure of human bone marrow CD34+ cells alone to the same radiation dosage caused a precipitous decline in haematopoietic progenitor cell which was not retrieved via cytokine treatment. Thus, vascular endothelial cells produce soluble factors that promote the repair and functional recovery of haematopoietic stem cells after radiation injury (Muramoto *et al.*, 2006). These results are in accordance with the secretory activity of human brain endothelial cells previously demonstrated to induce the expansion of purified bone marrow CD34+CD38– haematopoietic stem cells; additionally non-contact co-cultures with endothelial cells also increase to some extent the number of human cells capable of engrafting NOD/SCID mice at significantly higher rates than fresh CD34+ cells via the elaboration of soluble factors (Chute *et al.*, 2002, 2004, 2005).

Among nutrients, cell-to-cell signalling, soluble factors and vascular support, the extracellular matrix composition is another of the key factors that may promote stem cell phenotype retention and different cell type crosstalk. Adhering cells must attach to the appropriate matrix to enable survival and differentiation; surface molecules are important players in mediating cell–matrix interaction. Some ECM proteins are a prerequisite for matrix formation and mineralization, including alkaline phosphatase (ALP), osteocalcin, osteopontin, biglycan, and bone sialoprotein (BSP).

Bone sialoprotein, for example, is an acidic glycoprotein that plays an important role in cancer cell growth, migration and invasion (Kayed *et al.*, 2007). To date, BSP has emerged as the only *bona fide* candidate for nucleation of the mineralization process, displaying RGD motifs with the ability to bind hydroxyapatite and cell-surface integrins (Ganss *et al.*, 1999; Tye *et al.*, 2003). Interestingly BSP enhances osteogenic cell migration through basement membrane and collagen matrices *in vitro* by localizing matrix metalloprotease 2 (MMP-2) on the cell surface through alpha(v)beta(3)-integrin (Karadag and Fisher, 2006), possibly with the involvement of specific factors (Zha *et al.*, 2007). An environmental feedback, such as binding to a specific scaffold substrate, may induce BMSC responses promoting further cell differentiation; for example, Mg-modified alumina surfaces are more efficient in maintaining high transcript levels for BSP in human bone-derived cells than regular alumina (Zreiqat *et al.*, 1999), thus prompting cells to a faster and more efficient matrix deposition, a possible advantage for clinical applications. Very low levels of BSP as a coating solution have already been proposed on a rough surface of titanium-hydroxyapatite dental implant materials (Hilbig *et al.*, 2007) or onto intramedullary femoral implants (O'Toole *et al.*, 2004), with the aim of reducing the healing time for the patient. BSP-coated implants demonstrated improvement in cell adhesion and osteoinduction, but the pull-out strength of the femoral implants remained unaffected.

12.5 Stem cell-induced immunomodulation: a shared characteristic?

Strategies to recruit adult stem cells from different tissue sources are being readily implemented. The use of allogenic stem cells can expand their therapeutic interest, provided that the grafted cells could be tolerated. The immunosuppressive properties of BMSC are being extensively characterized. Some studies with bulk populations of BMSC indicate that soluble factors such as PGE2 and TGFbeta are important, while others support a role for cell–cell contact. Cloned BMSC, in fact, exhibit strong inhibitory effects on T-cell proliferation *in vitro*, and injection of a small number of these cells promoted the survival of allogenic skin grafts in mice. In comparison, direct co-culture of BMSC with stimulated lymphocytes resulted in much stronger immunosuppressive effect. Interestingly, the suppression was bidirectional, as BMSC proliferation was also reduced in the presence of lymphocytes (Xu *et al.*, 2007a). Possibly immunomodulation acts via inhibition of naive and memory T-cell responses to their cognate antigens, eventually due to a physical hindrance of the T-cells from the contact with the antigen-presenting cells in a non-cognate fashion (Krampera *et al.*, 2003).

The immunomodulatory effects of alternative stem cells are less defined. However, in recent papers, ADSC cells were shown to be unable to provoke *in vitro* alloreactivity of incompatible lymphocytes. They also suppressed mixed lymphocyte reaction and lymphocyte proliferative response to mitogens; for a full inhibitory effect cell contact was necessary. These findings support the proposal that ADSC cells share immunosuppressive properties with BMSC (Puissant *et al.*, 2005). Passaged ADSCs were shown to significantly increase their secretion of prostaglandin E2. Furthermore, only PGE2-production inhibitor indomethacine counteracted ADSC-mediated suppression on allogeneic lymphocyte proliferation, indicating that PGE2 is the major soluble factor involved in immunosuppressive effect of these cells (Cui *et al.*, 2007). Most relevantly, the immunomodulatory properties are retained even after osteogenic induction *in vitro* by both BMSCs and ADSC. Therefore allogenic transplantation of both cell types is feasible and should be considered in the context of tissue engineering (Niemeyer *et al.*, 2007). Recently contact- and dose-dependent immunosuppression of mesenchymal and epithelial amniotic stem cell populations were also reported (Wolbank *et al.*, 2007). Results showed inhibition of the immune response of peripheral blood mononuclear cells in mixed lymphocyte reactions and in phytohaemagglutinin activation assay for both cell types, suggesting their possible use for tissue engineered-allogeneic applications.

12.6 Future trends

Following our considerations, completely 'open structure' based scaffolds with no physical constraints (i.e. pore size, interconnection size, total porosity) might be modelled and developed in order to provide a *permissive* scaffold with time- and space-unlimited *in vivo* blood vessel invasion and neo-bone tissue ingrowth. Although highly porous bioceramics still represent a promising approach for generating osteoconductive grafts (Boyde *et al.*, 1999; Chang *et al.*, 2000; Gauthier *et al.*, 1998; Karageorgiou and Kaplan, 2005; Mankani *et al.*, 2001) their small pore interconnection size and internal pore architecture represent a physical constraint for blood vessel size and *in vivo* bone ingrowth within pore cavities (Mastrogiacomo *et al.*, 2006). A possible alternative approach might be the development of granule based grafts, where no physical limits are present and therefore the *in vivo* bone-forming efficiency might be significantly higher.

Lastly, different technologies have been developed to design scaffolds, using the precision multi-scale control of the biomaterial architecture. This has become possible with the introduction of a wide array of manufacturing tissue scaffold techniques, which include electrospinning, self-assembly, phase separation, solvent-casting and particulate-leaching, freeze drying, melt moulding, template synthesis, drawing, gas foaming, and solid-free forming (Murugan and Ramakrishna, 2007). Among them, electrospinning is considered an effectual practical method in designing nano-fibrous scaffolds with fibre orientation (Huang *et al.*, 2002; Murugan and Ramakrishna, 2007). Scaffolds made of nanofibres (nano-fibrous scaffolds) play a key role in the success of TE, providing the structural support for the cells, accommodating and guiding their growth in the 3-D space into a specific tissue. The fibre orientation may greatly influence cell growth and its related functions. Therefore, engineering scaffolds with control over fibre orientation may become an essential prerequisite for controlling the whole tissue repair process. In this respect applications of nano-fibrous scaffolds may be envisaged in tendon and ligament repair, in vascular and nerve-regeneration applications as well as in the infarcted muscle cell-based therapeutic attempts.

12.7 Conclusions

Ideal skeletal reconstruction depends on regeneration of normal tissues that result from initiation of progenitor cell activity. However, knowledge of the origins and phenotypic characteristics of these progenitors and of the controlling factors governing bone formation and remodelling is still limited. In this chapter we have described the combination of stem cells and biomaterials as a powerful potential therapeutic tool for the regeneration and repair of large skeletal lesions. Several experimental preclinical models applying this approach have provided promising results of high interest.

The interaction between cells and scaffolds is a critical point in a regenerative medicine strategy. Biomaterial scaffolds in fact have to provide physical, chemical and biological information, in order to promote cell adhesion, proliferation and differentiation towards the needed cell types and proper tissue structural organization. Additionally, scaffolds should encourage vascular invasion and *mimic or reproduce the required microenvironment* for regenerating tissue, and they should disappear once their function is not necessary any longer.

No scaffold can be considered really inert since any material placed in contact with or within the body is going to be recognized and decoded by the cells. Indeed, all materials are able to induce changes in gene expression, and therefore influence cell behaviour (Zreiqat *et al.*, 1999). Cells interact with some materials with physiological mechanisms, for instance integrin-mediated adhesion. This is the case of collagen-based materials, whose major component is a protein represented in all tissues, which provide a bio-mimetic substrate. Conversely, the interaction between cells and synthetic materials is less intuitive, these materials being virtually molecularly unknown. The extreme of this process is represented by the foreign body reaction.

Stem cells represent an important novel tool in regenerative medicine and surgery and will surely find striking applications in the therapy of a variety of pathologies. However, there are still gaps in our knowledge of their biology and the standardization level required for clinical applications.

BMSCs have been reported to be responsible for exerting a long-lasting suppressant activity on the immune system of the recipient organism following engraftment. This controversial capability could have important applications in the treatment of immune-mediated disorders. On the other hand, it could also have negative consequences in terms of suppression of the antineoplastic surveillance of the immune system, thus potentially favouring the onset of neoplasias in the transplanted host (Djouad et al., 2003). Further studies are indeed needed to assess the real importance of this point. As a result of the potential consequences of this critical and previously unrecognized aspect, BMSC use for therapeutic application should be carefully evaluated.

Scaffolds will also have to evolve to provide higher standards of efficiency and efficacy; possibly future biomaterials will need to have a geometry specifically designed for hosting cells and allow their survival and a better vascularization once delivered at the implant site *in vivo*. Scaffolds will have to be optimized in order to have resorption rates adequate to the specific application and possibly to become, at least to some extent, more easily workable and to allow some movement (flexion, torsion, compression, etc.).

Some issues remain at the forefront of the controversy involving stem cell research – legislation, ethics and public opinion, cost and production methods. As with any new technology, the generated enthusiasm, having potential to influence virtually every orthopaedic case management, must be balanced by subjecting it to stringent clinical and basic research investigations and to the respect of the current laws and guidelines for the preparation and therapeutical application of cell-based devices for tissue repair. The scientific and clinical challenge remains: to perfect cell-based tissue-engineering protocols in order to utilize the body's own rejuvenation capabilities; this can be achieved by managing surgical implantations of scaffolds, bioactive factors, and reparative cells to regenerate damaged or diseased skeletal tissues.

12.8 References

Afizah, H., Yang, Z., Hui, J. H., Ouyang, H. W. and Lee, E. H. (2007) *Tissue Eng*, **13**, 659–66.
Awad, H. A., Butler, D. L., Boivin, G. P., Smith, F. N., Malaviya, P., Huibregtse, B. and Caplan, A. I. (1999) *Tissue Eng*, **5**, 267–77.

Bianco, P., Riminucci, M., Gronthos, S. and Robey, P. G. (2001) *Stem Cells*, **19**, 180–92.
Bonzani, I. C., Adhikari, R., Houshyar, S., Mayadunne, R., Gunatillake, P. and Stevens, M. M. (2007) *Biomaterials*, **28**, 423–33.
Boyde, A., Corsi, A., Quarto, R., Cancedda, R. and Bianco, P. (1999) *Bone*, **24**, 579–89.
Bruder, S. P., Kraus, K. H., Goldberg, V. M. and Kadiyala, S. (1998) *J Bone Joint Surg Am*, **80**, 985–96.
Cetrulo, C. L., Jr. (2006) *Stem Cell Rev*, **2**, 163–8.
Chang, B. S., Lee, C. K., Hong, K. S., Youn, H. J., Ryu, H. S., Chung, S. S. and Park, K. W. (2000) *Biomaterials*, **21**, 1291–8.
Chen, J., Wang, C., Lu, S., Wu, J., Guo, X., Duan, C., Dong, L., Song, Y., Zhang, J., Jing, D., Wu, L., Ding, J. and Li, D. (2005) *Cell Tissue Res*, **319**, 429–38.
Choong, C. S., Hutmacher, D. W. and Triffitt, J. T. (2006) *Tissue Eng*, **12**, 2521–31.
Chute, J. P., Saini, A. A., Chute, D. J., Wells, M. R., Clark, W. B., Harlan, D. M., Park, J., Stull, M. K., Civin, C. and Davis, T. A. (2002) *Blood*, **100**, 4433–9.
Chute, J. P., Muramoto, G., Fung, J. and Oxford, C. (2004) *Stem Cells*, **22**, 202–15.
Chute, J. P., Muramoto, G. G., Fung, J. and Oxford, C. (2005) *Blood*, **105**, 576–83.
Cui, L., Yin, S., Liu, W., Li, N., Zhang, W. and Cao, Y. (2007) *Tissue Eng*, **13**, 1185–95.
De Oliveira, J. F., De Aguiar, P. F., Rossi, A. M. and Soares, G. A. (2003) *Artif Organs*, **27**, 406–11.
Dellavalle, A., Sampaolesi, M., Tonlorenzi, R., Tagliafico, E., Sacchetti, B., Perani, L., Innocenzi, A., Galvez, B. G., Messina, G., Morosetti, R., Li, S., Belicchi, M., Peretti, G., Chamberlain, J. S., Wright, W. E., Torrente, Y., Ferrari, S., Bianco, P. and Cossu, G. (2007) *Nat Cell Biol*, **9**, 255–67.
Diascro, D. D., Jr., Vogel, R. L., Johnson, T. E., Witherup, K. M., Pitzenberger, S. M., Rutledge, S. J., Prescott, D. J., Rodan, G. A. and Schmidt, A. (1998) *J Bone Miner Res*, **13**, 96–106.
Digirolamo, C. M., Stokes, D., Colter, D., Phinney, D. G., Class, R. and Prockop, D. J. (1999) *Br J Haematol*, **107**, 275–81.
D'Ippolito, G., Diabira, S., Howard, G. A., Menei, P., Roos, B. A. and Schiller, P. C. (2004) *J Cell Sci*, **117**, 2971–81.
Djouad, F., Plence, P., Bony, C., Tropel, P., Apparailly, F., Sany, J., Noel, D. and Jorgensen, C. (2003) *Blood*, **102**, 3837–44.
Dong, J., Kojima, H., Uemura, T., Kikuchi, M., Tateishi, T. and Tanaka, J. (2001) *J Biomed Mater Res*, **57**, 208–16.
Dong, J., Uemura, T., Shirasaki, Y. and Tateishi, T. (2002) *Biomaterials*, **23**, 4493–502.
Dragoo, J. L., Carlson, G., McCormick, F., Khan-Farooqi, H., Zhu, M., Zuk, P. A. and Benham, P. (2007) *Tissue Eng*, **13**, 1615–21.
El-Amin, S. F., Botchwey, E., Tuli, R., Kofron, M. D., Mesfin, A., Sethuraman, S., Tuan, R. S. and Laurencin, C. T. (2006) *J Biomed Mater Res A*, **76**, 439–49.
El-Ghannam, A. (2005) *Expert Rev Med Devices*, **2**, 87–101.
Erbe, E. M., Marx, J. G., Clineff, T. D. and Bellincampi, L. D. (2001) *Eur Spine J*, **10 Suppl 2**, S141–6.
Ferrari, G., Cusella-De Angelis, G., Coletta, M., Paolucci, E., Stornaiuolo, A., Cossu, G. and Mavilio, F. (1998) *Science*, **279**, 1528–30.
Flautre, B., Anselme, K., Delecourt, C., Lu, J., Hardouin, P. and Descamps, M. (1999) *J Mater Sci Mater Med*, **10**, 811–4.

Flynn, L., Semple, J. L. and Woodhouse, K. A. (2006) *J Biomed Mater Res A*, **79**, 359–69.
Follmar, K. E., Prichard, H. L., DeCroos, F. C., Wang, H. T., Levin, L. S., Klitzman, B., Olbrich, K. C. and Erdmann, D. (2007) *Ann Plast Surg*, **58**, 561–5.
Friedenstein, A. J. (1976) *Int Rev Cytol*, **47**, 327–59.
Fukuda, K. (2001) *Artif Organs*, **25**, 187–93.
Ganss, B., Kim, R. H. and Sodek, J. (1999) *Crit Rev Oral Biol Med*, **10**, 79–98.
Gauthier, O., Bouler, J. M., Aguado, E., Pilet, P. and Daculsi, G. (1998) *Biomaterials*, **19**, 133–9.
Gauthier, O., Muller, R., von Stechow, D., Lamy, B., Weiss, P., Bouler, J. M., Aguado, E. and Daculsi, G. (2005) *Biomaterials*, **26**, 5444–53.
Goodwin, H. S., Bicknese, A. R., Chien, S. N., Bogucki, B. D., Quinn, C. O. and Wall, D. A. (2001) *Biol Blood Marrow Transplant*, **7**, 581–8.
Griffith, C. K., Miller, C., Sainson, R. C., Calvert, J. W., Jeon, N. L., Hughes, C. C. and George, S. C. (2005) *Tissue Eng*, **11**, 257–66.
Gronthos, S., Mankani, M., Brahim, J., Robey, P. G. and Shi, S. (2000) *Proc Natl Acad Sci U S A*, **97**, 13625–30.
Gronthos, S., Franklin, D. M., Leddy, H. A., Robey, P. G., Storms, R. W. and Gimble, J. M. (2001) *J Cell Physiol*, **189**, 54–63.
Gronthos, S., Zannettino, A. C., Hay, S. J., Shi, S., Graves, S. E., Kortesidis, A. and Simmons, P. J. (2003) *J Cell Sci*, **116**, 1827–35.
Hegewald, A. A., Ringe, J., Bartel, J., Kruger, I., Notter, M., Barnewitz, D., Kaps, C. and Sittinger, M. (2004) *Tissue Cell*, **36**, 431–8.
Hilbig, H., Kirsten, M., Rupietta, R., Graf, H. L., Thalhammer, S., Strasser, S. and Armbruster, F. P. (2007) *Eur J Med Res*, **12**, 6–12.
Huang, J. I., Beanes, S. R., Zhu, M., Lorenz, H. P., Hedrick, M. H. and Benhaim, P. (2002) *Plast Reconstr Surg*, **109**, 1033–41; discussion 1042–3.
Hutson, E. L., Boyer, S. and Genever, P. G. (2005) *Tissue Eng*, **11**, 1407–20.
Jager, M., Degistirici, O., Knipper, A., Fischer, J., Sager, M. and Krauspe, R. (2007) *J Bone Miner Res*, **22**, 1224–33.
Jiang, T., Abdel-Fattah, W. I. and Laurencin, C. T. (2006) *Biomaterials*, **27**, 4894–903.
Kakudo, N., Shimotsuma, A., Miyake, S., Kushida, S. and Kusumoto, K. (2008) *J Biomed Mater Res A*, **84**, 191–7.
Karadag, A. and Fisher, L. W. (2006) *J Bone Miner Res*, **21**, 1627–36.
Karageorgiou, V. and Kaplan, D. (2005) *Biomaterials*, **26**, 5474–91.
Kayed, H., Kleeff, J., Keleg, S., Felix, K., Giese, T., Berger, M. R., Buchler, M. W. and Friess, H. (2007) *Cancer Lett*, **245**, 171–83.
Khillan, J. S. (2006) *Methods Mol Biol*, **330**, 161–70.
Kim, S. S., Park, M. S., Gwak, S. J., Choi, C. Y. and Kim, B. S. (2006) *Tissue Eng*, **12**, 2997–3006.
Knippenberg, M., Helder, M. N., Doulabi, B. Z., Semeins, C. M., Wuisman, P. I. and Klein-Nulend, J. (2005) *Tissue Eng*, **11**, 1780–8.
Knippenberg, M., Helder, M. N., de Blieck-Hogervorst, J. M., Wuisman, P. I. and Klein-Nulend, J. (2007) *Tissue Eng*, **13**, 2495–2503.
Kon, E., Muraglia, A., Corsi, A., Bianco, P., Marcacci, M., Martin, I., Boyde, A., Ruspantini, I., Chistolini, P., Rocca, M., Giardino, R., Cancedda, R. and Quarto, R. (2000) *J Biomed Mater Res*, **49**, 328–37.
Krampera, M., Glennie, S., Dyson, J., Scott, D., Laylor, R., Simpson, E. and Dazzi, F. (2003) *Blood*, **101**, 3722–9.

Kretlow, J. D. and Mikos, A. G. (2007) *Tissue Eng*, **13**, 927–38.
Kuznetsov, S. A., Robey P.G. (2005) *Skeletal Stem Cells*, Eurekah.com Landes Bioscience, Georgetown, TX.
Kuznetsov, S. A., Mankani, M. H., Gronthos, S., Satomura, K., Bianco, P. and Robey, P. G. (2001) *J Cell Biol*, **153**, 1133–40.
Lee, H. S., Crane, G. G., Merok, J. R., Tunstead, J. R., Hatch, N. L., Panchalingam, K., Powers, M. J., Griffith, L. G. and Sherley, J. L. (2003) *Biotechnol Bioeng*, **83**, 760–71.
Leung, V. Y., Chan, D. and Cheung, K. M. (2006) *Eur Spine J*, **15 Suppl 3**, S406–13.
Li, H., Dai, K., Tang, T., Zhang, X., Yan, M. and Lou, J. (2007) *Biochem Biophys Res Commun*, **356**, 836–42.
Li, Y., Ma, T., Kniss, D. A., Yang, S. T. and Lasky, L. C. (2001) *J Hematother Stem Cell Res*, **10**, 355–68.
Liu, L., DiGirolamo, C. M., Navarro, P. A., Blasco, M. A. and Keefe, D. L. (2004) *Exp Cell Res*, **294**, 1–8.
Livingston, T., Ducheyne, P. and Garino, J. (2002) *J Biomed Mater Res*, **62**, 1–13.
Livingston, T. L., Gordon, S., Archambault, M., Kadiyala, S., McIntosh, K., Smith, A. and Peter, S. J. (2003) *J Mater Sci Mater Med*, **14**, 211–8.
Lu, J. X., Flautre, B., Anselme, K., Hardouin, P., Gallur, A., Descamps, M. and Thierry, B. (1999) *J Mater Sci Mater Med*, **10**, 111–20.
Mankani, M. H., Kuznetsov, S. A., Fowler, B., Kingman, A. and Robey, P. G. (2001) *Biotechnol Bioeng*, **72**, 96–107.
Marcacci, M., Kon, E., Zaffagnini, S., Giardino, R., Rocca, M., Corsi, A., Benvenuti, A., Bianco, P., Quarto, R., Martin, I., Muraglia, A. and Cancedda, R. (1999) *Calcif Tissue Int*, **64**, 83–90.
Martin, I., Miot, S., Barbero, A., Jakob, M. and Wendt, D. (2007) *J Biomech*, **40**, 750–65.
Mastrogiacomo, M., Scaglione, S., Martinetti, R., Dolcini, L., Beltrame, F., Cancedda, R. and Quarto, R. (2006) *Biomaterials*, **27**, 3230–7.
McGovern, J. J., Russell, P. S., Atkins, L. and Webster, E. W. (1959) *N Engl J Med*, **260**, 675–83.
Mehlhorn, A. T., Niemeyer, P., Kaiser, S., Finkenzeller, G., Stark, G. B., Sudkamp, N. P. and Schmal, H. (2006) *Tissue Eng*, **12**, 2853–62.
Miao, Z., Jin, J., Chen, L., Zhu, J., Huang, W., Zhao, J., Qian, H. and Zhang, X. (2006) *Cell Biol Int*, **30**, 681–7.
Muramoto, G. G., Chen, B., Cui, X., Chao, N. J. and Chute, J. P. (2006) *Biol Blood Marrow Transplant*, **12**, 530–40.
Murugan, R. and Ramakrishna, S. (2007) *Tissue Eng*, **13**, 1845–66.
Navarro, M., del Valle, S., Martinez, S., Zeppetelli, S., Ambrosio, L., Planell, J. A. and Ginebra, M. P. (2004) *Biomaterials*, **25**, 4233–41.
Niemeyer, P., Kornacker, M., Mehlhorn, A., Seckinger, A., Vohrer, J., Schmal, H., Kasten, P., Eckstein, V., Sudkamp, N. P. and Krause, U. (2007) *Tissue Eng*, **13**, 111–21.
O'Toole, G. C., Salih, E., Gallagher, C., FitzPatrick, D., O'Higgins, N. and O'Rourke, S. K. (2004) *J Orthop Res*, **22**, 641–6.
Puissant, B., Barreau, C., Bourin, P., Clavel, C., Corre, J., Bousquet, C., Taureau, C., Cousin, B., Abbal, M., Laharrague, P., Penicaud, L., Casteilla, L. and Blancher, A. (2005) *Br J Haematol*, **129**, 118–29.

Rambhatla, L., Bohn, S. A., Stadler, P. B., Boyd, J. T., Coss, R. A. and Sherley, J. L. (2001) *J Biomed Biotechnol*, **1**, 28–37.
Ren, J., Zhao, P., Ren, T., Gu, S. and Pan, K. (2008) *J Mater Sci Mater Med*, **19**, 1075–82.
Ren, T., Ren, J., Jia, X. and Pan, K. (2005) *J Biomed Mater Res A*, **74**, 562–9.
Reyes, M., Lund, T., Lenvik, T., Aguiar, D., Koodie, L. and Verfaillie, C. M. (2001) *Blood*, **98**, 2615–25.
Sacchetti, B., Funari, A., Michienzi, S., Di Cesare, S., Piersanti, S., Saggio, I., Tagliafico, E., Ferrari, S., Robey, P. G., Riminucci, M. and Bianco, P. (2007) *Cell*, **131**, 324–36.
Simmons, P. J. and Torok-Storb, B. (1991) *Blood*, **78**, 55–62.
Sui, Y., Clarke, T. and Khillan, J. S. (2003) *Differentiation*, **71**, 578–85.
Takezawa, T., Nitani, A., Shimo-Oka, T. and Takayama, Y. (2007) *Cells Tissues Organs*, **185**, 237–41.
Tye, C. E., Rattray, K. R., Warner, K. J., Gordon, J. A., Sodek, J., Hunter, G. K. and Goldberg, H. A. (2003) *J Biol Chem*, **278**, 7949–55.
Varma, M. J., Breuls, R. G., Schouten, T. E., Jurgens, W. J., Bontkes, H. J., Schuurhuis, G. J., van Ham, S. M. and van Milligen, F. J. (2007) *Stem Cells Dev*, **16**, 91–104.
Vats, A., Bielby, R. C., Tolley, N., Dickinson, S. C., Boccaccini, A. R., Hollander, A. P., Bishop, A. E. and Polak, J. M. (2006) *Tissue Eng*, **12**, 1687–97.
Wakitani, S., Saito, T. and Caplan, A. I. (1995) *Muscle Nerve*, **18**, 1417–26.
Wang, J. S., Shum-Tim, D., Galipeau, J., Chedrawy, E., Eliopoulos, N. and Chiu, R. C. (2000) *J Thorac Cardiovasc Surg*, **120**, 999–1005.
Williams, J. M., Adewunmi, A., Schek, R. M., Flanagan, C. L., Krebsbach, P. H., Feinberg, S. E., Hollister, S. J. and Das, S. (2005) *Biomaterials*, **26**, 4817–27.
Wolbank, S., Peterbauer, A., Fahrner, M., Hennerbichler, S., van Griensven, M., Stadler, G., Redl, H. and Gabriel, C. (2007) *Tissue Eng*, **13**, 1173–83.
Woodbury, D., Schwarz, E. J., Prockop, D. J. and Black, I. B. (2000) *J Neurosci Res*, **61**, 364–70.
Wu, Y. C., Shaw, S. Y., Lin, H. R., Lee, T. M. and Yang, C. Y. (2006) *Biomaterials*, **27**, 896–904.
Xu, G., Zhang, L., Ren, G., Yuan, Z., Zhang, Y., Zhao, R. C. and Shi, Y. (2007a) *Cell Res*, **17**, 240–8.
Xu, Y., Balooch, G., Chiou, M., Bekerman, E., Ritchie, R. O. and Longaker, M. T. (2007b) *Biochem Biophys Res Commun*, **359**, 311–6.
Yan, H. and Yu, C. (2007) *Arthroscopy*, **23**, 178–87.
Zha, Y. H., He, J. F., Mei, Y. W., Yin, T. and Mao, L. (2007) *Cell Biol Int*, **31**, 1089–96.
Zreiqat, H., Evans, P. and Howlett, C. R. (1999) *J Biomed Mater Res*, **44**, 389–96.

13
Cellular response to osteoinductive materials in orthopaedic surgery

L DI SILVIO and P JAYAKUMAR,
King's College London, UK

Abstract: Osteoinductive materials provide a biological stimulus for induction, recruitment, stimulation and differentiation of primitive, undifferentiated and pluripotent stromal cells into osteoblast or preosteoblasts, which are the initial cellular phase of a bone-forming lineage. This chapter discusses the importance of autografts, allografts and synthetic bone-graft substitutes (SBGSs) in reconstructive orthopaedic surgery. It reviews the biological and cellular response to osteoinductive materials and discusses how they can be further enhanced to optimise surgical application, in particular, the benefits of bone graft materials for bone tissue engineering techniques. Cellular responses relating to immunology, physical incorporation and the remodelling effect within the surrounding tissue environment are also discussed. The chapter presents current work utilising various porous cell seeded scaffolds, bioactive factors, recombinant signalling molecules, and stem cells, and their capacity for osseous regeneration. It also highlights the role of these materials in the future management of skeletal disorders and state-of-the-art trauma and reconstructive surgery.

Key words: osteoinductive materials, bone grafts, bone morphogenetic proteins, stem cells, tissue engineering.

13.1 Introduction

Skeletal degeneration, pathology and trauma leading to bone loss are major modern-day health concerns with massive socio-economic implications. The advances in surgical practice combined with a rise in active and ageing populations has led to an increase in trauma and reconstructive orthopaedic surgery. Fracture repair and implant technology continues to improve and develop; however, there are a significant proportion of patients commonly encountered with complications such as fracture non-union and bony defects not amenable to healing by direct fracture fixation alone. Consequently there is a greater demand for bone graft and novel bone regeneration systems beyond the current conventional medical and surgical repair strategies.

Skeletal regeneration is essentially achieved by the interaction of three elements: cells, growth factors (GFs) and a scaffold material. The skeletal regenerative process requires these three elements and the facilitation of an appropriate cellular response. The materials triggering this cellular response should ideally be osteoinductive. Autografts, allografts and synthetic bone-graft substitutes (SBGSs) are important biomaterials in reconstructive orthopaedic surgery which have been used or developed as part of tissue engineered systems capable of regenerating bone with variable osteoinductive potentials.[1] Autografts and allografts currently play an important role in the treatment of non-union, bridging diaphyseal defects and filling metaphyseal defects. Vascularised and cancellous autografts are considered the 'gold standard' bone graft due to their excellent capacity for skeletal incorporation. However, they are limited in terms of supply when extensive grafting is required, such as in spinal arthrodesis and repair of large bony defects, and they are associated with post-operative pain and host donor site morbidity.[2,3] Allografts such as demineralised bone matrix demonstrate osteoinductive potential and prove useful alternatives to autografts, but the majority are fraught with problems of low osteogenicity, increased immunogenicity, high resorption rates compared with autogenous bone and risk of disease transmission.[4] Osteoinductive growth factors, such as bone morphogenetic proteins, have been discovered and continue to act as powerful elements in bone regeneration science. The general limitations of bone graft materials, however, necessitate the challenge to develop SBGSs with high biocompatibility, osteogenicity and osteoinductivity on demand.

Ultimately, understanding the biological and cellular response to osteoinductive materials will allow us to optimise this surgical technology and harness the benefits of bone graft materials and bone tissue engineering techniques. These responses include the cellular processes related to immunology, physical incorporation and the remodelling effect within the surrounding tissue environment. Current work utilising bioresorbable materials, porous cell seeded scaffolds, bioactive factors, recombinant signalling molecules and stem cells has already shown the capacity for osseous regeneration. This biotechnology promises to play an integral role in the future management of skeletal disorders and state-of-the-art trauma and reconstructive surgery.

13.1.1 Definitions

Understanding the science of bone regeneration involves precise definition of important terminology.[5–8]

Osteoinduction is defined as the process by which osteogenesis (i.e., new bone formation from osteocompetent cells in connective tissue or cartilage) is induced.[9] In effect, this phenomenon features in most bone healing processes and osteoinductive materials provide a biological stimulus for induction, recruitment, stimulation and differentiation of primitive, undifferentiated and pluripotent stromal cells into osteoblast or preosteoblasts, the initial cellular phase of a bone-forming lineage. The original definition of osteoinductive materials by Marshal Urist states that these biomaterials are capable of inducing bone to form when placed into an extraskeletal site.[5] Osteoinductive materials include autografts, demineralised bone matrix (DBM) and specific bone morphogenetic proteins (BMPs) which naturally form bone within the skeleton as well as extraskeletally.[8,10,11] The ideal bone regeneration system will be fashioned from an osteoinductive replacement material that elicits an appropriate cellular response with the aid of bioactive factors to allow healing and osseous regeneration.

Osteogenic materials are defined as those which contain living cells and are capable of differentiation into bone.

Osteoconduction is defined as the process of bony ingrowth from local osseous tissue onto surfaces. The original definition was not strictly restricted to biomaterials;[12] however, the contemporary concept of an osteoconductive material is one where bone formation is promoted to appose and conform to its surface, when the material is placed into bone, by virtue of its composition, shape or surface texture.[9] In effect, these materials act as receptive scaffolds that facilitate enhanced bone formation. Purely osteoconductive biomaterials, e.g. hydroxyapatite (HA), are not usually associated with bone formation outside bone.[8]

Osseointegration is the process of achieving stable direct anchorage and contact between bone and implant. The phenomenon of osseointegration was first described by Brånemark *et al.* in 1977[13] and first defined by Albrektsson *et al.* in 1981.[14] The definition stated direct contact, at the light microscope level, between living bone and implant. It has also been defined at the histological level as the direct anchorage of an implant by formation of bony tissue around the implant without the growth of fibrous tissue at the bone–implant interface.[15] A biomechanical definition has also evolved stating that osseointegration is 'A process whereby clinically asymptomatic rigid fixation of alloplastic materials is achieved, and maintained, in bone during functional loading'.[16] This mode of anchorage between bone and implant has been highly successful and well demonstrated in the field of craniofacial implantology.[17]

Osteoinduction, osteoconduction and osseointegration are powerful interrelated phenomena in bone regeneration and are intrinsically related to bone morphogenetic proteins, bone growth factors and direct bone anchorage factors respectively. It should be noted that these are relative, not absolute, terms. A material's position on the scale of osteoinductivity and osteoconductivity is intrinsically dependent on processing, donor site, surface chemistry, geometry and texture, and extrinsically dependent on the biological factors, and the chemical and mechanical tissue environment.

13.2 Osteoinduction and bone healing

Osteoinduction is the basic biological mechanism of recruiting and stimulating immature cells to develop into preosteoblasts. This phenomenon forms part of the initial phase of the bone morphogenesis cascade and commences immediately after skeletal injury or biomaterial implantation, being particularly active during the first week post-trauma. Actions of the newly developed preosteoblasts are more clearly observed several weeks later in the callus stage of fracture healing.[18,19]

13.2.1 The bone morphogenesis cascade

Bone morphogenesis is a tri-phase sequential cascade: chemotaxis and mitosis of mesenchymal cells, differentiation of the mesenchymal cells initially into cartilage, and replacement of the cartilage by bone. The natural repair mechanism of bone utilises bone-forming cells, a cartilaginous scaffold upon which the new woven bone is formed, and bioactive molecules to direct the repair process.

13.2.2 The cells

Bone healing is primarily dependent on the osteoinductive potential of the tissue environment. The natural skeletal tissue environment consists of differentiated bone cells, including osteoblasts, osteoclasts and osteocytes, as well as undifferentiated mesenchymal cells. In the area of bone tissue engineering, four cell types have been used: unfractionated fresh bone marrow, purified culture expanded mesenchymal stem cells (MSCs), embryonic stem cells (ESCs) and differentiated osteoblasts.

The popular theory of response to musculoskeletal injury by Frost states that bone, marrow and soft tissue injury trigger repair and healing by early sensitisation of surviving cells simultaneously with the release of local, biochemical and biophysical messengers directing an appropriate cellular response in the first week post-injury. Some messengers guide cellular differentiation whilst others encourage mitogenesis. This primitive healing response involves the stimulation of a variety of cell types during weeks two to three, including fibroblasts, capillaries and mesenchyme.[20,21] However, the proper repair and healing of bone or anchorage of an implant is achieved by the injury-induced recruitment and stimulation of the undifferentiated mesenchymal cells to form osteoprogenitor cells and further development to differentiated bone cells.[22] Pre-existing, pre-injury differentiated cells, such as osteoblasts, express only a minor contribution towards new bone formation in fracture-healing situations.[20,21]

MSCs are immature and undifferentiated cells that can be isolated in bone marrow and periosteum. They have the capacity for extensive replication without differentiation and possess multilineage developmental potential.[23] Understanding the stimulation, induction and control of differentiation of these cells is the key to harnessing the regenerative power of musculoskeletal tissue.

MSCs are either pluripotent or multipotent depending on their lineage. They are capable of differentiation into fibroblastic, adipogenic, reticular and most osteogenic cells.[24,25] Stem cells can also be classified into adult and embryonic types. Embryonic stem cells (ESCs) are usually isolated from the inner wall of the pre-implantation blastocysts. These embryonic cells in the early developmental stage are more proliferative and pluripotent due to their indefinite amplification, without the risk of dedifferentiation.[26] ESCs are non-autologous as they contain a haploid set of chromosomes from a different non-self genetic parent, hence they can therefore be immunoreactive. However, recent advances in gene transfer, MHC manipulation and nuclear cloning allow for the formation of autologous-like ESCs.[25] This technology elicits a variety of ethical and social implications.[27] Future research may see stem cell generation of entire skeletal tissues without biomaterial scaffolds.[28]

Adult MSCs can also transform through multiple passages without loss of characteristics and tend not to dedifferentiate unless exposed to specific biochemical and mechanical cues where they can then be directionally differentiated.[29] Jaiswal *et al.* (1997) reported passing MSCs through 30 population doublings *in vitro* without loss of osteogenic potential.[30] They have been used extensively in experimental bone tissue engineering. A number of animal studies have demonstrated that tissue engineering techniques using autologous MSCs, combined with a suitable scaffold, can create new bone in the defect site. A study by Korda *et al.* (2006) examined whether autologous MSCs seeded in an allograft were able to survive a range of normal clinical impaction forces. The study showed that MSCs survived normal impaction forces; this could have clinical significance in the treatement of revision hips.[31] However, the long-term biological effects of the stem cells at implant sites and interactions with biomaterials as issues concerning the phenomenon of cell plasticity, remain largely unknown.[32] Stem cell research and technology is still in its infancy.

Stem cells require osteoinductive and developmental cues which can be physical and chemical. Osteoinductive stimuli, such as mechanical stress and electrical signals, directly or indirectly influence bone induction.[33,34] Ultimately, osteoinductive agents are the instigators in transforming undifferentiated mesenchymal cells into preosteoblasts and subsequently differentiated bone cells. This classical cellular sequence is naturally observed in orthotopic environments but is also demonstrated experimentally in heterotopic bone formation, in sites such as muscle, to assess the osteoinductive potential of materials or agents.[35]

13.2.3 The osteoinductive agents

Growth factors (GFs) are proteins secreted by cells that act as signalling and regulation molecules triggering specific target cells to execute specific functions. They act as a cellular communications network and influence critical functions such as cell proliferation, matrix production and differentiation of tissues. A GF may affect multiple cell types and may induce several different cellular responses in a variety of tissues. Once a GF binds to a target cell receptor, it induces a ligand–receptor interaction. This induces an intracellular signal transduction system to the nucleus and results in a biological response. These interactions can be very specific, where a specific GF binds to a single cellular receptor, or complex, with one or more GFs binding to one or more receptors in order to produce an effect.

A number of GFs have been shown to be expressed and play significant roles in bone and cartilage formation, fracture healing and the repair of other musculoskeletal tissues. These include somatomedins, insulin-like growth factor I, II (IGF-1, IGF-II), fibroblast growth factor (FGF), transforming growth factor-β (TGF-β), platelet derived growth factor (PDGF), insulin-like growth factor (IGF) and bone morphogenentic proteins (BMPs).[36]

Osteoinduction is therefore mediated by numerous GFs and on this basis it is thought that these GFs may have potential use as therapeutic agents in bone healing and tissue engineering. Hauschka et al. (1988) showed that osteogenesis is in part due to the combination of actions of several growth factors acting at specific stages in different cells.[37] Bone is rich in many growth stimulating factors but the most important group is the transforming growth factor-beta (TGF-β) superfamily, of which the bone morphogenetic proteins (BMPs) are the most significant.

13.3 Bone morphogenetic proteins (BMPs)

In 1965 Urist was able to elicit new bone formation from an intramuscular injection of demineralised bone matrix.[10] In the late 1970s further research from this original work led to the isolation of the soluble glycoprotein bone morphogenetic protein (BMP) originating from the transforming growth factor-β (TGF-β) superfamily.[38,39] This superfamily of growth factors plays an important role during embryogenesis and postnatal tissue repair with BMPs specifically having the greatest osteogenic and osteoinductive potential.[40]

The process of osteoinduction was originally thought to be due to a single BMP; however, 15 BMPs (BMP 2–16) have been discovered to date, mostly distinguished by their potential to induce bone formation *de novo* via stem cell recruitment, proliferation and differentiation.[40–42] The evidence for proposed use in tissue engineered bone was provided by the study of Wang *et al.* (1990), who showed that BMPs caused commitment and differentiation of multipotential stem cells into osteoprogenitor cells.[43]

13.3.1 BMPs and basic science

BMPs are dimeric molecules with two polypeptide chains of over 400 amino acids linked by a single disulphide bond, exhibiting a characteristic cysteine knot on X-ray crystallography. Genes expressing BMP 2, 7 and 14 are located on chromosome 20.[44] BMPs act as cellular signalling agents crucial in signalling activity during bone growth and healing. The natural release of BMPs occurs after trauma or during bone remodelling.[36] The biological actions of BMPs are mediated via specific BMP surface receptors. To date, only two types have been defined. BMP binding to these serine/threonine protein kinase-based receptors and enzymes activates a signalling cascade and phosphorylation and activation of intracellular messenger proteins called *Smads*. The active *Smads* are then translocated to the cell nucleus to trigger and activate the transcriptional regulation of BMP responsive genes, resulting in transcription of macromolecules involved in bone and cartilage formation.[44,45] This process is responsible for the differentiation of primitive mesenchymal cells toward chondrocytic or osteoblastic phenotypes.[44]

BMP studies in animals and humans demonstrate a typical sequence of events in bone morphogenesis, chemotaxis, mitosis of mesenchymal cells and their differentiation into cartilage and subsequently bone.[44] The initiator of this cascade is unclear; however, studies have shown that the binding of plasma fibronectin to implanted demineralised matrix may trigger and facilitate the recruitment and proliferation of primitive mesenchymal cells (maximal on day 3).[46] These cells differentiate into chondrocytes (day 5–8), which subsequently hypertrophy, preceding the calcification of cartilage matrix (day 9). The differentiation of osteoblasts and bone formation occurs simultaneously with angiogenesis and vascular invasion (day 10–11). Newly formed woven bone remodels into trabecular and cortical bone alongside the formation of bone marrow. Remodelling in fracture callus occurs through zonal ossification where intercortical medullary areas create 'soft callus' undergoing endochondral ossification, and subperiosteal soft-tissue areas surrounding the fracture zone form 'hard callus' and bone via intramembranous ossification.[47] Classification as a BMP implies the substance is capable of inducing bone formation by endochondral ossification.

The expression of specific BMPs is highly variable throughout the bone morphogenesis cascade, involving variable and complex intra- and extracellular signalling processes. BMP-2, BMP-6, BMP-7 and BMP-9 are considered to have osteoinductive properties[48] and experimentally, a combination of BMP-2 with BMP-6 or BMP-7 has shown a 5–10-fold increase in bone formation.[49] BMP-2 and BMP-4 undergo early expression by primitive mesenchymal stem cells and throughout the cascade and a joint BMP-2/BMP-4 receptor has been discovered.[50] BMP-2, BMP-6 and BMP-9 are expressed in the early stages of differentiation of mesenchymal progenitor cells to preosteoblasts. BMPs 2, 4, 6, 7 and 9 increase osteocalcin expression and alkaline phosphatase expression in preosteoblasts, leading to mineralisation.[51] The osteoinductive capabilities of BMP-3 have been demonstrated alongside an inhibitory action in the presence of BMP-2 and BMP-7, acting as a negative regulator of bone formation, and BMP-7 is expressed by osteogenic cells from day 7, peaking at 2–4 weeks.[51,52] Interestingly, BMPs may also demonstrate a pleiotropic characteristic, having actions upon development of non-musculoskeletal tissue. A study demonstrated that knockout of BMP genes in mice resulted in developmental defects in the heart (BMP-2), eye and kidney (BMP-7).[53,54] It is still not known which molecules are absolutely required for the osteoinductive process.

13.3.2 BMP and clinical applications

BMPs have been extensively studied and applied to a variety of applications and clinical trials in the field of trauma and orthopaedic surgery, in particular the challenge of impaired fracture healing and non-union.[41,55,56] Recombinant human (rh) BMP-2 and rhBMP-7, known as Osteogenic Protein-1 (OP-1), have been shown to be the most effective types and are the only forms of BMP developed for clinical use.[40,42,57–60] Both have stimulated new bone formation in diaphyseal defects in the rat, rabbit, dog, sheep and non-human primates.[58–64]

A randomised controlled trial in 2001 showed that rhBMP-7 (or OP-1) is at least as effective as autograft in treatment of tibial non-union.[65] A study by Muschler et al. (2004) reported similar results with BMP-7 delivered in a collagen carrier.[66] Other studies have also demonstrated the successful use of rhBMP-7 in a variety of non-union scenarios, in particular the reconstruction of femoral non-unions using cortical bone allograft used as a delivery system for human BMP.[67–69] BMP-7 has also been evaluated in the treatment of degenerative intervertebral disc disease with studies showing enhanced repair of discal extracellular matrix and inhibition of pain-related behaviour in rat models and improvements in repair of disc height in rabbit models.[70] Extensive work by Ripamonti et al. using BMP-7 has shown that it elicits heterotopic bone formation and complete healing of 25 mm-diameter critical-sized defects in primates in 90 days.[71] BMP-7 has also recently been implicated in impaction allografting for the biological augmentation of deficient bone revision arthroplasty. A recent study by Tsiridis et al. (2007) showed that partial dimineralisation and addition of BMP-7 enhances the osteoinductivity of fresh allograft for impaction grafting.[72]

The importance of BMPs in fracture healing was demonstrated by Bostrom et al. (1995) when BMP-2 and BMP-4 were localised to fracture callus.[73] Einhorn et al. (1997) demonstrated the injection of rhBMP-2 into a standard fresh fracture model in a rat accelerated healing and achieved healed state two weeks earlier than controls.[74] Similar findings were observed in rabbit and goat fracture models.[75,76] Furthermore, a prospective randomised controlled trial in 2002 showed that rhBMP-2 significantly enhanced the healing of open tibial fractures, with fewer infections and a lower rate of non-union rate compared to controls.[77] Other studies have shown rhBMP-2 to be at least as effective clinically as autograft in anterior interbody spinal fusions and better in terms of fusion rates in degenerative lumbar disc disease, again avoiding the morbidity associated with autograft harvesting.[78–80] rhBMP-2 has also shown an enhanced ability to form bone in impaction grafting of filling defects in canine models with uncemented total hip arthroplasties and canine models with titanium implants in the proximal humerus.[81] The latter study showed that the greatest implant strength was obtained using combined rhBMP-2 and rhTGF-(β2). rhBMP-2 has been extensively used for *in vitro* and *in vivo* studies and has been shown to induce cell differentiation and form endochondral bone in ectopic and heterotopic locations.[82]

BMPs have also been investigated in cartilage repair, with BMP-4 showing enhanced chondrogenesis and improved repair of articular cartilage in rat models.[83] rhBMP-7 showed similar chondrogenetic potential by inducing full-thickness osteochondral defects in a canine model.[84] Studies have also shown the requirement of BMPs for chondrocyte differentiation and type II collagen expression during chondrogenesis.[85]

Both rhBMP-2 and rhBMP-7 were licensed for use by the United States FDA in 2001 for the treatment of single-level interbody fusions of the lumbar spine and non-union surgery respectively. Other BMPs have not been subjected to this level of evidence testing.

13.3.3 BMP and BMP-specific antagonists

BMP-specific antagonists have been isolated (e.g., 'noggin', 'gremlin').[86,87] The under-expression of 'noggin', an extracellular BMP antagonist which binds to BMP directly, has been shown to result in heterotopic ossification and fibrodysplasia ossificans progressiva (FOP) in *in vivo* animal studies.[88]

This work may lead to testing of BMP-specific antagonists for therapeutic challenges such as treating excessive bone-forming pathologies.

13.3.4 BMPs and dose delivery

GFs in general are species- and dose-dependent, requiring careful targeting within the host system. It has also been shown that the osteoinductive potential of individual rhBMPs follows a dose–response ratio where a threshold value must be reached before triggering osteoinduction.[89] The challenge is the carriage and delivery of safe but effective doses and local concentrations of GFs such as BMPs in a controlled manner to trigger this regenerative response, especially as systemic clearance of BMPs is high.[89] Kaplan et al. (1998) showed that too low a dose resulted in poor bone formation with reduced mechanical strength, while high doses may inhibit osteogenesis or lead to bone growth at sites outside the graft boundaries.[90]

Using extrapolations from primate models, the concentrations of BMPs used in current clinical applications are several orders of magnitude greater than BMPs occurring naturally in the body.[91] Thus, the development of efficient, cost-effective delivery systems is essential alongside the further understanding of the fracture healing cascade. Several BMP delivery methods have been described, including direct application of rhBMP to the regeneration site by injection, delivery with a carrier vehicle such as bone matrix, hydrogels, alginates, gelatin foams and microspheres, collagen, or calcium phosphate.[92] The incorporation of BMPs into a drug delivery system poses a challenge as purified BMPs are hydrophilic and hence, when incorporated in the body, tend to disperse rather than act locally. Thus, in order to slow the release down, they need to be complexed to a carrier, adding a further problem in that proteins are immunogenic.

Gene therapy has also been utilised with BMP being delivered directly, resulting in transfection of cells and protein expression (*in vivo* transduction), or indirectly through transfection of cultured cells (*in vitro* transduction) which are then implanted and express the required protein via its encoding gene.[93,94]

13.3.5 BMPs and safety

The increasing developments in the potential clinical uses of BMP will warrant a careful consideration of safety issues such as dose concentrations, immunogenicity, inflammatory response and long-term genetic effects in humans.

Therapeutic concentrations of BMP are approximately 4–6 orders of magnitude greater than naturally occurring concentrations and carry a relatively small risk of excessive bone formation.[95] However, cases of ectopic bone formation and overgrowth have been reported after use of high BMP doses.[89] Regulatory factors include the number of mesenchymal stem cells which express BMP receptors, as well as osteoclastic activity stimulated by high doses of BMPs and the systemic clearance of rhBMPs, which has been shown to be rapid.[96] Human studies have not demonstrated any systemic toxicity to date; however, collagen carrier materials and BMPs have been shown to induce antibody formation.[65,77]

13.4 Bone graft biomaterials and bone healing

Understanding and controlling the cellular response to bone graft materials will achieve the positive desired biological interaction between graft material and host site. The processes involved include the phenomena of osteoinduction, osteoconduction and osseointegration, resulting in new bone formation, union and incorporation between graft and host, leading to a skeletal tissue environment which is biologically and mechanically matched to the original.

The healing response triggered by placement of bone grafts in skeletal tissue include the host inflammatory reaction to trauma associated with surgical preparation and implantation of the graft, the inflammatory and immune reaction to the graft material itself, and the cellular processes such as proliferation, migration and differentiation. The speed and degree of this response and thus of graft incorporation into host skeletal tissue is dependent on local and systemic factors, namely type of graft material and surrounding local tissue integrity, and the systemic physiological state of the host, respectively.[8]

The cellular events occurring at the graft–tissue interface and graft material itself during its incorporation involve five phases, namely haematoma formation, inflammation, vascularisation, resorption, and regeneration and remodelling. Haematoma formation occurs on trauma related to host site preparation and graft implantation which also triggers the release of cytokines and growth factors. The inflammatory process involves proliferation, differentiation and migration of mesenchymal cells. Neutrophils, lymphocytes and monocytes are also attracted and recruited to the zone of trauma, migrating to the site and into organising haematoma. This process also involves platelet adherence and degranulation with release of several growth factors (e.g., FGF-2, PDGF, TGF-β) into a fibrin meshwork. Fibroblasts also produce collagen which occurs in response to various stimulatory factors (e.g., TGF-β) combined with the response to extent of collagen degradation by factors such as metalloproteinases also released by fibroblasts, as well as macrophages.[97] The vascularisation process involves the development of fibrovascular tissue within the graft and in the immediate tissue periphery. Moreover, there is vascular invasion into the graft which may also originate from pre-existing Haversian and Volkmann canal systems. This phase is essential for provision of the graft environment with nutrients and cell populations. The resorptive process involves focal osteoclasts acting on graft surfaces before bone regeneration is achieved via intramembranous or endochondral ossification. Once mechanically stable, the remodelling process occurs to optimise functionality. The rate of skeletal remodelling, dependent on age, ethnicity, genetics, metabolic and pathological conditions, will influence long-term graft turnover and integrity. The most important factor, however, is mechanical load. Mechanical loads, specifically strain, influence bone remodelling and cause bone to physically adapt in relation to the mechanical loads placed across it.[98] Experimental studies have suggested that no strain leads to bone resorption and increasing strain positively alters the anabolic activity of osteocytes directly or indirectly.[99] Direct effects have been stated to involve strain-induced changes in the shape of processes within osseous cannaliculi. Indirect effects are considered to involve strain-induced alterations in hydrostatic pressures which may subsequently exert anabolic or catabolic effects on cells, specifically osteoblasts.[100]

In general, most of these phases are orchestrated by a variety of inflammatory mediators such as kinins, prostaglandins, vasoactive amines, nitric oxide, interleukins, complement, angiogenic and growth factors. In association with the appropriate cellular response, a state of mechanical stability needs to be achieved in order to prevent granulation and fibrosis at the graft–host interface allowing graft incorporation. The important integral factors within local tissues in the zone of implantation include the vascularity of the graft bed alongside the number and competence of the endothelial and osteo-progenitor cells. Limitations in these cellular ingredients will result in a poor response to cues from osteoinductive *and* osteoconductive materials. Environments deficient in progenitor stem cells limiting graft incorporation include scar or ischaemic tissue, large bony defects, acute or chronic infection, immunosuppresion and radiotherapy.

13.5 Osteoinductivity and bone healing

The osteoinductive potential of a biomaterial is the ability to induce osteogenesis. It is important to appreciate that injury caused by implantation of biomaterials is in itself a sufficient trigger for recruitment of previously undifferentiated bone cells.

The mid-1960s saw modern research into osteoinduction and experiments using demineralised bone as an osteoinductive material.[5] It is important to state that osteoinductive potential of a biomaterial is a composite of its true osteoinductivity and the osteoinductive effect of the injury at placement in recruiting previously undifferentiated bone cells. Early studies have also shown that physical stimuli including stress and electrical activity have direct or indirect effects on osteoinduction.[101,102] Vacanti and Bonassar (1999) reported that for cell colonies to grow into three-dimensional functional tissues, external cues in the form of mechanical, electrical, structural and chemical signalling mimicking the local ECM were required.[1]

From a materials perspective, it is important note that *bone growth* over surfaces is highly dependent on the *osteoconductive* potential of the material. Bone conduction is impossible on pure copper and silver, whilst it can occur with materials of low biocompatibility such as stainless steel and naturally with materials of high biocompatibility such as commercially pure (c.p.) titanium.[6,103,104]

The natural skeletal repair process of fracture healing utilises bone-forming cells, a cartilaginous scaffold upon which the new woven bone forms, and bioactive molecules to direct the repair sequence. The cartilaginous scaffold ossified by the body is porous, and allows vascularisation and cellular ingrowth. Biomaterials engineered for the regeneration of bone are often based on porous scaffolds. In bone tissue engineering, the roles of biomaterial scaffolds are to define the architecture of growing tissues and provide effective delivery of osteogenic cells in combination with growth factors to trigger signalling cues for osteoinduction, cell proliferation, differentiation and angiogenesis. It is important for bone graft materials to physically act as templates for the growth and sites for cell anchorage by initially providing biomechanical stability and structural guidance to the tissue environment until cells produce adequate extracellular matrix.

13.6 Osteoconduction and bone healing

Bone growth on implant surfaces is reliant on the action of 12 formed differentiated bone cells. The source of these may be pre-existing surviving preosteoblasts or osteoblasts activated by trauma but more often primitive mesenchymal cells recruited by osteoinduction.[20,21] Thus, in practical terms osteoconduction occurs after osteoinduction. The cellular components are further reliant on bone growth factors and an adequate blood supply for appropriate bone formation.[19] *In vivo* studies on osteoconduction and remodelling have shown that a fully vascularised environment is necessary for bone formation.[19] Interestingly, a variety of growth factors possess both mitogenic and angiogenic activity.[105]

In terms of the implantation of biomaterials, bone conduction is dependent on these conditions for bone repair as well as the osteoconductive potential of the biomaterial itself. Materials such as copper and silver are not osteoconductive, whereas it is observed on materials of low and high biocompatibility such as stainless steel and pure titanium respectively.[6,103,104]

13.7 Osseointegration and bone healing

Osseointegration is a phenomenon that effectively occurs after previous osteoinduction and osteoconduction. This 'bone anchorage' process may be enhanced or accelerated by variable surface modifications, such as hydroxyapatite coating[106] or roughened implants,[107] as well as hyperbaric oxygen treatment[108] or surface oxidisation.[109] These biomaterial optimisation techniques may also rely on the concurrent elimination or reduction of negative tissue conditions. Secondary failure of osseointegration is caused by poor contact factors such as cylindrical surfaces without threads, rough plasma-sprayed surfaces and overloading.[14,18] The ultrastructure of the bone–titanium interface demonstrates simultaneous direct bone contact, osteogenesis and bone resorption.[110] It has been shown that functioning osseointegrated implants demonstrate an interfacial bone density similar to that of the bone involved.[111] Osseointegrated implants are commonplace in oral and craniofacial surgery with excellent functional results.[13,112,113] They are also established in orthopaedic surgery, especially in screw-type implant systems such as in hip arthroplasties, interphalangeal implants and vertebral screws.[114]

13.8 Bone graft materials

Bone graft materials can be classified into autografts, allografts, and synthetic materials or synthetic bone graft substitutes (SBGSs)[8,11] (see Table 13.1). They can be sub-classified based on composition and biological or mechanical properties, and more importantly into whether they have primarily osteoinductive or osteoconductive activity.

13.9 Osteoinductive materials

13.9.1 Autografts

Autografts are the 'gold standard' for bone transplantation. The process involves harvesting viable tissue from the donor site and transplanting this to a recipient site in the same patient. Autografts can be aspirated (e.g., bone marrow), cortical or cancellous and vascularised or devascularised. Vascularised and cancellous autografts may be osteogenic (due to viable cells), osteoinductive (due to matrix proteins), osteoconductive (due to bony matrix) and completely biocompatible. Devascularised autografts may also provide an osteoinductive response via the small number of surviving cells transplanted. Cancellous and cortical autografts have been widely used over decades in orthopaedic surgery to treat skeletal defects. Spinal fusions and fracture non-unions have utilised autograft bone excavated from the

Table 13.1 Bone graft and bone graft substitutes – a classification based on composition

Type	Characteristics	Examples
I	Autograft A. Iliac crest B. Locally harvested C. Vascularised or non-vascularised cortical bone D. Aspirated and/or enriched bone marrow stromal cells	
II	Allograft A. Mineralised B. Demineralised (demineralised bone matrix, DBM)	Grafton (DBM)
Synthetic bone graft substitutes		
III	Hydroxyapatite (HA) blocks and granules A. Formed by sintering or precipitation B. Formed by conversion of calcium carbonate	Apapore (HA) Endobon (HA) Pro Osteon® (HA) Pro Osteon-R® (CaCO₃ HA coated) Bonesave (HA/TCP; 60/40) OsSatura (HA/TCP; 80/20)
IV	Soluble calcium-based granules A. Tricalcium phosphate (TCP) B. Calcium sulphate	Vitoss (TCP) Skelite (CaPO₄) Osteoset® (CaSO₄)
V	Silicon-containing calcium phosphates	
VI	Bone morphogenetic proteins (BMPs) A. BMP-2 B. BMP-7 C. Growth differentiation factor-5 (GDF-5)	InFuse® (BMP-2) OssiGraft® (OP-1)
VII	Injectable cements A. Polymethyl methacrylate B. Calcium phosphate cements C. Silica-containing cements	Norian (CO₃ apatite)

Source: modified from Hing[7] and Bauer.[8,11]

iliac crest donor site and translocated to stabilise the defect at the recipient site. Autograft bone transplantation is highly successful and advantageous in terms of excellent immunological compatibility, and the direct transfer of osteogenic cells and osteoinductive cues. However, its donor tissue supply and thus therapeutic potential is limited. Further disadvantages include the complications of donor site morbidity including chronic pain, infection and revision surgery.

Vascularised autografts

Vascularised autografts demonstrate an optimal host cellular response allowing excellent graft incorporation. The inflammatory reaction is limited and the graft–host interaction is rarely complicated by ischaemic necrosis of cells in the graft, nor a significant immuno-reactive host response versus the graft.[8,11] A study in 1977 suggested that more than 90% of osteocytes may survive the transplantation process if adequate vascular anastomosis and graft stability is achieved.[115] The high osteoinductive potential of vascularised autografts allows new bone formation and subsequent incorporation via osteoconduction and osseointegration at a rapid rate. Mechanical loading over time determines the dynamics of skeletal remodelling and the amount of graft remaining over time.

Cancellous autografts

Implantation of cancellous autografts results in a significant loss of transplanted cells due to cellular ischaemic necrosis or the induction of apoptosis irrespective of the integrity of the host site. However, the 'osteoinductive strength' is maintained by the primitive mesenchymal stem cells present in bone marrow and endothelial progenitors, which are highly resistant to ischaemia post-transplantation. Survival of these cell lines may lead to the stimulation of cellular proliferation, differentiation and migration in response to variations in the local biochemical environment such as oxygen tension, pH and chemical factors present during transplantation. Cancellous autografts allow rapid invasion of their graft matrices by granulation tissue and studies have shown that osteoclastic bone resorption followed by new bone formation occurs after a few weeks.[116] It was shown that radiodensity levels changed from high to normal levels as the initial new bone formation onto cancellous autograft trabeculae progressed to a state of resorption of the graft matrix.[116] Cancellous autografts are considered to have a greater efficacy and rate of bone formation compared with cancellous allografts. This has been shown experimentally by a controlled study using rats where cancellous autografts demonstrated maximal genetic expression of type I and type III collagen genes one week before cancellous allograft and earlier bone formation, graft invasion by mesenchymal cells and remodelling.[117]

Non-vascularised cortical autograft

Non-vascular cortical autograft or vascularised autografts with compromise of their vascular anastomosis result in a graft of necrotic bone that is resistant to an immunological reaction by the host. This autograft subtype is strong and provides good mechanical support at graft sites. However, the structural make-up of the cortical bone matrix prevents useful diffusion activity to support survival of any significant populations of osteocytes post-transplantation. Moreover, the high density and low surface area prove a physical barrier to vascular ingrowth and the process of revascularisation. Overall, non-vascular cortical autografts are not significantly osteogenic and, although they present an osteoconductive potential, confer little benefit over cortical allograft.

Autologous bone marrow

Autologous bone marrow cells have been shown to induce bone formation at extraskeletal sites.[118] In 1968, Friendenstein *et al.* demonstrated the proliferation of fibroblasts and stromal components associated with bone formation after transplanting bone marrow cells. This work suggested the presence of undifferentiated precursor cells in bone marrow capable of osteoinduction and the formation of osteoblasts.[119] Several studies have since been conducted to characterise the osteo-progenitor populations in bone marrow whilst concurrently defining growth factors involved in regulating the differentiation of osteoblasts.[120] However, only a few studies have shown clinical application using aspirated bone marrow autograft.[121] Further studies are also required to determine the effects of dilution with peripheral blood on the concentration of bone marrow-derived progenitors. Achieving control of the osteo-progenitor population by methods such as limiting dilution or *in vitro* cell culture expansion techniques may significantly improve the biological effect of the graft and velocity of incorporation into the host tissue.

13.9.2 Allografts

Allografts 'extend' the scope of autografts and involve harvesting tissue from an individual, the donor host, and transplanting this tissue into the recipient site in a patient of the same species. The main advantage is the abundant graft supply of cortical and trabecular bone. However, this graft type elicits an immune reaction, especially in response to cells of fresh allograft, and potentially transmits disease. Moreover, the cleaning and processing of allografts to negate this immunoreactive element and microbiological risk also tends to remove cells and bioactive molecules compromising its incorporation into existing host tissue and limits the graft's osteogenic potential. Osteoinductivity, osteoconductivity and mechanical strength of allograft materials vary greatly in relation to the method of processing.[8,11]

Allografts can be classified by anatomical and handling properties or methods of processing and sterilisation (see Table 13.2). They have been

Table 13.2 Allografts – a classification based on graft anatomy, processing, sterilisation and handling properties

I **Graft anatomy**
   A. Cortical
   B. Cancellous
   C. Osteochondral
II **Graft processing**
   A. Fresh
   B. Frozen
   C. Freeze-dried
   D. Demineralised
	III **Graft sterilisation**
	   A. Sterilely processed
	   B. Irradiated
	   C. Ethylene oxide
	IV **Graft handling properties**
	   A. Powder
	   B. Particulate
	   C. Gel
	   D. Paste/putty
	   E. Chips
	   F. Strips/blocks
	   G. Massive

Source: Bauer.[11]

extensively used as structural or morsellised grafts, and prepared in fresh, frozen, freeze-dried or demineralised forms. A variety of demineralised bone allograft preparations are available.

Fresh bone allograft transplants are associated with a significant immune response and a high risk of disease transmission. HIV and hepatitis C transmission has been described.[122,123] These immune responses to the graft cells are mediated by Class I and Class II HLA antigens which are also the antigen types commonly associated with organ transplantation. Thus, the cellular response to allograft and graft incorporation is optimised by minimising the differences in histocompatibility or treatment protocols that reduce immunogenicity and more accurately match tissue type.[124,125] Several animal studies have demonstrated allograft-induced cell-mediated immunity and graft-specific antibody production.[125,126] Matched and mismatched, fresh and frozen allografts in canine fibular models have demonstrated the immune response, with humoral immunity directed predominantly at Class I antigens and greater immunoreactivity towards fresh bone allograft.[126] Other studies have shown allogenic T-cell activation not only by bone marrow cells, but also by cells within the diaphyseal cortex.[127] The clinical significance of these animal experiments in humans is unclear. Graft-specific anti-HLA antibodies have been found in humans receiving frozen allograft, but there was no compromise in clinical outcome.[128]

Thus, mineralised frozen and freeze-dried allografts are considered less immunoreactive than fresh allografts, and generally osteoconductive and weakly osteoinductive.[48,129] An extensive study which retrieved human frozen or freeze-dried allografts demonstrated a variety of interesting features related to the incorporation of allografts in terms of union, graft surface chemistry and the type of bone formation.[130]

Allografts, in general, fuse to bone at junctions containing matching bone types, e.g. cortical to cortical or medullary to medullary junctions. At the cortical junctions, bony union occurs by intramembranous ossification. Moreover, bone is formed by reconstituted periosteum and by bone extension in from the periosteum and not the cut cortical ends. Junctions involving cancellous bone also tend to form more rapidly than those between cortical regions.

The majority of allograft surfaces have a bony layer deposited by intramembranous bone formation onto external surface with a few remodelling buds extending into Volkmann canals. In contrast, there were areas of allograft surfaces with unrepaired osteoclast-lined erosions, filled with loose fibrous tissue and inflammatory cells.

In terms of healing, allografts tend to involve intramembranous bone formation over endochondral ossification where constructs of mineralised allograft are mechanically stable. Movement or micromotion at the graft–host site creates cartilage. Based on these observations, frozen and freeze-dried allografts offer mostly osteoconductive surfaces rather than osteogenesis or significant osteoinductivity.

13.9.3 Demineralised bone matrix

In 1965, demineralised bone matrix (DBM) was reported to induce the formation of heterotopic bone when transplanted in soft tissue.[5] These early studies observed the infiltration of DBM with inflammatory cells and MSCs before causing endochondral ossification within 3 weeks.[5] Implantation of DBM in rat muscle resulted in the formation of bone locally in the soft tissues.[38] Thus, DBM was found to possess both osteoconductive and osteoinductive properties. This research was further developed to isolate the active components of DBM and define the factors responsible for its osteoinductive potential. A series of glycoproteins belonging to the TGF-β superfamily and known as BMPs were isolated.

In 1981, Urist *et al.* described the use of an autolysed, antigen-extracted allogenic bone (AAA bone) in postero-lateral lumbar spinal fusion.[131] A variety of DBM allografts are available and have been used in a variety of clinical applications alone or in combination with autografts.[92,132–134] The main clinical use of DBM has been in the management of fracture non-union.[48] It is limited in terms of strength, being prepared in putty or paste forms, making it unsuitable for bony structural support such as in metaphyseal fractures. Furthermore, there is a theoretical risk of disease transmission, being of allograft origin, and batch-to-batch variation in osteoinductive activity.

Overall, the cellular responses and processes involved with incorporation of allografts are similar to those with the non-vascularised autografts. The difference is primarily related to the slower rate of incorporation and increased levels of inflammatory and immune response elicited by allograft, especially fresh bone allograft. Theoretically, this immune activity could naturally lead to local graft rejection as well as adverse implications on the sensitisation of the host to an expanded pool of reactive HLA antigens. This obviously has serious consequences and enforces limitations upon options for subsequent organ transplants.[135] In general, graft processing to minimise the immune response of the host tends to adversely affect the structural integrity of the material and destroy viable cells, reducing the osteogenic potential and osteoinductive activity. Allograft bone is used primarily for bone grafting of non-unions, segmental bone defects and reconstruction after bone tumour surgery. However, allografts are not commonly used in managing orthopaedic trauma.[48]

13.10 Synthetic bone graft substitutes (SBGSs)

Synthetic bone graft substitutes are in high demand as they promise to contend with the limitations associated with autograft and allografts such as supply, complications of donor site morbidity and compromised osteogenicity respectively. In modern surgical practice, composite preparations involving bone marrow or cancellous autografts, demineralised allografts and synthetic grafts (e.g., osteoconductive calcium-based biomaterials, recombinant osteoinductive proteins) are more readily used. To date, no SBGS exists that encompasses all the qualities of autograft. There is a continuing challenge to produce SBGSs in combination with various factors to achieve these qualities, such as by incorporating extracted or synthesised protein growth factors.

A wide variety of these SBGS materials have been introduced which vary with respect to composition, biological properties and mechanical strength (see Table 13.1). There is also great variation in level of osteoconductivity and osteoinductivity. Current synthetic biomaterials have usually undergone extensive preclinical testing to negate a significant inflammatory or immune response. The inflammatory reactions to SBGSs are rather dependent on material composition, size and surface chemistry. However, the majority do not provoke a major inflammatory reaction at implantation. It is important to note that in the field of total joint arthroplasty there is a concern related to osteoconductive synthetic biomaterials causing a wear debris-induced inflammatory reaction and secondary bone resorptive process. Particulate debris, of the order of 2 µm, in contact with or phagocytosed by macrophages are thought to trigger the release of cytokines, such as TNF-α and IL-1β, which induce osteolysis. This bone resorptive process occurs by direct or indirect cytokine-induced osteoclastic chemotaxis, maturation, attachment and activation. The theoretical risk of SBGSs producing particulate matter, secondary to motion or degradation *in vivo*, and invoking this erosive response exists. Practically, however, the structural composition of SBGSs as large granules, blocks or cements does not appear to create a significant particulate load and the acid solubility of most of these materials implies that macrophages are able to contend with these particles without creating a sustained inflammatory response. At present, SBGSs do not appear to cause significant particle-induced bone resorption in clinical practice.

13.10.1 SBGSs and osteoconductivity

SBGSs are predominantly osteoconductive. Osteoconductive materials in general were developed before osteoinductive and osteogenic substitutes. These materials encompass the variety of synthetic bone analogue materials that resemble the mineral phase of bone and are biocompatible. Materials including ceramics (e.g., porous HA, coralline HA and TCP), degradable polymers (e.g., polyglycolic acid and polylactic acid) and composites aim to provide a structure or scaffold which has a close interface with adjacent bone. They demonstrate high osteoconductivity, allowing cellular elements to grow into the material, but lack osteogenic and regenerative power compared with the osteoinductive activity of autografts, DBM and BMP. These materials are generally used as fillers for bony defects requiring mechanical support such as metaphyseal fractures of the distal radius, tibial plateau and calcaneum. However, they have also demonstrated roles in extending autogenous bone graft and carrier materials for osteoinductive proteins.[48]

The specific cellular responses involved in the osteoconductive effect of SBGSs and the extensive bony apposition demonstrated by some SBGSs are not clearly understood. Mechanisms triggering osteoblast-mediated bone formation or extracellular mineral precipitation with crystal growth have been postulated.[8,11] It is clear, however, that appropriate incorporation of SBGSs depends upon host site integrity, its own composition, the biochemical tissue environment and the mechanical loading that influences remodelling over time.

Calcium-based ceramics such as hydroxyapaptite or tricalcium phosphate are structurally similar to bone and their osteoconductive properties have allowed them to be used extensively as coatings on implants. HA is highly biocompatible, but is limited mechanically as it is brittle, weak and unable to withstand significant compressive loads. More importantly, it lacks osteoinductivity and true bone regeneration capability. Ceramic-based cellular techniques, however, have produced HA-based constructs in the fabrication of sintered porous hydroxyapatite (SPHA).[136] This 'smart biomaterial' consists of a repetitive sequence of concavities. More importantly, it is a gene- and cell-activating material, capable of attracting BMPs and other osteogenic factors into the material cavities by adsorption, and is intrinsically osteoinductive.

Current trends in SBSGs include combination products. Collagen mineral composite grafts have been produced, combining bovine collagen with HA as a SBGS, that gain osteoinductive capability with the addition of bone marrow aspirated from the iliac crest applied to fracture sites. A study comparing this combination BGS system with autogenous bone graft showed no difference in union rates or functional outcome in managing acute long bone fractures with defects in 249 fractures.[137] Thus, osteoinductivity and osteoconductivity were achieved. Under development is BMP-2 mixed with injectable calcium phosphate cement that contains bicarbonate.[138] Preclinical studies have shown promising results in the treatment of critical-sized skeletal defects in primates. The authors have also suggested that the combination of DBM or mineralised bone with resorbable polymers may provide both optimal strength with osteoinductive and osteoconductive capability.[8] The development of future complex combination products is most likely to continue along the evolutionary line of bone tissue engineering involving cells, a geometrically optimal matrix and growth factors. This goal will provide surgeons with a powerful treatment tool in managing skeletal trauma and disease.

13.11 Bone graft biomaterials and remodelling

The adaptation of bone graft biomaterials to mechanical loads has been extensively observed in all major graft types. Vascular autografts, in the form of fibular shaft used to treat femoral head osteonecrosis, have shown evidence of architectural change in response to mechanical loading.[8] Allografts used in revision hip arthroplasty have been preserved in zones of loading and shown to resorb in non-load-bearing sites.[139] There is also evidence of remodelling occurring in SBGSs. Norian Skeletal Repair System (SRS®, Cupertino, CA), a calcium phosphate cement with osteoconductive properties, demonstrated early skeletal apposition, speedier cortical bone remodelling than medullary bone remodelling, focal osteoclastic resorption, vascular ingrowth, and formation of Haversian systems, when injected into transmural defects in canine tibia.[140] Thus, this cement material has been shown to cure *in vivo* and form a mineral network perceptibly identical to bone as well as become subject to appropriate remodelling patterns and rates in correlation with mechanical loading. Bauer and colleagues suggested the requirements for SBGSs to respond to Wolff's law include the material to be osteoconductive, to enable physical and mechanical incorporation for load transmission across the graft site, possess mechanical properties to prevent mechanical failure, such as fracture, deformation, particulate wear under in vivo loads, and that material composition, should facilitate osteoclastic resorption.[8,11] Features such as porosity may alter the effects of mechanical loading causing variations in hydrodynamic pressure, alter streaming potentials throughout the material and subsequent affect cellular functions such as proliferation, differentiation and migration.

A further area of interest is the cellular response and influence of implant fixation systems combined with the utilisation of bone graft materials. An obvious example is the use of cage systems for spinal fusions which incorporate and clinically used in conjunction with autograft, allograft and SBGSs. The general concensus is that a positive cellular response and hisologically viable bone is achieved. Further work is required to determine the influence of hardware design on incorporation and remodelling of bone graft materials.

13.12 Discussion

In the era of bone regeneration, the ultimate biomaterial should possess a high osteoinductive potential which triggers the appropriate cellular response and has a positive influence on the interrelated phenomena of osteoconduction and osseointegration. Practically, bone graft substitutes can be classified as predominantly osteoinductive or osteoconductive. The next generation of materials incorporating osteoinductive proteins with osteoconductive carriers is an area of great interest. Osteoinductive activity is demonstrated by autogenous bone graft, DBM and BMPs. The cellular responses to these implanted osteoinductive materials are observed naturally as osteoinduction is part of normal bone and fracture healing. The cellular responses include cell attachment, migration, proliferation and differentiation.

These are part of a complex healing cascade which involves a variety of associated factors including BMPs. Further research is required to fully understand this process. There is good evidence for the clinical application of BMPs in orthopaedic surgery, such as BMP-7 in the treatment of non-union and BMP-2 in fresh fracture management and interbody spinal fusion. However, further work is required, particularly in the areas of delivery vehicles, delivery times, dosages, combinations of growth factors and mesenchymal stem cells. The future development of BMP science alongside improvements in surgical techniques and gene therapy may prove a massive clinical benefit.[141,142]

A basic understanding of the spatial and temporal distribution of cells and growth factors necessary for osteogenesis remains to be determined. Modern technologies such as molecular cloning and gene therapy may further enhance the ability of BMPs to play an important role in the clinical practice of the twenty-first century. The clinical application of a bone tissue engineered construct, composed of intrinsically osteoinductive materials, has the potential to greatly improve the treatment of conditions requiring bone repair.

13.13 References

1. Vacanti CA, Bonassar LJ. An overview of tissue engineered bone. *Clin Orthop* 1999; 367: S375–81.
2. Summers BN, Eisenstein SM. Donor site pain from the ilium: a complication of lumbar spine fusion. *J Bone Joint Surg [Br]* 1989; 71-B: 677–80.
3. Younger EM, Chapman MW. Morbidity at bone graft donor site. *J Orthop Trauma* 1989; 3: 192–5.
4. Goldberg VM, Stevenson S, Shaffer JW. Biology of autografts and allografts. In: Friedlaender GE, Goldberg VM (eds) *Bone and Cartilage Allografts: Biology and Clinical Applications*. Park Ridge, IL: The American Academy of Orthopaedic Surgeons, 1991: 3–11.
5. Urist MR. Bone: formation by autoinduction. *Science* 1965; 150: 893–9.
6. Albrektsson T, Johansson C. Osteoinduction, osteoconduction, and osseointegration. *Eur Spine J* 2001; 10: S96–101.
7. Hing KA. Bone repair in the twenty-first century: biology, chemistry or engineering? *Phil Trans R Soc Lond* 2004; 362: 2821–50.
8. Bauer TW. Bone graft substitutes. *Skeletal Radiol* 2007; 36: 1105–7.
9. Wilson-Hench J. Osteoinduction. In: Williams DF (ed) *Progress in Biomedical Engineering. Vol 4. Definitions in Biomaterials*. Amsterdam: Elsevier, 1987: 29.
10. Urist MR, Silverman BF, Buring K, Dubuc FL, Rosenberg JM. The bone induction principle. *Clin Orthop* 1967; 53: 243–83.
11. Bauer TW. Bone graft materials: an overview of the basic science. *Clin Orthop* 2000; 371: 10–27.
12. Glantz PO. Comment. In: Williams DF (ed) *Progress in Biomedical Engineering. Vol 4. Definitions in Biomaterials*. Amsterdam: Elsevier, 1987: 24.
13. Brånemark PI, Hansson BO, Adell R, Breine U, Lindström J, Hallén O, Öhman A. Osseointegrated titanium implants in the treatment of the edentulous jaw. *Scand J Plast Reconstr Surg* 1977; 11 [Suppl 16]: 1–175.
14. Albrektsson T, Brånemark PI, Hansson HA, Lindström J. Osseointegrated titanium implants. Requirements for ensuring a long-lasting, direct bone anchorage in man. *Acta Orthop Scand* 1981; 52: 155–70.
15. *Merriam-Webster's Medical Dictionary*. Springfield, MA: Merriam-Webster. New Enlarged Print Edition (26 Oct 2007).
16. Zarb G, Albrektsson T. Osseointegration – a requiem for the periodontal ligament? – An editorial. *Int J Periodont Rest Dentistry* 1991; 11: 88–91.
17. Albrektsson T. Principles of osseointegration. In: Hobkirk JA, Watson K (eds) *Dental and Maxillofacial Implantology*. London: Mosby-Wolfe, 1995: 9–19.
18. Albrektsson T. On long-term maintenance of the osseointegrated response. *Aust Prosthodont* 1993; J7 [Suppl]: 15–24.
19. Albrekston T. The healing of autologous bone grafts after varying degrees of surgical trauma. *J Bone Joint Surg [Br]* 1980; 32: 403–10.

20. Frost HM. The biology of fracture healing. An overview for clinicians, part I. *Clin Orthop Rel Res* 1989; 248: 283–93.
21. Frost HM. The biology of fracture healing. An overview for clinicians, part II. *Clin Orthop Rel Res* 1989; 248: 294–309.
22. Young RW. Nucleic acids, protein synthesis and bone. *Clin Orthop Rel Res* 1963; 26: 147–56.
23. Kuehnle I, Goodell MA. The therapeutic potential of stem cells from adults. *BMJ* 2002; 325(7360): 372–6.
24. Bianco P, Riminucci M, Kuznetsov S, Robey PG. Multipotential cells in the bone marrow stroma: regulation in the context of organ physiology. *Crit Rev Eukaryot Gene Expr* 1999; 9(2): 159–73.
25. Vats A, Tolley NS, Polak JM, Buttery LD. Stem cells: sources and applications. *Clin Otolaryngol* 2002; 27(4): 227–32.
26. Keller GM. In vitro differentiation of embryonic stem cells. *Curr Opin Cell Biol* 1995; 7(6): 862–9.
27. McLaren A. Ethical and social considerations of stem cell research. *Nature* 2001; 414(6859): 129–31.
28. Oreffo RO, Triffitt JT. Future potentials for using osteogenic stem cells and biomaterials in orthopedics. *Bone* 1999; 25(2): 5S–9S.
29. Haynesworth SE, Carrino DA, Caplan AI. Characterization of the core protein of the large chondroitin sulfate proteoglycan synthesized by chondrocytes in chick limb bud cell cultures. *J Biol Chem* 1987; 262(22): 10574–81.
30. Jaiswal N, Haynesworth SE, Caplan AI, Bruder SP. Osteogenic differentiation of purified, culture-expanded human mesenchymal stem cells in vitro. *J Cell Biochem* 1997; 64(2): 295–312.
31. Korda M, Blunn G, Phipps K, Rust P, Di Silvio L, Coathup M, Goodship A, Hua J. Can mesenchymal stem cells survive under normal impaction force in revision total hip replacements? *Tissue Engineering* 2006; 12: 3.
32. Rose FR, Oreffo RO. Bone tissue engineering: hope vs hype. *Biochem Biophys Res Commun* 2002; 292(1): 1–7.
33. Dealler SF. Electrical phenomena associated with bones and fractures and the therapeutic use of electricity in fracture healing. *J Med Eng Tech* 1981; 5: 73–82.
34. Buch F. On electrical stimulation of bone tissue. PhD thesis, Dept of Biomaterials/Handicap Research, University of Göteborg, Sweden, 1985; 1–139.
35. Levander G. A study of bone regeneration. *Surg Gynae Obstet* 1938; 67: 705–14.
36. Lind M. Growth factors: possible new clinical tools. *Acta Orthop Scand* 1996; 67: 407–17.
37. Hauschka PV, Chen TL, Mavrakos AE. Polypeptide growth factors in bone matrix. *Ciba Found Symp* 1988; 136: 207–25.
38. Urist MR, Mikulski A, Lietze A. Solubilized bone morphogenetic protein. *Proc Natl Acad Sci USA* 1979; 76: 1828–32.
39. Solheim E. Growth factors in bone. *Int Orthop* 1998; 22: 410–16.
40. Cheng H, Jiang W, Phillips FM, Haydon RC, Peng Y, Zhou L et al. Osteogenic activity of the fourteen types of human bone morphogenetic proteins (BMPs). *J Bone Jt Surg Am* 2003; 85-A(8): 1544–52.
41. Wozney JM. Overview of bone morphogenetic proteins. *Spine* 2002; 27(16 Suppl 1): S2–8.

42. Matthews SJ. Biological activity of bone morphogenetic proteins (BMPs). *Injury* 2005; 36 (Suppl 3): S34–7.
43. Wang EA, Rosen V, D'Alessandro JS, Bauduy M, Cordes P, Harada T et al. Recombinant human bone morphogenetic protein induces bone formation. *Proc Natl Acad Sci USA* 1990; 87(6): 2220–4.
44. Reddi AH. Bone morphogenetic proteins: from basic science to clinical applications. *J Bone Jt Surg Am* 2001; 83-A (Suppl 1, Part 1): S1–6.
45. Heldin CH, Miyazone K, ten Dijke P. TGF-beta signalling from cell membrane to nucleus through SMAD proteins. *Nature* 1997; 390(6659): 465–71.
46. Weiss RE, Reddi AH. Synthesis and localization of fibronectin during collagenous matrix–mesenchymal cell interaction and differentiation of cartilage and bone in vivo. *Proc Natl Acad Sci USA* 1980; 77(4): 2074–8.
47. Gerstenfeld LC, Cullinane DM, Barnes GL, Graves DT, Emhorn TA. Fracture healing as a post-natal developmental process: molecular, spatial, and temporal aspects of its regulation. *J Cell Biochem* 2003; 88(5): 873–84.
48. Keating JF, McQueen MM. Review article: Substitutes for autologous bone graft in orthopaedic trauma. *J Bone Joint Surg [Br]* 2001; 83-B: 3–8.
49. Israel DI, Nove J, Kerns KM. Heterodimeric bone morphogenetic proteins show enhanced activity in vitro and in vivo. *Growth Factors* 1996; 13: 291–300.
50. Mayer H, Scutt AM, Ankenbauer T. Subtle differences in the mitogenic effects of recombinant human bone morphogenetic proteins -2 to -7 DNA synthesis on primary bone-forming cells and identification of BMP-2/4 receptor. *Calcif Tissue Int* 1996; 58: 249–55.
51. Phillips AM. Overview of the fracture healing cascade. *Injury* 2005; 36 (Suppl 3): S5–7.
52. Bahamonde ME, Lyons KM. BMP3: to be or not to be a BMP. *J Bone Jt Surg Am* 2001; 83A (Suppl 1, Part 1): S56–62.
53. Zhang H, Bradley A. Mice deficient for BMP2 are nonviable and have defects in amnion/chorion and cardiac development. *Development* 1996; 122(10): 2977–86.
54. Dudley AT, Lyons KM, Robertson EJ. A requirement for bone morphogenetic protein-7 during development of the mammalian kidney and eye. *Genes Dev* 1995; 9(22): 2795–807.
55. Centrella M, Horowitz MC, Wozney JM, McCarthy TL. Transforming growth factor-beta gene family members and bone. *Endocr Rev* 1994; 15(1): 27–39.
56. Einhorn TA, Trippel SB. Growth factor treatment of fractures. *Instr Course Lect* 1997; 46: 483–6.
57. Einhorn TA, Lee CA. Bone regeneration: new findings and potential clinical applications. *J Am Acad Orthop Surg* 2001; 9(3): 157–65.
58. Cook SD, Baffes GC, Wolfe MW et al. The effect of recombinant human osteogenic protein-1 on healing of large segmental bone defects. *J Bone Joint Surg [Am]* 1994; 76-A: 827–38.
59. Cook SD, Baffes GC, Wolfe MW, Sampath TK, Rueger DC. Recombinant human bone morphogenetic protein-7 induces healing in a canine long-bone segmental defect model. *Clin Orthop* 1994; 301: 302–12.
60. Cook SD, Wolfe MW, Salkeld SL, Rueger DC. Effect of recombinant human osteogenic protein-1 on healing of segmental defects in non-human primates. *J Bone Joint Surg [Am]* 1996; 77-A: 734–50.

Cellular response to osteoinductive materials 339

61. Yasko AW, Lane JM, Fellinger EJ et al. The healing of segmental bone defects, induced by recombinant human bone morphogenetic protein (rhBMP-2): a radiographic, histological and biomechanical study in rats. *J Bone Joint Surg [Am]* 1992; 74-A: 659–70.
62. Gerhart TN, Kirker-Head CA, Kriz MJ et al. Healing segmental femoral defects in sheep using recombinant human bone morphogenetic protein. *Clin Orthop* 1993; 293: 317–26.
63. Bostrom M, Lane JM, Tomin E et al. Use of bone morphogenetic protein-2 in the rabbit ulnar nonunion model. *Clin Orthop* 1996; 327: 272–82.
64. Itoh T, Mochizuki M, Nishimura R et al. Repair of ulnar segmental defect by recombinant human morphogenetic protein-2 in dogs. *J Vet Med Sci* 1998; 60: 451–8.
65. Friedlaender GE, Perry CR, Cole JD, Cook SD, Cierny G, Muschler GF et al. Osteogenic protein-1 (bone morphogenetic protein-7) in the treatment of tibial nonunions. *J Bone Jt Surg Am* 2001; 83-A (Suppl 1, Part 2): S151–8.
66. Muschler GF, Nakamoto C, Griffith LG. Engineering principles of clinical cell-based tissue engineering. *J Bone Jt Surg Am* 2004; 86-A(7): 1541–58.
67. Johnson EE, Urist MR. One-stage lengthening of femoral nonunion augmented with human bone morphogenetic protein. *Clin Orthop* 1998; 347: 105–16.
68. Johnson EE, Urist MR. Human bone morphogenetic protein allografting for reconstruction of femoral nonunion. *Clin Orthop Rel Res* 2000; 371: 61–74.
69. Giannoudis PV, Tzioupis C. Clinical applications of BMP-7: the UK perspective. *Injury* 2005; 36 (Suppl 3): S47–50.
70. Masuda K, Imai Y, Okuma M, Muehleman C, Nakagawa K, Akeda K et al. Osteogenic protein-1 injection into a degenerated disc induces the restoration of disc height and structural changes in the rabbit annular puncture model. *Spine* 2006; 31(7): 742–54.
71. Ripamonti U, Crooks J, Rueger DC. Induction of bone formation by recombinant human osteogenic protein-1 and sintered porous hydroxyapatite in adult primates. *Plast Reconstr Surg* 2001; 107(4): 977–88.
72. Tsiridis E, Zubier A, Amit B, Roushdi I, Heliotis M, Gurav N, Di Silvio L. In vitro and in vivo optimisation of impaction allografting by demineralisation and addition of rhOP-1. *J Orthop Res* 2007; 25(11): 1425–37.
73. Bostrom MP, Lane JM, Berberian WS, Missri AA, Tomin E, Weiland A, Doty SB, Glaser D, Rosen VM. Immunolocalization and expression of bone morphogenetic proteins 2 and 4 in fracture healing. *J Orthop Res* 1995; 13(3): 357–67.
74. Einhorn TA. Problems with delayed and impaired fracture healing remain a challenge to the orthopedic trauma surgeon. *Orthop Trauma* 1997; 11(4): 243.
75. Turek TJ, Bostrom MPG, Camacho NP et al. Acceleration of bone healing in a rabbit ulnar osteotomy model with rhBMP-2. *Trans Orthop Res Soc* 1997; 22: 526.
76. Welch RD, Jones AL, Bucholz RW et al. Effect of recombinant human bone morphogenetic protein-2 on fracture healing in a goat tibial fracture model. *J Bone Miner Res* 1998; 13: 1483–90.
77. Govender S, Csimma C, Genant HK, Valentin-Opran A, Amit Y, Arbel R et al. BMP-2 Evaluation in Surgery for Tibial Trauma (BESTT) Study Group. Recombinant human bone morphogenetic protein-2 for treatment of open

tibial fractures: a prospective, controlled, randomized study of four hundred and fifty patients. *J Bone Jt Surg Am* 2002; 84-A(12): 2123–34.
78. Burkus JK, Gornet MF, Dickman CA, Zdeblick TA. Anterior lumbar interbody fusion using rhBMP-2 with tapered interbody cages. *J Spinal Disord Tech* 2002; 15(5): 337–49.
79. Burkus JK, Sandhu HS, Gornet MF, Longley MC. Use of rhBMP-2 in combination with structural cortical allografts: clinical and radiographic outcomes in anterior lumbar spinal surgery. *J Bone Jt Surg Am* 2005; 87(6): 1205–12.
80. Simpson AH, Mills L, Noble B. The role of growth factors and related agents in accelerating fracture healing. *J Bone Jt Surg Br* 2006; 88(6): 701–5.
81. Sumner DR, Turner TM, Urban RM, Virdi AS, Inoue N. Additive enhancement of implant fixation following combined treatment with rhTGF-beta2 and rhBMP-2 in a canine model. *J Bone Jt Surg Am* 2006; 88(4): 806–17.
82. Aspenberg P, Turek T. BMP-2 for intramuscular bone induction: effect in squirrel monkeys is dependent on implantation site. *Acta Orthop Scand* 1996; 67(1): 3–6.
83. Kuroda R, Usas A, Kubo S, Corsi K, Peng H, Rose T et al. Cartilage repair using bone morphogenetic protein 4 and muscle-derived stem cells. *Arthritis Rheum* 2006; 54(2): 433–42.
84. Cook SD, Patron LP, Salkeld SL, Rueger DC. Repair of articular cartilage defects with osteogenic protein-1 (BMP-7) in dogs. *J Bone Jt Surg Am* 2003; 85-A (Suppl 3): 116–23.
85. Goldring MB. Are bone morphogenetic proteins effective inducers of cartilage repair? Ex vivo transduction of muscle-derived stem cells. *Arthritis Rheum* 2006; 54(2): 387–9.
86. Zimmerman LB, De Jesus-Escobar JM, Hariand RM. The Spemann organizer signal noggin binds and inactivates bone morphogenetic protein 4. *Cell* 1996; 86(4): 599–606.
87. Hsu DR, Economides AN, Wang X, Eimon PM, Hariand RM. The Xenopus dorsalizing factor Gremlin identifies a novel family of secreted proteins that antagonize BMP activities. *Mol Cell* 1998; 1(5): 673–83.
88. Glaser DL, Economides AN, Wang L, Liu X, Kimble RD, Fandl JP. In vivo somatic cell gene transfer of an engineered Noggin mutein prevents BMP4-induced heterotopic ossification. *J Bone Jt Surg Am* 2003; 85-A(12): 2332–42.
89. Valentin-Opran A, Wozney J, Csimma C, Lilly L, Riedel GE. Clinical evaluation of recombinant human bone morphogenetic protein-2. *Clin Orthop Rel Res* 2002; 395: 110–20.
90. Kaplan FS, Shore EM. Encrypted morphogens of skeletogenesis: biological errors and pharmacologic potentials. *Biochem Pharmacol* 1998; 55(4): 373–82.
91. Termaat MF, Den Boer FC, Bakker FC, Patka P, Haarman HJ. Bone morphogenetic proteins. Development and clinical efficacy in the treatment of fractures and bone defects. *J Bone Jt Surg Am* 2005; 87(6): 1367–78.
92. Eleftherios T, Amit B, Zubier A, Neelam G, Manolis H, Sanjukta D, Di Silvio L. Enhancing the osteoinductive properties of hydroxyapatite by the addition of human mesenchymal stem cells, and recombinant human osteogenic protein-1 (BMP-7) in vitro. *Injury Int J* 2006; 37S: S25–32.
93. Winn SR, Uludag H, Hollinger JO. Carrier systems for bone morphogenetic proteins. *Clin Orthop Rel Res* 1999; 367 Suppl: S95–106.

94. Betz OB, Betz VM, Nazarian A, Pilapil CG, Vrahas MS, Bouxsein ML. Direct percutaneous gene delivery to enhance healing of segmental bone defects. *J Bone Jt Surg Am* 2006; 88(2): 355–65.
95. Ripamonti U, Ramoshebi LN, Matsaba T, Tasker J, Crooks J, Teare J. Bone induction by BMPs/OPs and related family members in primates. *J Bone Joint Surg Am* 2001; 83-A (Suppl 1, Part 2): S116–27.
96. Itoh K, Udagawa N, Katagiri T, Lemura S, Ueno N, Yasuda H *et al.* Bone morphogenetic protein 2 stimulates osteoclast differentiation and survival supported by receptor activator of nuclear factor-kappa B ligand. *Endocrinology* 2001; 142(8): 3656–62.
97. Abbas AK. Diseases of immunity. In: Cotran R, Kumar V, Collins T (eds) *Pathologic Basis of Disease*. Philadelphia, PA: WB Saunders, 7th edition, 2004: Chapter 6.
98. Wolff J. *Über die Wechselbeziehungen Zwischen der Form und der Function der Einzelnen Gebilde des Organismus*. Leipzig: FCW Vogel, 1901.
99. Lanyon LE. Control of bone architecture by functional load bearing. *J Bone Miner Res* 1992; 7: 369–75.
100. Claes LE, Heigele CA, Neidlinger-Wilke C, Kaspar D, Seidl W, Margevicius KJ, Augat P. Effects of mechanical factors on the fracture healing process. *Clin Orthop Rel Res* 1998; Oct (355 Suppl): S132–47.
101. Yasuda I. Fundamental aspects of fracture treatment. *J Kyoto Med Soc* 1953; 4: 395–404.
102. Bassett CAL. Biological significance of piezo-electricity. *Calcif Tissue Res I* 1968: 252–61.
103. Johansson C, Albrektsson T, Roos A. A biomechanical and histomorphometric comparison between differentiated types of bone implants evaluated in a rabbit model. *Eur J Exp Musculoskel Research* 1992; 1: 51–61.
104. Johansson C, Han CH, Wennerberg A, Albrektsson T. A quantitative comparison of machined commercially pure titanium and titanium-aluminum–vanadium implants in rabbit bone. *Int J Oral Maxillofac Implants* 1998; 13: 315–21.
105. Trippel SE, Coutts RD, Einhorn TA, Mundy GR, Rosenfeld RG. Growth factors as therapeutic agents. *J Bone Joint Surg [Am]* 1996; 78: 1272–86.
106. Gottlander M. On hard-tissue reactions to hydroxyapatite-coated titanium implants. PhD thesis, Dept of Biomaterials/Handicap Research, University of Göteborg, Sweden, 1994; 1–192.
107. Wennerberg A. On surface roughness and implant incorporation. PhD thesis, Dept of Biomaterials/Handicap Research, University of Göteborg, Sweden, 1996; 1–212.
108. Nilsson LP, Granström G, Albrektsson T. The effect of hyperbaric oxygen treatment on bone regeneration. An experimental study in the rabbit. *Int J Oral Maxillofac Implants* 1987; 3: 43–8.
109. Albrektsson T, Johansson C, Lundgren AK, Sul YT, Gottlow J. Experimental studies on oxidized implants. A histomorphometrical and biomechanical analysis. *Appl Osseointegration Res* 2000; 1: 21–4.
110. Röser K, Johansson C, Donath K, Albrektsson T. A new approach to demonstrate cellular activity in bone formation adjacent to implants. *J Biomed Mater Res* 2000; 51: 280–91.

111. Tsuboi N, Sennerby L, Johansson C, Albrektsson T, Tsuboi Y, Iizika T. Histomorphometric analysis of bone–titanium interface in human retrieved implants. In: Ueda M (ed) *Proceedings of the Third International Congress on Tissue Integration in Oral and Maxillofacial Reconstruction*.Tokyo:Quintessence, 1996, 84–5.
112. Tjellström A, Lindström J, Hallén O, Albrektsson T, Brånemark PI. Osseointegrated titanium implants in the temporal bone. A clinical study on bone-anchored hearing aids. *Am J Otol* 1981; 2: 303–10.
113. Tjellström A, Lindström J, Nylén O, Albrektsson T, Brånemark P-I, Birgersson B, Nero H, Sylvén C. The bone-anchored auricular epithesis. *Laryngoscope* 1981; 91: 811–15.
114. Albrektsson T, Carlsson LV, Jacobsson M, Macdonald W. The Gothenburg osseointegrated hip arthroplasty: experience with a novel type of hip design. *Clin Orth Rel Res* 1998; 352: 81–94.
115. Doi K, Tominaga S, Shibata T. Bone grafts with microvascular anastomoses of vascular pedicles. *J Bone Joint Surg Am* 1977; 59A: 809–15.
116. Burchardt H. The biology of bone graft repair. *Clin Orthop* 1983; 174: 28–42.
117. Virolainen P, Perala M, Vuorio E, Aro HT. Expression of matrix genes during incorporation of cancellous bone allografts and autografts. *Clin Orthop* 1995; 317: 263–72.
118. Ashton BA, Allen TD, Howlet CR *et al*. Formation of bone and cartilage by marrow stromal cells in diffusion chambers in vivo. *Clin Orthop* 1980; 151: 294–9.
119. Friedenstein AJ, Petrakova KV, Kurolesova AI, Frolova GP. Heterotopic transplants of bone marrow: Analysis of precursor cells for osteogenic and hematopoietic tissues. *Transplantation* 1968; 6: 230–5.
120. Majors AK, Boehm CA, Nitto H, Midura RJ, Muschler GF. Characterization of human bone marrow stromal cells with respect to osteoblastic differentiation. *J Orthop Res* 1997; 15: 546–57.
121. Healey JH, Zimmerman PA, McDonnell JM, Lane JM. Percutaneous bone marrow grafting of delayed union and nonunion in cancer patients. *Clin Orthop* 1990; 256: 280–5.
122. Simonds RJ, Holmberg SD, Hurwitz RL. Transmission of human immunodeficiency virus type I from a seronegative organ and tissue donor. *N Engl J Med* 1992; 326: 726–32.
123. Conrad EU, Gretch DR, Obermeyer KR *et al*. Transmission of the hepatitis-C virus by tissue transplantation. *J Bone Joint Surg [Am]* 1995; 77-A: 214–24.
124. Bonfiglio M, Jeter WS. Immunological responses to bone. *Clin Orthop* 1972; 87: 19–27.
125. Stevenson S. The immune response to osteochondral allografts in dogs. *J Bone Joint Surg* 1987; 69A: 573–82.
126. Stevenson S, Shaffer JW, Goldberg VM. The humoral response to vascular and nonvascular allografts of bone. *Clin Orthop Rel Res* 1996; 326: 86–95.
127. Horowitz MC, Friedlaender GE. Induction of specific T-cell responsiveness to allogenic bone. *J Bone Joint Surg* 1991; 73A: 1157–68.
128. Friedlaender GE. Immune responses to osteochondral allografts. Current knowledge and future directions. *Clin Orthop* 1983; 174: 58–66.

129. Burwell RG. Studies in the transplantation of bone. V. The capacity of fresh and treated hornografts of bone to evoke transplantation immunity. *J Bone Joint Surg* 1963; 45B: 38, 401.
130. Enneking WF, Mindell ER. Observations on massive retrieved human allografts. *J Bone Joint Surg* 1991; 73A: 1123–42.
131. Urist MR, Dawson E. Intertransverse process fusion with the aid of chemosterilized autolyzed allogenic (AAA) bone. *Clin Orthop* 1981; 154: 97–113.
132. Einhorn TA, Lane JM, Burstein AH, Kopman CR, Vigorita VJ. The healing of segmental bone defects induced by demineralized bone matrix. *J Bone Joint Surg* 1984; 66A: 274–9.
133. Ragni P, Lindholm TS, Lindholm TC. Vertebral fusion dynamics in the thoracic and lumbar spine induced by allogenic demineralized bone matrix combined with autogenous bone marrow. An experimental study in rabbits. *Ital J Orthop Traumatol* 1987; 13: 241–51.
134. Wilkins RM, Stringer EA. Demineralized bone powder: Use in grafting space-occupying lesions of bone. *Int Orthop* 1994; 2: 71–8.
135. Lee MY, Finn HA, Lazda VA, Thistlethwaite JR, Simon MA. Bone allografts are immunogenic and may preclude subsequent organ transplants. *Clin Orthop* 1997; 340: 215–19.
136. Ripamonti U, Crooks J, Rueger DC. Sintered porous hydroxyapatite in adult primates. *Plast Reconstr Surg* 2001; 107(4): 977–88.
137. Chapman MW, Bucholz R, Cornell C. Treatment of acute fractures with a collagen–calcium phosphate graft material: a randomized clinical trial. *J Bone Joint Surg [Am]* 1997; 79-A: 495–502.
138. Li RH, Bouxsein ML, Blake CA, D'Augusta D, Kim H, Li XJ, Wozney JM, Seeherman HJ. rhBMP-2 injected in a calcium phosphate paste (alpha-BSM) accelerates healing in the rabbit ulnar osteotomy model. *J Orthop Res* 2003; 21(6): 997–1004.
139. Trancik TM, Stulberg BN, Wilde AH, Feiglin DH. Allograft reconstruction of the acetabulum during revision total hip arthroplasty. *J Bone Joint Surg* 1986; 68A: 527–33.
140. Frankenburg EP, Goldstein SA, Bauer TW, Harris SA, Poser RD. Biomechanical and histological evaluation of a calcium phosphate cement. *J Bone Joint Surg* 1998; 80A: 1112–24.
141. De Biase P, Capanna R. Clinical applications of BMPs. *Injury* 2005; 36 (Suppl 3): S43–6.
142. Muschler CF, Perry CR, Cole JD *et al*. Treatment of established tibial non-unions using human recombinant osteogenic protein-1. *Proc Am Acad Orthop Surg* 1998; 65: 218.

14
Cellular response to bone graft matrices

A B M RABIE and R W K WONG,
University of Hong Kong, Hong Kong

Abstract: Bone induction is needed in treating bone trauma and filling osseous defects often required for successful surgery. The development of a graft matrix that is osteogenic is highly desirable. This chapter summarizes some of the current research in our centre in this field. The DBM_{IM}, the HMGRI-collagen matrix, the polymethoxylated flavonoid-collagen matrix and the phytoestrogen-collagen matrix have the effect of increasing new bone formation locally and can be used for bone regeneration, bone grafting or bone induction. Further research is needed to optimize their use and to gain further understanding on their bone-forming mechanism.

Key words: cellular responses, bone induction, matrices.

14.1 Introduction

Bone induction is needed in treating bone trauma and filling osseous defects often required for successful surgery. Over the years autogenous cancellous bone grafting has been considered the gold standard in replacing bone.[1] The major limitations of using autogenous grafting are the inadequacy of supply and surgical morbidity, including donor site pain, paresthesia, and infection, which can approach 8% to 10%.[2] Moreover, graft resorption posed a severe problem. In an experimental study, endochondral (EC) bone grafts showed 65% volume loss.[3] Allografts, an alternative to autogenous grafting, seem to be biologically inferior and are associated with infection and inflammation.[4] Therefore, the development of a graft matrix that is osteogenic is highly desirable. It is important to understand the cell responses to the osteogenic matrix. This chapter attempts to summarize some of the current research in our centre in this field.

14.2 Demineralized intramembranous bone matrix (DBM_{IM})

To develop bone graft materials to replace lost bone, osteoinductive matrices such as demineralized bone matrix (DBM) have been extensively studied[5-9] since they were reported by Urist in 1965.[10] The osteoinductive component in the DBM was identified as a family of bone morphogenetic proteins (BMPs).[10,11]

Bone induced by BMP in postfetal life recapitulates the process of embryonic and endochondral ossification.[12] BMPs are important regulators in osteogenic differentiation during fracture repair.[13] Wang and coworkers[14] showed that BMP-2 caused commitment and differentiation of multipotential stem cell line into osteoblastlike cells. Clinically, native human BMP has been used successfully for the treatment of established nonunions and spinal fusions.[15]

In an effort to augment the healing of autogenous EC bone, Rabie and Lie[6] mixed the autogenous EC bone with DBM of EC in origin (DBM$_{EC}$). This composite bone graft (EC-DBM$_{EC}$) produced 47% more bone than the EC autogenous bone alone. The advantage of the DBM was ascribed to the content and diffusibility of the bone morphogenetic proteins (BMP)[10] and other cytokines. In the presence of DBM,[7] autogenous EC bone showed better integration and faster incorporation with the host bone. In the clinic, the composite bone graft EC-DBM$_{EC}$ was successfully used in maxillary and mandibular reconstruction.[8] In these cases the DBM$_{EC}$, prepared from human cadaver long bones (demineralised cortical powder, Musculoskeletal Transplant Foundation, USA), was mixed with patient's autogenous bone. The successful outcome in the clinic confirmed the experimental results which showed that the DBM$_{EC}$ augmented the healing of EC autogenous bone.

In a quantitative analysis of the amount of bone newly formed during the healing of intramembranous (IM) and EC bone grafts, we reported that the IM bone produced 166% more bone than EC bone[16] when transplanted into rabbit skull defects, thus, pointing to the advantages of using IM bone or DBM of IM origin (DBM$_{IM}$) in the repair of bony defects. In this study, we prepared DBM from IM bone (see Appendix) and we hypothesize that DBM$_{IM}$ greatly augments the healing of EC bone.

In the healing of bone grafts, a significant correlation exists between the temporal sequence of vascular re-establishment and the eventual viability and incorporation of a graft.[17] Microangiographic techniques were used to study the rates of vascularization of IM and EC bone grafts and it was concluded that IM bone vascularized earlier than EC bone.[18] In the earlier stages of angiogenesis, in response to angiogenic mediators such as TGF-α and TGF-β, endothelial cells express cell–cell and cell–matrix adhesion molecules in preparation for vascular invasion into the newly formed matrix.[19] Since IM vascularizes earlier than EC bone, then, IM bone probably expresses these angiogenic mediators earlier than EC bone. Demineralized bone matrix prepared from IM bone should possess the same characteristics. Our working hypothesis proposed that DBM$_{IM}$, when mixed with EC bone, expresses angiogenic mediators at an earlier stage and induces earlier vascularization which should greatly enhance the osteogenic potential of EC bone grafts. Therefore, we decided to study the correlation between the temporal sequence of neovascularization and osteogenesis induced by EC bone graft alone and by composite EC-DBM$_{IM}$ bone graft. Anti-human angiogenesis related endothelial cells antibodies (EN 7/44, Bachem Feinchemikalien AG), a monoclonal antibody which was found to recognize budding capillaries and proliferating and migrating cells,[20,21] was used in this study to monitor the vascularization during the early stages of angiogenesis associated with composite EC-DBM$_{IM}$ and EC bone graft-induced osteogenesis on days 1, 2, 3, 4, 5, 6, 7 and 14 post-grafting.

The aims of this study were to quantify the amount of newly formed bone during the healing of composite EC-DBM$_{IM}$ bone graft and to compare it to EC bone graft alone; to examine the temporal sequence of vascular invasion during the early stages of osteogenesis induced by EC-DBM$_{IM}$ and to compare it to EC bone alone; and to identify cells involved in the healing of EC bone in the presence of DBM$_{IM}$.

Twenty-eight adult New Zealand White rabbits with a mean weight of 3.8 kg (SD 0.3) were used in this study. Twenty-four 10 × 5 mm full-thickness bony defects were created in the parietal bones of 24 rabbits in the experimental group. The experimental groups were divided into two groups: (1) the EC group – transplantation of autogenous EC bone graft alone; and (2) the composite EC-DBM$_{IM}$ group – transplantation of composite graft of EC autogenous bone and rabbit DBM$_{IM}$ particle. One rabbit from each group was sacrificed on days 1, 2, 3, 4, 5, 6 and 7 post-grafting and the remaining five rabbits from each group were sacrificed at day 14 post-grafting.

The remaining four rabbits with eight 10 × 5 mm full-thickness bony defects (two bony defects were created in each rabbit) were used as control groups: (1) active control – implantation of rabbit skin collagen; and (2) passive control – left empty. They were all sacrificed at day 14 post-grafting.

The animals were premedicated 1 hour before surgery with oxytetracycline hydrochloride (200 mg/mL, 30 mg/kg body weight, Tetroxyla, Bimeda, Dublin, Ireland) and buprenorphine hydrochloride (0.3 mL/kg body weight, Hypnorm, Janssen Pharmaceutical, Beerse, Belgium), supplemented with diazepam (5 mg/mL, 1 mg/kg body weight, Valium 10, Roche, Switzerland). In order to maintain the level of neuroleptanalgesia, increments of Hypnorm (0.1 mL/kg) were given at 30-minute intervals during the operation.

The surgical procedure consisted of the creation of 10 × 5 mm^2 full-thickness (approximately 2 mm) cranial critical-size defects, devoid of periosteum, using templates, in the parietal bones. In the experimental EC bone graft group, a piece of identical size EC bone harvested from the diaphyseal tibial shaft was grafted to the defect. Holes (approximately 1 mm diameter) were drilled at opposite ends of the bone grafts and likewise in the parietal bone to allow for fixation of the bone grafts with stainless steel wires (0.3 mm diameter). In the experimental composite EC-DBM$_{IM}$ bone graft group, the DBM$_{IM}$ particles mixed with whole blood were placed in the defects with the EC bone grafts. In the control groups, the defects were either left empty or grafted with rabbit skin collagen.

All wounds were closed with interrupted 3/O nylon sutures. No attempt was made to approximate the periosteum. Postoperatively, the rabbits were given an antibiotic 30 mg/kg body-weight, i.m. (oxytetracycline hydrochloride) daily, and for pain relief, Temgesic 50 µg/kg s.c. daily.

At each of the time periods 1, 2, 3, 4, 5, 6, 7 and 14 days postsurgically, animals were sacrificed with pento-barbital sodium, i.v. 60 mg/kg body weight, and the defect areas, including the surround tissue, were harvested for histological preparation. The specimens were fixed in 10% neutral phosphate-buffered formalin, decalcified in K's Decal Fluid (sodium formate/formic acid) for 2 days and double embedded in celloidin–paraffin. The specimens were cut into both parasagittal and coronal serial sections 3 µm thick (Hypercut, Schandon), mounted on poly-L-lysine coated slides and stained routinely with haematoxylin and eosin. The specimens were qualitatively evaluated by means of an Olympus transmitting light microscope (BH-2, Tokyo, Japan).

14.2.1 Immunohistochemistry for anti-human angiogenesis-related endothelial cells

Anti-human angiogenesis-related endothelial cells antibodies (EN 7/44 Bachem Feinchemikalien AG) were used for staining by means of the avidin–biotin complex (ABC) method. The specimens were mounted on poly-L-lysine coated slides, dewaxed, rehydrated, digested with trypsin for 20 min at 37°C, and then blocked with 10% normal swine rabbit serum in TBS at room temperature for 30 min. Diluted EN 7/44 primary anti-serum 1 in 50 was incubated at 37°C for 60 min, washed in TBS, and swine anti-rabbit serum 1 in 100 was applied for 45 min, followed by application of ABC complex 1 in 100 for 60 min. After washing in TBS, sections were counterstained with 1% TritonX100/TBS for 30 seconds, developed in 0.05% DAB for 1–3 min, cleared and mounted with coverslips.

Quantitative analysis was carried out on serial sections of defects of the experimental groups that were sacrificed after 14 days. Defects were divided into five regions spaced 1500 µm apart. From among more than 10 sections from each region, two sections were randomly drawn, giving a total of 10 sections from each defect. These sections were then stained by the periodic acid-Schiff (PAS) reaction technique to show new bone and mineralizing cartilage, and cover-slipped in histological medium (Permount, Fisher Scientific, NJ). The total amount of new bone formed within the surgically created defect was quantified by one observer, RW. Quantitative analysis was performed with a semiautomatic image analysis computer software system (Videoplan, Kontron Image Analysis Division, Germany) through a light transmitting microscope (Axioskop, Carl Zeiss, Germany) fitted with a video camera (Colour Video Camera, TK-1080E, JVC, Japan). Differences in staining properties and morphology between newly formed bone, old bone and cartilage made identification of newly formed bone not difficult.

Data were analysed with statistical analysis computer software (Graphpad Instat, v.2.04a, 1994, San Diego, CA). The one-way analysis of variance (ANOVA) method was used to compare sections drawn from the five regions in each defect. The arithmetic mean, standard deviation (SD) and 95% confidence intervals were calculated for each experimental group. The two means were compared by the Welch unpaired t-test. The critical level of statistical significance chosen was $P < 0.05$.

The size of the method error in digitizing the areas of new bone was calculated by the formula $\pm\sqrt{\frac{\Sigma d^2}{2n}}$, where d was the difference between the two registrations of a pair and n was the number of double registrations. The size of the method error is 0.014 mm^2. Ten randomly drawn histological sections were digitized on two separate occasions at least 3 months apart by the same observer and also by an independent observer. Paired t-tests were also performed to compare the intra-observer and the inter-observer registrations. The two-tailed P value to compare the intra-observer registrations was 0.5652, that to compare the inter-observer registrations was 0.5911, both considered not significant.

All animals remained in excellent health throughout the course of the experiment and rapidly recovered after surgery. There was no evidence of infection in any of the animals. Two weeks postoperatively, the defects in the non-grafted group did not exhibit bone formation, as they were soft to palpation. Defects implanted with either EC bone grafts or composite EC-DBM$_{IM}$ bone grafts were hard to palpation.

14.2.2 EC bone graft groups

Histological evaluation of the healing process revealed that the new bone was localized to the host bone graft surface (day 14, Fig. 14.1). Evaluation of the new bone formation demonstrated a route of entry of chondroblasts and osteoblasts mirroring the endochondral bone development as expected. Healing was characterized by the presence of a cartilage intermediate concomitant with the resorption of the original EC bone graft. Chondrocytes

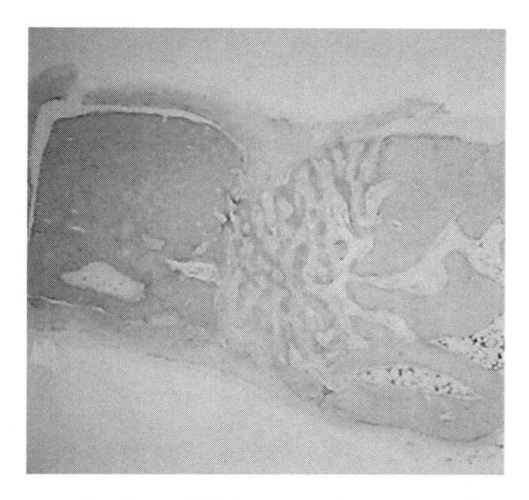

14.1 Healing of EC bone graft, 14 days after grafting, PAS stain, ×40.

with their typical territorial cartilage matrix were interposed between bone graft and host bone. Chondroblasts were first identified by day 6. Cartilage matrix with the typical array of chondrocytes was identified by day 7. Signs of osteogenesis accompanied vascular invasion at day 14.

14.2.3 Composite EC-DBM$_{IM}$ groups

Histological evaluation of the healing process revealed that healing of the composite EC-DBM$_{IM}$ bone grafts was characterized by endochondral ossification concomitant with resorption of the original DBM. New bone formation and vascularization were observed throughout the whole defect and the bone graft was being resorbed and remodelled; a wide bone marrow space was observed within the bone graft. The new bone formation demonstrated a route of entry of osteoblasts mirroring the endochondral bone development. Chondroblasts was first identified by day 4. Cartilage matrix with the typical array of chondrocytes was identified by day 5. Initial signs of osteogenesis accompanied vascular invasion at day 6. At the host bone–graft interface, high bone density with narrow bone marrow space was observed when compared to the same area in the EC group (Fig. 14.2).

14.2.4 Control groups

The defects were filled with fibrous connective tissue (Fig. 14.3). Occasionally, a narrow margin of new bone formation could be seen around the edges of the defects bridging the marrow space. The rabbit skin collagen did not induce bone formation across the defects.

350 Cellular response to biomaterials

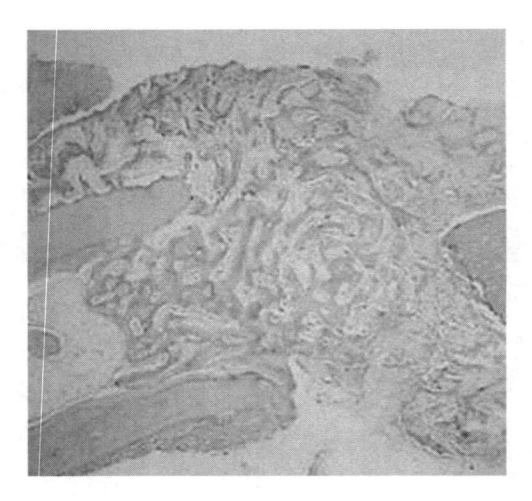

14.2 Healing of EC-DBM$_{IM}$ bone graft, 14 days after grafting, PAS stain, ×40.

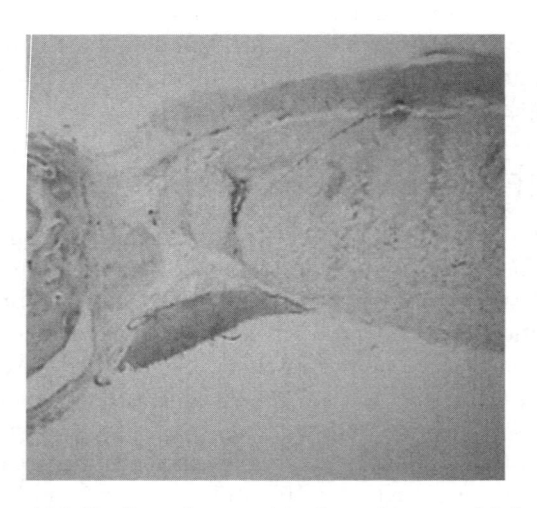

14.3 Healing of control (collagen) group, 14 days after grafting, PAS stain, ×40.

14.2.5 Immunohistochemical and angiogenic findings

In EC bone graft groups, immunohistochemical staining of the tissue collected 1–3 days after grafting revealed negative staining for EN7/44 antibodies. Positive staining was first detected at day 4 after implantation. By day 5, small blood vessels were first seen budding from the host bed towards the EC graft. Vascular invasion progressed towards the middle of the grafted

Table 14.1 Summary of the angiogenic cascade of endochondral bone induced osteogenesis

Number of days post EC bone implantation	Events related to osteogenesis	Events related to angiogenesis
Days 1–3	Inflammatory response	Blood clot formation
Day 4		Positive immunostaining for EN 7/44
Day 5		Budding of micro-vessels from host tissues
Day 6	Differentiation of chondroblasts	
Day 7	Chondrocytes and cartilage matrix synthesis	Appearance of small blood vessels into the newly formed matrix
Day 14	Osteoblasts, endochondral ossification, chondrocyte hypertrophy and signs of chondrolysis	Advent of vascular invasion towards one-third of the defect

area by day 14. In the composite EC-DBM$_{IM}$ group, immunohistochemical staining of the tissue collected 1 day after grafting revealed negative staining for EN 7/44 antibodies. Positive staining was only detected at day 2 after implantation. By day 3, small blood vessels were first seen budding from the host bed towards the DBM graft. Vascular invasion progressed towards the middle of the grafted area by day 6. A summary of the angiogenic cascade and its relation to the cellular events accruing from composite EC bone graft and EC-DBM$_{IM}$ bone graft-induced osteogenesis are discussed in Tables 14.1 and 14.2. Passive control revealed no background staining by using the secondary antibody.

In this study, we successfully prepared DBM from IM bone and showed that it greatly enhanced the osteogenic potential of EC bone grafts. One possible explanation of the enhanced osteogenesis seen in the composite bone graft was that DBM$_{IM}$ enhanced the bone induction ability of the host cells. Scott *et al.*[22] extracted an osteogenic factor from IM bone and showed that it was different from factors extracted from EC bone. This osteogenic factor produced bone directly, bypassing a cartilage intermediate stage. Therefore, the advantage of DBM$_{IM}$ was ascribed to the content and diffusibility of the osteoinductive factors, and other cytokines which interacted with undifferentiated osteogenic precursor cells in the host bed and caused them to differentiate into functionally active osteogenic elements.[22,23] The collagenase-resistant bone morphogenetic protein (BMP) was found to diffuse significant distances along surfaces of contact cells with bone matrix. These distances were measured on implants in diffusion chambers to be as

Table 14.2 Summary of the angiogenic cascade during composite EC-DBM$_{IM}$ induced osteogenesis

Number of days post EC-DBM$_{IM}$ implantation	Events related to osteogenesis	Events related to angiogenesis
Day 1	Inflammatory response	Blood clot formation
Day 2		Positive immunostaining for EN 7/44
Day 3		Budding of micro-vessels from host tissues
Day 4	Differentiation of chondroblasts	
Day 5	Chondrocytes and cartilage matrix synthesis	Appearance of small blood vessels into the newly formed matrix
Day 6	Osteoblasts, endochondral ossification, chondrocyte hypertrophy and signs of chondrolysis	Vascular invasion in the outer two-thirds of the defect
Day 7	New bone formation	
Day 14	Bone remodelling, dissolution of the implanted matrix	Advent of vascular invasion throughout the whole defect

great as 1000 μm.[24] The diffusion of these osteoinductive factors from the DBM$_{IM}$ induces new bone formation that helps in the integration of the EC autogenous bone and the recipient calvarial IM bone.

Integration of bone grafts is dependent on many interrelated processes including the nature of the host bone and its osteoprogenitor cells, osteoconduction, osteoinduction and the nature of the extracellular matrix of the bone graft. Therefore, in the present work, it was essential to identify the cells involved in the process of bone induction by the EC-DBM$_{IM}$ and to examine the effect of DBM$_{IM}$ on the integration of the bone graft with the host bone. In the current study, defects were created in the parietal bone of rabbit skulls. This makes the host bone of IM origin. The integration of EC bone grafted into calvarial IM bone defects was characterized by the presence of foci of cartilage, unlike the integration of IM bone graft which was characterized by the presence of osteogenic cells.[25] Osteogenesis induced by EC-DBM$_{IM}$ was characterized by the presence of both cartilage matrix and bone matrix (Table 14.2). Bone matrix with osteogenic cells was identified as early as day 4 next to the host and interposed between DBM$_{IM}$ particles in the composite EC-DBM$_{IM}$ group. Bone matrix in the case of the healing of EC autogenous bone was not identified until day 6. Therefore, the presence of bone at an earlier stage during osteogenesis induced by

EC-DBM$_{IM}$ is a direct indication that better amalgamation and integration of the EC bone graft with the host bone exists in the presence of DBM$_{IM}$.

Another factor that could contribute to the enhanced bone induction ability of EC-DBM$_{IM}$ is the ability of the DBM$_{IM}$ to induce earlier vascularization than EC bone. In this study, positive immunostaining of angiogenic mediators was first detected on day 4 in EC bone. In contrast, the first sign of positive immunostaining of the angiogenic mediators was detected on day 2 in the composite EC-DBM$_{IM}$. This is a direct indication that the presence of DBM$_{IM}$ has triggered the angiogenic cascade at an earlier stage when compared to the EC bone alone. In response to angiogenic mediators, 'primed' endothelial cells exhibit reorganization of cytoskeletal elements and prepare movement away from the parent venule.[26] In the current study, endothelial cells lining small vessels budding from the host bed tested positive on day 3 in the composite EC-DBM$_{IM}$ group and on day 6 in the EC group. The process of endothelial cell activation initiates the formation of capillary buds that continue to grow and mature unless the terminal steps in the angiogenic cascade are interrupted.[27,28] Therefore, the early invasion of new blood vessels into the grafted site is considered a source of potential osteoprogenitor cells. The appearance of small blood vessels budding from the host on day 3 and the identification of bone matrix interposed between the DBM$_{IM}$ particles on day 4 suggest that the osteoinductive BMPs present in the DBM$_{IM}$ diffused into the extracellular matrix and acted upon the nondifferentiated mesenchymal cells in the perivascular sites of the newly formed blood vessels and caused them to differentiate into osteogenic cells. The endochondral ossification seen in the healing of the EC bone graft was characterized by the presence of cartilage matrix on day 5. Bone matrix concomitant with neovascularization was evident on day 7. The wide spread of neovascularization seen in the composite EC-DBM$_{IM}$ group and the subsequent earlier bone formation are direct evidence of enhanced host responses in the presence of DBM$_{IM}$.

To conclude, DBM$_{IM}$ substantially increased the amount of new bone formation of autogenous EC bone grafts and induced earlier vascularization during osteogenesis. However, because of the potential risk of transmission of infection through the DBM, other possible sources of BMP were also actively investigated.

14.3 HMG-CoA reductase inhibitor: HMGRI-collagen matrix

To discover small molecules that induce BMP-2, Mundy and coworkers[29] examined more than 30,000 compounds from a natural products collection and tested the effects of compounds on BMP-2 gene expression. They identified the statin lovastatin, originally isolated as from *Penicillium citrinium* and *Penicillium brevicompactum*, an HMG-CoA reductase inhibitor (HMGRI), inhibitor of the rate-limiting enzyme in the mevalonate pathway in cholesterol synthesis, as the only natural product in the collection that specifically increased BMP-2 gene expression.[29]

Sugiyama et al.,[30] using real-time polymerized chain reaction analysis and alkaline phosphatase assay, revealed that the statin compactin induced an increase in the expression of BMP-2 mRNA and protein. Like compactin, simvastatin also activated the BMP-2 promoter, whereas pravastatin did not.[30] Oophorectomized rats given statins in oral dosages, comparable with those used in humans (that is, 1–10 mg/kg per day of simvastatin), had a 40% to 90% increase in the trabecular bone volume of the femur and lumber vertebrae within 35 days, relative to rats given placebo.[31] Skoglund et al.[32] showed that high-dose systemic treatment with simvastatin improved fracture healing in a mice femur fracture model.

Our laboratory decided to examine its bone-forming ability *in vivo* for the repair of bone defects by using a carrier to enable its use in the clinical setting. To achieve this, our laboratory measured the amount of new bone produced by simvastatin with collagen matrix carrier grafted into bony defects and compared with that of the collagen carrier alone.[33]

Eighteen 10 × 5 mm full-thickness bone defects were created in the parietal bones of nine New Zealand White rabbits from an inbred colony. The rabbits were 5 months old (adult stage) and weighed 3.5–4.0 kg. The handling of the animals and the experimental protocol were approved by the Committee on the Use of Live Animals in Teaching and Research, the University of Hong Kong. In the experimental group, six defects were grafted with simvastatin solution mixed with collagen matrix (statin group, see Appendix). In the control groups, six defects were grafted with collagen matrix alone (positive control) and six were left empty (negative control). The sample size was based on previous research using this model.[6] Our laboratory uses a small sample size because there was a large difference in bone formation between different groups so that a statistically significant difference can be detected with a minimal number of animals, as well as from the principle of the animal ethics committee that the smallest sample size that allows a significant difference to be detected should be used.

For the statin group and the control groups, two defects were created on the parietal bone of each rabbit. In each group six defects were created and surgery was performed, but after sacrifice only five were randomly drawn and prepared for analyses.

The details of the operation and the postoperative care of the animals were previously described in the last section. The surgical procedure consisted of the creation of one or two 10 × 5 mm full-thickness (approximately 2 mm) cranial defects, devoid of periosteum, using templates, in the parietal bones as described in the last section. For the statin group, the defects were filled with statin collagen grafts which consisted of 0.2 mL statin solution mixed with 0.02 g of collagen matrix (purified fibrillar collagen, Collagen Matrix, Inc., NJ). The grafts were prepared 15 minutes before grafting. In the control groups, the defects were either left alone (negative control) or grafted with 0.02 g of collagen matrix mixed with 0.2 mL water for injection (positive control).

All wounds were closed with interrupted 3/0 black silk sutures. No attempt was made to approximate the periosteum to prevent the barrier effect. Postoperatively, the rabbits were given oxytetracycline hydrochloride daily for 10 days and buprenorphine hydrochloride for 2 weeks. Two weeks after surgery, the animals were killed with sodium pentobarbitone. Immediately upon death, defects and surrounding tissue were removed for histological preparation. Tissues were fixed in 10% neutral buffered formal saline solution, demineralized with K's Decal Fluid (sodium formate/formic acid), and finally double embedded in celloidin/paraffin wax. Each tissue sample containing the defect was embedded intact. Serial, 5-μm-thick sections of the whole defect were cut perpendicular to the long axis.

Quantitative analysis was made on serial sections of defects in the experimental and the active control groups. Defects were divided into five regions spaced 1500 μm apart. From the serial sections in each region, two sections were selected randomly, giving a total of 10 sections from each defect. Therefore, the amount of new bone formation was assessed throughout the whole defect. The total amount of new bone formed within the surgically created defect was measured on 100 sections and the data were analysed with the same technique described in the previous section. All animals remained in excellent health throughout the course of the experiment and recovered rapidly after operation. There was no evidence of side effects or infection in any of the animals.

In the statin group, new bone was formed at the host bone–graft interface and tended to grow across the defect. Integration of the grafts with the recipient bed was characterized by the presence of new bone. No cartilage was found (Fig. 14.4). At higher magnification, new bone could be seen amalgamated with collagen matrix; bone cells were present showing that the collagen was not just calcified but rather that new bone was formed. The bone tended to grow towards the collagen matrix where statin was added (Fig. 14.5). In the active control group little new bone was formed at the host bone–graft interface. Some collagen fibres were present at the centre of the defects. In the passive control group, the defect was healed, with fibrous tissue bridging across the defect. Very little new bone had formed at the ends of the host bone, so no quantitative analysis was performed.

A total of 100 sections (50 sections for each group) of the statin group and the active control group were digitized and analysed. The amount of

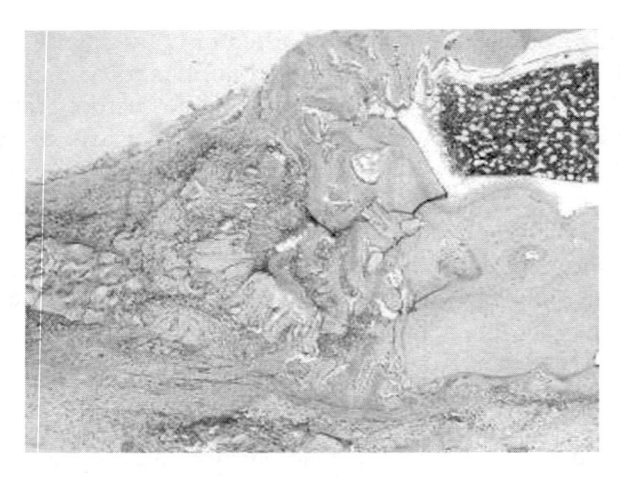

356   Cellular response to biomaterials

14.4 Healing of HMGRI-collagen matrix, 14 days after grafting, PAS stain, ×40.

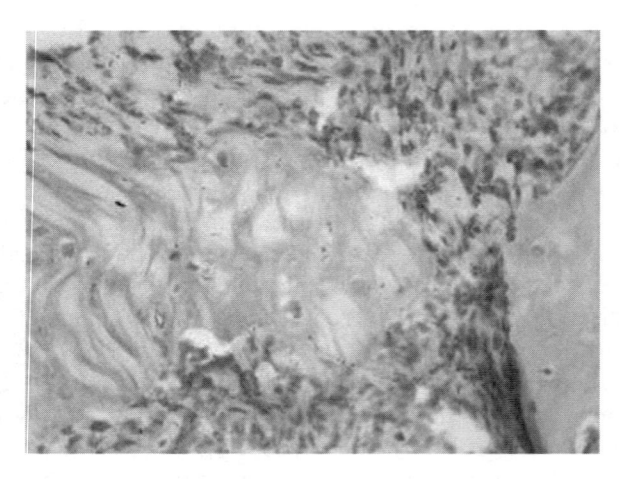

14.5 Healing of HMGRI-collagen matrix, 14 days after grafting, PAS stain, ×200.

newly formed bone was significantly greater in the defects of the experimental groups than in those of the active control. A total of 308% more new bone was present in defects grafted with statin in collagen matrix. Welch's unpaired t-test, which does not assume equal variances, was used to test the difference between each experimental group and the active control group; the two-tailed P value was <0.0001, considered significant.

In this study we demonstrated that purified fibrillar collagen is a good delivery vehicle to the statin and that the statin is osteoinductive when grafted into skull defects. *In vitro* experiments showed that when statin was added to cultured murine or human bone cells, it enhanced the expression of BMP-2 mRNA.[29] When statin was added to neonatal murine calvarial bone in organ culture it was shown that new bone formation increased two- to threefold.[29] Its direct use for the repair of bone defects was not carried out. Therefore in this study we decided to examine its osteoinductivity *in vivo* for the repair of bone defects and to find a carrier that enables its use in a clinical setting. This study showed that statin was osteoinductive when used with absorbable collagen sponge. This was probably caused by an increase in the expression of the BMP-2 gene in the bone-forming cells locally, causing a production of BMP-2, which in turn induced bone formation in the surrounding area. Mundy and coworkers showed that statins specifically enhance the expression of BMP-2 mRNA and failed to alter the expression of BMP-4, interleukin-6 and parathyroid hormone-related peptide.[29] Thus statin can be used as an osteoinductive agent by combining it with an absorbable carrier.

The bone inductive process involved here is basically bone formation induced by BMP which is similar to bone formation in the presence of DBM, which also releases BMP.[5-8] We have previously investigated ultrastructurally the early healing process of IM bone grafts in the presence of DBM of IM in origin.[34] In the presence of DBM, inflammatory response and formation of blood clot appeared on day 1–2 after surgery. Mesenchymal cell migration and proliferation appeared on day 3. Proliferation of preosteoblasts, osteoblasts, osteocytes and appearance of small blood vessels into the newly formed matrix appeared on day 4. New bone formation next to the host bone was observed on day 5, progress of osteogenesis towards the middle of the defect on day 6, and ongoing dissolution of the implanted matrix and newly formed bone filling across the bone defect at day 7. We compared this with the osteogenesis of autogenous IM bone grafts and found that the healing of bone graft in the presence of DBM presented a mirror image of the healing of autogenous bone graft except that the appearance of bone-forming cells and blood vessels was noted earlier.[34] It is possible that the route of healing of the statin collagen graft would be similar to that seen during bone induction using DBM since both enhance the expression of BMP-2.

In addition to DBM, we examined the early healing pattern of bone defects with statin, another HMG-CoA reductase inhibitor,[35] in a similar animal model with rabbits killed on days 1, 2, 3, 4, 5 and 6 postoperatively. Immunolocalization studies of the defects grafted with statin showed that VEGF was expressed on day 3 postoperatively, BMP-2 on day 4 and Cbfa1 on day 5 and that new bone was formed by day 5. These events occurred one day earlier than in the group grafted with the carrier alone.

Unfortunately, statin also has some side effects such as digestive disturbances, weakness, headache and possibility of serious muscle problems.[36] Therefore, other osteogenic agents like those shown below need to be developed.

14.4 Polymethoxylated flavonoid-collagen matrix

Further studies showed that the statin-mediated activation of BMP-2 promoter was completely inhibited by the downstream metabolite of HMG-CoA reductase, mevalonate, indicating that the activation was a result of the inhibition of that enzyme.[30] Therefore, it is possible that any agent that inhibits the HMG-CoA reductase may have the similar effect of statin in the activation of the BMP-2 promotor. One possible agent is naringin, which was also shown to have a HMG-CoA reductase inhibiting effect.[37] It may also activate the BMP-2 promotor and increase the bone formation.

Naringin is a polymethoxylated flavonoid that is also commonly found in citrus fruits. Research on it has been focused on its antioxidant and anticholesterol effect. The significance is that naringin is present in large amounts in common edible fruits like grapefruit. Therefore, if naringin can be shown to increase bone in an animal model of bone defect healing, it may be the long-sought-after safe agent for bone induction and bone defect repair. In addition, it also increased our understanding in the relationship between the mevalonate pathway and the bone-forming pathway. Therefore we carried out an experiment on naringin using the identical protocol of the previous statin study. The only difference was the experimental group.[38]

In the naringin group, six defects were grafted with naringin in collagen matrix which consisted of 0.2 mL naringin solution mixed with 0.02 g of collagen matrix (see Appendix). The other experimental details were the same as in the section above on lovastatin and compactin. All animals remained in excellent health throughout the course of the experiment and recovered rapidly after operation. There was no evidence of side effects or infection in any of the animals.

In the naringin group, new bone was formed at the host bone–graft interface and tended to grow across the defect. Integration of the grafts with the recipient bed was characterized by the presence of new bone. No cartilage was found (Fig. 14.6). At higher magnification, new bone could be seen amalgamated with collagen matrix; bone cells were present showing that the collagen was not just calcified but rather that new bone was formed. The bone tended to grow towards the collagen matrix where naringin was added (Fig. 14.7).

A total of 50 sections from five randomly drawn specimens of the naringin group were digitized and analysed. The amount of newly formed bone

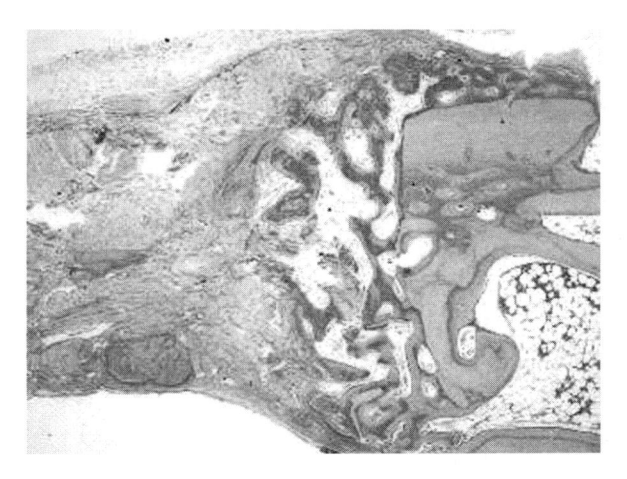

14.6 Healing of polymethoxylated flavonoid-collagen matrix, 14 days after grafting, PAS stain, ×40.

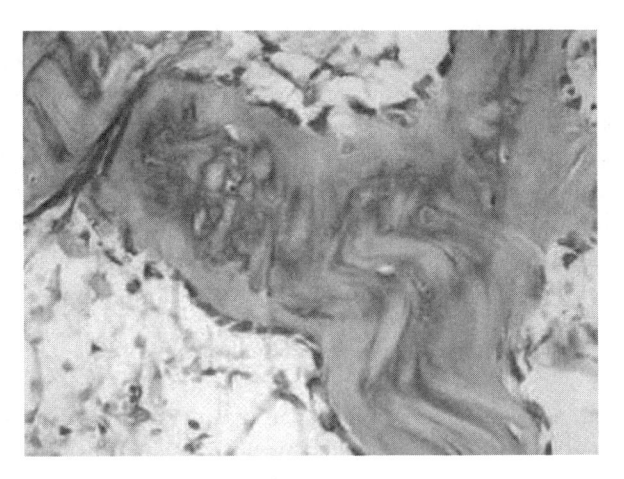

14.7 Healing of polymethoxylated flavonoid-collagen matrix, 14 days after grafting, PAS stain, ×200.

was significantly greater in the defects of the experimental groups than in those of the active control. A total of 490% more new bone was present in defects grafted with naringin in collagen matrix. Welch's unpaired *t*-test, which does not assume equal variances, was used to test the difference between each experimental group and the active control group; the two-tailed *P* value is <0.0001, considered significant.[38]

This study showed that naringin in collagen matrix significantly increased new bone formation locally when grafted into skull defects. This was probably caused by the HMG-CoA reductase inhibition effect causing an increase in the expression of the BMP-2 gene in the bone-forming cells locally, causing a production of BMP-2, which in turn induced bone formation in the surrounding area.

It is possible that the route of healing of the naringin in collagen graft would be similar to that seen during bone induction using DBM or statin since they enhance the expression of BMP-2. Further research on various gene expressions during the early healing of naringin in collagen matrix is needed to confirm this.

The dose for naringin in this study was estimated from a clinical study using naringin supplementation to lower the plasma lipids in hypercholesterolemic subjects.[39] This is because we hypothesized that naringin exerts its effect through blocking the cholesterol producing pathway. Therefore the dosage chosen had to have an effect on the plasma lipid levels. In that study the dose received per subject was 400 mg/day. We estimated the dose required for bone induction in rabbits, that is, the dose received after the grafting surgery, to be this dose multiplied by the body mass ratio. Using a pilot study this estimated dose did cause an increase in bone formation. In this study we have tried to demonstrate the osteogenic effect of naringin using one commonly used carrier for bone repair. Further research is needed to determine the optimum dose, the best choice of carrier and the release kinetics of naringin for bone grafting.

As mentioned in the introduction, naringin is a common polymethoxylated flavonoid that inhibits the HMG-CoA reductase other than statin. This study showed that it increased bone formation by 490% whereas statin increased bone formation by 308% as shown above. Therefore this provided further evidence for the hypothesis that inhibiting HMG-CoA reductase, in turn blocking the mevalonate pathway in cholesterol production, will increase bone formation. Further studies on various gene expressions of naringin on osteoblasts are needed to confirm that the actual bone-forming effect of naringin actually occurs through this mechanism to trigger BMP-2 formation, which in turn increases bone formation.

The significance of this study is that naringin is found in common edible fruit. It has recently gained interest in research by its antioxidant, anticholesterol and anticancer effect. Naringin has tremendous potential to be utilized in any application that requires increase in bone formation. Further studies are needed to determine the systemic effect of naringin on bone metabolism, for example on the prevention of osteoporosis.

14.5 Phytoestrogen-collagen matrix

The decrease in serum estrogen after the menopause is associated with bone loss and osteoporosis, and estrogen replacement therapy is considered to be effective in preventing bone loss.[40] It has been shown that estrogen enhances osteoblast differentiation and bone formation[41–43] and the conditioned medium of estrogen-treated osteoblast cultures inhibits osteoclast development.[44] Thus, estrogen is one of the most important sex steroids for the maintenance of bone balance.

Phytoestrogens are plant-derived non-steroidal compounds that bind to estrogen receptors (ERs) and have estrogen-like activity.[45] Phytoestrogens are divided into three classes: isoflavones, coumestans and lignans. In addition, some flavonoids, such as naringenin, are also classed as phytoestrogens. They have attracted much attention among public and medical communities because of their potential beneficial role in prevention and treatment of cardiovascular diseases, osteoporosis, diabetes and obesity, menopausal symptoms, renal diseases and various cancers.[46,47]

Kanno et al.[48] report the effects of phytoestrogens and environmental estrogens on osteoblast differentiation using MC3T3-E1 cells, a mouse calvaria osteoblast-like cell line. They increased alkaline phosphatase activity and enhanced bone mineralization in these cells.

Puerarin, 4H-1-benzopyran-4-one,8-β-D-glucopyranosyl-7-hydroxy-3-(4-hydroxy-phenyl), $C_{12}H_{20}C_9$, is one of the major phytoestrogens isolated from the root of a wild leguminous creeper, *Pueraria lobata* (Willd.) Ohwi.[49] This is a commonly used traditional Chinese medicine known as Gegen and has been demonstrated to have effects on decreasing loss in bone density in ovariectomized mice.[50] It also has other important uses on treatment of fever, liver diseases[51] and cardiovascular diseases.[52] In China, *P. lobata* is also used as a health supplement for reducing risk factors of cardiovascular diseases. There are wide application of *P. lobata* in clinical prescriptions and dietary supplements.[53]

It is possible that puerarin may also increase bone formation. The significance is that puerarin is commonly taken as a health supplement in dishes and soups in Asian meals. Therefore, it may be another agent for bone induction and bone defect repair. Therefore we carried out an experiment on puerarin using the identical protocol of the previous statin study. The only difference is the experimental group.[54] In the puerarin group, six defects were filled with puerarin in a collagen matrix which consisted of 0.2 mL puerarin solution mixed with 0.02 g of collagen matrix (see Appendix). The other experimental details were the same as in the earlier section on lovastatin and compactin. All animals remained in excellent health throughout the course of the experiment and recovered rapidly after operation. There was no evidence of side effects or infection in any of the animals.

In the puerarin group, new bone was formed at the host bone–graft interface and tended to grow across the defect. Integration of the grafts with the recipient bed was characterized by the presence of new bone. No cartilage was found (Fig. 14.8). At higher magnification, new bone could be seen amalgamated with collagen matrix; bone cells were present, showing that the collagen was not just calcified but rather that new bone was formed. The bone tended to grow towards the collagen matrix where statin was added (Fig. 14.9).

A total of 50 sections from five randomly drawn specimens of the puerarin group were digitized and analysed. The amount of newly formed bone was significantly greater in the defects of the puerarin group than in those

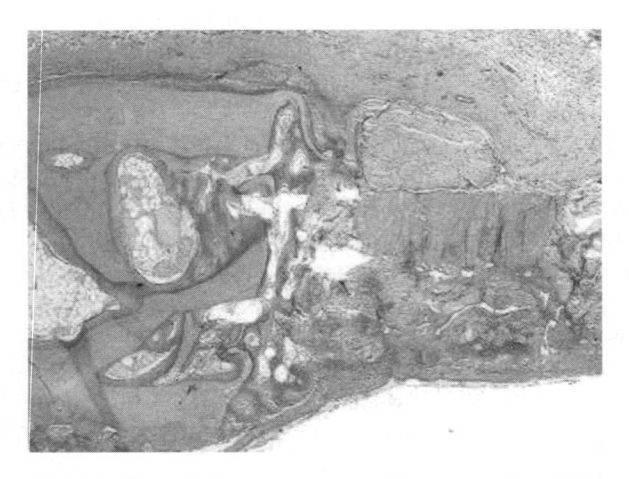

14.8 Healing of phytoestrogen-collagen matrix, 14 days after grafting, PAS stain, ×40.

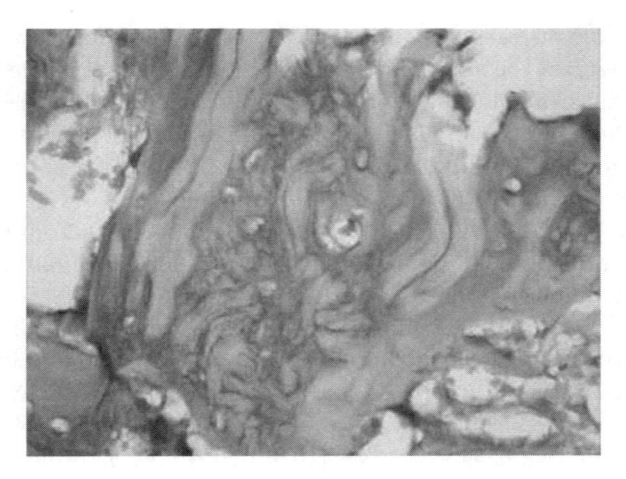

14.9 Healing of phytoestrogen-collagen matrix, 14 days after grafting, PAS stain, ×200.

of the active control. A total of 554% more new bone was present in defects grafted with puerarin in collagen matrix. Welch's unpaired t-test, which does not assume equal variances, was used to test the difference between each experimental group and the active control group; the two-tailed P value is <0.0001, considered significant.

This study demonstrated that puerarin in collagen matrix significantly increased new bone formation locally when grafted into skull defects. It is important to understand the underlying mechanisms of this effect. As mentioned, puerarin is an isoflavone phytoestrogen. Isoflavones stimulate osteogenesis at low concentrations and inhibit osteogenesis at high concentrations in osteoblasts and osteoprogenitor cells.[55,56] Isoflavones also influence the production of cytokines derived from osteoblasts and osteoprogenitor cells in a similar biphasic manner.[57] In addition, isoflavones inhibit osteoclast formation and activity.[58]

Pettersson et al.[59] and An et al.[60] proposed the molecular mechanisms of the osteogenic effects of phytoestrogens based on their estrogen receptors (ER)-mediated and/or enzyme-inhibiting effects. ER-mediated action has focused on the preferential binding of phytoestrogens to ERβ, which acts as a dominant-negative regulator of estrogen signalling. In addition, some phytoestrogens inhibit tyrosine kinases, and mitogen-activated protein kinases (MAPKs), which are crucial for cellular functions.[61] However, the biological effects of phytoestrogens cannot be fully explained by these mechanisms.[62]

Dang and coworkers identified peroxisome proliferator-activated receptors (PPARs) as additional molecular targets of phytoestrogens.[55,56] Regardless of their enzyme inhibitory actions, phytoestrogens can dose-dependently activate PPARs and induce divergent effects on osteogenesis and adipogenesis. As a result, the balance between concurrently activated ERs and PPARs determines the dose-dependent biological effects of phytoestrogens. As shown in their studies, dominant ER-mediated effects (an increase in osteogenesis and a decrease in adipogenesis) could only be seen at low concentrations of phytoestrogens, whereas dominant PPARγ-mediated effects (a decrease in osteogenesis and an increase in adipogenesis) were only evident at high concentrations.

Phytoestrogens can target Runx2 transcription factor and PPARγ, via binding and/or phosphorylation, resulting in the commitment of osteoprogenitor cells to differentiate into either osteoblasts or adipocytes.[61,63] Bone morphogenetic protein-2 (BMP-2), a potent inducer of osteogenic differentiation, is a target for estrogen and isoflavones.[64,65] Similar to estrogen, isoflavones increased the expression of mRNA encoding BMP-2 in mouse bone marrow mesenchymal stem cells, in C3H10T1/2 cells and in primary rat osteoblastic cells.[64,65] These effects are probably mediated by ERβ, and not ERα.[65] Further research on various gene expressions in osteoblasts, mesenchymal cells and endothelial cells during the early healing of puerarin in collagen matrix is needed to confirm the mechanism of its osteogenic effect.

The dose for puerarin in this study was estimated from a clinical study using puerarin injection,[66] by multiplying the dose for a human by the body mass ratio between rabbit and human. Using a pilot study, this estimated dose caused an increase in bone formation. In this study we have tried to demonstrate the osteogenic effect of puerarin using one commonly used carrier for bone repair. Further research is needed to determine the optimum dose, the best choice of carrier, and the release kinetics of puerarin in different carriers for bone grafting.

The significance of this study is that puerarin is a commonly used health supplement in Asia. It gained interest in research as a result of its multiple health maintenance (promoting) effects. This is the first study that demonstrated its local osteogenic effect. Puerarin has tremendous potential to be utilized in any application that requires an increase in bone formation. Further studies are needed to determine the systemic effect of puerarin on bone metabolism, for example on the prevention of osteoporosis.

14.6 Conclusions

The DBM_{IM}, the HMGRI-collagen matrix, the polymethoxylated flavonoid-collagen matrix and the phytoestrogen-collagen matrix have the effect of increasing new bone formation locally and can be used for bone regeneration, bone grafting or bone induction. Examples include repair of congenital bone defect as in cleft lip and palate; bone reconstruction after surgery, such as excision of tumours; trauma such as bone fracture; bone loss caused by other pathology such as periodontal disease; promotion of bone healing in compromised patients such as after radiotherapy, and diabetes mellitus or old age. Further research is needed to optimize their use and to gain further understanding of their bone-forming mechanism. In addition, they can be combined with other bone engineering techniques such as cell therapy, gene therapy or distraction osteogenesis, or other bone reconstruction materials such as calcium phosphate and hydroxyapatites, to optimize the treatment effects.

14.7 References

1. J.M. Lane, E. Tomin and M.P.G. Bostrom: Biosynthetic bone grafting. *Clin Orthop*, 367S (1999), S107.
2. E.M. Younger and M.W. Chapman: Morbidity at bone graft donor sites. *J Orthop Trauma*, 3 (1989), 192.
3. J.E. Zins and L.A. Whitaker: Membranous versus endochondral bone: implications for craniofacial reconstruction. *Plast Reconst Surg*, 72 (1983), 778.

4. D.M. Strong, G.E. Friedlaender, W.W. Tomford, D.S. Springfield, T.C. Shives, H. Burchardt, W.F. Enneking and H.J. Mankin: Immunologic responses in human recipients of osseous and osteochondral allografts. *Clin Orthop*, 326 (1996), 107.
5. A.B.M. Rabie, Y.M. Deng, N. Samman and U. Hägg: The effect of demineralized bone matrix on the healing of intramembranous bone grafts in rabbit skull defects. *J Dent Res*, 75(4) (1996), 1045.
6. A.B.M. Rabie and K.J. Lie: Integration of endochondral bone grafts in the presence of demineralized bone matrix. *Int J Oral Maxillofac Surg*, 25 (1996), 311.
7. A.B.M. Rabie, R.W.K. Wong and U. Hägg: Bone induction using autogenous bone mixed with demineralised bone matrices. *Aust J Orthod*, 15 (1999), 269.
8. A.B.M. Rabie and S.H. Chay: Clinical applications of composite intramembranous bone grafts. *Am J Orthod Dentofac Orthop*, 117 (2000), 375.
9. A.B.M. Rabie, R.W.K. Wong and U. Hägg: Composite autogenous bone and demineralised bone matrices used to repair defects in the parietal bone of rabbits. *Br J Oral Maxillofac Surg*, 38 (2000), 565.
10. M.R. Urist: Bone: formation by autoinduction. *Science*, 150 (1965), 839.
11. M.R. Urist, Y.K. Huo, A.G. Brownell *et al.*: Purification of bovine bone morphogenetic protein by hydroxyapatite chromatography. *Proc Natl Acad Sci USA*, 81 (1984), 371.
12. J. M. Wozney and V. Rosen: Bone morphogenetic protein and bone morphogenetic gene family in bone formation and repair. *Clin Orthop*, 346 (1998), 26.
13. T. Sakou: Bone morphogenetic proteins: from basic studies to clinical approaches. *Bone*, 22 (1998), 591.
14. E.A. Wang, D.I. Israel, S. Kelly *et al.*: Bone morphogenetic protein-2 causes commitment and differentiation in C3H10T1/2 and 3T3 cells. *Growth Factors*, 9 (1993), 57.
15. E.E. Johnson, M.R. Urist and G.A. Finerman: Resistant nonunions and partial or complete segmental defects of long bones. Treatment with implants of a composite of human bone morphogenetic protein (BMP) and autolyzed, antigen-extracted, allogeneic (AAA) bone. *Clin Orthop*, 277 (1992), 229.
16. R.W.K. Wong and A.B.M. Rabie: A quantitative assessment of the healing of intramembranous and endochondral autogenous bone grafts. *Eur J Orthod*, 21 (1999), 119.
17. B.L. Eppley, M. Doucet, D.T. Connolly and J. Feder: Enhancement of angiogenesis by bFGF in mandibular bone graft healing in the rabbit. *J Oral Maxillofac Surg*, 46 (1988), 391.
18. J.F. Kusiak, J.E. Zins and L.A. Whitaker: The early revascularization of membranous bone. *Plast Reconstr Surg*, 76 (1985), 510.
19. A.B.M. Rabie: Vascular endothelial growth pattern during demineralized bone matrix induced osteogenesis. *Con Tis Res*, 36 (1998), 337.
20. H.H. Hagenmeier, E. Volmer, S. Goerdt, K. Schulze-Osthoft and C. Sorg: A monoclonal antibody reacting with endothelial cells of budding capillaries in tumors and inflammatory tissues, and non-reactive with normal adult tissues. *Int J Cancer*, 38 (1986), 481.
21. K.M. Joos and A. Sandra: Microarterial synthetic graft repair: interstitial cellular components. *Microsurg*, 11 (1990), 268.
22. C.K. Scott, S.D. Bain and J.A. Hightower: Intramembranous bone matrix is osteoinductive. *Ana Res*, 238 (1994), 23.

23. J.M. Wozney, V. Rosen and A.J. Celeste: Novel regulators of bone formation: molecular clones and activities. *Science*, 242 (1988), 1528.
24. M.R. Urist, R. Granstein, H. Nogami, L. Stevenson and R. Murphy: Transmembrane bone morphogenesis across multiple-walled chambers: new evidence of a diffusible bone morphogenetic property. *Arch Surg*, 112 (1977), 612.
25. A.B.M. Rabie, D. Zhou and N. Samman: Ultrastructural identification of cells involved in the healing of intramembranous and endochondral bones. *Int J Oral and Maxillofac Surg*, 25 (1996), 383.
26. P.J. Polverini: The pathophysiology of angiogenesis. *Crit Rev Oral Biol Med*, 6 (1995), 230.
27. L. Liaw and S.M. Schwartz: Comparison of gene expression in bovine aortic endothelium in vivo versus in vitro. *Arteriosclerosis Thrombosis*, 13 (1993), 985.
28. M.M. Sholley, G.P. Fergusen, H.R. Seibel, J.L. Montour and J.D. Wilson: Mechanisms of neovascularization. Vascular sprouting can occur without proliferation of endothelial cells. *Lab Invest*, 51 (1984), 624.
29. G. Mundy, R. Garrett, S. Harris *et al.*: Stimulation of bone formation in vitro and in rodents by statins. *Science*, 286 (1999), 1946.
30. M. Sugiyama, T. Kodama and K. Konishi: Compactin and simvastatin, but not pravastatin, induce bone morphogenetic protein-2 in human osteosarcoma cells. *Biochem Biophys Res Comm*, 271 (2000), 688.
31. P.S. Wang, D.H. Solomon and H. Mogun: HMG-CoA reductase inhibitors and the risk of hip fractures in elderly patients. *J Am Med Assoc*, 283 (2000), 3211.
32. B. Skoglund, C. Forslund and P. Aspenberg: Simvastatin improves fracture healing in mice. *J Bone Miner Res*, 17 (2002), 2004.
33. R.W.K. Wong and A.B.M. Rabie: Statin collagen grafts used to repair defects in the parietal bone of rabbits. *Br J Oral Maxillofac Surg*, 41 (2003), 244.
34. S.H. Chay, A.B.M. Rabie and A. Itthagarun: Ultrastructural identification of cells involved in the healing of intramembranous bone grafts in both the presence and absence of demineralised intramembranous bone matrix. *Aust Orthod J*, 16 (2000), 88.
35. R.W.K. Wong and A.B.M. Wong: Early healing pattern of statin-induced osteogenesis. *Br J Oral Maxillofac Surg*, 43 (2005), 46.
36. D.W. Sifton: Zocor. In: D.W. Sifton (editor), *Physicians' Desk Reference*. Montvale, NJ: Medical Economics Company (2000), p. 1917.
37. Y.W. Shin, S.H. Bok and T.S. Jeong: Hypocholesterolaemic effect of naringin associated with hepatic cholesterol regulating enzyme changes in rats. *Int J Vit Nutr Res*, 69 (1999), 341.
38. R.W.K. Wong and A.B.M. Rabie: Effect of naringin collagen graft on bone formation. *Biomaterials*, 27 (2006), 1824.
39. H.J. Kim, G.T. Oh, Y.B. Park, M.K. Lee, H.J. Seo and M.S. Choi: Naringin alters the cholesterol biosynthesis and antioxidant enzyme activities in LDL receptor-knockout mice under cholesterol fed condition. *Life Sci*, 74 (2004), 1621.
40. R.T. Turner, B.L. Riggs and T.C. Spelsberg: Skeletal effects of estrogen. *Endocr Rev*, 15 (1994), 275.
41. J. Chow, J.H. Tobias, K.W. Colston and T.J. Chambers: Estrogen maintains trabecular bone volume in rats not only by suppression of bone-resorption but also by stimulation of bone-formation. *J Clin Invest*, 89 (1992), 74.

42. Q. Qu, M. Perala-Heape, A. Kapanen, J. Dahllund, J. Salo, H.K. Vaananen and P. Harkonen: Estrogen enhances differentiation of osteoblasts in mouse bone marrow culture. *Bone*, 22 (1998), 201.
43. T. Takano-Yamamoto and G.A. Rodan: Direct effects of 17-β-estradiol on trabecular bone in ovariectomized rats. *Proc Natl Acad Sci USA*, 21 (1990), 2172.
44. Q. Qu, P.L. Harkonen, J. Monkkonen and H.K. Vaananen: Conditioned medium of estrogen-treated osteoblasts inhibits osteoclast maturation and function in vitro. *Bone*, 25 (1999), 211.
45. F. Branca: Dietary phyto-oestrogens and bone health. *Proc Nutr Soc*, 62 (2003), 877.
46. S.J. Bhathena and M.T. Velasquez: Beneficial role of dietary phytoestrogens in obesity and diabetes. *Am J Clin Nutr*, 76 (2002), 1191.
47. A.M. Duncan, W.R. Phipps and M.S. Kurzer: Phyto-oestrogens. *Best Pract Res Clin Endocrinol Metab*, 17 (2003), 253.
48. S. Kanno, S. Hirano and F. Kayama: Effects of phytoestrogens and environmental estrogens on osteoblastic differentiation in MC3T3-E1 cells. *Toxicol*, 196 (2004), 137.
49. F.S. Chueh, C.P. Chang, C.C. Chio and M.T. Lin: Puerarin acts through brain serotonergic mechanisms to induce thermal effects. *J Pharmacological Sci*, 96 (2004), 420.
50. X. Wang, J. Wu, H. Chiba, K. Umegaki, K. Yamada and Y. Ishimi: Puerariae radix prevents bone loss in ovariectomized mice. *J Bone Miner Metab*, 21 (2003), 268.
51. D.H. Overstreet, Y.W. Lee, A.H. Rezvani, Y.H. Pei, H.E. Criswell and D.S. Janowsky: Suppression of alcohol intake after administration of the Chinese herbal medicine, NPI-028, and its derivatives. *Alcohol Clin Exp Res*, 20 (1996), 221.
52. L.Y. Wang, A.P. Zhao and X.S. Chai: Effects of puerarin on cat vascular smooth muscle in vitro. *Acta Pharmacol Sinica*, 15 (1994), 180.
53. R.W. Jiang, K.M. Lau, H.M. Lam, W.S. Yam, L.K. Leung, K.L. Choi, M.M. Waye, T.C. Mak, K.S. Woo and K.P. Fung: A comparative study on aqueous root extracts of *Pueraria thomsonii* and *Pueraria lobata* by antioxidant assay and HPLC fingerprint analysis. *J Ethnopharmacol*, 96 (2005), 133.
54. R.W.K. Wong and A.B.M. Rabie: Effect of puerarin on bone formation. *Osteoarthritis and Cartilage*, 15 (2007), 894.
55. Z.C. Dang, V. Audinot, S.E. Papapoulos, J.A. Boutin and C.W.G.M. Löwik: Peroxisome proliferator-activated receptor γ (PPARγ) as a molecular target for the soy phytoestrogen genistein. *J Biol Chem*, 278 (2003), 962.
56. Z.C. Dang and C.W. Löwik: The balance between concurrent activation of ERs and PPARs determines daidzein-induced osteogenesis and adipogenesis. *J Bone Miner Res*, 19 (2004), 853.
57. V. Viereck, C. Gründker, S. Blaschke, H. Siggelkow, G. Emons and L.C. Hofbauer: Phytoestrogen genistein stimulates the production of osteoprotegerin by human trabecular osteoblasts. *J Cell Biochem*, 84 (2002), 725.
58. K.D. Setchell and E. Lydeking-Olsen: Dietary phytoestrogens and their effect on bone: evidence from in vitro and in vivo, human observational, and dietary intervention studies. *Am J Clin Nutr*, 78 (2003), 593S.
59. K. Pettersson, F. Delaunay and J.A. Gustafsson: Estrogen receptor β acts as a dominant regulator of estrogen signaling. *Oncogene*, 19 (2000), 4970.

60. J. An, C. Tzagarakis-Foster, T.C. Scharschmidt, N. Lomri and D.C. Leitman: Estrogen receptor β-selective transcriptional activity and recruitment of coregulators by phytoestrogens. *J Biol Chem*, 276 (2001), 17808.
61. R. Agarwal: Cell signaling and regulators of cell cycle as molecular targets for prostate cancer prevention by dietary agents. *Biochem Pharmacol*, 60 (2000), 1051.
62. S. Barnes: Soy isoflavones-phytoestrogens and what else? *J Nutr*, 134 (2004), 1225S.
63. T. Akune, S. Ohba, S. Kamekura, M. Yamaguchi, U.I., Chung, N. Kubta, Y. Terauchi, H. Yoshifumi, Y. Azuma, K. Nakamura, T. Kadowaki and H. Kawaguchi: PPAR insufficiency enhances osteogenesis through osteoblast formation from bone marrow progenitors. *J Clin Invest*, 113 (2004), 846.
64. T.L. Jia, H.Z. Wang, L.P. Xie, X.Y. Wang and R.Q. Zhang: Daidzein enhances osteoblast growth that may be mediated by increased bone morphogenetic protein (BMP) production. *Biochem Pharmacol*, 65 (2003), 709.
65. S. Zhou, G. Turgeman, S.E. Harris, D.C. Leitman, B.S. Komm, P.V.N. Bodine and D. Gazit: Estrogens activate bone morphogenetic protein-2 gene transcription in mouse mesenchymal stem cells. *Mol Endocrinol*, 17 (2003), 56.
66. X. Chen: The clinical observation of puerarin injection on unstable angina pectoris. *Zhongyaocai*, 27 (2004), 77.

14.8 Appendix

Preparation of rabbit DBM$_{IM}$ particles

Fresh IM bone was excised from the mandibles and the parietal bones of New Zealand White rabbits, being careful to avoid the sutures and condyles to prevent endochondral contamination. Bone marrow, soft tissue and periosteum were removed and then the bones were cleaned in phosphate-buffered saline with protease inhibitors and defatted in diethyl ether. Cleaned bones were frozen in liquid nitrogen and ground using a pre-cooled analytical mill (IKA Labortechnik); the resulting powder was sieved to give a particle size of ≤0.25 mm. The bone powder was demineralized following a modified protocol of Reddi and Huggins.[67] To each gram of bone powder 50 mL of 0.5 M HCl (Merk, Germany) plus protease inhibitors were added and the resulting suspension was dialysed (Visking tubing, size 5: MW 12 kda, Medicell Internal Ltd, UK) against 0.5 M HCl for 3 hours and distilled in H_2O overnight to remove calcium hydroxyapatite from the bone matrix. The resulting powder was then dried before use.

Preparation of rabbit skin collagen

The collagen was prepared from the New Zealand White rabbit skin. The skin was cleaned with phosphate-buffered saline to remove soft tissue, minced and treated with a series of sodium chloride solutions, distilled water and 4% acetic acid, then lyophilized and dried before use.

Preparation of simvastatin solution

Zocor® tablet, 10 mg simvastatin, Merck & Co. Inc., NJ, dissolved in water for injection to the concentration of 2.5 mg/mL.

Preparation of naringin solution

Naringin, Sigma-Aldrich, MO, dissolved in water for injection to the concentration of 100 mg/mL.

Preparation of puerarin solution

Puerarin, Sigma-Aldrich, MO, dissolved in water for injection to the concentration of 100 mg/mL.

Rabbit model

The rabbit model used in this study was relevant because non-grafted control bone defects have been found not to heal with new bone formation within 14 days after their creation. In addition, there was minimal bone healing across the defect with collagen matrix as shown by the results of the control group. There was minimal morbidity due to this procedure as all the rabbits were in good health and condition after the surgery. Two weeks was chosen to examine the bone formation during the early healing of the bone defect, based on another study.[6,7] It gave better indication of the ability of new bone to grow across the bone defect. It was also the time span chosen for other studies on bone formation using the same animal model[6,7,9] so that comparisons can be made.

Collagen matrix (purified absorbable fibrillar collagen) was used in this study because it had been used successfully as a carrier for growth factors like BMP-2 to induce bone formation in animals and in humans.[68-70] It was derived from bovine tendon in the fibrillar form and was suggested from the Manufacturer (Collagen Matrix, Inc., NJ) to be useful for delivering cells and growth factors and for gene therapy. Recently it was successfully used with rhBMP-2 in the repair of alveolar clefts in humans.[71] Bouxsein and coworkers[72] assessed the retention time of ^{125}IrhBMP-2 in absorbable collagen sponge using gamma scintigraphy and showed that about 37% (±10%) of the initial dose remained at the site one week after surgery, and 8% (±7%) remained after two weeks. It is possible that the phytochemicals could have been retained similarly by the collagen matrix carrier and released over time to exert their effect that led to an increase in bone formation.

References

1. A.H. Reddi and C. Huggins: Biochemical sequences in the transformation of normal fibroblasts in adolescent rats. *Proc Natl Acad Sci USA*, 69 (1972), 1601.
2. A.B.M. Rabie and K.J. Lie: Integration of endochondral bone grafts in the presence of demineralized bone matrix. *Int J Oral Maxillofac Surg*, 25 (1996), 311.
3. A.B.M. Rabie, R.W.K. Wong and U. Hägg: Bone induction using autogenous bone mixed with demineralised bone matrices. *Aust J Orthod*, 15 (1999), 269.
4. A.B.M. Rabie, R.W.K. Wong and U. Hägg: Composite autogenous bone and demineralised bone matrices used to repair defects in the parietal bone of rabbits. *Br J Oral Maxillofac Surg*, 38 (2000), 565.
5. M. Nevins, C. Kirker-Head and M. Nevins: Bone formation in the goat maxillary sinus induced by absorbable collagen sponge implants impregnated with re-combinant human bone morphogenetic protein-2. *Int J Periodont Rest Dent*, 16 (1996), 9.
6. S.T. Li: Collagen as a delivery vehicle for bone morphogenetic protein (BMP). *Trans Orthop Res Soc*, 21 (1996), 647.
7. P.J. Boyne, R.E. Marx and M. Nevins: A feasibility study evaluating rhBMP-2/absorbable collagen sponge for maxillary sinus floor augmentation. *Int J Periodont Rest Dent*, 17 (1997), 11.
8. M. Chin, T. Ng, W.K. Torn and M. Carstens: Repair of alveolar clefts with recombinant human bone morphogenetic protein (rhBMP-2) in patients with clefts. *J Craniofac Surg*, 5 (2005), 778.
9. M.L. Bouxsein, T.J. Turek and C.A. Blake: Recombinant human bone morphogenetic protein-2 accelerates healing in a rabbit ulnar osteotomy model. *J Bone Joint Surg Am*, 83A (2001), 1219.

15
Cellular response to hydroxyapatite and Bioglass® in tissue engineering and regenerative medicine

J HUANG, University College London, UK

Abstract: This chapter discusses the biological responses of various human cells to surface reactive Bioglass®, hydroxyapatite (HA) and their composites for their potential applications in tissue engineering and regenerative medicine. Cytotoxicity and inflammatory response to Bioglass®, HA and substituted HA particles were investigated at first by indirect and direct contact tests using human osteoblast (HOB) and macrophage cell models; the attachment and proliferation of HOB cells in direct contact with nanometre-sized HA (nanoHA) and substituted HA were then compared. It was found that a high concentration of Bioglass® and nanoHA particles can reduce cell viability, but no cytotoxic effect was observed from their biocomposites where these bioactive particles were incorporated into polyethylene or polyhydroxyethylmethacrylate matrix, and a direct bonding between HOB cells and the composite was observed from electron microscopy examination. Release of Si from Bioglass® and the composite was found to increase the proliferation of HOB cells at a certain level. Therefore, incorporation of Si into the structure of HA to formulate silicon substituted HA (SiHA) is an attractive way to control release of Si. The optimum content of Si in nanoSiHA found in the study will be useful for its further applications.

Key words: Bioglass®, substituted hydroxyapatite, bioactive composite, human osteoblast, macrophage.

15.1 Introduction

Hydroxyapatite (HA) is the major inorganic constituent in bone, and synthetic HA has been successfully used as a bone graft and coatings on metallic implants in dental and orthopaedic applications due to its well-known biocompatibility, bioactivity and osteoconductivity [1]. Bioactive glasses, particularly Bioglass®, have been developed and investigated for a wide range of biomedical applications, more recently for use in tissue engineering and regenerative medicine [2, 3]. An important feature of bioactive glass and glass ceramics is the formation of a biologically active hydroxycarbonate apatite (HCA) layer on the surface both *in vitro* and *in vivo*.

This HCA phase is chemically and structurally equivalent to the mineral phase in bone, and it provides a direct bonding by bridging host tissue with implants, which results in the bioactivity of the materials. The desirable chemical reactivity of a bioactive material is conceptually different from the relative inertness possessed by conventional material, and is critical for the formation of direct bonding or replacement by host tissue. However, the control of the reactivity of bioactive material is highly important for the success of such implants.

A wide range of bioactive glasses and ceramics have been incorporated into bioinert polymeric matrix to formulate bioactive composites [4–8]. This is a logical approach from the point of view of matching the structure and properties of tissue to be replaced. At the ultramicrostructural level, bone can be considered as apatite or bone mineral reinforced collagen composite, where bone mineral crystals at the nanometre scale are embedded in the collagen matrix [4]. It is expected that the reactivity of these bioactive materials will provide and promote the direct bonding with host tissue and thus contribute to the functional longevity for the tissue repair and regeneration. A systematic investigation of these composites and their components is therefore required to test these hypotheses. In the development of a bioactive filler reinforced composite, optimizing the structure and biological response of the filler particles is essential as the structure of fillers (size, composition, crystallinity) can influence cell viability/activity [9, 10]. Recent studies have shown that the incorporation of nano-sized hydroxyapatite improved the mechanical properties and protein absorption of the composite [11, 12], thus encouraging various attempts at formulation of new bioactive nanocomposites. These nanocomposites closely resemble the bone structure, but their properties, particularly the releasing of nanoparticles to the biological systems, require thorough investigation.

In vitro cell culture, allowing the biological assessment of materials at a cellular level, is becoming increasingly useful for the evaluation of a new biomaterial, as it is possible to test any toxic effect on human cells. In this chapter, the biological responses of various human cells, such as primary human osteoblast (HOB) cells and human monocyte-derived macrophages, to surface reactive Bioglass®, substituted hydroxyapatite and their composites will be discussed.

15.2 Cellular responses to Bioglass® and hydroxyapatite particles

Bioglass® and HA are well known for their excellent bioactivity, but their relatively poor mechanical properties restrict their applications in the bulk form. The bioactive glass and ceramics particles have been found to have applications as reinforcement fillers in formulating bioactive composites with desirable properties; in particular, the nanocomposites (using nanoparticle fillers) have showed promising properties and therefore have attracted much attention recently. However, understanding of the biological responses to the bioactive particles in the micrometre to nanometre range is relatively limited. To build up a comprehensive knowledge of cellular responses of these particles, both direct and indirect testing and evaluations are required. In the direct testing, the test materials are brought into direct contact with the cells, i.e. macrophages. The damage to the cell membrane as the result of the contact is the sign of cytotoxic effects. In the indirect testing, cytotoxicity is measured by the exposure of the cells to the aqueous extract of the test materials for given time periods. Any potential toxic leachable into the surrounding environment/medium can therefore be detected. The assay is based either on the use of vital dyes, or by measuring the action of intracellular enzymes on tetrazolium salts, such as MTT.

15.2.1 Indirect test

To simulate the physiological environment and detect any potential toxic leachable from hydroxyapatite and Bioglass® particles, aqueous extracts from Bioglass® 45S5 (US Biomaterials) and HA powder (Plasma Biotal, UK) were made by adding 1 g of each powder into a sterile universal containing 10 ml ascorbate-free DMEM medium and rolling the container at 37°C for a certain period (i.e. 24 and 120 hours). The median particle sizes (d_{50}) of L-, SL-, S- and SS-Bioglass® and HA powder were 70, 35, 17, 6 and 5 µm respectively, measured from a Malvern Mastersizer X particle size analyser (detailed in Table 15.1).

Owing to its sensitivity, the MTT assay was chosen to detect the toxic leachables from the elution. This assay measures intracellular mitochondrial activity of the cells, which involves reduction of MTT (3-(4,5-dimethylthiazol-2-yl)-2,5-diphenyltetrazolium bromide) by intracellular dehydrogenase enzymes of viable cells to a blue formazan with an optical density of 570 nm. The number of surviving cells is directly proportional to the level of the formazan produced, which can be quantified using a plate reader.

Table 15.1 Particle size of HA and Bioglass® powders

Powder	d_{50} (µm)	d_{10} (µm)	d_{90} (µm)
HA	4.5	1.4	14.7
L-Bioglass®	70	10.8	110
SL-Bioglass®	35	7.0	90
S-Bioglass®	17	3.5	38
SS-Bioglass®	5.5	1.4	24.7

Tissue culture plastic (Tcp) and polyvinylchloride (PVC) + 0.3% tin were used as negative (non-toxic) and positive (toxic) controls, respectively. Various dilutions of the eluted media from Bioglass® powder and PVC control were made to determine the toxic range. All extracts were sterilized by filtration though 0.2 µm cellulose acetate filters prior to use.

Primary human osteoblast (HOB) cells were isolated from trabecular bone fragments [13], seeded at a density of 1×10^4 cells/well in DMEM medium and incubated to confluency. The medium was replaced with the 100% eluted extract from test and control samples ($n = 16$). The assay was performed after the culture plates were incubated at 37°C in humidified air with 5% CO_2 for 24 and 72 hours.

No toxic leachable was detected from elution of HA powder, as the viability of HOB cells exposed to the 100% eluted medium from HA powder was in the same range as that from the non-toxic control Tcp. However, a low cell viability was found on the culture in the 100% extract from Bioglass® powder, and a reduction (20–34%) in cell viability was observed on the 50% extract medium (dilution factor of 0.5), but this toxic effect was removed by further dilution (Fig. 15.1). An increase in relative cell viability (compared with Tcp control) was found on the culture in the 25% extract medium (dilution factor of 0.25) from the SL and L powders (large particle size). A transition from toxic to non-toxic was observed in 12.5–25% extract media from the smallest size powder (SS). A dramatic increase in cell viabil-

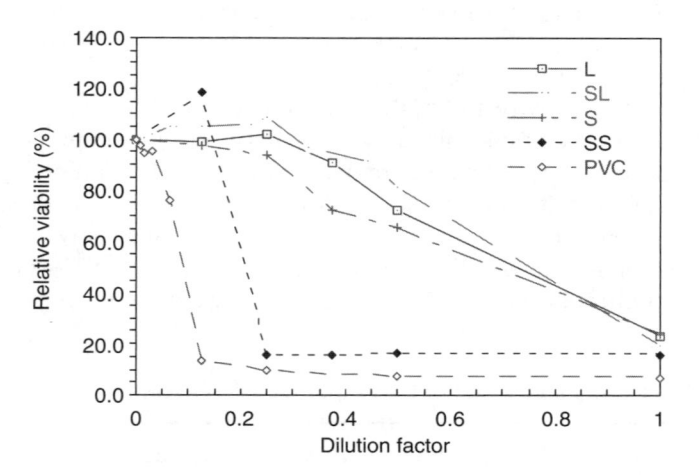

15.1 Relative viability (compared with Tcp control) of HOB cells following 24 hours' exposure to the extracts from Bioglass® powder and toxic control PVC. The median particle sizes of Bioglass® powders L-, SL-, S- and SS-Bioglass were 70, 35, 17 and 6 µm, respectively. A 'stimulatory' effect was observed in the elution from Bioglass® with a dilution factor from 0.25 to 0.125, depending on the particle size.

ity on the culture in the 12.5% extract medium indicated not only the non-toxicity of the material, but also the existence of a 'stimulatory' effect. For the toxic control PVC, the cells maintained 76% viability on the culture in the 6.25% extract, but no stimulatory effect was observed. In addition, an increase in HOB cell activity was found when incubated in eluted media for 72 hours, which indicated that the leachable released from Bioglass® powder could not be toxic for a longer exposure period.

15.2.2 Direct contact test

The toxic leachables from test materials can be detected from indirect testing. Cytotoxic responses of HOB cells were found to be related to the size of bioactive glass particles, but the micrometre-sized HA, the smallest size of particles in the test samples, was not cytotoxic to HOB cells. A nanometre-sized hydroxy-carbonate apatite (HCA) is known to form on the surface of Bioglass® both *in vitro* and *in vivo*. This phase is considered chemically and structurally equivalent to the mineral phase in bone, and provides a direct bonding to host tissue and bioactivity. Recently, nanometre-sized HA and calcium phosphate particles have been found applications in nanocomposites and non-viral gene delivery systems [11, 12, 14, 15]. Poor crystalline HA was found to stimulate the inflammatory cytokine release, and synthetic carbonate substituted HA (CHA), which more closely resembles bone minerals, is more soluble than HA. Therefore, the cellular response to nanosized HA and CHA when in direct contact with human cells needs to be further understood.

Human monocyte-derived macrophage

A human monocyte-derived macrophage model was used to study the cytotoxicity and inflammatory response of nanoHA and nanoCHA particles.

Cytotoxicity

Lactate dehydrogenase (LDH) is a cytosolic enzyme released when the cell membrane is damaged. It has been used to assess cell damage resulting from exposure to test substances [16]. The LDH released when human macrophages were in contact with nanoparticles, a marker indicating damage to the cell membrane, was determined as a measure of the cytotoxicity of nanoparticles.

NanoHA and nanoCHA with carbonate content of 1% and 2.5% (CHA1 and CHA2), respectively, were synthesized based on a precipitation reaction between calcium hydroxide ($Ca(OH)_2$) and orthophosphoric acid

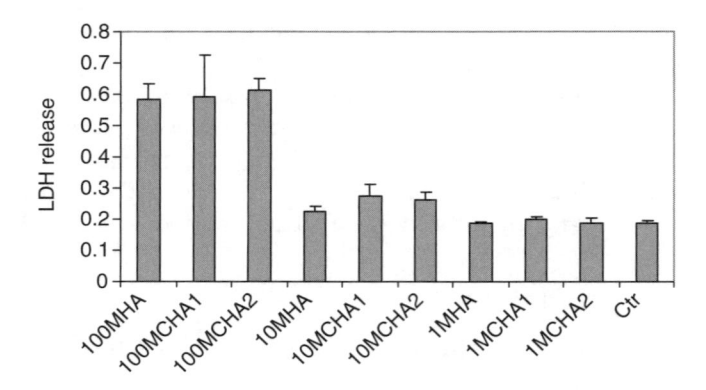

15.2 The release of LDH from human macrophage in contact with nanoHA, nanoCHA1 and nanoCHA2 particles with concentration from 1 to 100 million particles per test.

(H_3PO_4) (both AnalaR grade, BDH, UK). Carbonate substitution was achieved by dissolving CO_2 into the solution during the reaction process [17]. The concentration of nanoHA and nanoCHA particles in the culture medium was estimated by using a haemocytometer under a Leica microscope with frame grabber after staining nanoparticles with 1 wt% silver nitrate to increase the contrast.

NanoHA and nanoCHA particles with minimum numbers of 1 million to 100 million particles were added to each test well containing 0.5 million cells. After 24 hours of culturing, the release of LDH from the cells was measured (Cytotox 96, Promega, Southampton, UK). Macrophages not in contact with particles were used as the test control.

As shown in Fig. 15.2, there was a significant increase of LDH release from human macrophages when cells were in contact with high concentration of nanoHA and nanoCHA particles (100 million particles), indicating the cytotoxic effects, as the amount of enzyme activity correlates to the amount of cell membrane damage. There was no significant difference between nanoHA and nanoCHA. However, the levels of LDH release decreased significantly with a reduction in nanoparticle concentration from 100 million to 10 million and 1 million. No significant difference in LDH release was found between 10 million and 1 million particles with test control (no particles).

Inflammatory response

The inflammatory cytokine tumour necrosis factor alpha (TNF-α) is an important marker of the potential inflammatory response of cells in the presence of foreign particles [18, 19]. To measure the inflammatory response,

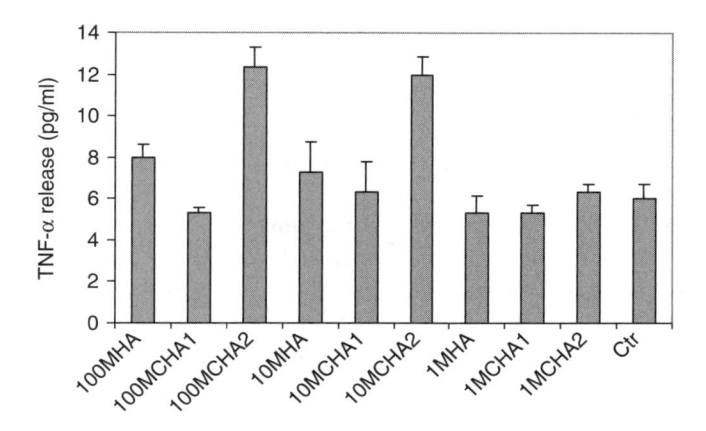

15.3 The release of TNF-α from human macrophage in contact with nanoHA, nanoCHA1 and nanoCHA2 particles with concentration from 1 to 100 million particles per test.

the release of the cytokine tumour necrosis factor alpha (TNF-α) from cells in the presence of nanoHA and nanoCHA particles was determined. Nanoparticles with minimum numbers of 1 million to 100 million particles were added to each test well containing 0.5 million cells. After 24 hours of culturing, the release of TNF-α was determined by an enzyme-linked immunosorbent assay (ELISA, Endogen, Woburn, MA).

The results showed that the level of cytokine TNF-α releases from human macrophages when in contact with nanoHA and nanoCHA1 particles was low, but there was a significant release of TNF-α when human macrophages were in contact with nanoCHA2 (Fig. 15.3), particularly at high concentrations (100 million and 10 million particles). At low concentration of nanoCHA2 (1 million particles), there was no significant release of TNF-α from the cells. It seems that the TNF-α release is correlated with the carbonate content, the higher carbonate content (2.5% in comparison with 1%), the greater the response.

Human osteoblast cells

For a bone replacement material, the osteoconductive potential is related to the biological responses from HOB cells, and much attention has been directed to the study of the behaviours of HOB cells, particularly the attachment, adhesion and spreading at the initial stage of cell/material interactions, as the quality of this interaction will influence subsequent cell proliferation and differentiation.

The extract from Bioglass® was found to be able to stimulate HOB cell activity and the dissolution products of bioactive glasses also had a positive effect on the expression of genes regulating osteogenesis [20, 21]. The structure of Bioglass® can be considered as a three-dimensional silica network, modified by incorporation of other oxides, (e.g. sodium, calcium and phosphorus oxides). Si is well known for its active roles in bone development [22], and soluble silicon was able to stimulate collagen type 1 synthesis in human osteoblast-like cells and enhance osteoblastic differentiation [23]. Formulation of silicon substituted hydroxyapatite (SiHA) by incorporation of Si into HA structure is an effective way to control the release of Si. It has recently been discovered that the *in vitro* and *in vivo* bioactivity of SiHA can be significantly improved [24–26], thus making SiHA an attractive alternative to conventional HA in bone replacement. In comparison with HA, dissolution on SiHA was more rapid, and the dissolution rate was increased with the concentration of Si content in SiHA [27]. The Si release as the result of dissolution plays an important part in the osteoconductivity of SiHA. The biological responses of HOB cells to nanoSiHA, in comparison with nanoHA and more resorbable nanoCHA, were studied in order to understand the role of Si in nanoSiHA.

NanoSiHA was chemically systhesized from the reaction between calcium hydroxide and phosphoric acid, similar to the production of nanoHA, with the addition of tetraethoxysilane as the source for the substitution of Si for P, and the content of Si in SiHA was 0.8 wt%, 1.2 wt% and 1.6 wt% respectively [28].

Cell proliferation

NanoSiHA, nanoHA and nanoCHA were deposited on glass discs, and the specimens were sterilized by dry heat at 600°C for 4 hours. The direct interaction of HOB cells with these nanoparticles as a function of time was determined using the alamarBlue™ assay (Serotec, Oxford, UK). This assay employs a redox indicator, which measures the response to chemical reduction of growth medium as a result of cell proliferation. The measured absorbance from alamarBlue™ assay is proportional to the cell density or cell number. An advantage of this assay is that it is not an end-point assay, thus it is very useful in monitoring a direct cellular response over a certain time period as required.

HOB cells (2×10^4) (Promocell, GmbH) were seeded and incubated at 37°C in a humidified air atmosphere of 5% CO_2. HOB cells cultured on tissue culture plastic (Tcp) were used as test controls. At days 2, 4 and 7, the reduction of alamarBlue™ during 4 hours of incubation with the cells was measured. The result showed that HOB cells were able to attach on the surfaces deposited with nanoHA, nanoCHA and nanoSiHA. There were more cells attached to the nanoSiHA surface in comparison to nanoHA and Tcp control at day 2 of culture; the HOB cell activity was also propor-

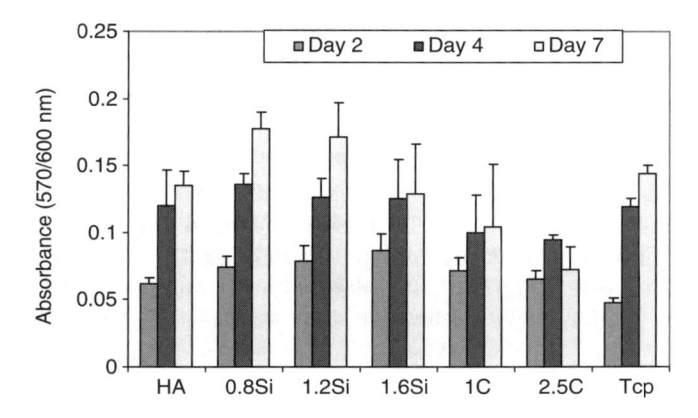

Cellular response to hydroxyapatite and Bioglass® 379

15.4 Metabolic activity of HOB cells in direct contact with nanoHA (HA), nanoSiHA (0.8Si, 1.2Si and 1.6Si) and nanoCHA (1C and 2.5C) during 7 days of culture measured from alamarBlue™ assay. A high cell activity was observed on nanoSiHA.

tional to the Si content in nanoSiHA; the activity on 0.8, 1.2 and 1.6 wt% nanoSiHA was significantly higher than that on nanoHA at day 2, indicating that nanoSiHA was able to promote cell attachment.

There was a significant increase in the absorbance with culture time, thus demonstrating that HOB cells were able to proliferate on the surfaces. The metabolic activity of HOB cells in direct contact with nanoSiHA increased with time during the 7 days of culture (Fig. 15.4) and in comparison to nanoHA the cell activity was significantly higher on 0.8 and 1.2 wt% nanoSiHA at day 7, but not on 1.6 wt% nanoSiHA, where a higher dissolution would be expected resulting from the high level of Si substitution.

In comparison, the initial cell attachment on nanoCHA was at the same level as that of nanoHA and more cells were attached than those on Tcp control, but the cell activity at day 4 and day 7 was lower than that on nanoHA. It is expected that the dissolution rate is higher on the nanosized ceramics particles, particularly on more resorbable carbonate substituted HA. The results showed that the chemistry of the materials influences their cellular response. The different responses observed in the study further demonstrate that the modifications of materials chemistry require careful controls and thorough investigation.

Immunofluorescence and scanning electron microscopy

Immunofluorescent staining of actin cytoskeletal proteins reveals the development of the cell cytoskeleton on the substrates, which will assist in the understanding of the quality of cell adhesion and cell/material interaction.

HOB cells (2×10^4 cells) (Promocell, GmbH) were cultured in direct contact with nanoparticles deposited on a glass surface and incubated at 37°C in a humidified air atmosphere of 5% CO_2. The development of actin cytoskeletal proteins of the HOB cells was observed by immunofluorescence study [29]. After three days in culture, the cells were fixed in 4% paraformaldehyde/PBS with 1% sucrose and permeabilized at 4°C for 5 minutes. The samples were incubated with 1% bovine serum albumin (BSA)/PBS at 37°C for 5 minutes to block the non-specific binding. This step was followed by the addition of FITC conjugated phalloidin (Sigma, Poole, UK) at 37°C for 1 hour. Nuclei were made visible by counterstaining with ToTo-3 (Molecular Probes Europe BV, Leiden, Netherlands) for 5 minutes. The samples were then mounted in a Vectorshield fluorescent mountant (Vector Laboratories, UK) and viewed by a Leica SP2 laser scanning confocal microscope.

Immunofluorescent staining of actin cytoskeletal proteins of HOB cells in contact with nanoparticles revealed the initial interactions with HOB cells. Cells could be seen to have actin filaments throughout the cells on all nanoHA, nanoSiHA and nanoCHA deposited surfaces, as shown in Fig. 15.5, where the actin stress fibres were more organized on the 0.8 wt% nanoSiHA surface.

The morphology of HOB cells in contact with nanoparticles was further studied by scanning electron microscopy. The cultures were fixed, stained with 1 wt% osmium tetroxide and dehydrated in a graduated series of alcohols and finally critical point dried (Polaron E3000 CPD). The sample surface was coated with a layer of carbon before it was examined using a JEOL 6340 field emission SEM.

SEM examination found that HOB cells were able to grow in the presence of nanoHA and nanoSiHA particles. HOB cells maintained the osteoblast morphology, with large areas of confluent cells after 7 days of culture, and fibre-like extracellular matrices (ECM) were produced (Fig. 15.6). There were larger amounts of ECM produced by the HOB cells on nanoSiHA deposits, the fibres were well organized, and plenty of spherical nodules were observed attached and/or embedded in the ECM. Energy dispersive X-ray (EDX) analysis revealed that these nanosized nodules contained Ca and P, which might be the calcification front for the further steps of bone mineralization.

In comparison, partial dissolution was found on nanoCHA and the changes of the morphology of nanoCHA particles were observed (Fig. 15.7); such high reactivity may affect the cell attachment and proliferation, thus partially explaining the low activity of HOB cells on nanoCHA at day 7 (from the alamarBlue™ assay).

The results obtained from the current study of cellular responses to Bioglass® and substituted HA further support the findings that Si plays a

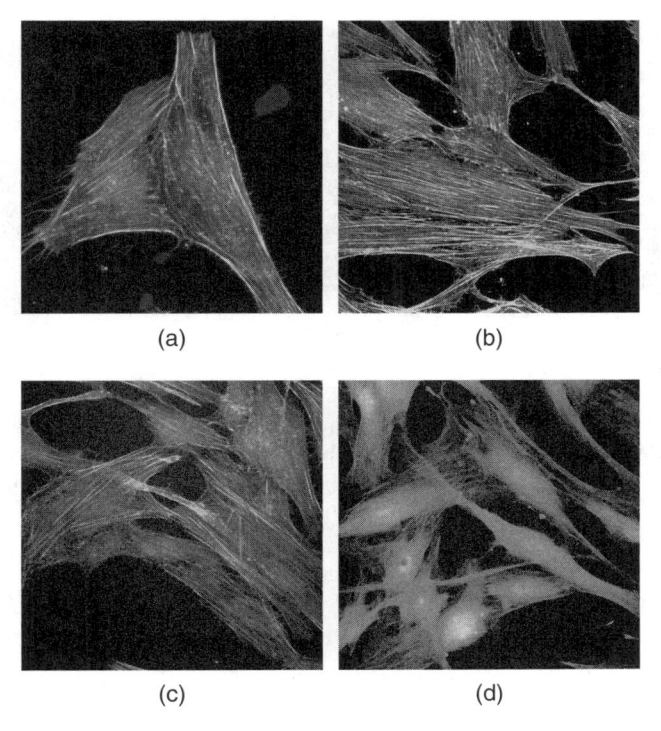

Cellular response to hydroxyapatite and Bioglass® 381

(a) (b)
(c) (d)

15.5 Actin cytoskeletal organization (green, labelled with phalloidin-FITC, counterstained with ToTo-3 for nuclei in blue) of HOB cells in contact with (a) nanoHA, (b) 0.8% nanoSiHA, (c) 1.6% nanoSiHA and (d) 1% nanoCHA, after 3 days of culture. The more organized actin stress fibres were observed on the 0.8% nanoSiHA surface.

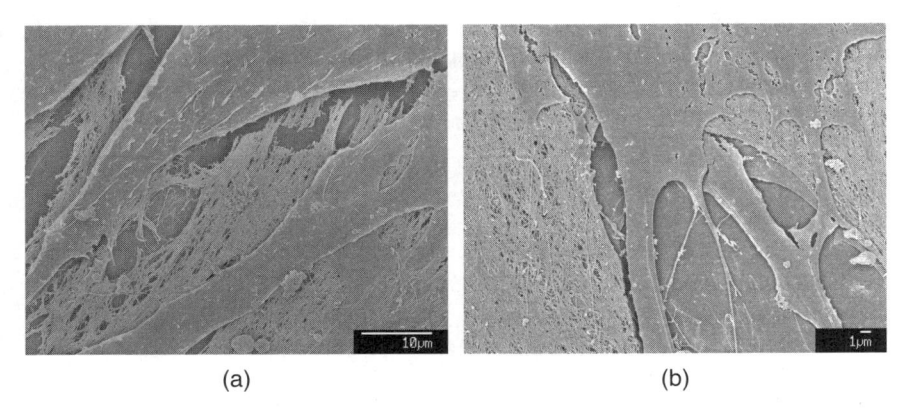

(a) (b)

15.6 SEM micrographs of HOB cells on (a) nanoHA and (b) nanoSiHA after 7 days of culture. Extracellular matrix was observed around the HOB cells.

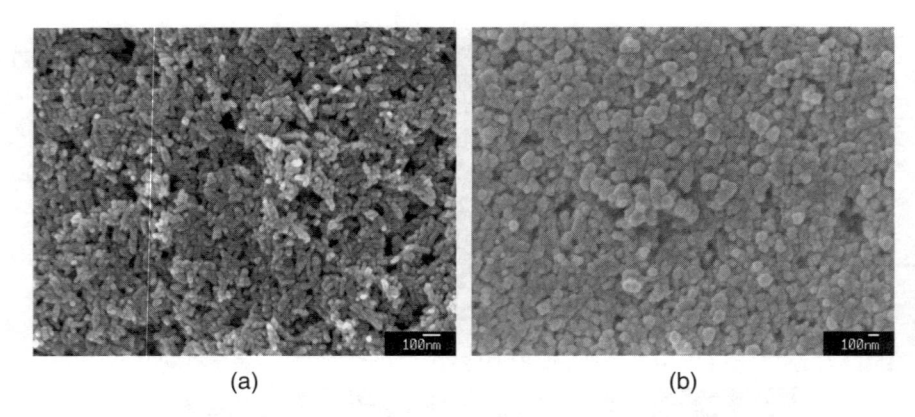

15.7 SEM micrographs of morphology of nanoCHA particles (a) before and (b) after 7 days of culture. The rod-like nanoCHA particles were changed to a more spherical shape, indicating their high surface reactivity.

critical role in the bone calcification process [30] and the incorporation of Si in SiHA is able to stimulate the HOB cell activity *in vitro* and enhance bone formation *in vivo* [24–27]. Incorporation of Si into the HA structure accelerated the dissolution process, which in turn contributed to the enhanced bioactivity of SiHA. The stimulatory effect was most likely correlated with the presence of Si, as it was only observed in the nanoSiHA deposited surface, not on more resorbable nanoCHA. At high Si concentration, 1.6 wt% nanoSiHA, the cell activity decreased, which might have resulted from the higher solubility observed, as a similar drop in cell activity was observed on soluble nanoCHA. It seems there is a threshold of the stimulatory effect: too much release of Si and other ions can have detrimental effect: as a transition from stimulating to reducing cell viability was also found from an elution study on Bioglass® particles. This indicates that the higher bioactivity is related to the concentration of Si. The level of Si required to stimulate bone cell activity and bone formation is still under further quantification, but an optimum composition of Si in nanoSiHA found in the study was between 0.8 and 1.2 wt%. The result demonstrated that the control of the chemistry of materials is important in determining the biological response.

15.3 Cellular responses to Bioglass® and hydroxyapatite reinforced biocomposites

Some cytotoxic effects were observed in the highly bioactive glass and nanoCHA particles, but the cytotoxicity could be reduced by dilution.

Therefore, an effective way to employ the bioactivity of these materials is to incorporate the stiff and bioactive particles into a tough and biocompatible matrix, thus reducing the risk of potential cytotoxicity dramatically. Extrusion and compression moulding were used to incorporate Bioglass® (BG) particles into polyethylene (PE) matrix [31] and produce Bioglass® reinforced polyethylene composite (BGPE). NanoCHA particles were mixed with 2-hydroxyethyl-methacrylate (HEMA, Aldrich) and polymerized at 80°C overnight to make nanoCHA/PHEMA composite.

15.3.1 Bioglass®/polyethylene composites

The specimens of unfilled polyethylene (PE), 20 vol% (a) and 40 vol% (b) BGPE composites were sterilized by gamma irradiation at a dose of 2.5 Mrad (Isotron, UK). HOB cells were seeded on the surface of gamma irradiated composites ($n = 3$) and incubated at 37°C in humidified air with 5% CO_2. Hydroxyapatite reinforced polyethylene composites, HAPEX™ [7] (HAPE), were used as test controls. Aqueous extracts from test composites were prepared by elution of one test specimen (10 × 10 mm) in a sterile universal containing 10 ml ascorbate-free DMEM medium at 37°C. All extracts were sterilized by filtration though 0.2 µm cellulose acetate filters prior to use.

No evidence of cytotoxicity was observed following incubation of cells in the eluted medium from HAPE and BGPE composites. Cell viability data from the MTT assay on 100% extract medium after 24 hours culture are presented in Fig. 15.8. None of the test materials released any 'toxic' leachables after 24 hours' elution. The cellular activity on HAPE was within the same range as in the non-toxic control Tcp, while a significant increase in cellular metabolic activity was found on both BGPEl (large particles) and BGPEs (small particles) composites ($p < 0.001$), particularly at higher filler contents (40 vol%), indicating the higher bioactivity of the composites. In contrast, the positive toxic control (PVC + 0.3% tin) showed considerable deterioration in cell viability.

The activity of HOB cells in direct contact with composites as a function of time was determined using the alamarBlue™ assay. HOB cells (5×10^4) were seeded and incubated at 37°C in a humidified air atmosphere of 5% CO_2 for up to 21 days. The alamarBlue™ assay showed that there was no cytotoxic effect when HOB cells were in direct contact with HAPE and BGPE composites. The metabolic activity of the cells on the test materials increased with time during the 21-day culture period (Fig. 15.9). A decrease in cell metabolic activity at day 2 was observed on all test materials, but an increase in cell activity was observed from day 4 to 21, with a higher activity seen on the HAPE and BGPEl (l = large particles) in comparison with BGPEs (s = small particles).

384 Cellular response to biomaterials

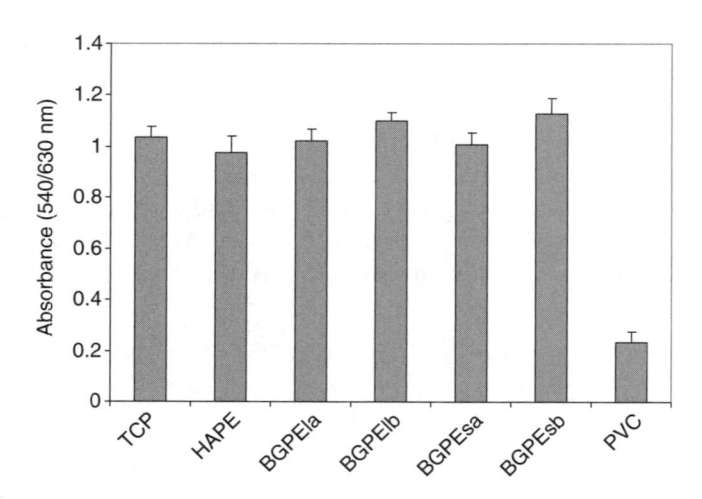

15.8 Viability of HOB cells following 24 hours' exposure to 24-hour eluted media from the BGPE composites and control materials, where (a) $V_p = 20\%$ and (b) $V_p = 40\%$. An increase in cellular activity was observed on both BGPEl (large particles) and BGPEs (small particles) composites, particularly at higher filler contents ($V_p = 40\%$).

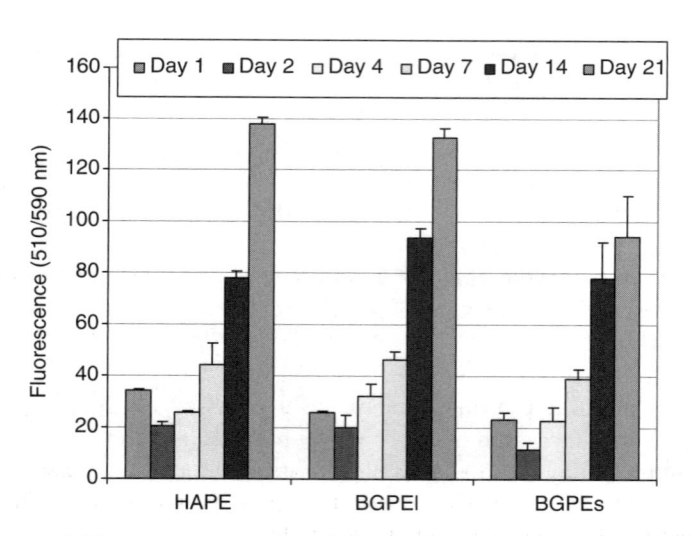

15.9 Metabolic activity of HOB cells on HAPE, BGPEl and BGPEs from day 1 to 21 measured from fluorescence in alamarBlue™ assay. An increase in cell activity with time was observed.

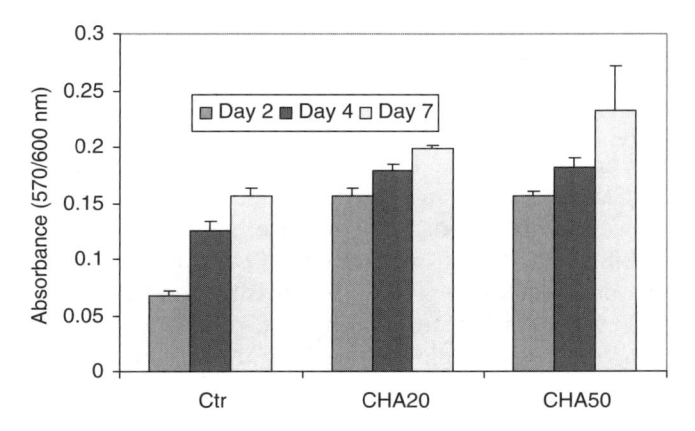

15.10 Metabolic activity of HOB cells on nanoCHA/PHEMA composites during 7 days of culture measured from alamarBlue™ assay. The cell activity on the composites was significantly higher than that on test controls.

15.3.2 NanoCHA/polyhydroxyethylmethacrylate composite

The attachment and growth of HOB cells on the 20 wt% and 50 wt% nanoCHA/PHEMA composites were also assessed using the alamarBlue™ assay. The measured absorbance showed that more cells were attached to the nanoCHA composite surfaces than on the test control at the earlier time point (day 2), as shown in Fig. 15.10. All the composites were able to support the growth and proliferation of HOB cells during 7 days of culture as the absorbance increased with culture time, thus indicating that the nanocomposites provided favourable surfaces for cell attachment and proliferation.

15.4 Interaction of human osteoblast (HOB) cells and biocomposites at the ultrastructural level

Bioglass® 45S5 (BG) has excellent bioactivity, and can bond directly to bone. It is expected that when highly bioactive BG particles are incorporated into a bioinert polyethylene (PE) matrix, the bioactivity will be retained in BGPE composites. The close examination of the interface between HOB cells and composite at ultra-microstructural level can reveal the interaction of HOB cells with the composite *in vitro*, thus providing further understanding of the biological response of cells to the composite *in vivo*. Transmission electron microscopy (TEM) was therefore used to study the interface of HOB cells with the composites after 6 weeks of *in vitro* culture.

386 Cellular response to biomaterials

HOB cells were seeded on the surface of the composites, as described earlier, and incubated at 37°C in humidified air with 5% CO_2 for 6 weeks. The cultures were then fixed, stained, dehydrated and embedded in the resin. Ultra-thin sections were produced by a diamond knife, stained with 2% aqueous uranyl acetate and Reynolds' lead citrate, and then examined on a Philips TEM-CM12 transmission electron microscope.

HOB cells were able to grow and produce extracelluar matrix on the surfaces of unfilled PE during the 6 weeks of culture, demonstrating that PE was biocompatible. However, it was not possible to examine the cell–PE interface as PE was detached from embedded resin during the sectioning stage, indicating the poor bonding between HOB cells and PE. In contrast, the multi-layers of HOB cells were retained on the HAPE and BGPE composite with their interface remaining intact during the whole period of processing for TEM examination (Fig. 15.11). An electron dense layer (dark region) was found around the BG particles of the composite. EDX analysis showed that the layer was rich in Ca and P. When examining the interface

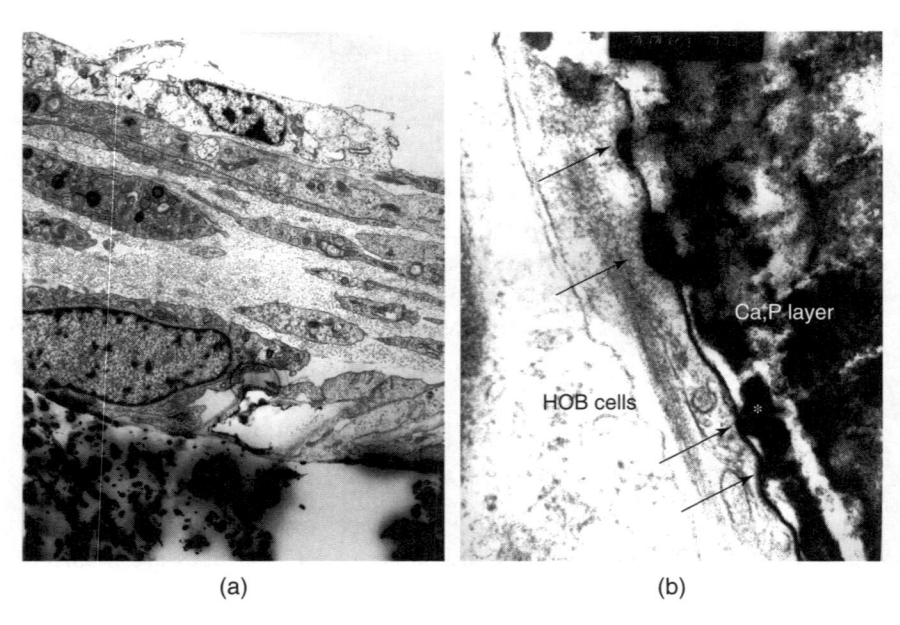

(a) (b)

15.11 TEM micrographs showing the interface between (a) HOB cells and HAPE composite, and (b) HOB cells and BGPE composite. A direct contact and firm bonding was found between HOB cells and composites, while this was absent from unfilled PE. An electron-dense layer (rich in Ca and P from EDX) was found around the BG particles in the BGPE composite, and intimate contacts (*) with HOB cells (arrows) were observed, thus showing the origin of the direct bonding between composite and cells.

between HOB cells and composites at high magnification, an intimate contact was found between this Ca, P layer with HOB cells (Fig. 15.11b), thus showing the origin of the strong bonding between composite and cells.

A hydroxy-carbonate apatite (HCA) layer was formed on the BG particles in the composite after immersion in SBF [8], thus indicating that the HCA formed in the composite provided sites favourable for cell attachment leading to the solid bonding between cells and composite. PE is biocompatible but a bioinert material; by introducing highly bioactive Bioglass® 45S5 particles into PE matrix, the bioactivity of Bioglass® was retained in the BGPE composites. The difference in bonding with HOB cells between the bioinert unfilled PE and bioactive BGPE and HAPE composite was clearly demonstrated from TEM investigation.

15.5 Summary

With recent advances in science and technology, particularly nanotechnology, a range of novel biomaterials has emerged. The traditional approach has changed from matching the properties of tissue to be replaced at macroscopic level to microscopic level and further down to recreating nanoscale details. This study has demonstrated the importance of the biological evaluation of cellular responses to a new material. Both indirect and direct tests are needed to establish the understanding of its potential medical applications.

Bioactive materials, such as Bioglass® 45S5, are highly surface reactive, and the release of ions, such as Si and Ca, into the surrounding environment has been reported to stimulate the biological responses [2], thus leading to formation of direct bonding with bone through a series of surface reactions. Carbonate substituted HA (CHA) more closely resembles bone minerals but is more soluble than HA, particularly at low crystallinity. The high surface activity can also have a detrimental effect as a result of high ion release and changes in local pH. Incorporation of bioactive glass and ceramics into a polymer matrix is an attractive way to effectively utilize their high bioactivity potential. No cytotoxic effect was observed on Bioglass®/PE and nanoCHA/PHEMA composites, and high-resolution TEM examination of the cell–material interface further demonstrated the strong bonding between HOB cells and the bioactive composites.

The role of Si in bone formation and calcification has long been recognized [20]. Silicon substituted HA (SiHA), incorporating Si into the HA structure, is an attractive way to control the release of Si. In this chapter, the stimulatory effect of Si on the cellular response was found not only in Bioglass® and the composites, but also in nanoSiHA with Si content in the range of 0.8 to 1.2 wt%, thus demonstrating that nanoSiHA is a very promising material with great potential. The mechanism of the stimulation function of SiHA requires further investigation, as the comprehensive understanding of the role of Si in stimulating the biological response will help to further application of nanoSiHA in tissue engineering and regenerative medicine.

15.6 References

1. LeGeros RZ, LeGeros JP (1993), 'Dense hydroxyapatite', in Hench LL and Wilson J (eds), *An Introduction to Bioceramics*, World Scientific, Singapore, 139–180.
2. Hench LL, Best SM (2004), 'Ceramics, glasses and glass-ceramics', in Ratner BD, Schoen FJ, Hoffman AS and Lemons JE, *Biomaterials Science: An Introduction to Materials in Medicine*, 2nd edition, Elsevier Academic Press, 153–170.
3. Polak J, Hench L (2005), 'Gene therapy progress and prospects: in tissue engineering', *Gene Therapy*, 12, 1725–1733. doi:10.1038/sj.gt.3302651
4. Bonfield W (1988), 'Composites for bone replacement', *J Biomed Eng*, 10, 522–526.
5. Juhasz JA, Best SM, Brooks R, Kawashita M, Miyata N, Kokubo T, Nakamura T, Bonfield W (2004), 'Mechanical properties of glass-ceramic A–W-polyethylene composites: effect of filler content and particle size', *Biomaterials*, 25, 949–955. doi:10.1016/j.biomaterials.2003.07.005
6. Silva GA, Pedro A, Costa FJ, Neves NM, Coutinho OP, Reis RL (2005), 'Soluble starch and composite starch Bioactive Glass 45S5 particles: Synthesis, bioactivity, and interaction with rat bone marrow cells', *Mater Sci Eng C*, 25, 237–246. doi:10.1016/j.msec.2005.01.012
7. Huang J, Di Silvio L, Wang M, Tanner KE, Bonfield W (1997), '*In vitro* mechanical and biological assessment of hydroxyapatite reinforced polyethylene composite', *J Mater Sci: Mater Med*, 8, 775–779. doi: 10.1023/A:1018516813604
8. Huang J, Di Silvio L, Wang M, Rehman I, Bonfield W (1997) 'Evaluation of *in vitro* biocompatibility of Bioglass® reinforced polyethylene composites', *J Mater Sci: Mater Med*, 8, 809–813. doi: 10.1023/A:1018581100400
9. Harada Y, Wang J, Doppalapudi V, Willis A, Jasty M, Harris W, Nagase M, Goldring S (1996), 'Differential effects of different forms of hydroxyapatite and hydroxyapatite tricalcium phosphate particulates on human monocyte macrophages in vitro', *J Biomed Mater Res*, 31, 19–26. doi:10.1002/(SICI)1097-4636(199605)31:1<19::AID-JBM3>3.0.CO;2-T
10. Sun J, Tsuang Y, Chang WH, Li J, Liu H, Lin F (1997), 'Effect of hydroxyapatite particle size on myoblasts and fibroblasts', *Biomaterials*, 18, 683–690. doi: 10.1016/S0142-9612(96)00183-4
11. Wang X, Li Y, Wei J, de Groot K (2002), 'Development of biomimetic nano-hydroxyapatite/poly(hexamethylene adipamide) composites', *Biomaterials*, 23, 4787–4791. doi:10.1016/S0142-9612(02)00229-6
12. Wei G, Ma PX (2005), 'Structure and properties of nano-hydroxyapatite/polymer composite scaffolds for bone tissue engineering', *Biomaterials*, 25, 4749–4757. doi:10.1016/j.biomaterials.2003.12.005

13. Di Silvio L (1995), 'A novel application of two biomaterials for the delivery of growth hormone and its effect on osteoblasts,' PhD thesis, University of London.
14. Chowdhury EH, Maruyama A, Kano A, Nagaoka M, Kotaka M, Hirose S, Kunou M, Akaike T (2006), 'pH-sensing nano-crystals of carbonate apatite: Effects on intracellular delivery and release of DNA for efficient expression into mammalian cell', *Gene*, 376(1), 87–94. doi:10.1016/j.gene.2006.02.028.
15. Olton D, Li J, Wilson ME, Rogers T, Close J, Huang L, Kumta PN, Sfeir C (2007), 'Nanostructured calcium phosphates (NanoCaPs) for non-viral gene delivery: Influence of the synthesis parameters on transfection efficiency,' *Biomaterials*, 28(6): 1267–1279. doi:10.1016/j.biomaterials.2006.10.026.
16. Suska F, Gretzer C, Esposito M, Tengvall P, Thomsen P (2005), 'Monocyte viability on titanium and copper coated titanium', *Biomaterials*, 26, 5942–5950. doi:10.1016/j.biomaterials.2005.03.017
17. Bonfield W, Gibson IR (1999), International Patent Publication No. WO 99/32401, 1999.
18. Wimhurst JA, Brooks RA, Rushton N (2001), 'Inflammatory responses of human primary macrophages to particulate bone cements *in vitro*', *J Bone Joint Surg*, 83-B, 40–44.
19. Osano E, Kishi JJ, Takahashi Y (2003), 'Phagocytosis of titanium particles and necrosis in TNF-α-resistant mouse sarcoma L929 cells', *Toxicology in Vitro*, 17, 41–47. doi:10.1016/S0887-2333(02)00127-3
20. Xynos ID, Edgar AJ, Buttery LDK, Hench LL, Polak JM (2001), 'Gene expression profiling of human osteoblasts following treatment with the ionic dissolution products of Bioglass® 45S5 dissolution', *J Biomed Mater Res*, 55, 151–157. doi: 10.1002/1097-4636(200105)55:2<151::AID-JBM1001>3.0.CO;2-D
21. Xynos ID, Edgar AJ, Buttery LD, Hench LL, Polak JM (2000), 'Ionic products of bioactive glass dissolution increase proliferation of human osteoblasts and induce insulin-like growth factor II mRNA expression and protein synthesis', *Biochem Biophys Res Commun*, 276, 461–465. doi: 10.1006/bbrc.2000.3503
22. Carlisle EM (1972), 'Silicon: an essential element for the chick', *Science*, 178, 619–621. doi: 10.1126/science.178.4061.619
23. Reffitt DM, Ogston N, Jugdaohsingh R, Cheung HFJ, Evans BAJ, Thompson RPH, Powell JJ, Hampson GN (2003), 'Orthosilicic acid stimulates collagen type 1 synthesis and osteoblastic differentiation in human osteoblast-like cells in vitro', *Bone*, 32(2), 127–135. doi: 10.1016/S8756-3282(02)00950-X
24. Patel N, Best SM, Bonfield W, Gibson IR, Hing KA, Damien E, Revell PA (2002), 'A comparative study on the in vivo behavior of hydroxyapatite and silicon substituted hydroxyapatite granules', *J Mater Sci: Mater Med*, 13, 1199–1206. doi: 10.1023/A:1021114710076
25. Patel N, Brooks RA, Clarke MT, Lee PMT, Rushton N, Gibson IR, Best SM, Bonfield W (2005), 'In vivo assessment of hydroxyapatite and silicate-substituted hydroxyapatite granules using an ovine defect model', *J Mater Sci: Mater Med*, 16, 429–440. doi: 10.1007/s10856-005-6983-6
26. Hing KA, Revell PA, Smith N, Buckland T (2006), 'Effect of silicon level on rate, quality and progression of bone healing within silicate-substituted porous hydroxyapatite scaffolds', *Biomaterials*, 27(29), 5014–5026. doi:10.1016/j.biomaterials.2006.05.039

27. Porter AE, Patel N, Skepper JN, Best SM, Bonfield W (2003), 'Comparison of in vivo dissolution processes in hydroxyapatite and silicon-substituted hydroxyapatite bioceramics', *Biomaterials*, 24, 4609–4620. doi:10.1016/S0142-9612(03)00355-7
28. Huang J, Jayasinghe SN, Best SM, Edirisinghe MJ, Brooks RA, Rushton N, Bonfield W (2005), 'Novel deposition of nano-sized silicon substituted hydroxyapatite by electrostatic spraying', *J Mater Sci: Mater Med*, 16, 1137–1142. doi: 10.1007/s10856-005-4720-9
29. Dalby MJ, Di Silvio L, Harper EJ, Bonfield W (2002), 'Increasing hydroxyapatite incorporation into poly(methylmethacrylate) cement increases osteoblast adhesion and response', *Biomaterials*, 23, 569–576. doi:10.1016/S0142-9612(01)00139-9
30. Carlisle EM (1970), 'Silicon: a possible factor in bone calcification', *Science*, 167, 279–280. doi: 10.1126/science.167.3916.279
31. Wang M, Hench LL, Bonfield W (1998), 'Bioglass®/high density polyethylene composite for soft tissue applications: Preparation and evaluation', *J Biomed Mater Res*, 42, 577–586. doi: 10.1002/(SICI)1097-4636(19981215)42:4<577::AID-JBM14>3.0.CO;2-2

16
Diamond-like carbon (DLC) as a biocompatible coating in orthopaedic and cardiac medicine

W MA, A J RUYS and H ZREIQAT,
The University of Sydney, Australia

Abstract: Diamond-like carbon (DLC) is a metastable form of amorphous carbon (a-C) containing a combination of four-fold coordinated sp^3 sites and three-fold coordinated sp^2 sites, with some of the bonds terminated by hydrogen (amorphous hydrogenated carbon, a-C:H). This chapter reviews the current progress in DLC coating deposition, coating characterization techniques, characteristics of various DLC coatings, and, in particular, the relevant biological and clinical advancements in the area of orthopaedic and blood-interfacing applications.

Key words: diamond-like carbon, biomaterials, deposition, biocompatibility, haemocompatibility.

16.1 Introduction

Biomaterials are used to replace damaged parts [1] in the human body where they will need to meet not only the demand of our biological systems but also its mechanical structure. This is particularly true for load-bearing applications, where the implants must withstand dynamic mechanical forces and the potentially damaging effects of the long-term biological interactions with the surrounding tissues. The physical characteristics of the implant are mainly controlled by its bulk properties, whereas the implant surface governs its interaction with surrounding tissue [2]. Most of the currently used biomaterials with excellent bulk properties do not possess the surface characteristic that meet the stringent biological requirement for tissue replacement [3]. Considerable efforts have been made to modify the surfaces of prostheses in an attempt to improve their initial interlocking to the surrounding tissue [4, 5]. Major causes of implant failure, both orthopaedic and blood-interfacing, are directly or indirectly related to the low wear and corrosion resistance of biomaterials, which can lead to deterioration of the implants, blood coagulation, and cellular damage [6, 7]. Therefore, a biocompatible, wear resistant, corrosion resistant coating is of significant interest to the biomedical industry.

16.1.1 Diamond-like carbon (DLC)

Carbon, the basic unit for organic molecules, forms a great variety of crystalline and disordered structures because it is able to exist in three hybridizations: sp^1 (carbon polymers), sp^2 (graphitic) and sp^3 (diamond-like). Diamond-like carbon (DLC) is a metastable form of amorphous carbon (a-C) containing a combination of four-fold coordinated sp^3 sites and three-fold coordinated sp^2 sites, with some of the bonds terminated by hydrogen (amorphous hydrogenated carbon, a-C:H). As a bulk material, DLC has high mechanical hardness [8], chemical inertness [9], wear resistance [10], corrosion resistance [11] and optical transparency [12], and it is a wide band gap semiconductor [13–18]. DLC coatings generally adhere well to various metallic and non-metallic substrates [9, 19–21]. The 'state of the art properties' [22] of DLC and its potential are well recognized by research communities from different fields of science [22].

DLC consists of a mixture of amorphous and crystalline phases, as shown in Fig. 16.1 [23]. Its properties, including physical properties [13], biological performance [2] and surface characteristics, vary considerably with composition. The fraction of sp^3 sites, and the hydrogen content, are the two key compositional parameters that determine the properties of DLC [15]. This is best represented by a ternary phase diagram, first constructed by Robertson [24].

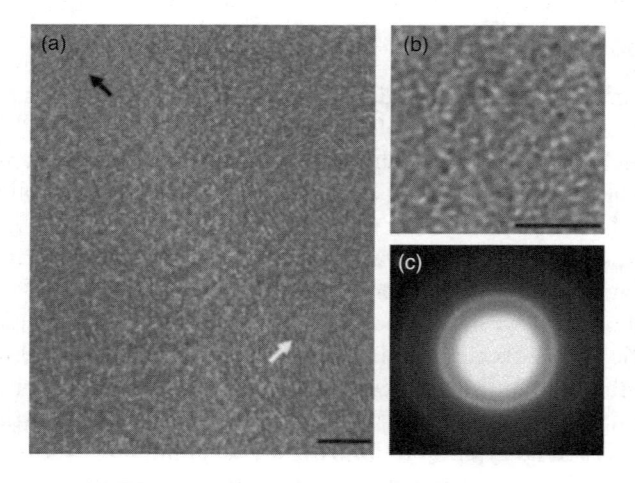

16.1 (a) TEM micrograph of the ta-C deposited by FAD showing both amorphous (black arrow) and nanocrystalline regions (white arrow). (b) Enlargement of the amorphous region. (c) Electron diffraction pattern of the nanocrystalline region. The brightest ring corresponds to the (0 0 2) plane reflection of graphite. Scale bar = 3 nm for both images A and B [23]. (Permission to reproduce images granted by authors).

The coating composition is controlled by the deposition process, and the effects of composition and deposition conditions on the mechanical and physical properties of DLC coating have been studied extensively. On the other hand, both *in vitro* and *in vivo* studies have often quoted DLC as a single compound [8], rather then a class of material with widely varying properties. Hence, few details with regard to the influence of coating composition on the coating biocompatibility have been reported. Results to date only amount to a database of evidence for DLC biocompatibility, rather than conclusive investigations, which detail the effect of physical and compositional characteristics of DLC on the corresponding biological response. This incompleteness in the literature has made it impossible to provide constructive feedback for coating evaluation, and thus hindered the further development and optimization of DLC coatings for a specific biological application. Here, we review the current progress in DLC coating deposition, coating characterization techniques and characteristics of various DLC coatings, and in particular the relevant biological and clinical advancements in the area of orthopaedic and blood-interfacing applications.

16.2 Deposition and characterization of DLC coatings

16.2.1 DLC deposition

DLC deposition methods can be divided into two categories, namely Chemical Vapour Deposition (CVD) using hydrocarbon gases and other Physical Vapour Deposition (PVD) methods using solid target materials. Whitmell and Williamson [25] were the first to produce hard carbon films by a chemical process, from hydrocarbon gas (ethylene–argon mixture) in a DC glow discharge. Aisenberg and Chabot [26] were the first to successfully prepare DLC films by a physical method – ion beam deposition.

In general, DLC films are formed when carbon or hydrocarbon radicals bombard a substrate with impact energies from 50 eV to several hundred electron volts. The influence of impact energy controls the sp^3 fraction, and thus determines the type of film obtained [27, 28]. Figure 16.2 shows the sp^3 fraction produced by several commonly used DLC deposition methods. A summary of the influence of impact energy on the type of DLC produced has been proposed by Angus *et al.* [29].

Ion beam deposition (IBD)

Ion beam deposition (IBD) is one of the most commonly used PVD deposition methods, in which the precursor (gas) ions are extracted from a plasma to form an ion beam. The beam can be directed at the substrate to provide

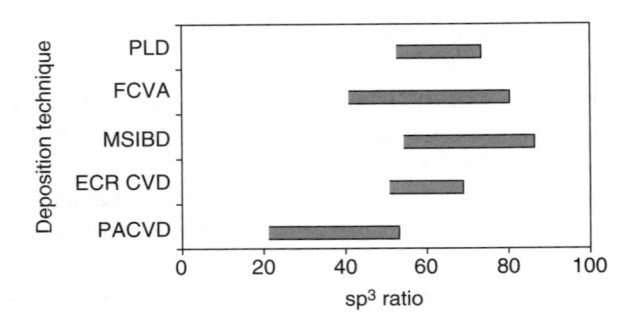

16.2 The sp³ fraction produced by various deposition techniques (PLD = pulsed laser deposition, FCVA = filtered cathodic vacuum arc, MSIBD = mass selective ion beam deposition, ECR CVD = electron cyclotron resonance chemical vapour deposition, PACVD = plasma assisted chemical vapour deposition).

ion bombardment of the growing film. IBD methods can be scaled up for manufacturing; the limitation with these methods is that the hardest films are obtained under conditions of low power and low gas pressure, which reduces the deposition rate.

In a typical ion beam deposition system, carbon ions are produced by the plasma sputtering of a graphite cathode by an ion source [26, 29–32]. However, a gas can also be used to produce the plasma [33]. An example of this is the ionization of methane in a plasma [34, 35]. In this scenario, a bias voltage can be used to extract the active carbon or hydrocarbon ions from the plasma, and these ions can then be accelerated to an appropriate high energy state in a vacuum chamber, where the deposition takes place. Regardless of whether the ion beam is formed by cathodic sputtering or gas ionization, the ion source should operate at a finite pressure. This means that the ion beam can also contain a significant quantity of un-ionized species. This can reduce the flux ratio of ions to neutral species to as low as 2–10%. Ion beam sources tend to run best at higher ion energies of 100–1000 eV [21]. A good deposition condition for DLC will provide a carbon ion flux at about 100 eV per carbon atom, with a narrow energy distribution, a single energetic species and a minimum number of non-energetic (generally neutral) species [36].

Mass selected ion beam deposition (MSIBD)

Mass selected ion beam deposition (MSIBD) is achieved by magnetically filtering out any electrically neutral species and ions with an e/m ratio equal to that of the C^+ ion [37–41]. It is the most controllable system for DLC deposition, but the flux is limited to approximately 10^{15}–10^{16} atoms $cm^{-2}s^{-1}$ [42]. MSIBD involves the selective utilization of carbon ions ejected from a graphite target. The selection criterion is generally ions that are within the energy range 1–10 eV. The ions are then accelerated to 5–40 kV and passed through a magnetic filter. The resultant ion beam is typically divergent as a result of electrical repulsion. This divergence is corrected for by means of an electrostatic lens. The beam impacts the substrate in a vacuum in the order of 1.3×10^{-5} Pa to produce a ta-C film [21, 43, 44]. Because MSIBD is highly controllable, it is frequently used in experimental work [24].

Sputtering

Sputtering is the preferred technique for industrial applications because of its versatility [45–48] and also because the deposition conditions can be controlled by the plasma power and gas pressure [21]. Direct current (DC) or Radio Frequency (RF) are the most common forms of sputtering of a graphite cathode by an argon plasma. Magnets are often placed behind the target to increase the degree of ionization of plasma. The 'unbalanced magnetron' with the influence of DC bias produces varied ion beam energies at the substrate [21].

Sputter deposition, similar to IBD, can product a-C with a relatively large sp^3 fraction, but at the expense of a slow growth rate [49, 50]. Compared to IBD, the sputtered carbon can be very pure, but has a very broad energy distribution depending on the energy and nature of the bombarding species [51].

Filtered cathodic vacuum arc deposition (FCVAD)

Filtered cathodic vacuum arc deposition (FCVAD) is a specific technique for the deposition of tetrahedrally coordinated amorphous carbon or ta-C [52–58]. The vacuum arc is a low-voltage, high-current plasma discharge that takes place between two metallic electrodes in high vacuum [59]. The two most important parameters influencing film properties are ion energy (which may be increased by applying a substrate bias) and substrate temperature. Ion deposition energy of 100 eV, with substrate temperature less than 300°C, is the optimum condition for ta-C formation, which produces a high ion density of up to 10^{-13} cm^3 [60], a sp^3 fraction of up to 85% and hardness of 85 GPa [61].

In FCVAD, because an arc is initiated in a high vacuum by touching a solid graphite cathode with a small striker electrode, it produces plasma with a large amount of carbon particles. These particles range in size from nanometres to micrometres, and are undesirable for most applications [61]. Thus the expanding plasma is guided into a curved toroidal magnetic filter in order to remove these particles.

The main advantages of the FCVA are that it produces DLC of high sp^3 fraction (ta-C) at high growth rate with a low capital cost, and that films can be deposited on an insulating substrate. The disadvantage of FCVA is incomplete filtering of carbon particles for some applications [62].

Pulsed laser deposition (PLD)

Pulsed laser deposition (PLD) is widely used in a laboratory environment, whereby a high power, pulsed beam such as ArF is focused inside an ultra-high vacuum chamber to strike and vaporize a graphite target to form an intense plasma that is used for deposition [63, 64]. During PLD, the laser pulse is absorbed by the target and the energy is first converted to electronic excitation and then into thermal, chemical and mechanical energy, resulting in evaporation, ablation, plasma formation and even exfoliation [65]. The properties of the thin films can be controlled by the laser intensity, substrate temperature, buffer gas pressure, and incident angle of the coating plume [64, 66–69].

Plasma assisted chemical vapour deposition (PACVD)

Plasma assisted chemical vapour deposition (PACVD) is often referred as glow discharge CVD. The reactor consists of two electrodes of different surface area, and both the DC and RF glow discharge techniques are used in the deposition of DLC [70–75]. Any hydrocarbon gas with sufficient vapour pressure can in principle be used as precursor gas for PACVD. Studies by Koidl and Wild [71] showed that the H/C ratio in the precursor gas strongly affects the H/C ratio of the resulting film.

In the DC discharge technique, a negative bias is applied to the substrate, which acts as an electrode. This bias voltage ionizes the hydrocarbon precursor producing plasma, thereby attracting the ions required for the growth of the film. Bias voltage and total pressure of the system are important processing parameters, which help in minimizing the sputtering of atoms from the film surface [21]. The processing pressure generally lies in the range 1.3 to 13 Pa [76]. The substrate must be electrically conductive for the deposition of films by DC glow discharge. Thus, it limits the choice of substrate and restricts the deposition of a-C:H films to conducting surfaces [21].

RF glow discharge overcomes the limitation of the DC glow discharge method as it can deposit a-C:H on insulating substrates. The RF power is usually capacitively coupled to the smaller electrode on which the substrate is mounted, and the other electrode (often including the reactor walls) is earthed. The RF frequency used in this process is generally greater than the ion plasma frequency (~2–5 MHz), thus electrons can follow the RF voltage

Table 16.1 The most commonly used sp³ fraction measurement methods

Method	Comments	References
Electron energy loss spectroscopy (EELS)	Destructive and time-consuming	[50, 82–84]
UV Raman	Fast and non-destructive, but not as accurate as EELS	[85–87]
Nuclear magnetic resonance (NMR)	Large samples needed for C¹³ dephasing	[88]
Spectroscopic ellipsometry	Not usable in situ; small spectral range	[46, 89]

but not the ions [72]. Therefore, the difference in electron and ion mobility produces a negative DC self bias on the powered electrode [76].

Many innovative PACVD techniques have been produced in recent years, such as the plasma beam source technique by Weiler et al. [77, 78], sheet-like plasma CVD by Nonogaki et al. [79] and the magnetron plasma technique by Nakamura et al. [80]. Perhaps the most compact, RF powered, high plasma-density is the Electron Cyclotron Wave Resonance (ECWR) source [81]. It produces an extremely high density plasma of 10^{12} cm^{-3} or over, with independent control of ion energy and ion current density [21].

16.2.2 Characterization of DLC

Owing to the amorphous nature of DLC, a great deal of complexity is involved in its structural characterization. Various modern characterization methods have been proposed to determine sp³ fraction, as compiled in Table 16.1.

16.3 Physical properties of DLC coatings

The physical properties of DLC are dependent on the coating structure and composition, and these factors are all related to the deposition process and the corresponding deposition parameters [20, 90–92]. Two major advantages of DLC compared to CVD polycrystalline diamond are superior smoothness and ability to be deposited at room temperature [92–94]. Some of the disadvantages of DLC films are the high intrinsic stress and thermal stability (the later is not a problem for ta-C). A complete understanding of structural and compositional effects on the coating physical properties is desirable for commercial applications of DLC films so as to circumvent potential problems, such as issues with implant–interface compatibility and to optimize the cost:benefit ratio.

Table 16.2 Comparison of elastic properties of various DLC films

	a-C:H	ta-C	Diamond
Density	2.35	3.26	3.515
H (at%)	30	0	0
sp^3 (%)	70	88	100
Young's modulus (E; GPa)	300	757	1144.6
Shear modulus (G; GPa)	115	227	534.3
Bulk modulus (B; GPa)	248	334	444.8
Poisson's ratio	0.3	0.12	0.07

Source: Robertson [21].

16.3.1 Elastic properties

Nanoindentation [95–97], laser induced surface acoustic wave (LISAW) [98, 99], and surface Brillouin scattering (SBS) are used to measure elastic properties of DLC coatings [100]. Although some of the methods offer better accuracy than others [101, 102], the general finding is that elastic properties of DLC are influenced by the sp^3 fraction. As shown in Table 16.2, the higher the sp^3 fraction, the higher the density, and the more elastic the coating.

16.3.2 Hardness and yield strength

The measurement of hardness by nanoindentation is related empirically to the yield stress Y and Young's modulus E [103], where the $H:E$ ratio (H stands for hardness) can be calculated by the Orowan approximation [104]:

$$\frac{H}{Y} = 0.07 + 0.06 \ln\left(\frac{H}{E}\right)$$

Resulting from a-C:H deposited by PACVD at a lower voltage, the hardness then rises to a maximum corresponding to the maximum diamond-like character (sp^3 bonds), and then it decreases as the films acquire a more graphitic bonding [16].

Individual measurements on the direct relationship between hardness and sp^3 fraction in ta-C made by Pharr *et al.* [95], Martinez *et al.* [105] and Friedmann *et al.* [106] showed that the sp^3 bond is responsible for the formation of hard coatings. This finding accords with Forrest *et al.* [107].

16.3.3 Coating-substrate adhesion strength

A thicker coating layer is preferred to maximize the wear life of DLC protective coating; however, the maximum thickness, h, is limited by the elastic energy per unit volume due to the compressive stress σ [108, 109]. Delamination generally occurs when σ exceeds the surface facture energy γ per surface [107, 110]. It has been reported by Tamor *et al.* [111] that the stress, Young's modulus and hardness are all proportional to each other. In other words, the most wear-resistant film only exists in thin layers [21]. Various strategies are used to maximize film thickness while maintaining coating–substrate adhesion strength [108, 112–117]. One approach to reduce stress is alloying DLC with, for example, Si [116, 118] or other metal elements. The process of alloying, also called doping, promotes sp^3 fraction via chemical means rather than ion bombardment as in undoped DLC coating. This approach increases the maximum coating thickness and thermal stability of hydrogenated amorphous carbon and improves the friction performance. One disadvantage of both Si and metal element doping is the reduction of the refractive index [21].

16.3.4 Tribological properties

DLC has been shown to possess very good tribological properties and there are a number of studies dealing with this issue [119, 120]. Friction and wear properties of DLC coatings reported in the literature are strongly affected by the following:

- The nature of the films (controlled by the deposition process)
- The tribological test conditions:
  - materials parameters: nature of substrate and pin material (for pin-on-disc wear tests)
  - mechanical parameters: type of contact and contact pressure
  - kinematic parameters: nature of motion, velocity
  - physical parameters: temperature during friction
  - chemical parameters: nature of environment (humidity, etc.).

Thus, a wide range of tribological properties have been reported in the literature for DLC films. The hypotheses proposed to justify the exhibited tribological behaviour are very specific to the DLC films synthesized and tested under specific conditions. Therefore, the proposed friction mechanisms should not be regarded as applicable to all types of DLC films [121–123].

Friction

The range of friction coefficients reported for DLC films is 0.007–0.04 in vacuum [128], below 10^{-4} Pa, and in ambient air at 20% < RH < 100% the range is 0.05–0.4 [126, 128]. The moisture sensitivity of DLC films is higher in the case of hydrogenated DLC than in hydrogen-free DLC films, whereby the friction coefficient dropped from 0.1 to 0.08 for ta-C, and increased from 0.02 to 0.16 when the relative humidity increased from 0% to 100% [124].

Gardos [129] suggested that the increase in friction coefficient of hydrogenated DLC films in humid air is due to the increase in van der Waals bond strength of hydrogen bonds to the adsorbed water molecules (~5 kcal/mol). However, inverse humidity sensitivity has been reported in ta-C (i.e., decreasing friction coefficient with increasing humidity) [124]. It is believed that the friction of DLC films is controlled by the formation of transfer layers of a graphitic nature [128] during wear on the sliding couple, which generally has a lubricating effect [130–133]. Kokaku and Kitoh [134] proposed that the tribo-chemical reactions leading to the oxidation of the DLC surface increase the friction coefficient in the presence of moisture, indicating that the presence of water inhibits the formation of a transfer layer. The friction coefficient decreased by etching the oxide layer by Ar ions.

Despite the environmental sensitivity, a series of experiments by Erdemir et al. [125, 135–137] at the same humidity level showed that hydrogen is a key element in determining the tribological behaviour, and the presence of hydrogen lowers the friction coefficient of DLC films [138]. It is believed that intense hydrogen ion bombardment prevents crosslinking or C=C double bonding in the growing DLC films and etches out the graphitic phases [136]. In addition, some of the carbon atoms at the surface can be dihydrated (i.e. two hydrogen atoms bonded to one carbon atom). The presence of dihydrated carbon atoms on DLC surfaces is expected to provide better shielding, or a higher degree of chemical passivation, and thus lower friction.

Wear

Generally, the wear rate for DLC varies depending on the hydrogen content as shown in Table 16.3; on average it can be as low as 10^{-6} mm^3/Nm [24]. Martinez et al. [105] demonstrated that the wear rate of ta-C had its minimum at the point of maximum hardness. Although the experiments were set up

Table 16.3 Comparison of wear rate for ta-C and a-C:H

	Wear rate (mm^3 N^{-1}m^{-1})
ta-C	10^{-9}
a-C:H	10^{-6} to 10^{-7}

Source: Voevodin et al. [124]

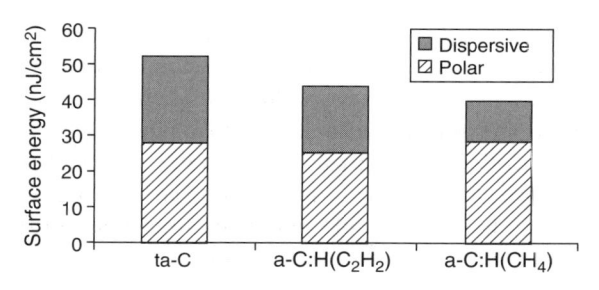

16.3 Surface energy (sessile drop test) on the various DLC surfaces [23]. (Permission to reproduce bar chart granted by authors.)

in different environments, all the above findings suggested that the higher the sp^3 fraction, the better the wear resistance of DLC coatings [121]. Addition of metals or other elements such as silicon into the film can lead to further reduction in wear rate. Other studies found that in some cases the failure of DLC during friction and wear is due to spalling, which is the result of poor adhesion onto the substrate [139]. The addition of certain elements reduced wear by improving the coating–substrate adhesion.

16.3.5 Surface properties

DLC characteristically has an inherently low surface energy [117]. In the case of amorphous hydrogenated DLC (a-C:H), this is typically 41 mN m^{-1}. Ion doping can be used to alter the surface energy. Oxygen doping can reportedly increase it to 52 mN m^{-1}, while silicon or fluorine doping can reduce it to 19–24 mN m^{-1} [140]. Indeed, it has been reported that fluorine doping can reduce surface energy to values approaching those of Teflon (18.5 mN m^{-1}) [141].

Tetrahedral hydrogen-free DLC (ta-C) with a high sp^3 fraction is hydrophobic, with a reported contact angle of 75–80° in water [141]. As is the case for amorphous hydrogenated DLC (a-C:H), doping can also reduce the surface energy of ta-C, thereby rendering it more hydrophobic. For example, aluminium or iron doping can reduce the surface energy of ta-C to 25 mN m^{-1} [136]. Figure 16.3 shows the surface energy (dispersive and polar) from the sessile drop test on various DLC surfaces [23].

16.4 Biocompatibility of DLC

16.4.1 Tissue-compatibility of DLC

Reports on tissue-compatibility of DLC coatings have contributed to a large collection of positive evidence confirming the biocompatibility of

DLC *in vitro* [23] and *in vivo* [142], identifying it as a desirable biomaterial for its long-term wear resistance and corrosion resistance and superior biocompatibility.

Mouse fibroblast and mouse peritoneal macrophages were grown on DLC coatings for 7 days and showed no biochemical and morphological signs of cellular damage and inflammatory reaction *in vitro* [9]. Follow-up studies using human osteoblast-like cells, macrophage and fibroblasts grown on DLC coating on a variety of substrates exhibited normal cellular growth, morphology and differentiation [143, 144]. Other experiments in tissue-compatibility of DLC surfaces by other groups have largely confirmed these results [145–147]. Human osteogenic sarcoma T385 cells [148], human 1BR3 fibroblasts, human myeloblastic ML-1 [148], human embryo kidney 293 cells and other cell types have been grown on DLC under different conditions [149], and cell responses such as proliferation rate, viability, cell adhesion, differentiation, cell morphology and cytoskeletal architecture have been monitored [9]. Taken together, these results confirmed that it is unlikely that contact with a DLC-coated surface in the human body will elicit inflammatory signals by these cells [150–156].

Further studies compared the osteoblast and fibroblast cell viability after 7 days in culture on DLC, amorphous carbon nitride film, stainless steel and medical-grade titanium surfaces. The results showed a significant up-regulation of cells on DLC surfaces [158]. *In vitro* comparison between DLC coated and uncoated CoCrMo surface for 48 hours using Z-coil neuron stem cells showed increased cell attachment on DLC surfaces [159]. Kumari *et al.* [160] proved that DLC is more biocompatible to mouse fibroblast cells (L-929), human osteoblast cells (HOS) and primary human umbilical cord vein endothelial cells (HUVEC) than medical-grade titanium. Moreover, Tang *et al.* [161] further confirmed experimentally both *in vitro* and *in vivo* that DLC is highly desirable for biomedical applications compared to stainless steel and medical-grade titanium.

In vivo studies confirmed the biocompatibility of the DLC-coated materials [162, 163]. Dowling *et al.* [164] inserted 12 mm × 4 mm diameter DLC-coated stainless steel cylinders into both the cortical bone and the muscular tissue of sheep for periods of 4 weeks and 12 weeks respectively. Both groups of tests demonstrated the biocompatibility of the coatings. In another *in vivo* study 5 × 1 mm diameter DLC-coated Ti–13Nb–13Zr cylinders were inserted into both muscular tissue and femoral condyles of rats for intervals of 4 and 12 weeks postoperatively. Histological analyses showed that the DLC coatings were well tolerated in both types of implantation [165]. Mohanty *et al.* [157] studied the long-term tissue response of DLC and DLC-coated titanium implants to the skeletal muscle of rabbits and confirmed the compatibility of the materials with skeletal muscle [166].

16.4.2 Haemocompatibility of DLC

It is generally believed that DLC-coated surfaces with appropriate structural and compositional characteristics display enhanced anti-coagulation characteristics and anti-thrombogenicity. However, a well-defined dependence model has not been proposed.

Albumin/fibrinogen adsorption ratio

A high albumin/fibrinogen protein ratio adsorbed on an implant surface prior to cell or platelet attachment correlates with a small number of adhering platelets, and therefore a low tendency of thrombus formation. A higher ratio of albumin/fibrinogen adsorption was observed on DLC surfaces, compared to Ti, TiN and TiC, indicating the ability of DLC to prevent thrombus formation [167]. Cui and Li [3] reported good tissue and blood-compatibility for DLC and amorphous CN films from *in vitro* experiments, with DLC also showing a high albumin/fibrinogen ratio. The albumin/fibrinogen ratios from the different research groups are summarized by Hauert [168], revealing that DLC has the highest albumin/fibrinogen ratio. Yang *et al.* [169–171] assessed the platelet adhesion and protein adsorption on various types of amorphous hydrogenated carbon, and showed that the biological coating was influenced by the coating structure. Ma *et al.* [23] studied the albumin/fibrinogen adsorption ratio on various DLC surfaces containing different hydrogen contents and confirmed that the albumin/fibrinogen adsorption ratio is inversely proportional to hydrogen content: the higher the hydrogen content, the lower the albumin/fibrinogen adsorption ratio. Figure 16.4 shows the ratio on various DLC surfaces [23].

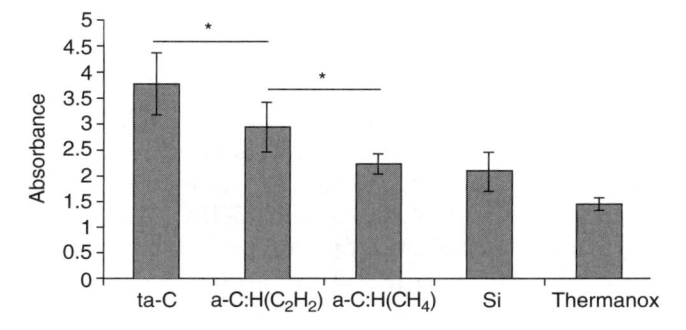

16.4 Albumin/fibrinogen adsorption ratio on various DLC surfaces [23]. (Permission to reproduce bar chart granted by authors.)

Platelet adhesion

The adhesion of platelets was lower on DLC-coated titanium than on uncoated titanium [167]. The haemocompatibility, thrombogenicity and contact surface morphology of DLC with rabbit blood platelet using the dynamic blood interaction method showed decreased platelet coverage area compared to titanium, TiN and TiC [172, 173]. A similar experiment was also conducted by Schaub *et al.* [174] who compared DLC with a range of commonly used blood interfacing materials including low temperature isotropic carbon (LTIC), Ti alloy, TiO, polycrystalline diamond (PCD), woven Dacron (WD), collagen-impregated Dacron, expanded polytetrafluoroethylene (ePTFE) and denucleated ePTFE, whereby the DLC-coated surface showed the least platelet adherence [174]. Weisenberg and Mooradian [175] showed DLC to be the most haemocompatible material for microelectromechanical systems, compared to silicon (Si), silicon dioxide (SiO_2), silicon nitride (Si_3N_4), low-stress silicon nitride ($Si_{1.0}N_{1.1}$), SU-8 photoresist, and Parylene thin films. Li and Gu [176] compared growth and morphology of fibroblasts, granulocytes and platelets on polymethylmethacrylate (PMMA) and two types of DLC prepared by different levels of bombarding energy. The results showed that both types of DLC outperformed PMMA in tissue and blood compatibility.

Platelet activation

Gutensohn *et al.* [177] analysed the intensity of platelet activation antigens CD62p and CD63 and showed that DLC-coated artery stainless steel stents resulted in a decrease of the CD62p and CD63 antigens, indicating lower platelet activation on DLC-coated stents, and thus a lower tendency for thrombus formation. Additionally, metal ions released from the stainless steel stents (which may negatively influence the haemocompatibility of the surface) could be suppressed by the DLC coating. Alanazi *et al.* [178], using a flow chamber and whole human blood, showed a low percentage of platelets adhering on the DLC surface, results confirmed by others [179–181] where DLC and ta-C demonstrated good biocompatibility and lower platelet adhesion compared with pyrolytic carbon, Ti and 316L stainless steel. Kwok *et al.* [169] found that both hydrogen-free and amorphous carbon nitride films are more biocompatible than low-temperature isotropic pyrolytic carbon (LTIPC); however, an optimized sp^3 fraction is desirable for optimized biocompatibility. Further reports by Yu *et al.* [180] investigated the haemocompatibility of DLC by haemolysis ratio measurements and platelet adhesion experiments and found that DLC films exhibit better anticoagulation properties than pyrolitic carbon.

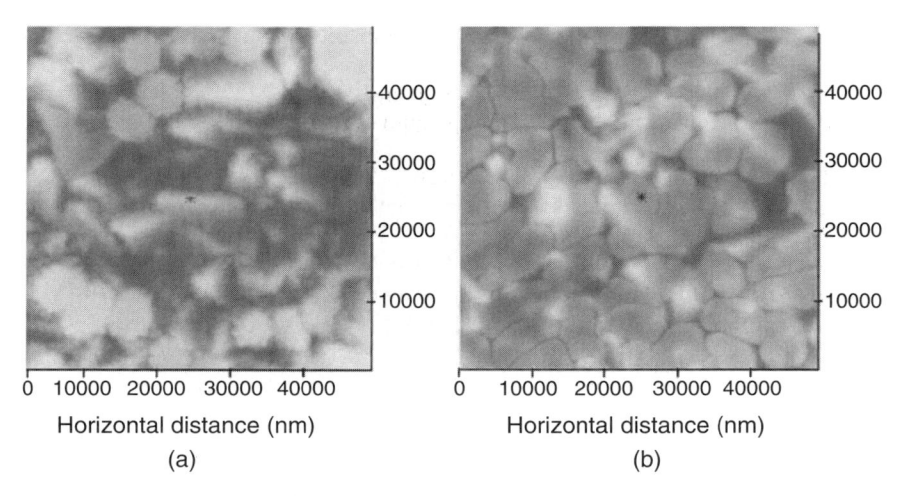

Diamond-like carbon (DLC) as a biocompatible coating 405

16.5 Smooth 'clotting' formed on DLC surface by (a) anti-coagulated blood and (b) fresh human blood [182]. (Permission to reproduce images granted by authors.)

Blood clotting

Baranauskas *et al.* [182] pioneered the study of blood clotting on DLC surfaces using high resolution atomic force microscopy (AFM). While a moderate amount of smooth fibre formed on the DLC surface, the AFM image indicates that this 'clot-like fibre' was induced by the lateral surface of the cells' membranes, rather than by the surface contact mechanism (Fig. 16.5).

In vivo *and clinical studies*

Yang [183] studied the haemocompatibility of different surfaces implanted for 2 hours into the intrathoracic venae cavae of Swedish native sheep. The results showed that there was significantly more thrombus on pyrolytic carbon and methylated titanium than on titanium, cobalt–chromium and DLC, and that the lowest coverage of thrombus was obtained for a TiN coated sample [184]. The DLC centre in the UK has reported on their home page on the use of DLC coating in a blood flow accelerator, whereby an inhibition of thrombus formation and platelet deposition was observed in the DLC-coated device *in vivo*.

16.5 Corrosion behaviour of DLC

The static corrosion performance of DLC was evaluated by Tiainen in 10 wt% HCl water solution for 45 days [185]. A lower corrosion rate was found with all DLC-coated samples. These results accord with those previously reported by others [8, 186].

Kim et al. [11] used electrochemical techniques (potentiodynamic polarization test, electrochemical impedance spectroscopy) and surface analysis (atomic force microscopy, scanning electron microscopy) and reported good corrosion resistance. Further research by others has confirmed that the corrosion resistance of DLC is excellent in acids, including HCl, HF, HNO_3 and H_2O_4, and in alkaline solutions, such as NaOH and KOH [187, 188]. Mitura et al. [152] carried out a 52-week *in vivo* test which showed that the implantation of DLC-coated discs subcutaneously, intramuscularly and in tibial bone resulted in the formation of a thin collagenous capsule, with no evidence of corrosion products or chronic inflammatory reactions.

16.6 DLC for orthopaedic applications

In 2003, Freitas stated that 'the largest anticipated biomedical use of diamond-like carbon is in orthopaedics and articulated prostheses' [189]. *In vivo* biocompatibility of DLC-coated cobalt–chromium cylinders implanted in intramuscular locations in rats and transcortical sites in sheep for 90 days demonstrated good tissue implant interaction [162]. DLC-coated zirconium exhibited enhanced *in vivo* osteointegration, compared to zirconium, titanium and aluminium, when implanted in the tibiae of Wistar rat *in situ* for a 30-day period [190]. Further research found DLC-coated artificial joints to have 'low immunoreactivity' [191]. *In vitro* comparisons of the cytotoxicity of possible intra-articular wear particles, HMWPE, bone cement, chromium–cobalt, DLC and SiC showed that DLC and Si are 'comparatively harmless'[192, 193]. Although the biocompatibility of DLC coating for orthopaedic applications is well established, inconsistent results have been found regarding the tribological aspects of the articulating DLC-coated surfaces [127].

16.6.1 DLC is not the best candidate for articulating implants

Scholes et al. [194, 195] compared the wear and frictional behaviour of five different hip prostheses: stainless steel/UHMWPE, zirconia/UHMWPE, DLC (on stainless steel)/UHMPW, alumina/alumina, and CoCrMo/CoCrMo. Carboxymethylcellulose solutions and bovine serum were used as the lubricant. Results showed that the performance of DLC/UHMPW was inconsistent; the tribological behaviour of DLC-coated hip prostheses is highly dependent on the surface roughness of the as-deposited surface, the surface chemistry and the lubricant condition. Gilmore and Hauert [196] and

Wimmer et al. [197] in the EMPA research group further revealed that the moisture and surface sensitivity of articulating DLC surfaces is affected by the generation and delamination of the tribochemical reaction layers. Saikko et al. [198] studied the wear of UHMWPE acetabular cups against alumina, CoCr and DLC-coated CoCr femoral heads, using a biaxial hip wear simulator with diluted calf serum as the lubricant. Although detailed studies were not made of the surface and compositional characteristics of the DLC coating used in their experiment, they reported that the DLC coatings did not markedly differ from alumina and CoCr, suggesting that DLC may not be the ideal coating for articulating implant parts.

Recently Taeger et al. [199] published data about the clinical failure rate of an 8-year follow-up on 101 patients with implanted DLC-coated femoral balls articulating against polyethylene. The DLC-coated femoral heads with the trade name Adamante™ (Biomécanique, France) consisted of a 2–3 nm thick DLC coating on Ti$_6$Al$_4$V alloy ball, made by ion-beam deposition. The DLC-coated implants showed no signs of complications within the first 1.5 years. Thereafter more and more DLC implants showed aseptic loosening, requiring revision of the implant. Within 8.5 years, 45% of the originally implanted DLC-coated joints had to be replaced. The DLC coating on the retrieved joint heads showed numerous, predominantly round-shaped pits, and the procedure was considered to be a 'clinical failure' despite the excellent experimental results for the coating's tribochemical and wear characteristics.

In 2001, a Diamond Rota Gliding™ (Implant Design, Switzerland) DLN/UHMWPE knee joint was marketed without sufficient preliminary tests. Within a short time, some of the approximately 190 implanted joints showed increased wear and partial coating delamination, and had to be replaced. Additionally, residual coating on the upper side of the implant was held responsible for the inadequate bone ingrowth. In July 2001, the implantation of this knee joint was forbidden by the Swiss Federal Office of Public Health [184].

16.6.2 DLC is proven superior for articulating implants

Wear resistance tests were conducted by Dowling et al. [164] on 22-mm diameter stainless steel (316L) femoral heads with DLC coatings, with an sp^3 fraction of 40–50%. This was done in a multi-station hip simulator in distilled water. Measurements in terms of volume changes of UHMWPE cups showed significant improvements in wear resistance compared to the uncoated surface. Lappalainen et al. [187] and Dearnaley and Lankford [200] compared polyethylene acetabular cups against CoCrMo alloy, stainless steel AISI316L, AISI420, Ti$_6$Al$_4$V alloy, alumina and DLC-coated CoCrMo specimens with sp^3 fraction of 80%. Results suggested that DLC coatings with an appropriate structure can potentially decrease the wear rate by a factor of 30–600. Affatato *et al.* [201] performed an *in vitro* test using a 10-station hip joint simulator for up to 6 million cycles and demonstrated that the wear rate of UHMWPE observed with the DLC-coated femoral heads was comparable to that for femoral heads made from zirconia.

Promising wear results were found from knee wear tests carried out by Oñate *et al.* [202] at 50 MPa using distilled water at 37°C as lubricant. Tests of 5 million cycles were carried out for DLC on CoCrMo/UHMWPE, TiN on CoCrMo/UHMWPE, and CoCrMo/UHMWPE. Dorner-Reisel *et al.* [203] also received positive results after 5 million cycles under a four-station knee simulator (Stanmore KS4) in bovine serum solution at 37°C for DLC on $Co_{28}Cr_6Mo$/UHMWPE and for $Co_{28}Cr_6Mo$/UHMWPE.

16.7 DLC for selective blood-interfacing applications

16.7.1 DLC for cardiac applications

In the late 1990s, all mechanical heart valves were still very thrombogenic, requiring mandatory high-dose warfarin treatment which led to the development of glassy carbon, a form of pyrolytic carbon used extensively for heart valves [204, 205]. However, several clinical failures unveiled the brittle nature of pyrolytic carbon [206]. *In vitro* studies proved that DLC-coated mechanical valves could further reduce the extent of valve-related thrombogensis compared to pyrolytic carbon [207, 208]. A compact (~6 × 6 cm, 280 gm) centrifugal blood pump has been developed as an implantable left ventricular assist system with the entire blood-contacting surface coated with diamond-like carbon (DLC) to improve blood biocompatibility [209].

Yamazaki and colleagues [209] at Sun Medical Technology Research Corporation (Suwa, Nagano, Japan) implanted a DLC-coated centrifugal ventricular blood pump device in calves, and found only minor thrombosis on DLC-coated surfaces, without post-operative anticoagulation. The Cardio Carbon Company (Swansea, UK) [210] developed two DLC-coated titanium implants that are currently the subject of clinical trials: an 'Angelini Laminaflo' mechanical heart valve and an 'Angelini Valvuloplasty' ring [210] used for heart valve repairs. The company Sorin Biomedica (Saluggia, Italy) produces heart valves and stents coated with 0.5 micron thick Carbofilm™ [210]. The film has a turbostatic structure equivalent to that of pyrolytic carbon. Clinical reports comparing the performance of various DLC-coated heart valves are rather inconsistent, due to the variations in leaflet design and inadequate information on the properties of the DLC coatings [211].

The VentraAssist™ left ventricular assist device by Ventracor Pty Ltd (Australia) is one of the latest FDA-approved and fully marketable DLC-coated cardiac devices. It is a new, third-generation cardiac assist system, primarily designed as a permanent alternative to heart transplant.

16.7.2 DLC for catheters and stents

Stents have traditionally been metallic because of the necessary mechanical requirements such as high expansibility with thin walls and high circumferential strength. However, metal surfaces are thrombogenic [212, 213]. Corrosion resistance is dependent upon the formation of a passive oxide film; if breached, metal ions are released, causing a foreign body inflammatory reaction [213–216] with a risk of tumour development [217]. The ideal stent should meet stringent requirements regarding thrombogenicity, biocompatibility and structure [218]. Histologically, in-stent restenosis appears to derive almost exclusively from neointimal hyperplasia [218]. Some improved prospects are reported for DLC-coated stents [177, 219, 220]. This is mainly due to the following advancements associated with DLC coatings [221]:

- Rapid proliferation of endothelial cells
- Non-activation of polymorphonuclear cells (no immune response)
- Strongly reduced activation of platelets and clotting factors (reduced risk of thrombosis)
- Strongly reduced serum proteins absorption, non-denaturation of proteins and enzymes (non-immune response, non-inhibition of enzymes, non-presentation of neoantigens)
- Non-recognition of foreign material, such as stainless steel, by biosystems
- No 'hyper-encapsulation' of foreign material (restenosis), as occurs with stainless steel 316 L.

At least four carbon-coated stents to the authors' knowledge are actively promoted for clinical use in Europe: the BioDiamond™ (Plasma Chem, Mainz, Germany), the Carbostent™ (Sorin, Saluggia, Italy), the Diamond Flex™ (Phytis, Dreieich, Germany), and the DLN or Dylyn™ (Bekaert, Kortrijk, Belgium). Despite the general consensus regarding the superiority of DLC-coated stents *in vitro* [221], the clinical reports on these stents are rather mixed [222, 223].

BioDiamond™

The first detailed studies on BioDiamond stents were performed by Barragan *et al.* [224] and suggested that BioDiamond stents helped to reduce haemostatic processes. Results from PTCA (percutaneous transluminal coronary angioplasty) indicated that the major adverse cardiac events are significantly lower with a DLC stent in comparison to other products [220]. The BioDiamond website claims that 'BioDiamond stents, without the use of any cytotoxic/cytostatic agent, demonstrate less than 4% restenosis for *de novo* lesions, and less than 10% for non-selected patients.' This implies that BioDiamond implants have the same or even better effectiveness than most modern drug eluting stents, for example the Cordis/Cypher.

Carbostent[TM]

Antoniucci et al. [210] demonstrated promising results in the prevention of restenosis using the Carbostent. The Carbostent was implanted in 112 patients, with 132 *in vivo* lesions. No stent deployment failure occurred, and no acute or subacute stent thrombosis. The 6-month event-free survival was 84 ± 4%. Bartorelli et al. [225] reported promising results based on non-randomized data from ANTARES (aspirin alone antiplatelet regimen after intracoronary placement of the carbostent) registry. This study indicated that the Carbostent may prevent stent thrombosis in selected patients treated with aspirin only. On the other hand, Colombo and Airoldi [226] randomly compared the Carbostent with different bare metal stents slightly differing in design and strut thickness (Tristar99, Tetra99, Penta99: Guidant, Santa Clara, CA). The authors reported similar incidence of binary angiographic restenosis at 6-month follow-up (18% in the Carbostent group, 21% in the stainless steel group; $p = 0.56$) as well as similar MACE (major adverse cardiac events) rates, suggesting that carbon coating does not provide clinical advantages in comparison to bare metal stents.

Three other studies comparing Carbostent with bare metal have recently been completed. To overcome the issue of possible bias due to different stent designs, these three studies compared stents with the same design, the only difference being the presence of the carbon coating. The main difference between these studies is in the type of carbon coating as well as clinical and angiographic characteristics. However, they all reached the same finding, that no significant difference in restenosis rates between carbostent and control stents was observed [226, 227].

Diamond Flex[TM]

Phytis L.D.A., a German stent-making company, has developed a stainless-steel stent 60–80 microns thick that is entirely coated with a diamond-like layer. Tests sponsored by the company showed that albumin adsorption was 20-fold less on the DLC-coated stents than on the SiO_2 and TiO_2 controls.

There was also a significant reduction of thrombogenic potential by the DLC stents compared to uncoated stents, which was further reduced for heparin-coated DLC stents [177].

Airoldi et al. [228] compared the performance of the stainless steel and the Diamond Flex stent in 347 patients (520 lesions) and found no significant improvement of the Diamond Flex over stainless steel stents with the same design.

Dylyn^{TM}

De Scheerder et al. [219] evaluated hydrogenated diamond-like carbon-coated (DLC, a-C:H) and diamond-like nanocomposite-coated (DLN or Dylyn, Bekaert, Kortrijk, Belgium) stents using a porcine scarified model, for 6 weeks. Results indicated that single-layered DLC stent coatings are compatible, resulting in decreased thrombogenicity and decreased neointimal hyperplasia. However, a double-layered DLC stent resulted in an increased inflammatory reaction and no additional advantage compared to the single-layer coating.

Clinical studies were performed by Riezebos et al. [229] on the composite incidence of death, non-fatal myocardial infarction, and revascularization, among 127 patients implanted with Dylyn-coated (DLC-coated) coronary stents. No episodes of acute stent thrombosis were reported. The number of patients with in-stent restenosis at 6 months was 29 (24%). The primary procedural, angiographical endpoint was achieved in 91 patients (70%). The clinical results at the 6-month follow-up were favourable with a MACE rate of 7.4%. However, the clinical results do not seem to outstrip traditional non-drug-eluting stents.

16.8 Conclusions

This detailed review has demonstrated that the scientific literature has not yet reached clear conclusions on some of the key claims made for the biomaterial benefits of DLC coatings. While there appears to be no doubt that DLC is highly biocompatible and non-thrombogenic, and can be produced with high hardness, the specific benefits of DLC wear coatings for orthopaedics and DLC anti-thrombogenic coatings for blood-interfacing implants are still inconclusive. Studies correlating the physicochemical properties of DLC coatings with *in vitro* and *in vivo* behaviour are scarce.

Reports over recent years have presented the dilemma between coating stability and tribological behaviour. In terms of biological evaluation of DLC coating, although significant data have been collected over the years, there is a lack of conclusive results in the literature concerning the effect of coating characteristics on biological performance. As a result, little progress can be made in evaluating clinical data for device improvement and optimization.

The fundamental issue for DLC in biomedical applications is to understand how the composition and structure of the coating influences the biocompatibility of the coating. Good basic-science studies in this area are the key to real progress. Therefore, while DLC has great potential as a biomaterial coating for improving wear resistance and reducing thrombogenicity, this potential has not yet been sufficiently explored or optimized.

16.9 Acknowledgements

The authors would like to acknowledge the Australia National Health, Medical Research Council and the Australian Research Council and AINSE postgraduate award for funding this research.

16.10 References

[1] Anderson JM, Miller KM. Biomaterial biocompatibility and the macrophage. *Biomaterials* 1984; 5(1): 5–10.
[2] Hauert R. A review of modified DLC coatings for biological applications. *Diamond and Related Materials* 2003; 12: 583–589.
[3] Cui FZ, Li DJ. A review of investigations on biocompatibility of diamond-like carbon and carbon nitride films. *Surface and Coatings Technology* 2000; 131(1–3): 481–487.
[4] Sioshansi P, Tobin EJ. Surface treatment of biomaterials by ion beam processes. *Surface and Coatings Technology: 9th International Conference on Surface Modification of Metals by Ion Beams* 1996; 83(1–3): 175–182.
[5] Ikada Y. Surface modification of polymers for medical applications. *Biomaterials* 1994; 15(10): 725–736.
[6] Courtney JM, Lamba NMK, Sundaram S, Forbes CD. Biomaterials for blood-contacting applications. *Biomaterials* 1994; 15: 737.
[7] Wolter JR. *Biomaterials in Ophthalmology*. Bologna: Academic Press, 1990.
[8] Maguire PD, McLaughlin JA, Okpalugo TIT, Lemoine P, Papakonstantinou P, McAdams ET, Needham M, Ogwu AA, Ball M, Abbas GA. Mechanical stability, corrosion performance and bioresponse of amorphous diamond-like carbon for medical stents and guidewires. *Diamond and Related Materials* 2005; 14(8): 1277–1288.
[9] Thomson LA, Law FC, Rushton N, Franks J. Biocompatibility of diamond-like carbon coating. *Biomaterials* 1991; 12(1): 37–40.
[10] Du C, Su XW, Cui FZ, Zhu XD. Morphological behaviour of osteoblasts on diamond-like carbon coating and amorphous C-N film in organ culture. *Biomaterials* 1998; 19(7–9): 651–658.
[11] Kim H-G, Ahn S-H, Kim J-G, Park SJ, Lee K-R. Corrosion performance of diamond-like carbon (DLC)-coated Ti alloy in the simulated body fluid environment. *Diamond and Related Materials* 2005; 14: 35–41.

[12] Pandey M, Bhatacharyya D, Patil DS, Ramachandran K, Venkatramani N, Dua AK. Structural and optical properties of diamond like carbon films. *Journal of Alloys and Compounds* 2005; 386(1–2): 296–302.
[13] Robertson J. Hard amorphous (diamond-like) carbons. *Progress in Solid State Chemistry* 1991; 21(4): 199–333.
[14] Robertson J. Properties of diamond-like carbon. *Surface and Coatings Technology* 1992; 50(3): 185–203.
[15] Roberston J. Amorphous carbon. *Advances in Physics* 1986; 35(4): 317–374.
[16] Koidl P, Wild C, Dischler B, Wagner J, Ramsteiner M. Plasma deposition, properties and structure of amorphous hydrogenated carbon films. *Material Science Forum* 1990; 52–53: 41–70.
[17] Balakrisnan B, Tomcik B, Blackwood DJ. Influence of carbon sputtering conditions on corrosion protection of magnetic layer by an electrochemical technique. *Journal of The Electrochemical Society* 2002; 149(3): B84–B88.
[18] McKenzie DR. Tetrahedral bonding in amorphous carbon. *Reports on Progress in Physics* 1996; 59(1996): 1611–1664.
[19] Franks J, Ng T, Wright A. Preparation and characteristics of diamond-like carbon films. *Vacuum* 1988; 38(8–10): 749–751.
[20] Matthews A, Eskildsen SS. Engineering applications for diamond-like carbon. *Diamond and Related Materials* 1994; 3(4–6): 902–911.
[21] Robertson J. Diamond-like amorphous carbon. *Materials Science and Engineering: R: Reports* 2002; 37(4–6): 129–281.
[22] Grill A. Diamond-like carbon: state of the art. *Diamond and Related Materials* 1999; 8(2–5): 428–434.
[23] Ma WJ, Ruys AJ, Mason RS, Martin PJ, Bendavid A, Liu Z, Ionescu M, Zreiqat H. DLC coatings: Effects of physical and chemical properties on biological response. *Biomaterials* 2007; 28(9): 1620–1628.
[24] Robertson J. Diamond-like carbon. *Pure and Applied Chemistry* 1994; 66(9): 1789–1796.
[25] Whitmell DS, Williamson R. Deposition of hard surface-layers by hydrocarbon cracking in a glow-discharge. *Thin Solid Films* 1976; 35: 255–261.
[26] Aisenberg S, Chabot R. Ion-beam deposition of thin films of diamondlike carbon. *Journal of Applied Physics* 1971; 42(7): 2953–2958.
[27] Kohary K, Kugler S. Growth of amorphous carbon: Low-energy molecular dynamics simulation of atomic bombardment. *Physical Review B (Condensed Matter and Materials Physics)* 2001; 63(19): 193–404.
[28] Marton D, Boyd KJ, Rabalais JW, Lifshitz Y. Semiquantitative subplantation model for low energy ion interactions with surfaces. II. Ion beam deposition of carbon and carbon nitride. *Journal of Vacuum Science and Technology A: Vacuum, Surfaces, and Films* 1998; 16(2): 455–462.
[29] Angus JC, Koidl P, Domitz S. *Plasma Deposited Thin Films*. Boca Raton, FL: CRC Press, 1986.
[30] Spencer EG, Schmidt PH, Joy DC, Sansalone FJ. Ion-beam-deposited polycrystalline diamondlike films. *Applied Physics Letters* 1976; 29(2): 118–120.
[31] Weissmantel C, Bewilogua K, Dietrich D, Erler H-J, Hinneberg H-J, Klose S, Nowick W, Reisse G. Structure and properties of quasi-amorphous films prepared by ion beam techniques. *Thin Solid Films* 1980; 72(1): 19–32.

[32] Lau WM, Bello I, Feng X, Huang LJ, Qin F, Yao Z, Ren Z, Lee ST. Direct ion beam deposition of carbon films on silicon in the ion energy range of 15–500 eV. *Journal of Applied Physics* 1991; 70(10): 5623–5628.
[33] Richter F. Formation of optical thin films by ion and plasma assisted techniques. *Contributions to Plasma Physics* 2001; 41: 549–561.
[34] Locher R, Wild C, Koidl P. Direct ion-beam deposition of amorphous hydrogenated carbon films. *Surface and Coatings Technology* 1991; 47(1–3): 426–432.
[35] Druz B, Ostan E, Distefano S, Hayes A, Kanarov V, Polyakov V, Rukovishnikov A, Khomich A, Rossukanyi N. Diamond-like carbon films deposited using a broad, uniform ion beam from an RF inductively coupled CH_4-plasma source. *Diamond and Related Materials* 1998; 7(7): 965–972.
[36] Weiler M, Sattel S, Jung K, Ehrhardt H, Veerasamy VS, Robertson J. Highly tetrahedral, diamond-like amorphous hydrogenated carbon prepared from a plasma beam source. *Applied Physics Letters* 1994; 64(21): 2797–2799.
[37] Martin PJ, Bendavid A. Review of the filtered vacuum arc process and materials deposition. *Thin Solid Films* 2001; 394: 1–15.
[38] Lifshitz Y, Lempert GD, Grossman E. Substantiation of subplantation model for diamondlike film growth by atomic force microscopy. *Physical Review Letters* 1994; 72: 2753–2756.
[39] Hofsäss H, Binder H, Klumpp T, Recknagel E. Doping and growth of diamond-like carbon films by ion beam deposition. *Diamond and Related Materials* 1994; 3(1–2): 137–142.
[40] Hofsäss H, Ronning C. *Beam Processing of Advanced Materials*. Cleveland, OH: ASME, 1995.
[41] Hirvonen JP, Koskinen J, Lappalainen R, Anttila A. In Pouch JJ, Alterovitz SA (eds), *Properties and Characterization of Amorphous Carbon Films*. Aedermannsdorf, Switzerland: Trans. Tech. Publications, 1990, p. 197.
[42] Chandra L, Allen M, Butter R, Rushton N, Hutchings IM, Clyne TW. Hydrogen-free amorphous carbon films: correlation between growth conditions and properties. *Diamond and Related Materials* 1996; 5(3–5): 388–400.
[43] Grill A, Meyerson BS. Development and status of diamond-like carbon. In Spear KE, Dismukes JP (eds), *Synthetic Diamond: Emerging CVD Science and Technology*. New York: Wiley, 1994, Chapter 6.
[44] Lifshitz Y. Diamond-like carbon – present status. *Diamond and Related Materials* 1999; 8(8–9): 1659–1676.
[45] Jansen F, Machonkin M, Kaplan S, Hark S. The effects of hydrogenation on the properties of ion beam sputter deposited amorphous carbon. *Journal of Vacuum Science and Technology A* 1985; 3: 605.
[46] Savvides N. Deposition parameters and film properties of hydrogenated amorphous silicon prepared by high rate dc planar magnetron reactive sputtering. *Journal of Applied Physics* 1984; 55(12): 4232–4238.
[47] Rossnagel SM, Russak MA, Cuomo JJ. Pressure and plasma effects on the properties of magnetron sputtered carbon films. *Journal of Vacuum Science and Technology A: Vacuum, Surfaces, and Films* 1987; 5(4): 2150–2153.
[48] Cho N-H, Krishnan KM, Viers DK, Rubin MD, Hopper CB, Bushan B, Bogy DB. Chemical structure and physical properties of diamond-like carbon films prepared by magnetron sputtering. *Journal of Materials Reseach* 1990; 5: 2543.

[49] Gielen JWAM, Van de Sanden MCM, Schram DC. Plasma beam deposited amorphous hydrogenated carbon: Improved film quality at higher growth rate. *Applied Physics Letters* 1996; 69: 152–154.
[50] Cuomo JJ, Doyle JP, Bruley J, Liu JC. Sputter deposition of dense diamond-like carbon films at low temperature. *Applied Physics Letters* 1991; 58: 466–468.
[51] Roberston J. Mechanical properties and coordinations of amorphous carbons. *Physical Review Letters* 1992; 68: 220–223.
[52] Kang GH, Uchida H, Koh ES. Macroparticle-free TiN films prepared by arc ion-plating process. *Surface and Coatings Technology* 1994; 68–69: 141–145.
[53] Aksenov II, Vakula SI, Padalka VG, Strel'nitskii VE, Khoroshikh VM. High-efficiency source of pure carbon plasma. *Soviet Physics – Technical Physics* 1980; 25(9): 1164–1166.
[54] McKenzie DR, Muller D, Pailthorpe BA. Compressive-stress-induced formation of thin-film tetrahedral amorphous carbon. *Physical Review Letters* 1991; 67(6): 773–776.
[55] Falabella S, Boercker DB, Sanders DM. Fabrication of amorphous diamond films. *Thin Solid Films* 1993; 236(1–2): 82–86.
[56] Schultrich B, Siemroth P, Scheibe H-J. High rate deposition by vacuum arc methods. *Surface and Coatings Technology* 1997; 93(1): 64–68.
[57] Tay BK, You GF, Lau SP, Shi X. Plasma flow simulation in an off-plane double bend magnetic filter. *Surface and Coatings Technology* 2000; 133–134: 593–597.
[58] Polo MC, Andujar JL, Hart A, Robertson J, Milne WI. Preparation of tetrahedral amorphous carbon films by filtered cathodic vacuum arc deposition. *Diamond and Related Materials* 2000; 9(3–6): 663–667.
[59] Bendavid A, Martin PJ, Kinder TJ, Preston EW. The deposition of NbN and NbC thin films by filtered vacuum cathodic arc deposition. *Surface and Coatings Technology* 2003; 163–164: 347–352.
[60] Coll BF, Chhowalla M. Modelization of reaction kinetics of nitrogen and titanium during TiN arc deposition. *Surface and Coatings Technology* 1994; 68–69: 131–140.
[61] Brown IG. Cathodic arc depositon of films. *Annual Review of Materials Science* 1998; 28: 243–269.
[62] Robertson J. Requirements of ultrathin carbon coatings for magnetic storage technology. *Tribology International* 2003; 36: 405–415.
[63] Voevodin AA, Donley MS. Preparation of amorphous diamond-like carbon by pulsed laser deposition: a critical review. *Surface and Coatings Technology* 1996; 82(3): 199–213.
[64] Siegal MP, Barbour JC, Provencio PN, Tallant DR, Friedmann TA. Amorphous-tetrahedral diamondlike carbon layered structures resulting from film growth energetics. *Applied Physics Letters* 1998; 73(6): 759–761.
[65] Venkatesan T. In Chrisey DB, Hubler GK (eds), *Pulsed Laser Deposition of Thin Films*. New York: John Wiley & Sons, 1994.
[66] Davanloo F, Juengerman EM, Jander DR, Lee TJ, Collins CB. Amorphic diamond film produced by a laser plasma source. *Journal of Applied Physics* 1990; 67: 2081–2087.
[67] Davanloo F, Juengerman EM, Jander DR, Lee TJ, Collins CB. Laser plasma diamond. *Journal of Materials Research* 1990; 5: 2398–2404.

[68] Collins CB, Davanloo F, Juengerman EM, Jander DR, Lee TJ. Preparation and study of laser plasma diamond. *Surface and Coatings Technology* 1991; 47: 244–251.
[69] Xiong F, Wang YY, Chang RPH. Complex dielectric function of amorphous diamond films deposited by pulsed-excimer-laser ablation of graphite. *Physics Review B* 1993; 48: 8016–8023.
[70] Catherine Y. Preparation techniques for diamond-like carbon. In Clausing RE et al. (ed), *Diamond and Diamond-like Films and Coatings*. NATO ASI 1991; 266: 193.
[71] Koidl P, Wild C. Structured ion energy distribution in radio frequency glow-discharge systems. *Applied Physics Letters* 1989; 54(6): 505–507.
[72] Zou JW, Reichelt K, Schmidt K, Dischler B. The deposition and study of hard carbon films. *Journal of Applied Physics* 1989; 65(10): 3914–3918.
[73] Zou JW, Schmidt K, Reichelt K, Dischler B. The properties of a-C:H films deposited by plasma decomposition of C_2H_2. *Journal of Applied Physics* 1990; 67(1): 487–494.
[74] Martinu L, Raveh A, Boutard D, Houle S, Poitras D, Vella N, Wertheimer MR. Properties and stability of diamond-like carbon films related to bonded and unbonded hydrogen. *Diamond and Related Materials* 1993; 2(5–7): 673–677.
[75] Zarrabian M, Fourches-Coulon N, Turban G, Marhic C, Lancin M. Observation of nanocrystalline diamond in diamondlike carbon films deposited at room temperature in electron cyclotron resonance plasma. *Applied Physics Letters* 1997; 70(19): 2535–2537.
[76] Celii FG, Butler JE. Diamond chemical vapor deposition. *Annual Review of Physical Chemistry* 1991; 42: 643–684.
[77] Weiler M, Sattel S, Jung K, Ehrhardt H. Highly tetrahedral, diamond-like amorphous hydrogenated carbon prepared from a plasma beam source. *Applied Physics Letters* 1994; 64: 2797–2799.
[78] Weiler M, Sattel S, Giessen T, Jung K, Ehrhardt H. Preparation and properties of highly tetrahedral hydrogenated amorphous carbon. *Physics Review B* 1996; 53: 1594–1608.
[79] Nonogaki R, Yamada S, Araki T, Wada T. High rate deposition of diamond-like carbon films by sheet-like plasma chemical vapor deposition. *Journal of Vacuum Science and Technology A: Vacuum, Surfaces, and Films* 1998; 17: 731–734.
[80] Nakamura K, Akaike H, Ninomiya Y, Tate Y, Fujimaki A, Hayakawa H. NbN/Nb/AlO$_x$/Nb/NbN junctions fabricated using an ultra-high vacuum dc-magnetron sputtering system. *Superconductor Science and Technology* 2001; 14: 1144–1147.
[81] Weiler M, Lang K, Li E, Robertson J. Deposition of tetrahedral hydrogenated amorphous carbon using a novel electron cyclotron wave resonance reactor. *Applied Physics Letters* 1998; 72: 1214–1316.
[82] Fallon PJ, Brown LM. Analysis of chemical-vapour-deposited diamond grain boundaries using transmission electron microscopy and parallel electron energy loss spectroscopy in a scanning transmission electron microscope. *Diamond and Related Materials* 1993; 2(5–7): 1004–1011.
[83] Chhowalla M, Robertson J, Chen CW, Silva SRP, Davis CA, Amaratunga GAJ, Milne WI. Influence of ion energy and substrate temperature on the optical

and electronic properties of tetrahedral amorphous carbon (ta-C) films. *Journal of Applied Physics* 1997; 81: 139–145.
[84] Egerton RF. *Electron Energy-loss Spectroscopy in the Electron Microscopy.* New York: Plenum Press, 1986.
[85] Gilkes KWR, Sands HS, Batchelder DN, Robertson J, Milne WI. Direct observation of sp^3 bonding in tetrahedral amorphous carbon using ultraviolet Raman spectroscopy. *Applied Physics Letters* 1997; 70(15): 1980–1982.
[86] Merkulov VI, Lannin JS, Munro CH, Asher SA, Veerasamy VS, Milne WI. UV Studies of tetrahedral bonding in diamondlike amorphous carbon. *Physical Review Letters* 1997; 78: 4869–4872.
[87] Ferrari AC, Robertson J. Resonant Raman spectroscopy of disordered, amorphous and diamond-like carbon. *Physics Review B* 2001; 64: 75414.
[88] Pan H, Pruski M, Gerstein BC, Li F, Lannin JS. Local coordination of carbon atoms in amorphous carbon. *Physics Review B* 1991; 44: 6741–6745.
[89] Savvides N. Optical constants and associated functions of metastable diamondlike amorphous carbon films in the energy range 0.5–7.3 eV. *Journal of Applied Physics* 1986; 59(12): 4133–4145.
[90] Kimock FM, Knapp BJ. Commercial applications of ion beam deposited diamond-like carbon (DLC) coatings. *Surface and Coatings Technology* 1993; 56(3): 273–279.
[91] Bull SJ. Tribology of carbon coatings: DLC, diamond and beyond. *Diamond and Related Materials* 1995; 4(5–6): 827–836.
[92] Bhushan B. Chemical, mechanical and tribological characterization of ultrathin and hard amorphous carbon coatings as thin as 3.5 nm: recent developments. *Diamond and Related Materials* 1999; 8(11): 1985–2015.
[93] Robertson J. Ultrathin carbon coatings for magnetic storage technology. *Thin Solid Films* 2001; 383(1–2): 81–88.
[94] Goglia PR, Berkowitz J, Hoehn J, Xidis A, Stover L. Diamond-like carbon applications in high density hard disc recording heads. *Diamond and Related Materials* 2001; 10(2): 271–277.
[95] Pharr GM, Callahan DL, McAdams SD, Tsui TY, Anders S, Anders A, Ager JW, Brown IG, Bhatia CS, Silva SRP, Robertson J. Hardness, elastic modulus, and structure of very hard carbon films produced by cathodic-arc deposition with substrate pulse biasing. *Applied Physics Letters* 1996; 68(6): 779–781.
[96] Jiang X, Reichelt K, Stritzker B. The hardness and Young's modulus of amorphous hydrogenated carbon and silicon films measured with an ultralow load indenter. *Journal of Applied Physics* 1989; 66(12): 5805–5808.
[97] Jiang X, Zou JW, Reichelt K, Grunberg P. The study of mechanical properties of a-C:H films by Brillouin scattering and ultralow load indentation. *Journal of Applied Physics* 1989; 66(10): 4729–4735.
[98] Schneider D, Meyer CF, Mai H, Schoneich B, Ziegele H, Scheibe HJ, Lifshitz Y. Non-destructive characterization of mechanical and structural properties of amorphous diamond-like carbon films. *Diamond and Related Materials* 1998; 7(7): 973–980.
[99] Nowack H, Muhlhan C. Plasma pretreatment of polypropylene for improved adhesive bonding. *Surface and Coatings Technology* 1998; 98(1–3): 1107–1111.

[100] Beghi MG, Bottani CE, Ossi PM, Lafford TA, Tanner BK. Combined surface Brillouin scattering and X-ray reflectivity characterization of thin metallic films. *Journal of Applied Physics* 1997; 81(2): 672–678.
[101] Moylan SP, Kompella S, Chandrasekar S, Faris TN. A new approach for studying mechanical properties of thin surface layers affected by manufacturing processes. *Journal of Manufacturing Science and Engineering* 2000; 125: 310–335.
[102] Pethica JB, Hutchings R, Oliver WC. Hardness measurement at penetration depth as small as 20 nm. *Philosophical Magazine A* 1983; 48(4): 593–606.
[103] Tabor D. The hardness of solids. *Review of Physics in Technology* 1970; 1: 145–179.
[104] Kelly A, Macmillan NH. *Strong Solid*. Oxford: Oxford University Press, 1986.
[105] Martinez E, Andujar JL, Polo MC, Esteve J, Robertson J, Milne WI. Study of the mechanical properties of tetrahedral amorphous carbon films by nanoindentation and nanowear measurements. *Diamond and Related Materials* 2001; 10(2): 145–152.
[106] Friedmann TA, McCarty KF, Barbour JC, Siegal MP, Dibble DC. Thermal stability of amorphous carbon films grown by pulsed laser deposition. *Applied Physics Letters* 1996; 68(12): 1643–1645.
[107] Forrest RD, Burden AP, Silva SRP, Cheah LK, Shi X. A study of electron field emission as a function of film thickness from amorphous carbon films. *Applied Physics Letters* 1998; 73(25): 3784–3786.
[108] Chhowalla M. Thick, well-adhered, highly stressed tetrahedral amorphous carbon. *Diamond and Related Materials* 2001; 10(3–7): 1011–1016.
[109] Grill A, Patel V. Stresses in diamond-like carbon films. *Diamond and Related Materials* 1993; 2(12): 1519–1524.
[110] Crouse PL. The effect of deposition parameters on the compressive stress in a-C:H thin films. *Diamond and Related Materials* 1993; 2(5–7): 885–889.
[111] Tamor MA, Vassell WC, Carduner KR. Atomic constraint in hydrogenated 'diamond-like' carbon. *Applied Physics Letters* 1991; 58(6): 592–594.
[112] Yan XB, Xu T, Yang SR, Liu HW, Xue QJ. Characterization of hydrogenated diamond-like carbon films electrochemically deposited on a silicon substrate. *Journal of Physics D: Applied Physics* 2004; 37: 2416.
[113] Meneve J, Dekempeneer E, Wegener W, Smeets J. Low friction and wear resistant a-C:H/a-Si$_{1-x}$C$_x$:H multilayer coatings. *Surface and Coatings Technology* 1996; 86–87(Part 2): 617–621.
[114] Sellers JC. The disappearing anode myth: Strategies and solutions for reactive PVD from single magnetrons. *Surface and Coatings Technology* 1997; 94–95(1–3): 184–188.
[115] Hirvonen J-P, Koskinen J, Jervis JR, Nastasi M. Present progress in the development of low friction coatings. *Surface and Coatings Technology* 1996; 80(1–2): 139–150.
[116] Oguri K, Arai T. Tribological properties and characterization of diamond-like carbon coatings with silicon prepared by plasma-assisted chemical vapour deposition. *Surface and Coatings Technology* 1991; 47(1–3): 710–721.
[117] Memming R, Tolle HJ, Wierenga PE. Properties of polymeric layers of hydrogenated amorphous carbon produced by a plasma-activated chemical vapour

deposition process. II: Tribological and mechanical properties. *Thin Solid Films* 1986; 143(1): 31–41.
[118] Oguri K, Arai T. Two different low friction mechanisms of diamond-like carbon with silicon coatings formed by plasma-assisted chemical vapor deposition. *Journal of Materials Research* 1992; 7(6): 1313.
[119] Ronkainen H, Koskinen J, Varjus S, Holmberg K. Experimental design and modelling in the investigation of process parameter effects on the tribological and mechanical properties of r.f.-plasma-deposited a-C:H films. *Surface and Coatings Technology* 1999; 122(2–3): 150–160.
[120] Krokoszinski H-J. Abrasion resistance of various thin film coatings on thick film resistor materials. *Surface and Coatings Technology* 1991; 49(1–3): 451–456.
[121] Grill A. Tribology of diamondlike carbon and related materials: an updated review. *Surface and Coatings Technology* 1997; 94–95(1–3): 507–513.
[122] Koskinen J, Schneider D, Ronkainen H, Muukkonen T, Varjus S, Burck P, Holmberg K, Scheibe H-J. Microstructural changes in DLC films due to tribological contact. *Surface and Coatings Technology* 1998; 108–109(1–3): 385–390.
[123] Donnet C. Recent progress on the tribology of doped diamond-like and carbon alloy coatings: a review. *Surface and Coatings Technology* 1998; 100–101(1–3): 180–186.
[124] Voevodin AA, Zabinski JS, Donley MS, Phelps AW. Friction induced phase transformation of pulsed laser deposited diamond-like carbon. *Diamond and Related Materials* 1996; 5(11): 1264–1269.
[125] Erdemir A, Eryilmaz OL, Nilufer IB, Fenske GR. Effect of source gas chemistry on tribological performance of diamond-like carbon films. *Diamond and Related Materials* 2000; 9(3–6): 632–637.
[126] Enke K. Some new results on the fabrication of and the mechanical, electrical and optical properties of i-carbon layers. *Thin Solid Films* 1981; 80(1–3): 227–234.
[127] Hauert R. An overview on the tribological behavior of diamond-like carbon in technical and medical applications. *Tribology International* 2004; 37: 991–1003.
[128] Zhang W, Tanaka A, Wazumi K, Koga Y. Effect of environment on friction and wear properties of diamond-like carbon film. *Thin Solid Films* 2002; 413(1–2): 104–109.
[129] Gardos MN. In Spear KE, Dismukes JP (eds), *Synthetic Diamond: Emerging CVD Science and Technology*. New York: John Wiley and Sons, 1994.
[130] Meletis EI, Erdemir A, Fenske GR. Tribological characteristics of DLC films and duplex plasma nitriding/DLC coating treatments. *Surface and Coatings Technology* 1995; 73(1–2): 39–45.
[131] Liu Y, Meletis EI, Erdemir A. A study of the wear mechanism of diamond-like carbon films. *Surface and Coatings Technology* 1996; 82(1–2): 48–56.
[132] Liu Y, Erdemir A, Meletis EI. An investigation of the relationship between graphitization and frictional behavior of DLC coatings. *Surface and Coatings Technology* 1996; 86–87 (Part 2): 564–568.
[133] Meletis EI, Liu Y, Erdemir A. Influence of environmental parameters on the frictional behavior of DLC coatings. *Surface and Coatings Technology* 1997; 94–95(1–3): 463–468.

[134] Kokaku Y, Kitoh M. Influence of exposure to an atmosphere of high relative humidity on tribological properties of diamond-like carbon films. *Journal of Vacuum Science & Technology A* 1989; 3: 2311–2314.
[135] Erdemir A, Nilufer IB, Eryilmaz OL, Beschliesser M, Fenske GR. Friction and wear performance of diamond-like carbon films grown in various source gas plasmas. *Surface and Coatings Technology* 1999; 120–121: 589–593.
[136] Erdemir A. The role of hydrogen in tribological properties of diamond-like carbon films. *Surface and Coatings Technology* 2001; 146–147: 292–297.
[137] Erdemir A, Eryilmaz OL, Fenske G. Synthesis of diamondlike carbon films with superlow friction and wear properties. *Journal of Vacuum Science and Technology A* 2000; 18: 1987–1992.
[138] Donnet C, Fontaine J, Grill A, Le Mogne T. The role of hydrogen on the friction mechanism of diamond-like carbon films. *Tribology Letters* 2000; 9: 137–142.
[139] Mori H, Tachikawa H. Increased adhesion of diamond-like carbon-Si coatings and its tribological properties. *Surface and Coatings Technology* 2002; 149(2–3): 224–229.
[140] Grischke M, Hieke A, Morgenweck F, Dimigen H. Variation of the wettability of DLC-coatings by network modification using silicon and oxygen. *Diamond and Related Materials* 1998; 7(2–5): 454–458.
[141] Han H, Ryan F, McClure M. Ultra-thin tetrahedral amorphous carbon film as slider overcoat for high areal density magnetic recording. *Surface and Coatings Technology* 1999; 120–121: 579–584.
[142] Terumitsu H, Atsushi S, Tetsuya S, Yoshiaki M, Toshiya S, Satoshi Y, Aki K, Nobuyuki S, Mutsumi H, Kanako K, Hirokuni Y, Sachio K. Fluorinated diamond-like carbon as antithrombogenic coating for blood-contacting devices. *Journal of Biomedical Materials Research Part A* 2006; 76(1): 86–94.
[143] Allen M, Law F, Rushton N. The effects of diamond-like carbon coatings on macrophages, fibroblasts and osteoblast-like cells in vitro. *Clinical Materials* 1994; 17(1): 1–10.
[144] Allen MJ, Myer BJ, Law FC, Rushton N. The growth of osteoblast-like cells on diamond-like carbon (DLC) coatings in vitro. *Transactions of the Orthopaedic Research Society* 1995; 20: 489.
[145] O'Leary A, Dowling DP, Donnelly K, O'Brien TP, Kelly TC, Weill N, Eloy R. Diamond-like carbon coatings for biomedical applications. *Key Engineering Materials* 1995; 99–100: 301–308.
[146] Singh A, Ehteshami G, Massia S, He J, Storer RG, Raupp G. Glial cell and fibroblast cytotoxicity study on plasma-deposited diamond-like carbon coatings. *Biomaterials* 2003; 24(28): 5083–5089.
[147] Linder S, Pinkowski W, Aepfelbacher M. Adhesion, cytoskeletal architecture and activation status of primary human macrophages on a diamond-like carbon coated surface. *Biomaterials* 2002; 23: 767–773.
[148] Franks J, Finch D. Medical applications of diamond-like carbon coatings. In Coombs RHR, Robinson DW (eds), *Nanotechnology in Medicine and the Biosciences*. Amsterdam: Gordon and Breach, 1996, pp. 133–138.
[149] Lu L, Jones MW, Wu RL. Diamond-like carbon as biological compatible material for cell culture and medical application. *Biomedical Materials and Engineering* 1993; 3(4): 223–228.

[150] Linder S, Pinkowski W, Aepfelbacher M. Adhesion, cytoskeletal architecture and activation status of primary human macrophages on a diamond-like carbon coated surface. *Biomaterials* 2002; 23(3): 767–773.
[151] Watanabe H, Awazu K, Yoshida H, Takahashi K, Iwaki M. Effect of ion implantation on ion-plated diamond-like carbon films. *Diamond and Related Materials* 1994; 3(8): 1117–1119.
[152] Mitura E, Wawrzyniak P, Rogacki G, Szmidt J, Jakubowski A. The properties of diamond-like carbon layers deposited onto SiO_2 aerogel. *Diamond and Related Materials* 1994; 3(4–6): 868–870.
[153] Olborska A, Swider M, Wolowiec R, Niedzielski P, Rylski A, Mitura S. Amorphous carbon – Biomaterial for implant coatings. *Diamond and Related Materials* 1994; 3(4–6): 899–901.
[154] Zolynski K, Witkowski P, Kaluzny A, Has Z, Niedzielski P, Mitura S. Implants with hard carbon layers for application in Pseudoarthrosis femoris sin. Ostitis post fracturam apertam olim factam. *Journal of Chemical Vapor Deposition* 1996; 4: 232–239.
[155] Heinrich G, Grogler T, Singer RF, Rosiwal SM. CVD diamond coated titanium alloys for biomedical and aerospace applications. *Surface and Coatings Technology* 1997; 94–95(1–3): 514–520.
[156] Dorner-Reisel A, Schürer C, Irmer G, Simon F, Nischan C, Müller E. Diamond-like carbon coatings with Ca-O-incorporation for improved biological acceptance. *Analytical and Bioanalytical Chemistry* 2002; 374: 753–755.
[157] Mohanty M, Anilkumar TV, Mohanan PV, Muraleedharan CV, Bhuvaneshwar GS, Derangere F, Sampeur Y, Suryanarayanan R. Long term tissue response to titanium coated with diamond like carbon. *Biomolecular Engineering* 2002; 19: 125.
[158] Rodil SE, Olivares R, Arzate H, Muhl S. Properties of carbon films and their biocompatibility using in-vitro tests. *Diamond and Related Materials* 2003; 12: 931–937.
[159] Sheeja D, Tay BK, Nung LN. Feasibility of diamond-like carbon coatings for orthopaedic applications. *Diamond and Related Materials* 2004; 13: 184–190.
[160] Kumari TV, Kumar PRA, Muraleedharan CV, Bhuvaneshwar GS, Sampeur Y, Derangere F, Suryanarayanan R. In vitro cytocompatibility studies of diamond like carbon coatings on titanium. *Bio-Medical Materials and Engineering* 2002; 12(4): 329–338.
[161] Tang L, Tsai C, Gerberich WW, Kruckeberg L, Kania DR. Biocompatibility of chemical-vapour-deposited diamond. *Biomaterials* 1995; 16(6): 483–488.
[162] Allen M, Myer B, Rushton N. In vitro and in vivo investigations into the biocompatibility of diamond-like carbon (DLC) coatings for orthopedic applications. *Journal of Biomedical Materials Research (Applied Biomaterials)* 2001; 58: 319–328.
[163] Thamaraiselvi TV, Rajeswari S. Biological evaluation of bioceramic materials – a review. *Trends in Biomaterials and Artificial Organs* 2004; 18(1): 9–17.
[164] Dowling DP, Kola PV, Donnelly K, Kelly TC, Brumitt K, Lloyd L, Eloy R, Therin M, Weill N. Evaluation of diamond-like carbon-coated orthopaedic implants. *Diamond and Related Materials* 1997; 6(2–4): 390–393.

[165] Uzumaki ET, Lambert CS, Belangero WD, Freire CMA, Zavaglia CAC. Evaluation of diamond-like carbon coatings produced by plasma immersion for orthopaedic applications. *Diamond and Related Materials* 2006; 15(4–8): 982–988.
[166] Geoffrey D, James H. Biomedical applications of diamond-like carbon (DLC) coatings: a review. *Surface and Coatings Technology* 2005; 200(7): 2518–2524.
[167] Krishnan LK, Varghese N, Muraleedharan CV, Bhuvaneshwar GS, Derangere F, Sampeur Y, Suryanarayanan R. Quantitation of platelet adhesion to Ti and DLC-coated Ti in vitro using 125I-labeled platelets. *Biomolecular Engineering* 2002; 19(2–6): 251–253.
[168] Hauert R. A review of modified DLC coatings for biological applications. *Diamond and Related Materials, 13th European Conference on Diamond, Diamond-Like Materials, Carbon Nanotubes, Nitrides and Silicon Carbide* 2003; 12(3–7): 583–589.
[169] Kwok SCH, Yang P, Wang J, Liu X, Chu PK. Hemocompatibility of nitrogen-doped, hydrogen-free diamond-like carbon prepared by nitrogen plasma immersion ion implantation–deposition. *Journal of Biomedical Materials Research Part A* 2004; 70A(1): 107–114.
[170] Yang P, Huang N, Leng YX, Chen JY, Fu RKY, Kwok SCH, Leng Y, Chu PK. Activation of platelets adhered on amorphous hydrogenated carbon (a-C:H) films synthesized by plasma immersion ion implantation-deposition (PIII-D). *Biomaterials* 2003; 24: 2821–2829.
[171] Yang P, Kwok SCH, Chu PK, Leng YX, Chen JY, Wang J, Huang N. Haemocompatibility of hydrogenated amorphous carbon (a-C:H) films synthesized by plasma immersion ion implantation-deposition. *Nuclear Instruments and Methods in Physics Research B* 2003; 206: 721–725.
[172] Jones MI, McColl IR, Grant DM, Parker KG, Parker TL. Protein adsorption and platelet attachment and activation, on TiN, TiC, and DLC coatings on titanium for cardiovascular applications. *Biomedical Materials Research (Applied Biomaterials)* 2000; 52: 413–421.
[173] Jones MI, McColl IR, Grant DM, Parker KG, Parker TL. Haemocompatibility of DLC and TiC–TiN interlayers on titanium. *Diamond and Related Materials* 1999; 8(2–5): 457–462.
[174] Schaub RD, Kameneva MV, Borovetz HS, Wagner WR. Assessing acute platelet adhesion on opaque metallic and polymeric biomaterials with fiber optic microscopy. *Journal of Biomedical Materials Research Part A* 2000; 49(4): 460–468.
[175] Weisenberg BA, Mooradian DL. Hemocompatibility of materials used in microelectromechanical systems: Platelet adhesion and morphology in vitro. *Journal of Biomedical Materials Research Part A* 2002; 60(2): 283–291.
[176] Li DJ, Gu HQ. Cell attachment on diamond-like carbon coating. *Bulletin of Materials Science* 2002; 25: 7–13.
[177] Gutensohn K, Beythien C, Bau J, Fenner T, Grewe P, Koester R, Padmanaban K, Kuehnl P. In vitro analyses of diamond-like carbon coated stents: Reduction of metal ion release, platelet activation, and thrombogenicity. *Thrombosis Research* 2000; 99(6): 577–585.
[178] Alanazi A, Nojiri C, Kido T, Noguchi T, Ohgoe Y, Matsuda T, Hirakuri K, Funakubo A, Sakai K, Fukui Y. Engineering analysis of diamond-like carbon

coated polymeric materials for biomedical applications. *Artificial Organs* 2000; 24: 624–627.
[179] Butter R, Allen M, Chandra L, Lettington AH, Rushton N. In vitro studies of DLC coatings with silicon intermediate layer. *Diamond and Related Materials* 1995; 4(5–6): 857–861.
[180] Yu LJ, Wang X, Wang XH, Liu XH. Haemocompatibility of tetrahedral amorphous carbon films. *Surface and Coatings Technology* 2000; 128–129: 484–488.
[181] Beck RM. PhD thesis. University of Tubingen, Germany, 2001.
[182] Baranauskas V, Fontana M, Guo ZJ, Ceragioli HJ, Peterlevitz AC. Analysis of the coagulation of human blood cells on diamond surfaces by atomic force microscopy. *Nanotechnology* 2004; 15: 1661–1664.
[183] Yang Y. PhD thesis. Linköping University, Sweden, 1997.
[184] Hauert R, Muller U. An overview on tailored tribological and biological behavior of diamond-like carbon. *Diamond and Related Materials* 2003; 12: 171–177.
[185] Tiainen V-M. Amorphous carbon as a bio-mechanical coating: mechanical properties and biological applications. *Diamond and Related Materials* 2001; 10: 153–160.
[186] Mudali UK, Sridhar TM, Raj B. Corrosion of bio implants. Sādhanā 2003; 28 (Parts 3 and 4): 601–637.
[187] Lappalainen R, Heinonen H, Anttila A, Santavirta S. Some relevant issues related to the use of amorphous diamond coatings for medical applications. *Diamond and Related Materials* 1998; 7: 482–485.
[188] Okpalugo TIT, Ogwu AA, Maguire PD, McLaughlin JAD, Hirst DG. In-vitro blood compatibility of a-C:H:Si and a-C:H thin films. *Diamond and Related Materials* 2004; 13: 1088–1092.
[189] Freitas RA. Cell response to diamond surfaces. In *Nanomedicine. Volume IIA: Biocompatibility*. Georgetown, TX: Landes Bioscience, 2003.
[190] Guglielmotti MB, Renou S, Cabrini RL. A histomorphometric study of tissue interface by laminar implant test in rats. *International Journal of Oral and Maxillofacial Implants* 1999; 14(4): 565–570.
[191] Pinneo M. Diamond growth: today and tomorrow. In Krummenacker M, Lewis J (eds), *Prospects in Nanotechnology: Toward Molecular Manufacturing, Proceedings of the First General Conference on Nanotechnology: Development, Applications, and Opportunities*. New York: John Wiley & Sons, 1992, pp. 147–172.
[192] Aspenberg P, Anttila A, Konttinen YT, Lappalainen R, Goodman SB, Nordsletten L, Santavirta S. Benign response to particles of diamond and SiC: Bone chamber studies of new joint replacement coating materials in rabbits. *Biomaterials* 1996; 17(8): 807–812.
[193] Santavirtas S, Takagi M, Gomez-Barrena E, Nevalainen J, Lassus J, Salo J, Konttinen YT. Studies of host response to orthopedic implants and biomaterials. *Journal of Long-term Effects of Medical Implants* 1999; 9(1–2): 67–76.
[194] Scholes SC, Unsworth A, Goldsmith AAJ. A frictional study of total hip joint replacments. *Physics in Medicine and Biology* 2000; 45: 3721–3735.
[195] Scholes SC, Unsworth A, Hall RM, Scott R. The effects of material combination and lubricant on the friction of total hip prostheses. *Wear* 2000; 241: 209–213.

[196] Gilmore R, Hauert R. Control of the tribological moisture sensitivity of diamond-like carbon films by alloying with F, Ti or Si. *Thin Solid Films* 2001; 398–399: 199–204.

[197] Wimmer MA, Sprecher C, Hauert R, Tager G, Fischer A. Tribochemical reaction on metal-on-metal hip joint bearings – a comparison between in-vitro and in-vivo results. *Wear* 2003; 225: 1007–1014.

[198] Saikko V, Ahlroos T, Calonius O, Keranen J. Wear simulation of total hip prostheses with polyethylene against CoCr, alumina and diamond-like carbon. *Biomaterials* 2001; 22: 1507–1514.

[199] Taeger T, Podleska LE, Schmidt B, Ziegler M, Nast-Kolb D. Comparision of diamond-like carbon and alumina-oxide with polyethylene in total hip arthroplasty. *Materialwissenschaft und Werkstofftechnik* 2003; 34(12): 1094–1100.

[200] Dearnaley G, Lankford J; Southwest Research Institute, assignee. Medical implants made of metal alloys bearing cohesive diamond like carbon coatings. US Patent 5725573, 1996.

[201] Affatato S, Frigo M, Toni A. An in vitro investigation of diamond-like carbon as a femoral head coating. *Biomedical Materials Research (Applied Biomaterials)* 2000; 53: 221–226.

[202] Oñate JI, Comin M, Braceras I, Garcia A, Viviente JL, Brizuela M, Garagorri N, Peris JL, Alava JI. Wear reduction effect on ultra-high-molecular-weight polyethylene by application of hard coatings and ion implantation on cobalt chromium alloy, as measured in a knee wear simulation machine. *Surface and Coatings Technology* 2001; 142–144: 1056–1062.

[203] Dorner-Reisel A, Schürer C, Muller E. The wear resistance of diamond-like carbon coated and uncoated Co28Cr6Mo knee prostheses. *Diamond and Related Materials* 2004; 13: 823–827.

[204] Okpalugo TIT, Ogwu AA, Maguire PD, McLaughlin JAD, Hirst DG. In-vitro blood compatibility of a-C:H:Si and a-C:H thin films. *Diamond and Related Materials* 2004; 13(4–8): 1088–1092.

[205] Ohgoe Y, Hirakuri KK, Tsuchimoto K, Friedbacher G, Miyashita O. Uniform deposition of diamond-like carbon films on polymeric materials for biomedical applications. *Surface and Coatings Technology* 2004; 184(2–3): 263–269.

[206] Richard G, Cao H. Structural failure of pyrolytic carbon heart valves. *Journal of Heart Valve Diseases* 1996; 5(S1): S79–S85.

[207] Dion I, Baquey C, Monties JR. Diamond: the biomaterial of the 21st century? *International Journal of Artificial Organs* 1993; 16: 623–627.

[208] Tran HS, Puc MM, Hewitt CW, Soll DB, Marra SW, Simonetti VA, Cilley JH, DelRossi AJ. Diamond-like carbon coating and plasma or glow discharge treatment of mechanical heart valves. *Journal of Investigative Surgery* 1999; 12(3): 133–140.

[209] Yamazaki K, Litwak P, Tagusari O, Mori T, Kono K, Kameneva M, Watach M, Gordon L, Miyagishima M, Tomioka J, Umezu M, Outa E, Antaki JF, Kormos RL, Koyanagi H, Griffith BP. An implantable centrifugal blood pump with a recirculating purge system (cool-seal system). *Artificial Organs* 1998; 22(6): 466–474.

[210] Antoniucci D, Bartorelli A, Valenti R, Montorsi P, Santoro GM, Fabbiocchi F, Bolognese L, Loaldi A, Trapani M, Trabattoni D. Clinical and angiographic outcome after coronary arterial stenting with the carbostent. *American Journal of Cardiology* 2000; 85(7): 821–825.

[211] Campbell A, Baldwin T, Peterson G, Bryant J, Ryder K. Pitfalls and outcomes from accelerated wear testing of mechanical heart valves. *Journal of Heart Valve Diseases* 1996; 5(s1): s124–s132.
[212] Pearce BJ, McKinsey MD, McKinsey JF. Current status of intravascular stents as delivery devices to prevent restenosis. *Vascular and Endovascular Surgery* 2003; 37(4): 231–237.
[213] Emneus H, Stenram U. Metal implants in the human body. A histopathological study. *Acta Orthopaedica Scandinavica* 1965; 36(2): 115–126.
[214] Emneus H. Experimental investigation of corrosion of stainless steels used in bone surgery. *Acta Orthopaedica Scandinavica* 1961; 44: 1–62.
[215] Hunt JA, Remes A, Williams DF. Stimulation of neutrophil movement by metal ions. *Journal of Biomedical Materials Research* 2004; 26(6): 819–828.
[216] Tracana RB, Sousa JP, Carvalho GS. Mouse inflammatory response to stainless steel corrosion products. *Journal of Materials Science: Materials in Medicine* 1994; 5(9–10): 596–600.
[217] Holmes SA, Cheng C, Whitfield HN. The development of synthetic polymers that resist encrustation on exposure to urine. *British Journal of Urology* 1992; 69: 651–655.
[218] Becker GJ. Intravascular stents. General principles and status of lower-extremity arterial applications. *Circulation* 1991; 83(2): 1122–1136.
[219] De Scheerder I, Szilard M, Yanming H, Ping XB, Verbeken E, Neerinck D, Demeyere E, Coppens W. Evaluation of the biocompatibility of two new diamond-like stent coatings (Dylyn) in a porcine coronary sent model. *Journal of Invasive Cardiology* 2000; 12(8): 389–394.
[220] Batyraliev TA. Medium and long-term effects of coronary stent BioDiamond implanation. *Terapevticheskii Arkhiv* 2002; 2: 57–60.
[221] Santin M, Mikhalovska L, Lloyd AW, Mikhalovsky S, Sigfrid L, Denyer SP, Field S, Teer D. In vitro host response assessment of biomaterials for cardiovascular stent manufacture. *Journal of Materials Science: Materials in Medicine* 2004; 15(4): 473–477.
[222] Monika W. The diamond prostheses. *Oberflaechen/Polysurfaces* 2006; 47(4): 16–17.
[223] Terumitsu H, Aki K, Atsushi H, Koki T, Tetsuya S. Application of DLC to medical apparatus. *Kagaku to kogyo* 2006; 59(10): 1064–1068.
[224] Barragan P, Herbst F, Kalachev A, Nader WF, Roquebert PO, Silvestri M, Siméoni JB. The BioDiamond and BioDiamond F stents. In Serruys PW, Kutryk MJB (eds), *Handbook of Coronary Stents*, 3rd edn. London: Dunitz, 2000, pp. 29–39.
[225] Bartorelli AL, Trabattoni D, Montorsi P, Fabbiocchi F, Galli S, Ravagnani P. Aspirin alone antiplatelet regimen after intracoronary placement of the Carbostent: the Antares study. *Catheterization and Cardiovascular Interventions* 2002; 55(2): 150–156.
[226] Colombo A, Airoldi F. Passive coating: the dream does not come true. *Journal of Invasive Cardiology* 2003; 15: 566–567.
[227] Edward A, Evans UH, Wusirika R, Morrison PW. Diamond like carbon coating for rhenium wires and foils. In *Proceedings of the MRS Fall Meeting* 1999, Boston, MA.

[228] Airoldi F, Colombo A, Tavano D, Stankovic G, Klugmann S, Paolillo V, Bonizzoni E, Briguori C, Carlino M, Montorfano M. Comparison of diamond-like carbon-coated stents versus uncoated stainless steel stents in coronary artery disease. *American Journal of Cardiology* 2004; 93(4): 474–477.

[229] Riezebos R, Ronner E, Kiemeneij F, Laarman GJ. A prospective multicentre registry of the procedural and long-term clinical and angiographic results of stent implantation using a Dylyn coated coronary stent system. *International Journal of Cardiovascular Intervention* 2004; 6(3–4): 137–141.

Part III
The effect of surfaces and proteins on cell response

17
Cell response to nanofeatures in biomaterials

A CURTIS and M DALBY, Glasgow University, UK

Abstract: Methods of making nanofeatures including nanotopography and nanoprinting are described and discussed. The range of cell reactions to these features is described in detail and the origins and possible alternative sources of these reactions is considered. Reactions considered include cell adhesion, cell spreading, cell orientation, mechanotransduction and gene expression. Symmetry, spacing and pattern on the nanofeatures are important in these reactions and the reasons for these are discussed. The involvement of mechanosensor systems and reactions at the genome are described. Finally possible uses nanofeaturing in medical devices is considered.

Key words: nanofeature, nanotopography, nanofabrication, cell shape, cell movement, mechanoreceptors. cytoskeleton, chromosome packing, gene expression, medical devices.

17.1 Introduction: why should cells respond to nanofeatures?

Though cells are appreciably larger than the range of sizes classed as nanometric, the irreducible size of many cellular components is nanometric in at least one dimension. For instance, actin fibres are about 6 nm in diameter, myosin filaments are 9 nm in diameter, the plasma membrane with its embedded mechanosensitive proteins is less than 10 nm thick, and attachment points of cells (the focal contacts) are down to 10 nm diameter and probably less than 3 nm high. Thus, it seems likely that cells would possess mechanisms for interacting with nanometric features of the surrounding environment. The smaller nanoparticles (<100 nm diameter) would have insufficiently large electrostatic forces of repulsion to prevent Brownian motion bringing them into contact with almost any surface, so that cells would be constantly experiencing encounters with these small particles. Cells would also be likely to interact with nanometric detail on extended surfaces, such as those on neighbouring cells or biomaterials.

In many ways the effective system for the lives of cells will be one that can react to nanometric dimensions or at least to objects that are of nanometric size on such surfaces. Perhaps even more important is the fact that many signalling molecules on the plasma membrane are individually nanometric and can be present with average molecule-to-molecule separation distances in the order of nanometres. One important matter that will be considered later is whether a cell can make some sort of reaction to a single nanometric feature or whether the reactions have to be to assemblies of such features and if so at what spacings. Once we consider single nanofeatures we are closing in on the realms of studies of nanoparticles, which is outside the areas considered in this chapter.

Studies of the effects of microtopography, mostly made in the 1980s and 1990s though dating back as far as Weiss (1945) at least, were a major origin for the work we now describe on nanotopography. At first these microscale studies were mostly interpreted as chemical effects on cell adhesion, etc., but the development of better and a wider range of fabrication methods allowed tests to be made of different topographies in the same material without local chemical differences and of the same topography in differing materials. Though surface chemical effects were clearly shown, topography seemed to be even more important: see, for example, Clark *et al.* (1987, 1990), Kasemo and Lausmaa (1994) and especially Britland *et al.* (1996) who competed chemical effects against topographic. There seems to be an almost seamless junction between micro- and nano-topographic effects without particular effects being limited to one or other scale (Clark *et al.* 1991, Brunette *et al.* 2001).

It is also noteworthy that many protozoa and similarly sized organisms bear complex surface nanostructures, e.g. ciliate protozoa, with which they interact with each other (e.g. in mating) or with other organisms, e.g. prey: see Yahel *et al.* (2006) and Hartmann *et al.* (2007).

Signalling between nanofeatured systems has only come to our attention with explorations of nanobioscience but clearly occurs, and this chapter will give special attention to that aspect.

Some major reviews have been published in or just before the first years of this century, for instance Wilkinson and Curtis (1996), Wilkinson *et al.* (1998), Curtis and Wilkinson (1997, 1999, 2001) and Desai (2000), but progress has been so rapid in recent years that the reader is advised to check up on the more recent literature, using databases and perhaps this review as sources.

17.2 Making artificial nanofeatures: fabrication methods including embossing

17.2.1 Top-down methods

The main methods of nanofabrication of topographies onto extended surfaces are divided into *top down*, e.g. electron beam lithography and holographic (direct writing) methods, and *bottom up*, which are self-assembling methods, e.g. colloidal lithography and polymer demixing.

Writing a pattern can be done by e-beam, X-ray beam and perhaps with lower resolution by ion beam. The writing is usually done with a resist coating to the surface to be treated. Some resists are positive ones, and others negative. The positive ones respond to beam irradiation by becoming more susceptible to chemical erosion during a developing stage so that the resist is removed. The negative ones become less susceptible so the exposed areas resist removal. Then the silicon, silica or other wafer material is etched by wet or dry processes – see Elwenspoek and Jansen (1998) and Madore and Landolt (1997). Wet processes (such as the use of CsOH) have a tendency to undercut vertical features.

Both the surface details of the dies used to make surfaces for cell attachment and also the cell attachment to replicas in polymer (p-caprolactone) made from these dies are shown in Fig. 17.1.

Once a topographically featured surface has been made it is common to wish to replicate this many times or to transfer the topography to a different material or to make an inverse copy of the original (so that pillars become pits and vice versa, for example).

Two main methods are available. The first consists of building up a metal (usually nickel) layer on the master by chemical deposition of a very thin metal film (often palladium) to provide an electrically conducting surface, followed by electroplating of about 280 microns of nickel. This value for thickness is related to a particular embossing machine. Usually detachment of the nickel shim is done by mechanical prising, but heat or the use of mould-release agents is possible. Ideally several shims can be made from one master.

Copies of masters or shims can also be made in polymer. Embossing or casting (see Casey *et al.* 1997, 1999) is done by pressing the shim into molten (casting) or near-melted polymer (above the glass transition temperature (T_g) embossing). Polymer casts from monomer plus polymerising catalyst, e.g. dimethylsiloxane and proprietary catalyst and epoxy polymers, or polymer in solution or melted polymer, are all routes to further copies (now the 'right-way round'). Polymer solutions are likely to be less satisfactory than other materials because the evaporation of the solvent is likely to distort the product. On some occasions the copy in polymer can be used for further castings or embossings. The silicone polymer polydimethylsiloxane (pDMS) is especially suitable for this.

Two points deserve special attention:

1. What sort of fidelity can we expect from these techniques? In part this depends upon the number of times a master or submaster is used and on the materials of which it is made. Other sources of deterioration are accumulation of dirt and damage in separating the master and its copy. Obviously overhanging and re-entrant masters are likely to give rise to

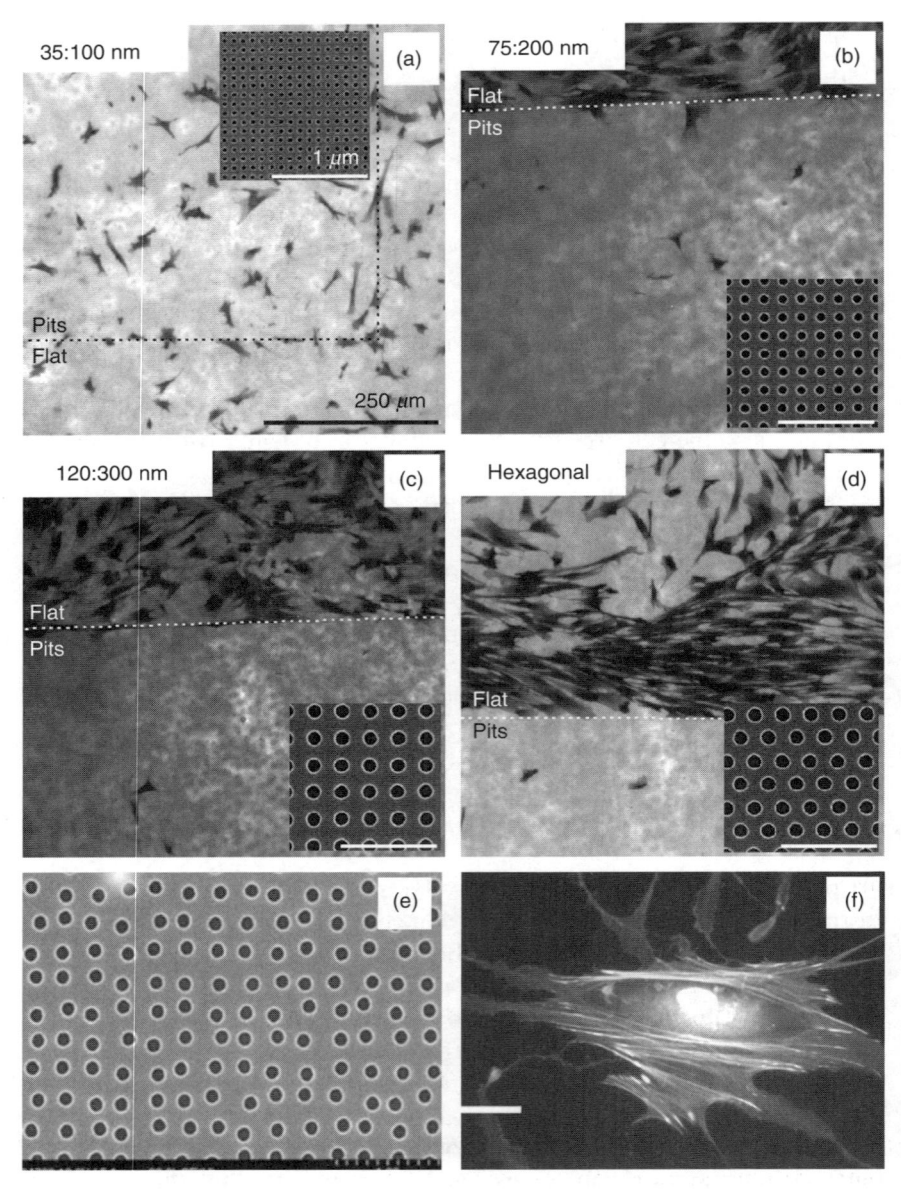

17.1 Growth and attachment reactions of *h-tert* (human immortalised) fibroblasts to four different nanopatterned surfaces (a to d) showing unpatterned (flat) and nanopitted areas in each image. Detail of the die shown as an insert in each image of cell attachment. Pit diameter and spacings shown in nm at the top left-hand corner for each image along with scale bars for both the detail of the die and of the cell attachment image. Cells were stained with Coomassie Blue. Note nearly complete absence of cells from the nanopitted areas of (b), (c) and (d). (e) The near-orthogonal pattern of pits (100 nm diameter; average spacing 300 nm) which is strongly adhesive for the *h-tert* cells compared with (c). (f) shows a cell growing on the surface made from die shown in (e). ((e) and (f) courtesy and permission of M. Biggs.)

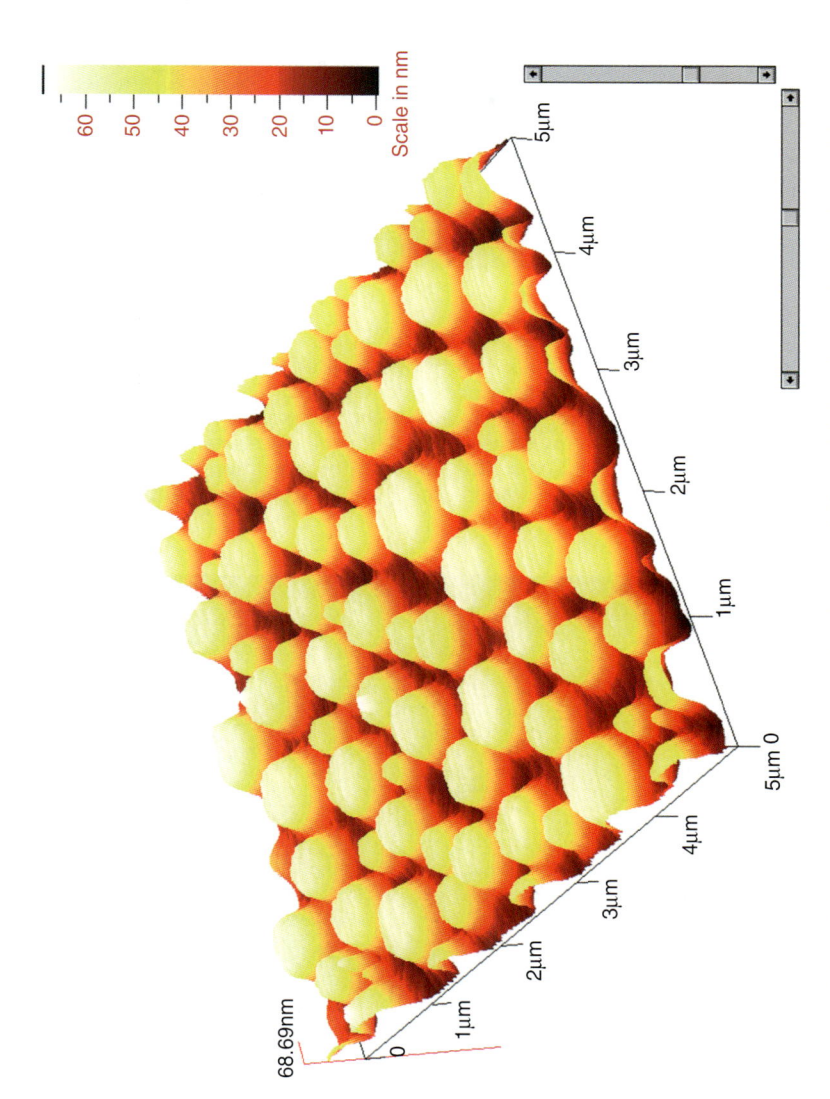

Plate 1 Atomic force microscopical image of 45 nm high polymer demixed nanoislands. Image courtesy of Dr Stanley Affrossman, University of Strathclyde.

(a) (b)

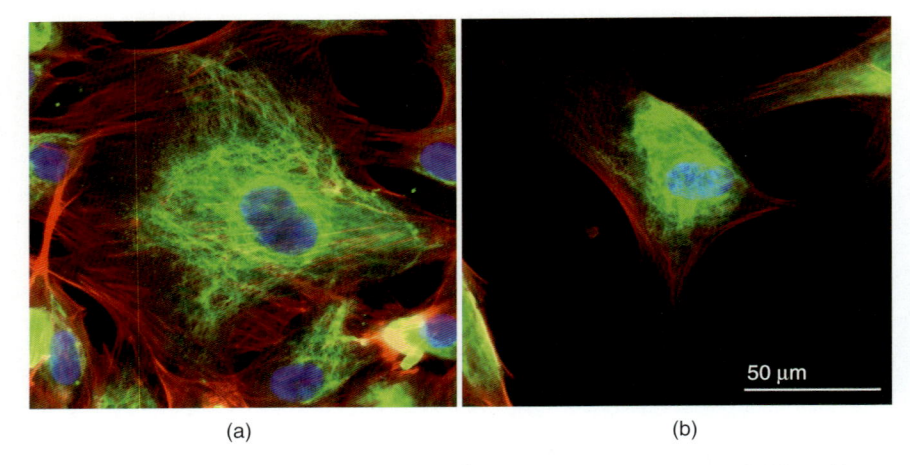

Plate 2 Human fibroblast cytoskeletons (actin and vimentin) in cells cultured on planar and nanopatterned (colloidal lithography – pillars 160 nm high, 100 nm diameter, fabricated by Dr Duncan Sutherland, Aarhus). (a) Well-spread cells with well-organised vimentin cytoskeleton and many stress fibres cultured on planar control. (b) Poorly spread cell with little vimentin organisation and few stress fibres cultured on nanoislands. Red = actin, green = vimentin, blue = nucleus.

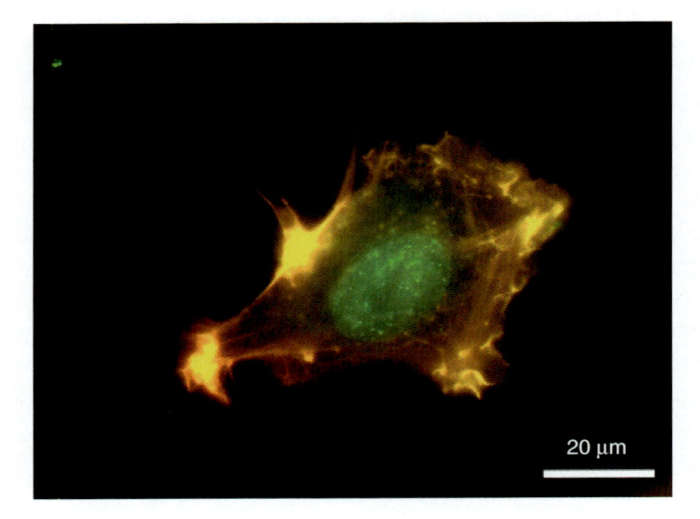

Plate 3 Cytoskeleton and nucleoskeleton of cells on nanostructures. Organisation of cytoskeleton appears to be linked to organisation of the nucleoskeleton – i.e. if the cell is well spread and the cytoskeleton well organised, the nuclear lamins appear diffuse. If, as in this case with a fibroblast cultured on a low-adhesion electron beam fabricated nanostructure, the cell is poorly spread and the cytoskeleton less well organised, the lamins appear dense and the nuclear morphology is clearly visible. yellow = actin, green = lamin.

difficulty in separating master and replicate. Good practice suggests that errors of 1–2 nm in dimension might be expected.
2. How do we best ensure that master and copy separate? Surface treatments such as Teflon sprays or fluorination may help. Swelling of the polymer by absorption of a swelling solvent such as heptane for pDMS may release the master from the embossed material or cast. Later the swelling solvent can be allowed to evaporate. Temperature change shock may help as well.

17.2.2 'Bottom-up' methods. Self-assembly and colloidal lithography

In these the nanostructure is assembled or self-assembled from nanometric-sized components, for example from gold colloids (size range 2 nm to thousands of nanometres): see especially the review by Wood *et al.* (2002a) and also Hanarp *et al.* (1999) and Cavalcanti-Adam *et al.* (2004). If the colloid suspension is at low concentration and its volume is small in relation to the area to be covered, the resulting dispersion of the particles will be essentially random, but with more concentrated colloids electrostatic and other forces may well induce a more ordered structure.

The nanoparticles can then be used as the topographic feature themselves, though there will of course be chemical differences between the particles and the intervening areas. Alternatively they can be used as a resist for some form of beam etch or as a master for embossing, etc., or they can be coated with a chemically homogeneous layer, for example of sputtered metal or vacuum-deposited materials from vapour.

A more controlled type of system is those where the components self-assemble into a nanostructure (Wood, 2002b). An early example of these is the phase separation systems which generate various types of nanotopography as the polymer mixtures cool and solidify. One rather widely used system has been the polystyrene/polybromostyrene (pS/pSBr) systems (Affrossman *et al.* 1993, 1994) which has the useful feature that after annealing the whole surface becomes a thin layer of pS so that the topography does not show surface variations in chemistry. Different proportions of the two polymers lead to different but reproducible topographies (see below).

Many techniques for preparing self-assembled nanometrically featured materials have been published in recent years and many of these will be suitable for making nanotopographic surfaces. A small selection of potentially useful techniques are given by Huck (2007), Nagai *et al.* (2007), Reches and Gazit (2007) and Lena *et al.* (2007), but here we concentrate on the polymer demixing strategy, which is inexpensive and capable of producing large areas of nanotopography, often with uniform surface chemistry.

Polymer demixing: 'islands'

The surface of a solid amorphous polymer, or polymer mixture, might be expected to be smooth and featureless. However, under certain circumstances marked topography can be obtained from a spontaneous demixing of the components of a binary blend, i.e. in thin films of certain polymer mixtures on selected substrates (see Plate 1 between pages 432–3).

The factors that determine demixing have been extensively studied. Principally, the mutual compatibility of the polymers controls the behaviour. The relationship between polymer compatibility and generation of topography in thin films of the blend is illustrated by examining a blend of poly(styrene) with a series of brominated polymers, poly(p-bromo$_x$-styrene), where x is the fraction of aromatic rings that are brominated, $1 \geq x \geq 0$. The compatibility of the brominated polymer with poly(styrene) depends on the extent of bromination (Affrossman et al. 1996). When $x = 1$ the polymers are incompatible and the compatibility increases as $x \to 0$. A measure of the compatibility is the thickness of the interfacial boundary layer between a poly(styrene) film and a poly(p-bromo$_x$-styrene) overlayer. Neutron reflectivity data showed a monotonic correlation between the interfacial layer thickness and the extent of bromination of the bromopolymer (Affrossman et al. 1998). Thin films of blends of the bromopolymer and polystyrene have smooth surfaces at low values of x. As x increases topography appears, becoming more marked with increase in x, corresponding to the decreasing compatibility.

The type of topography depends on the relative amounts of the polymer components of the blend (Affrossman et al. 1994, 1996). In the p(styrene)–p(p-bromostyrene) system, islands of the brominated polymer protrude from a sea of the p(styrene) when the weight fraction of the bromopolymer is less than ca. 60%. At higher weight fractions the bromopolymer islands link up to form a sheet with holes that expose the underlying poly(styrene). Similar changes in topography have been observed with other binary polymer mixtures.

The topography can be controlled further by changing the overall thickness of the polymer film. With spin-cast samples, this is readily accomplished by varying the concentration of the polymer blend solution and/or the spin speed. In general, increase in thickness of the cast films results in islands that are higher and wider. A concomitant change in the number density of the features is observed. More subtle control of the topography may be obtained by varying the molecular weights of the polymer components (Affrossman and Stamm 2000).

From the point of view of producing materials for study of cell interactions, it is desirable to reduce the experimental parameters by having the cells exposed to only one chemistry; the p(styrene)–p(p-bromostyrene) system produces a surface with patches of p(styrene) and patches of pp-bromostyrene. In this case, the surface chemistry can be changed by annealing the films (Affrossman et al. 2000). Heating the film to the glass transition temperature of the poly(styrene) causes the poly(styrene) to migrate over the surface, forming an overlayer, thus achieving topography with one chemistry at the surface. The topography is altered by this process, which typically causes the islands to grow in height and width. However, a minimum annealing gives a thin overlayer of p(styrene) with little change in the topography, (Affrossman et al. 1993).

Another polymer mixture, which gives similar island topography to the above, is poly(styrene)-poly(n-butylmethacrylate) (pnBMA) (Affrossman et al. 2000). In this system an overlayer is formed at room temperature. The cast film develops topography that is covered spontaneously by a thin overlayer of the acrylate.

The process of formation of the demixed polymer film is complex, involving the compatibility of the polymer components, their air–polymer and polymer–substrate surface tensions, the polymer component molecular weight, the solvent used, the rate of solvent evaporation, etc. The film is generally also metastable and can change if the temperature is sufficient to allow movement of one, or both, polymer components. These factors have been considered in the papers of Affrossman (Affrossman et al. 1998; Affrossman and Stamm 2000) and references therein, and the relevant paper for the PS–PnBMA system is Affrossman et al. (2000) so the reader can delve further into the subject if interested. It is not possible, however, to give a simple answer and a detailed discussion of polymer demixing/overlayer formation would be out of place in this review.

17.2.3 Surface treatment of nanofeatured materials

We can alter the surface chemistry of the whole of the nanostructured surface by choosing which polymer is used to make replicates or by an overall chemical treatment such as treatment with a plasma with an appropriate reactive ion (Favia et al. 2001), or a reactive wet chemical treatment which either oxidises, reduces or derivatises (e.g. silanes), etc., the surface or adsorbs strongly to the surface (e.g. polylysine). The silanes and some sulphydryl reagents, e.g. sulpho-SANPAH reagent (N-sulphosuccinimidyl-6-[4′-azido-2′-nitrophenylamino] hexanoate (photoactivatable)) are especially useful because further chemistries can be attached to the already derivatised surface. There are, however, possible problems because these derivatisations and other chemical reactions may show a preference for selective reaction sites determined by the topography. It may be remarked that little evidence for this has yet been found, and further remarked that not many people have tried to test this.

17.2.4 Printing and patterning

It may be more interesting to form chemical patterns on the surface. Printing can be done by using the 'stamp' system (Braun and Meyer 1999; Ostuni *et al.* 1999; Lee *et al.* 2002) in which the molecule to be printed is inked onto a stamp which carries the pattern upon it. This stamp is then contacted onto the surface to be printed. Usually the molecule is bound to the final substrate through reactive groups already bound to the substrate, but on occasion physical adsorption may work well enough to allow transfer and good final immobilisation. Usually pDMS stamps are used but that system does not really have nanometric resolution yet. However, colloidal lithography (see above) with gold nanoparticles allows greater resolution.

Natural structures as masters

It is appropriate to make mention of the fact that many natural structures provide regular master shapes that can be replicated by the methods mentioned above into similar structures in a variety of materials such as silica, pDMS and various organic polymers. With the exception of collagen (Gadegaard *et al.* 2003a) little use has been made of such possibilities.

17.3 The range of reactions to nanotopography

The data on the changes in expression of genes resulting from culture of cells on nanofeatured surfaces provides a logical way of introducing the wide variety of effects that have been reported. It is most apposite, partly from the history of research in this area and partly because interesting theories and explanations have their roots in that history, to start by describing the phenomena. These are widespread and at present our knowledge is too incomplete to make many useful generalisations, though some of the effects have been observed for quite a range of cell types and for ranges of feature size and spacing. Despite this caution it may be useful for the reader to have the data, which is summarised in Table 17.1.

For details of effects on shape, adhesion (including focal contacts), motility, cytoskeleton, phagocytosis, gene expression, apoptosis, endocytosis and activation, see Table 17.1 and Curtis *et al.* (2001, 2004). The results in Table 17.1 are not expressed in quantitative terms because this would occupy too much space. It is, however, clear that results can depend on the exact type of topography, spacings and sizes of nanotopography, as well as on the cell types. It seems probable that though the magnitude of the effect also depends on the chemistry of the substratum there is no known case where a change of substratum chemistry completely reverses the effect of the topography. Plate 2 (between pages 432–3) illustrates the changes that may

Table 17.1 The range of reactions to nanotopography

Property	Cell type	Topography	Effect	References
Cell adhesion	Fibroblast	Nanopit in pS	Decrease	Casey et al. 1997
	Endothelia	In silica and in pMMA	Decrease	Curtis et al. 2001, 2004
	Fibroblast	Demixed polymer islands pS	Reduction	Dalby et al. 2003a
	Fibroblast	Colloidal lith	Decrease	Dalby et al. 2004f
	Fibroblast	27 nm high nanoislands	Increase	Dalby et al. 2004g
	3T3	Complex in Ti nitride surface	Complex	Cyster et al. 2003
	3T3	Random nanotubes	Decrease	Lenardi et al. 2006
	Fibroblast	Ti and TAN surfaces	Depends on spacing	Meredith et al. 2005, 2007b
	Fibroblast	Nanopillar	Depends on spacing	Curtis et al. 2001, 2004
	Endothelia	Nanopillar	Raising	Cavalcanti-Adam et al. 2004
	Osteoblast	Nanopits near-square Ordered both in pMMA and silica	Decreased raised	Biggs et al. 2007a
	Osteoblast	Pyramids in Ti		
	Bladder smooth muscle	3-D nanostructured pEU		Pattison et al. 2006
Cell activation	Monocytes/macrophages	Silica nanodeep grooves	Increase	Wojciak-Stothard et al. 1995a, 1995b
	Platelets and monocytes	pCL/pEG mixtures demix to random hills	Increase	Hsu et al. 2004

Table 17.1 Continued

Property	Cell type	Topography	Effect	References
Cell movement	Fibroblast	Nanopillar	Changes related to type of nanotopography	Wood et al. 2006
	SMC	Nanogrooves	Increase	Yim et al. 2005
Morphology	Smooth muscle cells	pMMA, pDMS Nano G/R	Alignment	Yim et al. 2005
	Epithelia	Nano G/R on silicon oxide	Alignment	Teixeira et al. 2003
	Uroepithelia	Polystyrene nano G/R Colloidal lith	Alignment Alignment and shape	Zhu et al. 2004 Andersson et al. 2003a, 2003b
Neurite shape	Glioma	Nano G/R	Contact guidance	Johansson et al. 2006
	Various neurons	Nano G/R in poly(glycerol)sebacate	Alignment	Bettinger et al. 2006
Cell shape	Endothelia	pS, pSBr demix	Circularity	Dalby et al. 2002a, 2002b
	Human fibroblasts	Colloidal lithography	Spreading	Dalby et al. 2004e
Proliferation	Corneal cells	Polyurethane nano G/R	Corneal epithelia reduced; corneal fibroblasts little effect	Liliensiek et al. 2006
	Human fibroblasts	Polymer pS, pSBr demix	Increase	Dalby et al. 2002a, 2002b, 2002d
	Human fibroblasts	TAN	Reduction	Meredith et al. 2005

Cytoskeletal changes	Human fibroblasts	Polymer demix	Depends on topography	Dalby et al. 2002a, 2002b
Apoptosis	No record traceable			
Endocytosis	Human fibroblasts	Polymer demix	Enhanced attempts at endocytosis	Dalby et al. 2004b
Osteogenesis	Osteoblasts	Random nanoislands	Osteogenesis raised	Lim et al. 2005
	Osteoblasts	Nanopits/grooves	Some raised osteophenotype	Dalby et al. 2006a, 2006b
	Bone cells	Oxidised Ti surfaces, pattern not fully described	Osteogenesis	de Oliveira et al. 2007
Tissue rebuilding	Urinary epithelium	3-D nanoscaffold	Enhanced repair	Pattison et al. 2006
	Human fibroblasts	3-D nanoprojections on silicon	Lower proliferation, shape change, cell alignment	Choi et al. 2007
	Chondrocytes	3-D pLL nanofibres	Cell phenotype better preserved	Li et al. 2006
	Various	Semi-3-D electrospun polyamide nanofibres	Tissue-like growth	Schindler et al. 2005
General reviews				Curtis and Wilkinson 2001; Kriparamanan et al. 2006; Schindler et al. 2006

SMC = smooth muscle cell; G/R = grooved; TAN = titanium-aluminium-niobium alley; pEU = polyether urethane appear in a cell on being cultured (a) on a planar surface and (b) on a nanoisland surface.

Although no studies appear to have been carried out on the rate of change of a cell property in relation to a change of substratum and its nanofeatures, the impression is that such changes are well under way within a few hours. One easy way of investigating this is to provide the cell with a substratum composed of areas of differing nanotopography and to observe changes when a cell moves from one type of area to another. Explanations of these phenomena lie in two main areas, cell signalling and downstream from there the resulting changes in gene activity.

17.4 Cell signalling in response to nanofeatures

17.4.1 Introduction

This section will discuss the main features of nanostructure signalling. This section is divided into (i) properties of the nanofeatured surface and (ii) the signalling systems of the cell responsive to nanofeatures. The second of these topics is subdivided again.

It may seem obvious, but contact of the cell with the nanotopography is essential. But in saying this we do not know exactly how close that contact must be. Possibly any contact leads to signalling but duration of contact, area of contact and perhaps geometry all are likely to play a role. For many cells focal contacts appear to be involved (e.g., Meredith *et al.* 2007a, 2007b). Biggs *et al.* (2007b) have shown that changing the symmetry of a pattern (retaining nanofeature size and shape, changing only placing) can dramatically alter the types of adhesion cells form. For example, one symmetry may induce a prevalence of fibrillar adhesion formation (adhesions >5 μm long) whilst another may result in a shift to focal complex formation (adhesions <2 μm). Such adhesions have different roles with focal complexes acting as transient, motile structures and fibrillar adhesion resulting from alignment of focal adhesions (2–5 μm) to endogenous matrix proteins.

Sensitivity to size, spacing and pattern of array

One matter that has not been adequately studied is the proportion of the contact area of a cell that has to be in contact with a particular nanotopography for the cell to respond. However, the fact that cells often align to single cliffs of nanometric height though micrometric lateral extent (Andersson *et al.* 2003b) suggests that contact areas can be quite small, for example a few microns length of a single cliff.

17.4.2 Do effects arise solely from special features of nanofeatured surfaces?

One important and unresolved question is whether these effects of nanotopography can be explained solely in terms of the nanofeatures of the surface with which the cells are interacting or whether we have to bring in contributions from cellular mechanisms. Nanofeatured surfaces can be of low or high adhesion – correlated with superhydrophobicity and superhydrophilicity respectively – as well as of some intermediate value. In our work, surfaces patterned with random nanodots or pillars tended to be of intermediate adhesiveness for cells, but a high degree of order in the nanotopography (Curtis *et al.* 2001, 2004; Gadegaard *et al.* 2003b) tended to lower adhesion. We would expect quite a large effect of packing density simply because of increases in surface area due to the nanotopography, but conversely increased packing density will have the effect of increasing the average distance from surface to surface.

Certainly surface hydrophobicity and hydrophilicity are generally associated with adhesion processes and changes in surface nanotopography are reflected by changes in hydrophobicity and hydrophilicity (see Martines *et al.* 2005). But such findings are difficult to associate with some of the very precise reactions to feature pattern, e.g. high adhesiveness associated with 'near square' topography while perfect orthogonal patterns are of low adhesiveness. Such findings suggest that the cells may also be playing an important role in determining the reactions and perhaps in general, though we can make quite good predictions as to the adhesion and spreading of a fluid drop on such surfaces, predictions about cell behaviour are less reliable. If the surface adsorbs macromolecules the outcome depends in general terms on packing and conformation of the adsorbate (see Zhou *et al.* 2003) leading to enthalpic or entropic effects.

17.4.3 The involvement of cellular structures and processes

It is important to appreciate that though mechanical and cytoskeletal effects may be most important in the reaction of cells to nanofeatures, the whole process must start with the initial adhesion, which may take place in a very short time span, perhaps much less than seconds. See Curtis and Clark (1990) for some evidence on this. But it should also be borne in mind that the process of adhesion may continue in other parts of the cell while these other effects have started to come into play in the areas that were first to adhere.

Mechanotransduction

When we turn to consideration of the mechanisms that may be involved, one set of mechanisms seems especially relevant. This is that mechanotransduction is involved. The reasons for choosing this set are as follows:

- Externally applied mechanical forces produce many of the same results as growing cells on nanostructures, for instance cell alignment and shape change, and changes in gene expression.
- Cells growing on planar and on microstructured surfaces exert mechanical forces (Curtis 2005a; Curtis *et al.* 2007).
- Both *in vivo* and *in vitro*, many types of cells can strain their environment mechanically.

However, alternative explanations are possible and indeed their mechanisms may be combined with mechanotransduction. First, the cell may respond to very precise patterns of chemistry on the surfaces on the micro- or nanoscale. Evidence for this is not unambiguous because those chemical patterns may simply alter the attachments of the cells and thus the strengths and dispositions of the mechanical forces. Second, slow release of those surface chemicals might alter 'hormonal' pathways in the cells leading to more conventional signalling responses. Third, a more improbable idea is that the nanostructured surface interacts directly with the surface receptors in a manner sensitive to the exact nano-pattern. Little evidence for this is yet known. If we start to analyse the ways in which mechanotransduction is taking place we are again faced with two alternatives. The historically older explanation is that receptors such as stretch receptors, especially calcium channels, are being activated or shut down. These are mostly located in or near the cell surface. A more recent theory has its origin in the work of Heslop-Harrison *et al.* (1993) (see also Mosgoller *et al.* 1991). He and his colleagues observed the non-random placing of the chromosomes in the interphase nuclei of several organisms. Mechanical forces acting on the cell change the positions of the centromeres in the nucleus (Curtis and Seehar 1978, Curtis 1994) and of telomeric regions of the chromosomes (Dalby *et al.* 2007). (See also Section 17.5 on reactions at the genome, below.)

Thus mechanical forces could be changing chromosome packing, which would in turn alter the accessibility of various parts of the chromosome so that gene expression is altered (see below).

Mechanical transducers: filopodia

It seems likely that filopodia are one of the cells main sensory tools. Gustafson and Wolpert (1961) first described filopodia in living cells. They observed mesenchymal cells migrating up the interior wall of the blasto-

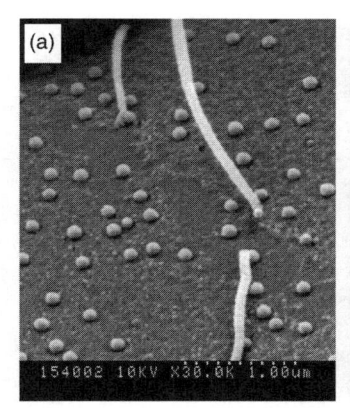

Cell response to nanofeatures in biomaterials 443

17.2 Fibroblast filopodial sensing of nanotopography (160 nm high, 100 nm in diameter) fabricated by colloidal lithography (fabricated by Dr D. Sutherland, iNano, Aarhus, Denmark): (a) low magnification SEM of several filopodia interacting with the nanocolumns; (b) high magnification SEM of a filopodium pushing against a nanocolumn.

coelic cavity in sea urchins and noted that the filopodia produced appeared to explore the substrate. This led them to speculate that they were being used to gather spatial information by the cells. This theme has been developed by Dalby *et al.* (2004a, 2004c).

When considering filopodial sensing of topography (Fig. 17.2), fibroblasts have been described as using filopodia to sense and align the cells to microgrooves (Clark *et al.* 1991). Macrophages have been reported to sense grooves down to a depth of 71 nm by actively producing many filopodia and elongating in response to the shallow topography (Wojciak-Stothard *et al.* 1995a, 1995b, 1996). Whilst distinctly different from the filopodia of the aforementioned cell types, neuronal growth cone filopodia have been described as firstly sensing microgrooves and then aligning neurons to the groove (Rajnicek *et al.* 1997, Rajnicek and McGaig 1997).

Cytoskeletal actin bundles drive the filopodia, and as the filopodia encounter a favourable guidance cue, they become stabilised following the recruitment of microtubules and accumulation of actin in a direction predictive of the future turn if a cell is to experience contact guidance.

Once cells locate a suitable feature using the filopodia presented on the cell's leading edge, lamellipodia are formed which move the cell to the desired site (Dalby *et al.* 2004c). These actions require G-protein signalling and actin cytoskeleton. Specifically of interest are Rho, Rac and Cdc42. Rho induces actin contractile stress fibre assembly to allow the cell to pull against the substrate, Rac induces lamellipodium formation, and Cdc42 activation is required for filopodial assembly (Schmitz *et al.* 2000). Rho and Rac are both required for cell locomotion, but cells can translocate when

Cdc42 is knocked out. Cells lacking Cdc42 cannot, however, sense chemotactic gradients and simply migrate in a random manner (Jones et al. 1998). This, again, presents compelling evidence for filopodial involvement in cell sensing.

17.4.4 Mechanosensor systems: ion channels

A considerable number of mechanosensitive ion channels have been described, for example calcium channels, potassium channels and chloride channels. Many of them seem to be involved in volume regulation in cells and in recent years the number of different proteins, known or putative, has increased considerably (see Hamill 2006, Matthews et al. 2007 and White 2006). But it is unclear in many cases whether the channel is directly linked to the internal cytoskeleton and how comparison of the strength of signal reception is handled between one part of the cell surface and another.

17.4.5 Importance of symmetry, spacing and pattern in nanofeatures

We have provided an introduction to these topics in the above sections. Here we deal with the question of cells measuring the substrate and possible implication of a near-surface structure in the cell cortex.

The remarkable sensitivity of cells to the dimensions and regularity of the features of the substrate on which they are growing or moving is such that feature spacings (x and y dimensions) or z-axis heights down to a few nanometres in some cases are detected by the cell. Cavalcanti-Adam et al. (2004, 2006) and Cavalcanti-Adam and Spatz (2006) found sharp cut-offs in cell adhesion once adhesion site arginine-glycine-aspartic acid tripeptide (RGD) patches fell below about 60 nm in patch-to-patch distance. Rather similar findings came from our own work with nanometric pillars or pits. In the first case the easy interpretation is that it is a chemical spacing, though the RGD patches will have physical height. In our own work there are strong arguments that it is simply physical dimension. The reason for that conclusion is that much the same effect is found with the same topography on a variety of different surfaces for a given cell type. Surfaces used have included silica, polycaprolactone (pCL), pDMS, polymethylmethacrylate (pMMA) (Curtis et al. 2001; Casey et al. 1997), etc., and a variety of different culture media with different macromolecular components that probably adsorb differentially to the surfaces. Cavalcanti-Adam et al. (2007) showed that cells on a 108 nm spaced pattern showed reduced spreading and poor persistence of contact related proteins. We discuss this in general terms in this section under the pattern matching concept.

It is tempting to consider that the cell uses some sort of measuring device to monitor the substratum. Several mechanisms can be envisaged, as follows.

Mechanical forces

The cell exerts mechanical forces on the substratum as a result of the actions of its cytoskeletal systems anchored in places onto the plasmalemma and thus through adhesion sites to the substratum. The pattern of forces applied to the substrate is affected by surface nanostructure. The forces arise from structures anchored to the cytoskeleton so that the interaction necessarily alters cell shape and movement (Safran *et al.* 2005). The spacing, symmetry, frequency and size of the nanofeatures would change the extent to which the cell was distorted by the nanofeatures. Such changes could be reflected in the stretch receptor reactions in the cell. The CLC-1 system is a pair of channels with three levels of response of the type that would be required for x and y sensing (Ludewig *et al.* 1997; Jentsch and Gunther 1997).

It is tempting to suggest that the transverse actin system of fibroblast-like cells (Albrecht-Buehler 1979; Lo and Gilula 1980; Curtis 2005b) is linked to this because it has rather regular spacing and could bring the adhesion receptors into good alignment with the intercellular adhesive proteins on the substratum, or very poor alignment if the spacing was wrong. One simple device would be a type of vernier device measuring distance by coincidences between substratum features and cytoskeletal ordering. This concept may have been foreshadowed by Bershadsky *et al.* (1985).

This concept can also be specifically set in the concept of the cell sending out microprobes (filopodia) (see Dalby *et al.* 2004c, 2004d); for filopodia which seem to explore the local environment, see above.

As the cell probes, the surroundings resist this (usually) and in turn generate forces within the cell. Safran and colleagues (Safran *et al.* 2005; Nicolas *et al.* 2004; Schwarz *et al.* 2002) have analysed the possible action of these forces in an interesting series of theoretical papers and have predicted polarising effects on the cells arising from interactions both with the substrate and with other nearby cells, as Curtis *et al.* (2007) have demonstrated by polarisation microscopy in cell cultures. A remarkable case of such interactions is shown in Fig. 17.3.

Cell fusion can be seen as an extreme aspect of this in which the cells pull themselves into so close an adhesion that collapse of the membrane structure leads to fusion.

The pattern matching concept

A development of the 'vernier' concept is that a 2-D pattern is established like a moiré pattern and that coincidences lead to signalling. There may be

446 Cellular response to biomaterials

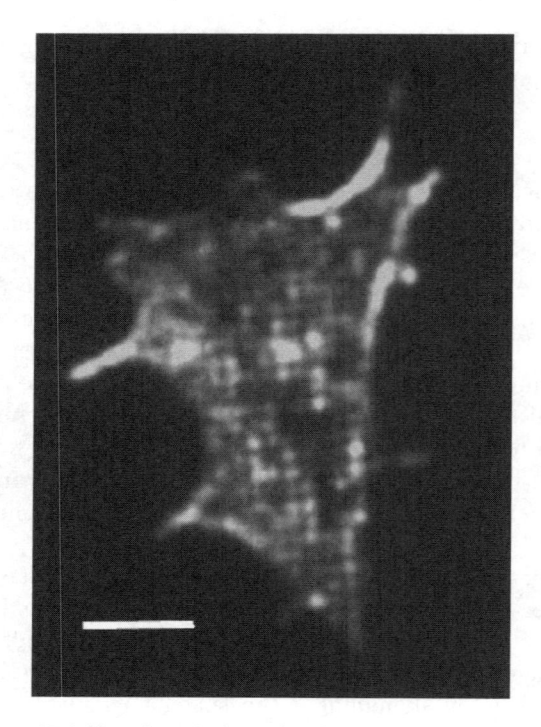

17.3 Nanoimprinting of substrate pattern into a cell and seen as pattern of actin. Note that the periodicity of the pattern in the cytoskeleton matches that of the nanotopography on the growth substrate. Scale bar is 2 µm.

interaction or attempted interaction between substratum and cell such that the nanopattern on the substratum imprints onto the cell. Curtis *et al.* (2006a) have seen this in platelets growing on a nanopatterned substratum (Fig. 17.3).

17.5 Reactions at the genome to morphological effects of topography on cells

17.5.1 Gene effects

It was considered that the morphological effects of topography on cells (adhesions, gross cell shape, cytoskeletal organisation, etc.) would result in downstream changes in the genome, but this, at the time, was theory rather than knowledge. The first studies to observe genome-wide responses utilised the then new technique of microarray and two types of topography, a grooved substrate fabricated by photolithography (Wilkinson *et al.* 2001, 2003) and a nanoisland substrate fabricated by polymer demixing (Affrossman *et al.* 1996, 1998, 2000; Affrossman and Stamm 2000).

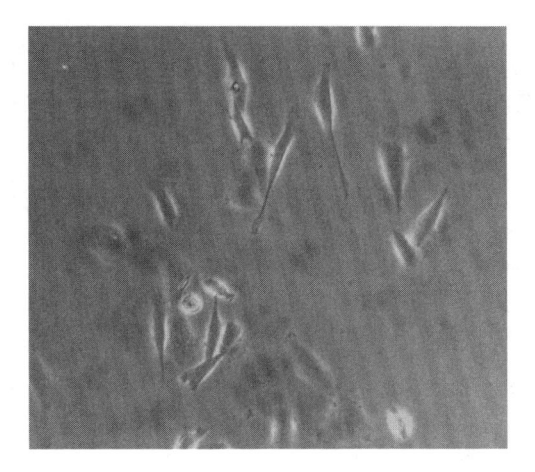

17.4 Capillary endothelial cells (mouse) responding by aligning to 120 nm deep grooves in silica buried under 6 μm of gelatin. Grooves are 5 μm wide.

The first studies (Dalby *et al.* 2002b, 2002c), were of 13 nm high islands, and were focused on just seeing if the genome was changed. The studies used 1.7 k arrays and grouped the changes into different gene functionalities: cytoskeleton, replication and cell function (transcription factors, etc.), extracellular matrix and signalling (G-proteins, kinase, ion signalling). It was seen that most of the major changes fitted into these categories and that these fitted well with the changes previously studied by immunohistochemistry and fluorescence microscopy of extracellular matrix, cytoskeletal staining and calcium transients (Wojciak-Stothard 1995a, 1995b) and more recently G-protein signalling in response to actin rearrangements in cell polarisation (Wojciak-Stothard and Ridley 2003).

The next study on grooves took a major step forward by relating changes in nucleus morphology alongside the expected changes in cell morphology (biopolarisation) associated with alignment to grooves (Dalby *et al.* 2003b). Again, the major gene regulatory changes fitted well into the aforementioned categories.

Of particular interest were the changes in cytoskeletal gene regulations (coupled with previous microscopical observations) and the changes in nuclear morphology resulting from contact guidance. The ability to change cytoskeletal organisation will lead to alterations in proliferation and differentiation as many signalling cascades are regulated by the cytoskeleton. These mechanotransductive signalling events may be chemical, e.g. kinase-based linked to focal adhesions influenced by cytoskeletal contraction; for example, integrin gathering as an adhesion is formed will activate myosin light chain kinase (MCLK) which will generate actin–myosin sliding (the key event in stress fibre contraction) and in turn will change focal adhesion kinase (FAK) activity (Burridge and Chrzanowska-Wodnicka 1996).

Another form of mechanotransduction, direct, is considered to be mediated by the cytoskeleton as an integrated unit. In such a case the cytoskeleton would have to act as an integrated unit applying tension to the nucleus and hence altering morphology (Hu et al. 2003; Wang and Suo 2005; Dalby et al. 2007). Several people have tried to describe the mechanical structure of the cytoskeleton needed to propagate the signals. Such theories include tensional integrity (Ingber 1993, 2003a, 2003b, 2003c; Charras and Horton 2002; Maniotis et al. 1997a, 1997b) and percolation theory (Forgacs 1995; Shafrir and Forgacs 2002; Forgacs et al. 2003).

It would also seem that an integrated cytoskeleton is important as actin microfilaments and tubulin microtubules are connected to focal adhesion, but only vimentin intermediate filaments are connected directly to the nucleus (Plate 2 between pages 432–3). They are, in fact, closely linked to the lamin intermediate filaments of the nucleoskeleton. It is further known that the telomeric ends of the interphase chromosomes are intimately linked to the lamins (Foster and Bridger 2005). Thus, tension directed through the cytoskeleton may be passed directly to the chromosomes during gene transcription. Changes in chromosomal three-dimensional arrangement may, in turn, affect transcriptional events such as access to the genes by transcription factors and polymerases. Also, changes in DNA tension can also cause polymerase enzymes to slow down, speed up or even stall completely (Bustamante et al. 2003). Further to this, if the theory of transcription factories (Osborne et al. 2004) is accounted for, perhaps changing nuclear morphology will move genes towards or away from the polymerase-rich factories and thus alter genomic regulation. We note that cytoplasmic hydrostatic pressure should also not be discounted when considering direct mechanotransduction (Charras et al. 2005, Plate 3 (between pages 432–3)).

These changes can be related to alterations in genome up and down regulations due in this case to growing cells on nanometrically deep groves (see Fig. 17.5).

These theories were first linked to topography through the study of the cytoskeleton, nuclear morphology and fluorescent in-situ hybridisation of centromeres in the interphase nuclei on 160 nm high columns in fibroblasts (Dalby 2005; Dalby et al. 2004a). These papers show that changes in nuclear morphology resulted in changes in centromere positioning and microarray output. This result was next repeated on 120 nm diameter pits fabricated by electron beam lithography and showed a similar result (Dalby et al. 2007). These studies both drew on the original ideas that interphase nuclei have a relative consistency of position (Heslop-Harrison et al. 1993). This

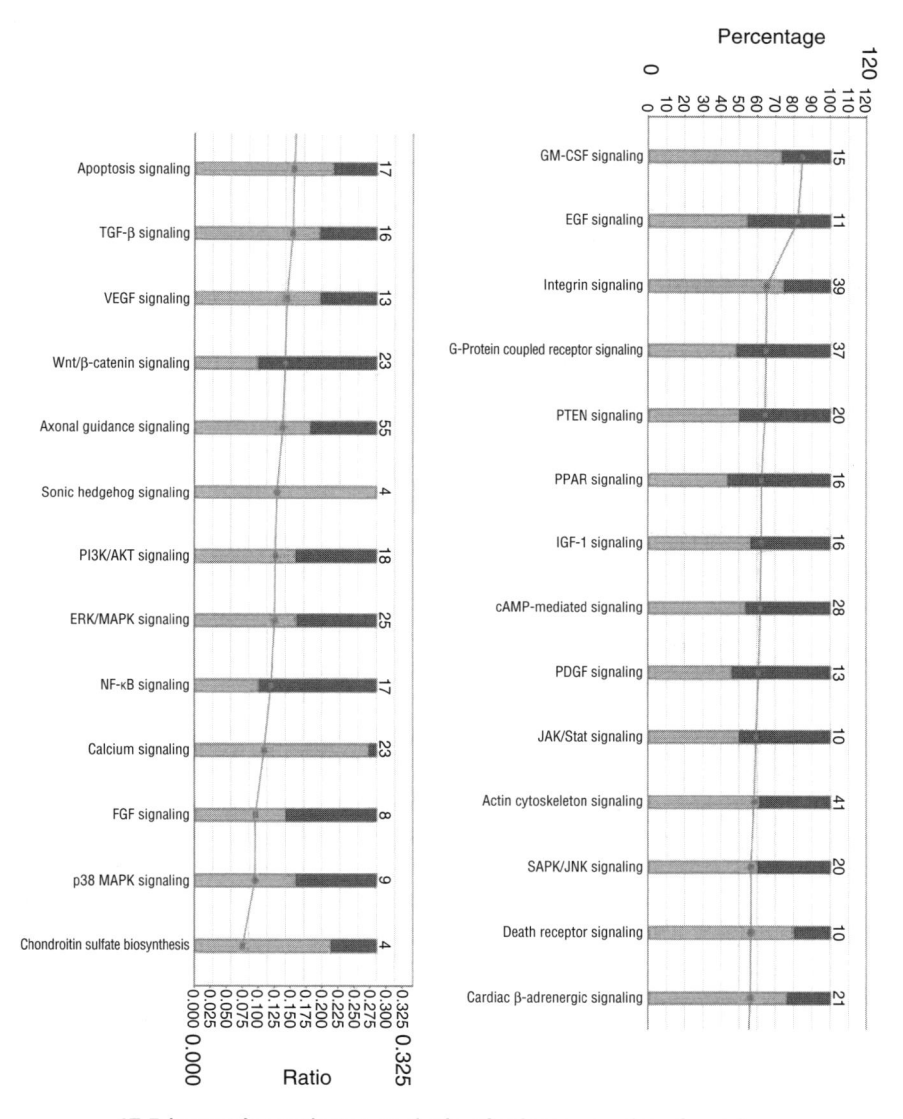

Cell response to nanofeatures in biomaterials 449

17.5 Ingenuity pathway analysis of microarray data for primary human osteoblasts cultured on 500 nm deep grooves. The figure shows canonical signalling pathway analysis, indicating that the culture of the cells on the grooves creates a large number of genome-wide up- and downregulations. Reproduced by courtesy of Mr M. Biggs.

concept has now developed into descriptions of chromosomal interphase territories (Kurz et al. 1996; Cremer and Cremer 2001).

A recent development regarding topography and genomic regulation has been the observation that changing nucleus morphology and centromere positioning leads to changes in transcription at different band positions along the chromosomal arms (Dalby et al. 2007). This study speculates that the chromosomes act rather like a collapsing net. As tension is released (due to the use of low-adhesion topographies) the first effects are felt near the centromeres (non-lamin associated) and then at the telomeres (lamin associated) as the collapse is more complete. Another version of this concept is that tension makes regions near the telomeres more accessible for transcription as the chromosome network is disentangled by tension.

An interesting thought is about cellular differentiation. We know that topography can alter stem cell differentiation (Dalby et al. 2006a, 2006b). Could this be by movement of non-protein coding genes for microRNA into areas of increased transcription? It has been postulated that these microRNAs have roles in differentiation (Bentwich 2005). Another discrete theory of how nanotopography may alter cytoskeletal mechanotransductive events has recently been mooted, that of nanoimprinting into cells by nanofeatures (Curtis et al. 2006a, 2006b). These papers describe a phenomenon that has been clearly seen in platelets and for which some evidence has been provided in more complex cell types. For nanoimprinting to occur, the pattern of the topography must be transferred to the cytoskeletal filaments, i.e. the topographies produce a template that is favourable or unfavourable to condensation of cytoskeletal polymer chains through invagination (embossing?) of the basal membrane against the topography. As with filopodia, the dimensions of the cytoskeletal elements are small enough to warrant consideration of influence by nanoscale features. There is evidence that this leads to increased 'attempted' endocytosis, i.e. the cells recognise the features as being in the correct size range to try to endocytose and to form clathrin coated pits (Dalby et al. 2004b). Such endocytotic vesicles are moved by actin cables, and perhaps it is this mechanism that is causing the topography to mimic the actin patterning described by Curtis et al. (2006a).

Three-dimensional nanostuctures and cells

Pattison et al. (2006) reported that the adhesion of bladder smooth muscle cells. Growth and extracellular material (ECM) production were enhanced by growth on nanofeatured scaffolds (see Schindler et al. 2006).

Concerning nanopatterning of 3-D scaffolds, the closest that has been achieved to date with controlled topography and cell testing is patterning of glass (Gadegaard et al. 2004) and nylon tubes (Berry et al. 2005, 2006) using polymer demixing. In essence, this is similar to polymer demixing on planar substrates, as has been described, but the polymers and solvent are blown up the tube rather than spun on the substrates. In the first study using glass tubes and fibroblasts, it was shown that varying the polymer blend could alter feature size (height of islands in this case from around 20 to 250 nm) and altering the pressure could change feature shape (from round islands at 1 psi to long, dash-shaped, islands at higher pressure, up to 10 psi tested). As with demixing on plane substrates, it was observed that varying the feature dimensions altered cell growth from low adhesion on larger islands (230 nm high) to enhanced adhesion on islands 67 nm high (Gadegaard et al. 2004). In nylon tubes, islands with a height of 90 nm were shown to inhibit both fibroblast (Berry et al. 2005) and osteoprogenitor (Berry et al. 2006) adhesion.

17.6 Related topics

17.6.1 Are micro-scale and nano-scale reactions fundamentally different?

Though we are still fairly ignorant about the mechanisms by which cells react to micrometric and nanometric features, no clear differences in effects on the cells between the two situations are yet known. In both cases groove–ridge (G/R) topographies align cells, though of course it is only in the z axis that they are likely to differ in dimension. Clark et al. (1991) did show that grooves nanometrically wide but micrometrically deep also aligned cells. In both cases cytoskeletal alignment is one of the obvious reactions of the cells. Changes in cell motility and morphology occur on both scales (Tan et al. 2000, Clark et al. 1991).

17.6.2 Relevance of nanotopography to making medical devices and other useful things

The simple deduction to be made from this review is that surface topography can be a powerful tool for selecting a specific set of cell responses. The precision with which these effects are achieved seems to depend on the precision and accuracy with which the structure is made. This leads to the conjecture that random arrangements and types of topography are likely to offer the cells a whole set of conflicting instructions. A slightly less obvious deduction is that it seems probable that different cell types will respond differently both to topography and to the dimensions of that topography. From this we can envisage a large range of materials with defined surface features that will either confer desirable properties on the materials or avoid undesirable consequences. These properties will be chosen from a range which includes cell adhesiveness, cell morphology and orientation, and cellular processes such as proliferation, movement, and various synthetic and degradative processes. All these will be underlain by changes in gene expression. Relevant reviews have been published by Wilkinson *et al.* (1998), Curtis (2005a, 2005b) and Curtis and Wilkinson (2001).

It is not clear whether there will be problems because a certain material surface offers some advantages along with some disadvantages. It should also be borne in mind that if biodegradable materials are used, a surface conferring certain properties can be removed as a result of biodegradation. This may be especially useful for tissue regeneration when stimulus of proliferation and of certain gene expressions may only be needed for a limited time.

17.6.3 Some practical aspects of nanotopography

Surface roughness is a deceitful friend. Though very many reports have stated that increasing surface roughness aids adhesion of cells and also of many non-living materials to each other, this is not a universal finding (Castellani *et al.* 1999; ter Brugge *et al.* 2002). For example, scratching (as in polishing) aids adhesion, while peening (shot blasting), which tends to produce relatively wide shallow pits, will frequently result in reduced cell adhesion. A cautionary tale which emerges from this review is that surface roughness is not sufficiently well characterised or defined to be of real utility is assessing or designing surfaces. Differing nanotopographies can have the same average surface roughness yet parallel grooves are likely to be far more adhesive for cells than other ones such as nanopits. The further implication is that techniques that allow the manufacture of surfaces of precise nanotopography will permit the manufacture of better products for nanomedicine and also probably for many other areas.

17.6.4 Mixed micro–nano scale biomaterials

These have only just begun to appear, presumably both because they represent major problems in fabrication and because we do not know enough about how cells respond to mixed scale cues to plan structures with any real confidence. However, they clearly might give us even more control over cell responses. Birch and Clayton (2007) describe a complex tube structure involving both micro- and nano-fabrication made by Meredith and colleagues which should be close to a capillary mimic.

17.7 Future trends

It should be clear by now that the studies on the reactions of cells to nanotopography impinge on a wide range of areas of research endeavour. These include fundamental questions about the control of gene expression, the mechanisms of interaction with nanotopography (and other topographies), whether the cell can operate a system to measure dimensional properties of the environment, mechanotransduction, whether some or all of the effects are the result of direct interactions and do not have to be routed through the genome, and how the cell compares different contacts on differing parts of the cell. All these are fundamental questions of cell biology. Then at the other end of our endeavours are questions and speculations about whether we can make use of any of this information in biomedicine and in the design and making of prostheses. Clearly there is much to be done but it may open up exciting vistas into parts of biology we have ignored because of the small scale of nanofeatures.

17.8 References

Affrossman, S. and Stamm, M. (2000). 'The effect of molecular weight on the topography of thin films of blends of poly(4-bromostyrene) and polystyrene'. *Colloid Polymer Science*, 278, 888–893.

Affrossman, S., Hartshorne, M., Jerome, R., Munro, H., Pethrick, R. A., Petitjean, S. and Vilar, M. R. (1993). 'Surface-composition of poly(styrene-D8-styrene) random copolymers studied by static secondary-ion mass-spectroscopy'. *Macromolecules*, 26(20), 5400–5404.

Affrossman, S., Hartshorne, M., Kiff, T., Pethrick, R. A. and Richards, R. W. (1994). 'Surface segregation in blends of hydrogenous polystyrene and perfluorohexane end-cappped deuterated polystyrene, studied by SSIMs and XPS'. *Macromolecules*, 27(6), 1588–1591.

Affrossman, S., Henn, G., O'Neill, S. A., Pethrick, R. A. and Stamm, M. (1996). 'Surface topography and composition of deuterated polystyrene–poly(bromostyrene) blends'. *Macromolecules*, 29, 5010–5015.

Affrossman, S., O'Neill, S. A. and Stamm, M. (1998). 'Topography and surface composition of thin films of blends of polystyrene with brominated polystyrenes: effects of varying the degree of bromination and annealing'. *Macromolecules* 31, 6280–6288.

Affrossman, S., Jerome, R., O'Neill, S. A., Schmitt, T. and Stamm, M. (2000). 'Surface structure of thin film blends of polystyrene and poly(n-butyl methacrylate)'. *Colloid Polymer Science*, 278(10), 993–999.

Albrecht-Buehler, G. (1979). 'Group locomotion of PtK1 cells'. *Exp. Cell Res.*, 122, 402–407.

Andersson, A.-S., Blackhed, F., von Euler, A., Richter-Dahlfors, A., Sutherland, D. and Kasemo, B. (2003a). 'Nanoscale features influence epithelial cell morphology and cytokine production'. *Biomaterials*, 24, 3427–3436.

Andersson, A. S., Olsson, P., Lidberg, U. and Sutherland, D. (2003b). 'The effects of continuous and discontinuous groove edges on cell shape and alignment'. *Exp. Cell Res.*, 288(1), 177–188.

Bentwich, I. (2005). 'A postulated role for microRNA in cellular differentiation'. *FASEB J.*, 19, 875–879.

Berry, C. C., Dalby, M. J., McCloy, D. and Affrossman, S. (2005). 'The fibroblast response to tubes exhibiting internal nanotopography'. *Biomaterials*, 26, 4985–4992.

Berry, C. C., Dalby, M. J., Oreffo, R.O.C., McCloy, D. and Affrossman, S. (2006). 'The interaction of human bone marrow cells with nanotopographical features in three dimensional networks'. *J. Biomed. Mater. Res.*, 79A, 431–439.

Bershadsky, A. D., Tint, I. S., Neyfakh, A. A. and Vasiliev, J. M. (1985). 'Focal contacts of normal and RSV-transformed quail cells. Hypothesis of the transformation-induced deficient maturation of focal contacts'. *Exp. Cell Res.*, 158, 433–444.

Bettinger, C. J., Orrick, B., Misra, A., Langer, R. and Borenstein, J. T. (2006). 'Micro fabrication of poly (glycerol-sebacate) for contact guidance applications'. *Biomaterials*, 27(12), 2558–2565.

Biggs, M. J. P., Richards, R. G., Gadegaard, N., Wilkinson, C. D. W. and Dalby, M. J. (2007a). 'The effects of nanoscale pits on primary human osteoblast adhesion formation and cellular spreading'. *J. Mater. Sci. – Mater. Med.*, 18(2), 399–404.

Biggs, M. J. P., Richards, R. G., Gadegaard, N., Wilkinson, C. D. W. and Dalby, M. J. (2007b). 'Regulation of implant surface cell adhesion: Characterization and quantification of S-phase primary osteoblast adhesions on biomimetic nanoscale substrates'. *J. Orthop. Res.*, 25(2), 273–282.

Birch, H. M. and Clayton, J. (2007). 'Close-up on cell biology'. *Nature*, 446, 937.

Braun, H. G. and Meyer, E. (1999). 'Microprinting – a new approach to study competitive structure formation on surfaces'. *Macromol. Rapid Commun.*, 20(6), 325–327.

Britland, S. I., Perridge, C., Denyer, M., Morgan, H., Curtis, A. and Wilkinson, C. (1996). 'Morphogenetic guidance cues can interact synergistically and hierarchically in steering nerve cell growth'. *Exp. Biol. Online – EBO*, 1:2.

Brunette, D. M., Tengvall, P., Textor, M. and Thomsen, P. (2001). *Titanium in Medicine*. Berlin: Springer.

Burridge, K. and Chrzanowska-Wodnicka, M. (1996). 'Focal adhesions, contractility, and signaling'. *Ann. Rev. Cell Dev. Biol.*, 12, 463–518.

Bustamante, C., Bryant, Z. and Smith, S. B. (2003). 'Ten years of tension: single-molecule DNA mechanics'. *Nature*, 421, 423–427.

Casey, B. G., Monaghan, W. and Wilkinson, C. D. W. (1997). 'Embossing of nanoscale features and environments'. *Microelectronic Engineering*, 35(1–4), 393–396.

Casey, B. G., Cumming, D. R. S., Khandaker, I. I., Curtis, A. S. G. and Wilkinson, C. D. W. (1999). 'Nanoscale embossing of polymers using a thermoplastic die'. *Microelectronic Engineering*, 46, 125–128.

Castellani, R., de Ruijter, A., Renggli, H. and Jansen, J. (1999). 'Response of rat bone marrow cells to differently roughened titanium discs'. *Clin. Oral Implants Res.*, 10, 369–378.

Cavalcanti-Adam, E. A. and Spatz, J. (2006). 'Regulation of adhesion dynamics, motility and survival by spacing of extracellular matrix peptides'. *J. Bone Mineral Research*, 21, S134.

Cavalcanti-Adam, E. A., Bezler, M., Tomakidi, P. and Spatz, J. P. (2004). 'Integrin ligands arranged in nanopatterns modulate focal adhesion assembly and signaling in osteoblasts'. *J. Bone Mineral Research*, 19, S64.

Cavalcanti-Adam, E. A., Micoulet, A., Blummel, J., Auernheimer, J., Kessler, H. and Spatz, J. P. (2006). 'Lateral spacing of integrin ligands influences cell spreading and focal adhesion assembly'. *Eur. J. Cell Biol.*, 85(3–4), 219–224.

Cavalcanti-Adam, E. A., Volberg, T., Micoulet, A., Kessler, H., Geiger, B. and Spatz, J. P. (2007). 'Cell spreading and focal adhesion dynamics are regulated by spacing of integrin ligands'. *Biophys. J.*, 92, 2964–2974.

Charras, G. T. and Horton, M. A. (2002). 'Single cell mechanotransduction and its modulation analyzed by atomic force microscope indentation'. *Biophys. J.*, 82, 2970–2981.

Charras, G. T., Yarrow, J. C., Horton, M. A., Mahadevan, L. and Mitchison, T. J. (2005). 'Non-equilibration of hydrostatic pressure in blebbing cells'. *Nature*, 435, 365–369.

Choi, C. H., Hagvall, S. H., Wu, B. M., Dunn, J. C. Y., Beygui, R. E. and Kim, C. J. (2007). 'Cell interaction with three-dimensional sharp-tip nanotopography'. *Biomaterials*, 28(9), 1672–1679.

Clark, P., Connolly, P., Curtis, A. S. G., Dow, J. A. T. and Wilkinson, C. D. W. (1987). 'Topographical control of cell behaviour. I. Simple step cues'. *Development*, 99, 439–448.

Clark, P., Connolly, P., Curtis, A. S. G., Dow, J. A. T. and Wilkinson, C. D. W. (1990). 'Topographical control of cell behaviour. II. Multiple grooved substrata'. *Development*, 108, 635–644.

Clark, P., Connolly, P., Curtis, A., Dow, J. A. and Wilkinson, C. (1991). 'Cell guidance by ultrafine topography in vitro'. *J. Cell Sci.*, 73–77.

Cremer, T. and Cremer, C. (2001). 'Chromosome territories, nuclear architecture and gene regulation in mammalian cells'. *Nat. Rev. Genet.*, 2(4), 292–301.

Curtis, A. S. G. (1994). 'Mechanical tensing of cells and chromosome arrangement'. *Biomechanics and Cells* (Soc. for Exp. Biol. Seminar Series), 54, 121–130.

Curtis, A. (2005a). 'The potential for the use of nanofeaturing in medical devices'. *Expert Rev. Med. Devices*, 2, 293–301.

Curtis, A. (2005b). 'Nanofeaturing materials for specific cell responses'. In *Nanoscale Materials Science in Biology and Medicine*, C. T. Laurencin and E. A. Botchwey (eds), Warrendale, PA: Materials Research Society, 354, pp. 175–179.

Curtis, A. S. G. and Clark, P. (1990). 'The effects of topographic and mechanical properties of materials on cell behaviour'. *Crit. Rev. Biocompatibility*, 5, 343–362.

Curtis, A. S. G. and Seehar, G. M. (1978). 'The control of cell division by tension or diffusion'. *Nature*, 274, 52–53.

Curtis, A. and Wilkinson, C. (1997). 'Reactions of cells to nanotopography'. In *Cellular and Molecular Biology Letters: Biophysics of Membrane Transport: XIII School Proceedings Pt. 1 Membrane Transport*, K. K. J. Kuczera, B. Rozycka-Roszak, and S. Przestalski (eds), Ladekzdroj, Poland: J. Szopa, A. Kozubek, A.F. Sikorski, pp. 9–18.

Curtis, A. and Wilkinson, C. (1999). 'Reactions of cells to nanotopography'. In *Cell Behaviour: Control and Mechanism of Motility*, J. M. Lackie, G.A. Dunn and G.E. Jones (eds), London: Portland Press, pp. 15–26.

Curtis, A. S. G and Wilkinson, C. D. W. (2001). 'Nanotechniques and approaches in biotechnology'. *Trends Biotechnol.*, 19, 97–101.

Curtis, A. S. G., Casey, B., Gallagher, J. D., Pasqui, D., Wood, M.A.W. and Wilkinson, C. D. W. (2001). 'Substratum nanotopography and the adhesion of biological cells. Are symmetry or regularity of nanotopography important?' *Biophys. Chem.*, 94, 275–283.

Curtis, A. S. G., Gadegaard, N., Dalby, M. J., Riehle, M. O., Wilkinson, C. D. W. and Aitchison, G. (2004). 'Cells react to nanoscale order and symmetry in their surroundings'. *IEEE Trans. Nanobioscience*, 3(1), 61–65.

Curtis, A. S. G., Dalby, M. J. and Gadegaard, N. (2006a). 'Nanoprinting onto cells'. *J. Roy. Soc. Interface*, 3, 393–398.

Curtis, A. S. G., Dalby, M. and Gadegaard, N. (2006b). 'Cell signaling arising from nanotopography: implications for nanomedical devices: a review'. *Future Medicine*, 1, 1–5.

Curtis, A., Sokolikova-Csaderova, L. and Aitchison, G. (2007). 'Measuring cell forces by a photoelastic method'. *Biophys. J.*, 92, 2255–2261.

Cyster, L. A., Parker, K. G., Parker, T. L. and Grant, D. M. (2003). 'The effect of surface chemistry and nanotopography of titanium nitride (TiN) films on 3T3-L1 fibroblasts'. *J. Biomed. Mater. Res. Part A*, 67A(1), 138–147.

Dalby, M. J. (2005). 'Topographically induced direct cell mechanotransduction'. *Medical Engineering Physics*, 27, 730–742.

Dalby, M. J., Childs, S., Riehle, M. O., Johnstone, H. J. H., Affrossman, S. and Curtis, A. S. G. (2002a). 'Fibroblast reaction to island topography: changes in cytoskeleton and morphology with time'. *Biomaterials*, 24, 927–935.

Dalby, M. J., Yarwood, S. J., Riehle, M. O., Johnstone, H. J., Affrossman, S. and Curtis, A. S. (2002b). 'Increasing fibroblast response to materials using nanotopography: morphological and genetic measurements of cell response to 13-nm-high polymer demixed islands'. *Exp. Cell Res.*, 276(1), 1–9.

Dalby, M. J., Yarwood, S. J., Johnstone, H., Affrossman, S. and Riehle, M. (2002c). 'Fibroblast signalling events in response to nanotopography: a gene array study'. *IEEE Trans. Nanobioscience*, 1, 12–17.

Dalby, M. J., Riehle, M. O., Johnstone, H. J. H., Affrossman, S. and Curtis, A. S. G. (2002d). 'Polymer-demixed nanotopography: control of fibroblast spreading and proliferation'. *Tissue Engineering*, 8(6), 1099–1107.

Dalby, M. J., Riehle, M. O., Johnstone, H. J. H., Affrossman, S. and Curtis, A. S. G. (2003a). 'Nonadhesive nanotopography: Fibroblast response to poly(n-butyl methacrylate)–poly(styrene) demixed surface features'. *J. Biomed. Mater. Res. Part A*, 67A(3), 1025–1032.

Dalby, M. J., Riehle, M. O., Yarwood, S. J., Wilkinson, C. D. and Curtis, A. S. (2003b). 'Nucleus alignment and cell signaling in fibroblasts: response to a micro-grooved topography'. *Exp. Cell Res.*, 284, 274–282.

Dalby, M. J., Riehle, M.O., Sutherland, D. S., Agheli, H. and Curtis, A. S. G. (2004a). 'Use of nanotopography to study mechanotransduction in fibroblasts – methods and perspectives'. *Eur. J. Cell Biol.*, 83(4), 159–169.

Dalby, M. J., Berry, C. C., Riehle, M. O., Sutherland, D. S., Agheli, H. and Curtis, A. S. G. (2004b). 'Attempted endocytosis of nano-environment produced by colloidal lithography by human fibroblasts'. *Exp. Cell Res.*, 295(2), 387–394.

Dalby, M. J., Riehle, M.O., Johnstone, H., Affrossman, S. and Curtis, A. S. G. (2004c). 'Investigating the limits of filopodial sensing: a brief report using SEM to image the interaction between 10 nm high nano-topography and fibroblast filopodia'. *Cell Biol. Int.*, 28(3), 229–236.

Dalby, M. J., Gadegaard, N., Riehle, M. O., Wilkinson, C. D. W. and Curtis, A. S. G. (2004d). 'Investigating filopodia sensing using arrays of defined nano-pits down to 35 nm diameter in size'. *Int. J. Biochem. Cell Biol.*, 36(10), 2005–2015.

Dalby, M. J., Riehle, M. O., Sutherland, D. S., Agheli, H. and Curtis, A. S. G. (2004e). 'Changes in fibroblast morphology in response to nano-columns produced by colloidal lithography'. *Biomaterials*, 25(23), 5415–5422.

Dalby, M. J., Riehle, M. O., Sutherland, D. S., Agheli, H. and Curtis, A. S. G. (2004f). 'Fibroblast response to a controlled nanoenvironment produced by colloidal lithography'. *J. Biomed. Mater. Res. Part A*, 69A(2), 314–322.

Dalby, M. J., Giannaras, D., Riehle, M. O., Gadegaard, N., Affrossman, S. and Curtis, A. S. G. (2004g). 'Rapid fibroblast adhesion to 27 nm high polymer demixed nano-topography'. *Biomaterials*, 25(1), 77–83.

Dalby, M. J., McCloy, D., Robertson, M., Wilkinson, C. D. W. and Oreffo, R. O. C. (2006a). 'Osteoprogenitor response to defined topographies with nanoscale depths'. *Biomaterials*, 27(8), 1306–1315.

Dalby, M. J., McCloy, D., Robertson, M., Agheli, H., Sutherland, D., Affrossman, S. and Oreffo, R. O. C. (2006b). 'Osteoprogenitor response to semi-ordered and random nanotopographies'. *Biomaterials*, 27(15), 2980–2987.

Dalby, M. J., Biggs, M. J. P., Gadegaard, N., Kalna, G., Wilkinson, C. D. W. and Curtis, A. S. G. (2007). 'Nanotopographical stimulation of mechanotransduction and changes in interphase centromere positioning'. *J. Cellular Biochem.*, 100, 326–338.

de Oliveira, P. T., Zalzal, S. F., Beloti, M. M., Rosa, A. L. and Nanci, A. (2007). 'Enhancement of in vitro osteogenesis on titanium by chemically produced nanotopography'. *J. Biomed. Mater. Res. Part A*, 80A(3), 554–564.

Desai, T. A. (2000). 'Micro- and nanoscale structures for tissue engineering constructs'. *Medical Engineering and Physics*, 22(9), 595–606.

Elwenspoek, M. and Jansen, H. V. (1998). *Silicon Micromachining*. Cambridge: Cambridge University Press.

Favia, P., Creatore, M., Palumbo, F., Colaprico, V. and d'Agostino, R. (2001). 'Process control for plasma processing of polymers'. *Surf. Coat. Technol.*, 142, 1–6.

Forgacs, G. (1995). 'On the possible role of cytoskeletal filamentous networks in intracellular signaling: an approach based on percolation'. *J. Cell. Sci.*, 108 (Pt 6), 2131–2143.

Forgacs, G., Newman, S. A., Hinner, B., Maier, C. W. and Sackmann, E. (2003). 'Assembly of collagen matrices as a phase transition revealed by structural and rheologic studies'. *Biophys. J.*, 84, 1272–1280.

Foster, H. A. and Bridger, J. M. (2005). 'The genome and the nucleus: a marriage made by evolution. Genome organisation and nuclear architecture'. *Chromosoma*, 114, 212–229.

Gadegaard, N., Mosler, S. and Larsen, N. B. (2003a). 'Biomimetic polymer nanostructures by injection moulding'. *Macromolecular Materials Engineering*, 288, 76–83.

Gadegaard, N., Thoms, S., MacIntyre, D. S., McGhee, K., Gallagher, J., Casey, B. and Wilkinson, C. D. W. (2003b). 'Arrays of nano-dots for cellular engineering'. *Microelectronic Engineering*, 67–68, 162–168.

Gadegaard, N., Dalby, M. J., Riehle, M. O., Curtis, A. S. G. and Affrossman, S. (2004). 'Tubes with controllable internal nanotopography'. *Advanced Materials*, 16, 1857–1860.

Gustafson, T. and Wolpert, L. (1961). 'Studies on the cellular basis of morphogenesis in the sea urchin embryo. Directed movements of primary mesenchyme cells in normal and vegetalized larvae'. *Exp. Cell Res.*, 24, 64–79.

Hamill, O. P. (2006). 'Twenty odd years of stretch-sensitive channels'. *Pflugers Archiv – Eur. J. Physiol.*, 453(3), 333–351.
Hanarp, P., Sutherland, D., Gold, J. and Kasemo, B. (1999). 'Nanostructured model biomaterial surfaces prepared by colloidal lithography'. *Nanostructured Materials*, 12(1–4), 429–432.
Hartmann, C., Ozmutlu, O., Petermeier, H., Fried, J. and Delgado, A. (2007). 'Analysis of the flow field induced by the sessile peritrichous ciliate *Opercularia asymmetrica*'. *J. Biomechanics*, 40(1), 137–148.
Heslop-Harrison, J. S., Leitch, A. R. and Schwarzacher, T. (1993). 'The physical organisation of interphase nuclei'. In *The Chromosome*, J. S. Heslop-Harrison and R. B. Flavell (eds), Oxford: Bios, pp. 221–232.
Hsu, S. H., Tang, C. M. and Lin, C. C. (2004). 'Biocompatibility of poly(epsilon-caprolactone)/poly(ethylene glycol) diblock copolymers with nanophase separation'. *Biomaterials*, 25(25), 5593–5601.
Hu, S., Chen, J., Fabry, B., Numaguchi, Y., Gouldstone, A., Ingber, D. E., Fredberg, J. J., Butler, J. P. and Wang, N. (2003). 'Intracellular stress tomography reveals stress focusing and structural anisotropy in cytoskeleton of living cells'. *Am. J. Physiol. Cell Physiol.*, 285(5), C1082–C1090.
Huck, W. T. S. (2007). 'Self-assembly meets nanofabrication: Recent developments in microcontact printing and dip-pen nanolithography'. *Ange. Chemie*, 46(16), 2754–2757.
Ingber, D. E. (1993). 'Cellular tensegrity: defining new rules of biological design that govern the cytoskeleton'. *J. Cell Sci.*, 104 (Pt 3), 613–627.
Ingber, D. E. (2003a). 'Tensegrity I. Cell structure and hierarchical systems biology'. *J Cell Sci.*, 116, 1157–1173.
Ingber, D. E. (2003b). 'Tensegrity II. How structural networks influence cellular information processing networks'. *J. Cell Sci.*, 116, 1397–1408.
Ingber, D. E. (2003c). 'Mechanosensation through integrins: cells act locally but think globally'. *Proc. Natl Acad. Sci. USA*, 100, 1472–1474.
Jentsch, T. J. and Gunther, W. (1997). 'Chloride channels: an emerging molecular picture'. *Bioessays*, 19(2), 117–126.
Johansson, F., Carlberg, P., Danielsen, N., Montelius, L. and Kanje, M. (2006). 'Axonal outgrowth on nano-imprinted patterns'. *Biomaterials*, 27(8), 1251–1258.
Jones, G. E., Allen, W. E. and Ridley, A. J. (1998). 'The Rho GTPases in macrophage motility and chemotaxis'. *Cell Adhes. Commun.* 6, 237–245.
Kasemo, B. and Lausmaa, J. (1994). 'Material–tissue interfaces: the role of surface properties and processes'. *Environ. Health Perspect.*, 102 Suppl. 5, 41–45.
Kriparamanan, R., Aswath, P., Zhou, A., Tang, L. P. and Nguyen, K. T. (2006). 'Nanotopography: cellular responses to nanostructured materials'. *J. Nanosci. Nanotechnol.*, 6(7), 1905–1919.
Kurz, A., Lampel, S., Nickolenko, J. E., Bradl, J., Benner, A., Zirbel, R. M., Cremer, T. and Lichter, P. (1996). 'Active and inactive genes localize preferentially in the periphery of chromosome territories'. *J. Cell Biol.*, 135(5), 1195–1205.
Lee, C. J., Huie, P., Leng, T., Peterman, M. C., Marmor, M. F., Blumenkranz, M. S., Bent, S. F. and Fishman, H. A. (2002). 'Microcontact printing on human tissue for retinal cell transplantation'. *Arch. Ophthalmology*, 120(12), 1714–1718.
Lena, S., Brancolini, G., Gottarelli, G., Mariani, P., Masiero, S., Venturini, A., Palermo, V., Pandoli, O., Pieraccini, S., Samori, P. and Spada, G. P. (2007). 'Self-assembly of an alkylated guanosine derivative into ordered supramolecular nanoribbons in

solution and on solid surfaces'. *Chemistry – a European Journal*, 13(13), 3757–3764.
Lenardi, C., Perego, C., Cassina, V., Podesta, A., D'Amico, A., Gualandris, D., Vinati, S., Fiorentini, F., Bongiorno, G., Piseri, P., Sacchi, F. V. and Milani, P. (2006). 'Adhesion and proliferation of fibroblasts on cluster-assembled nanostructured carbon films: the role of surface morphology'. *J. Nanosci. Nanotechnol.*, 6(12), 3718–3730.
Li, W. J., Jiang, Y. J. and Tuan, R. S. (2006). 'Chondrocyte phenotype in engineered fibrous matrix is regulated by fiber size'. *Tissue Engineering*, 12(7), 1775–1785.
Liliensiek, S. J., Campbell, S., Nealey, P. F. and Murphy, C. J. (2006). 'The scale of substratum topographic features modulates proliferation of corneal epithelial cells and corneal fibroblasts'. *J. Biomed. Mater. Res. Part A*, 79A(1), 185–192.
Lim, J. Y., Hansen, J. C., Siedlecki, C. A., Runt, J. and Donahue, H. J. (2005). 'Human foetal osteoblastic cell response to polymer-demixed nanotopographic interfaces'. *J. Roy. Soc. Interface*, 2(2), 97–108.
Lo, C. W. and Gilula, N. B. (1980). 'CC4 azal teratocarcinoma stem cell differentiation in culture. II. Morphological characterization'. *Devel. Biol.*, 75, 93–11.
Ludewig, U., Pusch, M. and Jentsch, T. J. (1997). 'Each subunit forms an independent pore in the dimeric CLC-0 chloride channel'. Paper M-AM-C6. *Biophys. J.* 72(2 Pt 2), A5.
Madore, C. and Landolt, D. (1997). 'Electrochemical micromachining of controlled topographies in titanium for biological applications'. *J. Micromech. Microeng.*, 7, 270–275.
Maniotis, A. J., Chen, C. S. and Ingber, D. E. (1997a). 'Demonstration of mechanical connections between integrins, cytoskeletal filaments, and nucleoplasm that stabilize nuclear structure'. *Proc. Natl Acad. Sci. USA*, 94, 849–854.
Maniotis, A. J., Bojanowski, K. and Ingber, D. E. (1997b). 'Mechanical continuity and reversible chromosome disassembly within intact genomes removed from living cells'. *J. Cell Biochem.*, 65, 114–130.
Martines, E., Seunarine, K., Morgan, H., Gadegaard, N., Wilkinson, C. D. W. and Riehle, M. O. (2005). 'Superhydrophobicity and superhydrophilicity of regular nanopatterns'. *Nanoletters*, 5, 2097–2103.
Matthews, B. D., Thodeti, C. K. and Ingber, D. E. (2007). 'Activation of mechanosensitive ion channels by forces transmitted through integrins and the cytoskeleton'. *Mechanosensitive Ion Channels, Part A*, 59–85.
Meredith, D. O., Eschbach, L., Wood, M. A., Riehle, M. O., Curtis, A. S. G. and Richards, R. G. (2005). 'Human fibroblast reactions to standard and electropolished titanium and Ti-6Al-7Nb, and electropolished stainless steel'. *J. Biomed. Mater. Res. Part A*, 75A(3), 541–555.
Meredith, D. O., Eschbach, L., Riehle, M., Curtis, A. and Richards, R. G. (2007a). 'Microtopography of metal surfaces influence fibroblast growth by modifying cell shape, cytoskeleton and adhesion'. *J. Orthop. Res.*, 25(11), 1523–1533.
Meredith, D. O., Riehle, M., Curtis, A. and Richards, R. G. (2007b). 'Is surface chemical composition important for orthopaedic implant materials?' *J. Mater. Sci. – Mater. Med.*, 18, 405–413.
Mosgoller, W., Leitch, A. R., Brown, J. K. M. and Heslop-Harrison, J. S. (1991). 'Chromosome arrangements in human fibroblasts at mitosis'. *Human Genetics*, 88, 27–33.

Nagai, A., Nagai, Y., Qu, H. J. and Zhang, S. G. (2007). 'Dynamic behaviors of lipid-like self-assembling peptide A(6)D and A(6)K nanotubes'. *J. Nanosci. Nanotechnol.*, 7(7), 2246–2252.

Nicolas, A., Geiger, B. and Safran, S. A. (2004). 'Cell mechanosensitivity controls the anisotropy of focal adhesions'. *Proc. Natl Acad. Sci. USA*, 101(34), 12520–12525.

Osborne, C. S., Chakalova, L., Brown, K. E., Carter, D., Horton, A., Debrand, E., Goyenechea, B., Mitchell, J. A., Lopes, S., Reik, W. and Fraser, P. (2004). 'Active genes dynamically colocalize to shared sites of ongoing transcription'. *Nature Genetics*, 36(10), 1065–1071.

Ostuni, E., Yan, L. and Whitesides, G. M. (1999). 'The interaction of proteins and cells with self-assembled monolayers of alkanethiolates on gold and silver'. *Colloids Surfaces B – Biointerfaces*, 15(1), 3–30.

Pattison, M. A., Webster, T. J. and Haberstroh, K. M. (2006). 'Select bladder smooth muscle cell functions were enhanced on three-dimensional, nano-structured poly(ether urethane) scaffolds'. *J. Biomater. Sci. – Polym. Ed.*, 17, 1317–1332.

Rajnicek, A. and McCaig, C. (1997). 'Guidance of CNS growth cones by substratum grooves and ridges: effects of inhibitors of the cytoskeleton, calcium channels and signal transduction pathways'. *J. Cell Sci.*, 110 (Pt 23), 2915–2924.

Rajnicek, A., Britland, S. and McCaig, C. (1997). 'Contact guidance of CNS neurites on grooved quartz: influence of groove dimensions, neuronal age and cell type'. *J. Cell Sci.*, 110 (Pt 23), 2905–2913.

Reches, M. and Gazit, E. (2007). 'Biological and chemical decoration of peptide nanostructures via biotin–avidin interactions'. *J. Nanosci. Nanotechnol.*, 7(7), 2239–2245.

Safran, S. A., Gov, N., Nicolas, A., Schwarz, U. S. and Tlusty, T. (2005). 'Physics of cell elasticity, shape and adhesion'. *Physica A – Statistical Mechanics Applications*, 352(1), 171–201.

Schindler, M., Ahmed, I., Kamal, J., Nur-E-Kamal, A., Grafe, T. H., Chung, H. Y. and Meiners, S. (2005). 'A synthetic nanofibrillar matrix promotes in vivo-like organization and morphogenesis for cells in culture'. *Biomaterials*, 26(28), 5624–5631.

Schindler, M., Nur-E-Kamal, A., Ahmed, I., Kamal, J., Liu, H. Y., Amor, N., Ponery, A. S., Crockett, D. P., Grafe, T. H., Chung, H. Y., Weik, T., Jones, E. and Meiners, S. (2006). 'Living in three dimensions – 3D nanostructured environments for cell culture and regenerative medicine'. *Cell Biochem. Biophys.*, 45(2), 215–227.

Schmitz, A. A., Govek, E. E., Bottner, B. and Van Aelst, L. (2000). 'Rho GTPases: signaling, migration, and invasion'. *Exp. Cell Res.*, 261, 1–12.

Schwarz, U. S., Balaban, N. Q., Riveline, D., Bershadsky, A., Geiger, B. and Safran, S. A. (2002). 'Calculation of forces at focal adhesions from elastic substrate data: the effect of localized force and the need for regularization'. *Biophys. J.*, 83(3), 1380–1394.

Shafrir, Y. and Forgacs, G. (2002). 'Mechanotransduction through the cytoskeleton'. *Am. J. Physiol. Cell Physiol.*, 282, C479–C486.

Tan, J., Shen, H., Carter, K. L. and Saltzman, W. M. (2000). 'Controlling human polymorphonuclear leukocytes motility using microfabrication technology'. *J. Biomed. Mater. Res.*, 51, 694–702.

Teixeira, A. I., Abrams, G. A., Bertics, P. J., Murphy, C. J. and Nealey, P. F. (2003). 'Epithelial contact guidance on well-defined micro- and nanostructured substrates'. *J. Cell Sci.*, 116, 1881–1892.

ter Brugge, P. J., Wolke, J. G. and Jansen, J. A. (2002). 'Effect of calcium phosphate coating crystallinity and implant surface roughness on differentiation of rat bone marrow cells'. *J. Biomed. Mater. Res.*, 60, 70–78.

Wang, N. and Suo, Z. (2005). 'Long-distance propagation of forces in a cell'. *Biochem. Biophys. Res. Commun.*, 328, 1133–1138.

Weiss, P. (1945). 'Experiments on cell and axon orientation in vitro: the role of colloidal exudates in tissue organization'. *J. Exp. Zool.*, 100, 353–386.

White, E. (2006). 'Mechanosensitive channels: therapeutic targets in the myocardium?' *Current Pharmaceutical Design*, 12(28), 3645–3663.

Wilkinson, C. D. W. and Curtis, A. S. G. (1996). 'Nanofabrication and its applications in medicine and biology'. *Dev. Nanotechnol.*, 3, 19–31.

Wilkinson, C. D. W., Curtis, A. S. G. and Crossan, J. (1998). 'Nanofabrication in cellular engineering'. *J. Vac. Sci. Technol.*, B, 16(6), 3132–3136.

Wilkinson, C. D. W., Riehle, M., Wood, M., Gallacher, J. and Curtis, A. S. G. (2001). 'The use of materials patterned on a nano- and micro-metric scale in cellular engineering'. *Mater. Sci. Engng. C*, 19(1–2), 263–269.

Wilkinson, C. D. W., Thoms, S., Macintyre, D. S. and Curtis, A. S. G. (2003). 'Nanofabrication of structures for cell engineering'. *Proc. SPIE*, 5220, 1–9.

Wojciak-Stothard, B. and Ridley, A. J. (2003). 'Shear stress-induced endothelial cell polarization is mediated by Rho and Rac but not Cdc42 or PI 3-kinases'. *J. Cell Biol.*, 161, 429–439.

Wojciak-Stothard, B., Curtis, A. S. G., Monaghan, W., McGrath, M., Sommer, I. and Wilkinson, C. D. W. (1995a). 'Role of the cytoskeleton in the reaction of fibroblasts to multiple grooved substrata'. *Cell Motil. Cytoskeleton*, 31, 147–158.

Wojciak-Stothard, B., Madeja, Z., Korohoda, W., Curtis, A. and Wilkinson, C. (1995b). 'Activation of macrophage-like cells by multiple grooved substrata – topographical control of cell behavior'. *Cell Biol. Int.*, 485–490.

Wojciak-Stothard, B., Curtis, A., Monaghan, W., Macdonald, K. and Wilkinson, C. (1995). 'Guidance and activation of murine macrophages by nanometric scale topography'. *Exp. Cell Res.*, 223, 426–435.

Wood, M. A., Riehle, M. and Wilkinson, C. D. W. (2002a). 'Patterning colloidal nanotopographies'. *Nanotechnology*, 13, 605–609.

Wood, M. A., Meredith, D. O. and Owen, G. R. (2002b). 'Steps toward a model nanotopography'. *IEEE Trans. Nanobioscience*, 1(4), 133–140.

Wood, M. A., Wilkinson, C. D. N. and Curtis, A. S. G. (2006). 'The effects of colloidal nanotopography on initial fibroblast adhesion and morphology.' *IEEE Trans. Nanobioscience*, 5(1), 20–31.

Yahel, G., Eerkes-Medrano, D. I. and Leys, S. P. (2006). 'Size independent selective filtration of ultraplankton by hexactinellid glass sponges'. *Aquatic Microbial Ecology*, 45(2), 181–194.

Yim, E. K. F., Reano, R. M., Pang, S. W., Yee, A. F., Chen, C. S. and Leong, K. W. (2005). 'Nanopattern-induced changes in morphology and motility of smooth muscle cells'. *Biomaterials*, 26(26), 5405–5413.

Zhou, D. J., Wang, X. Z., Birch, L., Rayment, T. and Abell, C. (2003). 'AFM study on protein immobilization on charged surfaces at the nanoscale: toward the fabrication of three-dimensional protein nanostructures'. *Langmuir*, 19(25), 10557–10562.

Zhu, B. S., Zhang, Q. Q., Lu, Q. H., Xu, Y. H., Yin, J., Hu, J. and Wang, Z. (2004). 'Nanotopographical guidance of C6 glioma cell alignment and oriented growth'. *Biomaterials*, 25(18), 4215–4223.

18
Cell response to surface chemistry in biomaterials

C A SCOTCHFORD, University of Nottingham, UK

Abstract: This chapter starts by identifying the importance of biomaterial surface chemistry in the resultant response of cells exposed to such substrates. It then reviews the role of adsorbed proteins in modulating the cellular response, before describing the role of model surfaces in elucidating mechanisms of the cell response at the molecular level. Elements of these mechanisms are described with respect to osteoblast–substrate interaction. Finally a selection of methods for surface chemical modification of titanium, titanium alloy and bioceramics are presented in the context of bone and dental biomaterials.

Key words: cell–material interaction, biomaterial surface chemistry, biomaterial surface chemical modification.

18.1 Introduction

Before considering cell responses to surface chemistry it may be useful to establish a working definition of surface chemistry. A surface may be defined as the boundary that separates an object from another object, substance or space. Given the increasingly sophisticated surface analytical tools available in recent years, the volume of material comprising the surface may be resolved to the single atomic level. Practically, however, the surface chemistry of a material may be interpreted according to its means of interrogation. For example, the response of a protein adsorbed to a material surface may be dictated by the chemistry of the outermost molecular layer; indeed Kasemo and Lausmaa,[1] among others, have postulated that biological tissues interact with mainly the outermost atomic layers of an implant. Although secondary and other by-product reactions will occur, the primary interaction zone is 0.1 to 1 nm into the surface. X-ray photoelectron spectroscopic surface chemical analysis, however, samples a depth of material from which photoelectrons can escape, typically two to five atomic layers or approximately 10 nm . For the purpose of this chapter, surface chemistry will be taken as indicating the depth responsible for the biological response whilst recognising that surface analytical techniques may sample deeper than this.

As detailed in other chapters of this book, cells have been shown to respond to several surface properties. These include surface roughness, surface topography, including nanotopography, surface chemistry and surface energy. While the influences of these properties on cell responses are dealt with in separate chapters, in practice it has proven difficult to investigate cell responses to changes on one surface property in isolation. Typically a change in one property may alter a second property inadvertently. Some studies have shown one property to be dominant over another in terms of cell response; however, control over any or all of these properties provides means for control and engineering of selected cell and tissue responses.

Surface topography, including nanotopography, has often been claimed as the dominant surface property, with several studies being carried out to assess the relative importance of topography and chemistry in influencing cell responses (Britland et al.,[2] Wong et al.[3]). While some of these studies do indeed demonstrate a dominant influence of topography over chemistry (Britland et al.[2]) Other studies have shown that chemistry may dominate over topography. For example, a study by Tan and Saltzman[4] demonstrated a variable response of neutrophils to grooved/ridged glass substrate (3 μm height, 6–14 μm spacing) dependent on the surface chemistry of the coating applied to this substrate. With gold–palladium coatings cells located within the grooves, whereas titanium-coated substrates demonstrated a predominance of cells on ridges. Further, the chemical composition of the substrate influenced the rate of cell motility, with significantly reduced cell motility on titanium-coated surfaces compared to gold–palladium substrates. This observation was attributed to a stronger adhesion of cells to the substrate with the higher wettability. Such results indicate that both surface topography and chemistry play important roles in the regulation of cell migration and the effect of microgeometry can be overridden by a strong chemical property.

Many studies have demonstrated diverse cellular responses to substrates with different surface chemistries, for example differential gene expression for several cell types on surfaces with varying hydrophobicity (Allen et al.[5]) and increased in vivo apoptosis and reduced foreign body giant cell formation on hydrophilic and anionic surfaces compared to hydrophobic and cationic surfaces (Brodbeck et al.[6]). The strong dependency of cell adhesion on the substrate chemistry has been demonstrated in numerous studies; for example on the water wettability and surface charge (Ratner et al.,[7] Lydon et al.[8]).

It is worth differentiating between surface chemistry and surface biochemistry at this point. This chapter will mainly focus on surface chemistry effects. For a review of surface biochemistry influences on cell responses see Healy.[9] Biochemical modification of surfaces has been investigated widely, with the emphasis on targeting receptor mediated cellular functions. For example, covalent immobilisation of adhesive peptides to material surfaces has been shown to modify cellular behaviour beyond initial adhesion and spreading, fibroblast motility being shown to vary as a function of substrate adhesiveness (Olbrich et al.[10]). The manner in which such peptides are tethered to substrates plays a key role in determining the binding affinity of peptide for receptor (Mrksich[11]). One approach to surface biochemical modification utilises cell adhesion molecules or adhesive ligands from cell adhesion molecules, for example the RGD peptide. However, recent work has shown reduced osseointegration to hydroxyapatite implants functionalised with the RGD peptide. It has been suggested that, for biomaterials that are highly interactive with the tissue microenvironment such as hydroxyapatite, the ultimate effects of RGD peptide functionalisation will depend on how signalling from the peptide integrates with endogenous processes such as protein adsorption (Hennessy et al.[12]). While polymeric biomaterials present a ready opportunity for surface biochemical functionalisation, for orthopaedic and dental applications metal surfaces possess a relative paucity of functional groups needed for immobilising such molecules. The passivating oxide film on the surfaces of these materials does, however, possess surface hydroxyl groups that provide locations for bonding using silane chemistry. This approach has been used to immobilise peptides and more complex biological molecules (Pegg[13]).

Clearly for materials that do not show a cytotoxic effect on cells and tissues, the structure and properties of the material surface play a key role in determining the observed outcome of material–biological interactions. Given this and recognising the level of sophistication available for both surface characterisation and surface fabrication down to the molecular level, great effort has been made in recent years to elucidate the interplay between surface chemistry and cell response.

18.2 The role of proteins in the cell response to surface chemistry

It is well established that the surface chemistry of a material dictates the characteristics of the adsorbed protein layer, which inevitably forms within seconds of a material being placed within a biological fluid, and that these characteristics determine the cellular response to that material. While these principles have been established for some time, based on early work by Leo Vroman and contemporaries, it is only in the last 10–15 years that the mechanisms of surface chemistry modulated cell behaviour have been studied at the molecular level. Central to this have been the use of well-defined model substrates, sophisticated surface chemical analysis and application of biological analytical tools such as PCR and gene arrays.

Protein adsorption is a complex dynamic process involving non-covalent interactions, including hydrophobic interactions, electrostatic forces, hydrogen bonding and van der Waals forces. It is known that proteins will undergo conformational change upon adsorption to most materials, and cellular interactions will greatly depend on the nature of this conformational change. Not all changes will be beneficial for cell attachment; for example, surface-denatured fibronectin will no longer support the adhesion and growth of cells (Zeng et al.[14]).

Many studies have focused on identifying the type, conformation and density of the adsorbed protein layer that dictates cell adhesion (Singhvi et al.,[15] Lewandowska et al.[16]). For example, adsorption of the cell adhesion protein fibronectin on different surfaces alters protein structure and modulates such alterations in cell adhesion, spreading and migration (Grinnell and Feld,[17] Pettit et al.,[18] Garcia et al.[19]). Michael and colleagues[20] demonstrated significant surface-chemistry dependent structural changes in fibronectin binding domains and adhesive activity. The study established a relationship between surface dependent changes in structural domains of fibronectin and functional activity.

Adsorbed proteins may provide binding ligands for cell-surface receptor molecules such as the integrin family, which in turn modulate the cell response to the surface. Integrin binding to adsorbed proteins occurs via defined ligands, a widely studied example being the interaction with the RGD peptide sequence found in a number of extracellular matrix (ECM) proteins such as fibronectin, vitronectin, osteopontin, bone sialoprotein and the collagens.

Protein adsorption experiments have identified fibronectin and vitronectin as primary proteins leading to endothelial, fibroblastic and bone-derived cell adhesion (Steele et al.,[21] Howlett et al.,[22] Underwood and Bennett[23]). These studies reported that, in vitro, vitronectin outcompetes fibronectin to be the major adhesive protein.

Taking the points made above, it is clear that a fundamental understanding of substrate-directed control of cell function is critical to the rational design of surfaces relevant to biomaterials, tissue engineering and in vitro culture supports for biotechnological applications. In the following sections attention will be focused on the role played by model surfaces in developing our understanding of cell responses to surface chemistry; some mechanisms implicated in such responses, and selected surface modifications designed to exploit such phenomena, will be considered in turn.

18.3 Cell response to model surfaces

The use of model surfaces based on technologies such as photolithography and self-assembly has provided a platform for the investigation of

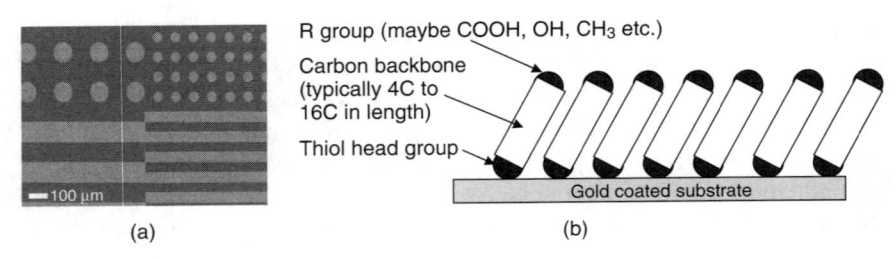

18.1 Model substrates for studying cell responses: (a) SEM micrographs of vanadium/niobium chemically patterned surfaces produced using photolithography methods, with dots and stripes in the range of 50 to 100 μm. (Reproduced with permission from Biomaterials. Copyright 2002 Elsevier Science Ltd.) (b) Schematic representation of a generalised alkanethiol self-assembled monolayer.

changes in protein adsorption and cell behaviour due to controlled variation in surface chemistry (Fig. 18.1). For example, recent studies using self-assembled monolayers (SAMs) have shown that surface chemistry can modulate cell adhesion, spreading and adhesion strength (Tegoulia and Cooper,[24] Singhvi et al.,[15] McClary et al.,[25] Scotchford et al.,[26] Tidwell et al.[27]), and self-assembled monolayers terminated in short oligomers of ethylene glycol have been effective in controlling non-specific protein adhesion.

Studies have been performed using surfaces of uniform chemistry, both single and mixed, chemically patterned surfaces to control cell behaviour with adhesive and non-adhesive regions and even chemical gradients. Patterned surfaces allow for control of size and chemistry of the distinct regions within patterns, giving scope for differential cell responses on the same substrate. While providing useful tools for fundamental studies, such surfaces have attracted interest for the generation of 2-D tissue/organ constructs. Indeed, spatial control of cell distribution on solid substrates *in vitro* is crucial in cell and tissue engineering, affecting cell guidance (Soekarno et al.[28]), cell shape effects on cell function, migration and proliferation (Singhvi et al.[15]). Regulation of cell spatial distribution has been expanded into three dimensions in an effort to promote cell migration throughout tissue engineering scaffolds. For example, Barry and colleagues employed plasma polymer deposition technology to produce polylactide tissue engineering scaffolds with distinct macroscopic core and sheath zones in terms of surface chemistry. The two zones were designed to both promote (core) and retard (sheath) cell adhesion in an attempt to overcome the common problem of preferential cell colonisation of the periphery of such scaffolds (Barry et al.[29]). The contrasting surface chemistries exhibited variation in fibronectin adsorption which provided the differential cell adhesion characteristics. The ability to control cell shape also allows

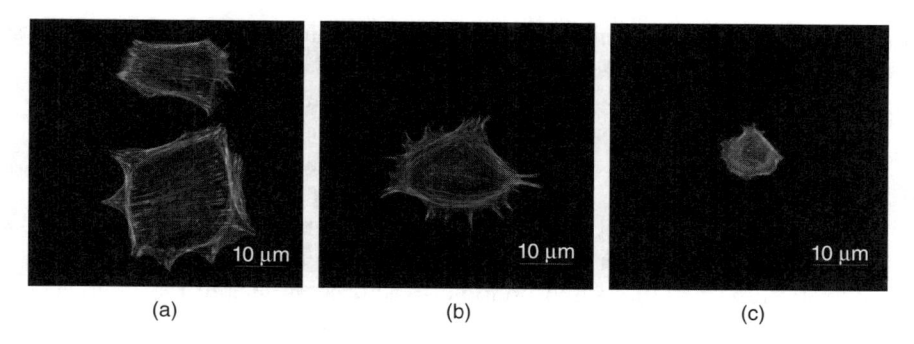

18.2 Confocal micrographs of osteoblasts cultured on SAMs after a 90-minute incubation showing filamentous actin and vinculin. (a) COOH-terminated SAM, (b) OH-terminated SAM, (c) CH$_3$-terminated SAM. From Scotchford et al.[40] (Reproduced with permission from *J. Biomed. Mater. Res.* Copyright 2001, John Wiley & Sons, Inc).

possibilities for enhanced function, differentiation and/or proliferation of anchorage-dependent cells (Healy et al.,[30] Folkman and Moscona,[31] Opas,[32] Chen et al.[33]).

The potential for patterning cells in response to 2-D chemically patterned substrate, whilst useful for studying cell behaviour, organisation and cellular interactions, also has application in biomedical device construction in areas such as wound and fracture healing implants, drug delivery, dental implants, biosensors and biochips (Mitchell et al.,[34] Hubbell,[35] Kwok et al.,[36] Kasemo and Gold,[37] Sanders et al.,[38] Gopel[39]).

Early studies using SAMs to investigate the influence of surface chemistry on cell adhesion and growth demonstrated differences in cell adhesion, morphology and proliferation as well as differences in relative protein adsorption (Fig. 18.2) (Scotchford et al.,[40] Tidwell et al.[27]). Subsequently, model surfaces have been used to examine the influence of surface chemistry on the conformation and level of organisation of adsorbed proteins and the interaction between this protein layer and adherent cells.

For example, it has been shown that hydrophobic surfaces strongly denature adsorbed proteins (Grinnell and Feld,[41] Sigal et al.,[42] Wertz and Santore,[43] Underwood et al.[44]) and have been shown to completely block the formation of an adsorbed fibronectin matrix (Altankov et al.[45]). Several studies have shown the effect of surface chemistry on the adsorption of ECM proteins such as collagen and fibronectin (Dupont-Gillain et al.[46]) and how alterations in protein adsorption can influence cellular signalling characteristics (Elliot et al.,[47] Altankov et al.[45]).

An example of how the surface chemistry of a material can have a significant effect not only on the structure of individual proteins but also on the supramolecular assembly of adsorbed proteins is reported by Elliott and colleagues.[48] This study used alkylthiol SAMs with the terminal chemistries CH_3, $COOH$, NH_2 and OH presented in the form of homogeneous, gradient and patterned substrates to examine the effect on formation of thin films of native fibrillar collagen. Supramolecular collagen structures assembled on CH_3 surfaces only. When mixtures of CH_3 and OH were used, structures assembled only when the measured contact angle was above 83°. Morphology studies with vascular smooth muscle cells indicated that only collagen films formed on hydrophobic structures mimicked the biological properties of fibrillar collagen gels. On hydrophobic surfaces, collagen adsorbed in a manner that initiated the assembly of large fibrils. This study also demonstrated that substrate surface chemistry, which influences the supramolecular structure of adsorbed collagen, can also influence cellular phenotype. Similar effects have been seen with fibrinogen adsorption (Evans-Nguyen and Schoenfisch[49]). Fibrinogen adsorbed on hydrophobic substrates forms extensive fibrin networks not seen on hydrophilic surfaces. This may play a significant role in determining inflammatory response to a biomaterial.

Therefore in addition to composition and conformation of adsorbed protein layers, the supramolecular structure of adsorbed proteins plays a role in the specific substrate–protein–cell interactions that control cell responses to surface chemistry.

Other studies (Singhvi et al.[15]) have demonstrated that the rate of endogenous protein synthesis and maintenance of cell phenotype can be controlled by constraining cell shape and/or size on chemically patterned substrates that provide regions of adhesive and non-adhesive domains. Such spatial differences may be maintained for extended periods on appropriate model substrates. Healy and colleagues, for example, using self-assembled silane patterns (Healy et al.[30]), demonstrated that the spatial distribution of mineralised tissue formation on the patterned substrates was the result of the initial spatial distribution of bone-derived cells on amine-terminated regions of the substrates.

18.4 Mechanisms for cell responses to surfaces

Recently, model surfaces have been used in studies attempting to elucidate the mechanisms by which surface chemistry influences the cell response. These, combined with investigations using modified biomaterial surfaces, have given some insight into molecular-scale events involved in the observed cell response. This section focuses on literature relevant to mechanisms of osteoblast responses to surface chemistry.

It is established that cell adhesion to proteins, mediated by integrins, adsorbed onto synthetic surfaces anchors cells and triggers signals that direct cell function. Fibronectin has been widely studied in this role. Adsorption of fibronectin onto substrates of varying surface chemistry alters its conformation/structure and therefore the ability of the molecule to support cell adhesion. Keselowsky and colleagues[50] used alkanethiol SAMs to examine the effect of surface chemistry on fibronectin adsorption and subsequent cellular interactions. CH_3, OH, COOH and NH_2 terminated SAM substrates modulated adsorbed fibronectin conformation as determined through differences in the binding affinities of monoclonal antibodies raised against the central cell binding domain (spanning the ninth and tenth type III repeats of the molecule and containing PHSRN and RGD peptide sequences, both of which are required for $\alpha_5\beta_1$ binding (Aota et al.[51])). The variation in binding affinity was in the order OH > COOH = NH_2 > CH_3. It is known that increases in relative distance between PHSRN and RGD sites nullify $\alpha_5\beta_1$ integrin binding, cell spreading and integrin mediated signalling (Grant et al.[52]). Keselowsky and colleagues[50] suggested that adsorption-induced changes in structural orientation of these binding domains may explain the observed substrate dependent differences in integrin binding and cell adhesion.

The importance of this PHSRN RGD peptide synergy and its conformational specificity is demonstrated by Petrie and colleagues.[53] This study used an alkanethiol SAM-based series of well-defined biointerfaces containing a non-fouling, protein adsorption-resistant background with recombinant fibronectin fragments containing the RGD and PHSRN adhesive motifs or the RGD peptide alone or RGD-PHSRN peptide separated by a polyglycine linker. When MC3T3-E1 cells were presented to these substrates, the recombinant fragment containing substrate showed higher adhesion strength, focal adhesion kinase (FAK) activation and cell proliferation rate than the other two substrates; also greater $\alpha_5\beta_1$ specificity was shown by the recombinant fragment substrate compared to the other substrates.

In addition to surface chemistry-driven variation in cell responses to bound fibronectin, variation in specific integrin binding within such cells has also been reported. Keselowsky et al.[50] demonstrated that preference of $\alpha_5\beta_1$ integrin binding to fibronectin adsorbed to alkanethiol SAMs was consistent with antibody binding affinities described above; however, α_v integrin binding followed a different trend, COOH >> OH = NH_2 = CH_3, demonstrating $\alpha_5\beta_1$ integrin specificity for Fn adsorbed onto the NH_2 and OH SAMs.

As well as having a role in fibronectin–integrin selectivity and binding strength, it appears that surface chemistry may influence the type of adhesive structure formed by integrins. For a review of adhesive structures formed by cells on material surfaces see Owen et al.[54] Faucheux and colleagues[55] used organosilane SAMs to investigate the effect of substrate chemistry on the ability of cells to translocate $\alpha_5\beta_1$ integrins and to form fibrillar adhesions. They demonstrated that the presence of fibrillar adhesions can be modulated by surface chemistry and this affects downstream events such as fibronectin fibrillogenesis and cell migration. Fibronectin was shown to bind more strongly to NH_2 than COOH-terminated SAMs. This observed difference in binding may have implications for fibrillar adhesion formation. α_5 and α_v subunits were shown to become segregated after 2 hours on COOH SAMs but not on the NH_2 SAM substrate, in a manner characteristic of early stage fibrillar adhesion. Further, the NH_2 SAM appeared to block fibronectin fibrillogenesis. It is known that the mechanism of fibronectin fibrillogenesis involves tension exerted by integrins to expose cryptic self-association sites within the fibronectin molecule required for polymerisation (Geiger et al.[56]). The strong electrostatic interaction of the fibronectin molecule with the NH_2-terminated SAM may prevent movement of the molecule on this substrate to form fibrils due to insufficient integrin traction. The result of these processes suggests that cells on the two different chemistries are effectively on different fibronectin matrices, with cell migration being faster on the COOH-terminated substrate.

It has been reported that surface chemistry modification of biomaterials modulates osteoblastic cell responses through the Erk/MAPK pathway (Zreiqat et al.,[57] Krause et al.[58]). Integrin-mediated adhesion to extracellular proteins activates multiple cytoskeletal-associated and intracellular signalling proteins such as FAK and Shc. FAK associates with Shc protein creating a Grb2 binding site. The Shc-Grb2-complex activates Ras, leading to stimulation of the mitogen-activated protein kinase (MAPK) signalling cascade (Marshall[59]). The phosphorylated extracellular regulated kinases 1/2 (Erk 1/2) translocate to the nucleus, providing a link between cytoplasmic signalling molecules and nuclear proteins. The nuclear target proteins of MAPK include c fos and c jun (members of the activated protein 1 family). AP-1 sites are located in the promoters of several genes expressed by osteoblasts such as osteocalcin and Type one collagen, hence AP-1 plays an important role in bone development (Angel and Karin,[60] Grigoriadis et al.[61]).

Surface chemistry has been reported to modulate the structure and molecular composition of cell–matrix adhesions as well as FAK signalling (Keselowsky et al.[62]). In such studies neutral hydrophilic OH-terminated SAMs supported the highest levels of recruitment of talin, α-actinin paxillin and tyrosine phosphorylated proteins to adhesive structures. Positively charged NH_2 and negatively charged COOH terminated SAMs exhibited intermediate levels of recruitment of these focal adhesion components, while hydrophobic CH_3 substrates displayed the lowest levels. These variations correlated well with $\alpha_5\beta_1$ binding. Phosphorylation of specific tyrosine residues in FAK also showed differential sensitivity to surface chemistry, suggesting a possible mechanism for differential activation of signalling cascades as a function of surface chemistry. This study further suggested that the surface chemistry-dependent differences observed in adhesive interactions modulated osteoblast differentiation.

Differences in integrin binding, specificity and localisation between varying surface chemistries are particularly important to understanding of cell–substrate interactions and hence engineering of surfaces to control cell function. Differential integrin binding between $\alpha_5\beta_1$ and $\alpha_v\beta_3$ triggers intracellular signals that control switching between proliferation and differentiation (Garcia et al.[63]). As indicated earlier, linear RGD peptides show little specificity among particular integrin receptors (Garcia and Reyes[64]). It has been suggested that $\alpha_v\beta_3$ may act to inhibit pro-differentiation signals from, for example, $\alpha_5\beta_1$ signalling on COOH substrates preadsorbed with fibronectin (Keselowsky et al.[62]). Based on such observations, Keselowsky and colleagues[62] proposed a model in which surface chemistry-dependent differences in integrin binding, in terms of bound numbers as well as integrin specificity, differentially regulate focal adhesion assembly and signalling which in turn modulate cellular functions.

18.5 Exploitation of cell response to surface chemistry

Whilst model surfaces have been used to investigate cell response to surface chemistry, research has also focused on surface modification of materials with current medical device applications to promote desired cell responses, typically in the form of coatings or chemical modification of existing surfaces. This section will focus on selected techniques applied to the modification of titanium and titanium alloys and bioceramics as examples of such research.

Ion implantation is a technique that has been applied to these groups of materials with the aim of modifying the cell response. Ion beam implantation, a powerful technique to modify surface chemistry and nanotopography, has been applied to medical devices to improve wear resistance and provide antimicrobial treatment for dental implants (Yoshinari et al.[65]). More recently a number of studies have investigated ion implantation as a means of modifying cell responses and are reviewed by Braceras and colleagues.[66] Ions of interest have included Ca, K, Mg, Zn, NH_2 and CO. Nayab and colleagues[67,68] studied the in vitro effects of Ca, K and Ar ion implantation on human osteoblast-like cell interactions with titanium. They found that high-dose Ca ion implantation enhanced spreading of MG-63, while strongly inhibiting cell attachment in early stages. After prolonged culture, cell adhesion was significantly greater; this was accompanied by up-regulation of $\alpha_5\beta_1$ integrin and vinculin positive adhesion plaques. A positive influence of Ca ion implantation in titanium on bone formation has also been reported in vivo. These studies are in agreement with earlier work by Hanawa and colleagues,[69] who subsequently reported that titanium plates implanted with Ca ions performed better than unmodified titanium plates in terms of osteoconduction. (Hanawa et al.[70]). These authors, following surface analysis of their modified surfaces, speculated that the presence of calcium titanate may be responsible for the observed biological responses.

Zreiqat and colleagues[57] investigated the effect of Mg and Zn ion implantation on signalling pathways in human osteoblasts. These ions were selected because bone mineral is known to contain trace levels of zinc and magnesium, and deficiencies in these elements have been associated with bone pathologies (Smith and Nisbet,[71] Herzberg et al.[72]). Modification of Ti alloy surface chemistry with these ions may have an enhancing effect on the skeletal tissue–device interface. Their data indicated that Mg may contribute to successful osteoblast function and differentiation at the skeletal tissue–device interface. Braceras and colleagues[66] investigated the effect of implanting Ti6Al4V with CO^+ ions. In vitro results with human bone cells in contact with implanted surfaces demonstrated enhanced cell spreading and proliferation with lower evidence of apoptosis. When implanted into New Zealand albino rabbit mandibles the treated surfaces supported higher and faster levels of osteintegration. Other ion implantation studies involving titanium substrates include the use of NH^{2+} ions (Yang et al.[73]). Here ion implantation formed a thicker surface oxide layer containing small amounts of nitrides. The authors claimed that the implanted surfaces supported enhanced osteoblast-like cell responses in vitro.

These reports suggest that surface chemistry modification of titanium or Ti6Al4V with divalent cations or bioactive bone-like ceramic is likely to have a role in modulating the extracellular signals in the bone microenvironment, thereby potentially contributing to successful osteoblast function and differentiation at the skeletal tissue–device interface.

In addition to ion implantation studies, Zreiqat and colleagues used sol-gel coating technology to modify Ti6Al4V surfaces with an alkoxide-derived hydroxy carbonate apatite (CHAP). This work was based on a biomimetic premise, as bone mineral is composed of nanocrystalline platelets of carbonate hydroxyapatite. This study demonstrated that Ti6Al4V modified with CHAP enhanced activation of Shc, a common point of integration between integrins and the RAS/Mapkinase pathway. The Mapkinase pathway was also upregulated, supporting a role for this pathway in mediating osteoblastic cell interactions with biomaterials (Zreiqat et al.[57]).

Surface modifications of titanium without calcium phosphate coatings or ion implantation have also been shown to activate osteoconduction. Based on the premise that titania and calcium titanate were the chemical species responsible for the enhanced bone responses associated with Ca ion implanted titanium, Hanawa and colleagues[70] tried simple methods to induce titanium surface modifications to produce calcium titanate or

Cell response to surface chemistry in biomaterials   473

enhance titanium oxide. Titanium surfaces were exposed to calcium nitrate (pH 3.9), calcium chloride (pH 7.4) and calcium oxide (pH 12.6) solutions. This study used the formation of apatite on the modified surfaces when exposed to Hank's solution as an indication for enhanced bone conductivity; this occurred on all modified surfaces but not on native titanium. The amount of apatite formed corresponded to the amount of calcium in the modified surface.

Micro arc oxidation (MAO) is another technique that has been used to modify titanium substrate surface chemistry in attempt to promote osseointegration. Micro arc oxidation refers to the process of applying a positive voltage to a Ti specimen immersed in an electrolyte; anodic oxidation of Ti occurs to form a TiO_2 layer on the surface. When the applied voltage is increased to a certain point, a micro arc occurs as a result of the dielectric breakdown of the TiO_2 layer. At the moment the dielectric breakdown occurs, Ti ions in the implant and OH ions in the electrolyte move in opposite directions very quickly to form TiO_2 again; during this process it is possible to incorporate Ca and P ions into the surface layer (Li et al.[74]). Li and colleagues[74] subjected titanium surfaces modified by MAO to *in vitro* and *in vivo* assessment of bone response. Increasing MAO voltage increased the roughness and thickness of the titanium oxide layer as well as the concentration of Ca and P ions in the oxide layer. Associated with these increases was an increase in alkaline phosphatase activity and a decrease in proliferation of osteoblast-like cells *in vitro*. The *in vivo* study was assessed in terms of interfacial strength between bone and implant. The screw-shaped implants demonstrated a threefold increase in removal torque after MAO treatment.

The methods described above relate to promotion of tissue formation, specifically bone. Titanium has also been surface modified to minimise biological activity, for example in blood-contacting applications. Diamond-like carbon (DLC) coating has proven to be effective in this respect and found wide commercial application. The compatibility of DLC coatings with blood-interfacing devices has been demonstrated to be associated with its surface hydrophobicity modulating a favourable ratio of adsorbed albumin to fibrinogen (Jones et al.[75]). For a review of biological applications of DLC coatings see Hauert.[76]

18.6 Future trends

Despite progress in understanding cell responses to surface chemistry, examples of which have been indicated, the full mechanisms by which surface chemistry modulates cell responses remains to be fully elucidated down to the molecular level. Whilst empirical approaches to controlling cell and tissue responses by surface chemistry modification continue to produce clinical benefits, future research will benefit from the use of model systems for addressing mechanistic processes, such as the role of focal adhesions and dynamic processes by which focal adhesions assemble and evolve. Beyond this novel model systems in which several material properties can be independently controlled would prove extremely valuable, recognising interplay between important substrate properties. These approaches will constitute important tools in the development of elusive design principles and materials selection rules to control cell response.

18.7 References

1. Kasemo B, Lausmaa J. Surface science aspects of inorganic biomaterials. *CRC Crit Rev Biocomp* 1986, 2: 335–380.
2. Britland S, Clark P, Connolly P, Moores G. Micropatterned substratum adhesiveness: A model for morphogenetic cues controlling cell behaviour. *Exp Cell Res* 1992, 198: 124–129.
3. Wong JY, Leach JB, Brown XQ. Balance of chemistry, topography and mechanics at the cell–biomaterial interface: Issues and challenges for assessing the role of substrate mechanics on cell response. *Surface Science* 2004, 570: 119–133.
4. Tan J, Saltzman WM. Topographical control of human neutrophil motility on micropatterned materials with various surface chemistry. *Biomaterials* 2002, 23: 3215–3225.
5. Allen LT, Fox EJ, Blute I, Kelly ZD, Rochev Y, Keenan AK, Dawson KA, Gallagher WM. Interaction of soft condensed materials with living cells: phenotype/transcriptome correlations for the hydrophobic effect. *Proc Natl Acad Sci USA* 2003, 100: 6331–6336.
6. Brodbeck WG, Patel JV, Voskerician G, Christenson E, Shive MS, Nakayama Y, Matsuda T, Ziats NP, Anderson JM. Biomaterial adherent macrophage apoptosis is increased by hydrophilic and anionic substrates in vivo. *Proc Natl Acad Sci USA* 2003, 99: 10287–10292.
7. Ratner BD, Horbett TA, Hoffman AS, Hauschka SD. Cell adhesion to polymeric materials: Implications with respect to biocompatibility. *J Biomed Mater Res* 1975, 9: 407–422.
8. Lydon MJ, Minett TW, Tighe BT. Cellular interactions with synthetic polymer surfaces in culture. *Biomaterials* 1985, 6: 396–402.
9. Healy KE. Molecular engineering of materials for bioreactivity. *Curr Opin Solid State Mater Sci* 1999, 4: 381–387.
10. Olbrich KC, Andersen TT, Blumenstock FA, Bizios R. Surfaces modified with covalently immobilised adhesive peptides affect fibroblast population motility. *Biomaterials* 1996, 17: 759–764.
11. Mrksich M. What can surface chemistry do for cell biology? *Curr Opin Chem Biol* 2002, 6: 794–797.
12. Hennessy HM, Clem WC, Phipps MC, Sawyer AA, Shaikh FM, Bellis SL. The effect of RGD peptides on osseointegration of hydroxyapatite biomaterials. *Biomaterials* 2008, 29: 3075–3083.
13. Pegg E. Organic functionalisation of titania. PhD thesis, University of Nottingham, 2008.

14. Zeng H, Chittur KK, Lacefield WR. Analysis of bovine serum albumin adsorption on calcium phosphate and titanium surfaces. *Biomaterials* 1999, 20: 377–384.
15. Singhvi R, Kumar A, Lopez GP, Stephanopoulos GN, Wang DIC, Whitesides GM, Ingber DE. Engineering cell shape and function. *Science* 1994, 264: 696–698.
16. Lewandowska K, Balachander N, Sukenik CN, Culp LA. Modulation of fibronectin adhesive functions for fibroblasts and neural cells by chemically derivatised substrata. *J Cell Physiol* 1989, 141: 334–345.
17. Grinnell F, Feld MK. Adsorption characteristics of plasma fibronectin in relationship to biological activity. *J Biomed Mater Res* 1981, 15: 363–381.
18. Pettit DK, Hoffman AS, Horbett TA. Correlation between corneal epithelial cell outgrowth and monoclonal antibody binding to the cell binding domain of adsorbed fibronectin. *J Biomed Mater Sci* 1994, 28: 685–691.
19. Garcia AJ, Vega MD, Boettiger D. Modulation of cell proliferation and differentiation through substrate-dependent changes in fibronectin conformation. *Mol Biol Cell* 1999, 10: 785–798.
20. Michael KE, Vernekar VN, Keselowsky BG, Meredith JC, Latour RA, Garcia AJ. Adsorption-induced conformational changes in fibronectin due to interactions with well-defined surface chemistries. *Langmuir* 2003, 19: 8033–8040.
21. Steele JG, McFarland C, Dalton BA, Johnson G, Evans MDM, Howlett CR, Underwood PA. Attachment of human bone cells to tissue culture polystyrene and to unmodified polystyrene: the effect of surface chemistry upon initial cell attachment. *J Biomater Sci Polym Ed* 1993, 5: 245–257.
22. Howlett CR, Evans MDM, Walsh WR, Johnson G, Steele JG. Mechanism of initial attachment of cells derived from human bone to commonly used prosthetic materials during cell culture. *Biomaterials* 1994, 15: 213–222.
23. Underwood PA, Bennett FA. A comparison of the biological activities of the cell-adhesive proteins vitronectin and fibronectin. *J Cell Sci* 1989, 93: 641–649.
24. Tegoulia VA, Cooper SL. Leucocyte adhesion on model surfaces under flow: effects of surface chemistry, protein adsorption and shear rate. *J Biomed Mater Res* 2000, 50: 291–301.
25. McClary KB, Ugarova T, Grainger DW. Modulating fibroblast adhesion, spreading and proliferation using self assembled monolayer films of alkylthiolates on gold. *J Biomed Mater Res* 2000, 50: 428–439.
26. Scotchford CA, Cooper E, Leggett GJ, Downes S. Growth of human osteoblast-like cells on alkanethiol on gold self-assembled monolayers: the effect of surface chemistry. *J Biomed Mater Res* 1998, 41: 431–442.
27. Tidwell CD, Ertel SI, Ratner BD. Endothelial cell growth and protein adsorption on terminally functionalised, self assembled monolayers of alkanethiolates on gold. *Langmuir* 1997, 13: 3404–3413.
28. Soekarno A, Lom B, Hockberger PE. Pathfinding by neuroblastoma cells in culture is directed by preferential adhesion to positively charged surfaces. *Neuroimage* 1993, 1: 129–144.
29. Barry JJA, Howard D, Shakesheff KM, Howdle SM, Alexander MR. Using a core-sheath distribution of surface chemistry through 3D tissue engineering scaffolds to control cell ingress. *Adv Mater* 2006, 18: 1406–1410.

30. Healy KE, Thomas CH, Rezania A, Kim JE, McKeown PJ, Lom B, Hockberger PE. Kinetics of bone cell organisation and mineralization on materials with patterned surface chemistry. *Biomaterials* 1996, 17: 95–108.
31. Folkman J, Moscona A. Role of cell shape in growth control. *Nature* 1978, 273: 345–349.
32. Opas M. Expression of the differentiated phenotype by epithelial cells in vitro is regulated by both biochemistry and mechanics of the substratum. *Dev Biol* 1989, 131: 281–293.
33. Chen CS, Mrksich M, Haung S, Whitesides GM, Ingber DE. Geometric control of cell life and death. *Science* 1997, 276: 1425–1428.
34. Mitchell SA, Poulsson AHC, Davidson MR, Bradley RH. Orientation and confinement of cells on chemically patterned polystyrene surfaces. *Colloids and Surfaces B: Biointerfaces* 2005, 46: 108–116.
35. Hubbell JA. Synthetic biodegradable polymers for tissue engineering and drug delivery. *Curr Opin Solid State Mater Sci* 1998, 3: 246.
36. Kwok CS, Horbett TA, Ratner BD. Design of infection-resistant antibiotic-releasing polymers II. Controlled release of antibiotics through a plasma-deposited thin film barrier. *J Controlled Release* 1999, 62: 301–311.
37. Kasemo B, Gold J. Implant surfaces and interface processes. *Adv Dental Res* 1999, 13: 8–20.
38. Sanders CA, Rodreiguez Jr M, Greenbaum E. Stand-off tissue-based biosensors for the detection of chemical warfare agents using photosynthetic fluorescence induction. *Biosens Bioelectron* 2001, 16: 439–446.
39. Gopel W. Bioelectronics and nanotechnologies. *Biosens Bioelectron* 1998, 13: 723–728.
40. Scotchford CA, Gilmore CP, Cooper E, Leggett GJ, Downes S. Protein adsorption and human osteoblast-like cell attachment and growth on alkylthiol on gold self-assembled monolayers. *J Biomed Mater Res* 2002, 59: 84–99.
41. Grinnell F, Feld MK. Fibronectin adsorption on hydrophilic and hydrophobic surfaces detected by antibody binding and analysed during cell adhesion in serum containing medium. *J Biol Chem* 1982, 257: 4888–4893.
42. Sigal GB, Mrksich M, Whitesides GM. Effect of surface wettability on the adsorption of proteins and detergents. *J Am Chem Soc* 1998, 120: 3464–3473.
43. Wertz CF, Santore MM. Effect of surface hydrophobicity on adsorption and relaxation kinetics in albumin and fibrinogen: single-species and competitive behaviour. *Langmuir* 2001, 17: 3006–3016.
44. Underwood PA, Steele JG, Dalton BA. Effects of polystryene surface chemistry on the biological activity of solid phase fibronectin and vitronectin analysed with monoclonal antibodies. *J Cell Sci* 1993, 104(3): 793–803.
45. Altankov G, Grinnell F, Groth T. Studies on the biocompatibility of materials: fibroblast reorganisation of substratum-bound fibronectin on surfaces varying in wettability. *J Biomed Mater Res* 1996, 30(3): 385–391.
46. Dupont-Gillain CC, Pamula E, Denis FA, De Cupere VM, Dufrene YF, Rouxhet PG. Controlling the supramolecular organisation of adsorbed collagen layers. *J Mater Sci Mater M* 2004, 15(4): 347–353.
47. Elliot JT, Tona A, Woodward J, Jones P, Plant A. Thin films of collagen affect smooth muscle cell morphology. *Langmuir* 2003, 19(5): 1506–1514.
48. Elliot JT, Woodward JT, Umarji A, Mei Y, Tona A. The effect of surface chemistry on the formation of thin films of native fibrillar collagen. *Biomaterials* 2007, 28: 576–585.

49. Evans-Nguyen KM, Schoenfisch MH. Fibrin proliferation and model surfaces: influence of surface properties. *Langmuir* 2005, 21(5): 1691–1694.
50. Keselowsky BG, Collard DM, Garcia AJ. Surface chemistry modulates fibronectin conformation and directs integrin binding and specificity to control cell adhesion. *J Biomed Mater Res* 2003, 66A: 247–259.
51. Aota S, Nomizu M, Yamada KM. The short amino acid sequence Pro-His-Ser-Arg-Asn in human fibronectin enhances cell-adhesive function. *J Biol Chem* 1994, 269: 24756–24761.
52. Grant RP, Spitzfaden C, Altroff H, Campbell ID, Mardon HJ. Structural requirements for biological activity of the ninth and tenth FIII domains of human fibronectin. *J Biol Chem* 1997, 272: 6159–6166.
53. Petrie TA, Capadona JR, Reyes CD, Garcia AJ. Integrin specificity and enhanced cellular activities associated with surfaces presenting a recombinant fibronectin fragment compared to RGD supports. *Biomaterials* 2006, 27: 5459–5470.
54. Owen GRh, Meredith DO, ap Gwynn I, Richards RG. Focal adhesion quantification – A new assay for material biocompatibility? *Review European Cells and Materials* 2005, 9: 85.
55. Faucheux N, Tzoneva R, Nagel MD, Groth T. The dependence of fibrillar adhesions in human fibroblasts on substratum chemistry. *Biomaterials* 2006, 27: 234–245.
56. Geiger B, Bershadsky A, Pankov R, Yamada KM. Transmembrane extracellular matrix–cytoskeleton crosstalk. *Nat Rev Mol Cell Biol* 2001, 2: 793–805.
57. Zreiqat H, Valenzuela SM, Nissan BB, Roest R, Knabe C, Radlanski RJ, Renz H, Evans PJ. The effect of surface chemistry modification of titanium alloy on signalling pathways in human osteoblasts. *Biomaterials* 2005, 26: 7579–7586.
58. Krause A, Cowles EA, Gronowicz G. Integrin-mediated signalling in osteoblasts on titanium implant materials. *J Biomed Mater Res* 2000, 52(4): 738–747.
59. Marshall CJ. MAP kinase kinase kinase, MAP kinase kinase and MAP kinase. *Curr Opin Genet Dev* 1994, 4(1): 82–89.
60. Angel P, Karin M. The role of Jun, Fos and AP-1 complex in cell proliferation and transformation. *Biochim Biophys Acta* 1991, 1072(2–3): 129–157.
61. Grigoriadis AE, Wang ZQ, Wagner EF. Fos and bone cell development: lessons from a nuclear omcogene. *Trends Genet* 1995, 11(11): 436–441.
62. Keselowsky BG, Collard DM, Garcia AJ. Surface chemistry modulates focal adhesion composition and signalling through changes in integrin binding. *Biomaterials* 2004, 25: 5947–5954.
63. Garcia AJ, Vega MD, Boettiger D. Modulation of cell proloferation and differentiation through substrate dependent changes in fibronectin conformation. *Mol Biol Cell* 1999, 10: 785–798.
64. Garcia AJ, Reyes CD. Bio-adhesive surfaces to promote osteoblast differentiation and bone formation. *J Dent Res* 2005, 84: 407–413.
65. Yoshinari M, Oda Y, Kato T, Okuda K. Influence of surface modifications to titanium on antibacterial activity in vitro. *Biomaterials* 2001, 22(14): 2043–2048.
66. Braceras I, Alava JI, Goikoetxea L, de Maeztu MA, Onate JI. Interaction of engineered surfaces with the living world: ion implantation vs. osseointegration. *Surf Coat Technol* 2007, 201: 8091–8098.
67. Nayab SN, Jones FH, Olsen I. Effect of calcium ion-implantation of titanium on bone cell function *in vitro*. *J Biomed Mater Res – Part A* 2007, 83: 296–302.

68. Nayab SN, Jones FH, Olsen I. Effects of calcium ion implantationon human bone cell interaction with titanium. *Biomaterials* 2005, 26: 4717–4727.
69. Hanawa T, Nodasaka Y, Ukai H, Murakami K, Asaoka K. Compatibility of MC3T3-E1 cells with calcium-ion-implanted titanium. *J Jpn Soc Biomater* 1994, 12: 209–216.
70. Hanawa T, Kamiura Y, Yamamoto S, Kohgo T, Amemiya A, Ukai H, Murakami K, Asaoka K. Early bone formation around calcium-ion-implanted titanium inserted into rat tibia. *J Biomed Mater Res* 1997, 36: 131–136.
71. Smith BS, Nisbet DI. Biochemical and pathological studies on magnesium deficiency in rat. II. Adult animals. *J Comp Pathol* 1972, 82(1): 37–46.
72. Herzberg M, Foldes J, Steinberg R, Menczel J. Zinc excretin in osteoporotic women. *J Bone Miner Res* 1990, 5(3): 251–257.
73. Yang Y, Tian J, Deng L, Ong JL. Morphological behaviour of osteoblast-like cells on surface modified titanium in vitro. *Biomaterials* 2002, 23: 1383–1389.
74. Li L-H, Kong Y-M, Kim H-W, Kim Y-W, Kim H-E, Heo S-J, Koak J-Y. Improved biological performance of Ti implants due to surface modification by micro-arc oxidation. *Biomaterials* 2004, 25: 2867–2875.
75. Jones MI, McColl IR, Grant DM, Parker KG, Parker TL. Protein adsorption and platelet attachment and activation, on TiN, TiC and DLC coatings on titanium for cardiovascular applications. *J Biomed Mater Res* 2000, 52: 413–421.
76. Hauert R. A review of modified DLC coatings for biological applications. *Diamond and Related Materials* 2003, 12: 583–589.

19
Bioactive surfaces using peptide grafting in tissue engineering

M DETTIN, University of Padova, Italy

Abstract: The design of bioactive surfaces consists in writing a message on the surface of a biomaterial in the biochemical language of the target cells. This chapter first discusses the possibility of using synthetic peptides instead of native proteins, then it presents the different means by which bioactive molecules can be delivered to tissue-implant interfaces and the criteria for bioactive surface design. The chapter reviews biomimetic material applications in the engineering of bone, neural and cardiovascular tissues.

Key words: biomimetic material, bioactive surface, peptide mimicry, covalent grafting.

19.1 Biomimetic materials and bioactive surfaces

While significant progress has been made in the reconstruction of organs or tissues damaged by trauma or diseases such as cancer and aging, organ or tissue transplant is still largely accepted as the gold therapy in treating patients. An autologous transplant is limited by the high possibility of infection, pain at the donor site and the necessity of additional surgery. Using tissues from other patients or animals is problematic due to the immunological response that can arise following transplantation, and by the difficulty in conserving donated organs and tissues. The technique of tissue engineering applies methods of engineering and medical sciences to create artificial constructs that can be used for tissue regeneration [1]. This field has opened new scenarios for non-invasive, less painful therapies that are now defined as *regenerative medicine*. The paradigm of tissue engineering consists in isolating specific cells by performing a small biopsy on the patient, growing these cells on a biomimetic three-dimensional (3-D) scaffold in carefully controlled culture conditions, inserting the construct into the body of the patient, and guiding the formation of new tissue within the 3-D scaffold that will eventually biodegrade over time. These protocols offer incredible possibilities for studying complex biological processes such as tissue growth *in vitro*, through analysis of multiple aspects of structure–function relationships associated with new tissue formation.

480 Cellular response to biomaterials

In order for tissue engineering to completely accomplish the regeneration of damaged organs and tissues, many factors need to be considered relative to the scaffold and the characteristics of the biomaterials used. The development of biomaterials for application in tissue engineering is currently focused on identifying biomimetic materials that can interact with the surrounding tissues biochemically [2, 3]. To better understand these molecular mechanisms, a brief summary of the events that take place in the physiological environment at the tissue–implant interface will be discussed (Fig. 19.1). Once inserted into the anatomical site, the implant surface is in close contact with exposed tissues. Consequently, a series of natural physical, chemical and biochemical interactions occurs in which macromolecules present in tissues and biological fluids are involved. Studying the tissue–biomaterial interface is crucial to understanding the complexity of these events, which modify the biomaterial via interaction with cells and biological fluids as well as modifying cell behavior via interaction with the biomaterial.

In 1999, Kasemo and Gold [4] suggested a plausible description of the events that take place over time and space on a cellular and molecular scale, following surgical implantation of an endosseous implant. In particular, the sequence of phenomena that modifies the interface appears to be rapid in

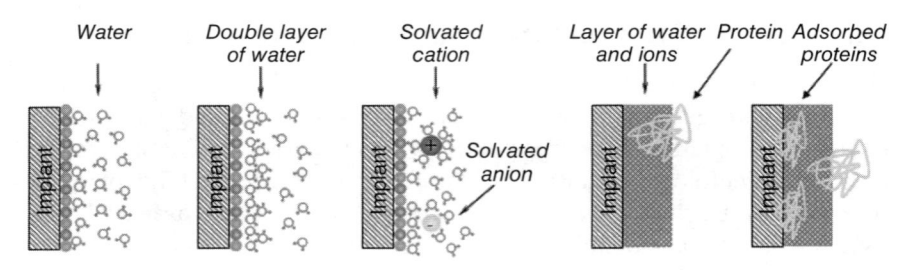

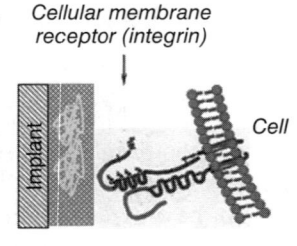

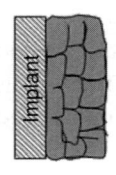

19.1 Schematic description of the events that take place following surgical implantation of an endosseous device.

the immediate phase after implant insertion. Once inserted, the implant's surface is exposed to the biological environment which prevails in the presence of water molecules: these bind quickly to the surface, forming within a matter of a few nanoseconds a simple or double layer whose structure is very different from that of liquid water and which depends strongly on the hydrophilic/hydrophobic properties of the surface. Once this aqueous film is formed, ions present in the environment (for example, Na^+ and Cl^-) are first solvated and then incorporated into the structure according to a spatial disposition that depends on surface characteristics. Subsequently, biomolecules surrounding the implant come into contact with its surface, where they are involved in a series of phenomena that vary from adsorption to conformational variations to their eventual denaturation.

Surface properties that influence these protein interactions are usually grouped into three categories: geometric, chemical and electrical [5]. Rougher surfaces expose a greater surface area for interactions with proteins than do smooth surfaces. The composition of the surface material determines which functional groups will be available for interaction with proteins. In any case, it should be noted that on a microscopic scale, the surface of the material is not homogeneous. Proteins adhere to the solid surface by adsorption: in implants that are exposed to blood, surface concentrations of proteins are thousands of times greater than those present in biological fluids. Proteins can be driven to the surface by the following transport systems: diffusion, thermal convention, flow and coupled transport. When in the vicinity of the scaffold surface, the protein can interact by forming polar, ionic or hydrophobic bonds or charge transfer. Even in the presence of only one type of protein, the layer will be heterogeneous. When a protein interacts with a surface on which another protein molecule is already bound, this secondary bond with the surface will not have the same orientation as presented by the first absorbed protein. The desorption of such proteins, or the breaking away of proteins from the surface and their return into solution, requires the simultaneous breaking of all protein–surface bonds, a very slow process [5].

Proteins should not be considered rigid structures: during the formation of bonds with the surface, they undergo alterations that change their conformation with partial or complete denaturation. Different levels of denaturation can also depend on the concentration of the proteins in solution: in low concentrations, proteins are bound to the surface in multiple locations, while in high concentrations, the proteins are bound by a lesser number of contacts and interactions among different protein chains can be activated.

At the physiological level, one cannot analyze just one protein since biological fluids contain many components. More than 150 proteins, as well as lipids, carbohydrates and hormones, are present in blood. When a multi-component solution interacts with a scaffold surface, certain proteins are deposited before others, with variations in the protein layer composition taking place over time. Many kinetic factors (i.e., concentration and size) and different affinities (i.e., size, charge, conformational stability) determine the composition of this protein layer, which is, broadly speaking, deposited chronologically. Considering a simple diffusion-limited situation at the interface, the molecules present in the solution at high concentration and with small size will arrive quickly and will be adsorbed from the surface. With time, molecules having greater affinity can approach the surface: exchange between proteins, resulting from competition for surface binding sites, takes place.

The protein layer composition, as well as the conformation of the proteins present, inevitably are influenced by the original characteristics of the biomaterial surface. According to its hydrophilic/hydrophobic properties, the proteins attach to the surface with their respective hydrophobic or hydrophilic domains, thus maintaining the water molecules of hydration in the first case. Hydrophilic domains are often located on the protein's surface and are thus available for interaction with the scaffold's surface. The implant's surface charge and its distribution can thus influence adhesion. However, charge is not the only determining factor. It should be noted that proteins bind more to surfaces that are closer to their isoelectric point (pI). There are two possible explanations for this. First, lateral interactions occur, since the proteins do not interact only with the surface but with the many other proteins present in solution. At the pI, reduced repulsions among molecules could favor interactions with the implant surface [5]. Second, if protein conformation is altered by surface–protein interactions, the surface composition of the protein itself would change and thus its means of binding to a substrate. The protein layer is known to be dynamic, in the sense that constant contact with biological fluids causes variations in the protein composition and conformation of the layer for the entire time necessary for healing. Therefore, the specificity of interactions between implant and biological tissue seems to be due principally to the way in which the protein layer is formed and organized on the implant surface: the characteristics of the protein layer depend, in turn, on how the implant surface is able to bind water molecules, ions, and other biomolecules transported by biological fluids.

Once the protein layer is formed, interface phenomena occur at an ever expanding scale. Cells, whose size is from 10^2 to 10^4 fold greater, and whose structural and functional complexity is enormously higher than that of proteins [4], begin to adhere. Proteins translate the surface characteristics into the biochemical language understandable to cells which interact with adsorbed proteins through their membrane receptors. The specificity of these interactions depends on how the protein layer is organized, in addition to temperature and surface texture. Cells adhere to the implant surface and, by activation of sequential biological cascades, grow, migrate, differentiate, and synthesize extracellular matrix (ECM), ultimately producing new tissue [6]. Attachment, adhesion and spreading are the first phases of cellular/biomaterial interaction and the quality of these phases will influence the capacity of cells to proliferate and differentiate on contact with the implant. The necessity of a complete fusion between the scaffold's surface and the tissue is particularly evident in determining the efficacy of osseous implants, avoiding when possible the formation of interface fibrosis. New-generation surfaces must therefore orchestrate the events that lead to an optimal integration between new tissue and implant, communicating with cells through the biochemical language of proteins.

These biomimetic materials must promote specific cellular responses that can direct the formation of new tissue. Biomolecular recognition by cells can be achieved by two methods: the first consists in providing the materials with bioactivity via incorporation of soluble molecules such as growth factors or plasmidic DNA in carriers that, upon release, promote or modulate the formation of new tissue [7–9]. Another approach involves the incorporation of proteins or peptides to the surface that can bind to cells by chemical or physical means. Molecules that interact in a specific way with receptors on the cell surface include entire or small sequences of proteins derived from the ECM or growth factors that promote specific cellular functions. Among these tasks, cellular adhesion is the most critical prerequisite to all vital functions of anchoring-dependent cells. This not only occurs through a simple suctioning effect, but is the initial event in a biochemical cascade that regulates activation of cytoskeleton proteins, the activity of transcription factors and gene expression. In the development of a biomimetic material, great importance is given to parameters such as density, spatial distribution and orientation of the biomolecule. The use of whole proteins is limiting due to their instability, high cost and often scarce availability. To avoid these problems, a strategy known as *peptide mimicry* is often employed.

19.2 Peptide mimicry for bioactive surface design

Peptide mimicry involves the identification of native protein fragments that retain at least in part one of the biological activities of the whole protein (Fig. 19.2). Short peptide sequences can be synthesized chemically [10], enzymatically or by recombinant DNA technology, and then purified to a high level of homogeneity. Subsequent studies using peptidic libraries have allowed for identifying peptidomimetic analogues that can resolve the problem of availability, potential immunogenicity and instability in physiological conditions that characterize native peptides and that have

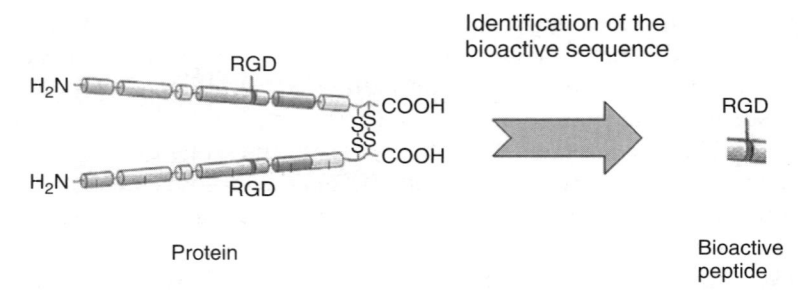

19.2 Schematic description of peptide mimicry strategy.

until today limited their use regardless of their considerable clinical potential [11].

Peptidomimetics can be defined as all peptide analogues obtained by post-translational modification: changes in the amino or carboxyl terminals; insertion of residues with inverted configurations (D-amino acids); the addition of isoteric groups or β-amino acids; substitutions of secondary proteic structures; and changes in peptide bonds and lipidization and conjugation with polymers (Fig. 19.3). Using brief amino acid sequences instead of native proteins in the preparation of bioactive surfaces offers many advantages. Native proteins tend to denature the moment they are adsorbed by the surface of the material such that subsequent bonding with the cellular membrane receptor is not always sterically possible. In comparison, synthetic peptides are very stable during the process of surface or material modification, and thus almost all sequences are available for interaction with cell receptors. In addition to being more easily characterized, peptides also have greater stability than proteins with regard to temperature, pH and storage. Peptides can be synthesized in the laboratory at competitive cost, avoiding the problems of harvesting that exist for some native proteins.

The addition of peptide sequences to biomimetic surfaces offers the advantage of introducing innovative biomaterials that can guide the formation of new tissue. This simple system can be used to study the complex mechanisms that direct tissue regeneration. For example, the effects of density and spatial distribution of bioactive peptides on specific cell functions such as adhesion, proliferation and differentiation can be studied using modified surfaces. Another aspect that must be emphasized is the greater versatility of peptides with respect to proteins in their anchoring to surfaces. Since they are obtained by chemical synthesis, peptides can be condensed via a particular introduced group or via only one functional group left available while other potentially reactive groups are reversibly protected. This versatility allows for the synthesis of biomimetic surfaces with peptides organized in a specific orientation.

Bioactive surfaces using peptide grafting in tissue engineering 485

Constrained amino acid residues

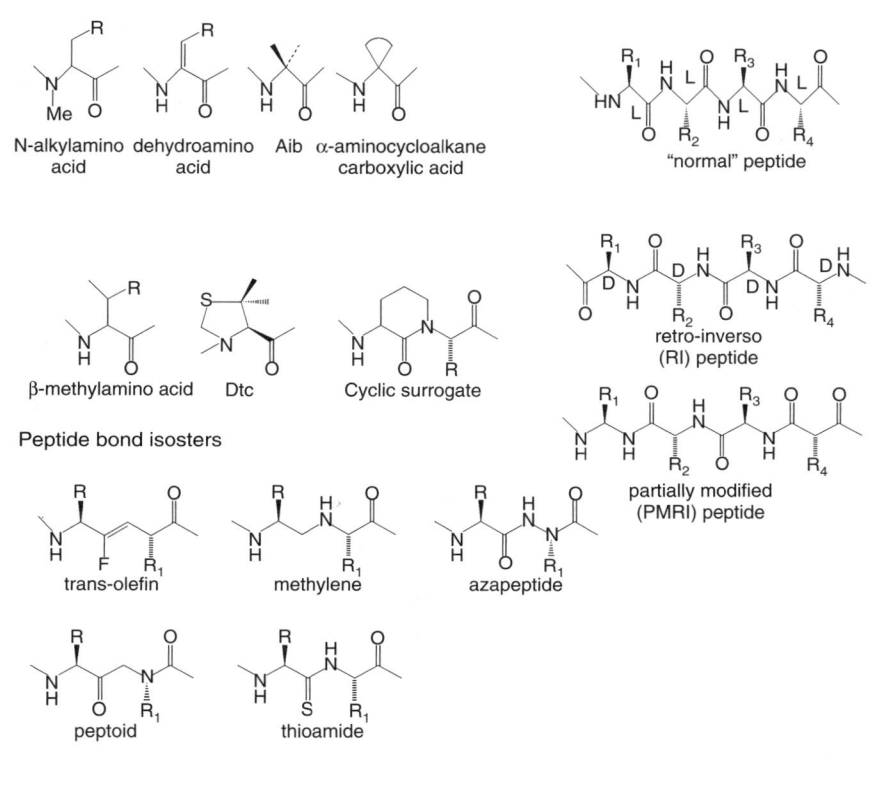

N-alkylamino acid   dehydroamino acid   Aib   α-aminocycloalkane carboxylic acid

"normal" peptide

β-methylamino acid   Dtc   Cyclic surrogate

retro-inverso (RI) peptide

Peptide bond isosters

trans-olefin   methylene   azapeptide

partially modified (PMRI) peptide

peptoid   thioamide

Some β-turn of peptidomimetics

19.3 Examples of the most common strategies for the peptidomimetic approach (adapted from Gentilucci *et al.*, 2006, ref. 11).

The most commonly utilized molecule is the RGD sequence (arginine-glycine-aspartic acid): a peptide signal for the adhesion of cells identified in 1984 and present in fibronectin, laminin, vitronectin, type I collagen, fibrinogen, von Willebrand factor, osteopontin, tenascin, bone sialoprotein, membrane proteins, bacterial and viral proteins, neurotoxins and disintegrins [12]. The great number of studies performed using this sequence derives from its vast distribution in every part of the organism, from its ability to activate multiple cell receptors for adhesion (binding to the integrin superfamily) and from its biological impact on the adhesion, behavior and survival of anchorage-dependent cells. At least 12 of the 24 known integrins adhere to molecules of the extracellular matrix by RGD-dependent mechanisms. The loop conformation contained in RGD and the sequences adjacent to it in diverse proteins are responsible for the different selectivity for specific integrins. In addition, the activity of this tripeptide is tied to blocking the carboxyl terminal. RGD activity decreases in the following order: GRGDSP > RGDS > RGD amide terminal > RGD carboxyl terminal [12]. Cyclic analogues of RGD are more active in promoting attachment if compared with linear sequences.

RGD as a soluble factor inhibits cellular adhesion to fibronectin, and as an anchoring agent it promotes cellular surface adhesion. Other signaling sequences utilized for the functionalization of surface models to promote cell adhesion are KQAGDV from gamma fibrinogen [13], the PHSRN motif of the FIII9 domain of fibronectin [14, 15], REDV from the III-CS domain of human plasma fibronectin [16], VAPG from elastin [17], and the sequences GYIGSR, SIKVAV [18], IKVAV, YIGSR [19], RNIAEIIKDI [20], YFQRYLI and PDSGR [21] from laminin. In promoting an adhesion mechanism specific only to bone cells, consensus sequences have been proposed such as XBBXBX or XBBBXXBX (B for a basic amino acid and X for a hydropathic amino acid) as binding domains for heparin. Among these signaling motifs are the peptides FHRRIKA [22], PRRARV [23], WQPPRARI [24], KRSR [25], polyarginine, polylysine [26], KKQRFRHRNRKG [27] and FRHRNRKGY [28] from vitronectin, and RKKRRQRRR from the Tat protein of Human Immunodeficiency Virus (HIV) [27].

19.3 Delivery of bioactive peptides to tissue–surface interface

Addition of bioactive molecules to material surfaces is the simplest way to render them biomimetic. Once the desired bioactive peptides that promote adhesion and cell growth are identified, strategies to tether them to a material are needed. The following are potential methods for the addition of bioactive molecules to surfaces:

- Simple adsorption
- Chemical anchoring of the molecule to the implant surface with covalent bonds
- Inclusion of the molecule in a coating material that allows the release of the active molecule at the implant–tissue interface [29] (Fig. 19.4).

Each of these methods presents advantages and disadvantages that will be briefly illustrated below.

Bioactive surfaces using peptide grafting in tissue engineering 487

Methods for delivering biomolecules
at tissue–implant interface

1. Adsorption

2. Release from coating

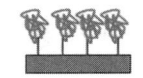

3. Immobilization

19.4 Different methods for preparing a biomimetic surface (adapted from Puleo and Nanci, 1999, ref. 29).

19.3.1 Adsorption

Adsorption takes advantage of the simple interactions between peptides and/or proteins and the material surface. It is accomplished by dipping the implant in a solution containing the protein or peptide for a predetermined time under controlled conditions. It is an extremely simple, inexpensive procedure, but it has the significant disadvantage of not allowing for a controlled deposition on the surface, and thus the precise quantity of biomolecule deposited is unknown. Furthermore, adsorption may cause a change in the conformation of the protein with subsequent denaturation and associated loss of biological activity.

19.3.2 Covalent grafting

Many techniques have been studied involving the immobilization of proteins on solid supports by covalent grafting. For example, derivatives of carboxyl-activated polymers have been coated with growth factors, fibronectin and collagen using covalent bonding of carbodiimide as a condensing agent [30]. This method allows for the use of existing reactive groups on the material surface for covalent bonding with the molecules of interest, theoretically one for every bonding site. The bond can be direct or with the introduction of another molecule that functions as a spacer. This approach has been widely used on polymeric surfaces, glass and SiO_2-based substances, as well as metal and oxide surfaces [31]. In orthopedic and dental applications, metal surfaces possess relatively few functional sites that can be used for anchoring of biomolecules, thus often these surfaces are oxidized by passivation. Silane derivatives on oxidized surfaces immobilize peptides, enzymes and adhesive proteins on different materials, including titanium and NiTi, Ti6Al4V and Co–Cr–Mo alloys. Biomaterial function depends on the type of silane, the experimental conditions and the substrate and biomolecule used. In past experiments, biomolecules have been successfully anchored to metallic surfaces for many days under simulated physiological conditions. The scarcity of functional groups can be compensated by including plasma treatments and deposition of self-assembled monolayers (SAMs). Plasma treatments can be used not only to increase the number of hydroxylic groups, but also to deposit different functional amino and carboxylic groups that offer great versatility in condensation reactions.

Deposition of SAMs involves both the process of silanization and gold coating the metal. Organosilanes can be used to bond an amino function to a surface oxide (i.e., SiO_2 or TiO_2), providing a second functional amino group which is anchored to the peptide or alternatively a bifunctional linker that can bind the peptide. The monomolecular layer of organosilanes modifies surface characteristics. An alternative chemical treatment can be used if the material is first gold coated [32]. In this case, the affinities of the thiol functions of gold are exploited; when using peptides that contain a cysteine residue, a simple immersion of the gold-coated material in the peptide solution is performed. The bonding that occurs is usually stable for at least 30 days. A significant number of materials such as glass [25], quartz [33], metallic oxides [34] and synthetic polymers [35] have been modified with bioactive peptide and their interaction with cells evaluated. For polymeric substrates that do not have functional groups for anchorage, a photochemical method of immobilization has been used [36].

19.3.3 Release from coating

This method involves the use of a coating for delivering adhesion or growth factors to the implant surface. This strategy has benefits since release is controlled, and concentration and exposure of the molecule to the material surface can be modulated to promote a tissue and cellular response. The peptide or protein is incorporated within the material, generally polymeric, allowing for its transport and release with kinetics that vary in function to the characteristics of the material and the method used for inclusion: it can be simply dissolved in the carrier or chemically bound to it. Many commercial products are used for this: co-polymers of lactic and glycolic acid [37], hyaluronic acid [38] and alginate hydrogels [39]. Naturally, these products must have the necessary properties of biocompatability, biodegradability and bioreabsorption. They also must not inhibit or mask the biological activity of the transported molecule. In addition the enriched coating must adhere to the implant surface even during surgical insertion in the anatomical site. Coatings made from collagen have also been proposed, as this fiber protein mimics the natural support onto which the biomolecules are adsorbed in the bone matrix. With this approach, biomolecule release is biphasic: an initial peak during the first four days, and a slow sustained release thereafter depending on the initial quantity of coating used and the biomolecule concentration within the coating. Ultimately, the collagen matrix is reabsorbed and replaced by new tissue.

19.4 Criteria for bioactive surfaces design

Biomimetic materials allow for the regulation and control of cellular interactions at a molecular level. Cell receptor binding to the ligand (bioactive peptide) on the biomimetic material determines the strength of adhesion of the cell to the surface [40], the velocity of cell migration on or through the material [41] and the extension and organization of the cytoskeleton [42]. Since biological responses are dependent on multiple variables such as receptor affinity and density and spatial distribution of the ligand, these must be considered the fundamental parameters for the development of a biomimetic material. The concentration of ligand and spacer used for separating the ligand from the surface or site of attachment is also important. Studies by Humphries and coworkers have identified the minimal surface concentration of fibronectin to be 1.3 pmoles/cm^2 for promoting cellular spreading [43]. Other studies indicate that the minimal surface concentration needed to obtain fibroblast spreading is 0.4 pmoles/cm^2 for vitronectin and 0.1 pmoles/cm^2 for fibronectin [44]. These values are low compared to those needed of peptides to induce sufficient spreading of fibroblasts. The concentration needed to produce adherence using an RGD peptide bonded to a hydrogel substrate of polyacrylic acid with a polyethyleneglycol (PEG) spacer was 12 pmoles/cm^2 [45]. Differences in the material used, the molecule delivery on the surface, the type of cell and the methods of assaying ligand concentrations and availability all contribute to variability in these results. For example, Massia and Hubbell indicated that 1 fmol/cm^2 of GRGDY covalently bound to glass was needed to obtain maximum spreading of fibroblasts, while 10 fmol/cm^2 of GRGDY was needed to assure focal contacts and stress fibers [46]. Nevertheless, surface peptide densities optimal for adhesion are not always optimal for subsequent phases of migration and proliferation. Neff and coworkers [47] observed that optimal proliferation of fibroblasts on RGD-functionalized polystyrene occurred at lower concentrations than those needed for optimal adhesion (1.33 pmoles/cm^2), indicating that concentrations for promoting cell proliferation could be different from those that maximize adhesion. In order to maintain biological activity of the peptide after immobilization, flexibility and minimal steric hindrance must be assured to allow for a seamless interaction with cell receptors. To this end, PEG spacers and non-specific sequences (GGGG or PVELP) have been employed.

19.5 In vitro and in vivo applications of bioactive surfaces

To demonstrate that the cellular response to the biomimetic surface is induced exclusively by a peptidic sequence such as the RGD motif, usually *in vitro* experiments are performed in serum-free media. When there is serum present in the media, a protein layer is formed that contains both the RGD motif and other binding sites that can compete for the anchored peptides, making it difficult to interpret data. Nevertheless, it is also rational to conduct these experiments in the presence of serum given the ultimate clinical objectives of this research. An improved cellular response in such a situation, compared with a non-biomimetic surface, could predict good performance in subsequent *in vivo* experiments.

The choice of time points and length of incubation times are also important. In the case of osteoblasts, cells are known to secrete collagen and non-collagen protein after 30–60 minutes [48, 49]. A shorter incubation time would allow the researcher to study the system in a time period when there is less interference between the surface-bound peptide and the proteins that are adsorbed on the surface from adherent cells in a serum-free environment. Data comparison from these types of *in vitro* experiments is very complex since the number of variables is extremely high: there is much heterogeneity in the cellular systems utilized (human, murine, rat), with regard to the extraction site (in the case of osteoblasts, calvaria versus femora) and in terms of donor age (from neonate to adult). It is thus quite evident that these multi-disciplinary studies do not always fully define all aspects relative to the physico-chemical and biological characteristics of bioactive surface materials.

19.5.1 Bone tissue engineering

Bone tissue formation is the result of a series of sequential events that begin with the recruitment and proliferation of bone progenitor cells from surrounding tissues, followed by differentiation, matrix formation and mineralization. A typical tissue engineering approach to treating skeletal defects involves the use of scaffolds containing osteo-promoting factors that promote repopulation of osteogenic cells. In particular, the ability of a biomimetic material to promote adhesion and migration of bone progenitor cells in the earliest stages of regeneration and healing is crucial for subsequent phases of new tissue formation.

Physiologically, many ECM proteins are expressed during the formation of new bone. Among these, osteopontin, thrombospondin and the sialoproteins have been identified to play a fundamental role in migration, proliferation, matrix deposition and mineralization of bone cells. Since integration of biomimetic scaffolds within the bone tissue takes place at the tissue–material interface, and is often determined by initial cellular interactions with the material surface, it is critical to develop functionalized scaffolds with peptides of the bone matrix.

The first studies in this field evaluated the RGD sequence. Adhesion of neonatal rat calvarial osteoblasts was tested on quartz supports treated with aminosilane and maleimide and then functionalized through Cys with an RGD-containing, 15-mer peptide (CGGNGE**RGD**TYRAY) derived from bone sialoprotein. The presence of the anchored peptide allowed for a stronger adhesion at 20 minutes compared to the control surface modified with CGGNGE**RGE**TYRAY. Adhesion strength was estimated by a radial flow apparatus that determined resistance to detachment. After 2 hours, a vinculin dye visualized a greater number of focal contacts with respect to controls [40]. Spreading, measured by the mean area of cells in contact with the surface, was higher in the RGD sample compared to the control RGE sample.

Studies on heparin binding sequences present on vitronectin, apolipoprotein E and B-100, fibronectin, thrombospondin, bone sialoprotein, osteopontin and platelet factor IV have identified the sequence signal for cellular adhesion to be of the type X-B-B-X-B-X and X-B-B-B-X-X-B-X, where B is a basic amino acid and X is a non-basic residue. Based on these studies, Dee and coworkers [25] have proposed the peptide KRSR as a new sequence signal for the adhesion of osteoblasts. This peptide, covalently bound through carbodiimide to a borosilicate glass treated with aminosilane (surface density ~80 pmol/cm^2), increased neonatal rat calvarial osteoblasts adhesion at 4 hours, in a serum-free media, without producing any effect on rat skin fibroblast adhesion. These results highlight the specificity of the molecular mechanisms involved in peptide activity. In this same study, insertion of a spacer between the bioactive sequence and the surface did not influence the interaction of osteoblasts and proteoglycans. The authors also demonstrated that the highest level of adhesion was achieved when the glass surface was bound to both RGD- and KRSR-containing peptides in a molar ratio of 1:1. Analogous studies in which both an RGD sequence (AcCGGNGE**RGD**TYRAY) and a binding site for heparin (i.e., a peptide containing the sequence FHRRIKA: AcCGGFHRRIKA) were added [42] demonstrated that (i) simultaneous use of two types of peptides anchored to quartz via silanization, functionalization with maleimide and interaction with thiol functional groups of Cys (surface density 4–6 pmol/cm^2) induced maximal adhesion of primary neonatal rat calvarial osteoblasts at 30 minutes in serum-free media, as well as optimizing cellular spreading at 4 hours, and mineralization at 24 days; (ii) surfaces containing both peptides in different proportions and surfaces functionalized with only the RGD peptide promoted the formation of focal contacts and stress fibers; while (iii) proliferation was no different in the presence of surfaces functionalized with one or both peptides. It was also demonstrated that, in this context, covalent functionalization of the surface is advantageous compared to adsorption [42].

Particular interest is now focused on biomimetic titanium, since this material is used in dental and orthopedic devices due to its exceptional weight-bearing capacity. Notwithstanding the current success of this biomaterial and the fact that its surface topography has achieved a high level of optimization, there is still room for improvement, above all in the case of implants of poor quality or limited quantity bone.

One of the first examples of a titanium biomimetic surface was presented by Xiao et al. [34]. He used a heterobifunctional reagent (N-succinimidyl-3-maleimidopropionate) to bond the silanized surface's amino group through 3-aminopropyltriethyloxysilane (APTES) with the biomimetic peptide to which a cysteine residue was added (RGDC). The surface peptide density, determined with X-ray photoelectron spectroscopy (XPS) and radiolabeling, was estimated to be approximately 30 pmol/cm^2 [34, 50].

Utilizing this strategy, Rezania and coworkers [33] have functionalized quartz disks, silicon wafers and sputter-deposited titanium films with the sequences AcCGGNGEPRGDTYRAY and AcCGGFHRRIKA, at a peptide density of 0.01 to 4 pmol/cm^2. The adhesion of neonatal rat calvarial osteoblasts and their spreading after 4 hours were excellent using RGD-functionalized surfaces with peptide densities greater than or equal to 0.6 pmol/cm^2. Rough, silanized titanium oxide was functionalized through its amino groups and glutaraldehyde [51] with two sequences, (GRGDSP)$_4$K and FHRNRKGY. This biomimetic surface increased not only cell adhesion, but also the strength of adhesion. Since this method of anchorage is non-specific (in addition to the amino terminal, other amino groups in lateral chains can react), these same researchers are working on new, more specific methods of anchorage [52]. In vivo implantation of titanium disks RGDC-functionalized with gold-thiol chemistry was carried out in bone defects induced in the tibia of Sprague Dawley rats. After four weeks, the regenerating bone adjacent to the peptide-coated surface was thicker than controls [35].

Polymeric materials (poly-α-hydroxyesters; polylactic acid (PLA); copolymers of polylactic and polyglycolic acid (PLGA) polypirrole (Ppy) and Interpenetrating Polymer Network [53]) have also been modified with the RGD peptide to create biomimetic surfaces or scaffolds. Often these types of polymers are used for coating titanium substrates to provide a very thin layer that does not modify titanium surface roughness (20 nm [54] or 270 nm

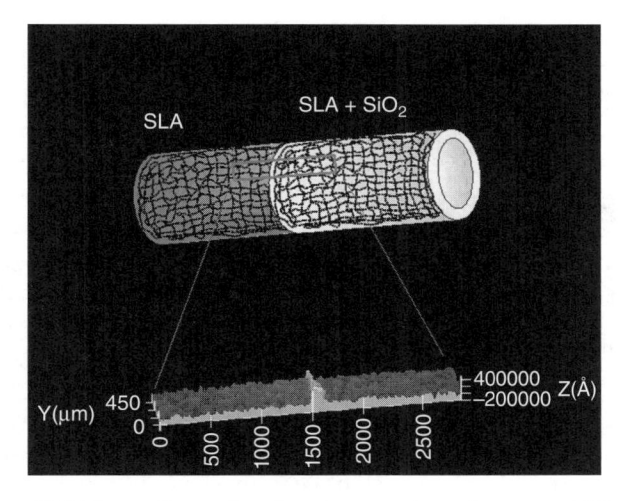

Bioactive surfaces using peptide grafting in tissue engineering 493

19.5 Three-dimensional reconstruction of an area (450 × 3000 μm²) of the surface sand-blasted and acid attached (on the right) partially coated with a silicon dioxide film (on the left). The peak in the middle is due to an artifact caused by the dip-coating procedure. From a morphological viewpoint there is no significant difference between the coated surface and the uncoated one (adapted from Dettin *et al.*, 2006, ref. 55).

[55]) (Fig. 19.5). To this aim, sequences have been covalently bound to coatings [54], thus assuring greater mineralization of neonatal rat calvarial osteoblasts, as well as including sequences in sol-gel films [55] that can promote adult rat bone marrow osteoblast adhesion and bone integration *in vivo* in the rabbit.

Recently, some researchers have attempted to identify surface treatments that render biomimetic surfaces resistant to non-specific protein absorption. Among these, Houseman and Mrksich [56] propose the SAMs on gold approach with GRGDS peptides and oligoethyleneglycols. Varying the layer composition, cellular adhesion of Swiss 3T3 fibroblasts was decreased as the length of the oligoethyleneglycol was increased, to a surface peptide density between 17 and 0.7 pmol/cm². This study clearly identified that the microenvironment influences the affinity of the ligand for cell integrin receptors. In the work of Schuler *et al.* [57], the results in different cell systems are presented (porcine epithelium, embryonal mouse fibroblasts and neonatal rat calvarial ostoblasts) seeded on both smooth and rough biomimetic titanium surfaces. As in the preceding case, these surfaces are called non-fouling and can be used in the presence of serum. They are obtained by conjugating the peptide Ac-GCRGYGRGDSPG-NH$_2$ with polymers PLL-g-PEG (poly-L-lysine-graft-PEG) via the vinylsulfone group on PEG (surface density between 0.001 pmol/cm² and 3 pmol/cm²)

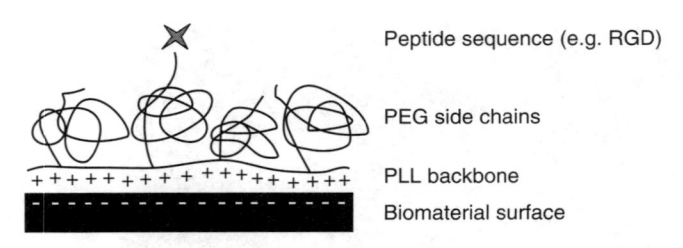

19.6 Schematic description of a non-fouling biomimetic surface: a PLL-g-PEG polymer is adsorbed on a negatively charged metal oxide surface. The sequence RGD is covalently attached to a fraction of the PEG-chains (adapted from Schuler et al., 2006, ref. 57).

(Fig. 19.6). Results showed that both adhesion and spreading were influenced by topography and surface chemical composition. Fibroblasts preferred smooth surfaces while osteoblasts preferred rough surfaces. Osteoblast adhesion increased with the density of surface peptides. In these studies, there was no synergism of effect between peptide density and surface roughness with regard to osteoblast adhesion.

Recently, peptide sequences that specifically bind $\alpha_2\beta_1$ integrin have been identified in type I collagen and have been utilized to obtain adhesive biomimetic surfaces. These sequences spontaneously configure into a collagen-mimetic triple helix that contains the sequence GFOGER (O is hydroxyproline) [58, 59]. Three different methods have been successfully used to functionalize polystyrene: two involve anchoring the peptides at the carboxyl terminal with biotin to surfaces treated with avidin or anti-avidin antibodies; another involves the reaction between the peptide and the carboxylic groups of adsorbed Bovine Serum Albumin (BSA) to the surfaces using 1-ethyl-3-[3-dimethylaminopropyl]carbodiimide (EDC) and N-hydroxysulfosuccinimide (sulfo-NHS) (Fig. 19.7). The most common osteogenic growth factors are members of the Transforming Growth Factor-β (TGF-β) family and the Bone Morphogenetic Proteins (BMP) [9]. Generally, the growth factors are incorporated into a carrier and released at the specific site to promote the formation of new bone tissue, or they are covalently bound to the biomaterial surface to increase the osteointegration of the implant. For example, a BMP-2 fragment has been used to functionalize an alginate hydrogel with osteoinductive potential [32].

Cartilage is a fibrous connective tissue composed of water and ECM containing type II collagen and a great quantity of proteoglycans, particularly chondroitin sulfate. Cartilage has many functions, including absorbing collisions, transferring load to the underlying bones, and allowing for movements of low friction in the articular junctions. Degeneration of cartilage can be caused by disease, congenital anomalies or trauma. This tissue does not have the capacity to repair itself since it is completely without blood

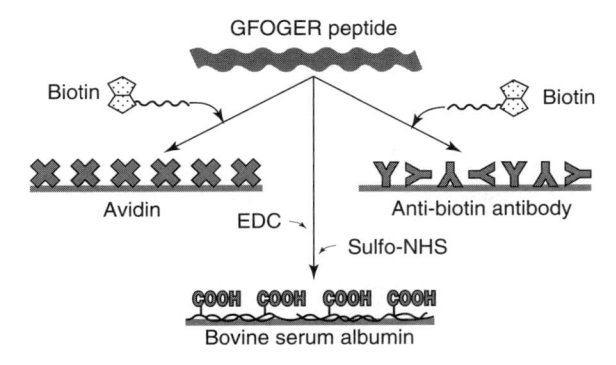

19.7 Different methods for delivering the collagen-mimetic GFOGER peptide to a surface (adapted from Reyes, ref. 86).

vessels. In the last 10 years, the use of cells incorporated into scaffolds has been widely investigated. Among these, the monolayer culture of chondrocytes is widely used. These cells easily expand but do not maintain their chondrogenic phenotype. To overcome this limitation, three-dimensional cultures have been investigated (pellet, bioreactor and scaffold cultures) using primary cell cultures and immortalized cell lines, as well as embryonic and adult stem cells. Molecular signals such as growth factors (TGF, FGF, IGF, Platelet-Derived Growth Factor), cytokines and non-protein molecules have been employed to promote scaffold colonization. Jeschke and coworkers have demonstrated that human and porcine chondrocytes adhere preferentially to coated surfaces with tailor-made cyclic RGD peptides even after numerous passages in culture monolayer [60]. In other studies, RGD sequences were utilized to modify carriers for the transport or differentiation of cells. Chen and coworkers used the sequence GRGDSPK to covalently functionalize microspheres of PLLA via carbodiimide, increasing chondrocyte adhesion in a reactor with intermittent flow [61]. In 2006, Hwang *et al.* demonstrated that encapsulation of human embryonal stem cells in poly(ethyleneglycol)-diacrylate containing hyaluronic acid or type I collagen did not lead to cell growth or matrix production. Conversely, the use of RGD-modified hydrogels led to the formation of new collagen within 3 weeks [62].

19.5.2 Neural tissue engineering

Given the anatomical and functional complexity of the nervous system, development of scaffolds for the repair and regeneration of brain, spinal cord and other nervous system tissues is an enormous challenge to modern medical research. Many prerequisites must be met to create a biocompatible microenvironment that permits cell infiltration and repair of damaged neuronal connections in a controlled and localized manner. Scaffolds are needed for cerebral tissue both to repair damage caused by trauma or diseases such as Parkinson's and Alzheimer's and also to serve as coatings for brain implants to prevent or limit inflammation at the site of graft [63]. The development of these types of scaffolds is further complicated by the characteristics of central nervous tissue, which has much less regenerative capacity than other tissue, including the peripheral nervous system. In the case of the spinal cord, a functioning scaffold would resolve problems derived from trauma. It is known that fluid fills the spinal cavity formed after trauma, producing a dense glial scar. This filling fluid, rich in active astrocytes, glycosaminoglycans and other inhibitory molecules, prevents the infiltration of neurons and other cells into the damaged site, thus the damage causes a loss in connection of axons and motor function [63]. Introduction of a scaffold into the traumatized site could limit glial scar formation and, using sequence signals for neuronal cell adhesion, could allow for the regeneration of neurons and extension of axons into the damaged site.

Currently, peripheral nerves are repaired by explanting one from another site and implanting it at the site of damage, for the joining of the two disconnected stumps. Limitations of this method include the need for two incision sites, the pathology created at the donor site, and the often partial repair affected at the damaged site. Tissue engineering proposes an alternative involving the use of scaffolds. Some materials used in this field are PEG, PLA/PLG (polyglycolicacid)/PLGA, Ppy, alginate, chitosan, collagen, dextran, fibrin and hyaluronic acid. These scaffolds are enriched with growth factors such as Nerve Growth Factor (NGF), Neutrophin-3 (NT-3), Ciliary Neutrophic Factor (CNTF), TGF-β or anti-inflammatory substances. Another alternative is the drug delivery system called HBDS (Heparin Binding Delivery System), which utilizes non-covalent interactions to enrich the scaffolds with growth factors. Specifically, fibrin scaffolds are covalently functionalized with a peptide containing the sequence signal for heparin derived from anti-thrombin III. The scaffold sequesters heparin which in turn binds growth factors for nerve cells (i.e., NGF or NT-3) [64]. Tested for the delivery of NT-3, this system has been shown to promote sprouting of neuronal fibers compared to controls, but long-term studies have demonstrated that this approach did not provide a greater recovery in function [65]. The selection of peptidic sequences to utilize in such systems was studied by Maxwell and co-workers using a phage display library. By using peptides with different gradually increasing affinities for heparin, it was possible to modulate both the release of heparin and growth factor [66]. Even when the lesion is miminal, optimal regeneration of nervous tissue requires axon growth of the proximal stump without interference in order to create a new connection with the distal stump. Recently, a new strategy involves the guided acceleration of axon growth by providing it a simple tube-shaped conduit. However, nerve regeneration requires the use of a scaffold that gives not only mechanical support for neuron growth and the prevention of fibrous tissue growth, but one that contains molecular signals that can guide the axons towards the distal stump.

The use of biomimetic materials allows for either the *in loco* placement of neurotrophic growth factors and/or *in vitro* grown cells that can stimulate the regeneration of neurons, or the placement of ligands that reproduce ECM protein fragments that can then actively control the morphogenesis of neurites, which play a key role in neuron migration. Use of peptides in this field began with the preparation of surfaces for the culture of neurons. Charged amino groups, covalently bonded to surface groups by reaction with APTES or introduced by coating with polylysine, facilitate negatively charged cell adhesion. The RGD sequence present in laminin also promotes the adhesion and growth of neurites even at low concentrations [67, 68]. There are differences in cell form and motility when adhered to polylysine or the RGD sequence, indicating that different structural elements are involved in the interactions with different surface molecules.

In recent years, many new proteins have been studied as neuronal guides, such as neural cell adhesion molecule (N-CAM) and members of the immunoglobulin superfamily. These proteins participate not only in adhesion but also in pathfinding of neurites and in *in vivo* fasciculation. In particular, the sequence KHIFSDDSSE [69], which mimics a homophilic binding domain of N-CAM, has been immobilized on glass by aminosilane and condensation of the peptide's carboxylic terminals with carbodiimide. This peptide was shown to increase rat cortical astrocyte adhesion without influencing fibroblasts. Other peptides, YIGSR and IKVAV, derived from laminin, and immobilized on biomimetic surfaces up to a concentration of 0.45 pmol/cm^2 [70], promoted the adhesion and survival of neurons, as well as the outgrowth of axons.

In order to facilitate the *in vivo* regeneration of nervous tissue, a three-dimensional scaffold must be used that will initially guide axon growth through its intraluminal architecture. To this end, the peptides CQAASIKVAV and CDPGYIGSR are covalently bound via the bifunctional reagent sulfo-(*N*-maleimidomethyl)cycloesan-1-carboxylate to a co-polymer of poly(2-hydroxymethyl methacrylate) and 2-aminoethyl methacrylate that is designed to have (i) numerous channels fixed longitudinally with a mean diameter of approximately 190 µm to promote the fasciculation of the regenerative cables, and (ii) a compression module of approximately 190 kPa, similar to that of nervous tissue. The surface peptide density is approximately 100 micromol/cm^2. This strategy presents the indubitable advantage of creating a scaffold in an aqueous solution, combining physical and biochemical stimuli that promote the adhesion of neural cells, and guides the growth of neurites better than non-functionalized controls [71].

In general, ideal scaffolds for biomaterials must fulfill the following criteria: (i) have a base unit that can be designed and modified; (ii) have a controllable level of biodegradability; (iii) be without cytotoxicity; (iv) have characteristics that promote or inhibit in a specific way cell–material interactions; (v) induce minimal immune or inflammatory response; (vi) be easy to produce on different scales, be easy to purify and process; and (vii) be chemically compatible with aqueous solutions and physiological conditions. Among the most promising scaffolds that can potentially fulfill all these criteria are those composed of self-assembled peptides. These molecules allow for obtaining structures in two or three dimensions that can promote multi-dimensional cell–cell interactions and can increase cell density [72]. To this aim, many types of natural materials derived from animals, for example collagen matrices, can be used as supports on which cells grow. Nevertheless, a serious drawback to many of these naturally derived versus synthetic scaffolds is the risk of transmitting pathogens, such as transmittable spongiform encephalopathy (TSE).

Another important advantage to using synthetic scaffolds is that they can be designed to satisfy specific criteria. When synthetic materials contain amino acids, these monomers assure excellent physiological compatibility and minimum toxicity. These new biomaterials based on synthetic peptides are composed of self-complementary scaffolds in which amino acids with a positive charge and residues with a negative charge are separated by hydrophobic residues [72]. Such matrices are produced in various forms such as tapes, strings or sheets of different widths. The peptide and saline concentration and the processing apparatus then determine the ultimate geometry and macroscopic dimensions of the matrix. In particular, a 16-residue peptide (AEAEAKAK)$_2$ demonstrates a high propensity to form β-sheet structures [73] (Fig. 19.8). After addition of salt, the peptide spontaneously forms a membrane particularly resistant to proteolytic digestion (trypsin, alpha-chymotrypsin, papain, K proteases and pronases), of a simple and completely non-toxic composition. In particular, the type of salt utilized seems to play an important role in inducing the process of self-assembly: the order of efficacy in inducing membrane formation is $Li^+ > Na^+ > K^+ > Cs^+$. The addition of Cs^+ causes precipitation of aggregates more than the formation of a membrane. According to this proposed model [73], membranes are formed by the stacking of β-sheets, each bound to the preceding one by ionic bridges formed between charged groups of lateral chains (glutamic acid and lysine residues), and to the subsequent sheet via hydrophobic interactions between lateral chains of alanine residues (Fig. 19.8). These peptides are in fact called *molecular Lego* [74] since they are stacked by their charged and hydrophobic sides just as the famous blocks attach to one another via pegs and holes.

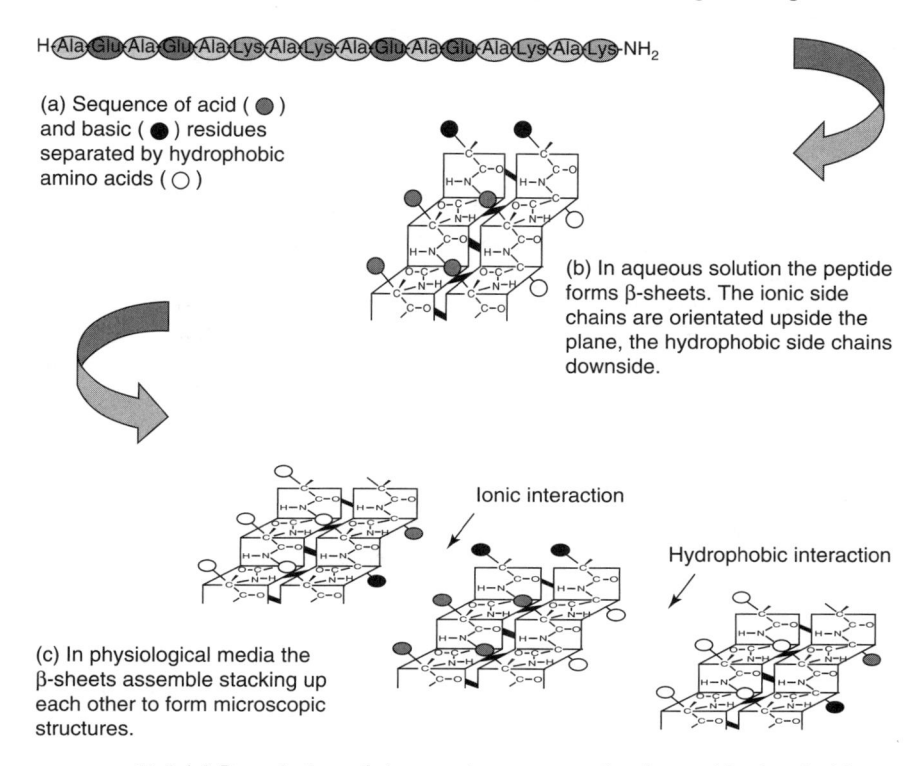

Bioactive surfaces using peptide grafting in tissue engineering 499

H-Ala-Glu-Ala-Glu-Ala-Lys-Ala-Lys-Ala-Glu-Ala-Glu-Ala-Lys-Ala-Lys-NH$_2$

(a) Sequence of acid (●) and basic (●) residues separated by hydrophobic amino acids (○)

(b) In aqueous solution the peptide forms β-sheets. The ionic side chains are orientated upside the plane, the hydrophobic side chains downside.

(c) In physiological media the β-sheets assemble stacking up each other to form microscopic structures.

Ionic interaction

Hydrophobic interaction

19.8 (a) Description of the regular pattern of polar and hydrophobic amino acids characteristic of self-assembling sequences. (b, c) Schematic view of the self-assembling process.

The new frontier for self-assembled peptide scaffolds involves modulating their capacity to induce specific cellular functions by addition of specific sequence signals that promote adhesion and/or growth of a particular cell type. Experiments conducted using neuronal cells from either cell lines or explants have demonstrated that self-assembled peptide matrices promote not only neuron growth but also the formation of functional synaptic connections. In particular, similar levels of synaptic activity and formation were obtained utilizing these self-assembled peptide matrices and Matrigel. This finding is extraordinary if one considers that commercial Matrigel contains laminin, collagen and other ECM proteins and growth factors [75]. In recent studies, self-assembled peptides were used to treat a severed hamster optic nerve [76]. Regeneration of axon connections was promoted to such a density to allow for recovery of vision. The results obtained with a single injection of peptide were similar to that achieved with surgical insertion of a piece of sciatic nerve, yet without the damage in motility provoked by the explantation. Axon growth through the scaffold-filled lesion was 100% in young and adult animals, while the recovery of vision was 75% in treated animals.

19.5.3 Cardiovascular tissue engineering

A synthetic cardiovascular graft can be used to repair tissues damaged during cardiac or bypass surgery when there is a limited quantity of autologous tissue. Synthetic vascular grafts generally take longer to achieve patency than natural tissues, primarily because of thrombi and hyperplasia of the intima following graft implantation. The differences in mechanical properties between the synthetic graft and surrounding tissues incite clinical complications. This slow achievement of patency of synthetic materials is even more of a problem in vessels with a diameter less than 6 mm. However, the design of cardiovascular grafts with high biocompatibility and mechanical stability is critical to the progress of cardiovascular tissue engineering.

The development of cardiovascular substitutes must consider the anatomical structure and biological function of blood vessels. These are formed in three layers of tissue composed of different cells that perform specific functions: (i) the intima, which is in contact with the blood and whose endothelial cell coating has anti-thrombotic properties; (ii) the middle tissue that possesses optimal mechanical properties due to the smooth muscle cells and elastin fibers; and (iii) the most external layer composed of fibroblasts and connective tissue. The ideal vascular graft must be designed to fulfill all these distinct functions. Currently, coronary artery bypass procedures are performed with autologous tracts of sapphic or mammary arterial veins. One approach to the engineering of vascular grafts involves the seeding of endothelial cells onto synthetic materials to assure that its antithrombogenetic properties are maintained. The success of these tissue engineering implants is tied to three principal aspects: (i) the type of cells and level of differentiation; (ii) the type of scaffold; and (iii) external stimuli, both mechanical and biochemical.

For cardiovascular systems, the general strategies of tissue engineering are complicated by the presence of circulating blood. Every attempt to modify a scaffold and to encourage re-endothelialization or cellular adhesion has an impact on blood components. Endothelialization through transplanted cells is problematic, since cells tend to fall off when the engineered surface is exposed to circulating blood. To strengthen adhesion of these cells to the support, ECM proteins or peptides derived from them have been used. This strategy was employed for polytetrafluorethylene supports using both fibronectin and RGD-containing peptides [77]. While the protein was simply adsorbed, the peptides were bound covalently: in both cases, a significant rise in cellular adhesion and stability between the support and cells was observed.

A possible complication arises from the fact that the RGD sequence is potentially capable of also increasing platelet adhesion, since these cells express integrin that recognizes the RGD sequence. Thus it is necessary to identify peptide sequences specific to endothelial cells that can promote their adhesion without creating surface thrombogenesis. To this aim, the effects of support-anchored adhesive peptides such as RGDS, KQAGDV and VAPG on vascular smooth muscle cell growth have been evaluated. Cells were shown to adhere better to functionalized surfaces; migration was greater compared to those not functionalized when the surface peptide density was 0.2 nmol/cm^2 while migration was lower if the surface density was as high as 2 nmol/cm^2. Proliferation was also less in treated samples, decreasing with increased surface peptide density, while the production of extracellular matrix was increased with respect to controls [78]. These complex findings lead one to think that it is not necessary to maximize a single cellular event, but rather adhesion in concert with proliferation, migration and matrix production. Other peptides that have been studied with success include YIGSR and REDV, the latter of which is derived from the III-CS domain of human plasma fibronectin [79].

Mann and coworkers [80] have demonstrated that ECM production due to endothelium and smooth muscle cells depends on the type and density of the peptide-signals present on the glass surface. It was also demonstrated that growth factors such as TGF-β1, in addition to sequence signals for adhesion, are necessary for optimal matrix production. Other studies on the grafting of cardiovascular tissue involve the inclusion of smooth muscle cells in biomimetic hydrogels.

Self-assembled peptides were successfully used to create a microenvironment rich in nanofibers that promotes the recruitment, survival and organization of vascular cells when injected into the myocardium [81]. Scaffolds of self-assembled peptides have been shown to promote adhesion of human microvascular endothelial cells. After adhesion, the cells were activated to produce a network completely analogous to what is obtained when cells are grown on Matrigel or on fibrin gels or collagen in the presence of growth factors such as VEGF (Vascular Endothelial Growth Factor) and beta-FGF (Fibroblast Growth Factor). These data indicate that the peptide scaffold supplies to the cells a 'fluffy' three-dimensional substrate that promotes the formation and survival of similar-capillary structures. While the cells included in fibrin or type I collagen gel activated proteases that degrade matrix and induce apoptosis, no such effects were observed for the self-assembled peptide scaffold over a two-week period. Furthermore, the peptide matrix seemed to promote the expression of VEGF [82]. Self-assembled peptides have also been used for prolonged drug delivery of IGF-1 (Insulin-like Growth Factor-1), and factors of growth and cardiomyolytic differentiation [83].

19.5.4 Other tissues

None of the models described herein is able to reproduce all the characteristics of natural, healthy skin. Tissue engineering has focused on skin, attempting to optimize not only the source of cells but also the matrix on which these cells will grow. As in other fields, natural polymer scaffolds such as collagen, fibronectin and hyaluronic acid, and synthetic polymers such as PLA, PGA, PLGA and polytetrafluoroethylene, have been investigated. The major disadvantage of synthetic polymers is that they do not contain cellular signals. To overcome this, strategies are being developed to incorporate adhesive peptides into synthetic materials [84]. Furthermore, PEG matrices have recently been functionalized with Matrix Metalloproteinase (MMP)-sensitive or plasmin-sensitive peptidic sequences to study proteolytic cellular migration that can be considered the starting point for the development of modifiable scaffolds [85].

19.6 Future trends

Future scenarios for the application of biomimetic peptides to surfaces or scaffolds will be characterized by: (i) identifying new biomimetic peptides that promote cellular events (adhesion rather than migration, growth or differentiation) of one specific cell type; (ii) effects of different signal sequences that promote diverse cellular functions or that induce multiple mechanisms that mediate one cell function; (iii) strategies for innovative functionalization, selective and of easy application, modulating the strength of binding between a ligand and the scaffold for different applications; (iv) strategies that optimize mechanical and biochemical stimuli; and (v) evaluation and optimization of all cellular aspects that ultimately determine the formation of new engineered tissue.

19.7 References

1. Vacanti, J.P., Langer R. Tissue engineering: the design and fabrication of living replacement devices for surgical reconstruction and transplantation, *Lancet*, 354, 32–34, 1999.
2. Sakiyama-Elbert, S.E., Hubbell, J.A. Functional biomaterials: design of novel biomaterials, *Annu Rev Mater Res*, 31, 183–201, 2001.
3. Shin, H., Jo, S.B., Mikos, A.G. Biomimetic materials for tissue engineering, *Biomaterials*, 24, 4353–4364, 2003.
4. Kasemo, B., Gold, J. Implant surfaces and interface processes, *Adv Dent Res*, 13, 8–20, 1999.
5. Dee, K.C., Puleo, D.A., Bizios, R. *An Introduction to Tissue–Biomaterial Interactions*, Wiley-Liss, Hoboken, NJ, 2002.
6. Ramires, P.A., Giuffrida, A., Milella, E. Three-dimensional reconstruction of confocal laser microscopy images to study the behavior of osteoblastic cells grown on biomaterials, *Biomaterials*, 23, 397–406, 2002.

7. Whitaker, M.J., Quirk, R.A., Howdle, S.M., Shakesheff, K.M. Growth factor release from tissue engineering scaffolds, *J Pharm Pharmacol*, 53, 1427–1437, 2001.
8. Richardson, T.P., Murphy, W.L., Mooney, D.J. Polymeric delivery of proteins and plasmid DNA for tissue engineering and gene therapy, *Crit Rev Eukaryot Gene Expr*, 11, 47–58, 2001.
9. Babensee, J.E., Mcintire, L.V., Mikos, A.G. Growth factor delivery for tissue engineering, *Pharm Res*, 17, 497–504, 2000.
10. Grant, G.A. *Synthetic Peptides: a User's Guide*, W.H. Freeman, New York, 1992.
11. Gentilucci, L., Tolomelli, A., Squassabia, F. Peptides and peptidomimetics in medicine, surgery and biotechnology, *Current Medicinal Chemistry*, 13, 2449–2466, 2006.
12. Hersel, U., Dahmen, C., Kessler, H. RGD modified polymers: biomaterials for stimulated cell adhesion and beyond, *Biomaterials*, 24, 4385–4415, 2003.
13. Hautanen, A., Gailit, J., Mann, D.M., Rouslahti, E. Effects of modifications of the RGD sequence and its context on recognition by the fibronectin receptor, *J Biol Chem*, 264, 1437–1442, 1989.
14. Kao, W.J., Lee, D. In vivo modulation of host response and macrophage behavior by polymer networks grafted with fibronectin-derived biomimetic oligopeptides: the role of RGD and PHSRN domain, *Biomaterials*, 22, 2901–2909, 2001.
15. Kao, W.J., Lee, D., Schense, J.C., Hubbell, J.A. Fibronectin modulates macrophage adhesion and FBGC formation: the role of RGD, PHSRN, and PRRARV domains, *J Biomed Mater Res*, 55, 79–88, 2001.
16. Hubbell, J.A., Massia, S.P., Desai, N.P., Drumheller, P.D. Endothelial cell-sensitive materials for tissue engineering in the vascular graft via new receptor, *Biotechnology* (NY), 9, 568–572, 1991.
17. Mann, K.B., West, J.L. Cell adhesion peptides alter smooth muscle cell adhesion, proliferation, migration, and matrix protein synthesis on modified surfaces and in polymer scaffolds, *J Biomed Mater Res*, 60, 86–93, 2002.
18. Tong, Y.W., Shoichet, M.S. Peptide surface modification of poly(tetrafluoroethylene-co-hexafluoropropylene) enhances its interaction with central nervous system neurons, *J Biomed Mater Res*, 42, 85–95, 1998.
19. Graf, J., Ogle, R.C., Robey, F.A., Sasaki, M., Martin, G.R., Yamada, Y., Kleinman, H.K. A peptapeptide from the laminin B1 chain mediates cell adhesion and binds the 67,000 laminin receptor, *Biochemistry*, 26, 6869–6900, 1987.
20. Schense, J.C., Bloch, J., Arbischer P., Hubbell, J.A. Enzymatic incorporation of bioactive peptides into fibrin matrices enhances neurite extension, *Nature Biotechnology*, 18, 415–419, 2000.
21. Huber, M., Heiduschka, P., Kienle, S., Pavlidis, C., Mack, J., Walk, T., Jung, G., Thanos, S. Modification of glassy carbon surfaces with synthetic laminin-derived peptides for nerve cell attachment and neurite growth, *J Biomed Mater Res*, 41, 278–288, 1998.
22. Healy, K.E., Rezania, A., Stile, R.A. Designing biomaterials to direct biological responses, *Ann N Y Acad Sci*, 875, 24–35, 1999.
23. Kao, W.J., Hubbell, J.A., Anderson, J.M. Protein-mediated macrophage adhesion and activation on biomaterials: a model for modulating cell behavior, *Mater Sci Mater Med*, 10, 601–605, 1999.

24. Woods, A., McCarthy, J.B., Furcht, L.T., Couchman, J.R. A synthetic peptide from the COOH-terminal heparin-binding domain of fibronectin promotes focal adhesion formation, *Mol Biol Cell*, 4, 605–613, 1993.
25. Dee, K.C., Andersen, T.T., Bizios, R. Design and function of novel osteoblast-adhesive peptides for chemical modification of biomaterials, *J Biomed Mater Res*, 40, 371–377, 1998.
26. Massia, S.P., Hubbell, J.A. Immobilized amines and basic amino acids as mimetic heparin-binding domains for cell surface proteoglycan-mediated adhesion, *J Biol Chem*, 267, 10133–10141, 1992.
27. Vogel, B.E., Lee, S.J., Hildebrand, A., Craig, W., Piershbacher, M.D., Wongstaal, F., Rouslahti, E. A novel integrin specificity exemplified by binding of the $\alpha V\beta 5$ integrin to the basic domain of the HIV tat protein and vitronectin. *J Cell Biol*, 121, 461–468, 1993.
28. Dettin, M., Conconi, M.T., Gambaretto, R., Bagno, A., Di Bello, C., Menti, A.M., Grandi, C., Parnigotto, P.P. Effect of synthetic peptides on osteoblast adhesion, *Biomaterials*, 26, 4507–4515, 2005.
29. Puleo, D.A., Nanci, A. Understanding and controlling the bone–implant interface, *Biomaterials*, 20, 2311–2321, 1999.
30. Ito, Y., Inoue, M., Liu, S.Q., Imanishi, Y. Cell growth on immobilized cell growth factor. VI. Enhancement of fibroblast cell growth by immobilized insulin and/or fibronectin, *J Biomed Mater Res*, 27, 901–907, 1993.
31. Puleo, D.A. Biochemical surface modification of Co–Cr–Mo, *Biomaterials*, 17, 217–222, 1996.
32. Ferris, D.M., Moodie, G.D., Dimond, P.M., Gioranni, C.W.D., Ehrlich, M.G., Valentini, R.F. RGD-coated titanium implants stimulate increased bone formation in vivo, *Biomaterials*, 20, 2323–2331, 1999.
33. Rezania, A., Johnson, R., Lefkow, A.R., Healy, K.E. Bioactivation of metal oxide surfaces. 1. Surface characterization and cell response, *Langmuir*, 15, 6931–6939, 1999.
34. Xiao, S.-J., Textir, M., Spencer, N.D. Covalent attachment of celladhesive (Arg-Gly-Asp)-containing peptides to titanium surface, *Langmuir*, 14, 5507–5516, 1998.
35. Massia, S.P., Hubbell, J.A. Covalently attached GRGD on polymer surfaces promotes biospecific adhesion of mammalian cells, *Ann N Y Acad Sci*, 589, 261–270, 1990.
36. Sugawara, T., Matsuda, T. Photochemical surface derivatization of a peptide containing Arg-Gly-Asp (RGD), *J Biomed Mater Res*, 29, 1047–1052, 1995.
37. Heckman, J.D., Ehler, W., Brookes, B.P., Aufdemorte, T.B., Lohmann, C.H., Morgan, T., Boyan, B.D. Bone Morphogenetic Protein but not Transforming Growth Factor-b enhances bone formation in canine diaphyseal nonunions implanted with a biodegradable composite polymer, *J Bone Joint Surg*, 81, 1717–1729, 1999.
38. Glass, J.R., Dickerson, K.T., Stecker, K., Polarek, J.W. Characterization of a hyaluronic acid-Arg-Gly-Asp peptide cell attachment matrix, *Biomaterials*, 17, 1101–1108, 1996.
39. Suzuki, Y., Tanihara, M., Suzuki, K., Saitou, A., Sufan, W., Nishimura, Y. Alginate hydrogel linked with synthetic oligopeptide derived from BMP-2 allows ectopic osteoinduction in vivo, *J Biomed Mater Res*, 50, 405–409, 2000.

40. Rezania, A., Thomas, C.H., Branger, A.B., Waters, C.M., Healy, K.E. The detachment strength and morphology of bone cells contacting materials modified with a peptide sequence found within bone sialoprotein, *J Biomed Mater Res*, 37, 9–19, 1997.
41. Kouvroukoglou, S., Dee, K.C., Bizios, R., Mcintire, L.V., Zygourakis, K. Endothelial cell migration on surfaces modified with immobilized adhesive peptides, *Biomaterials*, 21, 1725–1733, 2000.
42. Rezania, A., Healy, K.E. Biomimetic peptide surfaces that regulate adhesion, spreading, cytoskeletal organization, and mineralization of the matrix deposited by osteoblast-like cells, *Biotechnol Prog*, 15, 19–32, 1999.
43. Humphries, M.J., Akiyama, S.K., Komoriya, A., Olden, K., Yamada, K.M. Identification of an alternatively spliced site in human plasma fibronectin that mediates cell type specific adhesion, *J Cell Biol*, 103, 2637–2647, 1986.
44. Danilov, Y.N., Juliano, R.N. (Arg-Gly-Asp)n-albumin conjugates as a model substratum for integrin-mediated cell adhesion, *Exp Cell Res*, 182, 186–196, 1989.
45. Underwood, P.A., Bennett, F.A. A comparison of the biological activities of the cell-adhesive proteins vitronectin and fibronectin, *J Cell Sci*, 93, 641–649, 1989.
46. Massia, S.P., Hubbell, J.A. An RGD spacing of 440 nm is sufficient for integrin $\alpha_V\beta_3$-mediated fibroblast spreading and 140 nm for focal contact and stress fiber formation, *J Cell Biol*, 114, 1089–1100, 1991.
47. Neff J.A., Tresco, P.A., Caldwell, K.D. Surface modification for controlled studies of cell–ligand interactions, *Biomaterials*, 20, 2377–2393, 1999.
48. DiMuzio, M.T., Veis, A. The biosynthesis of phosphophoryns and dentin collagen in the continuously erupting rat incisor, *J Biol Chem*, 253, 6845–6852, 1978.
49. Vuust, J., Piez, K.A. A kinetic study of collagen biosynthesis, *J Biol Chem*, 247, 856–862, 1972.
50. Morra, M. Biochemical modification of titanium surfaces: peptides and ECM proteins, *European Cells and Materials*, 12, 1–15, 2006.
51. Bagno, A., Dettin, M., Chiarion, A., Brun, P., Gambaretto, R., Fontana, G., Di Bello, C., Palù, G., Castagliuolo, I. Human osteoblast-like cell adhesion on titanium substrates covalently functionalized with synthetic peptides, *Bone*, 40, 693–699, 2007.
52. Iucci, G., Dettin, M., Battocchio, C., Gambaretto, R., Di Bello, C., Polzonetti, G. Novel immobilizations of an adhesion peptide on the TiO_2 surface: a XPS investigation, *Materials Science and Engineering C*, 27, 1201–1206, 2007.
53. Bearinger, J.P., Castner, D.G., Healy, K.E. Biomolecular modification of p(AAm-co-EG/aa) IPNs supports osteoblast adhesion and phenotypic expression, *J Biomater Sci Polym Ed*, 9, 629–652, 1998.
54. Barber, T.A., Golledge, S.L., Castner, D.G., Healy, K.E. Peptide-modified p(AAm-co-EG/AAc) IPNs grafted to bulk titanium modulate osteoblast behaviour in vitro, *J Biomed Mater Res*, 64, 38–47, 2003.
55. Dettin, M., Bagno, A., Morpurgo, M., Cacchioli, A., Conconi, M.T., Di Bello, C., Gabbi, C., Gambaretto, R., Parnigotto, P.P., Pizzinato, S., Ravanetti, F., Guglielmi, M. Evaluation of SiO_2-based coating enriched with bioactive peptides mapped on human vitronectin and fibronectin: in vitro and in vivo assays, *Tissue Engineering*, 12, 3509–3523, 2006.

56. Houseman, B.T., Mrksich, M. The microenvironment of immobilized Arg-Gly-Asp peptides is an important determinant of cell adhesion, *Biomaterials*, 22, 943–955, 2001.
57. Schuler, M., Owen, G.R., Hamilton, D.W., deWild M., Textor, M., Brunette, D.M., Tosatti, S.G.P. Biomimetic modification of titanium dental implant model surfaces using the RGDSP-peptide sequence: A cell morphology study, *Biomaterials*, 27, 4003–4015, 2006.
58. Reyes, C.D., Garcia, A.J. Engineering integrin-specific surfaces with triple-helical collagen-mimetic peptide, *J Biomed Mater Res* A, 65, 511–523, 2003.
59. Reyes, C.D., Garcia, A.J. α2β1 integrin-specific collagen-mimetic surfaces support osteoblastic differentiation, *J Biomed Mater Res* A, 69, 591–600, 2004.
60. Jeschke, B., Meyer, J., Jonczyk, A., Kessler, H., Adamietz, P., Meenen, N.M., Kantlehner, M., Goepfert, C., Nies, B. RGD-peptides for tissue engineering of articular cartilage, *Biomaterials*, 23, 3455–3463, 2002.
61. Chen, R., Curran, S.J., Curran, J.M., Hunt, J.A. The use of poly(L-lactide) and RGD modified microspheres as cell carriers in a flow intermittency bioreactor for tissue engineering cartilage, *Biomaterials*, 27, 4453–4460, 2006.
62. Hwang, N.S., Varghese, S., Zhang, Z., Elisseff, J. Chondrogenic differentiation of human embryonic stem cell-derived cells in arginine-glycine-aspartate-modified hydrogels, *Tissue Eng*, 12, 2695–2706, 2006.
63. Willerth, S.M., Sakiyama-Elbert, S.E. Approaches to neural tissue engineering using scaffolds for drug delivery, *Advanced Drug Delivery Reviews*, 59, 325–338, 2007.
64. Sakiyama-Elbert, S.E., Hubbell, J.A. Controlled release of nerve growth factor from a heparin-containing fibrin-based cell ingrowth matrix, *J Control Release*, 69, 149–158, 2000.
65. Taylor, S.J., Sakiyama-Elbert, S.E. Effect of controlled delivery of neurotrophin-3 from fibrin on spinal cord injury in a long term model, *J Control Release*, 116, 204–210, 2006.
66. Maxwell, D.J., Hicks, B.C., Parsons, S., Sakiyama-Elbert, S.E. Development of rationally designed affinity-based drug delivery systems, *Acta Biomater*, 1, 101–113, 2005.
67. Massia, S.P., Hubbell, J.A. Covalent surface immobilization of Arg-Gly-Asp- and Tyr- Ile-Gly-Ser-Arg-containing peptides to obtain well-defined cell-adhesive substrates, *Anal Biochem*, 187, 292–301, 1990.
68. Xiao, S.J., Textor, M., Spencer, N.D., Wieland, M., Keller, B., Sigrist, H. Immobilization of the cell-adhesive peptide Arg-Gly-Asp-Cys (RGDC) on titanium surfaces by covalent chemical attachment, *J Mater Sci Mater Med*, 8, 867–872, 1997.
69. Kam, L., Shain, W., Turner, J.N., Bizios, R. Selective adhesion of astrocytes to surfaces modified with immobilized peptides, *Biomaterials*, 23, 511–555, 2002.
70. Saneinejad, S., Shoichet, M.S. Patterned glass surfaces direct cell adhesion and process outgrowth of primary neurons of the central nervous system, *J Biomed Mater Res*, 42, 13–19, 1998.
71. Yu, T.T., Shoichet, M.S. Guided cell adhesion and outgrowth in peptide-modified channels for neural tissue engineering, *Biomaterials*, 26, 1507–1514, 2005.
72. Holmes, T.H. Novel peptide-based biomaterial scaffolds for tissue engineering, *Trends in Biotechnology*, 20, 16–20, 2002.

73. Zhang, G.S., Holmes, T., Lockchin, C., Rich, A. Spontaneous assembly of a self-complementary oligopeptide to form a stable macroscopic membrane, *Proc Natl Acad Sci USA*, 90, 3334–3338, 1993.
74. Zhang, S. Emerging biological materials through molecular self-assembling, *Biotechnology Advances*, 20, 321–339, 2002.
75. Holmes, T.C., deLacalle, S., Su, X., Liu, G., Rich, A., Zhang, S. Extensive neurite outgrowth and active synapse formation on self-assembling peptide scaffolds, *Proc Natl Acad Sci USA*, 97, 67728–67733, 2000.
76. Ellis-Behnke, R.G., Liang, Y.X., You, S.W., Tay, D.K.C., Zhang, S., So, K.F., Schneider, G.E. Nano neuro knitting: Peptide nanofiber scaffold for brain repair and axon regeneration with functional return of vision, *Proc Natl Acad Sci USA*, 103, 5054–5059, 2006.
77. Walluscheck, K.P., Steinhoff, G., Kelm, S., Haverich, A. Improved endothelial cell attachment on ePTFE vascular grafts pretreated with synthetic RGD-containing peptides, *Eur J Vasc Endovasc Surg*, 12, 321–330, 1996.
78. Mann, B.K., West, J.L. Cell adhesion peptides alter smooth muscle cell adhesion, proliferation, migration, and matrix protein synthesis on modified surfaces and in polymer scaffolds, *J Biomed Mater Res*, 60, 86–93, 2002.
79. Hubbell, J.A., Massia, S.P., Drumheller, P.D. Surface-grafted cell binding peptide in tissue engineering of the vascular graft, *Ann N Y Acad Sci*, 665, 253–258, 1992.
80. Mann, B.K., Tsai, A.T., Scott-Burden, T., West, J.L. Modification of surfaces with cell adhesion peptides alters extracellular matrix deposition, *Biomaterials*, 20, 2281–2286, 1999.
81. Davis, M.E., Motion, M., Narmoneva, D.A., Takahasci, T., Hakuno, D., Kamm, R.D., Zhang, S., Lee, R.T. Injectable self-assembling peptide nanofibers create intramyocardial microenvironments for endothelial cells, *Circulation*, 111, 442–450, 2005.
82. Narmoneva, D.A., Oni, O., Sieminski, A.L., Zhang, S., Gertler, J.P., Kamm, R.D., Lee, R.T. Self-assembling short oligopeptides and the promotion of angiogenesis, *Biomaterials*, 26, 4837–4846, 2005.
83. Davis, M.E., Hsieh, P.C., Takahashi, T., Song, Q., Zhang, S., Kamm, R.D., Godzinsky, A.J., Anversa, P., Lee, R.T. Local myocardial insulin-like growth factor 1 (IGF-1) delivery with biotinylated peptide nanofibers improves cell therapy for myocardial infarction, *Proc Natl Acad Sci USA*, 103, 8155–8160, 2006.
84. Metcalfe, A.D., Ferguson, M.W.J. Tissue engineering of replacement skin: the crossroads of biomaterials, wound healing, embryonic development, stem cells and regeneration, *Journal of the Royal Society Interface*, 4, 413–437, 2006.
85. Raeber, G.P., Lutolf, M.P., Hubbell, J.A. Molecularly engineered PEG hydrogels: a novel model system for proteolytically mediated cell migration, *Biophys J*, 89, 1374–1388, 2005.
86. Reyes, C.D. Collagen- and fibronectin-mimetic integrin-specific surfaces that promote osseointegration, PhD Thesis in the School of Mechanical Engineering, Georgia Institute of Technology, 2006.

20
In vitro testing of biomaterials toxicity and biocompatibility

S INAYAT-HUSSAIN and N F RAJAB, Universiti Kebangsaan Malaysia, Malaysia and E L SIEW, Melaka Biotechnology Corporation, Malaysia

Abstract: New biomaterials will be required to undergo certain toxicity tests for biocompatibility and safety assessment. This chapter provides a broad overview of the *in vitro* toxicity tests, specifically on cytotoxicity and genotoxicity. The chapter further describes apoptosis as the mode of cytotoxicity and the relevant techniques to understand the effects of biomaterials on apoptosis.

Key words: apoptosis, biocompatibility, cytotoxicity, genotoxicity.

20.1 Introduction

Biomaterials have had a major impact in medicine to improve the quality of life of humans and animals. All materials used in the body or placed in contact with the body need to be assessed for toxicity. The use of *in vitro* techniques to study toxicity of various synthetic materials began some 30 years after tissue culture was first established as a technique (Dee *et al.*, 2002). Over the past decades, research and development of new biomaterials have changed from considering inactive materials to materials that are actively involved in biological activity. It is important to evaluate the safety of a biomaterial intended for human use by considering the historical description of its development. The characteristic of a biomaterial must be known prior to any medical application. The *in vitro* tests on biomaterials are performed to stimulate biological reactions to materials when they are in contact with the body.

Toxicology plays an important role in the development of biomaterial in order to ensure that the materials selected for use in devices are safe for their intended use. This chapter is divided into three sections providing important information on *in vitro* toxicity testing for assessment of cell death, namely cytotoxicity and mode of cell death followed by genotoxicity evaluation of biomaterials. The final intended use of a biomaterial must be considered prior to testing. International guidelines and standards such as those established by the American Society for Testing and Materials (ASTM) (www.astm.org) and the International Organization for Standardization (ISO) (www.iso.org) should be followed to ensure that the design of the study satisfies the relevant regulatory bodies.

20.2 Biocompatibility

Biocompatibility is defined as the ability of a material to perform an appropriate host response in a specific application (William, 1987). The biocompatibility of a test material can be examined either *in vitro* using appropriate cells that play an important role during wound healing and regeneration processes, or *in vivo* via implantation or injection of the test material. Previous studies have demonstrated a good correlation between *in vitro* and *in vivo* tests, therefore confirming the usefulness of *in vitro* tests as systems to select the materials (Cenni *et al.*, 1999). Although *in vitro* tests are much easier and cheaper to perform, where relevant *in vivo* tests should be performed for safety assessment.

New biomaterials may demonstrate harmful effects, therefore it is necessary to determine their biocompatibility by testing prior to contact with the human body. *In vitro* testing has been used successfully to screen biomaterial for biocompatibility. Advances in cell culture techniques have provided suitable *in vitro* models for evaluating certain aspects of biocompatibility.

The main question is 'how to design a test to study the biocompatibility of a biomaterial?'. Safety evaluation of medical devices is divided into a four-phase approach. This approach was outlined by the North American Material Science Association (NAMSA) as shown in Fig. 20.1. Basically, the first phase refers to the chemical, physical and biological characterization of material components. The second phase is the biocompatibility of materials and determining what tests are to be performed. Biocompatibility evaluation of biomaterials is based on guidelines provided by the International Organization for Standardization (ISO10993). Table 20.1 illustrates a testing matrix based on ISO10993-1 which is modified by the US Food and Drug Administration (FDA) to determine which tests should be conducted according to the intended use, type of tissue that will be in contact, and duration of contact time with the biomaterial (Hermansky, 2001). Each material or device should be considered individually with regard to the selection of evaluation tests. In addition, other information on determining the appropriate testing regime is also available from the American Society for Testing and Materials (ASTM) and the National Institutes of Health (NIH) (Dee *et al.*, 2002). This is followed by the third phase which covers the product and process validation and the last phase which relates to release and the audit testing procedure.

Phase	
I	Characterization of material • Chemical characterization • Physical characterization • Biological characterization
II	Biocompatibility of material
III	Product and process validation • Environment control • Manufacturing process control • Sterility Finished product quality
IV	Release and animal testing • Release testing • Periodic audit testing Product release

20.1 A four-phase approach for safety evaluation of medical devices (NAMSA).

In general, biomaterials should be tested in the final product form that will be used in the body. However, it should be emphasized that it is also important to test all components of the final product, as often they may individually be toxic. The toxic potential of materials and devices depends to a substantial degree on the leachability and toxicity of soluble components, thus extracts of biomaterial are usually used in the tests. In some tests, however, a material is used directly whereby an evaluation under normal-use conditions is mimicked. The most commonly used extraction media are physiological saline, vegetable oil, dimethylsulfoxide and ethanol. More commonly, *in vitro* cytotoxicity tests employ cell-culture medium containing serum (ISO10993-12). On the other hand, the amount of leachable compounds released in the extraction media is related to the surface area and thickness of the compound to be extracted (Table 20.2) and also at temperatures approximating the normal human body temperature (37°C). This temperature should be used when the body contact duration for the device is anticipated to be brief or when the integrity of the material may be compromised by using higher temperatures. The selection specific conditions to be used are dependent upon many factors such as the proposed end use of the materials, the degrading ability of the material at selected incubation temperature and the anticipated time of exposure under routine-use conditions of the device (Table 20.3).

Table 20.1 The ISO Standard 10993-1 Guidance for Selection of Biocompatibility Test with modifications: A = limited (24 hours); B = 24 hours to 30 days; C = permanent (>30 days); √ = ISO evaluation tests for consideration; O = additional test that may be applicable

Device categories		Toxicological effect									
Body contact	Contact duration	Cytotoxicity	Sensitization	Irritation or Intracutaneous reactivity	Systemic toxicity (acute)	Subchronic toxicity	Genotoxicity	Implantation	Hemocompatibility	Chronic toxicity	Carcinogenicity
Surface devices											
Skin	A	√	√	√							
	B	√	√	√							
	C	√	√	√							
Mucosal membrane	A	√	√	√							
	B	√	√	√	O	O		O			
	C	√	√	√	O	√	√	O		O	
Breached or compromised surfaces	A		√	√	O						
	B	√	√	√	O	O		O			
	C	√	√	√	O	√	√	O		O	
Blood path, indirect	A	√	√	√	√				√		
	B	√	√	√	√	O			√		
	C	√	√	O	√	√	√	O	√	√	√
External communicating devices											
Tissue, bone dentin communicating	A	√	√	√	O						
	B	√	√	O	O	O	√	√			
	C	√	√	O	√	O	√	√		O	√
Circulating blood	A	√	√	√	√		O		√		
	B	√	√	√	√	O	√	O	√		
	C	√	√	√	√	√	√	O	√	√	√
Implant devices											
Tissue/bone	A	√	√	√	O						
	B	√	√	O	O	O	√	√			
	C	√	√	O	O	O	√	√		√	√
Blood	A	√	√	√	√			√	√		
	B	√	√	√	√	O	√	√	√		
	C	√	√	√	√	√	√	√	√	√	√

Table 20.2 Common examples of extraction ratio (ISO10993-12)

Thickness (mm)	Extraction ratio (±10%)	Examples of materials
≤0.5	6 cm^2/ml	Metal, synthetic materials, ceramic, film, sheet, tubing wall
0.5	3 cm^2/ml	Metal, synthetic materials, ceramic, tubing wall, slab, moulded items
1.0	3 cm^2/ml	Elastomer
1.0	1.25 cm^2/ml	Elastomer
Irregular	0.12–0.2 g/ml: 6 cm^2/ml	Pellets, moulded parts

Table 20.3 Common examples of temperature and exposure time (ISO10993-12)

37 ± 1°C for 24 ± 2 h
37 ± 1°C for 72 ± 2 h
50 ± 2°C for 72 ± 2 h
70 ± 2°C for 24 ± 2 h
121 ± 2°C for 1.0 ± 0.2 h

20.3 Toxicity testing: *in vitro* methods

In vitro methods are used principally for screening purposes and for generating more comprehensive toxicology profiles. *In vitro* refers primarily to the handling of cells and tissue outside the body under conditions which support their growth, differentiation and stability. This section focuses on the importance of *in vitro* toxicity testing and methods used to determine the safety of a biomaterial.

Generally, *in vitro* toxicity testing is used to evaluate potential hazards of a variety of chemicals including pesticides, pharmaceutical, food additives and medical devices. Historically, this testing has relied on animal-based methods. As an alternative to animal testing, these methods hold the promise of not only reducing the usage of animals but also increasing the scientific sophistication of toxicity testing, as well as offering the practical advantages of being faster and cheaper than animal-based methods.

A prerequisite for the successful application of *in vitro* methods is the availability of appropriate validated test systems by an interagency committee, the Interagency Coordinating Committee on the Validation of Alternative Methods (ICCVAM). In 1982, the international harmonization of toxicity tests by the Organization for Economic Cooperation and Development (OECD) outlined the most effective steps in reducing duplication of testing in animals for regulatory purposes. Furthermore, the

In vitro testing of biomaterials toxicity and biocompatibility   513

International Conference on Harmonisation (ICH) has issued harmonized guidelines for efficacy and safety testing of pharmaceuticals including animal tests, which have subsequently been integrated into the respective national legislations. Consequently, the harmonization of test guidelines has led to significant reductions in animal testing, since regulatory agencies around the world now accept the results of tests conducted according to ICH guidelines (www.ich.org). *In vitro* models include a broad list of cell systems such as primary cells, genetically modified cells, immortalized cells and more recently stem cells, although the latter has not been accepted for regulatory submission.

In this chapter, cytotoxicity testing will be addressed where it is important and provides data of intrinsic value in defining toxic effects and also where it is used in designing more detailed *in vitro* studies. There are several test methods, some of which in our opinion are more innovative than others, which will be briefly outlined. In addition, the mode of cell death, especially apoptosis, and methods commonly used to determine cell death will also be discussed. This is followed by genotoxicity testing where such testing gives information on mechanisms of action, which is pivotal in the characterization of carcinogenic risk, supporting the use of non-threshold models for the estimation of low dose effects. For detail guidance on these areas, the reader is directed to the references at the end of this chapter.

20.4   Cytotoxicity

Cell cytotoxicity assays were among the first *in vitro* bioassay methods to be used to predict the toxicity of potential medical devices. These tests are recommended for all new biomaterials as they provide rapid evaluation, standardized protocols and comparative data, and have the ability to discard toxic materials even prior to testing in animals. Most importantly, cytotoxicity tests are necessary to define the concentration range for further and more in-depth *in vitro* testing to provide meaningful information on parameters such as genotoxicity, induction of mutation and apoptosis. Hartmann *et al.* (2001) have described the importance of cytotoxicity assessment of test articles prior to conducting genotoxicity testing. It is important to demonstrate that the DNA damage is the primary effect of the test article and is not due to secondary DNA damage during cytotoxicity. The ISO10993-5 provides guidelines for 'Tests for *in vitro* cytotoxicity' (1999). Guidelines include cell type, duration of the exposure and method of testing (Harmand, 1997). Many types of cells have been proposed and cell lines have been developed for growth *in vitro* because they maintain their genetic and morphological characteristics throughout an infinite lifespan. This is important for reproducibility of the test methods. For example, the L929 mouse fibroblast cells have been used most extensively for evaluating biomaterials

(You et al., 2007). This cell line is often selected because it is easy to maintain in culture and has produced results that have very good correlation with specific animal bioassay (Northup, 1986). However, cell lines from other tissues or species have also been used and the choice of cell types is usually dependent on the final application of the biomaterials. In the case of bone implants, osteoblasts and osteoclasts have been used to evaluate the effect of biomaterials on bone metabolism and mineralization (Sun et al., 1997). Other cell lines such as HeLa human cervical carcinoma epithelial cells (Chakrabarty et al., 2002) and human skin epithelial cells (Lestari et al., 2005) are also commonly used in cytotoxicity study.

The cytotoxicity endpoints are intended to show the degree of damage that may be caused by the biomaterial. In this case, many endpoints are possible, including (1) loss of membrane integrity, (2) release of cytosolic enzymes, (3) loss in metabolic process (e.g. ATP production) and (4) reduction of DNA synthesis or inability to continue cell replication. Several assays have been reported to determine the cytotoxic effect of biomaterials as presented in Table 20.4.

In this section, several frequently used methods for determining the cytotoxicity of biomaterials will be described, namely MTT (3-[4,5-dimethyl-2-thiazolyl]-2,5-diphenyl tetrazolium bromide) assay, neutral red uptake assay, direct contact test/colony forming assay and the agar diffusion test.

20.4.1 MTT assay

The well-documented MTT assay was introduced by Tim Mosmann in 1983. This colorimetric assay is used to determine the cell proliferation, viability and cytotoxicity. This assay is advantageous, as it can be rapidly performed on a microtitre plate assay and read on a spectrophotometer ELISA plate reader at the absorbance of 570 nm (Mosmann, 1983). Traditionally, the reduction of the tetrazolium salts to their equivalent formazan precipitates was used for the histochemical demonstration of the activities of oxidative and non-oxidative enzymes in mitochondrial using both light and electron microscopy (Altman, 1977). More recently, in MTT assay, the tetrazolium MTT (3-[4,5-dimethylthiazol-2-yl]-2,5-diphenyltetrazolium bromide) is reduced in a mitochondrial-dependent reaction to an insoluble purple formazan by cleavage of the tetrazolium ring by succinate dehydrogenase within the mitochondria (see Fig. 20.2). The formazan accumulates within the cell since it cannot pass through the cell membrane. By addition of spectrophotometric grade dimethylsulfoxide (DMSO), isopropanol or other suitable solvent, the formazan is solubilized and liberated and is readily quantified colorimetrically. Cytotoxic concentration is generally determined as the concentration that kills 50% of the cells, generally known as IC_{50}.

In vitro testing of biomaterials toxicity and biocompatibility   515

Table 20.4 Common examples of cytotoxicity tests

Test	Principle	Reference
MTT assay	Reduction of tetrazolium salts by succinate dehydrogenase	Mosmann, 1983 Wang *et al.*, 2004 Fotakis and Timbrell, 2006
Lactate dehydrogenase (LDH) leakage assay	Detection of LDH release into culture medium after cell membrane damage	Ohno *et al.*, 1995 Harbell *et al.*, 1997
Neutral red uptake (NRU) assay	Uptake of neutral red by lysosomes in viable cells	Borenfreund and Puerner, 1984 Clifford and Downes, 1996
Alamar blue assay	Reduction of resazurin (alamar blue) to resorufin (pink fluorescent)	Nakayama *et al.*, 1997 O'Brien *et al.*, 2000 Low *et al.*, 2006
Colony forming assay (CFA)	Measurement of cell viability by colony-forming efficiency (survival) and colony size (proliferation)	Puck and Marcus, 1956 Sasaki and Tanaka, 1991 Tsuchiya, 1994
Dye exclusion assay	Determine cell viability by dyes (e.g. trypan blue); cell with intact membranes exclude the dye and remain unstained	Weisenthal *et al.*, 1983 Harbell *et al.*, 1997
Sulforhodamine B (SRB) protein staining	Measures cell proliferation and chemosensitivity testing; the dye binds to basic amino acids of cellular proteins and colorimetric evaluation provides an estimate of total protein mass which is related to cell number	Skehan *et al.*, 1990 Papazisis *et al.*, 1997

20.2 Reduction of MTT salt to purple formazan.

The cytotoxic effects of several biomaterials had been studied previously by using MTT assay. Yang *et al.* (2002) demonstrated the cytotoxic effect of a polyhydroxyalkanoate (PHA), poly(hydroxybutyrate-*co*-hydroxyhexanoate) (PHBHHx), against L929 murine fibroblast cells where the MTT assay was employed to quantitatively assess the viable cell numbers of L929 attached and grown on polymer surfaces. Also in our previous work, the cytotoxicity of poly(3-hydroxybutyrate-*co*-4-hydroxybutyrate) copolymer was assessed using MTT assay in fibroblast cells and clearly showed there was slight decrease of MTT activity on the treated cells (Siew *et al.*, 2007). It has been reported that MTT assay appeared to be more sensitive in detecting early toxicity compared to LDH leakage assay and protein assay (Fotakis and Timbrell, 2006). XTT (2,3-bis[2-methyloxy-4-nitro-5-sulfophenyl]-2-tetrazolium-5-carboxanilide) assay, which is similar to MTT assay, gives greater sensitivity than LDH leakage assay (Mitsuyoshi *et al.*, 1999). Another similar assay is MTS, 3-(4,5-dimethylthiazol-2-yl)-5-(3-carboxymethoxyphenyl)-2-(4-sulfophenyl)-2H-tetrazolium. Based on an electron coupling reagent such as phenozine methosulfate (PMS), MTS produces a soluble product whose absorbance can be measured without a further dissolution step. However, sensitivity of the different salts has been reported and MTS was shown to be less sensitive than the MTT assay (Clifford and Downes, 1996).

20.4.2 Neutral red uptake assay

Neutral red uptake (NRU) assay has also been frequently used to measure cell viability and determination of biomaterial cytotoxicity. NRU assay, which was initially described by Finter in 1969, is being used commonly in biochemical and immunological studies and has been adopted as a recommended method for testing biomaterials. Neutral red is a vital dye, which is preferentially absorbed and endocytosed by viable cells and internalized inside lysosome. In this respect, it is considered as an indicator of lysosome and cell integrity. The basis of the test is that a cytotoxic material, regardless of site or mechanism of action, can interfere with this process and result in a reduction of the number of viable cells. NRU has been shown to correlate well with the number of cells in the culture and has also been found to be more accurate than MTT assay (Ciapetti *et al.*, 1996). In this respect, Ciapetti *et al.* (2002) utilized NRU to confirm that cement extracts are not cytotoxic on MG-63 osteoblast-like cells. One of the benefits of using NRU is that the measurement is not biased by occasional microbial contamination, which can lead to overestimation of cell viability (Ciapetti *et al.*, 1996). NRU is also used for evaluation of cytotoxicity of test articles by the National Toxicology Program Interagency Center for the Evaluation of Alternative Toxicological Methods (NICEATM) (2003).

20.4.3 Direct contact test

According to ISO10993-5, the direct contact test is useful to determine the cytotoxicity of biomaterial based on the morphological changes of the cells. Normally, fibroblast cells are used in this test. Direct contact tests are able to show the toxic effect of a material on adjacent cells. Dead cells lose their capability to remain attached to the culture plate and are frequently lost during the fixation process. Viable cells, however, adhere to the culture plate and are stained with cytochemical stain such as crystal violet. The toxic effect is evaluated by the absence of stained cells under and around the periphery of the test material.

In the direct contact test, the difference between living and dead cells is shown by an intermediate zone of damaged cells. The morphological changes such as vacuolization, rounding and swelling can be observed through an inverted microscope.

The direct contact test has the advantage that it (i) mimics the physiological conditions, (ii) has a zone of diffusion (a concentration gradient of toxic chemicals), and (iii) affects cellular trauma in accordance with the density of materials (Lee et al., 2000). Also this test has great advantages on testing biomaterials where it provides direct targeted cell contact with the material and does not need any extraction preparation. The biological reactivity is described and rated as 0–4 as illustrated in Table 20.5. The test material meets the requirements of the test if the response is not greater than grade 2 (mildly reactive), according to the US Pharmacopeia (USP26-NF21).

20.4.4 Colony-forming assay

The colony-forming assay (CFA) has been established for over 50 years (Puck and Marcus, 1956). It has been reported that CFA is one of the most useful cytotoxicity tests, with great sensitivity (Sasaki and Tanaka, 1991). CFA has been described and recommended by the Japanese standard test method for evaluating the cytotoxicity of biomaterials (Tsuchiya, 1994). In addition, CFA is used to determine the fraction of cells that survive after

Table 20.5 Reactivity grades for direct contact test and agar diffusion test (USP26-NF21)

Grade	Reactivity	Description of reactivity zone
0	None	No detectable zone around or under specimen
1	Slight	Some malformed or degenerated cells under specimen
2	Mild	Zone limited to area under specimen
3	Moderate	Zone extends 0.5 to 1.0 cm beyond specimen
4	Severe	Zone extends greater than 1.0 cm beyond specimen

treatment. In this respect, the cell survival is defined as the ratio of the plating efficiencies (PE), such as the percentage of seeded cells that grow into macro-colonies, between treated and untreated cells. When determining the PE, the assumption is made that the PE is independent of the number of cells seeded into culture, at least over a wide range of cell number. The cells are typically seeded between 20 and 200 colonies in a culture dish in order to obtain a relatively constant number of surviving macro-colonies. Cells are treated with any test materials for long-term incubation (6–8 days) until the colonies are formed and observed (Riddell et al., 1986; Garza-Ocañas et al., 1990). Subsequently the formed colonies are fixed with fixatives such as methanol to maintain the cell integrity and eliminate all the excessive water from the cells (dehydration). Lastly, the colonies are stained with Giemsa or crystal violet dye and the formation of a colony indicates that the treated cells are not cytotoxic.

In colony-forming assay, various types of cell lines are employed, including L929 murine fibroblast, Chinese hamster lung V79 fibroblast and mouse embryogenic Balb/3T3 cells. Studies reported to date demonstrate that V79 fibroblast cell is the most suitable cell line to be used in this assay due to its superiority in terms of colony-forming ability as compared to other cells (Tsuchiya, 1994).

20.4.5 Agar diffusion test

Agar diffusion assay is performed as recommended by ISO10993-5 and described by Schmalz (1988). The toxicity of the test substances is related to the zone of decolorization around the test material caused by unstained injured cells (zone index). The morphological signs of cell damage within the zone are expressed by the lysis index (Table 20.6) where cytotoxicity is estimated by the response index (zone index/lysis index) (Table 20.7). Thus, it is estimated as an average from replicated experiments. The toxicity of a test material refers to the scoring system as described by You et al. (2007) and shown in Table 20.7.

Table 20.6 Scoring of cell lysis using agar diffusion test (USP26-NF21)

Lysis index	Description of zone
0	No observable cytotoxicity
1	Less than 20% of zone affected
2	20 to 39% of zone affected
3	40 to 59% of zone affected
4	60 to 80% of zone affected
5	Greater than 80% of zone affected

Table 20.7 Scoring system for estimating cytotoxicity in agar diffusion test (USP26-NF21)

Response index	Interpretation of cytotoxicity
0/0–0.5/0.5	None
1/1–1.5/1.5	Mild
2/2–3/3	Moderate
>4/4	Marked

Originally, this test was designed for elastomeric biomaterials having a variety of shapes and uses (Ignatius and Claes, 1996; Ignatius *et al.*, 2001). Toxic substances leaching from the test material may depress the growth rate of the cells or damage them in various ways. In the agar diffusion assay, a piece of test material is placed on an agar layer covering a confluent monolayer of cells. Toxic substances leaching out from the test material can diffuse through the thin agar layer, killing or disrupting adjacent cells in the monolayer.

There is a caveat with using different test methods for cytotoxicity. It is possible that there are differences in the IC_{50} values generated with these different tests (Komissarova *et al.*, 2005; McKim *et al.*, 2005; Chan *et al.*, 2006). Therefore, for good in-house data, these tests can be coupled with understanding the mode of cell death.

20.5 Cell death

Cytotoxicity is synonymous with cell death for which a wealth of information has been generated on the modes of death. Typically, two types of cell death, namely necrosis and apoptosis, have been described in terms of their morphological and biochemical changes. Cells undergoing necrosis, which normally occurs by a variety of pathological and toxic stimuli, exhibit distinctive morphological and biochemical features (Trump *et al.*, 1984). Necrosis, which typically affects groups of contiguous cells, demonstrates early swelling of the cytoplasm and organelles, especially the mitochondria, known as high amplitude swelling. These changes result in organelle dissolution and rupture of the plasma membrane, allowing leakage of the cellular contents into the extracellular space and the influx of extracelullar Ca^{2+} into the cells. The Ca^{2+} influx results in cytoskeletal changes leading to membrane blebbing and activation of many enzymes, including proteases, phospholipases and nucleases (Orrenius *et al.*, 1989). Consequently, the cell contents leak out, leading to inflammatory reactions.

In contrast, apoptosis is a genetically orchestrated response for cells to 'commit suicide.' The first observation of apoptotic morphology, termed 'shrinkage necrosis', was discovered during normal liver and mild ischaemic damage after ligation of its portal blood supply (Kerr, 1971). Apoptosis is the Greek word for falling of the leaves from trees during autumn and this phenomenon can be initiated or inhibited by a flurry of stimuli, both physiological and pathological (Kerr et al., 1972). Apoptosis plays an important role in normal tissue homeostasis in addition to being involved in the pathophysiology of certain diseases. Typically, apoptosis is characterized by a marked condensation of chromatin, accompanied with shrinkage of the nucleus, compaction of cytoplasmic organelles, cell shrinkage and changes at the cell surface including loss of cell membrane phospholipid asymmetry. The plasma membrane remains intact throughout the process, avoiding leakage of the intracellular content, hence this type of cell death is not associated with inflammatory reactions. The apoptotic cells gradually separate apoptotic bodies, which are essentially membrane-bound fragments containing parts of the nucleus and other organelles. The apoptotic cells and bodies are phagocytosed by neighbouring cells or tissue macrophages. Apoptosis studies using *in vitro* cell culture models are generally without a phagocytosis system and therefore apoptotic cells will degenerate, leading to secondary necrosis (Wyllie, 1992; and Fig. 20.3). The major differences between apoptosis and necrosis can be listed as shown in Table 20.8.

20.5.1 Biochemical mechanisms of apoptosis

In the last two decades, a rapid growth in the understanding of the mechanisms of apoptosis in various models has been observed. The most commonly observed end-points during apoptosis are the condensation of chromatin and DNA fragmentation subsequent to a landmark report that

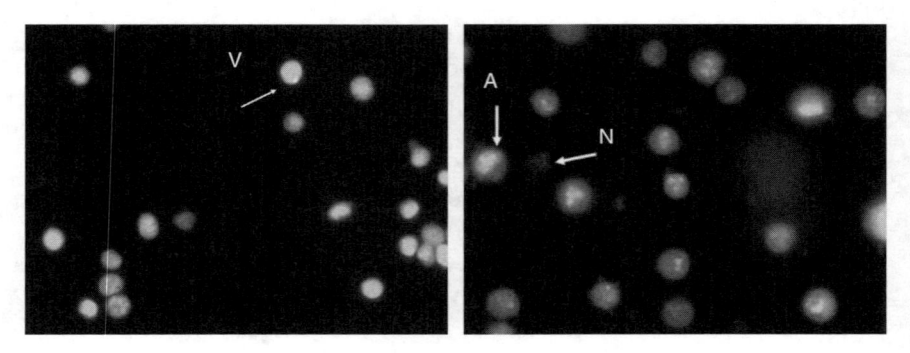

20.3 Representative images of apoptotic and necrotic cells induced by polyhydroxyalkanoate. V: viable cell, A: apoptotic cell, N: necrotic cells. Magnifications 200×.

Table 20.8 Characteristics of apoptosis and necrosis

Apoptosis	Necrosis
• Affects scattered individual cells	• Affects massive and contiguous cells
• Chromatin marginates as large crescent	• Chromatin marginates as small aggregates
• Internucleosomal cleavage	• Random DNA cleavage
• Organelles retain integrity	• Organelles swell (e.g. mitochondria) and cell contents are released
• Formation of apoptotic bodies	• Cell ruptures
• Presence of phosphatidylserine exposure	• Absence of phosphatidylserine exposure
• No inflammation (*in vivo*)	• Inflammation (*in vivo*)

an endonuclease was specifically activated during glucocorticoid induced thymocyte apoptosis (Wyllie, 1980). Cleavage of DNA at linker regions between nucleosomes by this enzyme resulted in DNA fragments of 180–200 bp and multiples thereof seen as a ladder pattern upon gel electrophoresis. This internucleosomal cleavage has been associated with the involvement of different cations including Mg^{2+} and Ca^{2+}/Mg^{2+}. Evidence from several studies has demonstrated that the cleavage of DNA occurs in a progressive and stepwise manner in which DNA is initially cleaved into ≥700, 200–250 and 30–50 kbp fragments prior to internucleosomal degradation (Cain *et al.*, 1994).

Apoptosis involves a series of degradative processes for which burgeoning evidence indicates that proteases, especially family members of the caspases, play a central role as regulators and 'executioners' in the cell death machinery (Budihardjo *et al.*, 1999). Caspase (cysteine *a*spartate prote*ase*) has a cysteine residue in the catalytic site and cleaves its substrates after aspartate residues. Caspases can generally be divided into three groups, namely cytokine activators, initiators and executioners of apoptosis (see Table 20.9).

These proteases exist as zymogens, i.e. inactive forms of procaspases, and are positioned to be activated upon cell death signalling. The recruitment of caspases in the apoptosis process can be the result of death signalling through either extrinsic or intrinsic pathways. The extrinsic pathway can be induced by members of the TNF family of cytokine receptors such as TNF1 and Fas. As a result of the interaction, these proteins recruit adapter proteins such as FADD or TRADD, resulting in the activation of initiator caspases 8 and 10 (see Fig. 20.4). In contrast, the intrinsic pathway of apoptosis occurs generally due to cellular stress, including cytotoxic drugs, ionizing radiation, oxidants and many apoptogenic toxicants. Activation of this pathway leads to the loss of mitochondrial membrane potential and release

Table 20.9 Characteristics of mammalian caspases

Caspase	Common name	Substrates	Functions	Size (kDa)
Caspase-1	ICE	Interleukins, pro-interleukin-1β, interleukin-18	Cytokine activator, inflammation	45
Caspase-2	ICH-1l, NEDD2	Golgin-160, lamins	Apoptosis initiator	48
Caspase-3	YAMA/Apopain/CPP32	PARP, topoisomerase I, lamins	Executioner	32
Caspase-4	Tx, Ich-2, ICErel-II	Caspase-1	Inflammation/apoptosis	43
Caspase-5	ICErel-III, Ty		Inflammation/apoptosis	46
Caspase-6	Mch-2	Lamins	Apoptosis executioner	34
Caspase-7	Mch3, CMH-1, ICE-LAP3	PARP, calpastatin	Apoptosis executioner	35
Caspase-8	Mch5, FLICE, MACH	Bid	Apoptosis initiator	55
Caspase-9	ICE-LAP6, Mch6	Caspase-3, caspase-7	Apoptosis initiator	48
Caspase-10	Mch4, FLICE-2	Caspase-3, caspase-4, caspase-6, caspase-7, caspase-8, caspase-9	Apoptosis initiator	55
Caspase-11	mICH-13		Inflammation/apoptosis	43
Caspase-12	mICH-4		Inflammation/ER stress	48
Caspase-13	ERICE	Caspase-3	Member of the ICE family of caspases that include caspase-1 and caspase-4, -5 and -11. Inflammation	43
Caspase-14	MICE		Keratinocyte differentiation	29.5

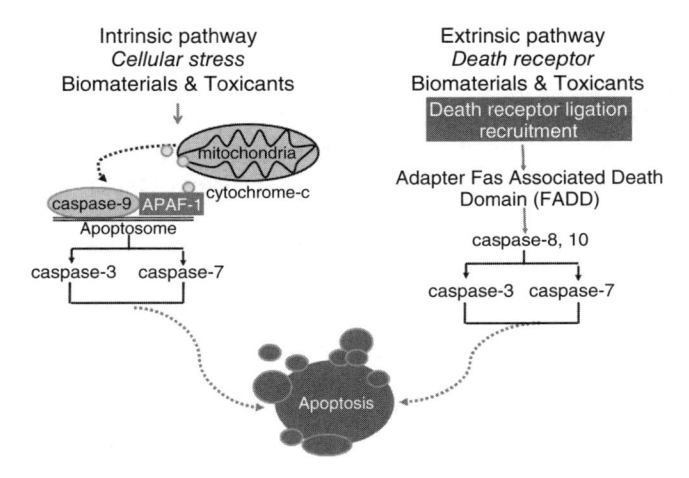

In vitro testing of biomaterials toxicity and biocompatibility 523

20.4 The intrinsic and extrinsic pathways of apoptosis.

of cytochrome from the mitochondrial intermembrane to the cytosol. Assembly of cytochrome c, apoptotic protease activating factor-1 (Apaf-1) and pro-caspase-9 forms the apoptosome complex, where activation of the initiator caspase-9 occurs, which in turn cleaves and activates the executioner caspases-3 and -7 (see Fig. 20.4)

Regardless of whether it is a death receptor pathway (extrinsic) or a mitochondrial pathway (intrinsic), there is a convergence of the downstream events where the executioner caspases (3, 6 and 7) are recruited, leading to further downstream events such as caspase substrate cleavage, lamin cleavage and DNA fragmentation. A complete list of the caspase substrates is beyond the scope of this chapter; however, Table 20.9 shows a selected list of substrates. Although apoptosis assessment is not required for the regulatory submission, it is a good endpoint to determine the mode of cytotoxicity and subsequently to predict whether cell death involves inflammation as the result of insult from the biomaterials. In addition, during early apoptosis, the cell membrane is still intact and the mitochondria and lysosome retain certain functions which may result in the underestimation of cytotoxicity of the biomaterials.

20.5.2 Assessment of apoptosis

Since the first description of apoptosis using electron microscopy, many methods have been utilized for the assessment of apoptosis. Although the methods described here are not exhaustive, they are sufficient for determining apoptosis induced by biomaterials.

Flow cytometry

Flow cytometry has been widely used in the assessment of apoptosis in both suspension and adherent cell lines (Givan, 2001). One of the events occurring during apoptosis is the flipping of phosphatidyl serine (PS) which is in the inner leaflet of the plasma membrane to the outer leaflet. This process is important as a phagocyte recognition signal on the surface of apoptotic cells. Annexin V belongs to a recently discovered family of proteins known as the annexins and has been demonstrated to be useful in detection of apoptosis, as it binds to negatively charged phospholipids like PS in the presence of Ca^{2+}. Annexin V conjugated with FITC (green fluorescence) can be used to detect early apoptosis, and simultaneous staining with a non-vital dye, propidium iodide (red fluorescence), allows a two-colour dot-plot analysis in the flow cytogram. Intact cells are unstained cells, known as double negatives, i.e. they neither express PS on the cell surface nor take up propidium iodide through leaky membranes. Cells that are stained with annexin V only are early apoptotic cells, and once these cells undergo the permeabilization of the cytoplasmic membrane, they are categorized as late apoptotic cells.

Another parameter often used in the study of apoptosis is the assessment of mitochondrial membrane potential (MMP) integrity. Dissipation of MMP occurs early during apoptosis and this event is associated with the release of cytochrome c and also apoptotic inducing factor (AIF). AIF has been demonstrated to be involved in proteolytic activation of the endonucleases involved in apoptosis, leading to chromatin condensation (Penninger and Kroemer, 2003). With the development and use of potentiometric fluorochromes, many models of apoptosis show the loss of the mitochondrial transmembrane potential (MTP) mediated by the opening of the megachannel (permeability transition pore), which precedes caspase activation (Zamzami et al., 2007). Several fluorochromes such as tetramethylrhodamine ethylester and 3,3'-dihexyloxacarbocyanine iodide have been utilized in flow cytometric assessment of mitochondrial involvement during apoptosis.

Morphological changes

During apoptosis, the salient feature is the condensation of chromatin in the nucleus which can be determined by assessment of the morphology. Basically the most common method to determine apoptosis is using fluorescence dyes such as Hoescht 33258, Hoescht 33242, acridine orange and propidium iodide. Fluorescence microscopy can be employed to detect these changes. In studies related to apoptosis, it is always recommended to determine the morphological changes using microscopic techniques. For a

In vitro testing of biomaterials toxicity and biocompatibility 525

detailed assessment of structural changes, transmission electron microscopy can be employed and ultrastructural changes can be observed (Inayat-Hussain *et al.*, 2001).

Caspases

The methods described above for apoptosis are mainly for the determination of apoptosis and hardly describe the mechanisms involved in apoptosis. For studies requiring a mechanistic approach, further assessment of caspase activation and cytochrome c release can be carried out. Proteolytic activities have been demonstrated to play an important role during apoptosis. The activities are related to cleavage of specific proteins and some of these substrates for caspases are shown in Table 20.9.

Current approaches to the assessment of caspase activation are immunoblotting and quantitation of the proteolytic activity using fluorogenic or chromogenic substrates in conjunction with fluorimeter or flow cytometry. All caspases are found in the form of zymogen, and activation of this proform will yield the active subunits which can be detected by immunoblotting (Inayat-Hussain *et al.*, 1997, 1999; Inayat-Hussain and Ross, 2005). A detailed description of the methods involved for assessment of caspases has been given by Kaufmann *et al.* (2001).

20.6 *In vitro* genotoxicity testing of biomaterial

Genotoxicity is a broad term that includes mutation as well as damage to DNA or the production of DNA adducts, by the chemical itself or its metabolites. Genotoxicity tests determine gene mutations, changes in chromosome structure and number, and other gene toxicities caused by medical devices, biomaterial or their extracts. This is described in ISO 10993-3:1992 Biological evaluation of medical devices – Part 3: Tests for genotoxicity, carcinogenicity and reproductive toxicity.

A number of different end-points can be used which may measure genetic changes or indicators for the potential to produce genetic change. Assays may be classified on the basis of these end-points (e.g., gene mutation, clastogenicity, aneugenicity and tests for DNA damage) and by consideration of the different phylogenetic levels represented. There are three major types of genotoxic effects: gene mutations, chromosomal aberrations, and DNA effects. No single *in vitro* assay is capable of detecting all three types of effects, therefore a battery of tests is recommended. Gene mutation and chromosomal aberration tests detect actual lesions in the DNA molecule, while DNA effects tests detect events that may lead to cell damage. *In vitro* tests in each category can be conducted using microorganisms or mammalian cells. Although ISO 10993-3 states that *in vivo*

Table 20.10 Principles of selected genotoxicity tests

Test or assay	Basic principles
Prokaryotic system	
1. Ames test	Exploitation of well-defined mutant bacteria at the histidine locus which lose the ability to grow in the absence of histidine
Eukaryotic system	
1. Cytogenetic assays	Examination of mammalian chromosomes for the presence of damage. Detection was based on either the numerical or structural aberration of the chromosome
2. Mammalian gene mutation assay	Based upon the detection of forward mutations which make used of genes that salvage nucleic acid breakdown, i.e. nucleotides. Loss of salvage enzyme through genotoxic damage will render cell resistant and survive as detected using colony formation technique
3. Alkaline Comet assay	Detection of alkali labile sites and overt strand breaks in the DNA of mammalian cells by utilizing microgel electrophoresis at the level of single cell

evaluations are required only if scientifically warranted or if the results of *in vitro* assays indicate a need for further testing, the International Conference on Harmonisation's document for pharmaceuticals recommends the inclusion of an *in vivo* model in the battery of genotoxicity tests. However, only an *in vitro* model will be discussed in this review as summarized in Table 20.10.

All assays were designed to provide the best chance of detecting potential activity, with respect to (i) the exogenous metabolic activation system; (ii) the ability of the compound or its metabolite(s) to reach the target DNA and/or targets such as the cell division apparatus, and (iii) the ability of the genetic test system to detect the given type of mutational event. Like several other sections of the international standards, ISO 10993-3 refers to the Organization for Economic Cooperation and Development (OECD) guidelines for specific test methods. As most biomaterials are insoluble and the OECD methods are designed to test soluble chemicals, the tests must be modified to accommodate the evaluation of fluid extracts. The United States Pharmacopeia (USP) (1995) has established standard preparation methods for material testing that can be used for genotoxicity testing, and ISO 10993-12, 'Sample Preparation and Reference Materials', also describes standard methods for the preparation of extracts of device materials. Selection of the appropriate extraction vehicle varies with the test system of choice.

20.6.1 Testing strategy

A range of tests has been developed which employ a wide variety of organisms, including bacteria, yeasts and other eukaryotic microorganisms, and mammalian cells studied *in vitro*, as well as whole mammals where effects in either somatic or germ cells can be measured. For the purpose of this chapter, only a representative selection of the assays will be discussed.

Bacterial reverse mutation assay (Ames test)

The most widely used *in vitro* test is the bacterial reverse mutation assay for gene mutations developed by Ames and his colleagues using *Salmonella typhimurium* (Gatehouse *et al.*, 1990). The very extensive database available for this assay justifies its inclusion in any initial testing package as outlined in ISO 10993-3, ICH S2A, ICH S2B and OECD 471. Several tester strains of bacteria capable of detecting both base-pair and frame-shift mutations must be included, the best validated strains being TA1535, TA1537 (or TA97 or TA97a), TA98 and TA100.

In general, the bacterial mutagenicity assays for direct-acting mutagens are the reverse-mutation plate test, the plate-incorporation test, or the agar overlay test which uses bacterial strains mixed with known concentrations of a test agent. The mixture is then overlaid onto the surface of an agar plate lacking the minimum essential medium component (minimal agar). On the agar plate, all auxotrophic organisms stop growing except those affected by the test agent that revert (reverse mutate) to prototrophic growth patterns.

Various studies have shown that this assay is capable of evaluating the potential mutagenic acting chemicals that were possibly present either directly or from eluted extracts in the test materials. Our own in-house data have demonstrated the ability of the assay to evaluate the mutagenic potential of two different types of biomaterial, namely hydroxyapatite and polyhydroxyalkanoate polymer (Siew *et al.*, 2008). Kaplan *et al.* (2004) have also shown the possible mutagenic activity of dental cements as detected using a few *Salmonella* strains with or without rat liver microsome S9 fraction. This fraction has the ability to detect any potential promutagenic chemicals that may be present in the treatment solution.

The *Salmonella* assay, whilst being an efficient primary screen for detecting compounds with inherent potential for inducing gene mutations, does not detect all compounds with mutagenic potential. Some compounds are clastogens but do not produce gene mutation in the *Salmonella* assay (e.g. inorganic arsenic compounds, IARC 1987). The second assay should therefore evaluate the potential of a chemical to produce both clastogenicity and aneugenicity, and it should use mammalian cells, either cell lines or primary human cultures such as fibroblasts or lymphocytes.

Cytogenetic analysis

A major development of novel techniques and methods is the use of chromosomal painting and *in vitro* micronucleus assay for the assessment of potential aneugenicity. It is now feasible to screen substances for their potential to induce aneuploidy in the initial testing stage. One approach is the *in vitro* cytogenetic assay for clastogenicity using metaphase analysis. Limited information can be obtained on potential aneugenicity by recording the incidence of hyperdiploidy, polyploidy and/or modification of mitotic index, etc. (Aardema *et al.*, 1998). If there are indicators of aneugenicity (e.g., induction of polyploidy) then this should be confirmed using appropriate staining procedures such as FISH (fluorescence in-situ hybridization) or chromosome painting to highlight alterations in the number of copies of selected chromosomes (reviewed by Parry, 1996). When cell lines are employed it is important that only those with a stable chromosome number are used. Reduced hypotonic treatment may be necessary to reduce artifactual changes in chromosome number. Only the detection of hyperploidy (gain in number) should be considered as a clear indication of induced aneuploidy.

The purpose of this test is to screen medical devices and materials to determine if they cause structural chromosome aberrations in Chinese Hamster Ovary (CHO) cells. This procedure allows for exposure with and without an exogenous source of metabolic activation. This procedure is designed to comply with the OECD 473, ISO 10993-3, ICH S2A, ICH S2B and OECD 471 as one of the three levels of *in vitro* tests for genotoxicity. There is evidence to support the proposal that chromosome mutation and related events that cause alteration in oncogenes and tumour suppressor genes of somatic cells are involved in cancer induction in humans and experimental animals. Structural chromosomal aberrations result from direct DNA breakage, replication on a damaged DNA template, inhibition of DNA synthesis, and other mechanisms such as topoisomerase inhibitors. Structural and numerical chromosomal aberrations are most commonly scored in proliferating cells arrested at metaphase using a tubulin polymerization inhibitor, e.g. Colcemid or colchicine.

Micronuclei (MN) arise in mitotic cells from chromosomal fragments or chromosomes that lag behind in anaphase and are not integrated into the daughter nuclei. Micronuclei harbouring chromosomal fragments result from direct DNA breakage, replication on a damaged DNA template and inhibition of DNA synthesis. MN harbouring whole chromosomes are primarily formed from failure of the mitotic spindle, kinetochore or other parts of the mitotic apparatus or by damage to chromosomal sub-structures, alterations in cellular physiology, and mechanical disruption. Thus, an increased frequency of micronucleated cells is a biomarker of genotoxic effects that can reflect exposure to agents with clastogenic chromosome breaking. The Committee on Mutagenicity of Chemicals in Food, Consumer Products and the Environment (COM) UK has reaffirmed the view stated in the 1989 guidelines that a combination of assays for gene mutation in bacteria and for chromosomal aberrations (plus aneuploidy) in mammalian cells may not detect a small proportion of agents with the potential for *in vitro* mutagenicity. Thus a third assay, comprising an additional gene mutation assay in mammalian cells, should be used, except for compounds for which there is little or no human exposure.

Mammalian gene mutation assay

Certain mammalian cell gene mutation protocols that have been widely employed, particularly some of those involving the use of Chinese hamster cells, have been considered to be insufficiently sensitive, predominantly on statistical grounds (UKEMS, 1989). Of the available systems, measuring mutations at the thymidine kinase (TK) locus in L5178Y mouse lymphoma cells has gained broad acceptance and has the advantage of detecting not only gene mutations but also various sizes of chromosome deletions.

COM has recommended the use of the mouse lymphoma assay (or an alternative test of equivalent statistical power) as the third *in vitro* test in Stage 1 genotoxicity testing. The use of the mouse lymphoma assay for the detection of all types of mutational end-point has been the subject of considerable debate, particularly by the International Conference on Harmonisation of Technical Requirements for Registration of Pharmaceuticals for Human Use (ICH, 1997). The ICH considers that the mouse lymphoma assay can be used on a routine basis as an alternative to clastogenicity tests that employ metaphase analysis (Müller *et al.*, 1999).

The mouse lymphoma assay identifies substances which induce gene mutations. In addition, there are some data to justify the use of the mouse lymphoma assay to identify potential clastogens. The HGPRT assay employs cells bearing genes that control the enzyme hypoxanthine-guanine phospho ribosyl transferase. HGPRT catalyses the reaction of hypoxanthine and guanine with phospho ribosyl pyro phosphate for purine salvage and it is a prime selection assay for determination of gene mutation in mammalian cells. In addition to its normal substrates, HGPRT catalyses the conversion of purine analogues such as 6-thioguanine (6-TG), rendering them cytotoxic to normal cells. Following mutation of the gene, cells deficient in HGPRT survive treatment with 6-TG as they cannot phosphoribosylate the analogue. A variant frequency (VF) is a measure of HGPRT variants as detected using auto radiography, whereas a mutant frequency (MF) is a measure of HGPRT mutants as determined by cloning assay.

Alkaline Comet assay

The Alkaline Comet assay or Single Cell Gel Electrophoresis (SCGE) assay, which is inceasingly being used for genotoxicity testing of pharmaceuticals and industrial chemicals, has now gained interest in the biomaterials field. The Comet assay was used to quantify DNA single-strand breaks, alkali labile and incomplete excision repair sites using cells exposed to genotoxins. Reviews of this assay have been published by McKelvey-Martin *et al.* (1993), Fairbairn *et al.* (1995), Tice *et al.* (2000), Olive *et al.* (1990), Rojas *et al.* (1999) and Speit and Hartmann (1999). Using this technique, virtually any accessible cell population can be evaluated for DNA damage. In this assay, cells suspended in molten agarose are layered onto a microscope slide, the cells lysed by detergents and high salt, and the liberated DNA electrophoresced under neutral or alkaline conditions.

Cells with increased levels of DNA damage display altered migration of the DNA towards the anode. 'Comets', i.e. individual cell DNA migration patterns, are viewed using fluorescence or non-fluorescence microscopy after staining with a suitable dye (see Fig. 20.5). Under neutral electrophoretic conditions, DNA migration is increased by double strand breaks (DSB). Under alkaline electrophoretic conditions, it appears that DNA migration is increased preferentially by strand breaks, both single and double, at a pH of 12.1 and by both strand breaks and alkali labile sites (ALS) at higher pH levels. One critically important advantage of this assay is the ability to detect increased levels of damage among subsets of cells within a larger population of apparently unaffected cells.

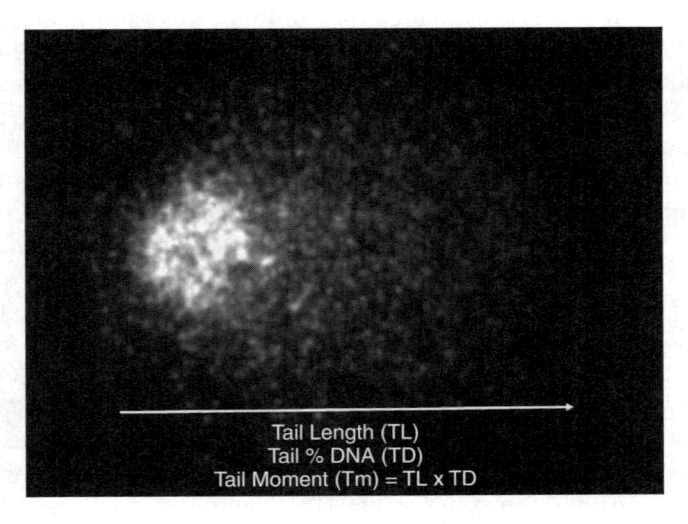

20.5 Representation of Comet image.

The level of damage can be evaluated using either the DNA tail length or tail moment values. Tail length is defined as the maximum length of DNA migration measured either from the estimated leading edge of the head or from the centre of the head, and is considered a measure of the smallest sized DNA fragments. The percentage of migrated DNA is the fraction of DNA in the tail as compared to the whole image. Tail moment is defined as some measure of tail length multiplied by the fraction of DNA in the tail. There are several methods for calculating tail moment and the choice of the method is at the discretion of the investigator (Olive and Banath, 1993). No matter what the cause of the genotoxicity detected by the SCGE assay, testing finished biomaterials using the comet assay makes it possible to evaluate interactions between biomaterials and living tissues that are much closer to actual application conditions.

Classical *in vitro* tests can be used to evaluate the genotoxicity of medical device materials. In all cases, adverse or equivocal findings warrant further investigation. Confirmation testing by dose–response relationship is the standard course of action. In addition, a presumptive positive finding in an *in vitro* assay can be confirmed by conducting an alternative *in vivo* model. Acceptable results from a battery of genotoxicity tests will not only go a long way towards ensuring the safety of a proposed biomaterial; in some cases such data can justify not pursuing *in vivo* carcinogenicity studies, particularly if there is existing information about the lack of genotoxicity of the material in question.

20.7 Future trends

There has been a significant shift towards alternatives to animal testing, resulting in the development of more sensitive and sophisticated *in vitro* tests. There is an extensive literature on the principles of validation and the various approaches that have been proposed (reviewed in Spielmann and Liebsch, 2002). Currently, the European Centre for the Validation of Alternative Methods (ECVAM) has applied many demanding procedures to assure the suitability of new *in vitro* tests.

To date, there are a number of accepted international *in vitro* test alternatives for local end-points, such as phototoxicity, skin penetration and skin corrosion. In general, phototoxicity can be divided into photocytotoxicity and photogenotoxicity. Photocytotoxicity is related to cell viability such as evaluation of *in vitro* 3T3 neutral red uptake test (Spielmann, 2001). On the other hand, photogenotoxicity refers to genotoxic effects on viable cells as can be determined by using alkaline comet assay and micronucleus assay where the procedures have been modified for use in 96 well plates (Kiskinis *et al.*, 2002). Both of these phototoxicity tests can be applied to compounds that absorb light energy within the sunlight spectrum and reach cells in the skin or the retina either by systemic distribution or through topical application (Spielmann et al., 2000).

High throughput screening (HTS) has been developed to enable testing of thousands of samples in a very short time. Although HTS is a useful tool for the pharmaceutical and biotechnology industries, this application should be expanded and adapted to identify the toxic compounds in the safety evaluation of new biomaterials. In addition, HTS is exemplified by one aspect of ECVAM's engagement in the A-Cute-Tox programme. For instance, yeast assay has been developed, in which upregulation of DNA damage responsive *RAD5A* gene is measured by using a promoter-GFP fusion (Billinton et al. 1998). Lately, the Green Screen HC *GADD45a-GFP* genotoxicity assay has been reported with high specificity and high sensitivity as a genotoxicity test. There are also a number of bacterial assays that have been developed for genotoxicity screening using microwell technology, such as the SOS chromotest (Quillardet et al., 1982; Xu et al. 2006) and the *umu* test (Oda et al., 2004). More recently, the SOS LUX test has been carried out in which the mutagenicity is determined by a chromogenic reporter system by expression of luciferase, an enzyme which catalyses ATP-dependent light emission from the substrate luciferin (Rettberg et al., 2000).

In summary, this chapter has attempted to highlight various methods available for testing biomaterials. Development of newer and more sensitive methods will ensure that the *in vitro* biocompatibility testing is more relevant to the *in vivo* application.

20.8 References

Aardema M J, Albertini S, Arni P, Henderson L M, Kirch-Volders M, Mackay J M, Sarrif D A, Stringer D A and Taalman R D F (1998), 'Aneuploidy: a report of an ECETOC task force', *Mutat Res*, 410, 3–79.

Altman F P (1977), 'Quantitative microscopy of enzyme reactions in tissue sections', *Microsc Acta*, 79, 327–338. Review.

Billinton N, Barker M G, Michael C E, Knight A W, Heyer W D, Goddard N J, Fielden P R and Walmsley R M (1998), 'Development of green fluorescent protein reporter for a yeast genotoxicity biosensor', *Biosens Bioelectron*, 13, 831–838.

Borenfreund E and Puerner J A (1984), 'A simple quantitative procedure using monolayer cultures for cytotoxicity assay', *J Tissue Culture Meth*, 9, 119–124.

Budihardjo I, Oliver H, Lutter M, Luo X and Wang X D (1999), 'Biochemical pathways of caspase activation during apoptosis', *Annu Rev Cell Dev Biol*, 15, 269–290.

Cain K, Inayat-Hussain S H, Wolfe J T and Cohen G M (1994), 'DNA fragmentation into 200–250 and/or 30–50 kilobase pair fragments in rat liver nuclei is stimulated by Mg^{2+} alone and Ca^{2+}/Mg^{2+} but not by Ca^{2+} alone', *FEBS Lett*, 349, 385–391.

Cenni E, Ciapetti G, Granchi D, Arciola C R, Savaring L, Stea S, Montanaro L and Pizzoferrato A (1999), 'Established cell lines and primary cultures in testing medical devices in vitro', *Toxicology in Vitro*, 13, 801–810.

Chakrabarty S, Roy M, Hazra B and Bhattacharya R K (2002), 'Induction of apoptosis in human cancer cell lines by diospyrin, a plant-derived bisnaphthoquinonoid, and its synthetic derivatives', *Cancer Lett*, 188, 85–93.
Chan K M, Rajab N F, Ishak M H A, Ali A M, Yusoff K, Din L B and Inayat-Hussain S H (2006), 'Goniothalamin induces apoptosis in vascular smooth muscle cells', *Chem Biol Interact*, 159 (2), 129–140.
Ciapetti G, Granchi D, Verri E, Savarino L, Cavedagna D and Pizzoferrato A (1996), 'Application of combination of neutral red and amido black staining for rapid, reliable cytotoxicity testing of biomaterials', *Biomaterials*, 17, 1259–1264.
Ciapetti G, Granchi D, Savarino E, Cenni E, Magrini E, Baldini N and Giunti A (2002), '*In-vitro* testing of the potential for orthopedic bone cements to cause apoptosis of osteoblast-like cells', *Biomaterials*, 23, 617–627.
Clifford C J and Downes S (1996), 'A comparative study of the use of colorimetric assays in the assessment of biocompatibility', *J Mater Med*, 7, 637–643.
Dee K C, Puleo D C and Bizios R (2002), *An Introduction to Tissue–Biomaterial Interactions*, 1st edn, Hoboken, NJ: John Wiley & Sons.
ESAC-ECVAM Scientific Advisory Committee (2000), 'Statement on the application of the EpiDerm™ human skin model for the skin corrosivity testing', *ATLA– Altern Lab Anim*, 28,365–367.
Fairbairn D W, Olive P L and O'Neill K L (1995), 'The Comet assay: a comprehensive review', *Mutat Res*, 339, 37–59.
Finter N B (1969), 'Dye uptake methods of assessing viral cytopathogenicity and their application to interferon asssays', *J Gen Virol*, 5, 419.
Fotakis G and Timbrell J A (2006), 'In vitro cytotoxicity assay: comparison of LDH, neutral red, MTT and protein assay in hepatoma cell lines following exposure to cadmium chloride', *Toxicol Lett*, 160, 171–177.
Garza-Ocañas L, Torres-Alanís O and Piñeyro-López A (1990), 'Evaluation of the cytotoxicity of 18 compounds in six rat cell lines', *ATLA*, 17, 246–249.
Gatehouse D G, Rowland I R, Wilcox P, Calander R D and Forster R (1990), 'Bacterial mutation assays in basic mutagenicity tests', in *UKEMS Recommended Procedures*, D Kirkland (ed.), UKEMS Report, University of Cambridge, 13–61.
Givan A L (2001), *Flow Cytometry: First Principles*, 2nd edn, Hoboken, NJ: Wiley-Liss. ISBN 0-471-38224-8 (paper); 0-471-22394-8 (electronic).
Harbell J W, Koontz S W, Lewis R W, Lovell D and Acosta D (1997), 'Cell cytotoxicity assays', *Food Chem Toxicol*, 35, 79–126.
Harmand M F (1997), 'Cytotoxicity – Part I: Toxicological risk evaluation using cell culture', in J H Braybrook, *Biocompatibility Assessment of Medical Devices and Materials*, Biomaterials Science and Engineering Series, Chichester: John Wiley & Sons, 119–124.
Hartmann A, Elhajouji A, Kiskinis F, Poetter H, Martus J, Fjällman A, Frieauff W and Suter W (2001), 'Use of the alkaline comet assay for industrial genotoxicity screening: comparative investigation with the micronucleus test', *Food Chem Toxicol*, 39, 843–858.
Hermansky S J (2001), 'Regulatory toxicology: medical devices', in M J Derelanko and M A Hollinger (eds), *Handbook of Toxicology*, 2nd edn, Boca Raton, FL: CRC Press, FL: 1173–1233.
IARC (1987), 'International Agency for Research in Cancer (IARC) monographs on the evaluation of the carcinogenic risks to humans', Supplement 6. Genetic

and related effects: an updating of selected IARC monographs from volume 1–42. Arsenic and Arsenic Compounds, IARC, Lyon, 71–76.

ICH International Conference on Harmonisation of Technical Requirements for Registration of Pharmaceuticals for Human Use (1997), 'Genotoxicity. A standard battery for genotoxicity testing of pharmaceuticals', in P F D'Arcy and D W G Harron (eds), *Proceedings of the Fourth International Conference on Harmonisation (ICH)*, Brussels, 1997, Belfast: Queens University Press, 987–997.

Ignatius A A and Claes L E (1996), 'In vitro biocompatibility of bioresorbable polymers: poly(L,DL-lactide) and poly(L-lactide-co-glycolide)', *Biomaterials*, 17, 831–839.

Ignatius A A, Schmidt C, Kaspar D and Claes L E (2001), 'In vitro biocompatibility of resorbable experimental glass ceramics for bone substitutes', *J Biomed Mater Res*, 55(3), 285–294.

Inayat-Hussain S H and Ross D (2005), 'Differential involvement of the mitochondrial/caspase-9 dependent apoptotic pathway in hydroquinone treated HL-60 and Jurkat leukemia cells', *Chem Res Toxicol*, 18(3), 420–427.

Inayat-Hussain S H, Couet C, Cohen G M and Cain K (1997), 'Processing/activation of CPP32-like proteases is involved in transforming growth factor-β_1 induced apoptosis in hepatocytes', *Hepatology*, 25, 1516–1526.

Inayat-Hussain S H, Osman A B, Din L B, Ali A M, Snowden R T, MacFarlane M and Cain K (1999), 'Caspases-3 and -7 are activated in goniothalamin-induced apoptosis in human Jurkat T-cells', *FEBS Lett*, 456(3), 379–383.

Inayat-Hussain S H, Winski S L and Ross D (2001), 'Differential involvement of caspases in hydroquinone-induced apoptosis in human leukemic HL-60 and Jurkat cells', *Toxicol Appl Pharmacol*, 175, 95–103.

ISO10993-1 (1998), International standards: Introduction to the standards.

ISO10993-5 (1999), International standards: Biological evaluation of medical device – test for in vitro cytotoxicity.

ISO10993-12 (2002), International standards: Biological evaluation of medical device – sample preparation and reference materials.

Kaplan C, Diril N, Sahin S and Cehreli M C (2004), 'Mutagenic potentials of dental cements as detected by the *Salmonella*/microsome test', *Biomaterials*, 25, 4019–4027.

Kaufmann S H, Kottke T J, Martins L M, Henzing A J and Earnshaw W C (2001), 'Analysis of caspase activation during apoptosis', *Curr Protoc Cell Biol*, 18.2.1–18.2.29.

Kerr J F R (1971), 'Shrinkage necrosis: a disctinct mode of cellular death', *J Path*, 105, 13–20.

Kerr J F R, Wyllie A H and Currie A R (1972), 'Apoptosis: a basic biological phenomenon with wide ranging implications in tissue kinetics', *Br J Cancer*, 26, 239–257.

Kiskinis E, Suter W and Hartmann A (2002), 'High throughput Comet assay using 96-well plates', *Mutagenesis*, 17, 37–43.

Komissarova E V, Saha S K and Rossman T G (2005), 'Dead or dying: the importance of time in cytotoxicity assays using arsenite as an example', *Toxicol Appl Pharmacol*, 202(1) 99–107.

Lee W K, Park K D, Han D K, Suh H, Park J C and Kim Y H (2000), 'Heparinized bovine pericardium as a novel cardiovascular bioprosthesis', *Biomaterials*, 21, 2323–2330.

Lestari F, Hayes A J, Green A R and Markovic B (2005), 'In vitro cytotoxicity of selected chemicals commonly produced during fire combustion using human cell lines', *Toxicology in Vitro*, 19, 653–663.

Low S P, Williams K A, Canham L T and Voelcker N H (2006), 'Evaluation of mammalian cell adhesion on surface-modified porous silicon', *Biomaterials*, 27, 4538–4546.

McKelvey-Martin V J, Green M H L, Schmezer P, Pool-Zobel B L, De Meo M P and Collins A (1993), 'The single cell gel electrophoresis assay (comet assay): a European review', *Mutat Res*, 288, 47–63.

McKim Jr. J M, Wilga P C, Pregenzer J F and Petrella D K (2005), 'A biochemical approach to *in vitro* toxicity testing', *Pharmaceutical Discovery*.

Mitsuyoshi H, Nakashima T, Sumida Y, Yoh T, Nakajima Y, Ishikawa H, Inaba K, Sakamoto Y, Okanoue T and Kashima K (1999), 'Ursodeoxycholic acid protects hepatocytes against oxidation injury via induction of antioxidants', *Biochem Biophys Res Commun*, 263, 537–542.

Mosmann T (1983), 'Rapid colorimetric assay for cellular growth and survival: application to proliferation and cytotoxicity assays', *J Immunol Meth*, 65, 55–63.

Müller L, Kikuchi Y, Probst G, Schechtman L, Shimada H, Sofuni T and Tweats D (1999), 'ICH-harmonised guidances on genotoxicity testing of pharmaceuticals: evolution, reasoning and impact', *Mutat Res*, 436, 195–225.

Nakayama G R, Caton M C, Nova M P and Parandosh Z (1997), 'Assessment of the alamar blue assay for cellular growth and viability in vitro', *J Immunol Meth*, 204, 205–208.

National Toxicology Program (NTP) Interagency Center for the Evaluation of Alternative Toxicological Methods (NICEATM) (2003), Test method protocol for NHK neutral red uptake cytotoxicity test.

Northup S J (1986), 'Testing biomaterials: in vitro assessment of tissue compatibility', in B D Ratner, A S Hoffman, F J Schoen and J E Lemons, (eds), *Biomaterials Science: An Introduction to Materials in Medicine*, San Diego, CA: Academic Press, 215–240.

O'Brien J, Wilson I, Orton T and Pognan F (2000), 'Investigation of the alamar blue (resazurin) fluorescent dye for the assessment of mammalian cell cytotoxicity', *Eur J Biochem*, 267, 5421–5426.

Oda Y, Funasaka K, Kitano M, Nakama A and Yoshikura T (2004), 'Use of a high throughput *umu*-microplate test system for rapid detection of genotoxicity produced by mutagenic carcinogens and airborne particulate matter', *Environ Mol Mutagen*, 43, 10–19.

Ohno T, Itagaki H, Tanaka N and Ono H (1995), 'Validation study on five different cytotoxicity assays in Japan – an intermediate report', *Toxicology in Vitro*, 4, 571–576.

Olive P L and Banath J P (1993), 'Induction and rejoining of radiation-induced DNA single-strand breaks – tail moment as a function of position in the cell-cycle', *Mutat Res*, 294, 275–283.

Olive P L, Banath J P and Durand R E (1990), 'Heterogeneity in radiation-induced DNA damage and repair in tumor and normal cells measured using the "comet" assay', *Radiat Res*, 122, 86–94.

Orrenius S, McConkey D J and Nicotera P (1989), 'The role of calcium in cytotoxicity', in G N Volans (ed.), *Basic Science in Toxicology*, New York: Taylor and Francis, 629–635.

Papazisis K T, Geromichalos G D, Dimitriadis K A and Kortsaris A H (1997), 'Optimization of the sulforhodamine B colorimetric assay', *J Immunol Meth*, 208, 151–158.

Parry J M (1996), 'Molecular cytogenetics', *Mutation Research*, Special Issue, 372, 151–294.

Penninger J M and Kroemer G (2003), 'Mitochondria, AIF and caspases rivaling for cell death execution', *Nature Cell Biology*, 5, 97–99.

Puck T T and Marcus P I (1956), 'Action of X-rays on mammalian cells', *J Exp Med*, 103, 653–666.

Quillardet P, Huisman O, D'ari R and Hofnung M (1982), 'SOS-Chromotest. a direct assay for induction of a SOS function in *E.coli* K-12 to measure genotoxicity', *Proc Natl Acad Sci*, 79, 5971–5975.

Rettberg P, Bandel K, Baumstark-Khan C and Horneck G (2000), 'Increased sensitivity of the SOS-LUX-test for the detection of hydrophobic genotoxic substances with *Salmonella typhimurium* TA1535 as host strain', *Anal Chim Acta*, 426, 167–173.

Riddell R J, Panacer D S, Wilde S M, Clothier R H and Balls M (1986), 'The importance of exposure period and cell type in in vitro cytotoxicity tests', *ATLA*, 14, 86–92.

Rojas E, Lopez M C and Valverde M (1999), 'Single cell gel electrophoresis assay: methodology and applications', *J Chromat B*, 722, 225–254.

Sasaki K and Tanaka N (1991), 'Chemical contamination suspected in cytotoxic effects of incubators on Balb 3T3 cells', *ATLA*, 19, 421–427.

Schmalz G (1988), 'Agar overlay method', *Int Endod J*, 21(2), 59–66.

Siew E L, Rajab N F, Osman A B, Sudesh K and Inayat-Hussain S H (2007), 'In vitro biocompatibility evaluation of poly(3-hydroxybutyrate-*co*-4-hydroxybutyrate) copolymer in fibroblast cells', *J Biomed Mater Res A*, 81(2), 317–325.

Siew E L, Rajab N F, Osman A B, Sudesh K and Inayat-Hussain S H (2008), 'Mutagenic and clastogenic characterisation of post-sterilised Poly(3-hydroxybutyrate co-4-hydroxybutyrate) copolymer biosynthesised by Deftia acidovoraus, *Journal of Biomedical Materials Research: Part A*. (in press).

Skehan P, Storeng R, Scudiero D, Monks A, McMahon J, Vistica D, Warren J T, Bokesch H, Kenney S and Boyd M R (1990), 'New colorimetric cytotoxicity assay for anticancer-drug screening', *J Natl Cancer Inst*, 82, 1107–1112.

Speit G and Hartmann A (1999), 'The comet assay (single-cell gel test): a sensitive genotoxicity test for the detection of DNA damage and repair', *DNA Repair Protocols*, 113, 203–212.

Spielmann H (2001), 'Acute phototoxicity testing', *Environmental Mutagen Research*, 23, 45–56.

Spielmann H and Liebsch M (2002), 'Validation successes: chemicals', *ATLA – Altern Lab Anim*, 30, 33–40. Review.

Spielmann H, Mueller L, Averbeck D, Balls M, Brendler-Schwaab S, Castell J V R, Curren de Silva O, Gibbs N K, Liebsch M, Lovell W W, Merk H F, Nash J F, Neumann N J, Pape W J W, Ulrich P and Vohr H W (2000), 'The Second ECVAM Workshop on phototoxicity testing – The report and recommendations of ECVAMWorkshop 42', *ATLA – Altern Lab Anim*, 28, 777–814.

Sun Z L, Wataha J C and Hanks C T (1997), 'Effects of metal ions on osteoblast-like cell metabolism and differentiation', *J Biomed Mater Res*, 34, 29–37.

Tice R R, Agurell E, Anderson D, Burlinson B, Hartmann A, Kobayashi H, Miyamae Y, Rojas E, Ryu J-C and Sasaki Y F (2000), 'Single cell gel/comet assay: Guidelines for in vitro and in vivo genetic toxicology testing', *Environ Mol Mutagen*, 35, 206–221.

Trump B F, Berezesky I K, Sato T, Laiho K U, Phelps P C and De Claris N (1984), 'Cell calcium, cell injury and cell death', *Environ Health Perpect*, 57, 281–287.

Tsuchiya T (1994), 'Studies on the standardization of cytotoxicity tests and new standard reference materials useful for evaluating the safety of biomaterials', *J Biomater Appl*, 9, 138–157.

UKEMS (1989), 'Statistical evaluation of mutagenicity test data. UKEMS subcommittee on guidelines for mutagenicity testing. Report. Part III', D J Kirkland (ed.), Cambridge: Cambridge University Press.

U.S. Pharmacopeia (1995), 'Biological reactivity tests, in-vitro,' in *U.S. Pharmacopeia 23*, United States Pharmacopeial Convention, Inc., Rockville, MD, 1697–1699.

U.S. Pharmacopeia (2003), USPXIII USP26-NF21. *The United States Pharmacopeia – The National Formulary*, United States Pharmacopeial Convention, Inc., Rockville, MD.

Wang Y W, Wu Q and Chen G Q (2004), 'Attachment, proliferation and differentiation of osteoblasts on random biopolyester poly(3-hydroxybutyrate-co-3-hydroxyhexanoate) scaffolds', *Biomaterials*, 25, 669–675.

Weisenthal L M, Dill P L, Kurnick N B and Lippman M E (1983), 'Comparison of dye exclusion assays with a clonogemic assay in the determination of drug-induced cytotoxicity', *Cancer Res*, 43, 258–264.

William D F (1987), 'Tissue–biomaterial interactions', *J Sci*, 22, 3421–3445.

Wyllie A H (1980), 'Glucocorticoid-induced apoptosis is associated with endogenous endonuclease activation', *Nature*, 284, 555–556.

Wyllie A H (1992), 'Apoptosis and the regulation of cell number in normal and neoplastic tissues', *Cancer Metastasis Rev*, 11, 95–103.

Xu H, Dutka B J and Schurr K (2006), 'Microtitration SOS chromotest: A new approach in genotoxicity testing', *Environ Toxicol Water Qual*, 4, 105–114.

Yang X, Zhao K and Chen G Q (2002), 'Effect of surface treatment on the biocompatibility of microbial polyhydroxyalkanoates', *Biomaterials*, 23, 1391–1397.

You E S, Jang H S, Ahn W S, Kang M I I, Jun M G, Kim Y C and Chun H J (2007), 'In vitro biocompatibility of surface-modified poly(DL-lactide-co-glycolide) scaffold with hydrophilic monomers', *J Ind Eng Chem*, 2, 219–224.

Zamzami N, Maisse C, Metivier D and Kroemer G (2007), 'Measurement of membrane permeability and the permeability transition of mitochondria', *Methods Cell Biol*, 80, 327–340.

21
The influence of plasma proteins on bone cell adhesion

Å ROSENGREN and S OSCARSSON,
Mälardalen University, Sweden

Abstract: This chapter gives an introduction to the tools and the background knowledge for further studies of cell–substrate interactions. A cell adhesion test is presented by which it is possible to study the functional behavior of specific proteins after adsorption. The test used model surfaces, cycloheximide treatment, fluorescence microscopy and divalent cations (Mg^{2+}/Ca^{2+}) for cell quantification, cell morphological studies and monitoring of specific/non-specific cell–substrate interactions.

Four plasma proteins, ceruloplasmin, prothrombin, α_2-HS-glycoprotein and α_1-antichymotrypsin, have been investigated. The goal was to monitor their orientation and (if possible) conformational changes after adsorption and to study their influence on bone cell adhesion as a result of the adsorption process.

The study was performed on ultraflat model surfaces (silicon wafers with a natural layer of silicon oxide/silica) to be able to characterize the adsorbed proteins with atomic force microscopy (AFM) and ellipsometry, and to single out the protein-induced effect on the cell activity by reducing interfering surface properties (e.g., topography, variations in surface chemistry). The dimensions of the adsorbed proteins on silicon wafers were measured with the AFM and the values were compared with available X-ray data. The molecules appeared to be oriented with their long axis parallel to the surface or, as in the case of ceruloplasmin, with one of its larger sides towards the surface.

Key words: cell–substrate interactions, AFM/proteins, cell–protein interactions.

21.1 Introduction

When a cell comes in contact with an artificial surface at an implant site, the cell response depends on the physiochemical nature of the surface (e.g., chemistry, crystallinity, topography, etc.) as well as possible deposits from the environment (e.g., proteins, lipids, saccharides, etc.). Proteins are strong cell guides and the purpose of this chapter is to describe how individual proteins affect the adhesion of cells (especially bone cells). To be able to do so, a high level of control is needed. Surface topography and chemistry have to be kept constant or eliminated and competition from other proteins and lipids from the environment must be regulated. This can be achieved by choosing appropriate experimental conditions.

The influence of topography was eliminated by investigating the interaction between proteins and MG63 cells on ultraflat silicon wafers with Thermanox® cell culture coverslips as reference surfaces. The divalent cations Mg^{2+} and Ca^{2+} were used to distinguish between specific integrin–ligand interactions and generic physicochemical interactions during serum-free conditions. It was clearly demonstrated that cell adhesion to the studied proteins was substrate-dependent and often involved receptor–ligand recognition, as judged by the divalent cation dependency.

Ceruloplasmin, prothrombin, α_1-antichymotrypsin and α_2-HS-glycoprotein were chosen for the study because in competition with other blood proteins they have shown differences in adsorption pattern to potential bone substitutes [1, 2]. Their influences on bone cells are therefore of interest.

Human plasma ceruloplasmin is a single-chain copper-containing glycoprotein with a molecular weight of ~132 kDa. Its precise function has not yet been defined, but it has been suggested to have a multifunctional role, including ferroxidase activity, antioxidant functions and involvement in the metabolism of copper, biogenic amines and nitric oxide [3–5]. Its involvement in cell adhesion is unclear, but there is *in vitro* evidence for the ability of ceruloplasmin to reduce polymorphonuclear leukocyte (PMN) and baby hamster kidney (BHK) cell adhesion to tissue culture plates [6, 7].

Prothrombin is a well-known coagulation factor consisting of a single polypeptide chain of ~72 kDa [8]. Its main function is to regulate blood clotting through conversion into the active component, thrombin, which cleaves fibrinogen to produce fibrinous clots [9]. Prothrombin contains an RGD sequence [8, 10] that could act as an adhesion ligand and, indeed, prothrombin has been found to support attachment and spreading of platelets, granulocyte, endothelial and smooth muscle cells [11–13].

α_1-Antichymotrypsin (ACT) is an acute-phase glycoprotein with a molecular weight of 64–68 kDa and an isoelectric point of ~3.9 [14–16]. ACT belongs to a large family of structural homologues, the serpins (*se*rine *p*roteinase *in*hibitor) and it is an effective inhibitor of chymotrypsin-like proteases [17, 18] and a deficiency is associated with emphysema [19, 20] and chronic obstructive pulmonary disease [21]. It is involved in anti-inflammatory actions by inhibiting cathepsin G, chymotrypsin, chymase, neutrophil chemotaxis and superoxide production [17, 18, 22, 23]. It is also found to play a role in Alzheimer's disease [24]. Previous cell adhesion studies have shown that ACT does not promote smooth muscle cell adhesion on ACT-coated plastic surfaces, but it supports cell spreading by inactivating proteases that degrade cell adhesion molecules.

α_2-HS-Glycoprotein (AHSG) is a negatively charged (pI ~4.3) disulfide-linked dimer with a molecular weight of 49–60 kDa [16]. Its crystal structure is not yet known, but according to comparative studies of deduced amino acid sequences, AHSG should be arranged into two N-terminal cystatin-like domains, D1 and D2, and a third, complex, proline-rich C-terminal domain D3 [25–27]. The molecule, which binds Ca^{2+} and calcium apatite [28], is greatly concentrated in bone [29] and serves thereby as an effective inhibitor of calcification [30]. Furthermore, lack of AHSG results in systemic calcification in mice and humans [31]. AHSG has also been implicated in other biological processes, such as the immune response, insulin signaling, brain development and transport of lipids [32–34]. In addition, it contains binding sites for TGF-β like growth factors [35] and lectins [30].

Fetuin, the bovine homologue (>70% sequence identity) of AHSG, stimulates the attachment, growth and differentiation of a variety of cell types in culture, for example epithelial cells, smooth muscle cells and fibroblasts [36]. However, it is unclear whether fetuin itself or some minor contaminant(s) is responsible for the growth-promoting activities. Fibronectin, albumin and human plasma were chosen as references. Fibronectin was used as adhesion promoter [37], albumin as adhesion rejecter [38] and human plasma to simulate the *in vivo* environment.

In this study, we sought to characterize the orientation and (if possible) conformational changes of ceruloplasmin, prothrombin, α_1-antichymotrypsin and α_2-HS-glycoprotein upon adsorption and to study their influence on bone-cell (MG63) adhesion as a result of the adsorption process. The characterization was performed by atomic force microscopy (AFM) and ellipsometry and the cell adhesion was monitored with fluorescence microscopy. Furthermore, we studied the influence of divalent cations (Mg^{2+} and Ca^{2+}) on the cell–substrate interactions. We found that cell adhesion to the four plasma proteins varied and that it was significantly influenced by the underlying material. The difference can be due to different molecular orientations and/or conformations on the two substrates.

The study of protein material interactions often involves materials which are available as planar surfaces or particles of different dimensions with or without pores. The large diversity of available materials requires many different analytical techniques for studies of protein material interactions. This chapter gives an introduction to the tools and the background knowledge for further studies of cell–substrate interactions.

21.2 Protein material interactions: flow-based protein adsorption analysis

The principle of flow-based adsorption analysis is based on the dynamic interaction of proteins with adsorbing surfaces under continuous flow. The experimental model relies on a traditional *chromatographic* setup consisting of a *mobile phase* and a *stationary phase* [39]. The stationary phase, the biomaterial, is packed into a glass column which is connected to a pump system thereby allowing a flow of mobile phase over the material surface. The sample, human blood plasma, is injected onto the column and the proteins are captured by the biomaterial. Proteins that do not possess any affinity for the biomaterial will either pass directly through the column or be eluted by a low-stringency washing step. The bound proteins are then recovered by washing the column with a solution that disrupts the interaction between the protein and the biomaterial (*desorption*). The proteins in the flow-through and the recovery liquids are readily collected in fractions for further analysis (Fig. 21.1).

21.2.1 Total protein binding capacity

The total protein binding capacity of the individual materials was calculated from the following equation:

$$\text{Total binding capacity (mg/m}^2) = (m_{\text{protein in}} - m_{\text{protein out}})/A_s = m_{\text{ads}}/A_s$$

where the amount of adsorbed plasma proteins was obtained by measuring the protein content in the flow-through during the injection phase and the low-stringency washing step, and subtracting it from the applied protein amount [1, 2]. The protein content was measured by BCA protein assay and the specific surface area, A_s, was determined by the BET method and Hg-porosimetry [40].

21.2.2 Dynamic binding capacity

The dynamic binding capacity, which is a relative measurement of available protein binding sites between different materials, was determined by comparing the flow-through volumes at the 50% saturation point of the breakthrough curves between the different materials.

21.2.3 Relative binding capacity for individual proteins

The relative binding capacity study is a purely qualitative comparison between materials regarding the presence of individual proteins in the flow-through. It is evaluated by identifying the proteins in the flow-through at 25%, 50% and 75% of saturation, i.e. when 25%, 50% and 75% of available protein binding sites have been occupied. The proteins were identified by 2-D gel electrophoresis.

21.3 Bicinchoninic acid (BCA) protein assay

The BCA assay [41] is a coloring assay that is used for protein quantification. The method is based on the Biuret reaction during which Cu^{2+} is

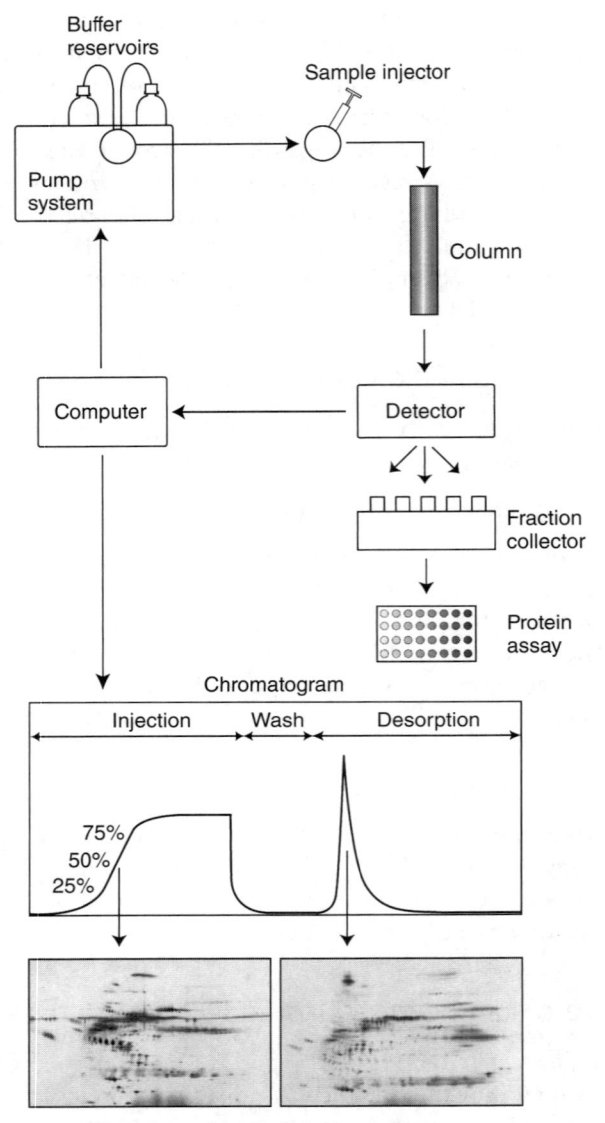

21.1 Schematic view showing the instrumental setup used for protein screening.

reduced to Cu^+ by peptide bonds in proteins under alkaline conditions. The Cu^{2+} is then detected by a chelating reaction with BCA yielding an intense purple color. The production of Cu^+ is a function of protein concentration and incubation time, allowing the protein content of unknown samples to be determined spectrophotometrically by comparison with known standards. The advantage of BCA assay over other assays is that it is relatively insensitive to detergents and denaturing agents. It is, however, more sensitive to sugars.

21.4 Two-dimensional gel electrophoresis

Two-dimensional (2-D) gel electrophoresis, first introduced by O'Farrell [42] and Klose [43], is a powerful and widely used method for the identification and analysis of proteins in complex protein mixtures. The technique, as in all electrophoresis techniques, relies on the migration of charged species (e.g., proteins) in electrical fields. The proteins are sorted according to two independent properties in two steps: the first dimension step, isoelectric focusing (IEF), separates proteins according to their *isoelectric points* (pI); the second dimension step, sodium dodecyl sulfate-polyacrylamide gel electrophoresis (SDS-PAGE), separates proteins according to their *molecular weight* (M_r, relative molecular weight). Each spot on the resulting 2-D array (Fig. 21.1) corresponds to a single protein species in the sample. Thousands of different proteins can thus be separated, and information such as the protein pI, the apparent molecular weight, and the amount of each protein is obtained.

A normal 2-D protocol involves thorough sample solubilization with highly concentrated urea, nonionic detergent, carrier ampholytes and reducing agent, IEF under denaturing conditions in a thin, nonsieving gel with an immobilized pH gradient, SDS-PAGE in homogeneous or gradient slab gels and protein staining with silver or Coomassie Blue™ followed by evaluation with an appropriate image evaluation software [44, 45].

In addition to its ability to resolve thousands of proteins, the technique allows rapid comparison of protein composition in different samples; it does not require any protein labeling which could modify the protein structure, it enables detection of post- and co-translational protein modifications which cannot be predicted from the genome sequence, and it is relatively economical compared to high-performance liquid chromatography (HPLC) and capillary electrophoresis (CE). The practical limitations include poor solubility of hydrophobic and membrane proteins, narrow dynamic range, difficulty in focusing highly basic and acidic proteins, inadequate sensitivity and poor quantitation, and it is also quite labor-intensive. Nevertheless it remains the technique of choice for separation of proteins in complex biological samples. Recently 2-D gel electrophoresis has experienced an enormous upswing when combined with mass spectrometry for sensitive and thorough protein identification.

21.5 Western immunoblotting

Immunoblotting [46] provides a simple and effective method for identifying *antigens* (proteins recognized by specific *antibodies*) in a complex mixture of proteins. Initially, the constituent proteins are separated by SDS-PAGE, or a similar technique, and thereafter are transferred electrophoretically or by diffusion onto a nitrocellulose filter. The immobilized proteins can then be identified using antibodies that bind the protein of interest and subsequent visualization of the resulting antibody–antigen complex. The visualization in this study relies on binding of an additional secondary antibody conjugated with a marker enzyme. The activity of the enzyme marker results in colored insoluble product when incubated with an appropriate chromogenic substrate. Immunoblotting is more sensitive than 2-D gel electrophoresis and is therefore used for identification of proteins in low concentrations.

21.6 Protein characterization

Atomic force microscopy (AFM) was used to study the orientation and conformational changes of single proteins after adsorption on model surfaces. Ellipsometry was used to monitor the formation of protein layers during adsorption.

21.6.1 Surface preparation

Surfaces are readily contaminated during regular storing. The contaminants occur in essentially three forms: (1) hydrocarbons, (2) metals, and (3) particles (macrocolloids, dust, inorganic debris) [47]. In the case of silicon wafers, the contaminants are generally removed with wet chemical cleaning such as etching in 'piranha' solution (2:1 H_2SO_4/H_2O_2) [48, 49]. The organic contaminants are removed by the oxidizing agent (H_2O_2) and the metals by the acid (H_2SO_4). The acidic and oxidative conditions increase the number of hydroxyl groups on the surface and enhance the growth of the natural oxide layer, yielding increased surface roughness. There have been reports of possible sulfur contamination on silica after piranha treatment [50]. The last steps in cleaning are rinsing and drying which should be done in ultrapure (low in carbon) deionized water and by physical removal (e.g., flow of N_2), respectively.

21.6.2 Atomic force microscopy in tapping mode™

Atomic force microscopy (AFM) or scanning force microscopy (SFM) is a non-optical method that employs mechanical sensing to image the topography of solid surfaces [51]. A sharp tip mounted at the end of a flexible cantilever is used to raster-scan the sample surface, registering the interaction force (typically in the order of 10^{-10} N for biomolecular interactions) between tip and sample at each position [57] (Fig. 21.2). During tapping mode™, this means registering the amplitude damping of an oscillating cantilever as it touches the surface [53]. The damping is monitored by an optical detector that measures the variation of the point of incidence of a laser beam reflected from the back of the cantilever. The optical signal alerts a feedback system which controls a piezoelectric positioner (*piezo*) on which the sample rests. The piezo either raises or lowers the sample (relative to the tip) to maintain a constant amplitude of the oscillating cantilever and thereby cancel any changes in the tip–sample interaction. The movement of the piezo, or more precisely the applied voltage, is proportional to the height of the sample and is used to reconstruct a pseudo-three-dimensional profile of the sample.

General guidelines for protein imaging are to use atomically flat surfaces and sharp tips with high aspect ratios [52]. The reason for the former is quite obvious since it is a prerequisite for the tip to 'see' the target molecules, and nothing but those molecules. The reason for the latter is that the tip imposes itself on the image, making every image a convolution of the actual molecule topography and the shape of the tip. At worst this will

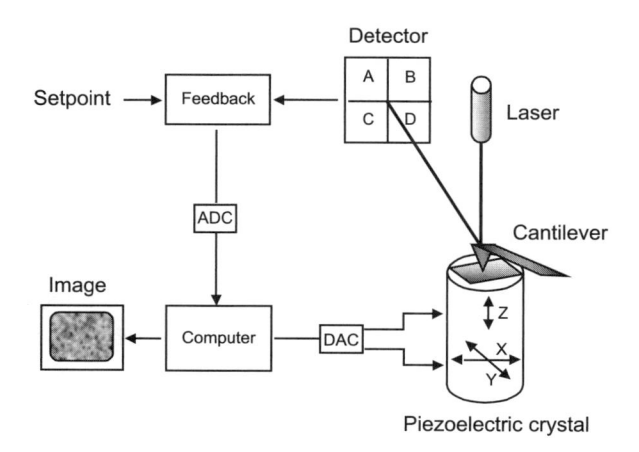

21.2 Principle of the AFM (reproduced with permission from Dr E Pavlovic).

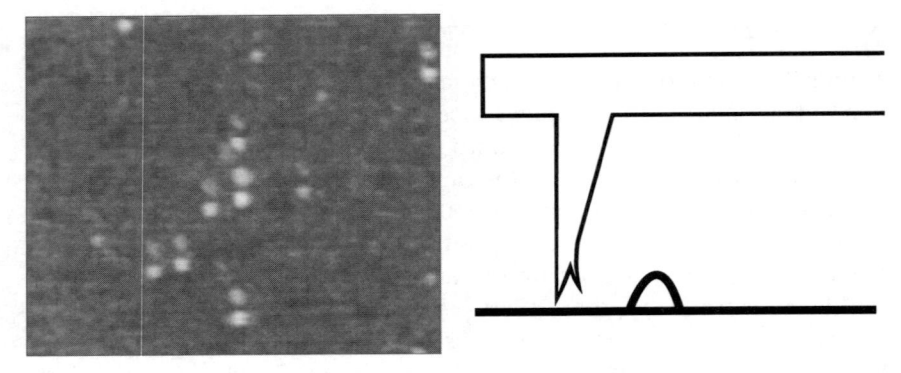

21.3 Tip shape induced artifacts: the surface was scanned with a blunt 'double' tip which magnifies and registers the same features twice.

generate a false appearance of the imaged object (Fig. 21.3) or at best a slight magnification. The latter always occurs and is known as *probe-broadening* [53]. There are methods to remove the 'tip effect' but these are usually quite time-consuming and require knowledge about the tip shape [54–56]. The displacement of the tip in the z direction is not affected by probe-broadening and, hence, the height of the (hard) object can be used for characterization. Nevertheless, the lateral dimensions work as guidelines to detect the orientation of objects, especially if the differences in dimensions between their axes are big enough [57, 58].

Protein imaging may be performed in air, liquid and vacuum [59, 60]. The advantage of liquid is that proteins can be imaged in their natural environment and that the solution conditions can be manipulated to minimize the force exerted by the tip on the sample [61]. Unfortunately, the AFM resolution is worse in liquid than in air/vacuum since the molecules tend to be more mobile. There is also a risk that the proteins are moved or removed by the tip, because proteins are less firmly attached in liquid than in air.

Imaging in air is easier but suffers from two main disadvantages. During drying of the samples, proteins lose most of their liquid, which might cause structural changes or even denaturation of the proteins. The liquid is, however, not completely removed, leading to strong capillary forces between the tip and the sample and possible protein deformation. This effect is more pronounced in other modes such as contact mode but still needs to be considered as a tapping mode™ artifact.

21.7 Ellipsometry

Ellipsometry is one of the oldest [62] and most frequently used techniques for direct monitoring of the adsorption of proteins on solid surfaces [63–66].

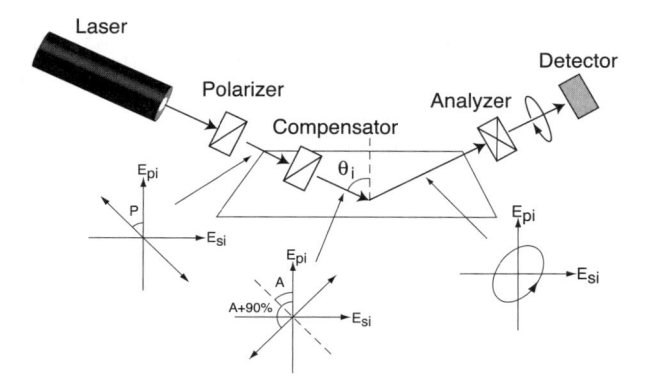

21.4 Principle of the rotating-analyzer ellipsometer.

It uses linear (circular) polarized light to determine the change of the polarization state by reflection at the sample surface. The polarization state is described by the two ellipsometric angles, Δ (relative phase change) and Ψ (relative amplitude change), from which surface properties such as (effective) refractive index ($N = n + ik$) and (mean) film thickness (d) can be calculated.

An ordinary rotating-analyzer ellipsometer is equipped with a light source (laser; λ: 623.8 nm), a polarizer, a compensator (quarter-wave plate), an analyzer (rotating polarizer) and a detector (Fig. 21.4). The laser sends out monochromatic light, which is linearly polarized by the polarizer and thereafter transformed into circularly polarized light by the compensator. The latter improves the sensitivity of the instrument. Upon reflection, the light is elliptically polarized and then directed through a rotating element generating a sinusoidal intensity-signal at the detector plate. The amplitude and phase of the exiting light is compared with the corresponding parameters for the entering light beam.

Ellipsometry is a good tool for protein adsorption analysis. It is sensitive (thickness resolution: ±1 Å), does not destroy the adsorbed proteins, does not require any protein labeling and can be performed *in situ* and in real-time. In addition it is robust and fairly simple to handle experimentally. The technique requires atomically flat and optically reflective surfaces, which limits the application. The main disadvantage is, however, its model dependency [63, 65]. To convert Δ and Ψ into effective refractive index (N) and mean film thickness (d), some assumptions must be made. We have to assume that the electrolyte–protein–substrate system can be described by a layer model with homogeneous optical parameters in different layers, meaning constant density and refractive index within the film, and discrete boundaries towards each layer. This is sometimes not true, especially not within submonolayers of protein. The latter result in a pseudo response that is an average of the optical properties of both surface and coating. However, keeping this in mind, ellipsometry can be used with confidence.

21.8 Cell–substrate interactions: study methods

A cell adhesion test was developed to study the functional behavior of specific proteins after adsorption. The test used model surfaces, cycloheximide treatment, fluorescence microscopy and divalent cations (Mg^{2+}/Ca^{2+}) for cell quantification, cell morphological studies and monitoring of specific/nonspecific cell–substrate interactions.

21.8.1 Cell adhesion test

Cell adhesion (attachment and spreading) can be used to evaluate the function of adsorbed proteins *in vitro* (Fig. 21.5). The number and morphology (structural appearance) of adhered cells gives an indication of the adsorbed proteins' potential to regulate cell establishment at surfaces. A protein that promotes cell establishment is expected to give high numbers of attached and spread cells and vice versa. To further classify the proteinaceous effect, reference proteins such as fibronectin (cell adhesion protein) and albumin (cell repellant) are often included in the experiments.

The nature of the cell–substrate interaction can be monitored by comparing the number of attached cells with the number of spread cells and/or by including or excluding divalent cations (e.g., Mg^{2+}/Ca^{2+}) from the cell suspension (Fig. 21.6). The first alternative relies on the assumption that cells need integrins for spreading but not for attachment; thus if the number of attached cells is higher than the number of spread cells then some of the

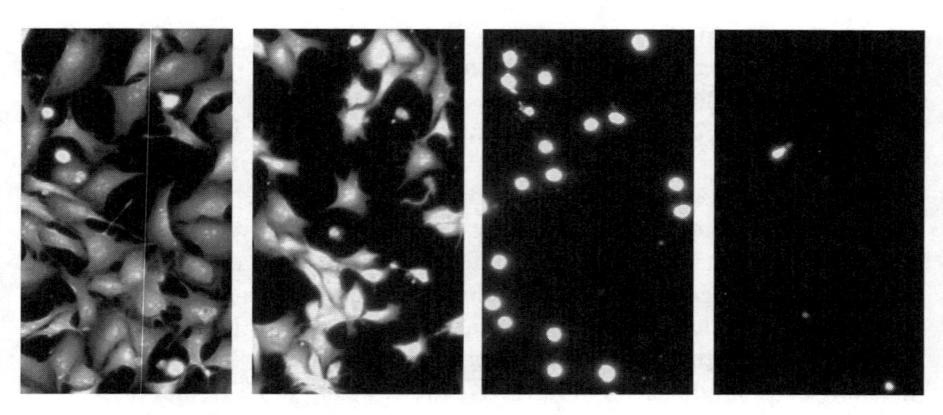

21.5 Fluorescence microscopy pictures of different levels of cell adhesion to surfaces.

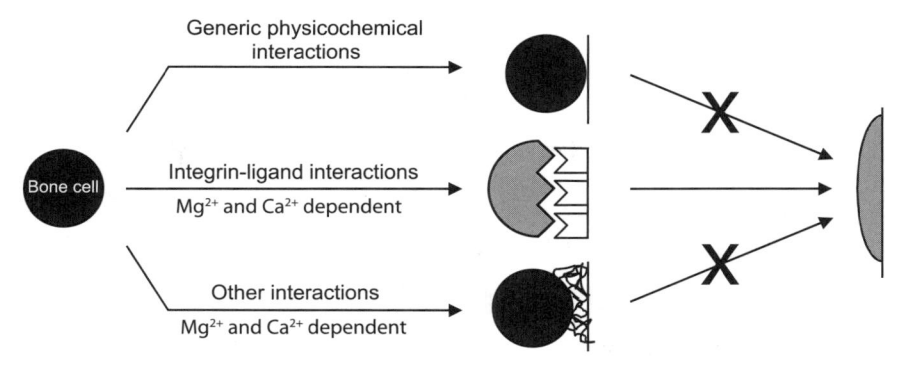

21.6 Principles of cell attachment and spreading.

attached cells use other interactions than ligand–integrin recognition for binding. The second alternative makes use of the basic condition that integrins require divalent cations to be able to bind ligands; thus if the number of attached cells dramatically decreases when Mg^{2+} and Ca^{2+} are excluded from the cell suspension then the cell attachment is integrin mediated. The test does not exclude the possibility that other receptors than integrins are involved, therefore studies with 'blocked' integrins are good complements.

A prerequisite for this type of experiment is that the system is kept free from other proteins and that the cells are not allowed to be influenced by the naked surfaces (topographical and physicochemical interactions) instead of the adsorbed proteins. The former is solved with cycloheximide treatment (see below) and careful washing of the cells after culturing, and the latter is settled by using flat and fully (protein) covered surfaces.

21.8.2 Cycloheximide treatment

In vivo (and *in vitro*), bone cells react on artificial surfaces by secreting proteins (part of the extra cellular matrix (ECM)) and thereby providing points of attachment. This secretion is undesirable when studying cell attachment to individual proteins. Cycloheximide (actidione) [67] is a glutarimide antibiotic that efficiently inhibits protein synthesis in culture. It prevents amino acids to be transferred from aminoacyl transfer RNA into nascent peptides on ribosomes of the 80 S type [68].

21.8.3 Fluorescence microscopy

Fluorescence microscopy provides a simple method to quantify cells and study their morphology on surfaces (Fig. 21.7). The technique relies on the

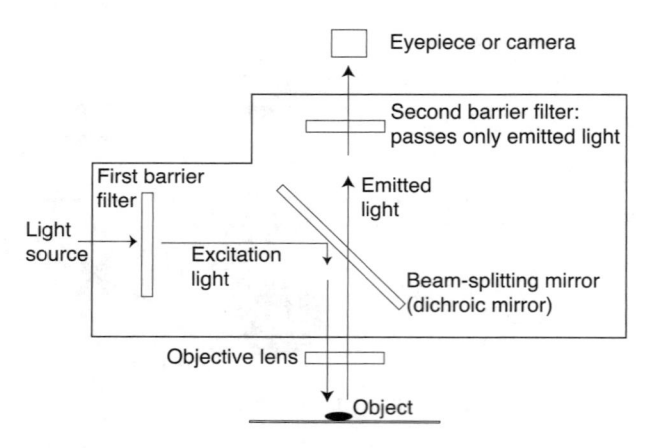

21.7 Principle of the fluorescence microscope.

phenomena that fluorescent molecules absorb light at one wavelength and emit it at another, longer wavelength [69]. If a cell, stained with such a dye, is illuminated at the dye's absorbing wavelength and then viewed through a filter that allows only light of the emitted wavelength to pass, it is seen to glow against a dark background.

The dye used is acridine orange (AO) which is a selective cationic stain for nucleic acids. It interacts with DNA and RNA by *intercalation* and electrostatic attraction, respectively. The absorption band of AO lies between 440 and 480 nm (blue light) and the emission occurs between 520 nm (green fluorescence for DNA) and 650 nm (orange fluorescence for RNA). Fluorescence microscopy combined with a video camera, computer and image software, has developed into a powerful tool for imaging of cells.

21.9 Protein–cell interactions

Four plasma proteins have been investigated. The goal was to monitor their orientation and (if possible) conformational changes after adsorption and to study their influence on bone cell adhesion as a result of the adsorption process. The choice fell on ceruloplasmin, prothrombin, α_2-HS-glycoprotein, and α_1-antichymotrypsin.

The study was performed on ultraflat model surfaces (silicon wafers with a natural layer of silicon oxide/silica) to be able to characterize the adsorbed proteins with atomic force microscopy (AFM) and ellipsometry, and to single out the protein-induced effect on the cell activity by reducing interfering surface properties (e.g., topography, variations in surface chemistry). Furthermore, we sought to distinguish between integrin-mediated cell adhe-

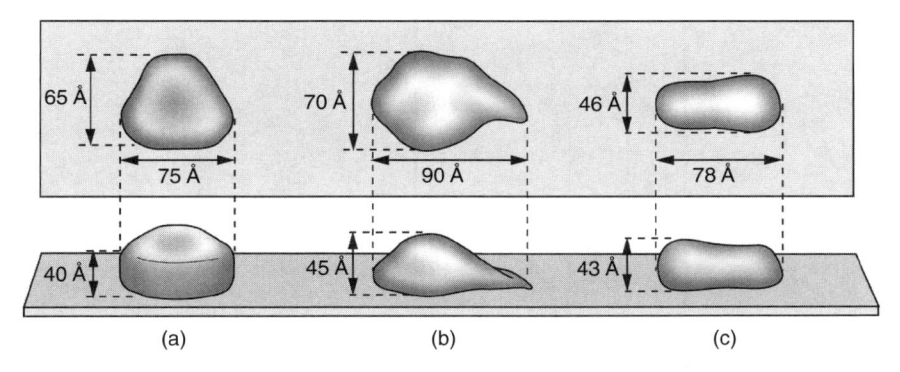

21.8 Schematic view, showing the orientations of (a) ceruloplasmin, (b) prothrombin, and (c) α1-antichymotrypsin.

sion and non-specific (i.e., divalent cation-independent) cell adhesion by regulating the integrin's ligand binding ability with divalent cations, Mg^{2+} and Ca^{2+}. The cell results were compared with equivalent experiments on a plasma-treated polyolefin surface, Thermanox®, which is frequently used as a positive cell adhesion reference.

The dimensions of the adsorbed proteins on silicon wafers were measured with the AFM and the values were compared with available X-ray data (Table 21.1). Most of the AFM values differed from the crystal data. This was expected, however, since AFM imaging always suffers from probe broadening and sometimes from probe compression (i.e., compression of the molecule by the probe) [57]. The former increases all lateral dimensions and the latter decreases the height. Nevertheless, by looking at the overall shapes of the molecules and comparing the AFM dimensions with the X-ray data, it was possible to speculate about the orientation of the adsorbed molecules. Accordingly, the molecules appeared to be oriented with their long axis parallel to the surface or, as in the case of ceruloplasmin, with one of its larger sides towards the surface (Fig. 21.8).

We could also see that the molecules had globular shapes but the resolution was insufficient to reveal any further conformational details. We could, however, conclude that prothrombin (but none of the others) formed multilayers at high protein concentrations and that the ceruloplasmin sample contained two different size populations. The latter gave us the idea that the AFM could be used to study the stability of protein solutions over time and consequently we found that ceruloplasmin gradually rearranges itself, probably due to loss of copper [72].

Another important issue when studying cell–protein interactions on surfaces is to minimize the influence of the underlying material. We therefore used ellipsometry to find out how well the surfaces could be covered with the individual proteins. The ellipsometric results (Fig. 21.9) together with

Table 21.1 Comparison of protein dimensions measured by AFM and X-ray crystallography

Protein	Fraction	AFM			X-ray		
		Height (Å)	Length[1] (Å)	Width[2] (Å)	Height (Å)	Length[1] (Å)	Width[2] (Å)
Ceruloplasmin[3]	100%	30 ± 9	165 ± 36	104 ± 28	~40/64[4]	75[3]	~65[3]
	67%	36 ± 3	178 ± 24	113 ± 21			
Prothrombin[5]	100%	19 ± 5	134 ± 29	86 ± 21	~45	~90	~70
α_2-HS-Glycoprotein	100%	8 ± 2	104 ± 31	63 ± 21	—[6]	—[6]	—[6]
α_1-Antichymotrypsin	100%	17 ± 4	117 ± 21	67 ± 21	~43[7]	~78[7]	~46[7]

[1] Length is the longest axis in the object.
[2] Width is the longest perpendicular bisector to the length.
[3] Ceruloplasmin is an equilateral triangle with side ~75 Å and triangular height ~65 Å.
[4] Without and with two extending loops.
[5] Prothrombin is a pseudoplanar 'drop-shaped' molecule.
[6] The crystal structure is not known.
[7] Dimensions obtained from native uncleaved structural homologue α_1-antitrypsin [70], cleaved α_1-antichymotrypsin [71] and α_1-antichymotrypsin with partial loop insertion [21].

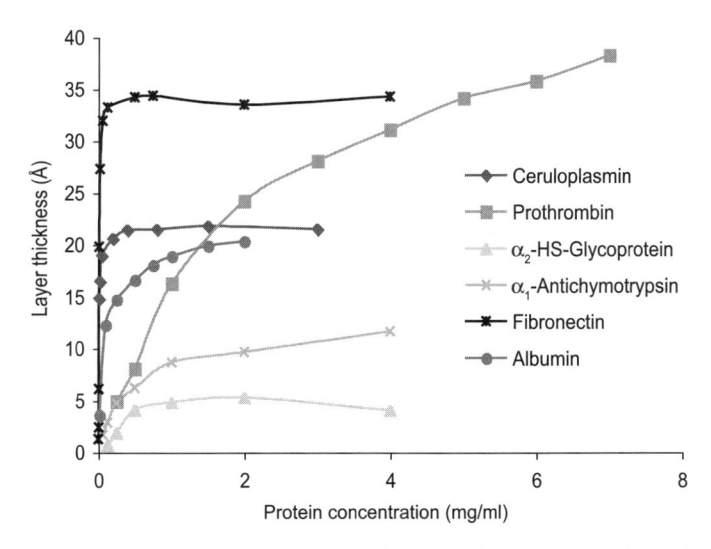

21.9 Deposition of proteins on silicon wafers as a function of applied protein concentration.

the AFM data showed that none of the protein coatings (except prothrombin) reached complete coverage. The prothrombin coating, however, consisted of at least two levels of protein molecules. In conclusion, the cells could be influenced by the material in nearly all cases but it is unclear to what degree.

An osteosarcoma cell line (MG63) was used to study the functional behavior (referring to cell binding and cell spreading capacity) of the proteins after adsorption. The cells were treated with cycloheximide and washed carefully to keep the system free from other proteins and cell-related contaminants. When applying the MG63 cells to the pre-coated silicon wafers and a pre-coated reference surface (Thermanox®), a clear difference in cell attachment (Fig. 21.10) and cell spreading (Table 21.2) could be seen as assessed by fluorescence microscopy. Ceruloplasmin, α_2-HS-glycoprotein and α_1-antichymotrypsin stimulated cell attachment to silica, but suppressed attachment to Thermanox®. The reason for this could be that the proteins adopted different conformations or orientations on the two materials and/or, as in the ceruloplasmin case, different protein moieties become adsorbed by the materials, and/or that the materials exposed different amounts of naked surface. In the Thermanox® case it is clear that most of the surface is covered with proteins, since very few cells attached compared to the naked surface. However, in the silica case we know from ellipsometry that some of the underlying surface might be available for the cells, so the difference in colonization could be explained by differences in the amount of

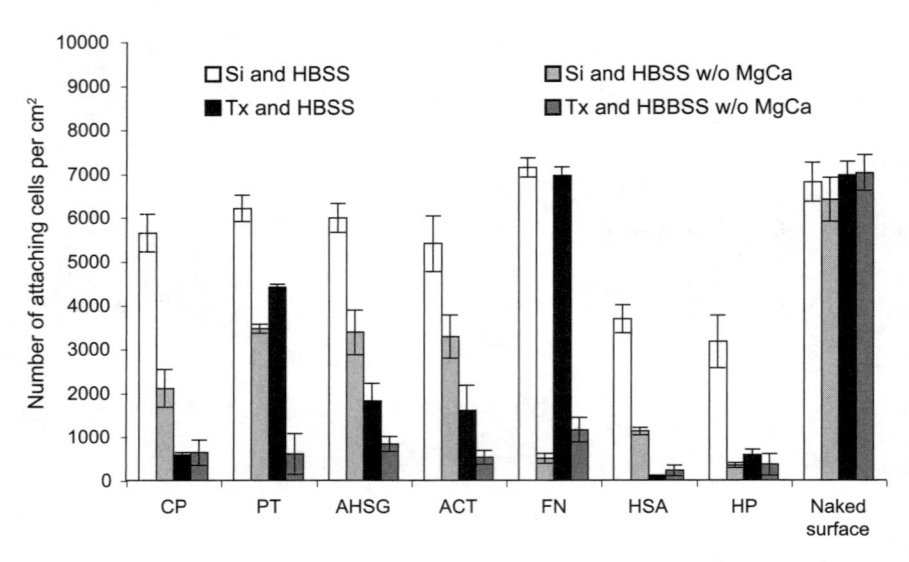

21.10 Comparison of number of attaching cells per cm^2 after 1.5 h on naked and protein coated silicon wafers and Thermanox® as a function of salt addition. The following proteins were used as coatings: ceruloplasmin (CP), prothrombin (PT), α_2-HS-glycoprotein (AHSG), α1-antichymotrypsin (ACT), fibronectin (FN), albumin (HSA) and human plasma (HP).

Table 21.2 Percentage of spread cells per cm^2 as a function of protein-, surface- and salt type

Surface	CP*	PT	AHSG	ACT	FN	HSA	HP	Naked surface
Si and HBSS	44	57	28	15	95	25	90	25
Si and HBSS without Mg and Ca	1–2	<1	<1	<1	1–2	<1	<1	1
Tmx and HBSS	75	53	8	10	95	<1	65	50
Tms and HBSS without Mg and Ca	<1	<1	<1	<1	1–2	<1	<1	5

*See Fig. 21.10 for meanings of abbreviations.

exposed surface. Nevertheless, we could conclude from the spreading behavior and the Mg^{2+}/Ca^{2+} dependency that at least some of the proteins must have a different conformation or orientation on silica than on Thermanox®, since 15–44% of the cells adhered through integrin-mediated interactions to the pre-coated silica while the integrin-mediated interactions and the cell spreading were negligible on pre-coated Thermanox®.

Prothrombin, on the other hand, stimulated cell attachment to both surfaces. The cells slightly preferred pre-coated silica over pre-coated Thermanox®. The adhesion to silica and Thermanox® was mediated both by integrins and by divalent cation-independent adhesion sites. In addition, the cells anchored to Thermanox® through an unidentified divalent cation-dependent interaction. We believe that these different adhesion mechanisms are due to different protein conformations and orientations, since the materials should be fully covered by proteins (as assessed by AFM, ellipsometry and Mg^{2+}/Ca^{2+} dependency) and are thereby unable to experience any direct cell–material interactions. Further studies are needed, however, to confirm this.

These studies demonstrated how important the adsorption of proteins at material interfaces is for cell adhesion. By manipulating the conformation and orientation of individual proteins as well as the degree of protein coverage, very different cell responses can be induced. Unfortunately, the studies also showed that the techniques today are insufficient to fully characterize a layer of adsorbed proteins. Even though we succeeded in monitoring the overall orientation of single proteins, we failed to see any conformational details. In addition it is highly speculative if the orientation and conformation of single proteins are the same for protein in layers. Nevertheless, we believe that with the development of new as well as old imaging techniques we will soon be able to relate protein configurations to their functional behavior.

In summary, the results obtained show how easily the composition of an adsorbed protein layer and the structure and functional behavior of individual proteins are manipulated by the properties of a material surface. By controlling these interactions it is possible to induce different cellular responses, a fact that might be put to use in the development of new biomaterials.

21.10 References

[1] Rosengren Å, Oscarsson S, Mazzocchi M, Krajewski A, Ravaglioli A. Protein adsorption onto two bioactive glass-ceramics. *Biomaterials* 2003; 1: 147–155
[2] Rosengren Å, Pavlovic E, Oscarsson S, Krajewski A, Ravaglioli A, Piancastelli A. Plasma protein adsorption pattern on characterized ceramic biomaterials. *Biomaterials* 2002; 4: 1237–1247
[3] Bianchini A, Musci G, Calabrese L. Inhibition of endothelial nitric-oxide synthase by ceruloplasmin. *J Biol Chem* 1999; 29: 20265–20270
[4] Lindley P, Card G, Zaitseva I, Zaitsev V. Ceruloplasmin: the beginning of the end of an enigma. *Perspect Bioinorg Chem* 1999; 51–89
[5] Saenko EL, Yaropolov AI, Harris ED. Biological functions of ceruloplasmin expressed through copper-binding sites and a cellular receptor. *J Trace Elem Exp Med* 1994; 2: 69–88

[6] Broadley C, Hoover RL. Ceruloplasmin reduces the adhesion and scavenges superoxide during the interaction of activated polymorphonuclear leukocytes with endothelial cells. *Am J Pathol* 1989; 4: 647–655

[7] Curtis ASG, Forrester JV. The competitive effects of serum proteins on cell adhesion. *J Cell Sci* 1984; 17–35

[8] Stubbs MT, Bode W. A player of many parts: the spotlight falls on thrombin's structure. *Thromb Res* 1993; 1: 1–58

[9] Ratner BD, Hoffman AS, Schoen FJ, Lemons JE, editors. *Biomaterials Science: an Introduction to Materials in Medicine*, 1st edn. San Diego, CA: Academic Press, 1996

[10] Furie B, Bing DH, Feldmann RJ, Robison DJ, Burnier JP, Furie BC. Computer-generated models of blood coagulation factor Xa, factor IXa, and thrombin based upon structural homology with other serine proteases. *J Biol Chem* 1982; 7: 3875–3882

[11] Byzova TV, Plow EF. Activation of $\alpha_v\beta_3$ on vascular cells controls recognition of prothrombin. *J Cell Biol* 1998; 7: 2081–2092

[12] Byzova TV, Plow EF. Networking in the hemostatic system. Integrin $\alpha_{IIb}\beta_3$ binds prothrombin and influences its activation. *J Biol Chem* 1997; 43: 27183–27188

[13] Chuang HYK, Mitra SP. Adsorption and reactivity of human prothrombin to artificial surfaces. *J Biomed Mater Res* 1984; 6: 695–705

[14] Travis J, Garner D, Bowen J. Human α_1-antichymotrypsin: purification and properties. *Biochemistry* 1978; 26: 5647–5651

[15] Berninger RW. α_1-Antichymotrypsin. *J Med* 1985; 16: 101–127

[16] Schwick HG, Haupt H. Purified human plasma proteins of unknown function. *Jpn J Med Sci Biol* 1981; 5: 299–327

[17] Beatty K, Bieth J, Travis J. Kinetics of association of serine proteinases with native and oxidized α_1-proteinase inhibitor and α_1-antichymotrypsin. *J Biol Chem* 1980; 9: 3931–3934

[18] Schechter NM, Jordan LM, James AM, Cooperman BS, Wang ZM, Rubin H. Reaction of human chymase with reactive site variants of α_1-antichymotrypsin. Modulation of inhibitor versus substrate properties. *J Biol Chem* 1993; 31: 23626–23633

[19] Faber JP, Poller W, Olek K, Baumann U, Carlson J, Lindmark B, Eriksson S. The molecular basis of α_1-antichymotrypsin deficiency in a heterozygote with liver and lung disease. *J Hepatol* 1993; 3: 313–321

[20] Poller W, Faber JP, Weidinger S, Tief K, Scholz S, Fischer M, Olek K, Kirchgesser M, Heidtmann HH. A leucine-to-proline substitution causes a defective α_1-antichymotrypsin allele associated with familial obstructive lung disease. *Genomics* 1993; 3: 740–743

[21] Gooptu B, Hazes B, Chang W-SW, Dafforn TR, Carrell RW, Read RJ, Lomas DA. Inactive conformation of the serpin α_1-antichymotrypsin indicates two-stage insertion of the reactive loop: implications for inhibitory function and conformational disease. *Proc Natl Acad Sci* 2000; 1: 67–72

[22] Kilpatrick L, Johnson JL, Nickbarg EB, Wang ZM, Clifford TF, Banach M, Cooperman BS, Douglas SD, Rubin H. Inhibition of human neutrophil superoxide generation by α_1-antichymotrypsin. *J Immunol* 1991; 7: 2388–2393

[23] Lomas DA, Stone SR, Llewellyn-Jones C, Keogan M-T, Wang Z-M, Rubin H, Carrell RW, Stockley RA. The control of neutrophil chemotaxis by inhibitors of cathepsin G and chymotrypsin. *J Biol Chem* 1995; 40: 23437–23443

[24] Abraham CR. Reactive astrocytes and α_1-antichymotrypsin in Alzheimer's disease. *Neurobiol Aging* 2001; 6: 931–936
[25] Brown WM, Dziegielewska KM, Saunders NR, Christie DL, Nawratil P, Mueller-Esterl W. The nucleotide and deduced amino acid structures of sheep and pig fetuin. Common structural features of the mammalian fetuin family. *Eur J Biochem* 1992; 1: 321–331
[26] http://www.expasy.ch/. PROSITE database
[27] Elzanowski A, Barker WC, Hunt LT, Seibel-Ross E. Cystatin domains in α_2-HS-glycoprotein and fetuin. *FEBS Lett* 1988; 2: 167–170
[28] Schinke T, Amendt C, Trindl A, Poeschke O, Mueller-Esterl W, Jahnen-Dechent W. The serum protein α_2-HS glycoprotein/fetuin inhibits apatite formation in vitro and in mineralizing calvaria cells. A possible role in mineralization and calcium homeostasis. *J Biol Chem* 1996; 34: 20789–20796
[29] Triffitt JT, Gebauer U, Ashton BA, Owen ME, Reynolds JJ. Origin of plasma α_2-HS-glycoprotein and its accumulation in bone. *Nature* 1976; 5565: 226–227
[30] Jahnen-Dechent W, Schafer C, Heiss A, Grotzinger J. Systemic inhibition of spontaneous calcification by the serum protein α_2-HS-glycoprotein/fetuin. *Z Kardiol* 2001; Suppl. 3: iii/47–iii/56
[31] Ketteler M, Bongartz P, Westenfeld R, Wildberger JE, Horst Mahnken AH, Bohm R, Metzger T, Wanner C, Jahnen-Dechent W, Floege J. Association of low fetuin-A (AHSG) concentrations in serum with cardiovascular mortality in patients on dialysis: a cross-sectional study. *Lancet* 2003; 9360: 827–833
[32] Dziegielewska KM, Brown WM. *Fetuin*. New York: Springer-Verlag, 1995
[33] Jersmann HPA, Dransfield I, Hart SP. Fetuin/α_2-HS-glycoprotein enhances phagocytosis of apoptotic cells and macropinocytosis by human macrophages. *Clin Sci* 2003; 3: 273–278
[34] Kalabay L, Cseh K, Pajor A, Baranyi E, Csakany GM, Melczer Z, Speer G, Kovacs M, Siller G, Karadi I, Winkler G. Correlation of maternal serum fetuin/α_2-HS-glycoprotein concentration with maternal insulin resistance and anthropometric parameters of neonates in normal pregnancy and gestational diabetes. *Eur J Endocrinol* 2002; 2: 243–248
[35] Demetriou M, Binkert C, Sukhu B, Tenenbaum HC, Dennis JW. Fetuin/α_2-HS-glycoprotein is a transforming growth factor-b type II receptor mimic and cytokine antagonist. *J Biol Chem* 1996; 22: 12755–12761
[36] Nie Z. Fetuin: its enigmatic property of growth promotion. *Am J Physiol* 1992; 3, Pt 1: C551–C562
[37] Hynes RO. *Fibronectins*. New York: Springer-Verlag, 1990
[38] Horbett TA. Protein adsorption on biomaterials. In: Cooper SL, Peppas NA. *Biomaterials: Interfacial Phenomena and Applications*. Washington, DC: ACS, 1982, 233–244
[39] Janson J-C, Ryden L, editors. *Protein Purification: Principles, High Resolution Methods and Applications*. New York: Wiley-VCH, 1998
[40] Allen T. *Surface Area and Pore Size Determination*. London: Chapman & Hall, 1997
[41] Smith PK, Krohn RI, Hermanson GT, Mallia AK, Gartner FH, Provenzano MD, Fujimoto EK, Goeke NM, Olson BJ, Klenk DC. Measurement of protein using bicinchoninic acid. *Anal Biochem* 1985; 1: 76–85
[42] O'Farrell PH. High resolution two-dimensional electrophoresis of proteins. *J Biol Chem* 1975; 10: 4007–4021

[43] Klose J. Protein mapping by combined isoelectric focusing and electrophoresis of mouse tissues. A novel approach to testing for induced point mutations in mammals. *Humangenetik* 1975; 3: 231–243
[44] Berkelman T, Stenstedt T. *2-D Electrophoresis: Principles and Methods*. San Francisco: Amersham Biosciences, Edition AB
[45] Issaq HJ, Conrads TP, Janini GM, Veenstra TD. Methods for fractionation, separation and profiling of proteins and peptides. *Electrophoresis* 2002; 17: 3048–3061
[46] Otto JJ. Immunoblotting. *Methods Cell Biol* 1993; 37: 105–117
[47] Mouche L, Tardif F, Derrien J. Particle deposition on silicon wafers during the wet cleaning processes. *J Electrochem Soc* 1994; 6: 1684–1691
[48] Pintchovski F, Price JB, Tobin PJ, Peavey J, Kobold K. Thermal characteristics of the sulfuric acid–hydrogen peroxide silicon wafer cleaning solution. *J Electrochem Soc* 1979; 8: 1428–1430
[49] Tamminen A, Anttila O. Cleaning processes for silicon wafers. *VTT Symp 1999*; 30th R3-Nordic Contamination Control Symposium: 431–438
[50] Amick JA. Cleanliness and the cleaning of silicon wafers. *Solid State Technology* 1976; 11: 47–52
[51] Binnig G, Quate C, Gerber C. Atomic force microscope. *Phys Rev Lett* 1986; 9: 930–933
[52] Morris VJ, Kirby AR, Gunning AP. *Atomic Force Microscopy for Biologists*. London: Imperial College Press, 1999
[53] Burnham N, Behrend O, Oulevey F, Gremaud G, Gallo P-J, Gourdon D, Dupas E, Kulik A, Pollock H, Briggs G. How does a tip tap? *Nanotechnology* 1997; 8: 67–75
[54] Nagase M, Namatsu H, Kurihara K, Makino T. Critical dimension measurement in nanometer scale by using scanning probe microscopy. *Jpn J Appl Phys, Part 1: Regular Papers, Short Notes & Review Papers* 1996; 7: 4166–4174
[55] Ramirez-Aguilar KA, Rowlen KL. Tip characterization from AFM images of nanometric spherical particles. *Langmuir* 1998; 9: 2562–2566
[56] Markiewicz P, Goh MC. Atomic force microscopy probe tip visualization and improvement of images using a simple deconvolution procedure. *Langmuir* 1994; 1: 5–7
[57] Bergkvist M, Carlsson J, Karlsson T, Oscarsson S. TM-AFM threshold analysis of macromolecular orientation: a study of the orientation of IgG and IgE on mica surfaces. *J Colloid Interface Sci* 1998; 2: 475–481
[58] Bergkvist M. Methods for studying orientation of surface adsorbed proteins. In: Hubbard AT, ed., *Encyclopedia of Surface and Colloid Science*. New York: Marcel Dekker, 2002, 3328–3341
[59] Haggerty L, Lenhoff AM. STM and AFM in biotechnology. *Biotechnol Prog* 1993; 1: 1–11
[60] Keller D. Scanning force microscopy in biology. In: Baszkin A, Norde W. *Physical Chemistry of Biological Interfaces*. New York: Marcel Dekker, 2000.
[61] Muller DJ, Fotiadis D, Scheuring S, Muller SA, Engel A. Electrostatically balanced subnanometer imaging of biological specimens by atomic force microscope. *Biophys J* 1999; 2: 1101–1111
[62] Rothen A. Measurements of the thickness of thin films by optical means, from Rayleigh and Drude to Langmuir, and the development of the present ellipsometer. In: Passaglia E, Stromberg RR, Kruger J, editors. *Symposium*

Proceedings on Ellipsometry in the Measurements of Surfaces and Thin Films, Volume 256. National Bureau of Standards, 1964, 7–21
[63] Arwin H. Ellipsometry. *Phys Chem Biol Interface* 2000; 577–607
[64] Elwing H. Protein adsorption and ellipsometry in biomaterial research. *Biomaterials* 1998; 4–5: 397–406
[65] Ivarsson B, Lundstroem I. Physical characterization of protein adsorption on metal and metal oxide surfaces. *Crit Rev Biocomp* 1986; 1: 1–96
[66] Langmuir I, Schaefer VJ. Optical measurements of the thickness of a film adsorbed from a solution. *J Am Chem Soc* 1937; 59: 1406
[67] Jost JL, Kominek LA, Hyatt GS, Wang HY. Cycloheximide: properties, biosynthesis, and fermentation. In Vandamme EJ, ed., *Biotechnology of Industrial Antibiotics.* New York: Marcel Dekker, 1984, 531–550
[68] Obrig TG, Culp WJ, McKeehan WL, Hardesty B. The mechanism by which cycloheximide and related glutarimide antibiotics inhibit peptide synthesis on reticulocyte ribosomes. *J Biol Chem* 1971; 1: 174–181
[69] Rost FWD. *Fluorescence Microscopy*. Cambridge University Press, 1995
[70] Kim S-J, Woo J-R, Seo EJ, Yu M-H, Ryu S-E. A 2.1. Å resolution structure of an uncleaved α_1-antitrypsin shows variability of the reactive center and other loops. *J Mol Biol* 2001; 1: 109–119
[71] Baumann U, Huber R, Bode W, Grosse D, Lesjak M, Laurell CB. Crystal structure of cleaved human α_1-antichymotrypsin at 2.7 Å resolution and its comparison with other serpins. *J Mol Biol* 1991; 3: 595–606
[72] Rydén L. Ceruloplasmin. In: Lontie R. *Copper Proteins and Copper Enzymes*. Boca Raton, FL: CRC Press, 1984, 37–100

22
Degradation of calcium phosphate coatings and bone substitutes

S OVERGAARD, Odense University Hospital,
Clinical Institute and University of Southern Denmark,
Denmark

Abstract: Calcium phosphate (Ca-P) bone substitutes and coatings on implants have been used clinically for many years. Hydroxyapatite (HA) coatings are well documented, though with little information on the long-term survival of the prosthetic components. There is a major need for bone transplantation today and Ca-P bone substitutes might be an alternative to bone allograft or autograft, either alone or in combination with osteogenic or inductive factors. There is a variety of requirements for Ca-P substitutes and coatings, and there are many factors that influence the degradation of Ca-P. There is a need for substitutes with various properties, sizes and shapes which can be applied with different methods. It is concluded that clinical use of HA-coated implants might be debated. There is a need for more controlled clinical trials for Ca-P bone substitutes, in order to validate each treatment application before a breakthrough in daily clinical use can take place.

Key words: calcium phosphate bone substitute, hydroxyapatite coating, calcium phosphate degradation, clinical.

22.1 Introduction and demand for coatings and bone substitutes

The present chapter will focus on degradation of calcium phosphate (Ca-P) bone substitutes and coatings and their clinical perspectives.

22.1.1 Coatings

The clinical use of hydroxyapatite (HA) coatings was motivated by the capability to enhance the early bone ingrowth to cementless prosthetic components.[1] The main purpose is to increase the long-term survival of joint replacements by increased implant fixation and reduced particle migration into the bone implant interface.[2,3]

22.1.2 Bone substitutes

Bone is among the most frequently transplanted tissue. Allograft and autograft account for the majority of bone graft procedures in revision surgery in joint replacements, spine fusion and trauma surgery. It has been estimated that 500,000 bone graft procedures are done yearly in Europe and it is expected that the number will increase with greater numbers of elderly people and longer lifetimes.

The use of autografts and allografts always requires a secondary surgical procedure, adding high costs to health services and increasing patient morbidity, and their availability may be a problem due to the limited quantities available. There is therefore universal interest in programmes to rebuild and restore the functions of degenerative tissue using artificial implant materials. Thus, the need for bone substitutes is obvious, and this has resulted in numerous bone substitutes being marketed during the last decade.

Although autografts and allografts are well validated clinically, it would be a major step forward in trauma and reconstructive surgery if they could be replaced by synthetically produced bone substitutes (Table 22.1). This is motivated by the risk of transmission of diseases with the use of allografts and by severe chronic donor site pain after harvesting of autografts. In addition, new EU regulations have resulted in more restrictive rules and a great increase in administrative work in order to improve the quality of bone banks and to reduce the risk for recipients. As a consequence, the costs of allografts have increased dramatically and fewer donors are found to be suitable. Thus, a number of factors favour the use of bone substitutes.

Table 22.1 Characteristics, advantages and disadvantages of Ca-P bone substitute, allograft and autograft

Type	Characteristics	Advantages	Disadvantages
Ca-P substitute	Osteoconductive	Porous structure, mimics trabecular bone Biocompatible Biodegradable	Brittle Poor mechanical properties Only serves as scaffold Poor clinical validation
Allograft	Osteoconductive Osteoinductive	Clinically validated No donor site pain 'Larger' amounts available	Possible immunological reactions Risk of transmission of disease, e.g. HIV Administrative costs
Autograft	Osteoconductive Osteoinductive Osteogenic	Clinically validated No immunological reactions No viral/bacterial recipient contamination	Donor site pain in 30–40% of cases Risk of infection Increased operation time, blood loss and complications

22.2 Requirements for coatings and bone substitutes

22.2.1 Coating

An ideal coating is bioactive in order to enhance bone ingrowth to the implant surface through osteoconductive properties. In addition, the mechanical properties of the coating ensure no cracks or delamination. Following bone ingrowth, the coating should be degraded along with bone formation, resulting in bone having direct contact with the implant surface.

22.2.2 Bone substitute

An ideal substitute is bioactive, and should provide an initial scaffold and support for bone formation followed by controlled degradation of the substitute in order to be replaced by newly formed bone. Thus, osteoconductive properties are mandatory, and these are created by the chemical composition and the porosity of the material. Porosity, pore size and interconnection of the pores are key factors and should be in the range 100–500 μm in order to get vascularisation followed by bone ingrowth. It might be questioned whether the osteoconductive properties alone can be sufficient for bone regeneration or that osteoinductive and/or osteogenic properties are needed together with the substitute. This property could be rendered by addition of growth factors and the use of bone marrow or stem cells.[4,5]

22.3 Mechanical properties of coatings and bone substitutes

22.3.1 Coating

The bonding strength between coating and metal substrate is very important for weight-bearing implants.[6] It has been demonstrated that HA reacts with titanium oxide at elevated temperatures (800–1000°C) creating a chemical bonding, whereas the reaction with Co-Cr alloy is less significant.[7] This might explain the fact that HA coating on titanium demonstrated higher bonding strength and better fatigue properties than on cobalt-chrome.[8] Tensile and shear strength of the coating are reduced significantly by increasing thickness.[9] Moreover, the underlying substrate surface roughness seems to be important.[10] It is assumed that the bonding strength is higher on a porous surface: this is supported by the observation that the tensile strength of an HA coating is lower on a polished implant surface than on a grit-blasted surface.

Key parameters for the mechanical properties and fixation of the coating are coating thickness, porosity, surface texture and the design of the pros-

Table 22.2 Typical clinical situations requiring bone transplantations

Speciality	Clinical procedures with bone transplantation	Warranted mechanical support
Trauma	Acute fractures with bone loss Delayed union Pseud-arthrosis	May be a requirement
Reconstructive orthopaedic surgery	Revision of joint replacements with bone loss	A necessary requirement
Spine fusion	Intercorporal and postero-lateral fusion	In most cases of minor importance
Tumours and infections	Sequelae from bone loss and resection	May be required
Maxillofacial surgery	Ridge augmentation procedures	Minor mechanical support is required

thetic component as mentioned above. Factors which influence coating resorption may also affect bonding strength, i.e. rapid resorption results in decreased bonding strength.[11]

22.3.2 Bone substitute

Mechanical properties of the substitute are important. However, the need for strength depends on the clinical situation and some transplanted defects and surgical application sites are not subjected to weight-loading, which reduces the demand for a strong structure (Table 22.2).

However, handling of the product should not result in damage to the material. Mechanically, porous structures have generally inferior properties than trabecular bone. Efforts have been put into optimising the porous structures and it might be possible to create a synthetic Ca-P porous structure with mechanical properties comparable to trabecular bone (Fig. 22.1). In a recent unpublished study, we have studied the effect of infiltrating the Ca-P porous structure with Poly-L-Acid (PLLA). *Ex vivo* analyses have shown that the failure energy of a PLA infiltrated Ca-P was equal to that of middle-aged human bone of the proximal tibia.

22.4 Degradation of coatings and bone substitutes

The question of when and how a degradable material degrades is a very important consideration in developing and selecting the right material for any specific clinical situation. Some degradation is needed in order to have a bioactive material; however, the degradation rate has to be controlled. A

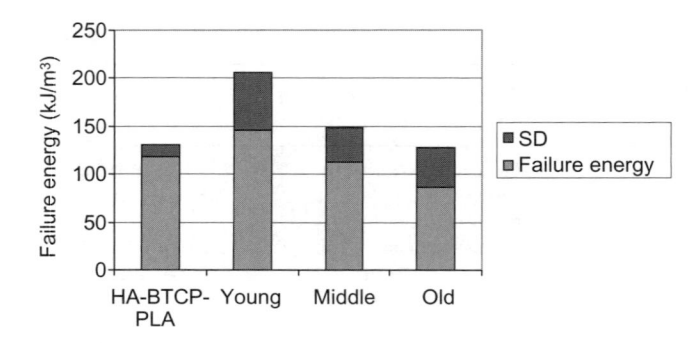

22.1 Compression test of a PLA-reinforced Ca-P bone substitute (unpublished results) compared to young, middle-aged and old human bone.[12] The bioactivity of the PLA Ca-P material has not been investigated, to date.

Table 22.3 Material properties that influence degradation *in vivo*

Property	Characteristics
Chemical composition	The lower the Ca/P ratio, the more rapid degradation of the material
Porosity and microstructure	The larger the porosity the faster degradation
Crystallinity and purity	Low crystalline Ca-P materials have greater solubility due to dissolution

balance must be achieved depending on the bone healing and degradation rate. Thus, the rate of bone healing has to keep up with the degradation rate of the material, otherwise a new defect will be created. Fast degradation is expected to take place when the local metabolic activity is high, depending on the material (Table 22.3).

22.4.1 Degradation mechanisms

There are several mechanisms involved in the degradation of bone substitutes *in vivo*: simple dissolution and cell-mediated resorption. In addition, the material is a key factor for degradation rate as listed in Table 22.3.

Simple dissolution

In the body environment, a Ca-P will dissolve under restricted conditions due to dissolution kinetics determined by ionic and solubility product.[13] Local undersaturation of the fluid with calcium and phosphate ions as determined by the thermodynamic solubility product leads to dissolution. Dissolution of Ca-P scaffolds is probably an important process in triggering bone formation. The process has been compared to that of dissolution of endogenous bone mineral.[14–17] In addition, dissolution is enhanced by increased impurity, low crystallinity and high porosity.[18–20]

Cell-mediated resorption

Cell-mediated bone resorption is caused by several cell types. However, the osteoclast is the only cell specialised for that function. Resorption lacunae are created at the ruffle border due to low pH.[18] The role of early postoperative osteoclastic resorption is most likely minor due to few osteoclasts at that stage; however, later, when bone ingrowth has occurred, osteoclastic resorption might take place. HA particles present in osteocytes, osteoclast-like cells, monocytes and fibroblasts have been demonstrated.[21–23] The significance of HA particles in osteocytes is not clear, but the particles might have been incorporated in the pre-osteoblast or osteoblast stage.[22]

22.4.2 Factors with influence on degradation

Material-related factors

Several material-related factors influence degradation as listed in Table 22.3. The Ca/P ratio and crystallinity together with porosity are suggested to be the main factors.[24]

Biological and mechanical factors

The significance of micromovements on degradation of coatings is well documented; however, it has not been evaluated in bone substitutes. It has been shown that continuous loading and micromotion of the implant accelerate resorption. In addition, coating degradation is enhanced on unstable fibrous anchored as compared with bony anchored implants. The difference between stable and unstable implants is most likely explained by the biological reactions. During micromotion a fibrous tissue membrane with high metabolic activities is developed. The fibrous tissue membrane is dominated by fibroblasts and macrophages able to phagocytose HA. Because of unstable conditions, low pH is maintained due to inhibited angiogenesis. In addition, fluid flow is increased along the interface, leading to accelerated dissolution due to changes in calcium and phosphate concentrations.[25,26] In contrast, the stable implant will result in rapid bone ingrowth without formation of fibrous tissue.

22.4.3 Experimental and clinical methods for estimation of degradation

Experimentally, evaluation of degradation of Ca-P materials, coating or bone substitutes is simply done by histology and histomorphometric methods, by either light microscopy or scanning electron microscopy.[23,27,28] Radioactively marked calcium has also been used to estimate loss of Ca-P coatings during implantation but without success.

Clinically, the difference in radiodensity between the bone substitute and the host bone might be easy to show on standard X-rays. However, it is not possible to quantify degradation of the bone substitute. Moreover, it is not possible to evaluate whether the substitute has been replaced by bone or just has been degraded on simple X-rays. In contrast, CT-scanning has been used to quantify degradation; however, in the early phase of degradation the reliability of the method is insufficient due to poor discrimination between new bone and substitute.[29] Thus the clinical methods for estimating degradation should be optimised. New CT-scanners might possibly give some improvement. Finally, magnetic resonance (MR) imaging has been used to study the degradation of PLA.[30]

22.5 Discussion

22.5.1 Risk factors in the clinical use of coatings and bone substitutes

Coatings

Several concerns regarding Ca-P coatings have been raised. First of all the risk of coating delamination by failure of the coating–implant interface has been suggested to cause implant loosening. Moreover, the effects of coating resorption have been advanced as a cause of reduced implant fixation. The degradation patterns of Ca-P coatings have been studied extensively in experimental settings over the years, but the longer-term fate clinically is insufficiently validated. It has been estimated that HA coatings are resorbed by approximately 20% yearly in humans.[28] With regard to the clinical use, HA-coated prostheses are well documented. Thus HA coating is capable of reducing the early migration of both femoral hip and tibial knee components as compared with uncoated implants when detected by radiostereometric analysis.[31,32] Moreover, several series have been published and 20 years of results are now available in the Norwegian Joint Arthroplasty Registry, showing very good survival of HA-coated femoral stems.[33] The results from this register are based mainly on the Corail stem.

However, the main object of reducing the revision rate of HA-coated prosthetic components has not yet been achieved. Results from the Danish

Hip Arthroplasty Registry showed that comparing identical brands of prosthetic components with and without HA coating had neither any overall additional nor a reduced effect of HA on implant survival.[34] A number of components have in retrieval studies showed failure of HA-coated prosthetic components exclusively on prosthetic components with a grit-blasted surface but not on a porous-coated surface.[35,36]

Bone substitutes

The majority of the synthetic materials have been validated both *in vitro* and *in vivo* using experimental animals for proof of concept. Moreover, numerous materials have been used clinically. However, few have been evaluated in proper design studies. For the end-users, i.e. the surgeons, there is an increasing request and demand for controlled clinical studies in order to have relevant clinical documentation of the substitute in question. A few studies have evaluated different materials as bone expanders together with autograft in spine surgery.[37] The concerns about bone substitutes involve adverse reactions and whether they will be inefficient.[38] Moreover, in weight-bearing situations mechanical failure is a concern. Finally, there is a concern that degradation of a specific material is too fast compared with bone formation. Bone formation has to keep up with the degraded material in order to avoid new defects.

22.6 Choice of bone substitute, and future trends

Owing to the fact that the clinical situations are very different, varying from small contained non-weight-bearing defects to large volume non-contained defects which might be weight-bearing, it seems logical that there is a need for substitutes with various properties, sizes and shapes which can be applied with different methods available (Table 22.2). Moreover, owing to the fact that Ca-P substitutes have only osteoconductive properties, they might be combined with materials which have osteoinductive properties like growth factors, and moreover the use of cells seems to be attractive as well. Therefore, bone tissue engineering might be the future choice for substituting bone in humans. However, growth factors such as BMP-2 and BMP-7 (Bone Morphogenetic Protein) are challenging, because the therapeutic window might be narrow. Too high a dose might result in accelerated degradation of the material, and even more serious, resorption of the host bone.[39] Regarding application of cells, the proof of concept has been performed in patients with a large defect in the proximal tibia.[40,41] The authors cultured autologous bone marrow stem cells *in vitro* and seeded them onto a porous HA scaffold. Complete fusion was achieved after 5–7 months, evaluated by X-rays and CT-scanning.

22.7 Conclusion

The clinical use of HA-coated prosthetic components is well validated and seems to be safe. However, the main purpose of HA coatings, to reduce the revision rate of the prosthetic components, has not been achieved. Thus, the use of HA-coated implants might be debated. It is of importance that HA coatings are applied on a rough or porous-coated surface, since failure of the HA coating has been shown on prosthetic components with a grit-blasted surface. The coatings should only be used on properly designed prosthetic components.

There is a major need for bone substitutes in the field of orthopaedic surgery, including spine fusion. Each specific application area such as spine fusion, fracture treatment and revision of joint replacement probably has to be approached differently with regard to the type of substitute, due to the very different clinical situations requiring various treatment strategies. Some of the clinical application sites might only require the osteoconductive substitute alone, whereas others might need addition of growth factors or cells in order to be successful.

There is a need for more controlled clinical trials in order to validate each treatment application before a breakthrough in daily clinical use can take place.

22.8 References

1. Geesink, R. G. T. Hydroxyapatite-coated total hip prostheses. Two-year clinical and roentgenographic results of 100 cases. *Clin. Orthop.* 261: 39–58, 1990.
2. Rahbek, O., Overgaard, S., Jensen, T. B., Bendix, K., and Søballe, K. Sealing effect of hydroxyapatite coating: a 12-month study in canines. *Acta Orthop. Scand.* 71(6): 563–573, 2000.
3. Rahbek, O., Overgaard, S., Lind, M., Bendix, K., Bünger, C., and Søballe, K. Sealing effect of hydroxyapatite coating on peri-implant migration of particles. An experimental study in dogs. *J. Bone Joint Surg. [Br.]* 83: 441–447, 2001.
4. Jensen, T. B., Overgaard, S., Lind, M., Rahbek, O., Bünger, C., and Søballe, K. Osteogenic protein 1 device increases bone formation and bone graft resorption around cementless implants. *Acta Orthop. Scand.* 73: 31–39, 2002.
5. Khan, Y., Yaszemski, M. J., Mikos, A. G., and Laurencin, C. T. Tissue engineering of bone: material and matrix considerations. *J. Bone Joint Surg. [Am.]* 90 Suppl 1: 36–42, 2008.
6. Jones, D. W. Coatings of ceramics on metals. In Ducheyne, P., and Lemons, J. E. (eds), *Annals of the New York Academy of Science: Bioceramics: Material Characteristics Versus in Vivo Behaviour*. New York: The New York Academy of Sciences, 1988, pp. 19–37.
7. Ducheyne, P., and Healy, K. E. The effect of plasma-sprayed calcium phosphate ceramic coatings on the metal ion release from porous titanium and cobalt-chromium alloys. *J. Biomed. Mater. Res.* 22: 1137–1163, 1988.

8. Kummer, F. J., and Jaffe, W. L. Stability of a cyclically loaded hydroxyapatite coating: effect of substrate material, surface preparation, and testing environment. *J. Appl. Biomater.* 3: 211–215, 1992.
9. Wang, B. C., Lee, T. M., Chang, E., and Yang, C. Y. The shear strength and the failure mode of plasma-sprayed hydroxyapatite coating to bone: the effect of coating thickness. *J. Biomed. Mater. Res.* 27: 1315–1327, 1993.
10. Filiaggi, M. J., Coombs, N. A., and Pilliar, R. M. Characterization of the interface in the plasma-sprayed HA coating/Ti-6Al-4V implant system. *J. Biomed. Mater. Res.* 25: 1211–1229, 1991.
11. Yang, C. Y., Lin, R. M., Wang, B. C., Lee, T. M., Chang, E., Hang, Y. S., and Chen, P. Q. In vitro and in vivo mechanical evaluations of plasma-sprayed hydroxyapatite coatings on titanium implants: the effect of coating characteristics. *J. Biomed. Mater. Res.* 37: 335–345, 1997.
12. Ding, M., Odgaard, A., Linde, F., and Hvid, I. Age-related variations in the microstructure of human tibial cancellous bone. *J. Orthop. Res.* 20: 615–621, 2002.
13. Christoffersen, J. The kinetics of dissolution of calcium hydroxyapatite. A contribution to understanding of biological demineralization. University of Copenhagen, Denmark, 1984, pp. 1–62.
14. Fallon, U. Alterations in the pH of osteoclast resorbing fluid reflects changes in bone degradative activity. *Calcif. Tissue Int.* 36: 458, 1984.
15. Klein, C. P. A. T., Wolke, J. G., de Blieck Hogervorst, J. M., and de Groot, K. Features of calcium phosphate plasma-sprayed coatings: an in vitro study. *J. Biomed. Mater. Res.* 28: 961–967, 1994.
16. Maxian, S. H., Zawadsky, J. P., and Dunn, M. G. In vitro evaluation of amorphous calcium phosphate and poorly crystallized hydroxyapatite coatings on titanium implants. *J. Biomed. Mater. Res.* 27: 111–117, 1993.
17. Vaes, G. Cellular biology and biomechanical mechanism of bone resorption. *Clin. Orthop.* 231: 239–271, 1988.
18. de Bruijn, J. D., Bovell, Y. P., Davies, J. E., and van Blitterswijk, C. A. Osteoclastic resorption of calcium phosphates is potentiated in postosteogenic culture conditions. *J. Biomed. Mater. Res.* 28: 105–112, 1994.
19. Klein, C. P., Driessen, A. A., de Groot, K., and van den Hooff, A. Biodegradation behavior of various calcium phosphate materials in bone tissue. *J. Biomed. Mater. Res.* 17: 769–784, 1983.
20. Maxian, S. H., Zawadsky, J. P., and Dunn, M. G. Effect of Ca/P coating resorption and surgical fit on the bone/implant interface. *J. Biomed. Mater. Res.* 28: 1311–1319, 1994.
21. Gomi, K., Lowenberg, B., Shapiro, G., and Davies, J. E. Resorption of sintered synthetic hydroxyapatite by osteoclasts in vitro. *Biomaterials* 14: 91–96, 1993.
22. Müller-Mai, C. M., Voigt, C., and Gross, U. Incorporation and degradation of hydroxyapatite implants of different surface roughness and surface structure in bone. *Scanning Microsc.* 4: 613–622, 1990.
23. Overgaard, S., Lind, M., Josephsen, K., Maunsbach, A. B., Bünger, C., and Søballe, K. Resorption of hydroxyapatite and fluorapatite ceramic coatings on weight-bearing implants: A quantitative and morphological study in dogs. *J. Biomed. Mater. Res.* 39: 141–152, 1998.
24. Overgaard, S. Calcium phosphate coatings for fixation of bone implants. *Acta Orthop. Scand.* 71 (Suppl. 297): 1–74, 2000.

25. Page, M., Hogg, J., and Ashhurst, D. E. The effects of mechanical stability on the macromolecules of the connective tissue matrices produced during fracture healing. I. The collagens. *Histochem. J.* 18: 251–265, 1986.
26. Prendergast, P. J., Huiskes, R., and Søballe, K. Systematic changes of biophysical stimuli occur in interfacial tissue during bone ingrowth at implant interfaces. *10th Conference of the ESB*, 151, 1996.
27. Overgaard, S., Søballe, K., Josephsen, K., Hansen, E. S., and Bünger, C. Role of different loading conditions on resorption of hydroxyapatite coating evaluated by histomorphometric and stereological methods. *J. Orthop. Res.* 14: 888–894, 1996.
28. Overgaard, S., Søballe, K., Lind, M., and Bünger, C. Resorption of hydroxyapatite and fluorapatite coatings in man. An experimental study in trabecular bone. *J. Bone Joint Surg. [Br.]* 79: 654–659, 1997.
29. Petruskevicius, J., Nielsen, S., Kaalund, S., Knudsen, P. R., and Overgaard, S. No effect of Osteoset, a bone graft substitute, on bone healing in humans: a prospective randomized double-blind study. *Acta Orthop. Scand.* 73: 575–578, 2002.
30. Pihlajamaki, H., Kinnunen, J., and Bostman, O. In vivo monitoring of the degradation process of bioresorbable polymeric implants using magnetic resonance imaging. *Biomaterials* 18: 1311–1315, 1997.
31. Karrholm, J., Malchau, H., Snorrason, F., and Herberts, P. Micromotion of femoral stems in total hip arthroplasty. A randomized study of cemented, hydroxyapatite-coated, and porous- coated stems with roentgen stereophotogrammetric analysis. *J. Bone Joint Surg. [Am.]* 76: 1692–1705, 1994.
32. Søballe, K., Toksvig-Larsen, S., Gelineck, J., Fruensgaard, S., Hansen, E. S., Ryd, L., Lucht, U., and Bünger, C. Migration of hydroxyapatite coated femoral prosthesis. A roentgen stereophotogrammetric study. *J. Bone Joint Surg. [Br.]* 75: 681–687, 1993.
33. Hallan, G., Lie, S. A., Furnes, O., Engesaeter, L. B., Vollset, S. E., and Havelin, L. I. Medium- and long-term performance of 11,516 uncemented primary femoral stems from the Norwegian arthroplasty register. *J. Bone Joint Surg. [Br.]* 89: 1574–1580, 2007.
34. Paulsen, A., Pedersen, A. B., Johnsen, S. P., Riis, A., Lucht, U., and Overgaard, S. Effect of hydroxyapatite coating on risk of revision after primary total hip arthroplasty in younger patients: findings from the Danish Hip Arthroplasty Registry. *Acta Orthop.* 78: 622–628, 2007.
35. Buma, P., and Gardeniers, J. W. Tissue reactions around a hydroxyapatite-coated hip prosthesis. Case report of a retrieved specimen. *J. Arthroplasty* 10: 389–395, 1995.
36. Nilsson, K. G., Cajander, S., and Kärrholm, J. Early failure of hydroxyapatite-coating in total knee arthroplasty. *Acta Orthop. Scand.* 65: 212–214, 1994.
37. Epstein, N. E. A preliminary study of the efficacy of beta tricalcium phosphate as a bone expander for instrumented posterolateral lumbar fusions. *J. Spinal Disord. Tech.* 19: 424–429, 2006.
38. Kanayama, M., Hashimoto, T., Shigenobu, K., Yamane, S., Bauer, T. W., and Togawa, D. A prospective randomized study of posterolateral lumbar fusion using osteogenic protein-1 (OP-1) versus local autograft with ceramic bone substitute: emphasis of surgical exploration and histologic assessment. *Spine* 31: 1067–1074, 2006.

39. Laursen, M., Hoy, K., Hansen, E. S., Gelineck, J., Christensen, F. B., and Bünger, C. E. Recombinant bone morphogenetic protein-7 as an intracorporal bone growth stimulator in unstable thoracolumbar burst fractures in humans: preliminary results. *Eur. Spine J.* 8: 485–490, 1999.
40. Marcacci, M., Kon, E., Moukhachev, V., Lavroukov, A., Kutepov, S., Quarto, R., Mastrogiacomo, M., and Cancedda, R. Stem cells associated with macroporous bioceramics for long bone repair: 6- to 7-year outcome of a pilot clinical study. *Tissue Eng.* 13: 947–955, 2007.
41. Quarto, R., Mastrogiacomo, M., Cancedda, R., Kutepov, S. M., Mukhachev, V., Lavroukov, A., Kon, E., and Marcacci, M. Repair of large bone defects with the use of autologous bone marrow stromal cells. *N. Engl. J. Med.* 344: 385–386, 2001.

23
Surface-modified titanium to enhance osseointegration in dental implants

L SARINNAPHAKORN and L DI SILVIO,
King's College London, UK

Abstract: Titanium is the preferred material for the fabrication of dental implants. The demand and need for reducing the healing time between placement and loading of dental implants, and the use of implants in sub-optimal bone conditions, have led to substantial research work on surface modified titanium. Since, surface modifications can play a significant role in the interaction and success of the implant to the adjacent tissue, a great deal of current research focuses on the effect of different treatments on the bioactivity of titanium. Hence, the aim of this chapter is to review and discuss all the relevant research work on modified titanium surfaces, surface topography and surface chemistry of titanium implants, in particular, and their effects on protein adsorption and subsequently cellular interaction. Present day surface modification technology and commercialised dental implant surfaces are summarised. This will be providing a useful update on the state of the art of contemporary dental implant surface technology. In addition, our own laboratory research work on *in vitro* studies using two human cell model types; human osteoblast (HOB) and mesenchymal stem cells (MSCs) is also presented and discussed. Finally, future trend in implant surface technology is given, based on present prediction for future application. Further development of this technology will assist in identifying ideal surface templates for tissue regeneration and engineering.

Key words: dental implant, surface modification, surface topography, surface roughness, surface chemistry, titanium surface, osteoblast, mesenchymal stem cell, osseointegration.

23.1 Introduction

With increasing life expectancy and the ever-growing number of the ageing population, it is not unusual that at some point in life, there may be the need to replace one or several parts of the body due to disease or a degenerative process. This health issue can affect any part of the body, including teeth. As a result, biomimetic materials are being developed and employed to alleviate this huge problem. In the past three decades, the number of dental implant procedures has increased steadily worldwide, predominantly due to the improvement in treatment concepts as well as the increased success rate of the treatment modalities. This has led to a new era in the history of dentistry.

The clinical success of oral implants is related to their early osseointegration of the titanium implant. Direct bone apposition onto the surface of titanium is critical for the rapid loading of dental implants. Following implantation, titanium implants interact with biological fluids and tissues. The surface of the implant becomes almost spontaneously coated with a protein conditioning layer. The composition and confirmation of this layer subsequently affect cellular responses such as cell adhesion, spreading and proliferation. Cells respond primarily to the protein layer, rather than the actual surface of the biomaterial. Since cells respond specifically to proteins, this interfacial protein film may be the event that controls subsequent bio-reaction to implants.

However, in a clinical scenario, there are two types of responses observed post-implantation. The first, involves the formation of a fibrous soft tissue capsule around the implant. This fibrous tissue capsule does not ensure proper biomechanical fixation and leads to clinical failure of the dental implant. The second type of bone response is related to direct bone–implant contact without an intervening connective tissue layer, this is what is known as osseointegration. This biological fixation is considered to be a prerequisite for implant-supported prostheses and their long-term success. The rate and quality of osseointegration in titanium implants are related to their surface properties. Surface topography and chemistry are two main parameters that may play a role in implant–tissue interaction and osseointegration.

23.2 Dental implant and osseointegration

In the early 1960s, Brånemark and co-workers at the University of Gothenburg started developing a novel implant which for clinical function depended on direct bone anchorage (Albrektsson et al. 2003). Since then, dental implants have gradually become one of the treatment modalities in modern dentistry, offered to patients who are fully or partially edentulous. The first part of the treatment involves a surgical procedure during which the implant, a screw-like titanium fixture, is installed in the jawbone, maxilla and/or mandible. The implant serves as the root and is integrated as new bone mass forms in contact with the surface of the titanium. A titanium or ceramic abutment is tightened to the implant and then a crown, which is the visible part of the reconstructed tooth, is then placed on the abutment. When multiple teeth are to be replaced, two or more implants are installed to build the foundation for either a fixed implant bridge reconstruction or a semi-fixed overdenture.

This scientific breakthrough, considered to be state-of-the-art in dental practice, depended upon one important phenomenon named 'osseointegration' which was partially defined at the time. However, to date, from a clinical point of view, osseointegration is a process whereby clinically asymptomatic rigid fixation of alloplastic materials is achieved, and maintained, in bone during functional loading (Zarb and Albrektsson 1998). This definition is based mainly on stability of implant materials instead of histological criteria, whereas osseointegration represents direct connection between bone and implant without an interposed soft tissue layer (Albrektsson et al. 2003).

23.3 Mechanism of peri-implant endosseous healing

The mechanism by which endosseous implants become integrated in bone can be subdivided into three separate phenomena: osteoconduction, *de novo* bone formation, and bone remodelling (Davies 1998, 2003). The first and most important healing phase, osteoconduction, relies on the recruitment and migration of osteogenic cells to the implant surface, through the residue of the peri-implant blood clot. Among the most important aspects of osteoconduction is the effect generated at the implant surface, by the initiation of platelet activation, which results in directed osteogenic cell migration. The second healing phase, *de novo* bone formation, results in a mineralized interfacial matrix equivalent to that seen in the cement line in natural bone tissue. The third healing phase, bone remodelling, relies on slower processes whereby the stability of the implant increases for long-term functional load.

23.4 The 'loss of osseointegration' phenomenon

As mentioned previously, one of the two interactions between implanted materials and tissue is fibrous encapsulation, and this is considered to be 'loss of osseointegration' or 'failed implant'. Although this phenomenon is considered to account for only a very small percentage of the cases, approximately about 5% of the machined implant surface placed in the clinical setting, it is generally unpredictable, and is considered to be unacceptable for standard medical treatment. In addition, the current trend of dental implant treatment is moving towards shorter treatment periods and earlier loading. Therefore, the precise prognosis and expected outcome is the most important factor to be considered in the treatment planning. The roughened surface which was introduced in the latter development of surface implant has increased the success outcome and this became the benchmark standard of contemporary dental implant surface finish.

23.5 Surface modification of dental implant

The original dental implant, manufactured by a machining process which had a previously called 'smooth' surface, had performed well in the human alveolar bone. However, the demand and request for reducing the time between placement and loading of dental implants and for the use of implants in sub-optimal bone has led to substantial interest and research on surface-modified implant materials. Therefore, they have gradually substituted the 'old-fashioned' machined implant with roughened surface implants. The advantages of surface-modified implants are that they (1) provide a better mechanical stability between bone and implant immediately following implantation, established by greater contact area, (2) provide a surface configuration that properly retains the blood clot, and (3) stimulate the bone healing process (Wennerberg *et al.* 2003).

23.5.1 Surface topography of dental implant

As previously stated, the traditional 'smooth' or 'machined' or 'tuned' titanium surface has been superseded by 'roughened' titanium surfaces. However, the conventional 'smooth' surface inevitably comes with ridges and grooves created by the machining process during manufacture. This can be observed by scanning electron microscopy (SEM). Therefore, it is reasonable to assume that all dental implant surfaces used have a certain degree of roughness, and this definitely has an effect on the cell–material interaction. The surface roughnesses of dental implants results in an increase in surface area which may influence the amount of protein adsorbed onto the implant surface, and subsequently influence the rate of cell adhesion and proliferation on the implanted titanium. It has also been shown that the surface topography (roughness), by itself, regulates the cell–material interaction, especially reported for dental implants at submicron level. Osteoprogenitor cells have been shown, in a number of studies, to be sensitive to topographical changes. Hence, surface topography can influence and regulate cell shape and cell behaviour, creating an enhanced cellular interaction, promoting cell integration with the titanium implant. Surface roughness can be divided into two levels depending on the scale of the features: micron and submicron or nanometre range topographies.

The micro level is defined for topographical features as being in the range of 1 to 10 µm. This scale is directly related to implant geometry, with small grooves or a micro-threaded design and macroporous surface treatments, normally from blasting methods. The roughening of the titanium surface consists in blasting the implants with hard ceramic particles. The ceramic particles are projected through a nozzle at high velocity by means of compressed air. Depending on the size of the ceramic particles, different surface roughness can be produced on titanium implants. The blasting material should be chemically stable and biocompatible and should not hamper the osseointegration of the titanium implants. Various ceramic particles have been used, such as alumina, titanium oxide and calcium phosphate particles.

Alumina (Al_2O_3) is frequently used as a blasting material and produces surface roughness varying with the shape and size of the blasting media. However, the blasting material is often embedded into the implant surface and a residue remains even after ultrasonic cleaning, acid passivation and sterilization. Alumina is insoluble in acid and is thus hard to remove from the titanium surface. In some cases, these particles have been released into the surrounding tissues and have interfered with the osseointegration of the implants.

Titanium oxide (TiO_2) is also used for blasting titanium dental implants. Titanium oxide particles with an average size of 25 μm produce a moderately rough surface in the 1–2 μm range on dental implants. An experimental study using microimplants in humans has shown a significant improvement in bone-to-implant contact (BIC) for the TiO_2 blasted implants in comparison with machined surfaces (Ivanoff *et al.* 2001). Wennerberg *et al.* (1996) demonstrated, using a rabbit model, that grit-blasting with TiO_2 or Al_2O_3 particles gave similar values of bone–implant contact, but drastically increased the biomechanical fixation of the implants when compared to smooth titanium. These studies have shown that the torque force increased with the surface roughness of the implants while comparable values in bone apposition were observed. These studies corroborate that roughening titanium dental implants increases their mechanical fixation to bone but not their biological fixation.

The high roughness resulted in mechanical interlocking between the implant surface and bone ongrowth when compared to the 'smooth' surface. Moderate roughness in the range of 1–2 μm has resulted in a favourable cellular adaptation. This range of roughness maximizes the interlocking between mineralized bone and the surface of the implant (Wennerberg *et al.* 1996).

The main clinical indication for using an implant with a rougher surface is where the quality or volume of the host bone is very poor. In these unfavourable clinical situations, early and high bone-to-implant contact would be beneficial in order to allow high levels of loading. Numerous studies have shown that surface roughness in this range resulted in greater bone-to-implant contact and higher resistance to torque removal than other types of surface topography. These reports have demonstrated that titanium implants with roughened surfaces have greater contact with bone than titanium implants with smoother surfaces (Cochran *et al.* 1998).

Surface profiles at the submicron or nanometre level are in the range 0.1–1.0 μm (100–1000 nm). This roughness generally results from either the chemically etched procedure (acid or alkali) or the electrochemical procedure. With the advance of surface technology, it has been shown that the creation of nano-sized surface structure or even a nanocrystal particle coating could be possible. These submicron and nanometre level surfaces may play an important role in the adsorption of proteins, the adhesion of osteoblastic cells and thus the rate of osseointegration (Brett *et al.* 2004). However, *reproducible* surface roughness in the nanometre range is still questionable; together with the methods used for measurement, which are quite variable, this makes it difficult to compare the surface finish of implant surfaces between one study and another. In addition, the optimal surface nanotopography for selective adsorption of proteins leading to the adhesion of osteoblastic cells and rapid bone apposition is still not conclusive and is under investigation.

23.5.2 The effect of surface topography on cell behaviour

The effect of topography on cellular behaviour has been extensively studied. The topography may be surrounding cells, intercellular materials or biomaterials, and the reactions include cell orientation, rate of movement, and activation of the cells (Curtis and Wilkinson 1997). The importance of topography in cell response and the underlying mechanisms have been described in wound repair. In theory, cells use the morphology of the substrate for orientation and migration (Boyan *et al.* 1996). Variations in surface texture can include microtopography and, more recently described, nanotopography, both of which can act as signalling cues and influence cellular response to an implant. It has been reported in cell adhesion studies that a rougher surface of cpTi is preferred by osteoblast-like cells, whereas a smoother surface is preferred by gingival and periodontal ligament fibroblast and epithelial cells. Further investigations have been performed to evaluate the effect of pre-treating surfaces with attachment proteins and examining their effects on fibroblasts and also on bone cells (Vrouwenvelder *et al.* 1994).

For implant materials that are to be anchored to bone, optimal osseointegration requires an optimal response of the bone cells and many studies have focused on the effects of surface roughness on osteoblast-like cells (Martin *et al.* 1995, 1996; Nishimura and Kawai 1998). Other studies have shown that not all responses are mediated by surface features alone, but systemic factors such as hormones also play a role, thus demonstrating the importance of understanding how osteogenesis at the bone–material interface is modulated not only by the material surface itself, but also by local cytokines, systemic hormones and pharmacological manipulation (Jones 2001).

23.5.3 Surface chemistry of dental implant

Dental implants are usually made from commercially pure titanium or titanium alloys, due to their excellent biocompatibility and mechanical properties. Commercially pure titanium (cpTi) has various degrees of purity (graded from 1 to 4). This purity is characterized by oxygen, carbon and iron content; this is also reflected in the mechanical properties among the different grades. Most dental implants are made from grade 4 cpTi as it is stronger than other grades. Titanium alloys are mainly composed of Ti6-Al4-V (grade 5 titanium alloy) with greater yield strength and fatigue properties than pure titanium.

With its highly active titanium dioxide layer, the material is well tolerated by local tissue. This oxide layer spontaneously forms (passively), when it is exposed to the oxygen, primarily titanium oxide, in the thickness range of 2–10 nm (Kasemo 1983), depending on the amount of oxygen bonded to Ti. A variety of different stoichiometrics of titanium oxide are known that cover a wide range of oxygen (O) to titanium (Ti) ratios: from Ti_3O to Ti_2O, Ti_3O_2, TiO, Ti_2O_3, Ti_3O_5 and TiO_2 (Textor *et al.* 2001). The most stable titanium oxide is TiO_2 with titanium in the preferred oxidation state +IV. It exists in three different crystallographic forms: rutile, anatase and brooklite, the first two being the most commons forms.

In a recent extensive commercial implant surface study (Jarmar *et al.* 2008) which used a variety of techniques – scanning electron microscopy (SEM), focused ion beam (FIB), high-resolution transmission electron microscopy (TEM) and X-ray photoelectron spectroscopy (XPS) – various different crystallographic and amorphous forms of titanium oxide layer were attributed to the different modification methods. In fact, the authors concluded that 'from a biological/chemical point of view, these manufacturers probably have different theories regarding osseointegration enhancement, while the topography, surface chemistry, and crystallinity are very different'.

Surface chemistry of the titanium involves various elements, the two most common being calcium (Ca) and phosphorus (P). According to their similarity to hydroxyapatite compositions (HA), they have long been accepted as the chemical compositions of choice, acting as osteoconductive materials. Titanium plasma spraying (TPS) is one of the techniques used to coat the dental implant surface for this purpose. Other methods have also been used, including electrochemical treatment, fluoride treatment and CaP solution treatment, the objective in all cases being that these deposited elements would be beneficial in the interaction between implant surface and surrounding tissue.

23.5.4 The effect of surface chemistry and cell behaviour

Although there is little doubt that surface topography is extremely important in governing interactions at biological interfaces, surface chemistry is also critical (Jones 2001). The cells would be sensitive to differences in chemistry of the outermost functional groups of the surface, although the exact mechanism is not clearly understood (Boyan *et al.* 1996). One of the theories proposed is that surface chemistry alteration may affect preferential protein adsorption and hence modulate cellular response. Furthermore, the issue of whether treatment to alter surface topography also influences surface chemistry, and vice versa, has also been addressed. This concern, however, not only makes this kind of experiment difficult to interpret, but also makes comparison difficult.

There have been many studies on titanium surfaces that have been chemically changed, or changed unintentionally. One such change is the sterilization procedure used for cleaning the titanium surface. The method may affect titanium surface composition and subsequent cellular response. For example, it has been reported that multiple sterilization of a titanium implant surface can have an inhibitory effect on cell spreading, while a single exposure did not appear to have the same effect.

The surface chemistry can also be intentionally modified for specific purposes with surface modification techniques such as different ion implantation and various coating methods. Surfaces can also be modified to enhance biomolecular adsorption. For example, titanium surfaces can be treated by $CaCl_2$ to give a calcium ion-rich surface and this surface has been shown to selectively adsorb serum protein (Ellingsen 1991). This study suggested that the surface characteristics of the TiO_2 surface layer were changed from anionic to cationic by calcium ion adsorption, allowing the adsorption of proteins via acidic functional groups. Several studies have investigated the influence of calcium ion and concluded that even very small amounts of calcium in the oxide layer on titanium surfaces may have a significant effect on the surface properties.

In summary, various methods have been developed and used in order to create a rough surface and improve the osseointegration of titanium for dental implants. The roughening process may also alter the surface chemistry of the outer surface layer of the implant material and vice versa. The main modification procedures are grit blasting, acid etching and the combination of the two. The electrochemical process is also one of the main modification processes. Manufacturers of dental implants have developed a variety of surfaces with different compositions and degrees of roughness. However, there is some controversy as to the optimal features for implant surfaces regarding osseointegration kinetics. Some of the methods are now obsolete and have been removed from the market, but others have been further developed and optimized to improve the long-term clinical outcome. The following section reviews all major implant products that are currently used worldwide.

23.6 Commercial surface-modified titanium implants

23.6.1 TiUnite®

TiUnite®, from Nobel Biocare AB, Sweden, is the titanium surface treated by an electrochemical method. This surface is a highly crystalline and phosphate-enriched titanium dioxide which is made by an electrochemical process. Its surface also presents a microstructured topography without sharp features and is characterized by the presence of uniformly distributed open pores in the low micrometre range. It has brought a new dimension to implant dentistry since its appearance on the market in 2000. A number of studies have reported that this modified surface dental implant integrated successfully and illustrated an excellent tissue–implant interface (Schupbach *et al.* 2005).

23.6.2 TiOblast™ and OsseoSpeed™

OsseoSpeed™, from AstraTech AB, Sweden, was launched in 2004 and is a further development of the moderately roughened (grit blasted with titanium dioxide particles) titanium surface TiOblast™. OsseoSpeed™ gains its additional surface characteristics via chemical (fluoride) treatment and a slight topographic modification of the TiOblast™ surface. The OsseoSpeed™ surface incorporates a small amount of fluoride ions in the oxide layer, and a slight decrease on the micrometre scale in the surface roughness and the appearance of nanometre-sized peaks.

23.6.3 SLA® and SLActive®

The SLA® surface, developed by Straumann AG, Switzerland, in 1994, is created by large-grit blasting and cleaned with an acid etching procedure (S = sand blasted, L = large grit, A = acid etched). The macro- and microstructure of this surface proved to be in closer contact with bone (Buser *et al.* 1991) and was deemed to be highly attractive as a rough surface dental implant in the commercial market. The SLActive® surface takes the proven concept of SLA one step further to set the new surface benchmark as a hydrophilic dental implant surface (Buser *et al.* 2004; Zhao *et al.* 2005).

23.6.4 OSSEOTITE® and Nanotite™

OSSEOTITE®, from 3i (Implant Innovation) which is now BIOMET 3i (Palm Beach Gardens, FL, USA), was introduced in the early 1990s, and has now been further developed as Nanotite™, which was launched at the beginning of 2007. The OSSEOTITE® surface is a combination of double surface-etched implant surface by HCl and H_2SO_4. For more than 10 years, with documentation from numerous global multicentre clinical evaluations, the OSSEOTITE® surface has proven to be one of the most predictable and well-researched surfaces. Clinical studies on the OSSEOTITE® surface continue to document the benefits of increased 'contact osteogenesis', especially in poor-quality bone.

The Nanotite™ surface, a further improved surface, is made by the discrete crystalline deposition (DCD™) process, an innovative implant surface technology. Traditionally, CaP has been plasma sprayed (Titanium Plasma Spraying, TPS) on the implant surface, creating a coating typically in the range of 50–100 μm. The nature of plasma-sprayed coatings makes them susceptible to delamination or dissolution of the amorphous content of the coating. Hence the positive attributes of CaP may be offset by certain risk factors. In contrast to the Nanotite™ implant, the CaP is not applied via a plasma-sprayed process but is rather a solution-based form of self-assembly. It is not a coating but consists of actual deposits of discrete crystals that occupy approximately 50% of the OSSEOTITE® surface, according to the manufacturer.

Furthermore, the dissolution of DCD™ on a Nanotite™ implant is extremely low in physiologically neutral pH given the highly crystalline nature of the CaP crystals. This provides implants with a more consistent and stable phase of CaP, allowing the implant site to capitalize on the positive attributes of this biomaterial. In addition, the DCD process increases the micro-surface area by 200%, providing greater micro-complexity. A recent *in vivo* study has shown that with this special surface complexity, there was evidence of bone-bonding which may be attributed to the increase of disruption force value compared to the control surface (Mendes *et al.* 2007).

23.6.5 FRIADENT® plus

The FRIADENT® plus surface, from DENTSPLY Friadent, GmbH, Germany, was one of the pioneer roughening titanium surfaces more than 15 years ago. Surface macro-roughness is achieved by grit-blasting the endosseous section of the implants. The implants then go through a unique thermal etching process called FRIADENT® BioPoreStructuring. The growth activating microporosity and surface morphology needed for osseointegration is achieved during this step. It was claimed that the specific acid used by the company creates the ideal physical, chemical and biological properties needed to attract bone-inducing cells to the implant surface.

In summary, from the dental implant surfaces presently on the market, all have one thing in common: surface roughness, since surface roughness plays a major role in both the quality and rate of osseointegration of titanium dental implants. Hence it can be concluded that highly roughened implants such as TPS or grit-blasted implants favour mechanical anchorage and primary fixation to bone. Topographies in the nanometre range have been used to promote protein adsorption, osteoblastic cell adhesion and the rate of bone tissue healing in the peri-implant region.

23.7 Current treatment modalities

Current research (Dental Institute, Guy's Hospital, King's College London) is focused on novel treatments of titanium surfaces, such as anodic spark deposition (ASD), sol-gel coating (SG), and electrohydrodynamic atomisation (EHDA). Examination using scanning electron microscopy (SEM), illustrated a sub-micron pore structure on an ASD surface (Fig. 23.1 (a)), a crystal-like, possibly hydroxyapatite crystal structure on a SG surface (Fig. 23.1 (b)), as well as nano-hydroxyapatite particles on an EHDA surface (Fig. 23.1 (c)). Chemical element analysis by energy dispersive X-ray spectroscopy (EDS) revealed a titanium dioxide layer surface doped with calcium (Ca) and phosphorous (P) in all three surfaces mentioned. The surface roughness was found to be in the nano-topographic range (Table 23.1). *In vitro* cellular response of titanium modified surface illustrated by cell adhesion suggested that these treated surfaces were favoured by primary human osteoblast cells (HOB) and human mesenchymal stem cells (MSCs)

Table 23.1 Surface roughness quantification from AFM data (mean ± SD)

Surface*	Line roughness; R_a (nm)	Surface roughness; S_a (nm)
a	76 ± 16	146 ± 18
b	253 ± 65	318 ± 27
c	468 ± 36	486 ± 185
d	203 ± 18	274 ± 18
e	30 ± 7	80 ± 29

*a = anodic spark deposition, b = sol gel, c = electrohydrodynamic atomization, d = chemically etched, e = machined (turned) surface.

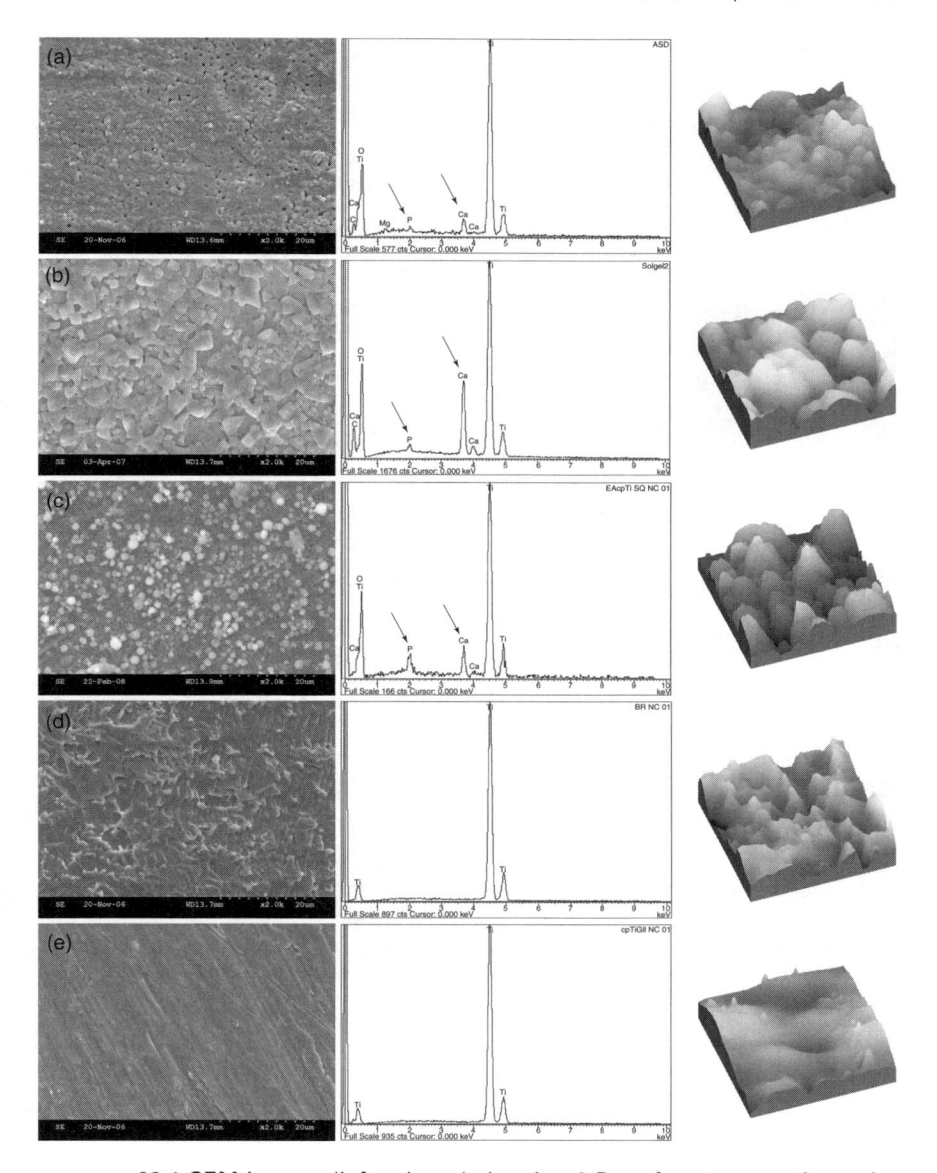

Surface-modified titanium in dental implants 583

23.1 SEM images (left column) showing 2-D surface topography and texture. EDS analysis (middle column) showing the surface chemical compositions added (arrows). Atomic force microscopy (AFM) images (right column) showing 3-D surface topography (10 × 10μm) and quantitative analysis for surface roughness (Table 23.1). (a) = anodic spark deposition, (b) = sol gel, (c) = electrohydrodynamic atomization, (d) = chemically etched, (e) = machined (turned) surface.

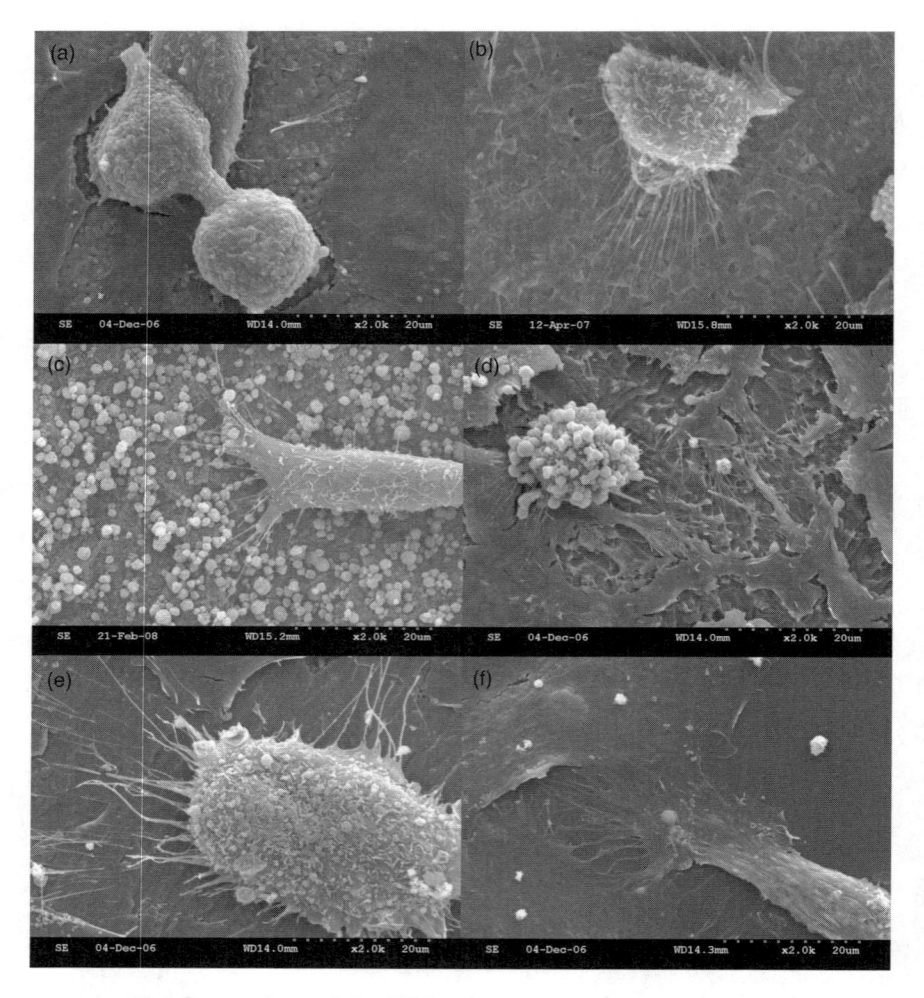

23.2 Comparison of the HOB cell morphology on the materials surfaces. (a) = anodic spark deposition, (b) = sol gel, (c) = electrohydrodynamic atomisation, (d) = chemically etched, (e) = machined (turned) surface, (f) = Thermanox™ (TMX) (standard plastic cover slip).

and our studies indicate that both cells types show good proliferation potential on these modified titanium surfaces (Fig. 23.2).

23.8 Future trends in dental implant surfaces

Whereas second-generation biomaterials were designed to be either resorbable or bioactive, the next generation of biomaterials is combining these two properties, with the aim of developing materials which, once implanted, will help the body heal itself (Hench and Polak 2002). As reviewed by Ratner and Bryant (2004), direct mimicry of biological processes represents an important strategy in modern biomaterial design. Normal tissues have a complex 3-D architecture important for the mechanics and functionality of the biological organism. However, synthetic materials offer the ability to generate many different kinds of 3-D structures with precise control over the final macroscopic properties and degradation profile by varying the chemistry and processing techniques. In order to generate a biomaterial for a specific cell type or tissue, efforts have focused on adding biological cues to synthetic materials in an attempt to mimic the native tissue while simultaneously maintaining control over the material properties.

The techniques that are currently available to impart implant surfaces with the desired osteoconductive properties are necessary, but still limited for early loading requirement. In order to overcome this problem, biomimetic coatings incorporating growth factors such as bone morphogenetic proteins (BMPs) are being investigated.

The theory of osseointegration is characterized by graft derived factors that actively stimulate new osteogenic activity. Urist (1965) first introduced this theory and demonstrated that protein extracts, later named 'bone morphogenetic proteins (BMP)', when implanted into animal at orthopic and ectopic sites, had the potential to induce the formation of new cartilage and bone tissues. Subsequently, many researchers have attempted to characterize BMPs. To date, more than 12 different types of BMP have been identified, of which only nine have been shown individually to induce bone in an *in vivo* assay system. Methods for delivering single BMP species to wound sites in order to promote bone formation have been described and are being focused upon by many researchers (Heckman *et al.* 1991). Recently, different types of BMPs have also been applied in conjunction with endosseous implants in animal models, and the presence of these proteins seems to induce faster bone formation around the implants as exhibited through histological examination (Xiang *et al.* 1994).

23.9 Conclusions

As reviewed above, there are five major commercially modified surfaces available for dental implants. Most of these surfaces have proven clinical efficacy (approximately 95% survival rate over 5 years follow-up). However, there are also a number of other small implant manufacturers who produce dental implants with some form of surface modification, and the advantage of one surface over another has been anecdotal, requiring unbiased *in vitro* and *in vivo* studies. In most cases tests were not standardized, using different surface sample sizes, different cell types or animal models. The exact role of surface chemistry and topography on the early events of osseointegration of dental implants still remains unclear. Furthermore, comparative clinical studies with different implant surfaces are rarely performed.

One of the main strategies for future dental implant material development should be aimed at developing surfaces with controlled and standardized topography or chemistry. By understanding more about the effect of protein adsorption, cell and tissue interactions with implant surfaces, we may be able to 'tailor-make' implants to match the individual patient's needs. The ability to control local release of bone-stimulating drugs or factors in the peri-implant region may significantly improve difficult clinical situations with poor bone quality and quantity, commonly found in ageing or compromised patients. Further development of these therapeutic strategies would ultimately result in enhancing the osseointegration process of dental implants for an expanding population of ageing patients, thus improving their quality of life and reducing socio-economic costs.

23.10 References

Albrektsson, T., Berglundh, T., & Lindhe, J. 2003, 'Osseointegration: Historic background and current concepts', in *Clinical Periodontology and Implant Dentistry*, 4th edn, J. Lindhe, T. Karring, & N. P. Lang, eds, Blackwell, Oxford, UK, pp. 809–820.

Boyan, B. D., Hummert, T. W., Dean, D. D., & Schwartz, Z. 1996, 'Role of material surfaces in regulating bone and cartilage cell response', *Biomaterials*, vol. 17, no. 2, pp. 137–146.

Brett, P. M., Harle, J., Salih, V., Mihoc, R., Olsen, I., Jones, F. H., & Tonetti, M. 2004, 'Roughness response genes in osteoblasts', *Bone*, vol. 35, no. 1, pp. 124–133.

Buser, D., Schenk, R. K., Steinemann, S., Fiorellini, J. P., Fox, C. H., & Stich, H. 1991, 'Influence of surface characteristics on bone integration of titanium implants. A histomorphometric study in miniature pigs', *Journal of Biomedical Materials Research*, vol. 25, no. 7, pp. 889–902.

Buser, D., Broggini, N., Wieland, M., Schenk, R. K., Denzer, A. J., Cochran, D. L., Hoffmann, B., Lussi, A., & Steinemann, S. G. 2004, 'Enhanced bone apposition to a chemically modified SLA titanium surface', *Journal of Dental Research*, vol. 83, no. 7, pp. 529–533.

Cochran, D. L., Schenk, R. K., Lussi, A., Higginbottom, F. L., & Buser, D. 1998, 'Bone response to unloaded and loaded titanium implants with a sandblasted and acid-etched surface: a histometric study in the canine mandible', *Journal of Biomedical Materials Research*, vol. 40, no. 1, pp. 1–11.

Curtis, A. & Wilkinson, C. 1997, 'Topographical control of cells', *Biomaterials*, vol. 18, no. 24, pp. 1573–1583.

Davies, J. E. 1998, 'Mechanisms of endosseous integration', *International Journal of Prosthodontics*, vol. 11, no. 5, pp. 391–401.

Davies, J. E. 2003, 'Understanding peri-implant endosseous healing', *Journal of Dental Education*, vol. 67, no. 8, pp. 932–949.

Ellingsen, J. E. 1991, 'A study on the mechanism of protein adsorption to TiO_2', *Biomaterials*, vol. 12, no. 6, pp. 593–596.

Heckman, J. D., Boyan, B. D., Aufdemorte, T. B., & Abbott, J. T. 1991, 'The use of bone morphogenetic protein in the treatment of non-union in a canine model', *Journal of Bone and Joint Surgery*, vol. 73, no. 5, pp. 750–764.

Hench, L. L. & Polak, J. M. 2002, 'Third-generation biomedical materials', *Science*, vol. 295, no. 5557, pp. 1014–1017.

Ivanoff, C. J., Widmark, G., Hallgren, C., Sennerby, L., & Wennerberg, A. 2001, 'Histologic evaluation of the bone integration of TiO_2 blasted and turned titanium microimplants in humans', *Clinical Oral Implants Research*, vol. 12, no. 2, pp. 128–134.

Jarmar, T., Palmquist, A., Branemark, R., Hermansson, L., Engqvist, H., & Thomsen, P. 2008, 'Characterization of the surface properties of commercially available dental implants using scanning electron microscopy, focused ion beam, and high-resolution transmission electron microscopy', *Clinical Implant Dentistry and Related Research*, vol. 10, no. 1, pp. 11–22.

Jones, F. H. 2001, 'Teeth and bones: Applications of surface science to dental materials and related biomaterials', *Surface Science Reports*, vol. 42, no. 3–5, pp. 75–205.

Kasemo, B. 1983, 'Biocompatibility of titanium implants: Surface science aspects', *The Journal of Prosthetic Dentistry*, vol. 49, no. 6, pp. 832–837.

Martin, J. Y., Schwartz, Z., Hummert, T. W., Schraub, D. M., Simpson, J., Lankford, J., Jr., Dean, D. D., Cochran, D. L., & Boyan, B. D. 1995, 'Effect of titanium surface roughness on proliferation, differentiation, and protein synthesis of human osteoblast-like cells (MG63)', *Journal of Biomedical Materials Research*, vol. 29, no. 3, pp. 389–401.

Martin, J. Y., Dean, D. D., Cochran, D. L., Simpson, J., Boyan, B. D., & Schwartz, Z. 1996, 'Proliferation, differentiation, and protein synthesis of human osteoblast-like cells (MG63) cultured on previously used titanium surfaces', *Clinical Oral Implants Research*, vol. 7, no. 1, pp. 27–37.

Mendes, V. C., Moineddin, R., & Davies, J. E. 2007, 'The effect of discrete calcium phosphate nanocrystals on bone-bonding to titanium surfaces', *Biomaterials*, vol. 28, no. 32, pp. 4748–4755.

Nishimura, N. & Kawai, T. 1998, 'Effect of microstructure of titanium surface on the behaviour of osteogenic cell line MC3T3-E1', *Journal of Materials Science: Materials in Medicine*, vol. 9, no. 2, pp. 99–102.

Ratner, B. D. & Bryant, S. J. 2004, 'Biomaterials: Where we have been and where we are going', *Annual Review of Biomedical Engineering*, vol. 6, pp. 41–75.

Schupbach, P., Glauser, R., Rocci, A., Martignoni, M., Sennerby, L., Lundgren, A., & Gottlow, J. 2005, 'The human bone-oxidized titanium implant interface: A light microscopic, scanning electron microscopic, back-scatter scanning electron microscopic, and energy-dispersive X-ray study of clinically retrieved dental implants', *Clinical Implant Dentistry and Related Research*, vol. 7 Suppl 1, pp. S36–S43.

Textor, M., Sitig C., Frauchiger, V., Tosatti, S., & Brunette, D. M. 2001, 'Properties and biological significance of natural oxide films on titanium and its alloys', in *Titanium in Medicine*, D. M. Brunette et al., eds, Springer, pp. 171–230.

Urist, M. R. 1965, 'Bone: formation by autoinduction', *Science*, vol. 150, no. 698, pp. 893–899.
Vrouwenvelder, W. C. A., Groot, C. G., & de Groot, K. 1994, 'Better histology and biochemistry for osteoblasts cultured on titanium-doped bioactive glass: Bioglass 45S5 compared with iron-, titanium-, fluorine- and boron-containing bioactive glasses', *Biomaterials*, vol. 15, no. 2, pp. 97–106.
Wennerberg, A., Albrektsson, T., Johansson, C., & Andersson, B. 1996, 'Experimental study of turned and grit-blasted screw-shaped implants with special emphasis on effects of blasting material and surface topography', *Biomaterials*, vol. 17, no. 1, pp. 15–22.
Wennerberg, A., Albrektsson, T., & Lindhe, J. 2003, 'Surface topography of titanium implants', in *Clinical Periodontology and Implant Dentistry*, 4th edn, J. Lindhe, T. Karring, & N. P. Lang, eds, Blackwell, Oxford, UK, pp. 821–840.
Xiang, W., Yan, J., Baolin, L., Shuxia, Z., Lianjia, Y., Yang, X., & White, F. H. 1994, 'Tissue reactions to titanium implants containing bovine bone morphogenetic protein: A scanning electron microscopic investigation', *International Journal of Oral and Maxillofacial Surgery*, vol. 23, no. 2, pp. 115–119.
Zarb, G. A. & Albrektsson, T. 1998, 'Consensus report: towards optimized treatment outcomes for dental implants', *Journal of Prosthetic Dentistry*, vol. 80, no. 6, p. 641.
Zhao, G., Schwartz, Z., Wieland, M., Rupp, F., Geis-Gerstorfer, J., Cochran, D. L., & Boyan, B. D. 2005, 'High surface energy enhances cell response to titanium substrate microstructure', *Journal of Biomedical Materials Research A*, vol. 74, no. 1, pp. 49–58.

24
Surface modification of titanium for the enhancement of cell response

R CHIESA, Politecnico di Milano, Italy

Abstract: Titanium is one of the most frequently used metallic materials for biomedical applications, due to its mechanical, chemical and biological properties. This chapter discusses the bulk and surface properties of titanium as implantable material, than considers surface modification treatments for enhancing its biological performance, and finally considers specific interface reactions of the metal with bone cells and tissues.

Key words: titanium, titanium mechanical properties, titanium surface modification treatments, titanium biocompatibility.

24.1 Titanium for hard and soft tissue applications

Titanium (Ti) and Ti alloys are the materials of choice for the fabrication and use of load-bearing implants in both dentistry and orthopaedics. They are biocompatible, have high strength and high stiffness, and in addition they have the potential for being further enhanced to promote osteointegration.

Unalloyed pure titanium is considered the best implantable metal in terms of biocompatibility. Titanium properties required for biomedical applications are described by International Standard ISO 5832 part 2, and American Standard ASTM F67. The ISO 5832 standard (from part 1 to part 12) describes the chemical, structural and mechanical properties of several metals required for long-term clinical applications. The ISO 5832 is divided into standard parts:

ISO 5832 – Part 1: Wrought stainless steel
ISO 5832 – Part 2: Unalloyed titanium
ISO 5832 – Part 3: Wrought titanium 6-aluminium 4-vanadium alloy
ISO 5832 – Part 4: Cobalt-chromium-molybdenum casting alloy
ISO 5832 – Part 5: Wrought cobalt-chromium-tungsten-nickel alloy
ISO 5832 – Part 6: Wrought cobalt-nickel-chromium-molybdenum alloy
ISO 5832 – Part 7: Forgeable and cold-formed cobalt-chromium-nickel-molybdenum-iron alloy

ISO 5832 – Part 8: Wrought cobalt-nickel-chromium-molybdenum-tungsten-iron alloy
ISO 5832 – Part 9: Wrought high nitrogen stainless steel
ISO 5832 – Part 10: Wrought titanium 5-aluminium 2,5-iron alloy
ISO 5832 – Part 11: Wrought titanium 6-aluminium 7-niobium alloy
ISO 5832 – Part 12: Wrought cobalt-chromium-molybdenum alloy

Before considering any implant for biomedical application, the introduction posted on each of the ISO 5832 standard parts should be considered: '*No known surgical implant material has ever been shown to cause absolutely no adverse reaction in the human body. However, long-term clinical experience of the use of the material referred to in this part of ISO 5832 has shown that an acceptable level of biological response can be expected when the material is used in appropriate applications.*' This point surely concerns titanium, and should always be taken into account when a metal, or more generically a material, is considered for any implantable application: the *appropriate application* is a major issue related to biocompatibility in the long-term success of any implanted biomedical device.

Titanium alloys used for biomedical applications are described in the standard parts ISO 5832-3, ISO 5832-10 and ISO 5832-11. In general, when higher strength is required for a specific biomedical application, titanium alloys can effectively substitute for unalloyed titanium grades. Compared to pure titanium, titanium alloys are more expensive and possess lower biocompatibility, generally due to the presence of the alloying elements, but have much higher mechanical strength. Table 24.1 reports the general chemical composition and mechanical properties of unalloyed titanium and titanium alloys as reported by ISO 5832 standard parts. It should also be noted that American standards specify some titanium alloys for biomedical applications. Ti6AL4V is specified by ASTM F1472 and ASTM F136, Ti6Al7Nb is specified by ASTM F1295. ASTM specifies also two other titanium alloys: Ti12Mo6Zr4Fe by ASTM F1813, and Ti13Nb13Zr by ASTM F1713 standard.

Titanium and its alloys have an elastic modulus around 110 GPa, half that of stainless steels and cobalt alloys, but much higher than any polymeric material and biological tissue. Such high mechanical properties are the key factors orienting the choice of titanium (alloys) for the manufacture of implantable devices.

Metals in general and titanium in particular are widely used for prosthetic applications and hard tissue repair and substitution. Titanium is widely used for the manufacture of dental implants, artificial joints (hip, knee and shoulder), spine and osteosynthesis components and many other implantable devices.

Surface modification of titanium for enhancement of cell response 591

Table 24.1 Chemical composition and mechanical properties in the annealed and in the work-hardened states of commercially pure titanium (Grades 1 to 4) and ISO standardized titanium alloys used for biomedical applications

Kind	ISO classification	Chemical composition (%)	UTS (MPa) Annealed – cold worked	TYS (MPa) Annealed – cold worked
Titanium (unalloyed)	ISO 5832-2 G1	Ti = balance, O < 0.18	240	170
	ISO 5832-2 G2	Ti = balance, O < 0.25	345	230
	ISO 5832-2 G3	Ti = balance, O < 0.35	450	300
	ISO 5832-2 G4	Ti = balance, O < 0.45	550–680	440–520
Ti6Al4V	ISO 5832-3	Ti = balance, Al = 5.5–6.75, V = 3.5–4.5	860	780
Ti5Al2.5Fe	ISO 5832-10	Ti = balance, Al = 4.5–5.5, Fe = 2.5–3	900	800
Ti7Al8Nb	ISO 5832-11	Ti = balance, Al = 5.5–6.75, Nb = 6.5–7.5	900	800

As a first analysis, the biocompatibility of metals can be associated with their corrosion resistance. Only inert metals, such as gold and platinum, are thermodynamically immune from corrosion in the human body. On the other hand, the unsatisfactory mechanical properties of these metals limit their use to specific biomedical applications. Titanium, as well as the other metals listed in the ISO 5832 standard, is not immune from corrosion in body fluids, but in normal situations the degradation and dissolution rate is very low, and the effect of ions released in the body fluids is negligible. Nevertheless, when localized corrosion phenomena, like fretting corrosion, take place, metal ions can be released at a high rate in biological systems, and biocompatibility problems may arise. In the case of high release of metal ions due to corrosion, the chemical compositions of the alloys play a major role: local or systemic reactions to some metal ions may impair cell and tissue response to biomaterials. With regard to titanium alloys, the role of vanadium, aluminium or molybdenum alloying element on biocompatibility is still a debated point (Howlett *et al.*, 1999; Scharnweber *et al.*, 2002).

The properties of titanium make it particularly suitable for applications where direct integration and binding with the bone are required. Hence, most current osseointegrated dental and orthopaedic implants are made of titanium or titanium alloys. Moreover, the excellent compatibility of titanium with biological tissues makes it suitable for many other biomedical applications. Titanium is used in cardiovascular and cardiac applications: heart valve components, pacemaker chassis, vascular stents and filters, etc. In these applications titanium devices are often in contact with soft tissues, blood or other biological fluids. The biological compatibility of titanium is related to its specific surface properties. The few nanometres-thick semiconducting titanium oxide TiO_2, spontaneously covering its surface, exhibits amphoteric behaviour in water solutions. In normal conditions the titanium surface shows weak positive or negative charge at different pH values. Depending on the surface conditions and properties, adsorption of different ions and proteins from the biological fluids may occur, interacting with many and complex biological and cellular factors such as inflammation and tissue repairing processes, complement activation, extracellular matrix formation, specific cellular type activation, etc.

Titanium surface properties give rise to specific interaction and adsorption processes, and a proper modification of its chemical, structural and morphological properties can be designed to properly inhibit or enhance specific adsorption processes. Particularly intriguing is the potential to modify the structure of the titanium oxide that naturally covers and protects the metal from oxidation and corrosion processes. Considering the titanium surface, proceeding from the macro scale to the micro and nanometric dimensions, the metal surface provides different interaction capabilities with the biological environment. When a titanium device is implanted, physiological solutions and proteins promptly interact with the metal surface. Depending on the pH of the solution, a pure untreated titanium surface, covered by few nanometres-thick TiO_2, exhibits a different charge. In physiological conditions, the weak negative charge on the titanium surface provides an interactive structure to the water dipoles, on which layer ions and proteins may interact and adsorb. Particularly interesting is the role of adsorbed proteins on the titanium surface. Albumin, fibronectin, fibrinogen, immunoglobulin G, and various complement factors when adsorbed on the surface are able to influence defined biological functions (Tengvall, 2001). Moreover, specific cellular types are recognized to discriminately interact with biomaterial surfaces through specific proteins; therefore the selective adsorption of proteins onto the surface may represent a key point for the development of new implantable devices.

Considering the osseointegration process of titanium dental implants, for example, the surface adsorbed proteins play a fundamental role for the integration of bone. The important role of fibronectin has been recognized; this glycoprotein is bonded by many cellular types through its RGD sequence. In particular, osteoblast activity, differentiation and adhesion is known to be positively regulated by fibronectin (Globus *et al.*, 1998). Moreover, fibronectin is always present in early bone formation (Ku *et al.*, 2002). Fibronectin binding on implant surfaces has also been shown to improve osteoblast gene expression and organic matrix mineralization (Stephansson *et al.*, 2002). Evaluation of fibronectin adsorption on the titanium surface can be assessed by SDA-PAGE and western blot techniques. The production and/or deposition of many other non-collagenous proteins can be assessed during evaluation of the properties of materials for osseointegration. From this perspective, interesting proteins are osteonectin, osteocalcin and osteopontin, whose role in mineralization of new bone has been widely recognized, but is still weakly understood (Ku *et al.*, 2002; Thompson and Puleo, 1996; Lincks *et al.*, 1998). Other proteins have been described that negatively affect the osseointegration process. For example, the negatively charged albumin and osteocalcin, freely dissolved in body fluids, are known to be strong inhibitors for the crystallization of new bone phases (Hlady and Furedi-Mihofer, 1979).

From this point of view, the possibility of effectively modifying the surface chemical and biological properties of titanium through several chemical and physical techniques is one of the main intriguing opportunities offered today to biomaterials scientists.

24.2 Mechanical, physical and thermal modification treatments of titanium

The performance of titanium as a biomaterial is related to its mechanical properties, to the biomechanical performance of the implanted device, and to its surface properties. Surface properties and characteristics of biomaterials are particularly important, as they are strictly related to the biological tissue response.

A wide variety of surface treatments have been applied to metal components in general, and to titanium and titanium alloy in particular, to improve specific properties. Main traditional industrial finishing processes, including sandblasting, pickling, etching, degreasing and polishing, have been commonly applied to titanium implants. Although in the past surface treatments were usually applied to titanium to improve specific mechanical properties (fatigue endurance, hardness, corrosion resistance), most of the efforts today are focused on the improvement of biocompatibility for triggering specific biological responses. Some of the surface treatments applied to titanium implantable devices are relatively simple. The surface roughness of osseointegrated dental implants and non-cemented hip prosthetic stems can be effectively increased by mechanical or physical finishing processes, such as sand and grit blasting or the deposition of a porous coating. It is generally recognized that the increase in roughness ensures a better load transfer between the metal component and the bone, providing a wider contact area and ensuring stronger mechanical interconnection with bone tissue that grows on the metal interface (Browne and Gregson, 1994; Hanawa, 1999).

Current research in the biomaterials field is focused on the modification of surface properties of titanium to address specific biological needs. One of the most used techniques to improve the biological performance of titanium implants, and in particular its osseointegration capability, is the deposition of calcium phosphate coatings (Fini et al., 1999; Ducheyne and Healy, 1988). The most commonly applied deposition technique is plasma spraying, which uses a thickness between 70 and 150 µm, and the most studied calcium phosphate coating is hydroxyapatite (HA) (Brossa et al., 1993). HA coating is very effective in improving the implant/bone bonding (Tisel et al., 1994; Chae et al., 1992). Hydroxyapatite plasma spray coatings are today widely used for the fixation of non-cemented orthopaedic implants, while their application for osseointegrated dental implants is more problematic. Hydroxyapatite coatings have shown high failure rates on dental implants, mainly related to the relatively weak adhesion of the HA coating to the metal substrate, associated with the specific use of these implantable devices. Especially considering the mechanical stresses associated with the insertion of screw-shaped dental implants, thick and relatively brittle bioceramic coatings such as plasma spray hydroxyapatite have shown limitations. Moreover, the dissolution rate of the calcium phosphate coatings, which depends on the degree of crystallinity and on the adopted deposition technique, is another debated and still unresolved problem.

Considering such problems, of particular interest, especially when considering applications in the dental field, are the various treatments for improving osseointegration, and in particular those using chemical and electrochemical modification of the titanium surface, that will be discussed in the following section.

24.3 Chemical and electrochemical modification treatments of titanium

There are many possible ways to modify the titanium surface through chemical methods, such as alkali or acid etching, oxidation with hydrogen peroxide, sol-gel routes, etc. Despite their differences, all these methods try to simulate a 'biologically friendly' environment on the titanium surface. In particular, using anodic oxidation and some chemical treatments, titanium surfaces can be effectively modified in order to achieve a surface bioactive layer able to provide biomimetic capability (Brunette et al., 2002).

Chemical and electrochemical modification of the surface for the improvement of the biocompatibility of titanium and titanium alloys are currently extensively studied and some applications are today clinically applied. Such treatments can be used to modify the surface morphology and roughness on different scales, as well as to change the chemical composition of the metal oxide surface for biological optimization and tissue integration improvement.

Osseointegration of metal implants always involves the *in vivo* enhancement of osteogenetic activity of bone cells, but the *in vitro* biological response of osteoblasts is a good provisional parameter to evaluate the performance of the metal surface (Deligianni *et al.*, 2001). Surface topography and chemistry significantly influence cell morphology, proliferation, differentiation and, more specifically, the gene expression activity.

Titanium and its alloys are known to be a suitable substrate for bone cell colonization (Brunette *et al.*, 2002). It was observed that osteoblasts improve proliferation when cultured on smooth titanium surfaces, while expressing higher differentiation growing on roughened surfaces (Boyan *et al.*, 1998). Surface modification at the micro- and nanoscale can guide direct osteoblast phenotype by altering adhesion, movement, morphology, apoptosis and, in general, protein production. A balance between the achievement of a high osteogenic cell colonization and the enhancement of their differentiation into mature osteoblasts is one of the key factors for *in vivo* implant osseointegration (Boyan *et al.*, 1998). Chemical and electrochemical etching processes using different acid solutions (hydrofluoric, oxalic, sulphuric, hydrochloric, etc.) at different temperatures and for different treatment times can be used to modify the titanium surface morphology (Fig. 24.1). In particular, the roughness modification at the sub-micrometric scale was

24.1 SEM image of the surface of an acid-etched dental implant thread. The etching provides a micro roughened surface with minimal modification of the sharpness of the thread's edge.

observed to play a fundamental role in the cell differentiation process and in the synthesis of osteogenic factors (Giordano et al., 2006).

24.3.1 Cathodic polarization

Along with surface morphology modification, electrochemical treatments can be used to obtain deposition of different crystalline phases. Cathodic polarization deposition was widely investigated in the 1990s (Shirkhanzadeh, 1991, 1995; Redepenning et al., 1996). Deposition by cathodic polarization can effectively be used to coat the surface of titanium and other metals with variable thickness layers of different calcium phosphate structures.

If the titanium component is placed in an electrochemical solution and connected to an electric circuit as cathode, many reactions may take place on its surface, depending on the pH of the solution, on the current density and on the voltage reached. This simple electrochemical technique can be used and optimized to cover the metal surface by different chemical compounds, precipitated by the solution. Some authors coated titanium and its alloys with calcium-phosphate brushite and apatite crystals by coupling cathodic deposition with further thermal or chemical treatments (Shirkhanzadeh, 1991, 1995; Redepenning et al., 1996; Ban and Maruno, 1995).

Shirkhanzadeh (1991) studied and obtained bioactive calcium phosphate ceramic coatings using cathodic deposition on titanium and titanium alloys. The electrolyte solution used by Shirkhanzadeh and colleagues was a solution prepared by dissolving tribasic calcium phosphate salt in a sodium chloride water solution. These coatings were achieved by potentiostatic polarization of the cathode with a potential in the range from −1500 to −1300 mV (vs. SCE, Standard Calomelan Electrode). The coatings obtained exhibited interesting features: they were micro-porous and showed an interlocking network of non-oriented plate-like apatite crystals (Fig. 24.2). Further studies carried out by the same team (Shirkhanzadeh, 1995) showed that the properties of the deposited phases were dependent on several parameters: electrolyte concentration, pH, applied voltage, ionic strength of the electrolyte, electrolyte temperature, cathode surface state, solution agitation and, of course, electrolyte composition. Shirkhanzadeh introduced a further stage for the coating technique by performing a hydrothermal treatment and a further thermal treatment to the cathodic deposition. The proposed sequence of treatments was able to produce a thin, relatively adherent and uniform hydroxyapatite coating.

Other authors (Redepenning et al., 1996), following Shirkhanzadeh's methods, developed a treatment to coat metals with pure brushite, and further transformed brushite into hydroxyapatite by solid state transforma-

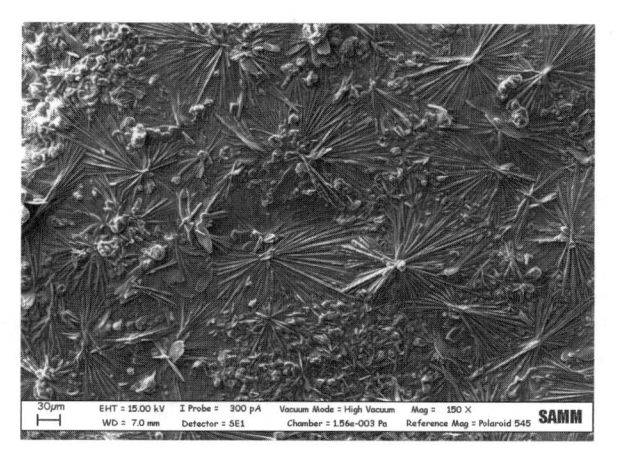

Surface modification of titanium for enhancement of cell response 597

24.2 SEM image of a brushite structured calcium phosphate coating obtained by cathodic polarization on titanium surface.

tion. These authors investigated and performed a cathodic polarization of the titanium substrate in an aqueous solution of $Ca(H_2PO_4)_2$, maintaining the current at the steady state by applying variable voltage in the range 0–100 V.

The electrochemical model proposed by these authors may also explain the Shirkhanzadeh deposition methods. The first step of the process is a preliminary cathodic water reduction with production of hydrogen and hydroxide ions. The hydroxide ions generated at the surface may react with di-hydrogen phosphate according to the equilibrium shown in reaction 24.1:

$$OH^- + H_2PO_4^- \leftrightarrow H_2O + HPO_4^{2-} \qquad 24.1$$

The precipitation of brushite ($CaHPO_4 \cdot 2H_2O$) finally happens in the presence of calcium ions (reaction 24.2):

$$Ca^{2+} + HPO_4^{2-} + 2H_2O \leftrightarrow CaHPO_4 \cdot 2H_2O \qquad 24.2$$

The deposition rate and the deposition morphology can be controlled by controlling the current density. Indeed the current density is associated with the pH at the surface; while the pH determines the reaction equilibrium of reaction 24.1, the HPO_4^{2-} concentration influences the equilibrium of reaction 24.2 and therefore the deposition rate of brushite.

Preliminary *in vivo* tests demonstrated the effectiveness of these methods to enhance osteointegration. The treated specimens implanted in rabbit exhibited higher push-out strength and bone apposition than pure titanium (Redepenning *et al.*, 1996). The low stability of the brushite structure of the calcium phosphate coating obtained can be improved by a chemical conversion to a more stable hydroxyapatite structure by chemical post-deposition treatment operated at high pH (Han *et al.*, 1999, 2001).

Current research on cathodic polarization coating methods is reaching new interest due to the possibilities offered by this technique to insert in the calcium phosphate coatings bioactive molecules and drugs, such as antibiotics or other drugs, during or after the electrochemical process.

24.3.2 Anodic oxidation

In anodic oxidation, electrode reactions in combination with electrical field-driven metal and oxygen ion diffusion lead to the formation of an oxide film at the surface of the anode. Protective films may be achieved by anodic oxidation on metals such as titanium and aluminium. Anodic oxidation might be used for producing increased oxide thickness, coloration and porous coatings. Physicochemical, morphological and structural features of anodic oxides on titanium can be varied by modifying process parameters such as the anode potential, the current density, the electrolyte composition and the process temperature.

The main reactions leading to anodic oxidation of titanium surface are the following:

At Ti/Ti oxide interface:

$$Ti \leftrightarrow Ti^{2+} + 2e^- \qquad 24.3$$

At Ti oxide/electrolyte interface:

$$2H_2O \leftrightarrow 2O^{2-} + 4H^+ \qquad 24.4$$

$$2H_2O \leftrightarrow O_2 \text{ (gas)} + 4H^+ + 4e^- \qquad 24.5$$

At both interfaces:

$$Ti^{2+} + O_2 \leftrightarrow TiO_2 + 2e^- \qquad 24.6$$

The externally applied electric field drives through the oxide the Ti and O ions formed by the redox reactions, leading to the oxide film growth. This obtained oxide film exhibits high resistivity, and the voltage drop will mainly occur over the anode oxide film. The oxide will continue to grow as long as the electric field is strong enough to drive the ions through the film. The oxide film will increase in thickness and its growth will be almost linearly dependent on the applied voltage until the breakdown of the film occurs. The dielectric breakdown, depending on the electrolyte nature, varies over a wide range, usually between 100 and 160 V.

The anodizing process may be carried out by both constant voltage and constant current control. When anodizing is carried out at voltages higher than the dielectric limit, the process will lead to increased gas evolution and often to the development of sparks, with higher film thickening and porous oxide morphology development.

24.3.3 Anodic oxidation of titanium below dielectric breakdown

Anodic oxide film achieved below the breakdown limit was extensively studied by many authors in the 1990s (Larsson *et al.*, 1994, 1996; Ask *et al.*, 1990; Lausmaa *et al.*, 1990). One interesting characteristic of anodized titanium is the bright colours produced by the treatment. These colours are due to light interference and diffraction phenomena through oxide of different thickness. The thickness can easily be varied almost linearly with the applied voltage, and many colours can be obtained in several solutions.

XPS analysis confirmed that anodic films are mainly composed of titanium dioxide (Lausmaa *et al.*, 1990). Many works reported the presence of electrolyte elements embedded in the titanium dioxide, along with oxygen enrichment (Delplancke and Winand, 1973; Aladjen, 1973; Lausmaa *et al.*, 1990). Ti6Al4V behaves very similarly to pure graded titanium, although some variability in the Al and V concentrations in the oxide with respect to the metal substrate have been observed.

Different kinds of structural defects, such as pits and sometimes pores, depending also on the solution, were often detected (Larsson *et al.*, 1994). Diffraction studies pointed out the oxide films to be either totally amorphous or in some cases partially crystalline. The most commonly observed phases of titanium oxide are anatase and rutile (both tetragonal), but also Ti_3O_5 and brookite (orthorhombic or tetragonal) have been detected (Aladjen, 1973; Delplancke and Winand, 1973). Besides the possibility of changing the colour and structure of the oxide film, another recent interesting application of anodic polarization below the dielectric breakdown concerns the immobilization of biomolecules on the titanium surface. Oligonucleotides can be immobilized on the titanium surface by a proper immobilization process, involving electrochemical anodic polarization (Michael *et al.*, 2007). Just to summarize such a technique, oligonucleotide anchor strand (AS) sequences can be oriented on the titanium surface due to weak electrostatic interactions (Fig. 24.3a) related to the amphoteric behaviour of titanium oxide, that is dependent on the isoelectric point (IEP), on the oligonucleotide pK_a values at a specific pH of the solution. The oriented strands may adsorb on the titanium surface (Fig. 24.3b), and anodic polarization can be used to grow the oxide film and to bind the AS previously oriented and adsorbed onto the (oxide) surface (Fig. 24.4a). The immobilized anchored strands can be finally used to bind complementary strands (Fig. 24.4b), perhaps carrying functional groups and biomolecules.

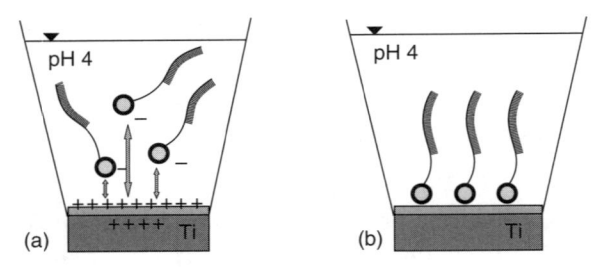

24.3 Oligonucleotide single strands: (a) orientation and (b) adsorption on titanium surface.

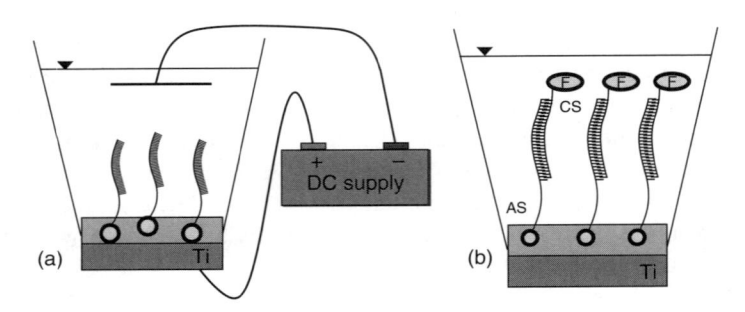

24.4 (a) Oligonucleotide immobilization by anodic polarization. (b) The titanium oxide growth binds the terminal group of the strand. Complementary oligonucleotide strands, carrying biomolecules or functional groups, can be finally bound to the anchored strands.

24.3.4 Anodic oxidation of titanium above dielectric breakdown: Anodic Spark Deposition

Anodic Spark Deposition (ASD) is an electrochemical technique inducing the formation of ceramic coatings on the metal surface. ASD is applicable to the so-called valve metal. A valve metal, such as titanium, aluminium and tantalum, is a material whose resulting thin film after anodization passes electrical current readily only in one direction.

During anodic oxidation of titanium in ionic solutions, the surface oxide layer increases in thickness. Increasing the electric potential, the breakdown of the dielectric film eventually occurs, and high-density current may generate high-temperature sparks, capable of melting the oxide layer surface and incorporating the metal ion species from the surrounding electrolytic solution. A ceramic coating will be formed with a typical microporous structure (Fig. 24.5a) composed of titanium oxide and the other elements (Fig. 24.5b) contained in the electrolytic solution. Many aqueous solutions of different electrolyte composition are suitable for ASD and the choice of the metal/

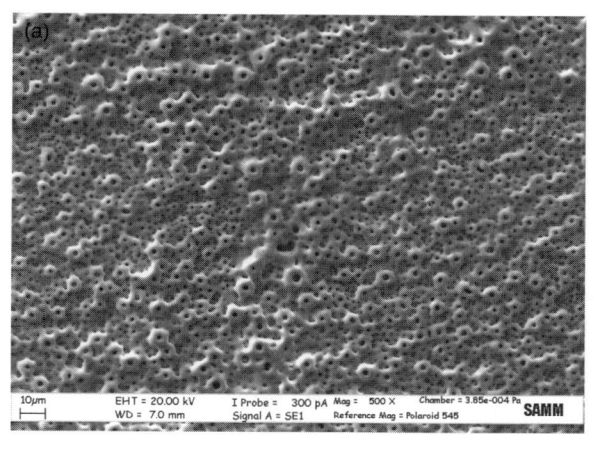

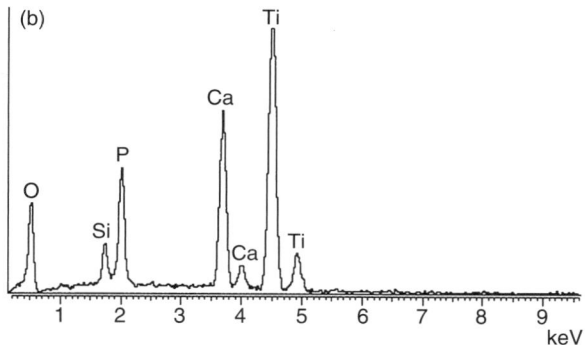

Surface modification of titanium for enhancement of cell response 601

24.5 (a) SEM image of a titanium surface treated by the Anodic Spark Deposition (ASD) process: a microporous titanium oxide layer covers the titanium surface. (b) EDS spectrum of the ASD coating of Fig. 24.5a: the coating is mainly composed of Ti, O, Ca, P and Si.

electrolyte couple is fundamental for designing a specific coating. One or more constituents of the anode usually take part in the ASD reaction and the final coating is composed of the metal alloy elements and the solution ionic species.

Among the many variables affecting the coating properties and characteristics, the most important are:

- composition of the electrolyte
- current density
- final voltage
- process temperature and time.

ASD treatments of titanium and its alloys ensure the formation of a thin superficial layer whose chemical composition and morphology can be controlled. Moreover, layers porous in morphology and containing calcium, phosphate, silicon, sodium and other elements which can promote tissue ingrowth can be designed and obtained (Sandrini et al., 2003). Since a great variety of results have been published by many authors concerning the osteointegration potential of ASD treated titanium implants (Chiesa et al., 2003), biocompatibility studies are necessary to understand the real benefit of this surface functionalization technology. The extremely interesting possibility of modifying the surface morphology and of enriching the titanium oxide layers with several ions, and specifically of controlling the thickness, chemical composition, coating stability and surface morphology and roughness, is currently the main area of investigation of many biomaterials groups in Europe and the United States, and new products are currently in clinical application or development.

24.4 The biocompatibility enhancement of titanium

24.4.1 Biocompatibility: material–host tissue interaction

A widely recognized definition of biocompatibility is '*the ability of a material to perform an appropriate host response in a specific application*' (Williams, 1991). This definition focuses on the idea of interaction between the biological environment, material and material function. After implantation, a biomaterial always induces a biological response in the surrounding biological environment. This response depends on the material properties, on the device biomechanics and function, as well as on the specificity of the tissue and on the host conditions. The host and the biomaterial always interact, and nowadays it is generally recognized that no biomaterial can be considered absolutely inert (Lemons, 1990).

Other important aspects concern the dynamic process of biocompatibility. After insertion of the implant, biomaterials are always subjected to changes caused by mechanical, chemical or structural modification. The biomaterial modification, induced by the surrounding biological environment and by the specific device function, especially concerns the material surface.

The interactions between material, function and host last throughout the implant positioning, but it is recognized that the initial interactions with the surrounding tissue will determine the success or failure of the implant. Therefore, it is not possible to define the biocompatibility of a material without framing it in the context of its localization and function inside the body. The implantation of any medical device generates a sequence of biological reactions, each playing a specific role in the biocompatibility process.

The first critical phase during surgical implantation of any biomedical device is the tissue damage; the damage leads to the formation of a wound

Surface modification of titanium for enhancement of cell response

where bleeding is gradually stopped by the formation of the clot. After clot formation, the inflammation process always starts with the involvement of humoral and cellular factors. Humoral factors include the complement system which, once activated, attracts inflammatory cells to the wound site, acting as cell receptors. Contemporarily, polymorphous nucleates leucocytes, lymphocytes and macrophages generate the cellular response. The activity of these cells is controlled by chemotactic and growth factors, which also regulate the generation and growth of new tissue. The implantation of a biomaterial may alter the balance of events involved in the repair process, inducing the so-called host response. Considering the osseointegration of a dental implant, a host response process may result in a chronic inflammation which leads to the formation of a fibrotic capsule around the implant (Silver and Doillon, 1989). The formation of this fibrotic tissue walls off the implant, thus preventing its integration in the surrounding tissue. As a consequence, the implant cannot form a structural continuum with the bone tissue, and its mechanical performance may be significantly impaired. It is widely recognized that this problem can be managed by modifying the biomaterial surface chemistry and topography by physical and chemical treatment modifications.

24.4.2 Osteointegration: material–host tissue interaction

Osteointegration is a process which allows the direct apposition of mature and mineralized bone tissue on the surface of an implanted material without interposition of soft tissue (Albrektsson et al., 1983). Indeed, one of the main reasons for the failure of dental implants is the deposition of fibrous tissue at the implant/bone interface. This kind of tissue, interrupting the structural continuity between bone and implant, may impair the optimal mechanical load transfer from the load-bearing implant to the surrounding tissue, and over time may lead to the implant mobilization and failure.

As a consequence, the American Academy of Implants Dentistry in 1996 defined osteointegration as the contact between an implant and the bone tissue that is shaped without a direct interposition of non-bone tissue; this contact must be able to transmit an ideal stress distribution at the interface.

Osteointegration, as well as any material–tissue interaction, depends on various and numerous factors related to the features of the implanted device (i.e., material, shape, mechanical properties, superficial treatments, topography and others), to the implantation technique (shape of the implantation site, cleaning of the surfaces, exact introduction of the implant, etc.), and to the general physiological condition of the patient (age, weight, height, hormonal balance, illnesses, etc.).

Many studies are oriented towards developing knowledge on bone biology in order to develop new biomaterials with an enhanced osteointegration potential; among them, surface treatments are aimed at guaranteeing a short healing time and longer durability of the implant.

In the specific case of bone/titanium implant interfaces, osseointegration is strictly related to the ability of the material to integrate in the surrounding bone tissues by forming a tight mechanical interconnection with the mineral phase produced by the bone remodelling process.

24.4.3 Bone/implant interface reactions: role of surface topography

The bone–implant interface plays a key role in enhancing the integration and fixation of the implanted devices. In particular, at a cellular level, it was shown that the phenotypic expression of osteoblasts is modulated by the substrate surface topography (Nanci and Poleo, 1999; Martin *et al.*, 1995; Deligianni *et al.*, 2001; Lincks *et al.*, 1998; Anselme *et al.*, 2000). Although the chemistry of a surface certainly plays an important role, when roughness increases ($R_a > 2$ μm), proliferation of cells may decrease, and their morphology becomes more typical of mature osteoblasts (Lincks *et al.*, 1998).

In vitro studies have provided some insights into the response of specific cell types to surface properties. In particular, osteoblast-like cells exhibit roughness-dependent phenotypic characteristics, and they tend to adhere more readily to surfaces with a rougher morphology. Moreover, they appear to be more differentiated on rougher surfaces with respect to morphology, extracellular matrix synthesis, alkaline phosphatase specific activity and osteocalcin production (Boyan *et al.*, 1998).

Dental implant roughness study is fundamental to understanding osseointegration properties of implant design. The two milestones of dental implant roughness modification history have been the titanium plasma-sprayed roughened surface and later sandblasting and chemical etching surface treatments. These surface modification treatments are today widely applied on dental implants, with a high rate of success.

24.4.4 Bone/implant interface reactions: role of surface chemistry

Once implanted *in vivo*, the implant surface will adsorb proteins from the plasma, and the cell/biomaterial interface reactions are surely dependent on the proteins adsorbed on the implant surface (Eckert *et al.*, 1997). Host response and cellular processes are consequently modulated and mediated by the protein-enriched biomaterial surface (Kirkpatrick *et al.*, 1997). When a titanium device is implanted, the surface of the titanium oxide layer undergoes important chemical changes, incorporating calcium, phosphorus and other ionic species from the plasma (Nanci and Poleo, 1999). The host response to titanium implants always involves chemicals, proteins and several biological and cell activating factors that lead to inflammation reactions and, in the normal and positive situation, to the apposition of new extracellular matrix from the mineralized tissue and, finally, to the implant osseointegration. Implant fixation is a dynamic process influenced also by mechanical stimulations from the new load transfer configuration, and the bone remodelling is a continuous and dynamic process both at the biomaterial interface and in bone sections distant from the implant. All mechanical, chemical and biological factors impairing the osseointegration process should be recognized and controlled, while the many factors recognized to improve biomaterial/tissue integration capability have to be studied and evaluated on the device. A greater understanding of the overall picture of the tissue integration process requires the effort of an interdisciplinary team: engineers, biologists, biochemists, cell biologists, material scientists (nanomedicine experts) and clinicians.

24.5 Conclusions

Titanium and its alloys have been successfully used in the biomedical field. In the last decade, researchers have started to understand the mechanisms involved that make this metal one of the best materials for load transfer implantable applications. The specific mechanical and chemical properties offer bioengineers the possibility of designing innovative devices in which properties such as high mechanical performance can be combined with the intriguing biological modifications possible for this metal and its alloys.

Surface modification of titanium implants aims to achieve a better, faster and more durable implant integration through the stimulation of many biological and cellular factors. Although research during the last decade has focused particularly on the understanding of the role of the many biological factors on the promotion or inhibition of biomaterial/tissue integration, the comprehension of the whole mechanism has still to be achieved. Nevertheless, the new class of physical, chemical and electrochemical surface modification technologies that can be effectively applied to titanium will help biomaterials scientists to better address the challenging requirements for new applications and improve the functionalities of the next generation of implantable medical devices.

24.6 References

Aladjen A. 1973. Anodic oxidation of titanium and its alloys. *Journal of Materials Science* 8: 688–704.

Albrektsson T, Hansonn H-A, Kasemo B, Larson K, Lundstrom I. 1983. The interface zone of inorganic implants in vivo: titanium implants in bone. *Annals of Biomedical Engineering* 11: 1–27

Anselme K, Linez P, Bigerelle M, Le Maguer D, Le Maguer A, Hardouin P, Hildebrand HF, Iost A, Leroy JM. 2000. The relative influence of the topography and chemistry of Ti6Al4V surfaces on osteoblast cell behaviour. *Biomaterials* 21: 1567–1577

Ask M, Lausmaa J, Kasemo B, Rolander U. 1990. Microsctructure and morphology of surface oxide films on Ti6Al4V. *Journal of Materials Research* 5: 1662–1667

Ban S, Maruno S. 1995. Effect of temperature on elecrochemical deposition of calcium phosphate coatings in a simulated body fluid. *Biomaterials*, 16: 977–981

Boyan BD, Batzer R, Kieswetter K, Liu Y, Cochran DL, Szmuckler-Moncler S, Dean DD, Schwartz Z. 1998. Titanium surface roughness alters responsiveness of MG63 osteoblast-like cells to 1 alpha, 25-(OH)2D3. *Journal of Biomedical Materials Research* 39(1): 77–85

Brossa F, Cigada A, Chiesa R, Paracchuni L, Consonni C. 1993. Adhesion properties of plasma sprayed hydroxyapatite coatings for orthopedic prostheses. *Biomedical Materials and Engineering* 3(3): 127–136

Browne M, Gregson PJ. 1994. Surface modification of titanium alloy implants. *Biomaterials* 15: 894–898

Brunette DM, Tengvall P, Textor M, Thompsen P. 2002. Metallurgy and technological properties of titanium and titanium alloys. *Titanium in Medicine* 15: 487–489

Chae JC, Collier JP, Mayer MB, Surprenant VA, Dauphinais LA. 1992. Enhanced ingrowth of porous-coated CoCr implants plasma spray with tricalcium phosphate. *Journal of Biomedical Materials Research* 56: 93–102

Chiesa R, Sandrini E, Santin M, Rondelli G, Cigada A. 2003. Osteointegration of titanium and its alloys by anodic spark deposition and other electrochemical techniques: a review. *Journal of Applied Biomaterials and Biomechanics* 1: 91–107

Deligianni DD, Katsale N, Ladas S, Sotiropolou D, Amebee J, Missirlis YF. 2001. Effect of surface roughness of the titanium alloy Ti-6Al-4V on human bone marrow cell response and on protein adsorption. *Biomaterials* 22: 1241–1251

Delplancke JL, Winand R. 1973. Galvanostatic anodization of titanium – I. Structures and composition of the anodic films. *Electrochimica Acta* 33(11): 1539–1549

Ducheyne P, Healy KE. 1988. The effect of plasma-sprayed calcium phosphate ceramic coatings on the metal ion release from porous titanium and cobalt–chromium alloys. *Journal of Biomedical Materials Research* 22: 1137–1163

Eckert R, Jeney S, Hörber JK. 1997. Understanding intracellular interactions and cell adhesion: lessons from studies on protein–metal interactions. *Cell Biology International* 21: 707–713

Fini M, Cigada A, Rondelli G, Chiesa R, Giordano R. 1999. In vitro and in vivo behaviour of Ca- and P-enriched anodized titanium. *Biomaterials* 20: 1587–1594

Giordano C, Sandrini E, Busini V, Chiesa R, Fumagalli G, Giavaresi G, Fini M, Giardino R, Cigada A. 2006. A new chemical etching process to improve endosseous implant osseointegration: in vitro evaluation on human osteoblast-like cells. *International Journal of Artificial Organs* 29: 772–780

Globus RK, Doty SB, Lull JC, Holmuhamedov E, Humphries MJ, Damsky CH. 1998. Fibronectin is a survival factor for differentiated osteoblasts. *Journal of Cell Science* 111: 1385–1393

Han Y, Xu K, Lu J. 1999. Morphology and composition of hydroxyapatite coatings prepared by hydrothermal treatment on electrodeposited brushite coatings. *Journal of Materials Science: Materials in Medicine* 10: 243–248

Han Y, Fu T, Lu, J, Xu, K. 2001. Characterization and stability of hydroxyapatite coatings prepared by an electrodeposition and alkaline-treatment process. *Journal of Biomedical Materials Research* 54: 96–101

Hanawa T. 1999. In vivo metallic biomaterials and surface modification. *Materials Science and Engineering* A267: 260–266

Hlady V, Furedi-Mihofer H. 1979. Adsorption of human serum albumin on precipitated hydroxy-apatite. *Journal of Colloid Interface Science* 69: 460–468

Howlett CR, Zreiqat H, Wu Y, McFall DW, McKenzie DR. 1999. Effect of ion modification of commonly used orthopedic materials on the attachment of human bone-derived cells. *Journal of Biomedical Materials Research* 45(4): 345–354

Kirkpatrick CJ, Wagner M, Kohler H, Bittinger F, Otto M, Klein CL. 1997. The cell and the molecular approach to biomaterial research: a perspective. *Journal of Materials Science: Materials in Medicine* 8: 131–141

Ku CH, Pioletti D, Browne M, Gregson PJ. 2002. Effect of different Ti-6Al-4V surface treatments on osteoblast behaviour. *Biomaterials* 23: 1447–1454

Larsson C, Thomsen P, Lausmaa J, Rodahl M, Kasemo B, Ericson LE. 1994. Bone response to surface modified titanium implants: studies on electropolished implants with different oxide thicknesses and morphology. *Biomaterials* 15: 1062–1074

Larsson C, Thomsen P, Lausmaa J, Rodahl M, Kasemo B, Ericson LE, Aronsson BO. 1996. Bone response to surface modified titanium implants: studies on the early tissue response to machined and electropolished implants with different oxide thicknesses. *Biomaterials* 17: 605–616

Lausmaa J, Kasemo B, Mattsson H, Odelius H. 1990. Multi-technique surface spectroscopic characterisation of electropolished and anodised Ti. *Applied Surface Science* 45: 189–200

Lemons J. 1990. Dental implant biomaterials. *Journal of the American Dental Association* 121: 716–719

Lincks J, Boyan BD, Blanchard CR, Lohmann CH, Liu Y, Cochran DL, Dean DD, Schwartz Z. 1998. Response of MG63 osteoblast-like cells to titanium and titanium alloy is dependent on surface roughness and composition. *Biomaterials* 19: 2219–2232

Martin JY, Schwartz Z, Hummert TW, Schraub DM, Simposon J, Lankford J, Dean DD, Cochran DL, Boyan BD. 1995. Effect of titanium surface roughness on proliferation, differentiation, and protein synthesis of human osteoblast-like cells (MG63). *Journal of Biomedical Materials Research* 29: 389–401

Michael J, Beutner R, Hempel U, Scharnweber D, Worch H, Schwenzer B. 2007. Surface modification of titanium-based alloys with bioactive molecules using electrochemically fixed nucleic acids. *Journal of Biomedical Materials Research Part B: Applied Biomaterials* 80: 146–155

Nanci A, Poleo DA. 1999. Understanding and controlling the bone–implant interface. *Biomaterials* 20: 2311–2321

Redepenning J, Schlessinger T, Burnham S, Lippiello L, Miyano J. 1996. Characterisation of electrochemically prepared brushite and hydroxyapatite coatings on orthopaedic alloys. *Journal of Biomedical Materials Research* 30: 287–294

Sandrini E, Chiesa R, Rondelli G, Santin M, Cigada A. 2003. A novel biomimetic treatment for an improved osteointegration of titanium. *Journal of Applied Biomaterials and Biomechanics* 1: 33–42

Scharnweber D, Beutner R, Rössler S, Worch H. 2002. Electrochemical behaviour of titanium-based materials – are there relations to biocompatibility? *Journal of Materials Science: Materials in Medicine* 13: pp. 1215–1220

Shirkhanzadeh M. 1991. Bioactive calcium phosphate coatings prepared by electrodeposition. *Journal of Materials Science Letters* 10: 1415–1417

Shirkhanzadeh M. 1995. Calcium phosphate coatings prepared by electrocrystallization from aqueous electrolytes. *Journal of Materials Science Letters* 6: 90–93

Silver F, Doillon C. 1989. *Biocompatibility: Interactions of Biological and Implantable Materials*. New York: VCH Publishers

Stephansson SN, Byers BA, Garcia AJ. 2002. Enhanced expression of the osteoblastic phenotype on substrates that modulate fibronectin conformation and integrin receptors. *Biomaterials* 23: 2527–2534

Tengvall P. 2001. Proteins at titanium interfaces. In Brunette DM, Tengvall P, Textor M and Thomsen P, *Titanium in Medicine: Materials Science, Surface Science, Engineering, Biological Responses and Medical Applications*. Berlin: Springer, Chapter 14

Thompson GJ, Puleo DA. 1996. Ti-6Al-4V ion solution inhibition of osteogenic cell phenotype as a function of differentiation time-course in vitro. *Biomaterials* 17: 1949–1954

Tisel CL, Goldberg VM, Parr JA. 1994. The influence of hydroxyapatite and tricalcium phosphate coating on bone growth into titanium fiber-metal implants. *Journal of Bone Joint Surgery* 76: 139–171

Williams DF. 1991. Definitions in biomaterials. *Proceedings of the 2nd Consensus Conference on Biomaterials*, Chester, UK

Index

acellular peripheral nerves 263
acellular scaffolds 190, 221–2
acid–base reaction 164
acid etching 595–6
acidic by-products 196
acridine orange (AO) 550
actin cytoskeletal proteins 380, 381
actuators 128–9
adhesion tension 94–5
adipose tissue-derived stem cells (ADSC) 298–9, 301
   immunomodulation 306
adsorption
   mechanisms of 93–7
   peptides 486, 487
   proteins *see* protein adsorption
adsorption isotherms 93
adult progenitor cells 230
adult stem cells, sources of 294–300, 301
adventitia 213, 214
affine hydrogel networks 115
agar diffusion test 517, 518–19
Alamar blue assay 378–9, 383, 384, 515
albumin 41
   attachment of endothelial cells to SAMs 105
   protein adsorption 96–7, 98, 99
albumin/fibrinogen adsorption ratio 403
alginates 39–40
   composites 47–8
   muscle regeneration 243–4
alignment, fibre 11, 12–13
Alkaline Comet assay 526, 530–1
alkanethiols on gold SAMs 90–2
   surface chemistry and cell response 102–6
allografts 262
   bone grafts 314, 327, 329–32, 344, 560–1
   DBM 331–2
α_1-antichymotrypsin (ACT) 539
   protein–cell interactions 550–5
α_2-HS-glycoprotein (AHSG) 539–40
   protein–cell interactions 550–5

alumina (aluminium oxide) 137, 143–4, 158, 576
American Society for Testing and Materials (ASTM) 509
   standards for titanium alloys 590
Ames test 526, 527
anchored strands 599–600
angiogenesis 345–53
anionic hydrogels 126
anodic oxidation 598–602
   above dielectric breakdown 600–2
   below dielectric breakdown 599–600
anodic spark deposition (ASD) 582, 583, 584, 600–2
antigens 544
anti-human angiogenesis-related endothelial cells 347–8
antimicrobial glasses 165–6, 170–2
   glass fibres 177
apatite-like layer 145–6, 147, 149
apatite/wollastonite (A/W) glass-ceramics 148–9
apoptosis 74–5, 438, 519–25
   assessment of 523–5
   biochemical mechanisms of 520–3
apoptotic inducing factor (AIF) 524
arterial grafts 216
artery structure 213
articulating implants 406–8
   see also hip prostheses
artificial degradation methods 72
artificial ECM (aECM) proteins 89
artificial muscles 128–9
atomic force microscopy (AFM) 544
   protein–cell interactions 551, 552
   in tapping mode 545–6
autografts 254
   bone grafts 314, 326–9, 344, 560–1
   peripheral nerves 262
autologous blood vessels 212, 215–16
autologous bone marrow 329

609

axons 255, 264–5, 267
  degradation 257, 258
  outgrowth 260–2

bacterial reverse mutation assay (Ames test) 526, 527
basal lamina tubes 272
β-sheets 498, 499
bicinchoninic acid (BCA) protein assay 541–3
binary phosphate glasses 166
bioabsorbable glass fibres 245
bioabsorbable polymers 29
bioactive glass-ceramics 147–9
bioactive glasses 53, 146–7
  see also Bioglass
bioactive materials 138, 139
  ceramics 45–6, 145–9
bioactive surfaces 479–507
  biomimetic materials and 479–83
  criteria for design 489–90
  delivery of bioactive peptides to tissue–surface interface 486–9
  future trends 502
  in vitro and in vivo applications 490–502
    bone tissue engineering 490–5
    cardiovascular tissue engineering 500–1
    neural tissue engineering 495–500
  peptide mimicry for bioactive surface design 483–6
bioactivity index 146
bio-artificial tendons (BATs) 241
bioceramics 45–6, 136–55
  biological and mechanical behaviour 138–41
  cell reactions 142–9
  dense vs porous 141–2
  future trends 149–50
  history of use 136–7
  types used in the body 137–8
biochemical modification of surfaces 463–4
biocompatibility 509
  biocompatible polymers 293
  DLC coatings 401–5
  enhancement of titanium 602–5
biocompatibility testing 61–84
  approach to 64–5
  degradation mechanisms and devising the appropriate tests 70
  immune response 76–9
  in vitro testing 65, 70–5, 509–12
  in vivo testing 65, 75–6
  practical problems 68–70
  standards 63–4, 509, 510, 511, 512
biodegradable composites 42–9, 53
  injectable composites 51–2
  scaffolds for tissue engineering 42–4
  types of composites 44–9

biodegradable polymers 28–60
  applications 66–8
  biocompatibility testing see biocompatibility testing
  degradation 36–9
  drug delivery from resorbable scaffolds 52
  fibre scaffolds 49–51
  muscle regeneration 243, 244–5, 247–8
  natural see natural polymers
  synthetic see synthetic polymers
biodegradable scaffolds
  composites 42–4
  vascular grafts 218–20
BioDiamond 409–10
Bioglass 45–6, 48, 137, 146–7, 371–90
  cellular responses to Bioglass reinforced composites 382–5
  cellular responses to particles 372–82
    direct contact test 375–82
    indirect test 373–5
  interaction of HOB cells and biocomposites at the ultrastructural level 385–7
Bioglass–polyethylene (BGPE) composites 383–4, 385–7
bioinert materials 137, 139, 144
biological fixation 141
biological/pharmaceutical agents 224–5
biomimetic materials
  and bioactive surfaces 479–83
  scaffolds 292–4
Bio-Oss 47
bioresorbable materials 138, 139, 143
biosafety 70–1
biostable polymers 29
biphasic calcium phosphate (BCP) 145
Biuret reaction 541–3
blasting 576
blood clotting 405
blood-interfacing applications 408–11
blood vessels 213
  vascular grafts see vascular grafts
BMP-specific antagonists 321–2
body water retainers 128
bone 144
  de novo bone formation 574
  remodelling 324, 334–5, 574
bone cell adhesion 538–59
bone graft materials
  and bone healing 323–4
  classification 326, 327
  and remodelling 334–5
  see also osteoinductive materials
bone graft matrices 344–70
  demineralised intramembranous bone matrix 344–53
  HMGRI–collagen matrix 353–8
  phytoestrogen–collagen matrix 360–4

Index 611

polymethoxylated flavonoid–collagen matrix 358–60
bone healing
　BMPs and fracture healing 321
　bone graft biomaterials and 323–4
　osseointegration and 326
　osteoconduction and 325
　osteoinduction and 316–18
　osteoinductivity and 324–5
bone marrow 316
　autologous 329
bone marrow stromal cells (BMSC) 294–6, 301, 308
　immunomodulation 305–6
　targeting stemness within the bone marrow niche 300–5
bone morphogenesis cascade 316
bone morphogenetic proteins (BMPs) 318–23, 335, 345, 567, 585
　and basic science 319–20
　and BMP-specific antagonists 321–2
　and clinical applications 320–1
　and dose delivery 322
　and safety 322–3
bone-related glasses 168–70
bone repair 291–312
　adult stem cell sources 294–300, 301
　biomimetic scaffolds 292–4
　future trends 306–7
　stem cell-induced immunomodulation 305–6
　targeting stemness within the bone marrow niche 300–5
bone sialoprotein (BSP) 304–5
bone substitutes 314, 327, 332–5
　calcium phosphate 560–71
　　choice and future trends 567
　　degradation 563–6
　　demand 560–1
　　mechanical properties 563, 564
　　requirements 562
　　risk factors in use 567
　phosphate-based glasses 165
bone tissue engineering 567
　bioactive surfaces 490–5
　electrospinning 51
bottom up nanofabrication 430, 433–5
bovine aortic endothelial cells (BAECs) 104–5
bridging oxygen 160

cage implant system 75
cage systems for spinal fusions 335
calcitonin 67
calcium 473
　ion implantation 471–2
　surface chemistry of dental implants 578–9
calcium chloride 579

calcium oxide 168
calcium phosphate 144–6, 292–3
　coatings and bone substitutes 560–71
　　degradation 563–6
　　demand for 560–1
　　future trends 567
　　mechanical properties 562–3
　　requirements for 562
　　risk factors 566–7
　coatings on titanium 594, 596–8
　co-precipitation of genes with 199–200
calcium phosphate cements (CPC) 52
calcium sulphate 136–7
calcium titanate 472–3
cancellous autografts 328
Candida albicans 171
carbon, diamond-like *see* diamond-like carbon
carbonate substituted HA (CHA) 375–85, 387
　nanoCHA 375–82, 385
Carbostent 409, 410
cardiac implant devices 408–9
cardiovascular tissue engineering 500–1
cartilage 49–50, 494–5
caspases 521–3, 525
casting 431–3
catheters 409–11
cathodic polarisation 596–8
cationic hydrogels 126
cell activation 437
cell adhesion 102, 437, 480, 482–3
　and cell signalling 441
　hydrogels 129, 130
　influence of plasma proteins on bone cell adhesion 538–59
　surface chemistry and nanostructural biomaterials 102–6
cell adhesion test 548–9
cell attachment 553–5
　on fibre bundles 21–2
cell-based scaffolds 190
　vascular grafts 222
cell culture media, immersion in 14
cell cultures 71
　3-D and muscle regeneration 245–6
cell death 74–5, 519–25
　see also apoptosis; necrosis
cell fusion 445
cell lines 72
cell-material interaction monitoring techniques 197–8
cell-mediated bone resorption 565
cell movement 438
cell proliferation 378–9, 438
cell shape 438
cell signalling 440–6, 447
cell spreading 72
　plasma proteins 553–5

612　Index

cell–substrate interactions 548–50
Cerabone A/W 148–9
ceramics
　blasting of dental implants 576
　ceramic–polymer composites 293
　for scaffolds 191–2, 197
　see also bioceramics
Ceravital 148
Cerosium 44
ceruloplasmin 539
　protein–cell interactions 550–5
chain scission 36, 37
chemical adsorption 93
chemical etching 582, 583
chemical gels 115, 116
　hydrogels 117–23, 124–6
chemical potential 122
chemical surface modification treatments 594–6
chemical vapour deposition (CVD) 393, 394, 396–7
chemically effective degree of crosslinking 123
chemically patterned surfaces 466–7
chemistry, surface *see* surface chemistry
chitin 40
chitin–polycaprolactone composite 239
chitosan 40–1
　composites 49
chondrocytes 302, 319
　bone tissue engineering 495
chromosomal aberration tests 525, 526, 528–9
chromosomes 442, 448–50
circulating stem cells 296–7, 301
clotting, blood 405
coating–substrate adhesion strength 398–9
coatings
　calcium phosphate 560–71
　　degradation 563–6
　　demand 560
　　mechanical properties 562–3
　　requirements 562
　　risk factors in use 566–7
　DLC *see* diamond-like carbon (DLC) coatings
　release of peptides from 486, 487, 488–9
　vascular grafts 224
co-cultures 300–5
collagen 41–2, 307
　bone graft matrices 353–64, 369–70
　　HMGRI–collagen matrix 353–8
　　phytoestrogen–collagen matrix 360–4
　　polymethoxylated flavonoid–collagen matrix 358–60
　bone tissue engineering 494–5
　composites 47–8
　functionalising electrospun fibres 15–17

muscle regeneration 243–4, 247
　surface chemistry and adsorption of 467–8
　tendons 18–19, 237
　　polypeptides 238
　　regeneration 241
　vascular grafts 220, 224
collagen–HA composite grafts 334
Collagraft 47
colloidal lithography 433–5
colony-forming assay (CFA) 515, 517–18
colony-forming unit-fibroblasts (CFU-f) 294
combination products 334
Comet assay 526, 530–1
Committee on Mutagenicity of Chemicals in Food, Consumer Products and the Environment (COM) 529
compactin 354
composites
　bioactive glasses and HA 372, 382–7
　　interaction of HOB cells and biocomposites at the ultrastructural level 385–7
　biodegradable *see* biodegradable composites
　phosphate-based glasses in 173
compressive strength 267, 268, 270
conditional glass formers 159
conductivity 6
connective tissue sheaths 254–5
contact angle 14, 15
　SAMs and surface wettability 91, 92
contact area (cell signalling) 440
continuous random network theory 157–8
controlled release glasses 164
copolymer combinations of biomaterials 222–3
copper ions 170–1
copper oxide 176–7
copper-releasing phosphate glasses 165
cord blood stem cells 299–300
corrosion 591
　DLC coatings 405–6
Cosecure 165
coumestans 361
counterions 124
covalent grafting 486, 487–8
crosslink density 120–1
　hydrogel scaffolds 130–1
crosslinking, degree of *see* degree of crosslinking
crystallinity, in polymers 32–3
crystallisation 163
crystallisation temperature 158, 159
CT-scanning 566
culture conditions 229–30
cycloheximide treatment 549
cytogenetic analysis 526, 528–9

Index   613

cytokine activators 521, 522
cytoskeleton 439, 447–8, Plate 2, Plate 3
cytotoxicity testing 513–19
  agar diffusion test 517, 518–19
  cellular response to HA 375–6
  colony-forming assay 515, 517–18
  direct contact test 517
  MTT assay 514–16
  neutral red uptake assay 515, 516

Dacron 212, 216, 226, 239
de novo bone formation 574
deacetylation, degree of 40
decellularised scaffolds 190, 221–2
degradable composites *see* biodegradable composites
degradable polymers *see* biodegradable polymers
degradation
  biodegradable polymers 36–9
    degradation mechanisms and devising biocompatibility tests 70
  calcium phosphate coatings and bone substitutes 563–6
    degradation mechanisms 564–5
    estimation of degradation 566
    factors influencing degradation 565
  phosphate-based glasses 163–4
DegraPol 244
degree of crosslinking 118, 119–23
  hydrogel scaffolds 130–1
  mechanical measurements 123
  swelling measurements 121–3
degree of deacetylation 40
degree of ionisation 125
demineralised bone matrix (DBM) 331–2, 344–5
  endochondral (DBM$_{EC}$) 345
  intramembranous (DBM$_{IM}$) 344–53, 357
    preparation of rabbit DBM$_{IM}$ particles 368–9
dense ceramics 141–2
dental implants 572–88, 603
  commercial surface-modified titanium implants 580–2
  current treatment modalities 582–4
  future trends 584–5
  'loss of osseointegration' phenomenon 574–5
  mechanism of peri-implant endosseous healing 574
  and osseointegration 573–4, 592–3, 603, 604
  surface chemistry 578–9
    and cell behaviour 579
  surface topography 575–8, 604
    and cell behaviour 577–8
depolymerisation model 160–2
deposition techniques 393–7

Dexon 66
Diamond Flex 409, 410–11
diamond-like carbon (DLC) coatings 391–426, 473
  biocompatibility 401–5
    haemocompatibility 403–5
    tissue-compatibility 401–2
  characterisation 397
  corrosion behaviour 405–6
  deposition 393–7
  for orthopaedic applications 406–8
  physical properties 397–401
  for selective blood-interfacing applications 408–11
dielectric breakdown 598
differentiation
  BMSC 295–6
  nanofeatures and 450
direct contact test
  cellular response to HA and Bioglass 375–82
  cytotoxicity 517
discrete crystalline deposition (DCD) 581
dissolution 564–5
DNA effects tests 525, 526, 530–1
DNA tail length 530, 531
DNA tail moment 530, 531
Donnan theory 124–5
doping 399
double bonded oxygens (DBOs) 159, 162
drug delivery
  BMPs and 322
  degradable polymers 66–8
  hydrogels 127–8
  resorbable scaffolds 52
dye exclusion assay 515
Dylyn-coated (DLN-coated) stents 409, 411
dynamic protein binding capacity 541

EAK16 peptide 88–9
elastic properties 398
elastically effective crosslink density 123
electrochemical surface modification treatments 594–602
  anodic oxidation 598–602
  cathodic polarisation 596–8
electrohydrodynamic atomisation 582, 583
electron beam lithography 431
electrospinning 3–27, 50–1, 306–7
  advantages 4
  applications in tissue engineering 8–10
  electrostatically spun scaffolds as vascular grafts 230–2
  fibre bundles 19–22
  fibre orientation 12–13
  functionalising fibres 15–17
  future trends 23
  porosity 11–12

process parameters 5–8
  ambient conditions 8
  polymeric solution properties 6–7
  setup 7–8
  surface wettability 14–15
  for tendon regeneration 17–19
ellipsometry 101, 546–8
  protein–cell interactions 551–3
embossing 431–3
embryonic stem (ES) cells 294, 302, 316, 317
endochondral (EC) bone grafts 344, 348–9
  EC-DBM$_{EC}$ composite 345
  EC-DBM$_{IM}$ composite 345–53
endocytosis 438, 450
endoneurium 254–6
endosseus implants 480–3
  mechanism of peri-implant endosseous healing 574
endotenon 18
endothelial cells 255
  attachment to SAMs 104–5
  microvascular 303
  and vascular grafts 213, 214, 217, 227, 230
entubulation 263
  see also nerve conduits/guides
enzyme degradation of polymers 68, 69, 70
epineurium 254–5
epitenon 18
epithelial amniotic stem cells 306
Escherichia coli 171
ethidium bromide (EB) 71
executioner caspases 521, 522, 523
expanded polytetrafluoroethylene (ePTFE) 212, 216, 226
exposure time 510, 512
extracellular matrix (ECM) 18, 86, 192–3, 242–3
  cardiovascular tissue engineering 500–1
  HOB cell responses to HA and Bioglass 380, 381
  self-assembling polymers and mimicking 87–90
  stem cells 304–5
  techniques to aid development of artificial ECM constructs 4
extraction ratio 510, 512
extrinsic pathway of apoptosis 521–3

FAK signalling 470–1
fascicles 254
fetuin 540
fibre alignment 11, 12–13
fibre bundles 19–22
  advantages and disadvantages 20–1
  cell attachment 21–2
  rope-like structures 22
fibre mesh bonding 194
fibre orientation 12–13
fibrillar adhesions 469–70

fibrin 224
  clot 78
fibrinogen 99–100
  albumin/fibrinogen adsorption ratio 403
fibroblasts 78, 227, 230, 255
  bone marrow-derived 303–4
  dermal 303
  electrospun fibres 15, 16
  tendons 237–8, 240
fibronectin 97, 99–100, 224, 246, 465
  adsorption on titanium implants 592–3
  attachment of endothelial cells to SAMs 104–5
  mechanisms of cell responses to surfaces 469–70
fibrous encapsulation 78, 146, 265–6, 573, 574, 603
fibrous scaffolds 49–51, 228
fibrous tissue membrane 565
filopodia 442–4
filtered cathodic vacuum arc deposition (FCVAD) 394, 395–6
fixed charges 118, 124–6
fixed point fibre bundles 19–21
flavonoids 361
  polymethoxylated flavonoid–collagen matrix 358–60
Flory–Huggins polymer–solvent interaction parameter 117–19
Flory–Rehner thermodynamic theory 121
flow-based protein adsorption analysis 540–1, 542
flow cytometry 524
flow rate 7–8
fluorescein diacetate (FDA) 71
fluorescence microscopy 524, 549–50
foam cells 259
Food and Drug Administration (FDA) 509
formazan 514, 515
Fourier-transform infrared spectroscopy 102
fracture healing see bone healing
free radicals 77
freeze drying 195
FRIADENT plus surface 581–2
friction 399–400, 406–7
frustrated phagocytosis 77
functionalised surfaces 97–100
functionalising fibres 15–17

G-protein signalling 443–4
gallium 172
gap method of alignment 12
gelatin
  coating of vascular grafts 224
  composites 47–8
gels 114–16
  see also hydrogels
gene mutation tests 525, 526, 527, 529
gene therapy 199–200

Index 615

genetically modified cells 230
genome, changes in 446–51
genotoxicity testing 525–31
   Alkaline Comet assay 526, 530–1
   Ames test 526, 527
   cytogenetic analysis 526, 528–9
   mammalian gene mutation assay 526, 529
GFOGER peptide 494, 495
glass-ceramics 147–9
glass fibres, phosphate-based 173–7, 243, 244, 245
glass formers (network formers) 159–62
glass modifiers 159–62
glass transition temperature 158, 159, 163
glasses 156
   bioactive *see* bioactive glasses; Bioglass
   formation 158–9
   glass structure theories 157–8
   patterning of 450–1
   phosphate-based *see* phosphate-based glasses
glial scar 496
glow discharge CVD (PACVD) 394, 396–7
glycosoaminoglycan (GAG) chains 89–90
good manufacturing practice (GMP) 201
Gore-Tex 238–9
Graham's salt 156–7
green compact 140
Green Screen HC *GADD45a–GFP* genotoxicity assay 532
grooves 447, 448, 449
growth cones 270–1
growth factors 318, 496
   see also bone morphogenetic proteins
guluronic acid 39

haematoma formation 323
haemocompatibility 403–5
HAPEX 44–5, 383–4, 385–7
'hard' proteins 95–6
hardness 398
healing process 78–9
Healos 47
heart valves 408–9
heparin 496
heparin binding delivery system (HBDS) 496
HGPRT assay 529
hierarchical tendon structure 18
high amplitude swelling 519
high-pressure gas foaming 194
high throughput screening (HTS) 532
hip prostheses 143
   DLC coatings 406–8
   HA-coated 566–7
HMG-CoA reductase inhibitor (HMGRI)–collagen matrix 353–8
host response 602–3
HT-29 cells 303

human bone marrow-derived fibroblasts (HBMFs) 303–4
human dermal fibroblasts (HDFs) 303
human microvascular endothelial cells (HMVECs) 303
human monocyte-derived macrophage 375–7
human osteoblast (HOB) cells
   responses to HA and Bioglass
   Bioglass and HA reinforced composites 382–5
   direct contact test 377–82
   indirect test 374–5
   interaction with biocomposites at the ultrastructural level 385–7
   surface-modified titanium implants 582–4
Human Tissue Act 2004 186
humidity 8
hyaluronic acid (hyaluronan) 42, 221
hydration 164
hydrogels 114–35
   in biomedical fields 126–31
     artificial muscles, actuators and valves 128–9
     body water retainers 128
     controlled drug delivery 127–8
     scaffolds 129–31
   definition and classification 114–16
   microstructural parameters/design variables 117–26
     degree of crosslinking 118, 119–23
     mesh size 118, 124
     micro- and macro-porosity 118, 126
     pendant fixed charges 118, 124–6
     polymer–solvent interaction 117–19
   nerve conduits 272
   swelling ratio 116–17
hydrogen bonds 87
hydrolytic degradation 36–7, 70, 164
hydrophilic surfaces 94, 95
hydrophobic surfaces 94, 95
hydroxyapatite (HA) 137, 144, 145–6, 292–3, 371–90
   cellular responses to HA particles 372–82
     direct contact test 375–82
     indirect test 373–5
   cellular responses to HA reinforced biocomposites 382–5
   coatings 560–71
     degradation 563–6
     demand 560
     mechanical properties 562–3
     requirements 562
     risk factors 566–7
     on titanium 594, 596–8
   composites with natural polymers 47–8, 334
   interaction of HOB cells and biocomposites at the ultrastructural level 385–7

Index

hydroxyapatite reinforced polyethylene (HAPE) composites 44–5, 383–4, 385–7
hydroxycarbonate apatite (HCA) 45–6, 371–2, 375, 387, 472

immune response 76–9
immunoblotting 525, 544
immunofluorescence 379–80, 381
immunoglobulin 99
immunomodulation, stem cell-induced 305–6
immunosuppression 262
in situ-forming hydrogels 128
in vitro preformed vascular network 303
in vitro studies
  phosphate-based glass fibres 175–7
  phosphate-based glasses 165–72
in vitro tests 508–37
  biocompatibility 65, 70–5, 509–12
  toxicity *see* toxicity testing
in vivo studies
  haemocompatibility of DLC coatings 405
  phosphate-based glasses 172
in vivo tests 509
  biocompatibility 65, 75–6
indirect test 373–4
inflammation 603
  biocompatibility testing 76
  bone graft materials 323
  immune response 77, 78–9
  inflammatory response to HA 376–7
infrared spectroscopy 102
initiator caspases 521, 522, 523
injectable composites 51–2
injection moulding 194–5
integrin binding 465, 469, 471
integrins 11, 548–9
  mechanisms of cell responses to surfaces 469–70
Interagency Coordinating Committee on the Validation of Alternative Methods (ICCVAM) 512
interfacial tension 93–4
internal scaffolds 268–74
International Conference on Harmonisation of Technical Requirements for Registration of Pharmaceuticals for Human Use (ICH) 513, 529
International Organization for Standardization (ISO) 61–2, 509
  biocompatibility testing 63–4, 75, 509, 510, 511, 512
  ISO 10993 standard 5832 589–90, 591
  ISO 63–4, 75, 525–6
interpenetrating polymer networks (IPNs) 127

inter-stump gap 263
  bridging a long gap 274
  bridging a short gap 276–7
  see also nerve conduits/guides
intramembranous (IM) bone matrix, demineralised 344–53, 357, 368–9
intrinsic pathway of apoptosis 521–3
ion beam deposition (IBD) 393–5
ion channels 444
ion implantation 471–2
ionic (polyelectrolyte) hydrogels 115, 116, 124–6
ionic strength 125
ionisation, degree of 125
ions, layer of 480, 481
ischaemia 257
islands, nano 434–5, 447, Plate 1, Plate 2
isoelectric focusing (IEF) 543
isoelectric point 126, 482
isoflavones 361, 363
  see also puerarin

keratin hydrogel 272
knee prostheses 407, 408
KRSR peptide sequence 491

lactate dehydrogenase (LDH) leakage assay 375–6, 515
lactic acid 196
lamellipodia 443
laminin 224, 243, 246
lamins 448, Plate 3
Langmuir adsorption isotherm 93
leukocytes 103–4
ligaments 49–50
lignans 361
linear elastic fracture mechanics 140–1
lipid bilayers 87
liquid reservoir fibre bundles 19–21
long gaps, nerve conduits across 274
lovastatin 353–4
lower critical solution temperature (LCST) 119
luciferase 532
lymphocytes 78
lysine 89
lysis index 518

machined (turned) surface 582, 583
macromolecular (polymer) gels 115, 116
  see also hydrogels
macrophages 77–8, 102, 256
  human monocyte-derived and response to HA 375–7
  Wallerian degeneration 258, 259
macro-porosity 118, 126
magnesium ion implantation 472
magnesium oxide 168–9
mammalian gene mutation assay 526, 529

Index   617

mammary artery 216
mannuronic acid 39
mass selected ion beam deposition (MSIBD) 394–5
mass swelling ratio 116–17
mast cells 256
Matrigel 499
mechanical loads
  bone graft materials 324, 334–5
  forces and cell signalling 445, 446
mechanical properties
  calcium phosphate bone substitute 563, 564
  HA coatings 562–3
  implant materials 139
  peripheral nerves 256–7
  scaffolds 189
  tendons 19–20, 237–8
mechanical surface modification treatments 593–4
mechanotransduction 442–4
Medicines and Healthcare Products Regulatory Agency 63
mesenchymal stem cells (MSCs) 306, 316, 317
  surface-modified titanium implants 582–4
mesh size 118, 124
metal element doping 399
metals, implant 138, 139, 186, 190–1, 197
metaphase analysis 528
metaphosphate glass structure 161–2
micro arc oxidation (MAO) 473
microarray technologies 79
microcontact printing 92
microenvironment
  axonal outgrowth 261
  stem cells for bone repair 300–5
micromotion 565
micronuclei 528–9
micro-porosity 118, 126
micro-scale 451
  mixed micro–nano scale biomaterials 200–1, 452
microtopography 430, 575–7
microtube formation 174–5, 177
microvascular-associated stem cells 296–7, 301
mineralisation 51
minifascicles 262, 273, 276
mitochondrial membrane potential (MMP) 524
mitogen-activated protein kinase (MAPK) signalling cascade 470
mixed micro–nano scale biomaterials 452
model surfaces 465–8
molecular Lego 498–9
monocytes 102
  human monocyte-derived macrophage model 375–7

morphology 438
  cells and *in vitro* biocompatibility testing 72–4
  changes and assessment of apoptosis 524–5
  polymers 32–3
mouse lymphoma assay 529
MTS assay 516
MTT assay 373–5, 514–16
multiarm PEG 90
muscles 242–8
  artificial 128–9
  biomaterials used in regeneration 243–8
  structure and function 242–3
myoblasts 245–6
myooids 246
myotendinous junction formation 247

nanoCHA 375–82
nanoCHA–polyhydroxyethylmethacrylate composite 385
nanofabrication 430–6
nanofeatures 429–61
  cell signalling in response to 440–6, 447
  difference between micro-scale and nano-scale reactions 451
  fabrication methods for artificial nanofeatures 430–6
  future trends 452–3
  mixed micro–nano scale biomaterials 452
  range of reactions to nanotopography 436–40
  reactions at the genome 446–51
  why cells should respond to 429–30
nanofibrous scaffolds 11, 307
nanoHA 375–82
nanoimprinting 450
nanoislands 434–5, 447, Plate 1, Plate 2
nanoparticles 149–50
  gene-containing calcium phosphate nanoparticles (GeneCaP) 200
nanoSiHA 378–82, 387–8
nanostructural biomaterials 86–113
  future trends 106
  influence of surface chemistry on cell response 102–6
  influence of surface chemistry on protein adsorption 93–102
  mixed micro–nano scale biomaterials 200–1, 452
  SAMs *see* self-assembled monolayers and scaffolds 198–9, 200–1, 228
  self-assembling polymers 87–90
  self-assembly in nature 86–7
Nanotite 581
nanotopography 430, 577
  practical aspects 452
  range of reactions to 436–40
  relevance to making medical devices 451–2

Index

naringin 358–60, 369
National Institutes of Health (NIH) 509
natural nerve guides 263–4
natural polymers 30, 39–42
  HA composites with 47–8, 334
  nerve guides 266
  for scaffolds 190, 191, 192–3
  scaffolds for vascular grafts 220–1
natural structures as masters 436
necrosis 74–5, 519, 520, 521
needle-tip to collector distance 7, 8
nerve conduits/guides 254, 263–77
  bridging a long gap 274
  bridging short gaps 276–7
  conduit design 267–74
  natural 263–4
  sieve electrodes 274–6
  'smart' 277
  synthetic 264–7
network formers (glass formers) 159–62
neural cell adhesion molecule (N-CAM) 497
neural tissue engineering 495–500
neurite outgrowth 105
neurite shape 438
Neurolac 263
neuroma 263
neurotrophic factors 264, 273
neutral red uptake (NRU) assay 515, 516
neutrophils 76–7
non-bridging oxygen (NBO) 160, 161
non-degradable synthetic polymers 238–9
non-fouling biomimetic surfaces 493–4
non-vascularised cortical autografts 328–9
Norian Skeletal Repair System (SRS) 334
North American Material Science Association (NAMSA) 509, 510
nucleoskeleton Plate 3
nylon tubes 450–1

oestrogen 360–1
oestrogen receptors 363
oligonucleotide immobilisation 599–600
organosilanes 488
  on hydroxylated surfaces 90–2
Organization for Economic Cooperation and Development (OECD) 512, 525
orientation, fibre 12–13
Orowan approximation 398
orthopaedics 71, 568
  cellular response to osteoinductive materials *see* osteoinductive materials
  DLC applications 406–8
  see also hip prostheses; knee prostheses
orthophosphate glass structure 161, 162
osmotic pressure 122, 124–5

osseointegration 315, 595, 603–4
  and bone healing 326
  titanium dental implants and 573–4, 592–3, 603, 604
  'loss of osseointegration' phenomenon 574–5
OsseoSpeed 580
OSSEOTITE 581
osteoblasts 316, 319, 490
  attachment to SAMs 104
  electrospun fibres 15, 16
  mechanisms of responses to surface chemistry 468–71
  surface-modified titanium 595
  see also human osteoblast (HOB) cells
osteoconduction 315, 574
  and bone healing 325
ostoeconductivity 138
  bone substitutes 333–4, 562
osteogenesis 314, 438
osteogenic materials 315
Osteogenic Protein-1 (OP-1) 320–1
osteoinduction 314–15
  and bone healing 316–18
osteoinductive materials 138, 313–43
  allografts 327, 329–32
  autografts 326–9
  BMPs 318–23
  bone graft materials
    and bone healing 323–4
    classification 326, 327
    and remodelling 334–5
  demineralised bone matrix 331–2
  osteoinductive agents 317, 318
  synthetic bone graft substitutes 314, 327, 332–5
osteoinductive potential 324–5
osteoinductivity 138
  and bone healing 324–5
osteointegration 603–4
osteopontin 105
oxide film 598–602

paraneural layer 254
paratenon 18
particulate gene-delivery systems 199
pattern matching 445–6, 447
patterns/patterning 436
  cell signalling in response to nanofeatures 440, 444–6, 447
  chemically patterned surfaces 466–7
pendant fixed charges 118, 124–6
peptide grafting 479–507
  criteria for bioactive surface design 489–90
  delivery of bioactive peptides to tissue–surface interface 486–9
  peptide mimicry 483–6
  see also bioactive surfaces

Index   619

peptides
  adsorption 486, 487
  immobilisation 464
  self-assembled 498–500, 501
pericytes 297
perineurium 254–5
peripheral nerves 252–90
  alternatives to autografts 262–3
  biomechanical properties 256–7
  crossing short gaps 276–7
  injury response in peripheral nervous system 257–62
  nerve guides 263–76
  neural tissue engineering 495–500
  'smart' conduit 277
  structure 254–7
peroxisome proliferator-activated receptors (PPARs) 363
phagocytosis 76–7
phantom hydrogel networks 115
pharmaceutical/biological agents 224–5
phase separation 4, 194
phosphate-based glasses 156–82
  biomedical applications 164–5
  chemistry of 159–62
  in composites 173
  future trends 177
  glass fibres 173–7, 243, 244, 245
  glass formation 158–9
  glass structure theories 157–8
  historical overview 156–7
  in vitro studies 165–72
  in vivo studies 172
  nanofibre 198–9
  properties 162–4
phosphatidyl serine (PS) 524
phosphorus 473
  surface chemistry of dental implants 578–9
photocytotoxicity 531–2
photogenotoxicity 531–2
photolithography 91–2
phototoxicity 531–2
PHSRN peptide 469
physical adsorption 93
physical gels 115, 116
  hydrogels 117, 118, 124, 126
physical surface modification treatments 593–4
physical vapour deposition (PVD) 393–6
phytoestrogen–collagen matrix 360–4
Piroxicam 67–8
placental stem cells 300
plaited electrospun fibre 22
plasma assisted chemical vapour deposition (PACVD) 394, 396–7
plasma proteins see proteins
plasma sprayed HA coatings 594

platelets
  activation 404
  adhesion 103, 404
Polyactive 35–6, 47
polyalkylcyanoacrylates 36
poly-α-hydroxyacids 31–4
  degradation 36–7, 38
  see also polyglycolic acid; polylactic acid
polyamino acids 36
polyampholyte hydrogels 126
polycaprolactone (PCL) 31, 32, 51, 240
  chitin–PCL composite 239
  degradation 36–7, 38
    in vitro 68, 69
  electrospinning 6, 7, 9, 14, 15
    collagen-coated 15–17
  muscle regeneration 244
  vascular grafts 220
polydimethylsiloxane (pDMS) 431, 436
polydioxanone 34
polyelectrolyte (ionic) hydrogels 115, 116, 124–6
polyethylene–Bioglass composites 383–4, 385–7
polyethylene glycol (PEG) 35–6
  multiarm 90
  spacers 489–90
polyethylene–hydroxyapatite composite (HAPEX) 44–5, 383–4, 385–7
polyethylene oxide (PEO) 35–6, 51
polyethylene terephthalate (PET) 212, 216, 226, 239
polyglycolic acid (PGA) 31–4, 240
  composites 46–7
  degradation 36–7, 38
  muscle regeneration 244
  vascular grafts 220
polyhydroxyalkanoates (PHAs) 35
polyhydroxybutyrate (PHB) 10, 35
polyhydroxyethylmethacrylate (PHEMA) 240
polyhydroxyethylmethacrylate (PHEMA)–nanoCHA composite 385
polyhydroxyvalerate (PHV) 35
polylactic acid (PLA) 31–4, 67–8, 239
  composites 46–7
  degradation 36–7, 38
  drug delivery of calcitonin 67
  vascular grafts 220
polylactide-co-glycolide (PLGA) 9–10, 67–8
  PLLA–PLGA composite 247–8
poly-L-lactic acid (PLLA) 32, 33–4, 240
  in vivo biocompatibility testing 76
  muscle regeneration 244
  PLLA–PLGA composite 247–8
polymer demixing 433, 434–5, 450–1
polymer–solvent interaction 117–19

620 Index

polymers 28–9, 139, 140
  bioactive surfaces 492–3
  biodegradable *see* biodegradable polymers
  electrospinning applications in tissue engineering 8–10
  properties of polymeric solutions for electrospinning 6–7
  self-assembling 87–90
  tissue engineering 502
polymethoxylated flavonoid–collagen matrix 358–60
polymorphonuclear leukocytes (PMNs) 77
polyphosphate glass structure 161, 162, 164
polysaccharides 221
polystyrene–polybromostyrene (pS–pSBr) systems 433, 434–5
polystyrene–poly-*n*-butylmethacrylate (pS–pnBMA) system 435
polyurethanes 218–19
porosity 10
  hydrogels 118, 126, 131
  scaffolds 11–12, 131, 188–9, 293–4, 306
porous ceramics 141–2
preosteoblasts 320
primary glass formers 159
printing 436
probe-broadening 546
progesterone 67
protein adsorption 93–102, 142, 480, 481–2
  albumin/fibrinogen adsorption ratio of DLC coatings 403
  cell response to surface chemistry 464–5
  characterisation techniques 100–2
  flow-based analysis 540–1, 542
  on functionalised surfaces 97–100
  mechanisms of adsorption 93–7
  selective 100
  titanium and titanium alloys 592–3
    bone–titanium implant interface reactions 604–5
protein scaffolds 238
proteins 11
  native proteins and tendon regeneration 241
  plasma proteins and bone cell adhesion 538–59
    BCA protein assay 541–3
    flow-based protein adsorption analysis 540–1, 542
    protein–cell interactions 550–5
    protein characterisation 544–8
    two-dimensional gel electrophoresis 542, 543–4
  uprated in acutely denervated Schwann cells 259, 260
proteoglycans 243
proteomics 79

prothrombin 539
  protein–cell interactions 550–5
pruning 261
pseudo-nerves 263
puerarin 361–4, 369
pulsed laser deposition (PLD) 394, 396
pyrolytic carbon 408
pyrophosphate glass structure 161, 162

Q^i terminology 160, 161
quartz crystal microbalance (QCM) 101
quaternary phosphate glass systems 168–72
  glass fibres 176–7

rabbit model 369–70
radiolabelling technique 101
radius ratio 157
rapid prototyping (RP) 195–6
recombinant human BMP-7 (rhBMP-7) 320–1
regeneration chambers 264
regeneration stagger 267
regenerative medicine 479
  approach to vascular grafts 217
relative protein binding capacity 541
remodelling, bone 324, 334–5, 574
repair phase (of healing) 78–9
repolymerisation model 162
resorbability 142
resorbable scaffolds 52
resorption, bone 324
  cell-mediated 565
response index 518, 519
RGD (arginine–glycine–aspartic acid) sequence 444, 464, 469
  bone tissue engineering 491–2
  cardiovascular tissue engineering 500–1
  peptide mimicry 485–6
  vascular grafts 225
rope-like structures 22
rotating analyser ellipsometer 547
rotating mandrel 12, 13

safety 322–3
Salmonella assay 527
saphenous vein 215–16
scaffolds 185–211
  advantages 196
  applications in gene therapy 199–200
  bone repair 306–7
    biomimetic scaffolds 292–4
    targeting stemness within the bone marrow niche 300–5
  degradable composites 42–4
  design and manufacture 188–9
  design and processing 193–6
  electrospinning *see* electrospinning
  fibrous 49–51, 228
  future trends 200–1

Index   621

hydrogels 129–31
initial scaffolds investigated for use as vascular grafts 217–22
　cell-based scaffolds 222
　decellularised scaffolds 221–2
　natural scaffolds 220–1
　synthetic scaffolds 217–20
internal (nerve guides) 268–74
limitations 196–7
monitoring cell behaviour in 197–8
nanofibrous 11, 307
nanostructural biomaterials 198–9, 200–1, 228
nature of 187–8
porosity 11–12, 131, 188–9, 293–4, 306
requirements 10
resorbable 52
types of material for 189–93
scanning electron microscopy 380–2
scanning force microscopy (SFM) see atomic force microscopy
Schwann cells 255, 264–5
　acutely denervated 258, 259–60
　bridging a long gap 274
Schwann tubes 259
seeding techniques 229
selective protein adsorption 100
self-assembled monolayers (SAMs) 90–2, 488
　protein adsorption on functionalised surfaces 98–100
　selective protein adsorption 100
　surface chemistry and cell response 102–6, 466, 467–8
self-assembled peptides
　cardiovascular tissue engineering 501
　neural tissue engineering scaffolds 498–500
self-assembling polypeptide hydrogels 129–30
self-assembly 4, 86–92
　nanofabrication 433–5
　in nature 86–7
　polymers 87–90
　SAMs see self-assembled monolayers
SEVAC–hydroxyapatite composite 48–9
shear modulus of polymer network 123
shear stresses 228–9
sheath 18–19
short gaps, nerve conduits across 276–7
sieve electrodes 274–6
silanes 488
silica 550–5
silicon doping 399
silicon substituted HA (SiHA) 378–82, 387–8
silicone nerve guides 264–6
silver ions 170–1
silver-releasing phosphate glasses 165

simulated body fluid (SBF) 46
simvastatin 354–8, 369
single cell gel electrophoresis (SCGE) 526, 530–1
sintered porous hydroxyapatite (SPHA) 333–4
sintering 140
skeletal muscle 242–3
SLA surface 580
SLActive surface 580–1
Smads 319
'smart' nerve conduit 277
smooth muscle cells 213, 214, 227, 230
sodding 229
sodium dodecyl sulphate–polyacrylamide gel electrophoresis (SDS–PAGE) 543
sodium hexametaphosphate (Graham's salt) 156–7
'soft' proteins 95–6
sol-gel coating technology 472–3, 582, 583
solid freeform fabrication (SFF) 195–6
solution, polymeric 6–7
solvent-casting particulate-leaching 193–4
solvent–polymer interaction 117–19
SOS chromotest 532
SOS LUX test 532
spacing 440, 444–6
spinal cord 496
sputtering 395
stamps 436
staphylococcal enterotoxin B toxoid 67
Staphylococcus aureus 171
starch based composites 48–9
statins 353–8, 369–70
stem cells 53, 79
　embryonic 294, 302, 316, 317
　mesenchymal see mesenchymal stem cells (MSCs)
　surface chemistry and adhesion 105–6
　and tissue scaffolds for bone repair 291–312
　　adult stem cell sources 294–300, 301
　　future trends 306–7
　　stem cell-induced immunomodulation 305–6
　　targeting stemness within the bone marrow niche 300–5
stents 409–11
sterilisation 579
strain 324, 334–5
strength, and porosity of ceramics 141
Streptococcus sanguis 170–1
subcutaneous fat 230
sulphorhodamine B (SRB) protein staining 515
superabsorbent hydrogels 128
supercooled liquid 158, 159
superporous hydrogels 127–8

622 Index

surface chemistry 462–78
  bone–titanium implant interface reactions 604–5
  cell response to model surfaces 465–8
  dental implants 578–9
    effect on cell behaviour 579
  exploitation of cell response to 471–3
  future trends 473–4
  mechanisms for cell responses to surfaces 468–71
  nanostructural biomaterials 85–113
    influence on cell response 102–6
    influence on protein adsorption 93–102
  role of proteins in cell response 464–5
surface energy 94, 401
surface modification
  biochemical modification 463–4
  exploitation of cell response to surface chemistry 471–3
  nanofeatured materials 435
  surface-modified titanium 589–608
    biocompatibility enhancement 602–5
    chemical and electrochemical modification treatments 594–602
      anodic oxidation 598–602
      cathodic polarisation 596–8
    dental implants see dental implants
    exploitation of cell response to surface chemistry 471–3
    mechanical, physical and thermal modification treatments 593–4
surface patterning techniques 91–2
surface plasmon resonance (SPR) 101
surface preparation 544
surface properties, of scaffolds 189
surface tension 6–7
surface topography 463
  bone–titanium implant interface reactions 604
  dental implants 575–8, 604
    effect on cell behaviour 577–8
  scaffolds 10, 11
  vascular grafts 226–7
  see also microtopography; nanotopography
surface wettability 14–15
swelling
  measurement and degree of crosslinking 121–3
  rate 126
  ratio 116–17
symmetry 444–6
synthetic bone graft substitutes (SBGSs) 314, 327, 332–4
  and osteoconductivity 333–4
  remodelling 334–5
  see also bone substitutes

synthetic nerve guides 264–7
synthetic polymers 502
  degradable 29–36, 190, 191, 193
    scaffolds for vascular grafts 217–20
    tendon regeneration 239–40
  nerve guides 266
  non-degradable 238–9
synthetic scaffolds
  porous 293–4
  vascular grafts 217–20

tail length 530, 531
tail moment 530, 531
tailoring of polymers 37–9
temperature
  crystallisation temperature 158, 159
  electrospinning 8
  glass transition temperature 158, 159, 163
  lower critical solution temperature (LCST) 119
  toxicity testing 510, 512
  upper critical solution temperature (UCST) 119
tendons 23, 49–50, 237–42
  biomaterials used in regeneration 238–42
  electrospinning and 17–23
    fibre bundles 19–22
    tendon regeneration 17–19
  and their mechanical properties 19–20, 237–8
  structure 18–19, 238
tenocytes 237–8, 240
  attachment of fibre bundles 21, 22
  electrospun fibres 15, 16
tension 450
termino-lateral neurorraphy (TLN) 263
ternary phosphate glasses 166–8
  glass fibres 175
thermal surface modification treatments 593–4
Thermanox 551–5
thin mandrel fibre bundles 19–21
three-dimensional nanofeatured scaffolds 450–1
three-dimensional stem cell cultures 304
thrombus formation 103
TiOblast 580
tip effect 545–6
tissue-compatibility 401–2
tissue engineering 3, 292
  bioactive surfaces using peptide grafting 479–507
  cardiovascular tissue engineering 500–1
  neural tissue engineering 495–500
  biodegradable composites for see biodegradable composites
  biodegradable polymers for see biodegradable polymers

bone tissue engineering 51, 490–5, 567
  electrospinning applications 8–10
  scaffolds for *see* scaffolds
tissue–implant interface 480–3
tissue rebuilding 438
titanium
  biomimetic 492
  surface-modified *see* surface-modified titanium
  and titanium alloys for hard and soft tissue applications 589–93
titanium oxide 158, 169–70, 172, 472–3, 592
  dental implants 576–7, 578
  protein adsorption 96–7
TiUnite 580
top down nanofabrication 430–3
topography *see* surface topography
total protein binding capacity 541
toxic materials 138
toxicity testing 508–37
  cell death 519–25
  cytotoxicity 513–19
  future trends 531–2
  genotoxicity 525–31
transforming growth factor-beta (TGF-β) superfamily 318
transmission electron microscopy (TEM) 385–7, 525
transparency factor 276
tribological properties 399–401, 406–8
  see also wear
tricalcium phosphate (TCP) 137, 144–5, 292–3
tricarboxylic acid (TCA) cycle 31
tumour necrosis factor-alpha (TNF-α) 376–7
twisted electrospun fibre 22
two-dimensional gel electrophoresis 542, 543–4

ultraphosphate glass structure 161, 162
ultrastructure 385–7
umbilical cord blood stem cells (UCBC) 299–300
umu test 532
United States Pharmacopeia (USP) 525
upper critical solution temperature (UCST) 119

vaccination 67
valve metals 600
valves 128–9
  heart valves 408–9

vascular grafts 212–36
  clinical use to date 215–16
  electrostatically spun scaffolds as 230–2
  further development and modification 222–30
    combinations of biomaterials and copolymers 222–3
    experimental conditions 228–30
    further modifications 223–5
    structures 225–8
  future trends 232
  initial scaffolds investigated for use 217–22
  regenerative medicine approach 217
  requirements of replacement vessels 214–15
vascularisation 43, 71
  bone graft materials 323–4
  stem cells and tissue scaffolds for bone repair 303–4
vascularised autografts 328
venous grafts 216
VentroAssist 409
via holes 276
vimentin 448, Plate 2
virus-derived gene carriers 199
viscosity 6, 7
vitronectin 465
voltage 7
volume swelling ratio 116–17
Vroman effect 96

Wallerian degeneration (WD) 257–60
water
  body water retainers 128
  interaction of endosseous implant with 480, 481
water-mediated hydrogen bonds 87
wear
  debris and SBGSs 332–3
  DLC coatings 400–1, 406–8
wettability, surface 14–15

X-rays 566

yeast assay 532
yield strength 398
Young equation 93–4

zinc ion implantation 472
zinc oxide 169, 170
zinc phosphate glasses 177
zymogens 521, 525